Free! Online PDF version of the
BNi FACILITIES MANAGER'S 2013 COSTBOOK
that you can download, print, share and *customize*!

Just go to www.ConstructionWorkZone.com/bnicostbooks

On ConstructionWorkZone.com you can quickly create a PDF version of this entire Costbook. You can even customize it! Just enter your own labor and markup rates and every cost item contained in this book will be instantly recalculated.

Your purchase of this BNi Costbook not only entitles you to full access to the interactive online PDF version, it also brings you a wide range of additional estimating tools available on ConstructionWorkZone.com, including a FREE Davis-Bacon wage rate database you can use to pinpoint the prevailing labor costs for over 400 metro areas throughout the country!

**ConstructionWorkZone.com …
your source for "all things construction"**

In addition to a wide range of online estimating tools, ConstructionWorkZone.com is your go-to source for current industry-specific news articles, search tools, manufacturer data, 3-part specifications, marketing resources and much more. And because you are a valued BNi customer, registration is free!

Questions?
Call us toll-free: 1.888.BNI.BOOK

BNI Building News

FACILITIES MANAGER'S

2013 COSTBOOK

NINETEENTH EDITION

BNi® Building News

• Anaheim • Vista

HUMBER LIBRARIES NORTH CAMPUS
205 Humber College Blvd
TORONTO, ON.   M9W 5L7

BNi Building News

EDITOR-IN-CHIEF
William D. Mahoney, P.E.

TECHNICAL SERVICES
Tony De Augustine
Anthony Jackson
Eric Mahoney, AIA
Ana Varela
Mathew Woolsey

GRAPHIC DESIGN
Robert O. Wright Jr.

BNi Publications, Inc.

ANAHEIM
1612 S. CLEMENTINE STREET
ANAHEIM, CA 92802

VISTA
990 PARK CENTER DRIVE, SUITE E
VISTA, CA 92081

1-888-BNI-BOOK (1-888-264-2665)
www.bnibooks.com

ISBN 978-1-55701-762-8

Copyright © 2012 by BNI Publications, Inc. All rights reserved. Printed in the United States of America. Except as permitted under the United States Copyright Act of 1976, no part of this publication may be reproduced or distributed in any form or by any means, or stored in a data base or retrieval system, without the prior written permission of the publisher.

While diligent effort is made to provide reliable, accurate and up-to-date information, neither BNI Publications Inc., nor its authors or editors, can place a guarantee on the correctness of the data or information contained in this book. BNI Publications Inc., and its authors and editors, do hereby disclaim any responsibility or liability in connection with the use of this book or of any data or other information contained therein.

Table of Contents

Format v
Features of This Book vii
Sample Costbook Page ix
Costbook Pages 1
Man-Hour Tables 453
Supporting Reference Data 675
Geographic Cost Modifiers 761
Square Foot Tables 773
Index 785

Preface

For over 65 years, BNi Building News has been dedicated to providing construction professionals with timely and reliable information. Based on this experience, our staff has researched and compiled thousands of up-to-the-minute costs for the **BNi Costbooks**. This book is an essential reference for contractors, engineers, architects, facility managers — any construction professional who must provide an estimate for any type of building project.

Whether working up a preliminary estimate or submitting a formal bid, the costs listed here can be quickly and easily tailored to your needs. All costs are based on prevailing labor rates. Allowances for overhead and profit are included in all costs. Man-hours are also provided.

All data is categorized according to the 16-division format. This industry standard provides an all-inclusive checklist to ensure that no element of a project is overlooked. In addition, to make specific items even easier to locate, there is a complete alphabetical index.

The "Features of this Book" section presents a clear overview of the many features of this book. Included is an explanation of the data, sample page layout and discussion of how to best use the information in the book.

Of course, all buildings and construction projects are unique. The information provided in this book is based on averages from well-managed projects with good labor productivity under normal working conditions (eight hours a day). Other circumstances affecting costs such as overtime, unusual working conditions, savings from buying bulk quantities for large projects, and unusual or hidden costs must be factored in as they arise.

The data provided in this book is for estimating purposes only. Check all applicable federal, state and local codes and regulations for local requirements.

Format

All data is categorized according to the 16-division format. This industry standard provides an all-inclusive checklist to ensure that no element of a project is overlooked.

GENERAL REQUIREMENTS .... DIVISION 1

- SUMMARY 1100
- PRICE AND PAYMENT PROCEDURES 1200
- ADMINISTRATIVE REQUIREMENTS 1300
- QUALITY REQUIREMENTS 1400
- TEMPORARY FACILITIES AND CONTROLS 1500
- PRODUCT REQUIREMENTS 1600
- EXECUTION REQUIREMENTS 1700
- FACILITY OPERATION 1800
- FACILITY DECOMMISSIONING 1900

SITE CONSTRUCTION .... DIVISION 2

- BASIC SITE MATERIALS AND METHODS 2050
- SITE REMEDIATION 2100
- SITE PREPARATION 2200
- EARTHWORK 2300
- TUNNELING, BORING, AND JACKING 2400
- FOUNDATION AND LOAD-BEARING ELEMENTS 2450
- UTILITY SERVICES 2500
- DRAINAGE AND CONTAINMENT 2600
- BASES, BALLASTS, PAVEMENTS, AND APPURTENANCES 2700
- SITE IMPROVEMENTS AND AMENITIES 2800
- PLANTING 2900
- SITE RESTORATION AND REHABILITATION 2950

CONCRETE .... DIVISION 3

- BASIC CONCRETE MATERIALS AND METHODS 3050
- CONCRETE FORMS AND ACCESSORIES 3100
- CONCRETE REINFORCEMENT 3200
- CAST-IN-PLACE CONCRETE 3300
- PRECAST CONCRETE 3400
- CEMENTITIOUS DECKS AND UNDERLAYMENT 3500
- GROUTS 3600
- MASS CONCRETE 3700
- CONCRETE RESTORATION AND CLEANING 3900

MASONRY .... DIVISION 4

- BASIC MASONRY MATERIALS AND METHODS 4050
- MASONRY UNITS 4200
- STONE 4400
- REFRACTORIES 4500
- CORROSION-RESISTANT MASONRY 4600
- SIMULATED MASONRY 4700
- MASONRY ASSEMBLIES 4800
- MASONRY RESTORATION AND CLEANING 4900

METALS .... DIVISION 5

- BASIC METAL MATERIALS AND METHODS 5050
- STRUCTURAL METAL FRAMING 5100
- METAL JOISTS 5200
- METAL DECK 5300
- COLD-FORMED METAL FRAMING 5400
- METAL FABRICATIONS 5500
- HYDRAULIC FABRICATIONS 5600
- RAILROAD TRACK AND ACCESSORIES 5650
- ORNAMENTAL METAL 5700
- EXPANSION CONTROL 5800
- METAL RESTORATION AND CLEANING 5900

WOOD AND PLASTICS .... DIVISION 6

- BASIC WOOD AND PLASTIC MATERIALS AND METHODS 6050
- ROUGH CARPENTRY 6100
- FINISH CARPENTRY 6200
- ARCHITECTURAL WOODWORK 6400
- STRUCTURAL PLASTICS 6500
- PLASTIC FABRICATIONS 6600
- WOOD AND PLASTIC RESTORATION AND CLEANING 6900

THERMAL AND MOISTURE PROTECTION .... DIVISION 7

- BASIC THERMAL AND MOISTURE PROTECTION MATERIALS AND METHODS 7050
- DAMPPROOFING AND WATERPROOFING 7100
- THERMAL PROTECTION 7200
- SHINGLES, ROOF TILES, AND ROOF COVERINGS 7300
- ROOFING AND SIDING PANELS 7400
- MEMBRANE ROOFING 7500
- FLASHING AND SHEET METAL 7600
- ROOF SPECIALTIES AND ACCESSORIES 7700
- FIRE AND SMOKE PROTECTION 7800
- JOINT SEALERS 7900

DOORS AND WINDOWS .... DIVISION 8

- BASIC DOOR AND WINDOW MATERIALS AND METHODS 8050
- METAL DOORS AND FRAMES 8100
- WOOD AND PLASTIC DOORS 8200
- SPECIALTY DOORS 8300
- ENTRANCES AND STOREFRONTS 8400
- WINDOWS 8500
- SKYLIGHTS 8600
- HARDWARE 8700
- GLAZING 8800
- GLAZED CURTAIN WALL 8900

FINISHES .... DIVISION 9

- BASIC FINISH MATERIALS AND METHODS 9050
- METAL SUPPORT ASSEMBLIES 9100
- PLASTER AND GYPSUM BOARD 9200
- TILE 9300
- TERRAZZO 9400
- CEILINGS 9500
- FLOORING 9600
- WALL FINISHES 9700
- ACOUSTICAL TREATMENT 9800
- PAINTS AND COATINGS 9900

Format (Continued)

SPECIALTIES .... DIVISION 10

VISUAL DISPLAY BOARDS	10100
COMPARTMENTS AND CUBICLES	10150
LOUVERS AND VENTS	10200
GRILLES AND SCREENS	10240
SERVICE WALLS	10250
WALL AND CORNER GUARDS	10260
ACCESS FLOORING	10270
PEST CONTROL	10290
FIREPLACES AND STOVES	10300
MANUFACTURED EXTERIOR SPECIALTIES	10340
FLAGPOLES	10350
IDENTIFICATION DEVICES	10400
PEDESTRIAN CONTROL DEVICES	10450
LOCKERS	10500
FIRE PROTECTION SPECIALTIES	10520
PROTECTIVE COVERS	10530
POSTAL SPECIALTIES	10550
PARTITIONS	10600
STORAGE SHELVING	10670
EXTERIOR PROTECTION	10700
TELEPHONE SPECIALTIES	10750
TOILET, BATH, AND LAUNDRY ACCESSORIES	10800
SCALES	10880
WARDROBE AND CLOSET SPECIALTIES	10900

EQUIPMENT .... DIVISION 11

MAINTENANCE EQUIPMENT	11010
SECURITY AND VAULT EQUIPMENT	11020
TELLER AND SERVICE EQUIPMENT	11030
ECCLESIASTICAL EQUIPMENT	11040
LIBRARY EQUIPMENT	11050
THEATER AND STAGE EQUIPMENT	11060
INSTRUMENTAL EQUIPMENT	11070
REGISTRATION EQUIPMENT	11080
CHECKROOM EQUIPMENT	11090
MERCANTILE EQUIPMENT	11100
COMMERCIAL LAUNDRY AND DRY CLEANING EQUIPMENT	11110
VENDING EQUIPMENT	11120
AUDIO-VISUAL EQUIPMENT	11130
VEHICLE SERVICE EQUIPMENT	11140
PARKING CONTROL EQUIPMENT	11150
LOADING DOCK EQUIPMENT	11160
SOLID WASTE HANDLING EQUIPMENT	11170
DETENTION EQUIPMENT	11190
WATER SUPPLY AND TREATMENT EQUIPMENT	11200
HYDRAULIC GATES AND VALVES	11280
FLUID WASTE TREATMENT AND DISPOSAL EQUIPMENT	11300
FOOD SERVICE EQUIPMENT	11400
RESIDENTIAL EQUIPMENT	11450
UNIT KITCHENS	11460
DARKROOM EQUIPMENT	11470
ATHLETIC, RECREATIONAL, AND THERAPEUTIC EQUIPMENT	11480
INDUSTRIAL AND PROCESS EQUIPMENT	11500
LABORATORY EQUIPMENT	11600
PLANETARIUM EQUIPMENT	11650
OBSERVATORY EQUIPMENT	11660
OFFICE EQUIPMENT	11680
MEDICAL EQUIPMENT	11700
MORTUARY EQUIPMENT	11780
NAVIGATION EQUIPMENT	11850
AGRICULTURAL EQUIPMENT	11870
EXHIBIT EQUIPMENT	11900

FURNISHINGS .... DIVISION 12

FABRICS	12050
ART	12100
MANUFACTURED CASEWORK	12300
FURNISHINGS AND ACCESSORIES	12400
FURNITURE	12500
MULTIPLE SEATING	12600
SYSTEMS FURNITURE	12700
INTERIOR PLANTS AND PLANTERS	12800
FURNISHINGS RESTORATION AND REPAIR	12900

SPECIAL CONSTRUCTION .... DIVISION 13

AIR-SUPPORTED STRUCTURES	13010
BUILDING MODULES	13020
SPECIAL PURPOSE ROOMS	13030
SOUND, VIBRATION, AND SEISMIC CONTROL	13080
RADIATION PROTECTION	13090
LIGHTING PROTECTION	13100
CATHODIC PROTECTION	13110
PRE-ENGINEERED STRUCTURES	13120
SWIMMING POOLS	13150
AQUARIUMS	13160
AQUATIC PARK FACILITIES	13165
TUBS AND POOLS	13170
ICE RINKS	13175
KENNELS AND ANIMAL SHELTERS	13185
SITE-CONSTRUCTED INCINERATORS	13190
STORAGE TANKS	13200
FILTER UNDERDRAINS AND MEDIA	13220
DIGESTER COVERS AND APPURTENANCES	13230
OXYGENATION SYSTEMS	13240
SLUDGE CONDITIONING SYSTEMS	13260
HAZARDOUS MATERIAL REMEDIATION	13280
MEASUREMENT AND CONTROL INSTRUMENTATION	13400
RECORDING INSTRUMENTATION	13500
TRANSPORTATION CONTROL INSTRUMENTATION	13550
SOLAR AND WIND ENERGY EQUIPMENT	13600
SECURITY ACCESS AND SURVEILLANCE	13700
BUILDING AUTOMATION AND CONTROL	13800
DETECTION AND ALARM	13850
FIRE SUPPRESSION	13900

CONVEYING SYSTEMS .... DIVISION 14

DUMBWAITERS	14100
ELEVATORS	14200
ESCALATORS AND MOVING WALKS	14300
LIFTS	14400
MATERIAL HANDLING	14500
HOISTS AND CRANES	14600
TURNTABLES	14700
SCAFFOLDING	14800
TRANSPORTATION	14900

MECHANICAL .... DIVISION 15

BASIC MECHANICAL MATERIALS AND METHODS	15050
BUILDING SERVICES PIPING	15100
PROCESS PIPING	15200
FIRE PROTECTION PIPING	15300
PLUMBING FIXTURES AND EQUIPMENT	15400
HEAT-GENERATION EQUIPMENT	15500
REFRIGERATION EQUIPMENT	15600
HEATING, VENTILATING, AND AIR CONDITIONING EQUIPMENT	15700
AIR DISTRIBUTION	15800
HVAC INSTRUMENTATION AND CONTROLS	15900
TESTING, ADJUSTING, AND BALANCING	15950

ELECTRICAL .... DIVISION 16

BASIC ELECTRICAL MATERIALS AND METHODS	16050
WIRING METHODS	16100
ELECTRICAL POWER	16200
TRANSMISSION AND DISTRIBUTION	16300
LOW-VOLTAGE DISTRIBUTION	16400
LIGHTING	16500
COMMUNICATIONS	16700
SOUND AND VIDEO	16800

Features of this Book

The construction estimating information in this book is divided into two main sections: Costbook Pages and Man-Hour Tables. Each section is organized according to the 16-division format. In addition there are extensive Supporting References.

Sample pages with graphic explanations are included before the Costbook pages. These explanations, along with the discussions below, will provide a good understanding of what is included in this book and how it can best be used in construction estimating.

Material Costs

The material costs used in this book represent national averages for prices that a contractor would expect to pay plus an allowance for freight (if applicable) and handling and storage. These costs reflect neither the lowest or highest prices, but rather a typical average cost over time. Periodic fluctuations in availability and in certain commodities (e.g. copper, conduit) can significantly affect local material pricing. In the final estimating and bidding stages of a project when the highest degree of accuracy is required, it is best to check local, current prices.

Labor Costs

Labor costs include the basic wage, plus commonly applicable taxes, insurance and markups for overhead and profit. The labor rates used here to develop the costs are typical average prevailing wage rates. Rates for different trades are used where appropriate for each type of work.

Fixed government rates and average allowances for taxes and insurance are included in the labor costs. These include employer-paid Social Security/Medicare taxes (FICA), Worker's Compensation insurance, state and federal unemployment taxes, and business insurance.

Please note, however, most of these items vary significantly from state to state and within states. For more specific data, local agencies and sources should be consulted.

Equipment Costs

Costs for various types and pieces of equipment are included in Division 1 - General Requirements and can be included in an estimate when required either as a total "Equipment" category or with specific appropriate trades. Costs for equipment are included when appropriate in the installation costs in the Costbook pages.

Overhead and Profit

Included in the labor costs are allowances for overhead and profit for the contractor/employer whose workers are performing the specific tasks. No cost allowances or fees are included for management of subcontractors by the general contractor or construction manager. These costs, where appropriate, must be added to the costs as listed in the book.

The allowance for overhead is included to account for office overhead, the contractors' typical costs of doing business. These costs normally include in-house office staff salaries and benefits, office rent and operating expenses, professional fees, vehicle costs and other operating costs which are not directly applicable to specific jobs. It should be noted for this book that office overhead as included should be distinguished from project overhead, the General Requirements (Division 1) which are specific to particular projects. Project overhead should be included on an item by item basis for each job.

Depending on the trade, an allowance of 10-15 percent is incorporated into the labor/installation costs to account for typical profit of the installing contractor. See Division 1, General Requirements, for a more detailed review of typical profit allowances.

Features of this Book *(Continued)*

Adjustments to Costs

The costs as presented in this book attempt to represent national averages. Costs, however, vary among regions, states and even between adjacent localities.

In order to more closely approximate the probable costs for specific locations throughout the U.S., a table of Geographic Multipliers is provided. These adjustment factors are used to modify costs obtained from this book to help account for regional variations of construction costs. Whenever local current costs are known, whether material or equipment prices or labor rates, they should be used if more accuracy is required.

Hours (Man-Hours)

These productivities represent typical installation labor for thousands of construction items. The data takes into account all activities involved in normal construction under commonly experienced working conditions such as site movement, material handling, start-up, etc.

Editor's Note: This **Costbook** is intended to provide accurate, reliable, average costs and typical productivities for thousands of common construction components. The data is developed and compiled from various industry sources, including government, manufacturers, suppliers and working professionals. The intent of the information is to provide assistance and guidance to construction professionals in estimating. The user should be aware that local conditions, material and labor availability and cost variations, economic considerations, weather, local codes and regulations, etc., all affect the actual cost of construction. These and other such factors must be considered and incorporated into any and all construction estimates.

Sample Costbook Page

In order to best use the information in this book, please review this sample page and read the "Features In This Book" section.

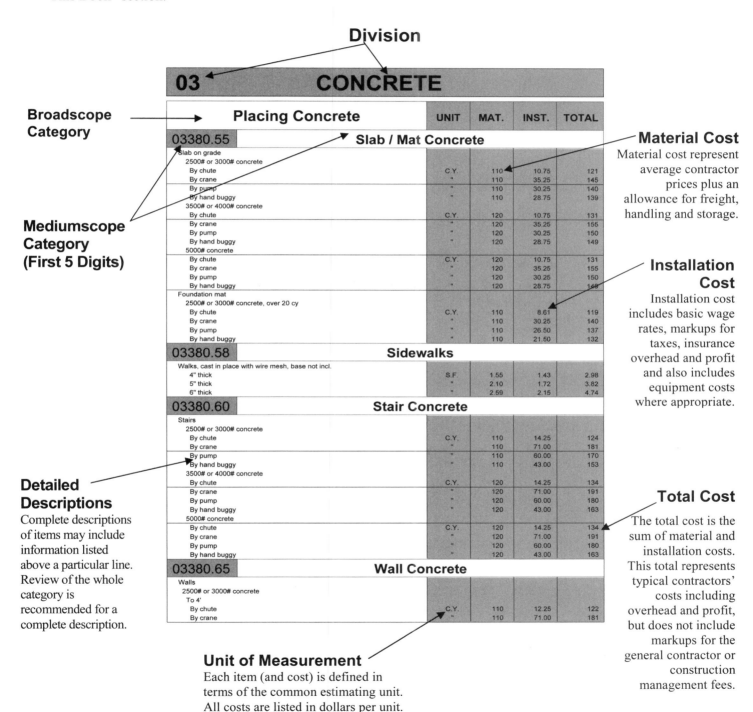

Division

Broadscope Category

Mediumscope Category (First 5 Digits)

Detailed Descriptions
Complete descriptions of items may include information listed above a particular line. Review of the whole category is recommended for a complete description.

Unit of Measurement
Each item (and cost) is defined in terms of the common estimating unit. All costs are listed in dollars per unit.

Material Cost
Material cost represent average contractor prices plus an allowance for freight, handling and storage.

Installation Cost
Installation cost includes basic wage rates, markups for taxes, insurance overhead and profit and also includes equipment costs where appropriate.

Total Cost
The total cost is the sum of material and installation costs. This total represents typical contractors' costs including overhead and profit, but does not include markups for the general contractor or construction management fees.

BNi Building News

01 GENERAL Requirements

	UNIT	MAT.	INST.	TOTAL
01020.10 Allowances				
Overhead				
$20,000 project				
Minimum	PCT.			15.00
Average	"			20.00
Maximum	"			40.00
$100,000 project				
Minimum	PCT.			12.00
Average	"			15.00
Maximum	"			25.00
$500,000 project				
Minimum	PCT.			10.00
Average	"			12.00
Maximum	"			20.00
$1,000,000 project				
Minimum	PCT.			6.00
Average	"			10.00
Maximum	"			12.00
$10,000,000 project				
Minimum	PCT.			1.50
Average	"			5.00
Maximum	"			8.00
Profit				
$20,000 project				
Minimum	PCT.			10.00
Average	"			15.00
Maximum	"			25.00
$100,000 project				
Minimum	PCT.			10.00
Average	"			12.00
Maximum	"			20.00
$500,000 project				
Minimum	PCT.			5.00
Average	"			10.00
Maximum	"			15.00
$1,000,000 project				
Minimum	PCT.			3.00
Average	"			8.00
Maximum	"			15.00
Professional fees				
Architectural				
$100,000 project				
Minimum	PCT.			5.00
Average	"			10.00
Maximum	"			20.00
$500,000 project				
Minimum	PCT.			5.00
Average	"			8.00
Maximum	"			12.00
$1,000,000 project				
Minimum	PCT.			3.50
Average	"			7.00
Maximum	"			10.00
Structural engineering				

01 GENERAL

Requirements	UNIT	MAT.	INST.	TOTAL
01020.10 Allowances *(Cont.)*				
Minimum	PCT.			2.00
Average	"			3.00
Maximum	"			5.00
Mechanical engineering				
Minimum	PCT.			4.00
Average	"			5.00
Maximum	"			15.00
Electrical engineering				
Minimum	PCT.			3.00
Average	"			5.00
Maximum	"			12.00
Taxes				
Sales tax				
Minimum	PCT.			4.00
Average	"			5.00
Maximum	"			10.00
Unemployment				
Minimum	PCT.			3.00
Average	"			6.50
Maximum	"			8.00
Social security (FICA)	"			7.85
01050.10 Field Staff				
Superintendent				
Minimum	YEAR			75,053
Average	"			93,100
Maximum	"			113,203
Field engineer				
Minimum	YEAR			73,868
Average	"			84,992
Maximum	"			97,053
Foreman				
Minimum	YEAR			49,762
Average	"			79,475
Maximum	"			93,687
Bookkeeper/timekeeper				
Minimum	YEAR			29,029
Average	"			37,921
Maximum	"			47,930
Watchman				
Minimum	YEAR			21,563
Average	"			28,590
Maximum	"			36,000
01310.10 Scheduling				
Scheduling for				
$100,000 project				
Minimum	PCT.			1.00
Average	"			2.00
Maximum	"			5.00
$500,000 project				
Minimum	PCT.			0.50
Average	"			1.00

01 GENERAL

Requirements	UNIT	MAT.	INST.	TOTAL
01310.10 Scheduling *(Cont.)*				
Maximum	PCT.			2.00
$1,000,000 project				
Minimum	PCT.			0.33
Average	"			0.75
Maximum	"			1.50
Scheduling software				
Minimum	EA.			550
Average	"			3,130
Maximum	"			62,600
01330.10 Surveying				
Surveying				
Small crew	DAY		950	950
Average crew	"		1,440	1,440
Large crew	"		1,900	1,900
Lot lines and boundaries				
Minimum	ACRE		680	680
Average	"		1,440	1,440
Maximum	"		2,370	2,370
01380.10 Job Requirements				
Job photographs, small jobs				
Minimum	EA.			140
Average	"			210
Maximum	"			500
Large projects				
Minimum	EA.			720
Average	"			1,070
Maximum	"			3,580
01410.10 Testing				
Testing concrete, per test				
Minimum	EA.			24.25
Average	"			40.50
Maximum	"			81.00
Soil, per test				
Minimum	EA.			50.00
Average	"			120
Maximum	"			330
Welding, per test				
Minimum	EA.			24.50
Average	"			40.75
Maximum	"			160
01500.10 Temporary Facilities				
Barricades, temporary				
Highway				
Concrete	L.F.	14.50	4.74	19.24
Wood	"	4.73	1.89	6.62
Steel	"	5.11	1.58	6.69
Pedestrian barricades				
Plywood	S.F.	3.80	1.58	5.38
Chain link fence	"	3.77	1.58	5.35

01 GENERAL Requirements

01500.10 Temporary Facilities (Cont.)

Item	UNIT	MAT.	INST.	TOTAL
Trailers, general office type, per month				
Minimum	EA.			240
Average	"			410
Maximum	"			810
Crew change trailers, per month				
Minimum	EA.			150
Average	"			160
Maximum	"			240

01505.10 Mobilization

Item	UNIT	MAT.	INST.	TOTAL
Equipment mobilization				
Bulldozer				
Minimum	EA.			230
Average	"			480
Maximum	"			810
Backhoe/front-end loader				
Minimum	EA.			140
Average	"			240
Maximum	"			530
Crane, crawler type				
Minimum	EA.			2,540
Average	"			6,240
Maximum	"			13,390
Truck crane				
Minimum	EA.			580
Average	"			900
Maximum	"			1,550
Pile driving rig				
Minimum	EA.			11,550
Average	"			23,100
Maximum	"			41,570

01525.10 Construction Aids

Item	UNIT	MAT.	INST.	TOTAL
Scaffolding/staging, rent per month				
Measured by lineal feet of base				
10' high	L.F.			14.75
20' high	"			26.50
30' high	"			37.25
40' high	"			42.75
50' high	"			51.00
Measured by square foot of surface				
Minimum	S.F.			0.65
Average	"			1.11
Maximum	"			2.00
Safety nets, heavy duty, per job				
Minimum	S.F.			0.42
Average	"			0.52
Maximum	"			1.13
Tarpaulins, fabric, per job				
Minimum	S.F.			0.29
Average	"			0.50
Maximum	"			1.31

01 GENERAL

Requirements	UNIT	MAT.	INST.	TOTAL
01570.10 Signs				
Construction signs, temporary				
Signs, 2' x 4'				
Minimum	EA.			42.75
Average	"			100
Maximum	"			360
Signs, 4' x 8'				
Minimum	EA.			90.00
Average	"			230
Maximum	"			1,000
Signs, 8' x 8'				
Minimum	EA.			120
Average	"			360
Maximum	"			3,640
01600.10 Equipment				
Air compressor				
60 cfm				
By day	EA.			110
By week	"			320
By month	"			970
300 cfm				
By day	EA.			220
By week	"			700
By month	"			2,120
600 cfm				
By day	EA.			620
By week	"			1,840
By month	"			5,590
Air tools, per compressor, per day				
Minimum	EA.			44.25
Average	"			55.00
Maximum	"			77.00
Generators, 5 kw				
By day	EA.			110
By week	"			330
By month	"			1,020
Heaters, salamander type, per week				
Minimum	EA.			130
Average	"			190
Maximum	"			400
Pumps, submersible				
50 gpm				
By day	EA.			89.00
By week	"			260
By month	"			790
100 gpm				
By day	EA.			110
By week	"			330
By month	"			990
500 gpm				
By day	EA.			180
By week	"			530
By month	"			1,590

01 GENERAL

Requirements

01600.10 Equipment (Cont.)

Requirements	UNIT	MAT.	INST.	TOTAL
Diaphragm pump, by week				
Minimum	EA.			150
Average	"			260
Maximum	"			550
Pickup truck				
By day	EA.			170
By week	"			490
By month	"			1,500
Dump truck				
6 cy truck				
By day	EA.			440
By week	"			1,330
By month	"			3,980
10 cy truck				
By day	EA.			550
By week	"			1,660
By month	"			4,980
16 cy truck				
By day	EA.			880
By week	"			2,650
By month	"			7,970
Backhoe, track mounted				
1/2 cy capacity				
By day	EA.			910
By week	"			2,770
By month	"			8,190
1 cy capacity				
By day	EA.			1,440
By week	"			4,310
By month	"			12,950
2 cy capacity				
By day	EA.			2,430
By week	"			7,300
By month	"			21,910
3 cy capacity				
By day	EA.			4,650
By week	"			13,950
By month	"			41,840
Backhoe/loader, rubber tired				
1/2 cy capacity				
By day	EA.			550
By week	"			1,660
By month	"			4,980
3/4 cy capacity				
By day	EA.			660
By week	"			1,990
By month	"			5,980
Bulldozer				
75 hp				
By day	EA.			770
By week	"			2,320
By month	"			6,970
200 hp				

01 GENERAL

01600.10 Equipment (Cont.)

Requirements	UNIT	MAT.	INST.	TOTAL
By day	EA.			2,210
By week	"			6,640
By month	"			19,920
400 hp				
By day	EA.			3,320
By week	"			9,960
By month	"			29,880
Cranes, crawler type				
15 ton capacity				
By day	EA.			990
By week	"			2,990
By month	"			8,960
25 ton capacity				
By day	EA.			1,220
By week	"			3,650
By month	"			10,950
50 ton capacity				
By day	EA.			2,210
By week	"			6,640
By month	"			19,920
100 ton capacity				
By day	EA.			3,320
By week	"			9,960
By month	"			29,980
Truck mounted, hydraulic				
15 ton capacity				
By day	EA.			940
By week	"			2,820
By month	"			8,140
Loader, rubber tired				
1 cy capacity				
By day	EA.			660
By week	"			1,990
By month	"			5,980
2 cy capacity				
By day	EA.			990
By week	"			3,870
By month	"			11,620
3 cy capacity				
By day	EA.			1,770
By week	"			5,310
By month	"			15,940

01740.10 Bonds

Requirements	UNIT	MAT.	INST.	TOTAL
Performance bonds				
Minimum	PCT.			0.64
Average	"			2.00
Maximum	"			3.18

02 SITE CONSTRUCTION

Site Remediation	UNIT	MAT.	INST.	TOTAL
02115.60 Underground Storage Tank Removal				
Remove underground storage tank, and backfill				
50 to 250 gals	EA.		820	820
600 gals	"		820	820
1000 gals	"		1,230	1,230
4000 gals	"		1,970	1,970
5000 gals	"		1,970	1,970
10,000 gals	"		3,280	3,280
12,000 gals	"		4,100	4,100
15,000 gals	"		4,920	4,920
20,000 gals	"		6,150	6,150
02115.66 Septic Tank Removal				
Remove septic tank				
1000 gals	EA.		200	200
2000 gals	"		250	250
5000 gals	"		310	310
15,000 gals	"		2,460	2,460
25,000 gals	"		3,280	3,280
40,000 gals	"		4,920	4,920

Site Preparation	UNIT	MAT.	INST.	TOTAL
02210.10 Soil Boring				
Borings, uncased, stable earth				
2-1/2" dia.				
Minimum	L.F.		20.50	20.50
Average	"		30.75	30.75
Maximum	"		49.25	49.25
4" dia.				
Minimum	L.F.		22.25	22.25
Average	"		35.25	35.25
Maximum	"		62.00	62.00
Cased, including samples				
2-1/2" dia.				
Minimum	L.F.		24.50	24.50
Average	"		41.00	41.00
Maximum	"		82.00	82.00
4" dia.				
Minumum	L.F.		49.25	49.25
Average	"		70.00	70.00
Maximum	"		98.00	98.00
Drilling in rock				
No sampling				
Minimum	L.F.		44.75	44.75
Average	"		65.00	65.00
Maximum	"		88.00	88.00

02 SITE CONSTRUCTION

Site Preparation	UNIT	MAT.	INST.	TOTAL
02210.10 **Soil Boring** *(Cont.)*				
With casing and sampling				
Minimum	L.F.		62.00	62.00
Average	"		82.00	82.00
Maximum	"		120	120
Test pits				
Light soil				
Minimum	EA.		310	310
Average	"		410	410
Maximum	"		820	820
Heavy soil				
Minimum	EA.		490	490
Average	"		620	620
Maximum	"		1,230	1,230

Demolition	UNIT	MAT.	INST.	TOTAL
02220.10 **Complete Building Demolition**				
Wood frame	C.F.		0.34	0.34
Concrete	"		0.52	0.52
Steel frame	"		0.69	0.69
02220.15 **Selective Building Demolition**				
Partition removal				
Concrete block partitions				
4" thick	S.F.		2.37	2.37
8" thick	"		3.16	3.16
12" thick	"		4.30	4.30
Brick masonry partitions				
4" thick	S.F.		2.37	2.37
8" thick	"		2.96	2.96
12" thick	"		3.95	3.95
16" thick	"		5.92	5.92
Cast in place concrete partitions				
Unreinforced				
6" thick	S.F.		16.50	16.50
8" thick	"		17.50	17.50
10" thick	"		20.50	20.50
12" thick	"		24.50	24.50
Reinforced				
6" thick	S.F.		19.00	19.00
8" thick	"		24.50	24.50
10" thick	"		27.25	27.25
12" thick	"		32.75	32.75
Terra cotta				
To 6" thick	S.F.		2.37	2.37
Stud partitions				

02 SITE CONSTRUCTION

02220.15 Selective Building Demolition (Cont.)

Demolition	UNIT	MAT.	INST.	TOTAL
Metal or wood, with drywall both sides	S.F.		2.37	2.37
Metal studs, both sides, lath and plaster	"		3.16	3.16
Door and frame removal				
Hollow metal in masonry wall				
Single				
2'6"x6'8"	EA.		59.00	59.00
3'x7'	"		79.00	79.00
Double				
3'x7'	EA.		95.00	95.00
4'x8'	"		95.00	95.00
Wood in framed wall				
Single				
2'6"x6'8"	EA.		33.75	33.75
3'x6'8"	"		39.50	39.50
Double				
2'6"x6'8"	EA.		47.50	47.50
3'x6'8"	"		53.00	53.00
Remove for re-use				
Hollow metal	EA.		120	120
Wood	"		79.00	79.00
Floor removal				
Brick flooring	S.F.		1.89	1.89
Ceramic or quarry tile	"		1.05	1.05
Terrazzo	"		2.10	2.10
Heavy wood	"		1.26	1.26
Residential wood	"		1.35	1.35
Resilient tile or linoleum	"		0.47	0.47
Ceiling removal				
Acoustical tile ceiling				
Adhesive fastened	S.F.		0.47	0.47
Furred and glued	"		0.39	0.39
Suspended grid	"		0.29	0.29
Drywall ceiling				
Furred and nailed	S.F.		0.52	0.52
Nailed to framing	"		0.47	0.47
Plastered ceiling				
Furred on framing	S.F.		1.18	1.18
Suspended system	"		1.58	1.58
Roofing removal				
Steel frame				
Corrugated metal roofing	S.F.		0.94	0.94
Built-up roof on metal deck	"		1.58	1.58
Wood frame				
Built up roof on wood deck	S.F.		1.45	1.45
Roof shingles	"		0.79	0.79
Roof tiles	"		1.58	1.58
Concrete frame	C.F.		3.16	3.16
Concrete plank	S.F.		2.37	2.37
Built-up roof on concrete	"		1.35	1.35
Cut-outs				
Concrete, elevated slabs, mesh reinforcing				
Under 5 cf	C.F.		47.50	47.50
Over 5 cf	"		39.50	39.50

02 SITE CONSTRUCTION

Demolition

02220.15 Selective Building Demolition *(Cont.)*

	UNIT	MAT.	INST.	TOTAL
Bar reinforcing				
Under 5 cf	C.F.		79.00	79.00
Over 5 cf	"		59.00	59.00
Window removal				
Metal windows, trim included				
2'x3'	EA.		47.50	47.50
2'x4'	"		53.00	53.00
2'x6'	"		59.00	59.00
3'x4'	"		59.00	59.00
3'x6'	"		68.00	68.00
3'x8'	"		79.00	79.00
4'x4'	"		79.00	79.00
4'x6'	"		95.00	95.00
4'x8'	"		120	120
Wood windows, trim included				
2'x3'	EA.		26.25	26.25
2'x4'	"		28.00	28.00
2'x6'	"		29.50	29.50
3'x4'	"		31.50	31.50
3'x6'	"		33.75	33.75
3'x8'	"		36.50	36.50
6'x4'	"		39.50	39.50
6'x6'	"		43.00	43.00
6'x8'	"		47.50	47.50
Walls, concrete, bar reinforcing				
Small jobs	C.F.		31.50	31.50
Large jobs	"		26.25	26.25
Brick walls, not including toothing				
4" thick	S.F.		2.37	2.37
8" thick	"		2.96	2.96
12" thick	"		3.95	3.95
16" thick	"		5.92	5.92
Concrete block walls, not including toothing				
4" thick	S.F.		2.63	2.63
6" thick	"		2.78	2.78
8" thick	"		2.96	2.96
10" thick	"		3.38	3.38
12" thick	"		3.95	3.95
Rubbish handling				
Load in dumpster or truck				
Minimum	C.F.		1.05	1.05
Maximum	"		1.58	1.58
For use of elevators, add				
Minimum	C.F.		0.23	0.23
Maximum	"		0.47	0.47
Rubbish hauling				
Hand loaded on trucks, 2 mile trip	C.Y.		37.50	37.50
Machine loaded on trucks, 2 mile trip	"		24.50	24.50

02 SITE CONSTRUCTION

Selective Site Demolition	UNIT	MAT.	INST.	TOTAL
02225.10 — Catch Basin / Manhole Demolition				
Abandon catch basin or manhole (fill with sand)				
Minimum	EA.		310	310
Average	"		490	490
Maximum	"		820	820
Remove and reset frame and cover				
Minimum	EA.		160	160
Average	"		250	250
Maximum	"		410	410
Remove catch basin, to 10' deep				
Masonry				
Minimum	EA.		490	490
Average	"		620	620
Maximum	"		820	820
Concrete				
Minimum	EA.		620	620
Average	"		820	820
Maximum	"		980	980
02225.13 — Core Drilling				
Concrete				
6" thick				
3" dia.	EA.		43.25	43.25
4" dia.	"		50.00	50.00
6" dia.	"		60.00	60.00
8" dia.	"		100	100
8" thick				
3" dia.	EA.		60.00	60.00
4" dia.	"		76.00	76.00
6" dia.	"		86.00	86.00
8" dia.	"		120	120
10" thick				
3" dia.	EA.		76.00	76.00
4" dia.	"		86.00	86.00
6" dia.	"		100	100
8" dia.	"		150	150
12" thick				
3" dia.	EA.		100	100
4" dia.	"		120	120
6" dia.	"		150	150
8" dia.	"		200	200
02225.15 — Curb & Gutter Demolition				
Curb removal				
Concrete, unreinforced				
Minimum	L.F.		4.92	4.92
Average	"		6.15	6.15
Maximum	"		7.68	7.68
Reinforced				
Minimum	L.F.		7.93	7.93
Average	"		8.78	8.78
Maximum	"		9.84	9.84
Combination curb and 2' gutter				
Unreinforced				

02 SITE CONSTRUCTION

Selective Site Demolition	UNIT	MAT.	INST.	TOTAL
02225.15 Curb & Gutter Demolition *(Cont.)*				
Minimum	L.F.		6.47	6.47
Average	"		8.48	8.48
Maximum	"		12.25	12.25
Reinforced				
Minimum	L.F.		10.25	10.25
Average	"		13.75	13.75
Maximum	"		24.50	24.50
Granite curb				
Minimum	L.F.		7.02	7.02
Average	"		8.20	8.20
Maximum	"		9.46	9.46
Asphalt curb				
Minimum	L.F.		4.10	4.10
Average	"		4.92	4.92
Maximum	"		5.85	5.85
02225.20 Fence Demolition				
Remove fencing				
Chain link, 8' high				
For disposal	L.F.		2.37	2.37
For reuse	"		5.92	5.92
Wood				
4' high	S.F.		1.58	1.58
6' high	"		1.89	1.89
8' high	"		2.37	2.37
Masonry				
8" thick				
4' high	S.F.		4.74	4.74
6' high	"		5.92	5.92
8' high	"		6.77	6.77
12" thick				
4' high	S.F.		7.90	7.90
6' high	"		9.48	9.48
8' high	"		11.75	11.75
12' high	"		15.75	15.75
02225.25 Guardrail Demolition				
Remove standard guardrail				
Steel				
Minimum	L.F.		6.15	6.15
Average	"		8.20	8.20
Maximum	"		12.25	12.25
Wood				
Minimum	L.F.		5.34	5.34
Average	"		6.30	6.30
Maximum	"		10.25	10.25
02225.30 Hydrant Demolition				
Remove fire hydrant				
Minimum	EA.		310	310
Average	"		410	410
Maximum	"		620	620
Remove and reset fire hydrant				

02 SITE CONSTRUCTION

Selective Site Demolition	UNIT	MAT.	INST.	TOTAL
02225.30 **Hydrant Demolition** *(Cont.)*				
Minimum	EA.		820	820
Average	"		1,230	1,230
Maximum	"		2,460	2,460
02225.40 **Pavement And Sidewalk Demolition**				
Bituminous pavement, up to 3" thick				
On streets				
Minimum	S.Y.		7.02	7.02
Average	"		9.84	9.84
Maximum	"		16.50	16.50
On pipe trench				
Minimum	S.Y.		9.84	9.84
Average	"		12.25	12.25
Maximum	"		24.50	24.50
Concrete pavement, 6" thick				
No reinforcement				
Minimum	S.Y.		12.25	12.25
Average	"		16.50	16.50
Maximum	"		24.50	24.50
With wire mesh				
Minimum	S.Y.		19.00	19.00
Average	"		24.50	24.50
Maximum	"		30.75	30.75
With rebars				
Minimum	S.Y.		24.50	24.50
Average	"		30.75	30.75
Maximum	"		41.00	41.00
9" thick				
No reinforcement				
Minimum	S.Y.		16.50	16.50
Average	"		20.50	20.50
Maximum	"		24.50	24.50
With wire mesh				
Minimum	S.Y.		26.00	26.00
Average	"		30.75	30.75
Maximum	"		37.75	37.75
With rebars				
Minimum	S.Y.		32.75	32.75
Average	"		41.00	41.00
Maximum	"		55.00	55.00
12" thick				
No reinforcement				
Minimum	S.Y.		20.50	20.50
Average	"		24.50	24.50
Maximum	"		30.75	30.75
With wire mesh				
Minimum	S.Y.		29.00	29.00
Average	"		35.25	35.25
Maximum	"		44.75	44.75
With rebars				
Minimum	S.Y.		41.00	41.00
Average	"		49.25	49.25
Maximum	"		62.00	62.00

02 SITE CONSTRUCTION

Selective Site Demolition	UNIT	MAT.	INST.	TOTAL
02225.40 **Pavement And Sidewalk Demolition** (Cont.)				
Sidewalk, 4" thick, with disposal				
Minimum	S.Y.		5.85	5.85
Average	"		8.20	8.20
Maximum	"		11.75	11.75
Removal of pavement markings by waterblasting				
Minimum	S.F.		0.19	0.19
Average	"		0.23	0.23
Maximum	"		0.47	0.47
02225.42 **Drainage Piping Demolition**				
Remove drainage pipe, not including excavation				
12" dia.				
Minimum	L.F.		8.20	8.20
Average	"		10.25	10.25
Maximum	"		13.00	13.00
18" dia.				
Minimum	L.F.		11.25	11.25
Average	"		13.00	13.00
Maximum	"		16.50	16.50
24" dia.				
Minimum	L.F.		13.75	13.75
Average	"		16.50	16.50
Maximum	"		20.50	20.50
36" dia.				
Minimum	L.F.		16.50	16.50
Average	"		20.50	20.50
Maximum	"		26.00	26.00
02225.43 **Gas Piping Demolition**				
Remove welded steel pipe, not including excavation				
4" dia.				
Minimum	L.F.		12.25	12.25
Average	"		15.25	15.25
Maximum	"		20.50	20.50
5" dia.				
Minimum	L.F.		20.50	20.50
Average	"		24.50	24.50
Maximum	"		30.75	30.75
6" dia.				
Minimum	L.F.		26.00	26.00
Average	"		30.75	30.75
Maximum	"		41.00	41.00
8" dia.				
Minimum	L.F.		37.75	37.75
Average	"		49.25	49.25
Maximum	"		65.00	65.00
10" dia.				
Minimum	L.F.		49.25	49.25
Average	"		62.00	62.00
Maximum	"		82.00	82.00

02 SITE CONSTRUCTION

Selective Site Demolition	UNIT	MAT.	INST.	TOTAL
02225.45 **Sanitary Piping Demolition**				
Remove sewer pipe, not including excavation				
4" dia.				
Minimum	L.F.		6.83	6.83
Average	"		9.84	9.84
Maximum	"		16.50	16.50
6" dia.				
Minimum	L.F.		7.68	7.68
Average	"		11.25	11.25
Maximum	"		20.50	20.50
8" dia.				
Minimum	L.F.		8.20	8.20
Average	"		12.25	12.25
Maximum	"		24.50	24.50
10" dia.				
Minimum	L.F.		8.78	8.78
Average	"		13.00	13.00
Maximum	"		27.25	27.25
12" dia.				
Minimum	L.F.		9.46	9.46
Average	"		13.75	13.75
Maximum	"		30.75	30.75
15" dia.				
Minimum	L.F.		10.25	10.25
Average	"		14.50	14.50
Maximum	"		35.25	35.25
18" dia.				
Minimum	L.F.		11.25	11.25
Average	"		16.50	16.50
Maximum	"		41.00	41.00
24" dia.				
Minimum	L.F.		12.25	12.25
Average	"		20.50	20.50
Maximum	"		49.25	49.25
30" dia.				
Minimum	L.F.		13.75	13.75
Average	"		24.50	24.50
Maximum	"		62.00	62.00
36" dia.				
Minimum	L.F.		16.50	16.50
Average	"		30.75	30.75
Maximum	"		82.00	82.00
02225.48 **Water Piping Demolition**				
Remove water pipe, not including excavation				
4" dia.				
Minimum	L.F.		9.84	9.84
Average	"		11.25	11.25
Maximum	"		13.00	13.00
6" dia.				
Minimum	L.F.		10.25	10.25
Average	"		11.75	11.75
Maximum	"		13.75	13.75
8" dia.				

02 SITE CONSTRUCTION

Selective Site Demolition

02225.48 Water Piping Demolition (Cont.)

	UNIT	MAT.	INST.	TOTAL
Minimum	L.F.		11.25	11.25
Average	"		13.00	13.00
Maximum	"		15.25	15.25
10" dia.				
Minimum	L.F.		11.75	11.75
Average	"		13.75	13.75
Maximum	"		16.50	16.50
12" dia.				
Minimum	L.F.		12.25	12.25
Average	"		14.50	14.50
Maximum	"		17.50	17.50
14" dia.				
Minimum	L.F.		13.00	13.00
Average	"		15.25	15.25
Maximum	"		19.00	19.00
16" dia.				
Minimum	L.F.		13.75	13.75
Average	"		16.50	16.50
Maximum	"		20.50	20.50
18" dia.				
Minimum	L.F.		14.50	14.50
Average	"		17.50	17.50
Maximum	"		22.25	22.25
20" dia.				
Minimum	L.F.		15.25	15.25
Average	"		19.00	19.00
Maximum	"		24.50	24.50
Remove valves				
6"	EA.		120	120
10"	"		140	140
14"	"		150	150
18"	"		200	200

02225.50 Saw Cutting Pavement

	UNIT	MAT.	INST.	TOTAL
Pavement, bituminous				
2" thick	L.F.		1.88	1.88
3" thick	"		2.35	2.35
4" thick	"		2.89	2.89
5" thick	"		3.13	3.13
6" thick	"		3.35	3.35
Concrete pavement, with wire mesh				
4" thick	L.F.		3.61	3.61
5" thick	"		3.91	3.91
6" thick	"		4.27	4.27
8" thick	"		4.70	4.70
10" thick	"		5.22	5.22
Plain concrete, unreinforced				
4" thick	L.F.		3.13	3.13
5" thick	"		3.61	3.61
6" thick	"		3.91	3.91
8" thick	"		4.27	4.27
10" thick	"		4.70	4.70

02 SITE CONSTRUCTION

Selective Site Demolition

	UNIT	MAT.	INST.	TOTAL
02225.80 Wall, Exterior, Demolition				
Concrete wall				
Light reinforcing				
6" thick	S.F.		12.25	12.25
8" thick	"		13.00	13.00
10" thick	"		13.75	13.75
12" thick	"		15.25	15.25
Medium reinforcing				
6" thick	S.F.		13.00	13.00
8" thick	"		13.75	13.75
10" thick	"		15.25	15.25
12" thick	"		17.50	17.50
Heavy reinforcing				
6" thick	S.F.		14.50	14.50
8" thick	"		15.25	15.25
10" thick	"		17.50	17.50
12" thick	"		20.50	20.50
Masonry				
No reinforcing				
8" thick	S.F.		5.46	5.46
12" thick	"		6.15	6.15
16" thick	"		7.02	7.02
Horizontal reinforcing				
8" thick	S.F.		6.15	6.15
12" thick	"		6.64	6.64
16" thick	"		7.93	7.93
Vertical reinforcing				
8" thick	S.F.		7.93	7.93
12" thick	"		9.11	9.11
16" thick	"		11.25	11.25
Remove concrete headwall				
15" pipe	EA.		180	180
18" pipe	"		200	200
24" pipe	"		220	220
30" pipe	"		250	250
36" pipe	"		270	270
48" pipe	"		350	350
60" pipe	"		490	490

Site Clearing

	UNIT	MAT.	INST.	TOTAL
02230.10 Clear Wooded Areas				
Clear wooded area				
Light density	ACRE		6,150	6,150
Medium density	"		8,200	8,200
Heavy density	"		9,840	9,840

02 SITE CONSTRUCTION

Site Clearing

02230.50 Tree Cutting & Clearing

	UNIT	MAT.	INST.	TOTAL
Cut trees and clear out stumps				
9" to 12" dia.	EA.		490	490
To 24" dia.	"		620	620
24" dia. and up	"		820	820
Loading and trucking				
For machine load, per load, round trip				
1 mile	EA.		98.00	98.00
3 mile	"		110	110
5 mile	"		120	120
10 mile	"		160	160
20 mile	"		250	250
Hand loaded, round trip				
1 mile	EA.		240	240
3 mile	"		270	270
5 mile	"		310	310
10 mile	"		380	380
20 mile	"		470	470
Tree trimming for pole line construction				
Light cutting	L.F.		1.23	1.23
Medium cutting	"		1.64	1.64
Heavy cutting	"		2.46	2.46

Dewatering

02240.10 Wellpoint Systems

	UNIT	MAT.	INST.	TOTAL
Pumping, gas driven, 50' hose				
3" header pipe	DAY		940	940
6" header pipe	"		1,180	1,180
Wellpoint system per job; 150' length of PVC header				
6" header pipe, 2"wellpoints, 5' centers	L.F.	62.00	3.76	65.76
8" header pipe	"	74.00	4.70	78.70
10" header pipe	"	110	6.26	116
Jetting wellpoint system				
14' long	EA.	84.00	63.00	147
18' long	"	96.00	78.00	174
Sand filter for wellpoints	L.F.	4.03	1.56	5.59
Replacement of wellpoint components	EA.		18.75	18.75

02 SITE CONSTRUCTION

Shoring And Underpinning	UNIT	MAT.	INST.	TOTAL
02250.10 — Trench Sheeting				
Closed timber, including pull and salvage, excavation				
8' deep	S.F.	3.27	8.39	11.66
10' deep	"	3.35	8.83	12.18
12' deep	"	3.43	9.32	12.75
14' deep	"	3.52	9.87	13.39
16' deep	"	3.61	10.50	14.11
18' deep	"	3.73	12.00	15.73
20' deep	"	3.86	13.00	16.86
02260.10 — Cofferdams				
Cofferdam, steel, driven from shore				
15' deep	S.F.	21.75	20.75	42.50
20' deep	"	21.75	19.50	41.25
25' deep	"	21.75	18.25	40.00
30' deep	"	21.75	17.00	38.75
40' deep	"	21.75	16.25	38.00
Driven from barge				
20' deep	S.F.	21.75	22.50	44.25
30' deep	"	21.75	20.75	42.50
40' deep	"	21.75	19.50	41.25
50' deep	"	21.75	18.25	40.00
02260.70 — Steel Sheet Piling				
Steel sheet piling, 12" wide				
20' long	S.F.	22.50	14.50	37.00
35' long	"	22.50	10.50	33.00
50' long	"	22.50	7.27	29.77
Over 50' long	"	22.50	6.61	29.11

Earthwork, Excavation & Fill	UNIT	MAT.	INST.	TOTAL
02315.10 — Base Course				
Base course, crushed stone				
3" thick	S.Y.	4.35	0.66	5.01
4" thick	"	5.78	0.71	6.49
6" thick	"	8.64	0.77	9.41
8" thick	"	11.50	0.88	12.38
10" thick	"	14.50	0.94	15.44
12" thick	"	17.25	1.10	18.35
Base course, bank run gravel				
4" deep	S.Y.	3.57	0.69	4.26
6" deep	"	5.46	0.75	6.21
8" deep	"	7.21	0.82	8.03
10" deep	"	9.03	0.88	9.91
12" deep	"	10.75	1.01	11.76
Prepare and roll sub base				

02 SITE CONSTRUCTION

Earthwork, Excavation & Fill	UNIT	MAT.	INST.	TOTAL
02315.10	**Base Course** *(Cont.)*			
Minimum	S.Y.		0.66	0.66
Average	"		0.82	0.82
Maximum	"		1.10	1.10
02315.20	**Borrow**			
Borrow fill, F.O.B. at pit				
Sand, haul to site, round trip				
10 mile	C.Y.	24.25	13.25	37.50
20 mile	"	24.25	22.00	46.25
30 mile	"	24.25	33.25	57.50
Place borrow fill and compact				
Less than 1 in 4 slope	C.Y.	24.25	6.62	30.87
Greater than 1 in 4 slope	"	24.25	8.83	33.08
02315.30	**Bulk Excavation**			
Excavation, by small dozer				
Large areas	C.Y.		1.88	1.88
Small areas	"		3.13	3.13
Trim banks	"		4.70	4.70
Drag line				
1-1/2 cy bucket				
Sand or gravel	C.Y.		4.10	4.10
Light clay	"		5.46	5.46
Heavy clay	"		6.15	6.15
Unclassified	"		6.56	6.56
2 cy bucket				
Sand or gravel	C.Y.		3.78	3.78
Light clay	"		4.92	4.92
Heavy clay	"		5.46	5.46
Unclassified	"		5.78	5.78
2-1/2 cy bucket				
Sand or gravel	C.Y.		3.51	3.51
Light clay	"		4.47	4.47
Heavy clay	"		4.92	4.92
Unclassified	"		5.17	5.17
3 cy bucket				
Sand or gravel	C.Y.		3.07	3.07
Light clay	"		4.10	4.10
Heavy clay	"		4.47	4.47
Unclassified	"		4.68	4.68
Hydraulic excavator				
1 cy capacity				
Light material	C.Y.		4.10	4.10
Medium material	"		4.92	4.92
Wet material	"		6.15	6.15
Blasted rock	"		7.02	7.02
1-1/2 cy capacity				
Light material	C.Y.		1.65	1.65
Medium material	"		2.20	2.20
Wet material	"		2.65	2.65
Blasted rock	"		3.31	3.31
2 cy capacity				
Light material	C.Y.		1.47	1.47

02 SITE CONSTRUCTION

Earthwork, Excavation & Fill

02315.30 Bulk Excavation (Cont.)

Description	UNIT	MAT.	INST.	TOTAL
Medium material	C.Y.		1.89	1.89
Wet material	"		2.20	2.20
Blasted rock	"		2.65	2.65
Wheel mounted front-end loader				
7/8 cy capacity				
Light material	C.Y.		3.31	3.31
Medium material	"		3.78	3.78
Wet material	"		4.41	4.41
Blasted rock	"		5.30	5.30
1-1/2 cy capacity				
Light material	C.Y.		1.89	1.89
Medium material	"		2.03	2.03
Wet material	"		2.20	2.20
Blasted rock	"		2.41	2.41
2-1/2 cy capacity				
Light material	C.Y.		1.55	1.55
Medium material	"		1.65	1.65
Wet material	"		1.76	1.76
Blasted rock	"		1.89	1.89
3-1/2 cy capacity				
Light material	C.Y.		1.47	1.47
Medium material	"		1.55	1.55
Wet material	"		1.65	1.65
Blasted rock	"		1.76	1.76
6 cy capacity				
Light material	C.Y.		0.88	0.88
Medium material	"		0.94	0.94
Wet material	"		1.01	1.01
Blasted rock	"		1.10	1.10
Track mounted front-end loader				
1-1/2 cy capacity				
Light material	C.Y.		2.20	2.20
Medium material	"		2.41	2.41
Wet material	"		2.65	2.65
Blasted rock	"		2.94	2.94
2-3/4 cy capacity				
Light material	C.Y.		1.32	1.32
Medium material	"		1.47	1.47
Wet material	"		1.65	1.65
Blasted rock	"		1.89	1.89

02315.40 Building Excavation

Description	UNIT	MAT.	INST.	TOTAL
Structural excavation, unclassified earth				
3/8 cy backhoe	C.Y.		17.75	17.75
3/4 cy backhoe	"		13.25	13.25
1 cy backhoe	"		11.00	11.00
Foundation backfill and compaction by machine	"		26.50	26.50

02 SITE CONSTRUCTION

Earthwork, Excavation & Fill	UNIT	MAT.	INST.	TOTAL
02315.45 — Hand Excavation				
Excavation				
To 2' deep				
Normal soil	C.Y.		53.00	53.00
Sand and gravel	"		47.50	47.50
Medium clay	"		59.00	59.00
Heavy clay	"		68.00	68.00
Loose rock	"		79.00	79.00
To 6' deep				
Normal soil	C.Y.		68.00	68.00
Sand and gravel	"		59.00	59.00
Medium clay	"		79.00	79.00
Heavy clay	"		95.00	95.00
Loose rock	"		120	120
Backfilling foundation without compaction, 6" lifts	"		29.50	29.50
Compaction of backfill around structures or in trench				
By hand with air tamper	C.Y.		33.75	33.75
By hand with vibrating plate tamper	"		31.50	31.50
1 ton roller	"		47.00	47.00
Miscellaneous hand labor				
Trim slopes, sides of excavation	S.F.		0.07	0.07
Trim bottom of excavation	"		0.09	0.09
Excavation around obstructions and services	C.Y.		160	160
02315.50 — Roadway Excavation				
Roadway excavation				
1/4 mile haul	C.Y.		2.65	2.65
2 mile haul	"		4.41	4.41
5 mile haul	"		6.62	6.62
Excavation of open ditches	"		1.89	1.89
Trim banks, swales or ditches	S.Y.		2.20	2.20
Bulk swale excavation by dragline				
Small jobs	C.Y.		6.15	6.15
Large jobs	"		3.51	3.51
Spread base course	"		3.31	3.31
Roll and compact	"		4.41	4.41
02315.60 — Trenching				
Trenching and continuous footing excavation				
By gradall				
1 cy capacity				
Light soil	C.Y.		3.78	3.78
Medium soil	"		4.07	4.07
Heavy/wet soil	"		4.41	4.41
Loose rock	"		4.82	4.82
Blasted rock	"		5.09	5.09
By hydraulic excavator				
1/2 cy capacity				
Light soil	C.Y.		4.41	4.41
Medium soil	"		4.82	4.82
Heavy/wet soil	"		5.30	5.30
Loose rock	"		5.89	5.89
Blasted rock	"		6.62	6.62
1 cy capacity				

02 SITE CONSTRUCTION

Earthwork, Excavation & Fill

02315.60 Trenching *(Cont.)*

Description	UNIT	MAT.	INST.	TOTAL
Light soil	C.Y.		3.11	3.11
Medium soil	"		3.31	3.31
Heavy/wet soil	"		3.53	3.53
Loose rock	"		3.78	3.78
Blasted rock	"		4.07	4.07
1-1/2 cy capacity				
Light soil	C.Y.		2.79	2.79
Medium soil	"		2.94	2.94
Heavy/wet soil	"		3.11	3.11
Loose rock	"		3.31	3.31
Blasted rock	"		3.53	3.53
2 cy capacity				
Light soil	C.Y.		2.65	2.65
Medium soil	"		2.79	2.79
Heavy/wet soil	"		2.94	2.94
Loose rock	"		3.11	3.11
Blasted rock	"		3.31	3.31
2-1/2 cy capacity				
Light soil	C.Y.		2.41	2.41
Medium soil	"		2.52	2.52
Heavy/wet soil	"		2.65	2.65
Loose rock	"		2.79	2.79
Blasted rock	"		2.94	2.94
Trencher, chain, 1' wide to 4' deep				
Light soil	C.Y.		2.35	2.35
Medium soil	"		2.68	2.68
Heavy soil	"		3.13	3.13
Hand excavation				
Bulk, wheeled 100'				
Normal soil	C.Y.		53.00	53.00
Sand or gravel	"		47.50	47.50
Medium clay	"		68.00	68.00
Heavy clay	"		95.00	95.00
Loose rock	"		120	120
Trenches, up to 2' deep				
Normal soil	C.Y.		59.00	59.00
Sand or gravel	"		53.00	53.00
Medium clay	"		79.00	79.00
Heavy clay	"		120	120
Loose rock	"		160	160
Trenches, to 6' deep				
Normal soil	C.Y.		68.00	68.00
Sand or gravel	"		59.00	59.00
Medium clay	"		95.00	95.00
Heavy clay	"		160	160
Loose rock	"		240	240
Backfill trenches				
With compaction				
By hand	C.Y.		39.50	39.50
By 60 hp tracked dozer	"		2.35	2.35
By 200 hp tracked dozer	"		1.47	1.47
By small front-end loader	"		2.68	2.68
Spread dumped fill or gravel, no compaction				

02 SITE CONSTRUCTION

Earthwork, Excavation & Fill	UNIT	MAT.	INST.	TOTAL
02315.60 Trenching *(Cont.)*				
6" layers	S.Y.		1.56	1.56
12" layers	"		1.88	1.88
Compaction in 6" layers				
By hand with air tamper	S.Y.		1.20	1.20
Backfill trenches, sand bedding, no compaction				
By hand	C.Y.	24.25	39.50	63.75
By small front-end loader	"	24.25	3.78	28.03
02315.70 Utility Excavation				
Trencher, sandy clay, 8" wide trench				
18" deep	L.F.		2.08	2.08
24" deep	"		2.35	2.35
36" deep	"		2.68	2.68
Trench backfill, 95% compaction				
Tamp by hand	C.Y.		29.50	29.50
Vibratory compaction	"		23.75	23.75
Trench backfilling, with borrow sand, place & compact	"	24.25	23.75	48.00
02315.75 Gravel And Stone				
F.O.B. PLANT				
No. 21 crusher run stone	C.Y.			52.00
No. 26 crusher run stone	"			52.00
No. 57 stone	"			52.00
No. 67 gravel	"			39.00
No. 68 stone	"			52.00
No. 78 stone	"			52.00
No. 78 gravel, (pea gravel)	"			39.00
No. 357 or B-3 stone	"			52.00
Structural & foundation backfill				
No. 21 crusher run stone	TON			41.50
No. 26 crusher run stone	"			41.50
No. 57 stone	"			41.50
No. 67 gravel	"			31.25
No. 68 stone	"			41.50
No. 78 stone	"			41.50
No. 78 gravel, (pea gravel)	"			31.25
No. 357 or B-3 stone	"			41.50
02315.80 Hauling Material				
Haul material by 10 cy dump truck, round trip distance				
1 mile	C.Y.		5.22	5.22
2 mile	"		6.26	6.26
5 mile	"		8.54	8.54
10 mile	"		9.40	9.40
20 mile	"		10.50	10.50
30 mile	"		12.50	12.50
Site grading, cut & fill, sandy clay, 200' haul, 75 hp dozer	"		3.76	3.76
Spread topsoil by equipment on site	"		4.17	4.17
Site grading (cut and fill to 6") less than 1 acre				
75 hp dozer	C.Y.		6.26	6.26
1.5 cy backhoe/loader	"		9.40	9.40

02 SITE CONSTRUCTION

Soil Stabilization & Treatment	UNIT	MAT.	INST.	TOTAL
02340.05 — Soil Stabilization				
Straw bale secured with rebar	L.F.	8.91	1.58	10.49
Filter barrier, 18" high filter fabric	"	2.15	4.74	6.89
Sediment fence, 36" fabric with 6" mesh	"	5.10	5.92	11.02
Soil stabilization with tar paper, burlap, straw and stakes	S.F.	0.42	0.06	0.48
02340.30 — Geotextile				
Filter cloth, light reinforcement				
Woven				
12'-6" wide x 50' long	S.F.	0.40	0.06	0.46
Various lengths	"	0.61	0.06	0.67
Non-woven				
14'-8" wide x 430' long	S.F.	0.22	0.06	0.28
Various lengths	"	0.31	0.06	0.37
02360.20 — Soil Treatment				
Soil treatment, termite control pretreatment				
Under slabs	S.F.	0.45	0.26	0.71
By walls	"	0.45	0.31	0.76
02370.10 — Slope Protection				
Gabions, stone filled				
6" deep	S.Y.	32.50	23.50	56.00
9" deep	"	39.75	26.75	66.50
12" deep	"	53.00	31.25	84.25
18" deep	"	67.00	37.50	105
36" deep	"	120	63.00	183
02370.40 — Riprap				
Riprap				
Crushed stone blanket, max size 2-1/2"	TON	41.75	70.00	112
Stone, quarry run, 300 lb. stones	"	52.00	65.00	117
400 lb. stones	"	54.00	60.00	114
500 lb. stones	"	57.00	56.00	113
750 lb. stones	"	59.00	52.00	111
Dry concrete riprap in bags 3" thick, 80 lb. per bag	BAG	7.04	3.49	10.53

Tunneling, Boring & Jacking	UNIT	MAT.	INST.	TOTAL
02445.10 — Pipe Jacking				
Pipe casing, horizontal jacking				
18" dia.	L.F.	140	93.00	233
21" dia.	"	160	100	260
24" dia.	"	170	100	270
27" dia.	"	190	100	290
30" dia.	"	210	110	320
36" dia.	"	240	120	360

02 SITE CONSTRUCTION

Tunneling, Boring & Jacking	UNIT	MAT.	INST.	TOTAL
02445.10 Pipe Jacking *(Cont.)*				
42" dia.	L.F.	280	130	410
48" dia.	"	340	140	480

Piles And Caissons	UNIT	MAT.	INST.	TOTAL
02455.60 Steel Piles				
H-section piles				
8x8				
36 lb/ft				
30' long	L.F.	19.00	12.00	31.00
40' long	"	19.00	9.70	28.70
50' long	"	19.00	8.08	27.08
10x10				
42 lb/ft				
30' long	L.F.	22.25	12.00	34.25
40' long	"	22.25	9.70	31.95
50' long	"	22.25	8.08	30.33
57 lb/ft				
30' long	L.F.	30.00	12.00	42.00
40' long	"	30.00	9.70	39.70
50' long	"	30.00	8.08	38.08
12x12				
53 lb/ft				
30' long	L.F.	28.25	13.25	41.50
40' long	"	28.25	10.50	38.75
50' long	"	28.25	8.08	36.33
74 lb/ft				
30' long	L.F.	39.25	13.25	52.50
40' long	"	39.25	10.50	49.75
50' long	"	39.25	8.08	47.33
14x14				
73 lb/ft				
40' long	L.F.	39.00	13.25	52.25
50' long	"	39.00	10.50	49.50
60' long	"	39.00	8.08	47.08
89 lb/ft				
40' long	L.F.	47.25	13.25	60.50
50' long	"	47.25	10.50	57.75
60' long	"	47.25	8.08	55.33
102 lb/ft				
40' long	L.F.	54.00	13.25	67.25
50' long	"	54.00	10.50	64.50
60' long	"	54.00	8.08	62.08
117 lb/ft				
40' long	L.F.	62.00	13.75	75.75
50' long	"	62.00	10.75	72.75

02 SITE CONSTRUCTION

Piles And Caissons	UNIT	MAT.	INST.	TOTAL
02455.60 — **Steel Piles** (Cont.)				
60' long	L.F.	62.00	8.31	70.31
Splice				
8"	EA.	110	79.00	189
10"	"	120	95.00	215
12"	"	160	95.00	255
14"	"	210	120	330
Driving cap				
8"	EA.	58.00	47.50	106
10"	"	58.00	59.00	117
12"	"	58.00	59.00	117
14"	"	58.00	68.00	126
Standard point				
8"	EA.	78.00	47.50	126
10"	"	93.00	59.00	152
12"	"	110	68.00	178
14"	"	120	79.00	199
Heavy duty point				
8"	EA.	66.00	53.00	119
10"	"	78.00	68.00	146
12"	"	100	79.00	179
14"	"	130	95.00	225
Tapered friction piles, fluted casing, up to 50'				
With 4000 psi concrete no reinforcing				
12" dia.	L.F.	21.25	7.27	28.52
14" dia.	"	24.50	7.46	31.96
16" dia.	"	29.25	7.65	36.90
18" dia.	"	33.00	8.56	41.56
02455.65 — **Steel Pipe Piles**				
Concrete filled, 3000# concrete, up to 40'				
8" dia.	L.F.	26.00	10.50	36.50
10" dia.	"	33.50	10.75	44.25
12" dia.	"	38.75	11.25	50.00
14" dia.	"	42.50	11.75	54.25
16" dia.	"	48.50	12.00	60.50
18" dia.	"	67.00	12.75	79.75
Pipe piles, non-filled				
8" dia.	L.F.	24.50	8.08	32.58
10" dia.	"	30.75	8.31	39.06
12" dia.	"	37.75	8.56	46.31
14" dia.	"	39.75	9.09	48.84
16" dia.	"	45.25	9.38	54.63
18" dia.	"	59.00	9.70	68.70
Splice				
8" dia.	EA.	110	95.00	205
10" dia.	"	120	95.00	215
12" dia.	"	130	120	250
14" dia.	"	140	120	260
16" dia.	"	180	160	340
18" dia.	"	230	160	390
Standard point				
8" dia.	EA.	150	95.00	245
10" dia.	"	200	95.00	295

02 SITE CONSTRUCTION

Piles And Caissons	UNIT	MAT.	INST.	TOTAL
02455.65 **Steel Pipe Piles** *(Cont.)*				
12" dia.	EA.	210	120	330
14" dia.	"	220	120	340
16" dia.	"	290	160	450
18" dia.	"	400	160	560
Heavy duty point				
8" dia.	EA.	260	120	380
10" dia.	"	360	120	480
12" dia.	"	390	160	550
14" dia.	"	530	160	690
16" dia.	"	530	190	720
18" dia.	"	580	190	770
02455.80 **Wood And Timber Piles**				
Treated wood piles, 12" butt, 8" tip				
25' long	L.F.	11.50	14.50	26.00
30' long	"	12.25	12.00	24.25
35' long	"	12.25	10.50	22.75
40' long	"	12.25	9.09	21.34
12" butt, 7" tip				
40' long	L.F.	13.75	9.09	22.84
45' long	"	13.75	8.08	21.83
50' long	"	15.50	7.27	22.77
55' long	"	15.50	6.61	22.11
60' long	"	15.50	6.06	21.56
02455.90 **Pile Testing**				
Pile test				
50 ton to 100 ton	EA.			23,270
To 200 ton	"			32,830
To 300 ton	"			37,960
To 400 ton	"			44,460
To 600 ton	"			55,580
02465.50 **Prestressed Piling**				
Prestressed concrete piling, less than 60' long				
10" sq.	L.F.	18.50	6.06	24.56
12" sq.	"	25.75	6.32	32.07
14" sq.	"	27.00	6.46	33.46
16" sq.	"	33.25	6.61	39.86
18" sq.	"	45.75	7.09	52.84
20" sq.	"	62.00	7.27	69.27
24" sq.	"	79.00	7.46	86.46
More than 60' long				
12" sq.	L.F.	26.50	5.19	31.69
14" sq.	"	29.00	5.29	34.29
16" sq.	"	35.00	5.38	40.38
18" sq.	"	46.00	5.49	51.49
20" sq.	"	62.00	5.59	67.59
24" sq.	"	74.00	5.70	79.70
Straight cylinder, less than 60' long				
12" dia.	L.F.	24.00	6.61	30.61
14" dia.	"	32.50	6.76	39.26
16" dia.	"	39.50	6.92	46.42

02 SITE CONSTRUCTION

Piles And Caissons	UNIT	MAT.	INST.	TOTAL
02465.50 **Prestressed Piling** *(Cont.)*				
18" dia.	L.F.	50.00	7.09	57.09
20" dia.	"	59.00	7.27	66.27
24" dia.	"	74.00	7.46	81.46
More than 60' long				
12" dia.	L.F.	24.00	5.29	29.29
14" dia.	"	32.50	5.38	37.88
16" dia.	"	39.50	5.49	44.99
18" dia.	"	50.00	5.59	55.59
20" dia.	"	59.00	5.70	64.70
24" dia.	"	75.00	5.82	80.82
Concrete sheet piling				
12" thick x 20' long	S.F.	27.50	14.50	42.00
25' long	"	27.50	13.25	40.75
30' long	"	27.50	12.00	39.50
35' long	"	27.50	11.25	38.75
40' long	"	27.50	10.50	38.00
16" thick x 40' long	"	37.75	8.08	45.83
45' long	"	37.75	7.65	45.40
50' long	"	37.75	7.27	45.02
55' long	"	37.75	6.92	44.67
60' long	"	37.75	6.61	44.36
02475.10 **Caissons (includes Casing)**				
Caisson, 3000# conc., 60 # reinf./CY, stable ground				
18" dia., 0.065 CY/ LF	L.F.	14.25	29.00	43.25
24" dia., 0.116 CY/ LF	"	23.00	30.25	53.25
30" dia., 0.182 CY/ LF	"	35.25	36.50	71.75
36" dia., 0.262 CY/ LF	"	49.00	41.50	90.50
48" dia., 0.465 CY/ LF	"	88.00	48.50	137
60" dia., 0.727 CY/ LF	"	140	66.00	206
72" dia., 1.05 CY/ LF	"	220	81.00	301
84" dia., 1.43 CY/ LF	"	260	100	360
Wet ground, casing required but pulled				
18" dia.	L.F.	14.25	36.50	50.75
24" dia.	"	23.00	40.50	63.50
30" dia.	"	35.25	45.50	80.75
36" dia.	"	49.00	48.50	97.50
48" dia.	"	88.00	61.00	149
60" dia.	"	140	81.00	221
72" dia.	"	220	120	340
84" dia.	"	260	180	440
Soft rock				
18" dia.	L.F.	14.25	100	114
24" dia.	"	23.00	180	203
30" dia.	"	35.25	240	275
36" dia.	"	49.00	360	409
48" dia.	"	88.00	490	578
60" dia.	"	140	730	870
72" dia.	"	220	810	1,030
84" dia.	"	260	910	1,170

02 SITE CONSTRUCTION

Utility Services	UNIT	MAT.	INST.	TOTAL
02510.10 Wells				
Domestic water, drilled and cased				
4" dia.	L.F.	32.75	73.00	106
6" dia.	"	36.00	81.00	117
8" dia.	"	42.50	91.00	134
02510.13 Gate Valves				
Gate valve, (AWWA) mechanical joint, with adjustable box				
4" valve	EA.	1,090	82.00	1,172
6" valve	"	1,230	98.00	1,328
8" valve	"	1,640	120	1,760
10" valve	"	2,460	140	2,600
12" valve	"	3,290	180	3,470
14" valve	"	8,220	200	8,420
16" valve	"	10,960	220	11,180
18" valve	"	13,700	250	13,950
Flanged, with box, post indicator (AWWA)				
4" valve	EA.	980	98.00	1,078
6" valve	"	1,150	110	1,260
8" valve	"	1,640	140	1,780
10" valve	"	2,460	160	2,620
12" valve	"	3,560	200	3,760
14" valve	"	8,220	250	8,470
16" valve	"	10,960	310	11,270
02510.15 Water Meters				
Water meter, displacement type				
1"	EA.	230	66.00	296
1-1/2"	"	810	74.00	884
2"	"	1,220	83.00	1,303
02510.17 Corporation Stops				
Stop for flared copper service pipe				
3/4"	EA.	49.75	33.25	83.00
1"	"	67.00	37.00	104
1-1/4"	"	180	44.25	224
1-1/2"	"	220	55.00	275
2"	"	300	66.00	366
02510.19 Thrust Blocks				
Thrust block, 3000# concrete				
1/4 c.y.	EA.	140	100	240
1/2 c.y.	"	200	120	320
3/4 c.y.	"	250	200	450
1 c.y.	"	340	400	740
02510.20 Tapping Saddles & Sleeves				
Tapping saddle, tap size to 2"				
4" saddle	EA.	84.00	23.75	108
6" saddle	"	98.00	29.50	128
8" saddle	"	110	39.50	150
10" saddle	"	130	47.50	178
12" saddle	"	160	68.00	228

02 SITE CONSTRUCTION

Utility Services

02510.20 Tapping Saddles & Sleeves (Cont.)

	UNIT	MAT.	INST.	TOTAL
14" saddle	EA.	180	95.00	275
Tapping sleeve				
4x4	EA.	880	31.50	912
6x4	"	1,140	36.50	1,177
6x6	"	1,150	36.50	1,187
8x4	"	1,180	47.50	1,228
8x6	"	1,210	47.50	1,258
10x4	"	1,920	98.00	2,018
10x6	"	2,790	98.00	2,888
10x8	"	2,900	98.00	2,998
10x10	"	2,970	100	3,070
12x4	"	2,990	100	3,090
12x6	"	3,020	110	3,130
12x8	"	3,090	120	3,210
12x10	"	3,250	140	3,390
12x12	"	3,370	150	3,520
Tapping valve, mechanical joint				
4" valve	EA.	830	310	1,140
6" valve	"	1,000	410	1,410
8" valve	"	1,480	620	2,100
10" valve	"	2,370	820	3,190
12" valve	"	4,180	1,230	5,410
Tap hole in pipe				
4" hole	EA.		59.00	59.00
6" hole	"		95.00	95.00
8" hole	"		160	160
10" hole	"		190	190
12" hole	"		240	240

02510.25 Valve Boxes

	UNIT	MAT.	INST.	TOTAL
Valve box, adjustable, for valves up to 20"				
3' deep	EA.	230	15.75	246
4' deep	"	230	19.00	249
5' deep	"	230	23.75	254

02510.30 Fire Hydrants

	UNIT	MAT.	INST.	TOTAL
Standard, 3 way post, 6" mechanical joint				
2' deep	EA.	2,200	820	3,020
4' deep	"	2,370	980	3,350
6' deep	"	2,630	1,230	3,860
8' deep	"	2,960	1,410	4,370

02510.35 Chilled Water Systems

	UNIT	MAT.	INST.	TOTAL
Chilled water pipe, 2" thick insulation, w/casing				
Align and tack weld on sleepers				
1-1/2" dia.	L.F.	27.50	2.23	29.73
3" dia.	"	44.00	3.51	47.51
4" dia.	"	51.00	4.92	55.92
6" dia.	"	58.00	6.15	64.15
8" dia.	"	81.00	7.02	88.02
10" dia.	"	100	8.20	108
12" dia.	"	120	9.84	130
14" dia.	"	170	10.75	181

02 SITE CONSTRUCTION

Utility Services

02510.35 Chilled Water Systems *(Cont.)*

	UNIT	MAT.	INST.	TOTAL
16" dia.	L.F.	210	12.25	222
Align and tack weld on trench bottom				
18" dia.	L.F.	220	13.75	234
20" dia.	"	290	15.25	305
Preinsulated fittings				
Align and tack weld on sleepers				
Elbows				
1-1/2"	EA.	570	41.50	612
3"	"	720	66.00	786
4"	"	930	83.00	1,013
6"	"	1,290	110	1,400
8"	"	1,830	130	1,960
Tees				
1-1/2"	EA.	870	44.25	914
3"	"	1,230	74.00	1,304
4"	"	930	95.00	1,025
6"	"	2,010	130	2,140
8"	"	2,680	170	2,850
Reducers				
3"	EA.	930	55.00	985
4"	"	1,470	66.00	1,536
6"	"	1,980	83.00	2,063
8"	"	2,150	110	2,260
Anchors, not including concrete				
4"	EA.	400	83.00	483
6"	"	580	83.00	663
Align and tack weld on trench bottom				
Elbows				
10"	EA.	2,410	150	2,560
12"	"	2,880	180	3,060
14"	"	3,630	190	3,820
16"	"	3,910	200	4,110
18"	"	4,360	220	4,580
20"	"	5,260	250	5,510
Tees				
10"	EA.	3,910	150	4,060
12"	"	4,960	180	5,140
14"	"	5,260	190	5,450
16"	"	5,560	200	5,760
18"	"	6,320	220	6,540
20"	"	7,220	250	7,470
Reducers				
10"	EA.	2,980	100	3,080
12"	"	3,610	110	3,720
14"	"	3,910	120	4,030
16"	"	4,810	140	4,950
18"	"	5,260	150	5,410
20"	"	5,560	180	5,740
Anchors, not including concrete				
10"	EA.	800	100	900
12"	"	880	110	990
14"	"	1,010	120	1,130
16"	"	1,420	140	1,560

02 SITE CONSTRUCTION

Utility Services	UNIT	MAT.	INST.	TOTAL
02510.35 — Chilled Water Systems *(Cont.)*				
18"	EA.	1,970	150	2,120
20"	"	2,680	180	2,860
02510.40 — Ductile Iron Pipe				
Ductile iron pipe, cement lined, slip-on joints				
4"	L.F.	15.75	6.83	22.58
6"	"	19.25	7.23	26.48
8"	"	25.00	7.68	32.68
10"	"	34.50	8.20	42.70
12"	"	42.50	9.84	52.34
14"	"	54.00	12.25	66.25
16"	"	66.00	13.75	79.75
18"	"	74.00	15.25	89.25
20"	"	84.00	17.50	102
Mechanical joint pipe				
4"	L.F.	18.00	9.46	27.46
6"	"	21.25	10.25	31.50
8"	"	28.25	11.25	39.50
10"	"	36.75	12.25	49.00
12"	"	46.75	16.50	63.25
14"	"	59.00	19.00	78.00
16"	"	64.00	22.25	86.25
18"	"	73.00	24.50	97.50
20"	"	84.00	27.25	111
Fittings, mechanical joint				
90 degree elbow				
4"	EA.	240	31.50	272
6"	"	310	36.50	347
8"	"	450	47.50	498
10"	"	650	68.00	718
12"	"	880	95.00	975
14"	"	1,360	120	1,480
16"	"	1,700	160	1,860
18"	"	2,550	190	2,740
20"	"	2,840	240	3,080
45 degree elbow				
4"	EA.	210	31.50	242
6"	"	280	36.50	317
8"	"	400	47.50	448
10"	"	570	68.00	638
12"	"	680	95.00	775
14"	"	1,130	120	1,250
16"	"	1,360	160	1,520
18"	"	1,980	240	2,220
20"	"	2,350	240	2,590
Tee				
4"x3"	EA.	340	59.00	399
4"x4"	"	370	59.00	429
6"x3"	"	420	68.00	488
6"x4"	"	440	68.00	508
6"x6"	"	480	68.00	548
8"x4"	"	590	79.00	669
8"x6"	"	680	79.00	759

02 SITE CONSTRUCTION

Utility Services	UNIT	MAT.	INST.	TOTAL
02510.40 **Ductile Iron Pipe** *(Cont.)*				
8"x8"	EA.	630	79.00	709
10"x4"	"	790	95.00	885
10"x6"	"	880	95.00	975
10"x8"	"	910	95.00	1,005
10"x10"	"	1,020	95.00	1,115
12"x4"	"	890	120	1,010
12"x6"	"	960	120	1,080
12"x8"	"	1,050	120	1,170
12"x10"	"	1,190	120	1,310
12"x12"	"	1,280	130	1,410
14"x4"	"	1,560	140	1,700
14"x6"	"	1,660	140	1,800
14"x8"	"	1,690	140	1,830
14"x10"	"	1,740	140	1,880
14"x12"	"	1,790	150	1,940
14"x14"	"	1,770	150	1,920
16"x4"	"	2,010	160	2,170
16"x6"	"	2,050	160	2,210
16"x8"	"	1,830	160	1,990
16"x10"	"	1,860	160	2,020
16"x12"	"	1,810	160	1,970
16"x14"	"	1,920	160	2,080
16"x16"	"	1,960	160	2,120
18"x6"	"	2,340	170	2,510
18"x8"	"	2,380	170	2,550
18"x10"	"	2,430	170	2,600
18"x12"	"	2,470	170	2,640
18"x14"	"	2,750	170	2,920
18"x16"	"	2,720	170	2,890
18"x18"	"	3,090	170	3,260
20"x6"	"	2,980	190	3,170
20"x8"	"	3,010	190	3,200
20"x10"	"	3,070	190	3,260
20"x12"	"	3,120	190	3,310
20"x14"	"	3,210	190	3,400
20"x16"	"	3,750	190	3,940
20"x18"	"	3,920	190	4,110
20"x20"	"	4,000	190	4,190
Cross				
4"x3"	EA.	350	79.00	429
4"x4"	"	390	79.00	469
6"x3"	"	400	95.00	495
6"x4"	"	420	95.00	515
6"x6"	"	470	95.00	565
8"x4"	"	650	110	760
8"x6"	"	710	110	820
8"x8"	"	780	110	890
10"x4"	"	910	120	1,030
10"x6"	"	960	120	1,080
10"x8"	"	1,050	120	1,170
10"x10"	"	1,240	120	1,360
12"x4"	"	1,160	140	1,300
12"x6"	"	1,310	140	1,450

02 SITE CONSTRUCTION

Utility Services	UNIT	MAT.	INST.	TOTAL
02510.40 **Ductile Iron Pipe** *(Cont.)*				
12"x8"	EA.	1,270	140	1,410
12"x10"	"	1,470	150	1,620
12"x12"	"	1,590	150	1,740
14"x4"	"	1,500	160	1,660
14"x6"	"	1,700	160	1,860
14"x8"	"	1,770	160	1,930
14"x10"	"	1,900	160	2,060
14"x12"	"	2,040	170	2,210
14"x14"	"	2,240	170	2,410
16"x4"	"	1,960	190	2,150
16"x6"	"	2,010	190	2,200
16"x8"	"	2,130	190	2,320
16"x10"	"	2,270	190	2,460
16"x12"	"	2,370	190	2,560
16"x14"	"	2,570	190	2,760
16"x16"	"	2,720	190	2,910
18"x6"	"	2,530	210	2,740
18"x8"	"	2,610	210	2,820
18"x10"	"	2,720	210	2,930
18"x12"	"	2,860	210	3,070
18"x14"	"	3,400	210	3,610
18"x16"	"	3,630	210	3,840
18"x18"	"	3,830	210	4,040
20"x6"	"	3,040	230	3,270
20"x8"	"	3,120	230	3,350
20"x10"	"	3,260	230	3,490
20"x12"	"	3,400	230	3,630
20"x14"	"	3,580	230	3,810
20"x16"	"	4,140	230	4,370
20"x18"	"	4,430	240	4,670
20"x20"	"	4,680	240	4,920
02510.60 **Plastic Pipe**				
PVC, class 150 pipe				
4" dia.	L.F.	4.94	6.15	11.09
6" dia.	"	9.35	6.64	15.99
8" dia.	"	14.75	7.02	21.77
10" dia.	"	21.00	7.68	28.68
12" dia.	"	31.25	8.20	39.45
Schedule 40 pipe				
1-1/2" dia.	L.F.	1.24	2.78	4.02
2" dia.	"	1.84	2.96	4.80
2-1/2" dia.	"	2.79	3.16	5.95
3" dia.	"	3.80	3.38	7.18
4" dia.	"	5.37	3.95	9.32
6" dia.	"	10.00	4.74	14.74
90 degree elbows				
1"	EA.	1.21	7.90	9.11
1-1/2"	"	2.29	7.90	10.19
2"	"	3.60	8.61	12.21
2-1/2"	"	11.00	9.48	20.48
3"	"	13.00	10.50	23.50
4"	"	23.50	11.75	35.25

02 SITE CONSTRUCTION

Utility Services

02510.60 Plastic Pipe (Cont.)

	UNIT	MAT.	INST.	TOTAL
6"	EA.	74.00	15.75	89.75
45 degree elbows				
1"	EA.	1.85	7.90	9.75
1-1/2"	"	3.24	7.90	11.14
2"	"	4.21	8.61	12.82
2-1/2"	"	11.00	9.48	20.48
3"	"	17.00	10.50	27.50
4"	"	30.50	11.75	42.25
6"	"	76.00	15.75	91.75
Tees				
1"	EA.	1.60	9.48	11.08
1-1/2"	"	3.06	9.48	12.54
2"	"	4.43	10.50	14.93
2-1/2"	"	14.50	11.75	26.25
3"	"	19.25	13.50	32.75
4"	"	34.75	15.75	50.50
6"	"	120	19.00	139
Couplings				
1"	EA.	0.97	7.90	8.87
1-1/2"	"	1.40	7.90	9.30
2"	"	2.15	8.61	10.76
2-1/2"	"	4.75	9.48	14.23
3"	"	7.44	10.50	17.94
4"	"	10.75	11.75	22.50
6"	"	34.00	15.75	49.75
Drainage pipe				
PVC schedule 80				
1" dia.	L.F.	2.97	2.78	5.75
1-1/2" dia.	"	3.58	2.78	6.36
ABS, 2" dia.	"	4.59	2.96	7.55
2-1/2" dia.	"	6.53	3.16	9.69
3" dia.	"	7.69	3.38	11.07
4" dia.	"	10.50	3.95	14.45
6" dia.	"	17.25	4.74	21.99
8" dia.	"	23.25	6.47	29.72
10" dia.	"	31.00	7.68	38.68
12" dia.	"	51.00	8.20	59.20
90 degree elbows				
1"	EA.	5.22	7.90	13.12
1-1/2"	"	6.53	7.90	14.43
2"	"	7.86	8.61	16.47
2-1/2"	"	18.75	9.48	28.23
3"	"	19.50	10.50	30.00
4"	"	34.25	11.75	46.00
6"	"	75.00	15.75	90.75
45 degree elbows				
1"	EA.	8.49	7.90	16.39
1-1/2"	"	10.75	7.90	18.65
2"	"	13.50	8.61	22.11
2-1/2"	"	25.25	9.48	34.73
3"	"	26.75	10.50	37.25
4"	"	51.00	11.75	62.75
6"	"	120	15.75	136

02 SITE CONSTRUCTION

Utility Services

02510.60 Plastic Pipe (Cont.)

	UNIT	MAT.	INST.	TOTAL
Tees				
1"	EA.	5.52	9.48	15.00
1-1/2"	"	17.75	9.48	27.23
2"	"	21.50	10.50	32.00
2-1/2"	"	25.25	11.75	37.00
3"	"	27.50	13.50	41.00
4"	"	52.00	15.75	67.75
6"	"	100	19.00	119
Couplings				
1"	EA.	4.59	7.90	12.49
1-1/2"	"	7.86	7.90	15.76
2"	"	11.50	8.61	20.11
2-1/2"	"	24.50	9.48	33.98
3"	"	25.25	10.50	35.75
4"	"	26.25	11.75	38.00
6"	"	44.00	15.75	59.75
Pressure pipe				
PVC, class 200 pipe				
3/4"	L.F.	0.24	2.37	2.61
1"	"	0.35	2.49	2.84
1-1/4"	"	0.60	2.63	3.23
1-1/2"	"	0.71	2.78	3.49
2"	"	1.18	2.96	4.14
2-1/2"	"	1.78	3.16	4.94
3"	"	2.73	3.38	6.11
4"	"	4.76	3.95	8.71
6"	"	9.52	4.74	14.26
8"	"	18.25	7.02	25.27
90 degree elbows				
3/4"	EA.	1.00	7.90	8.90
1"	"	1.10	7.90	9.00
1-1/4"	"	1.64	7.90	9.54
1-1/2"	"	2.08	7.90	9.98
2"	"	3.27	8.61	11.88
2-1/2"	"	9.98	9.48	19.46
3"	"	15.50	10.50	26.00
4"	"	27.75	11.75	39.50
6"	"	69.00	15.75	84.75
8"	"	120	23.75	144
45 degree elbows				
3/4"	EA.	1.13	7.90	9.03
1"	"	1.44	7.90	9.34
1-1/4"	"	2.07	7.90	9.97
1-1/2"	"	2.52	7.90	10.42
2"	"	3.55	8.61	12.16
2-1/2"	"	5.82	9.48	15.30
3"	"	13.25	10.50	23.75
4"	"	25.00	11.75	36.75
6"	"	62.00	15.75	77.75
8"	"	130	23.75	154
Tees				
3/4"	EA.	0.95	9.48	10.43
1"	"	1.24	9.48	10.72

02 SITE CONSTRUCTION

Utility Services	UNIT	MAT.	INST.	TOTAL
02510.60 Plastic Pipe *(Cont.)*				
1-1/4"	EA.	1.78	9.48	11.26
1-1/2"	"	2.48	9.48	11.96
2"	"	3.64	10.50	14.14
2-1/2"	"	5.74	11.75	17.49
3"	"	17.75	13.50	31.25
4"	"	25.25	15.75	41.00
6"	"	88.00	19.00	107
8"	"	180	26.25	206
Couplings				
3/4"	EA.	0.58	7.90	8.48
1"	"	0.86	7.90	8.76
1-1/4"	"	1.13	7.90	9.03
1-1/2"	"	1.22	7.90	9.12
2"	"	1.72	8.61	10.33
2-1/2"	"	3.64	9.48	13.12
3"	"	5.74	10.50	16.24
4"	"	8.07	10.50	18.57
6"	"	25.25	11.75	37.00
8"	"	45.25	15.75	61.00

Sanitary Sewer	UNIT	MAT.	INST.	TOTAL
02530.10 Cast Iron Flanged Pipe				
Cast iron flanged sections				
4" pipe, with one bolt set				
3' section	EA.	49.25	22.25	71.50
4' section	"	69.00	24.50	93.50
5' section	"	88.00	27.25	115
6' section	"	100	30.75	131
8' section	"	130	35.25	165
10' section	"	190	49.25	239
12' section	"	200	82.00	282
15' section	"	250	120	370
18' section	"	300	160	460
6" pipe, with one bolt set				
3' section	EA.	86.00	24.50	111
4' section	"	120	29.00	149
5' section	"	150	32.75	183
6' section	"	190	37.75	228
8' section	"	220	55.00	275
10' section	"	320	62.00	382
12' section	"	340	82.00	422
15' section	"	430	120	550
18' section	"	510	180	690
8" pipe, with one bolt set				
3' section	EA.	130	30.75	161

02 SITE CONSTRUCTION

Sanitary Sewer	UNIT	MAT.	INST.	TOTAL
02530.10 **Cast Iron Flanged Pipe** *(Cont.)*				
4' section	EA.	190	35.25	225
5' section	"	230	41.00	271
6' section	"	280	49.25	329
8' section	"	360	70.00	430
10' section	"	490	82.00	572
12' section	"	530	120	650
15' section	"	670	160	830
18' section	"	800	200	1,000
10" pipe, with one bolt set				
3' section	EA.	250	31.50	282
4' section	"	400	36.25	436
5' section	"	470	42.50	513
6' section	"	560	51.00	611
8' section	"	750	75.00	825
10' section	"	820	88.00	908
12' section	"	890	140	1,030
15' section	"	1,110	180	1,290
18' section	"	1,330	250	1,580
12" pipe, with one bolt set				
3' section	EA.	300	34.25	334
4' section	"	450	39.75	490
5' section	"	560	47.25	607
6' section	"	670	56.00	726
8' section	"	900	82.00	982
10' section	"	1,040	95.00	1,135
12' section	"	1,320	150	1,470
15' section	"	1,500	200	1,700
18' section	"	1,720	270	1,990
02530.11 **Cast Iron Fittings**				
Mechanical joint, with 2 bolt kits				
90 deg bend				
4"	EA.	78.00	31.50	110
6"	"	130	36.50	167
8"	"	300	47.50	348
10"	"	450	68.00	518
12"	"	680	95.00	775
14"	"	930	120	1,050
16"	"	1,070	160	1,230
45 deg bend				
4"	EA.	65.00	31.50	96.50
6"	"	100	36.50	137
8"	"	230	47.50	278
10"	"	340	68.00	408
12"	"	590	95.00	685
14"	"	720	120	840
16"	"	940	160	1,100
Tee, with 3 bolt kits				
4" x 4"	EA.	140	47.50	188
6" x 6"	"	230	59.00	289
8" x 8"	"	650	79.00	729
10" x 10"	"	810	120	930
12" x 12"	"	1,440	160	1,600

02 SITE CONSTRUCTION

Sanitary Sewer	UNIT	MAT.	INST.	TOTAL
02530.11 **Cast Iron Fittings** *(Cont.)*				
Wye, with 3 bolt kits				
6" x 6"	EA.	280	59.00	339
8" x 8"	"	590	79.00	669
10" x 10"	"	850	120	970
12" x 12"	"	1,680	160	1,840
Reducer, with 2 bolt kits				
6" x 4"	EA.	130	59.00	189
8" x 6"	"	200	79.00	279
10" x 8"	"	570	120	690
12" x 10"	"	690	160	850
Flanged, 90 deg bend, 125 lb.				
4"	EA.	160	39.50	200
6"	"	200	47.50	248
8"	"	270	59.00	329
10"	"	490	79.00	569
12"	"	680	120	800
14"	"	1,350	160	1,510
16"	"	2,020	160	2,180
Tee				
4"	EA.	250	59.00	309
6"	"	340	68.00	408
8"	"	530	79.00	609
10"	"	980	95.00	1,075
12"	"	1,320	120	1,440
14"	"	2,930	160	3,090
16"	"	4,410	240	4,650
02530.20 **Vitrified Clay Pipe**				
Vitrified clay pipe, extra strength				
6" dia.	L.F.	6.16	11.25	17.41
8" dia.	"	7.37	11.75	19.12
10" dia.	"	11.25	12.25	23.50
12" dia.	"	16.25	16.50	32.75
15" dia.	"	29.50	24.50	54.00
18" dia.	"	44.25	27.25	71.50
24" dia.	"	81.00	35.25	116
30" dia.	"	140	49.25	189
36" dia.	"	200	70.00	270
02530.30 **Manholes**				
Precast sections, 48" dia.				
Base section	EA.	390	200	590
1'0" riser	"	110	160	270
1'4" riser	"	130	180	310
2'8" riser	"	200	190	390
4'0" riser	"	370	200	570
2'8" cone top	"	240	250	490
Precast manholes, 48" dia.				
4' deep	EA.	760	490	1,250
6' deep	"	1,150	620	1,770
7' deep	"	1,310	700	2,010
8' deep	"	1,480	820	2,300
10' deep	"	1,660	980	2,640

02 SITE CONSTRUCTION

Sanitary Sewer	UNIT	MAT.	INST.	TOTAL
02530.30 **Manholes** (Cont.)				
Cast-in-place, 48" dia., with frame and cover				
5' deep	EA.	710	1,230	1,940
6' deep	"	930	1,410	2,340
8' deep	"	1,360	1,640	3,000
10' deep	"	1,580	1,970	3,550
Brick manholes, 48" dia. with cover, 8" thick				
4' deep	EA.	750	580	1,330
6' deep	"	940	640	1,580
8' deep	"	1,210	720	1,930
10' deep	"	1,510	820	2,330
12' deep	"	1,890	960	2,850
14' deep	"	2,290	1,150	3,440
Inverts for manholes				
Single channel	EA.	130	230	360
Triple channel	"	150	290	440
Frames and covers, 24" diameter				
300 lb	EA.	440	47.50	488
400 lb	"	460	53.00	513
500 lb	"	530	68.00	598
Watertight, 350 lb	"	550	160	710
For heavy equipment, 1200 lb	"	1,210	240	1,450
Steps for manholes				
7" x 9"	EA.	15.00	9.48	24.48
8" x 9"	"	18.75	10.50	29.25
Curb inlet, 4' throat, cast in place				
12"-30" pipe	EA.	410	1,230	1,640
36"-48" pipe	"	450	1,410	1,860
Raise exist frame and cover, when repaving	"		490	490
02530.40 **Sanitary Sewers**				
Clay				
6" pipe	L.F.	7.07	8.20	15.27
8" pipe	"	9.43	8.78	18.21
10" pipe	"	11.75	9.46	21.21
12" pipe	"	18.75	10.25	29.00
PVC				
4" pipe	L.F.	3.57	6.15	9.72
6" pipe	"	7.14	6.47	13.61
8" pipe	"	10.75	6.83	17.58
10" pipe	"	14.25	7.23	21.48
12" pipe	"	21.50	7.68	29.18
Cleanout				
4" pipe	EA.	13.50	59.00	72.50
6" pipe	"	29.75	59.00	88.75
8" pipe	"	88.00	59.00	147
Connect new sewer line				
To existing manhole	EA.	110	160	270
To new manhole	"	80.00	95.00	175

02 SITE CONSTRUCTION

Sanitary Sewer	UNIT	MAT.	INST.	TOTAL
02540.10 — **Drainage Fields**				
Perforated PVC pipe, for drain field				
4" pipe	L.F.	2.43	5.46	7.89
6" pipe	"	4.55	5.85	10.40
02540.50 — **Septic Tanks**				
Septic tank, precast concrete				
1000 gals	EA.	880	410	1,290
2000 gals	"	2,860	620	3,480
5000 gals	"	9,750	1,230	10,980
25,000 gals	"	55,900	4,920	60,820
40,000 gals	"	66,300	8,200	74,500
Leaching pit, precast concrete, 72" diameter				
3' deep	EA.	810	310	1,120
6' deep	"	1,430	350	1,780
8' deep	"	1,820	410	2,230

Piped Energy Distribution	UNIT	MAT.	INST.	TOTAL
02550.10 — **Gas Distribution**				
Gas distribution lines				
Polyethylene, 60 psi coils				
1-1/4" dia.	L.F.	2.16	4.42	6.58
1-1/2" dia.	"	2.94	4.74	7.68
2" dia.	"	3.71	5.53	9.24
3" dia.	"	7.87	6.64	14.51
30' pipe lengths				
3" dia.	L.F.	6.44	7.37	13.81
4" dia.	"	10.00	8.30	18.30
6" dia.	"	16.00	11.00	27.00
8" dia.	"	29.50	13.25	42.75
Steel, schedule 40, plain end				
1" dia.	L.F.	9.53	5.53	15.06
2" dia.	"	15.75	6.03	21.78
3" dia.	"	22.75	6.64	29.39
4" dia.	"	28.00	16.50	44.50
5" dia.	"	52.00	17.50	69.50
6" dia.	"	70.00	20.50	90.50
8" dia.	"	87.00	22.25	109
Natural gas meters, direct digital reading, threaded				
250 cfh @ 5 lbs	EA.	220	130	350
425 cfh @ 10 lbs	"	570	130	700
800 cfh @ 20 lbs	"	790	170	960
1,000 cfh @ 25 lbs	"	2,350	170	2,520
1,400 cfh @ 100 lbs	"	5,430	220	5,650
2,300 cfh @ 100 lbs	"	7,560	330	7,890
5,000 cfh @ 100 lbs	"	11,100	660	11,760

02 SITE CONSTRUCTION

Piped Energy Distribution

	UNIT	MAT.	INST.	TOTAL
02550.10 Gas Distribution *(Cont.)*				
Gas pressure regulators				
Threaded				
3/4"	EA.	98.00	83.00	181
1"	"	100	110	210
1-1/4"	"	110	110	220
1-1/2"	"	700	110	810
2"	"	720	130	850
Flanged				
3"	EA.	2,030	170	2,200
4"	"	3,040	220	3,260
02550.40 Steel Pipe				
Steel pipe, extra heavy, A 53, grade B, seamless				
1/2" dia.	L.F.	7.83	6.64	14.47
3/4" dia.	"	9.54	6.98	16.52
1" dia.	"	11.50	7.37	18.87
1-1/4" dia.	"	14.75	8.30	23.05
1-1/2" dia.	"	17.75	9.48	27.23
2" dia.	"	21.50	11.00	32.50
3" dia.	"	23.75	12.25	36.00
4" dia.	"	35.50	13.75	49.25
6" dia.	"	87.00	15.25	102
8" dia.	"	110	17.50	128
10" dia.	"	160	20.50	181
12" dia.	"	200	24.50	225
02550.80 Steam Meters				
In-line turbine, direct reading, 300 lb, flanged				
2"	EA.	4,370	83.00	4,453
3"	"	4,680	110	4,790
4"	"	5,150	130	5,280
Threaded, 2"				
5" line	EA.	8,270	660	8,930
6" line	"	8,430	660	9,090
8" line	"	8,580	660	9,240
10" line	"	9,050	660	9,710
12" line	"	9,200	660	9,860
14" line	"	9,520	660	10,180
16" line	"	10,140	660	10,800

02 SITE CONSTRUCTION

Power & Communications

02580.20 High Voltage Cable

	UNIT	MAT.	INST.	TOTAL
High voltage XLP copper cable, shielded, 5000v				
#6 awg	L.F.	3.23	1.00	4.23
#4 awg	"	3.71	1.23	4.94
#2 awg	"	4.84	1.46	6.30
#1 awg	"	5.11	1.62	6.73
#1/0 awg	"	6.06	1.84	7.90
#2/0 awg	"	10.25	2.24	12.49
#3/0 awg	"	10.25	2.62	12.87
#4/0 awg	"	11.00	2.80	13.80
#250 awg	"	12.75	3.33	16.08
#300 awg	"	14.50	3.73	18.23
#350 awg	"	16.25	4.10	20.35
#500 awg	"	23.25	5.60	28.85
#750 awg	"	34.25	6.16	40.41
Ungrounded, 15,000v				
#1 awg	L.F.	7.73	2.37	10.10
#1/0 awg	"	9.14	2.62	11.76
#2/0 awg	"	10.50	2.80	13.30
#3/0 awg	"	12.00	3.08	15.08
#4/0 awg	"	13.25	3.52	16.77
#250 awg	"	15.00	3.73	18.73
#300 awg	"	17.00	4.10	21.10
#350 awg	"	18.75	4.74	23.49
#500 awg	"	24.75	6.16	30.91
#750 awg	"	36.75	7.51	44.26
#1000 awg	"	54.00	9.48	63.48
Aluminum cable, shielded, 5000v				
#6 awg	L.F.	1.95	0.84	2.79
#4 awg	"	2.11	1.00	3.11
#2 awg	"	2.35	1.15	3.50
#1 awg	"	2.58	1.31	3.89
#1/0 awg	"	2.83	1.46	4.29
#2/0 awg	"	3.14	1.54	4.68
#3/0 awg	"	3.58	1.62	5.20
#4/0 awg	"	3.96	1.84	5.80
#250 awg	"	4.22	1.98	6.20
#300 awg	"	4.60	2.37	6.97
#350 awg	"	4.99	2.62	7.61
#500 awg	"	6.01	2.80	8.81
#750 awg	"	7.78	3.42	11.20
#1000 awg	"	8.77	3.85	12.62
Ungrounded, 15,000v				
#1 awg	L.F.	3.53	1.62	5.15
#1/0 awg	"	3.73	1.92	5.65
#2/0 awg	"	4.02	2.08	6.10
#3/0 awg	"	4.39	2.16	6.55
#4/0 awg	"	4.77	2.24	7.01
#250 awg	"	5.11	2.37	7.48
#300 awg	"	5.53	2.46	7.99
#350 awg	"	5.95	2.80	8.75
#500 awg	"	6.98	3.33	10.31
#750 awg	"	9.14	3.97	13.11
#1000 awg	"	12.00	4.93	16.93

02 SITE CONSTRUCTION

Power & Communications	UNIT	MAT.	INST.	TOTAL
02580.20 High Voltage Cable *(Cont.)*				
Indoor terminations, 5000v				
#6 - #4	EA.	74.00	12.00	86.00
#2 - #2/0	"	86.00	12.00	98.00
#3/0 - #250	"	110	12.00	122
#300 - #750	"	120	210	330
#1000	"	150	290	440
In-line splice, 5000v				
#6 - #4/0	EA.	180	290	470
#250 - #500	"	190	770	960
#750 - #1000	"	260	1,000	1,260
T-splice, 5000v				
#2 - #4/0	EA.	180	920	1,100
#250 - #500	"	190	1,540	1,730
#750 - #1000	"	260	1,930	2,190
Indoor terminations, 15,000v				
#2 - #2/0	EA.	98.00	270	368
#3/0 - #500	"	120	410	530
#750 - #1000	"	150	470	620
In-line splice, 15,000v				
#2 - #4/0	EA.	170	690	860
#250 - #500	"	240	920	1,160
#750 - #1000	"	290	1,390	1,680
T-splice, 15,000v				
#4	EA.	170	1,390	1,560
#250 - #500	"	230	2,310	2,540
#750 - #1000	"	290	3,460	3,750
Compression lugs, 15,000v				
#4	EA.	9.26	30.75	40.01
#2	"	10.75	41.00	51.75
#1	"	11.75	41.00	52.75
#1/0	"	17.75	51.00	68.75
#2/0	"	18.75	51.00	69.75
#3/0	"	21.00	66.00	87.00
#4/0	"	23.25	66.00	89.25
#250	"	27.00	73.00	100
#300	"	31.50	73.00	105
#350	"	32.50	89.00	122
#500	"	49.00	96.00	145
#750	"	78.00	120	198
#1000	"	110	150	260
Compression splices, 15,000v				
#4	EA.	10.50	51.00	61.50
#2	"	11.50	56.00	67.50
#1	"	13.00	69.00	82.00
#1/0	"	13.75	77.00	90.75
#2/0	"	14.75	89.00	104
#3/0	"	16.25	96.00	112
#4/0	"	17.75	110	128
#250	"	19.25	120	139
#350	"	22.00	130	152
#500	"	32.50	150	183
#750	"	53.00	190	243

02 SITE CONSTRUCTION

Power & Communications

02580.40 Supports & Connectors	UNIT	MAT.	INST.	TOTAL
Cable supports for conduit				
1-1/2"	EA.	91.00	26.75	118
2"	"	130	26.75	157
2-1/2"	"	140	30.75	171
3"	"	180	30.75	211
3-1/2"	"	240	38.50	279
4"	"	290	38.50	329
5"	"	520	51.00	571
6"	"	1,090	56.00	1,146
Split bolt connectors				
#10	EA.	3.20	15.50	18.70
#8	"	3.78	15.50	19.28
#6	"	4.17	15.50	19.67
#4	"	4.86	30.75	35.61
#3	"	6.95	30.75	37.70
#2	"	7.87	30.75	38.62
#1/0	"	10.25	51.00	61.25
#2/0	"	16.25	51.00	67.25
#3/0	"	24.75	51.00	75.75
#4/0	"	28.00	51.00	79.00
#250	"	29.00	77.00	106
#350	"	51.00	77.00	128
#500	"	67.00	77.00	144
#750	"	110	120	230
#1000	"	160	120	280
Single barrel lugs				
#6	EA.	1.13	19.25	20.38
#1/0	"	2.22	38.50	40.72
#250	"	5.36	51.00	56.36
#350	"	6.95	51.00	57.95
#500	"	13.50	51.00	64.50
#600	"	14.25	69.00	83.25
#800	"	16.25	69.00	85.25
#1000	"	19.50	69.00	88.50
Double barrel lugs				
#1/0	EA.	4.37	69.00	73.37
#250	"	12.75	99.00	112
#350	"	18.25	99.00	117
#600	"	27.50	150	178
#800	"	31.50	150	182
#1000	"	32.25	150	182
Three barrel lugs				
#2/0	EA.	35.25	99.00	134
#250	"	67.00	150	217
#350	"	110	150	260
#600	"	120	210	330
#800	"	190	210	400
#1000	"	270	210	480
Four barrel lugs				
#250	EA.	75.00	210	285
#350	"	120	210	330
#600	"	140	270	410
#800	"	210	270	480

02 SITE CONSTRUCTION

Power & Communications	UNIT	MAT.	INST.	TOTAL
02580.40 Supports & Connectors *(Cont.)*				
Compression conductor adapters				
#6	EA.	9.70	22.75	32.45
#4	"	10.25	26.75	37.00
#2	"	10.75	34.25	45.00
#1	"	12.25	34.25	46.50
#1/0	"	12.75	41.00	53.75
#250	"	24.25	62.00	86.25
#350	"	29.00	66.00	95.00
#500	"	37.50	84.00	122
#750	"	51.00	88.00	139
Terminal blocks, 2 screw				
3 circuit	EA.	22.00	15.50	37.50
6 circuit	"	30.50	15.50	46.00
8 circuit	"	35.50	15.50	51.00
10 circuit	"	41.00	22.75	63.75
12 circuit	"	46.50	22.75	69.25
18 circuit	"	63.00	22.75	85.75
24 circuit	"	79.00	26.75	106
36 circuit	"	110	26.75	137
Compression splice				
#8 awg	EA.	3.68	29.25	32.93
#6 awg	"	4.37	21.25	25.62
#4 awg	"	4.62	21.25	25.87
#2 awg	"	7.16	41.00	48.16
#1 awg	"	10.00	41.00	51.00
#1/0 awg	"	12.25	41.00	53.25
#2/0 awg	"	13.00	66.00	79.00
#3/0 awg	"	15.25	66.00	81.25
#4/0 awg	"	16.25	66.00	82.25
#250 awg	"	17.25	100	117
#300 awg	"	18.75	100	119
#350 awg	"	19.25	110	129
#400 awg	"	26.00	110	136
#500 awg	"	30.50	120	151
#600 awg	"	46.00	120	166
#750 awg	"	48.75	130	179
#1000 awg	"	62.00	130	192

Drainage And Containment	UNIT	MAT.	INST.	TOTAL
02630.10 Catch Basins				
Standard concrete catch basin				
Cast in place, 3'8" x 3'8", 6" thick wall				
2' deep	EA.	370	620	990
3' deep	"	500	620	1,120
4' deep	"	650	820	1,470

02 SITE CONSTRUCTION

Drainage And Containment

	UNIT	MAT.	INST.	TOTAL
02630.10 Catch Basins *(Cont.)*				
5' deep	EA.	770	820	1,590
6' deep	"	860	980	1,840
4'x4', 8" thick wall, cast in place				
2' deep	EA.	400	620	1,020
3' deep	"	560	620	1,180
4' deep	"	740	820	1,560
5' deep	"	850	820	1,670
6' deep	"	940	980	1,920
Frames and covers, cast iron				
Round				
24" dia.	EA.	340	120	460
26" dia.	"	370	120	490
28" dia.	"	450	120	570
Rectangular				
23"x23"	EA.	300	120	420
27"x20"	"	360	120	480
24"x24"	"	350	120	470
26"x26"	"	380	120	500
Curb inlet frames and covers				
27"x27"	EA.	600	120	720
24"x36"	"	440	120	560
24"x25"	"	410	120	530
24"x22"	"	350	120	470
20"x22"	"	460	120	580
Airfield catch basin frame and grating, galvanized				
2'x4'	EA.	840	120	960
2'x2'	"	590	120	710
02630.40 Storm Drainage				
Concrete pipe				
Plain, bell and spigot joint, Class II				
6" pipe	L.F.	8.93	11.25	20.18
8" pipe	"	9.33	12.25	21.58
10" pipe	"	9.49	13.00	22.49
12" pipe	"	12.75	13.75	26.50
15" pipe	"	17.25	14.50	31.75
18" pipe	"	21.25	15.25	36.50
21" pipe	"	25.75	16.50	42.25
24" pipe	"	32.50	17.50	50.00
Reinforced, class III, tongue and groove joint				
12" pipe	L.F.	18.25	13.75	32.00
15" pipe	"	20.50	14.50	35.00
18" pipe	"	22.75	15.25	38.00
21" pipe	"	29.75	16.50	46.25
24" pipe	"	38.75	17.50	56.25
27" pipe	"	45.75	19.00	64.75
30" pipe	"	50.00	20.50	70.50
36" pipe	"	75.00	22.25	97.25
42" pipe	"	100	24.50	125
48" pipe	"	140	27.25	167
54" pipe	"	150	30.75	181
60" pipe	"	210	35.25	245
66" pipe	"	260	41.00	301

02 SITE CONSTRUCTION

Drainage And Containment

	UNIT	MAT.	INST.	TOTAL
02630.40 Storm Drainage *(Cont.)*				
72" pipe	L.F.	280	49.25	329
Flared end-section, concrete				
12" pipe	L.F.	83.00	13.75	96.75
15" pipe	"	98.00	14.50	113
18" pipe	"	120	15.25	135
24" pipe	"	140	17.50	158
30" pipe	"	170	20.50	191
36" pipe	"	230	22.25	252
42" pipe	"	240	24.50	265
48" pipe	"	270	27.25	297
54" pipe	"	290	30.75	321
Porous concrete pipe standard strength				
4" pipe	L.F.	5.29	9.46	14.75
6" pipe	"	5.72	9.84	15.56
8" pipe	"	6.57	10.25	16.82
10" pipe	"	13.75	10.75	24.50
12" pipe	"	18.25	11.25	29.50
Corrugated metal pipe, coated, paved invert				
16 ga.				
8" pipe	L.F.	13.00	8.20	21.20
10" pipe	"	17.50	8.48	25.98
12" pipe	"	19.75	8.78	28.53
15" pipe	"	24.00	9.46	33.46
18" pipe	"	28.25	10.25	38.50
21" pipe	"	34.75	11.25	46.00
24" pipe	"	41.50	12.25	53.75
30" pipe	"	54.00	13.75	67.75
36" pipe	"	74.00	15.25	89.25
12 ga., 48" pipe	"	130	17.50	148
10 ga.				
60" pipe	L.F.	170	20.50	191
72" pipe	"	220	24.50	245
Galvanized or aluminum, plain				
16 ga.				
8" pipe	L.F.	11.00	8.20	19.20
10" pipe	"	15.25	8.48	23.73
12" pipe	"	17.50	8.78	26.28
15" pipe	"	21.75	9.46	31.21
18" pipe	"	26.25	10.25	36.50
24" pipe	"	39.25	12.25	51.50
30" pipe	"	52.00	13.75	65.75
36" pipe	"	65.00	15.25	80.25
12 ga., 48" pipe	"	120	17.50	138
10 ga., 60" pipe	"	160	20.50	181
Galvanized or aluminum, coated oval arch				
16 ga.				
17" x 13"	L.F.	40.25	11.25	51.50
21" x 15"	"	54.00	12.25	66.25
14 ga.				
28" x 20"	L.F.	76.00	13.75	89.75
35" x 24"	"	110	17.50	128
12 ga.				
42" x 29"	L.F.	130	20.50	151

02 SITE CONSTRUCTION

Drainage And Containment	UNIT	MAT.	INST.	TOTAL
02630.40 Storm Drainage *(Cont.)*				
57" x 38"	L.F.	200	24.50	225
64" x 43"	"	240	26.00	266
Oval arch culverts, plain				
16 ga.				
17" x 13"	L.F.	18.75	11.25	30.00
21" x 15"	"	27.25	12.25	39.50
14 ga.				
28" x 20"	L.F.	51.00	13.75	64.75
35" x 24"	"	64.00	17.50	81.50
12 ga.				
57" x 38"	L.F.	100	20.50	121
64" x 43"	"	120	24.50	145
71" x 47"	"	180	26.00	206
Nestable corrugated metal pipe				
16 ga.				
10" pipe	L.F.	15.25	8.48	23.73
12" pipe	"	19.00	8.78	27.78
15" pipe	"	24.50	9.46	33.96
18" pipe	"	28.50	10.25	38.75
24" pipe	"	39.75	12.25	52.00
30" pipe	"	49.25	13.75	63.00
14 ga., 36" pipe	"	57.00	15.25	72.25
Headwalls, cast in place, 30 deg wingwall				
12" pipe	EA.	490	150	640
15" pipe	"	600	150	750
18" pipe	"	730	170	900
24" pipe	"	1,090	170	1,260
30" pipe	"	1,310	200	1,510
36" pipe	"	1,420	300	1,720
42" pipe	"	1,770	300	2,070
48" pipe	"	1,880	400	2,280
54" pipe	"	2,130	500	2,630
60" pipe	"	2,550	600	3,150
4" cleanout for storm drain				
4" pipe	EA.	710	59.00	769
6" pipe	"	870	59.00	929
8" pipe	"	1,210	59.00	1,269
Connect new drain line				
To existing manhole	EA.	160	160	320
To new manhole	"	140	95.00	235
02630.70 Underdrain				
Drain tile, clay				
6" pipe	L.F.	5.34	5.46	10.80
8" pipe	"	8.52	5.72	14.24
12" pipe	"	17.25	6.15	23.40
Porous concrete, standard strength				
6" pipe	L.F.	4.81	5.46	10.27
8" pipe	"	5.20	5.72	10.92
12" pipe	"	6.87	6.15	13.02
15" pipe	"	12.50	6.83	19.33
18" pipe	"	16.75	8.20	24.95
Corrugated metal pipe, perforated type				

02 SITE CONSTRUCTION

Drainage And Containment

02630.70 Underdrain *(Cont.)*

	UNIT	MAT.	INST.	TOTAL
6" pipe	L.F.	6.66	6.15	12.81
8" pipe	"	7.87	6.47	14.34
10" pipe	"	9.54	6.83	16.37
12" pipe	"	13.50	7.23	20.73
18" pipe	"	16.75	7.68	24.43
Perforated clay pipe				
6" pipe	L.F.	6.43	7.02	13.45
8" pipe	"	8.62	7.23	15.85
12" pipe	"	15.00	7.45	22.45
Drain tile, concrete				
6" pipe	L.F.	3.82	5.46	9.28
8" pipe	"	5.94	5.72	11.66
12" pipe	"	11.75	6.15	17.90
Perforated rigid PVC underdrain pipe				
4" pipe	L.F.	2.26	4.10	6.36
6" pipe	"	4.33	4.92	9.25
8" pipe	"	6.62	5.46	12.08
10" pipe	"	10.25	6.15	16.40
12" pipe	"	15.75	7.02	22.77
Underslab drainage, crushed stone				
3" thick	S.F.	0.29	0.82	1.11
4" thick	"	0.40	0.94	1.34
6" thick	"	0.61	1.02	1.63
8" thick	"	0.80	1.06	1.86
Plastic filter fabric for drain lines	"	0.18	0.47	0.65
Gravel fill in trench, crushed or bank run, 1/2" to 3/4"	C.Y.	38.00	62.00	100

Base Courses And Ballasts

02720.10 Railroad Ballast, Rail, Appurtenances

	UNIT	MAT.	INST.	TOTAL
Rail				
90 lb	L.F.	32.75	0.98	33.73
100 lb	"	37.25	0.98	38.23
115 lb	"	42.00	0.98	42.98
132 lb	"	46.75	0.98	47.73
Rail relay				
90 lb	L.F.	15.00	0.98	15.98
100 lb	"	16.75	0.98	17.73
115 lb	"	20.25	0.98	21.23
132 lb	"	24.50	0.98	25.48
New angle bars, per pair				
90 lb	EA.	120	1.23	121
100 lb	"	130	1.23	131
115 lb	"	170	1.23	171
132 lb	"	190	1.23	191
Angle bar relay				

02 SITE CONSTRUCTION

Base Courses And Ballasts

02720.10 Railroad Ballast, Rail, Appurtenances *(Cont.)*

	UNIT	MAT.	INST.	TOTAL
90 lb	EA.	50.00	1.23	51.23
100 lb	"	51.00	1.23	52.23
115 lb	"	54.00	1.23	55.23
132 lb	"	57.00	1.23	58.23
New tie plates				
90 lb	EA.	16.25	0.87	17.12
100 lb	"	17.00	0.87	17.87
115 lb	"	18.75	0.87	19.62
132 lb	"	19.75	0.87	20.62
Tie plate relay				
90 lb	EA.	5.13	0.87	6.00
100 lb	"	7.21	0.87	8.08
115 lb	"	7.21	0.87	8.08
132 lb	"	8.90	0.87	9.77
Track accessories				
Wooden cross ties, 8'	EA.	57.00	6.15	63.15
Concrete cross ties, 8'	"	300	12.25	312
Tie plugs, 5"	"	24.00	0.61	24.61
Track bolts and nuts, 1"	"	6.30	0.61	6.91
Lockwashers, 1"	"	1.65	0.41	2.06
Track spikes, 6"	"	1.56	2.46	4.02
Wooden switch ties	B.F.	2.67	0.61	3.28
Rail anchors	EA.	6.39	2.23	8.62
Ballast	TON	20.00	12.25	32.25
Gauge rods	EA.	48.25	9.84	58.09
Compromise splice bars	"	610	16.50	627
Turnout				
90 lb	EA.	17,340	2,460	19,800
100 lb	"	18,270	2,460	20,730
110 lb	"	19,650	2,460	22,110
115 lb	"	20,110	2,460	22,570
132 lb	"	21,960	2,460	24,420
Turnout relay				
90 lb	EA.	11,100	2,460	13,560
100 lb	"	12,250	2,460	14,710
110 lb	"	12,710	2,460	15,170
115 lb	"	13,400	2,460	15,860
132 lb	"	14,560	2,460	17,020
Railroad track in place, complete				
New rail				
90 lb	L.F.	220	24.50	245
100 lb	"	220	24.50	245
110 lb	"	210	24.50	235
115 lb	"	230	24.50	255
132 lb	"	230	24.50	255
Rail relay				
90 lb	L.F.	120	24.50	145
100 lb	"	140	24.50	165
110 lb	"	140	24.50	165
115 lb	"	150	24.50	175
132 lb	"	150	24.50	175
No. 8 turnout				
90 lb	EA.	43,460	3,280	46,740

02 SITE CONSTRUCTION

Base Courses And Ballasts	UNIT	MAT.	INST.	TOTAL
02720.10 **Railroad Ballast, Rail, Appurtenances** *(Cont.)*				
100 lb	EA.	48,310	3,280	51,590
110 lb	"	52,940	3,280	56,220
115 lb	"	54,090	3,280	57,370
132 lb	"	55,250	3,280	58,530
No. 8 turnout relay				
90 lb	EA.	31,600	3,280	34,880
100 lb	"	33,230	3,280	36,510
110 lb	"	34,920	3,280	38,200
115 lb	"	38,850	3,280	42,130
132 lb	"	41,320	3,280	44,600
Railroad crossings, asphalt, based on 8" thick x 20'				
Including track and approach				
12' roadway	EA.	1,080	620	1,700
15' roadway	"	1,270	700	1,970
18' roadway	"	1,480	820	2,300
21' roadway	"	1,650	980	2,630
24' roadway	"	1,870	1,230	3,100
Precast concrete inserts				
12' roadway	EA.	1,560	250	1,810
15' roadway	"	1,870	310	2,180
18' roadway	"	2,230	410	2,640
21' roadway	"	2,870	490	3,360
24' roadway	"	3,480	550	4,030
Molded rubber, with headers				
12' roadway	EA.	9,390	250	9,640
15' roadway	"	11,750	310	12,060
18' roadway	"	13,910	410	14,320
21' roadway	"	15,340	490	15,830
24' roadway	"	18,700	550	19,250

Flexible Surfaces	UNIT	MAT.	INST.	TOTAL
02740.10 **Asphalt Repair**				
Coal tar seal coat, rubber add., fuel resist.	S.Y.	2.75	0.67	3.42
Bituminous surface treatment, single	"	2.53	0.47	3.00
Double	"	3.36	0.05	3.41
Bituminous prime coat	"	1.70	0.05	1.75
Tack coat	"	0.83	0.04	0.87
Crack sealing, concrete paving	L.F.	1.41	0.31	1.72
Bituminous paving for pipe trench, 4" thick	S.Y.	16.25	16.50	32.75
Polypropylene, nonwoven paving fabric	"	2.08	0.23	2.31
Rubberized asphalt	"	3.36	4.30	7.66
Asphalt slurry seal	"	8.30	2.78	11.08

02 SITE CONSTRUCTION

Flexible Surfaces

	UNIT	MAT.	INST.	TOTAL
02740.20 Asphalt Surfaces				
Asphalt wearing surface, flexible pavement				
1" thick	S.Y.	5.08	2.42	7.50
1-1/2" thick	"	7.67	2.91	10.58
2" thick	"	10.25	3.63	13.88
3" thick	"	15.25	4.85	20.10
Binder course				
1-1/2" thick	S.Y.	7.26	2.69	9.95
2" thick	"	9.65	3.30	12.95
3" thick	"	14.50	4.40	18.90
4" thick	"	19.25	4.85	24.10
5" thick	"	24.00	5.38	29.38
6" thick	"	29.00	6.06	35.06
Bituminous sidewalk, no base				
2" thick	S.Y.	11.00	2.89	13.89
3" thick	"	16.50	3.07	19.57

Rigid Pavement

	UNIT	MAT.	INST.	TOTAL
02750.10 Concrete Paving				
Concrete paving, reinforced, 5000 psi concrete				
6" thick	S.Y.	29.50	22.75	52.25
7" thick	"	34.50	24.25	58.75
8" thick	"	39.50	26.00	65.50
9" thick	"	44.25	28.00	72.25
10" thick	"	49.25	30.25	79.50
11" thick	"	54.00	33.00	87.00
12" thick	"	59.00	36.50	95.50
15" thick	"	74.00	45.50	120
Concrete paving, for pipe trench, reinforced				
7" thick	S.Y.	59.00	24.50	83.50
8" thick	"	63.00	27.25	90.25
9" thick	"	68.00	30.75	98.75
10" thick	"	73.00	35.25	108
Fibrous concrete				
5" thick	S.Y.	30.25	28.00	58.25
8" thick	"	38.50	30.25	68.75
Roller comp.conc., (RCC), place and compact				
8" thick	S.Y.	37.25	36.50	73.75
12" thick	"	57.00	45.50	103
Steel edge forms up to				
12" deep	L.F.	0.96	1.58	2.54
15" deep	"	1.24	1.89	3.13
Paving finishes				
Belt dragged	S.Y.		2.37	2.37
Curing	"	0.42	0.47	0.89

02 SITE CONSTRUCTION

Rigid Pavement	UNIT	MAT.	INST.	TOTAL
02760.10 Pavement Markings				
Pavement line marking, paint				
4" wide	L.F.	0.29	0.11	0.40
6" wide	"	0.29	0.26	0.55
8" wide	"	0.45	0.39	0.84
Reflective paint, 4" wide	"	0.62	0.39	1.01
Airfield markings, retro-reflective				
White	L.F.	1.01	0.39	1.40
Yellow	"	1.09	0.39	1.48
Preformed tape, 4" wide				
Inlaid reflective	L.F.	2.73	0.06	2.79
Reflective paint	"	1.87	0.11	1.98
Thermoplastic				
White	L.F.	1.15	0.23	1.38
Yellow	"	1.15	0.23	1.38
12" wide, thermoplastic, white	"	3.25	0.67	3.92
Directional arrows, reflective preformed tape	EA.	160	47.50	208
Messages, reflective preformed tape (per letter)	"	82.00	23.75	106
Handicap symbol, preformed tape	"	33.75	47.50	81.25
Parking stall painting	"	7.80	9.48	17.28

Site Improvements	UNIT	MAT.	INST.	TOTAL
02810.40 Lawn Irrigation				
Residential system, complete				
Minimum	ACRE			18,840
Maximum	"			35,860
Commercial system, complete				
Minimum	ACRE			28,600
Maximum	"			45,210
02820.10 Chain Link Fence				
Chain link fence, 9 ga., galvanized, with posts 10' o.c.				
4' high	L.F.	7.93	3.38	11.31
5' high	"	10.50	4.30	14.80
6' high	"	12.00	5.92	17.92
7' high	"	13.75	7.29	21.04
8' high	"	15.75	9.48	25.23
For barbed wire with hangers, add				
3 strand	L.F.	2.89	2.37	5.26
6 strand	"	4.90	3.95	8.85
Corner or gate post, 3" post				
4' high	EA.	92.00	15.75	108
5' high	"	100	17.50	118
6' high	"	110	20.50	131
7' high	"	140	23.75	164
8' high	"	140	26.25	166

02 SITE CONSTRUCTION

Site Improvements

02820.10 Chain Link Fence (Cont.)

	UNIT	MAT.	INST.	TOTAL
4" post				
4' high	EA.	160	17.50	178
5' high	"	190	20.50	211
6' high	"	210	23.75	234
7' high	"	220	26.25	246
8' high	"	250	29.50	280
Gate with gate posts, galvanized, 3' wide				
4' high	EA.	100	120	220
5' high	"	130	160	290
6' high	"	160	160	320
7' high	"	190	240	430
8' high	"	200	240	440
Fabric, galvanized chain link, 2" mesh, 9 ga.				
4' high	L.F.	4.32	1.58	5.90
5' high	"	5.27	1.89	7.16
6' high	"	7.38	2.37	9.75
8' high	"	12.25	3.16	15.41
Line post, no rail fitting, galvanized, 2-1/2" dia.				
4' high	EA.	30.00	13.50	43.50
5' high	"	32.75	14.75	47.50
6' high	"	35.75	15.75	51.50
7' high	"	40.75	19.00	59.75
8' high	"	45.25	23.75	69.00
1-7/8" H beam				
4' high	EA.	37.50	13.50	51.00
5' high	"	41.75	14.75	56.50
6' high	"	50.00	15.75	65.75
7' high	"	57.00	19.00	76.00
8' high	"	61.00	23.75	84.75
2-1/4" H beam				
4' high	EA.	27.25	13.50	40.75
5' high	"	34.00	14.75	48.75
6' high	"	39.00	15.75	54.75
7' high	"	45.25	19.00	64.25
8' high	"	53.00	23.75	76.75
Vinyl coated, 9 ga., with posts 10' o.c.				
4' high	L.F.	8.58	3.38	11.96
5' high	"	10.25	4.30	14.55
6' high	"	12.25	5.92	18.17
7' high	"	13.25	7.29	20.54
8' high	"	15.25	9.48	24.73
For barbed wire w/hangers, add				
3 strand	L.F.	3.12	2.37	5.49
6 Strand	"	5.03	3.95	8.98
Corner, or gate post, 4' high				
3" dia.	EA.	110	15.75	126
4" dia.	"	170	15.75	186
6" dia.	"	200	19.00	219
Gate, with posts, 3' wide				
4' high	EA.	120	120	240
5' high	"	150	160	310
6' high	"	170	160	330
7' high	"	200	240	440

02 SITE CONSTRUCTION

Site Improvements

02820.10 Chain Link Fence (Cont.)

	UNIT	MAT.	INST.	TOTAL
8' high	EA.	220	240	460
Line post, no rail fitting, 2-1/2" dia.				
4' high	EA.	30.00	13.50	43.50
5' high	"	32.75	14.75	47.50
6' high	"	35.75	15.75	51.50
7' high	"	40.75	19.00	59.75
8' high	"	45.25	23.75	69.00
Corner post, no top rail fitting, 4" dia.				
4' high	EA.	160	15.75	176
5' high	"	190	17.50	208
6' high	"	210	20.50	231
7' high	"	220	23.75	244
8' high	"	250	26.25	276
Fabric, vinyl, chain link, 2" mesh, 9 ga.				
4' high	L.F.	4.32	1.58	5.90
5' high	"	5.27	1.89	7.16
6' high	"	7.38	2.37	9.75
8' high	"	12.25	3.16	15.41
Swing gates, galvanized, 4' high				
Single gate				
3' wide	EA.	230	120	350
4' wide	"	260	120	380
Double gate				
10' wide	EA.	610	190	800
12' wide	"	660	190	850
14' wide	"	670	190	860
16' wide	"	750	190	940
18' wide	"	810	270	1,080
20' wide	"	860	270	1,130
22' wide	"	960	270	1,230
24' wide	"	980	320	1,300
26' wide	"	1,020	320	1,340
28' wide	"	1,090	380	1,470
30' wide	"	1,150	380	1,530
5' high				
Single gate				
3' wide	EA.	250	160	410
4' wide	"	280	160	440
Double gate				
10' wide	EA.	660	240	900
12' wide	"	700	240	940
14' wide	"	1,210	240	1,450
16' wide	"	790	240	1,030
18' wide	"	810	270	1,080
20' wide	"	920	270	1,190
22' wide	"	960	270	1,230
24' wide	"	1,010	320	1,330
26' wide	"	1,050	320	1,370
28' wide	"	1,170	380	1,550
30' wide	"	1,220	380	1,600
6' high				
Single gate				
3' wide	EA.	310	160	470

02 SITE CONSTRUCTION

Site Improvements

02820.10 Chain Link Fence (Cont.)

	UNIT	MAT.	INST.	TOTAL
4' wide	EA.	340	160	500
Double gate				
10' wide	EA.	750	240	990
12' wide	"	850	240	1,090
14' wide	"	900	240	1,140
16' wide	"	970	240	1,210
18' wide	"	1,030	270	1,300
20' wide	"	1,070	270	1,340
22' wide	"	1,150	270	1,420
24' wide	"	1,230	320	1,550
26' wide	"	1,270	320	1,590
28' wide	"	1,380	380	1,760
30' wide	"	1,450	380	1,830
7' high				
Single gate				
3' wide	EA.	380	240	620
4' wide	"	420	240	660
Double gate				
10' wide	EA.	970	320	1,290
12' wide	"	1,070	320	1,390
14' wide	"	1,150	320	1,470
16' wide	"	1,230	320	1,550
18' wide	"	1,300	380	1,680
20' wide	"	1,380	380	1,760
22' wide	"	1,460	380	1,840
24' wide	"	1,570	470	2,040
26' wide	"	1,670	470	2,140
28' wide	"	1,760	590	2,350
30' wide	"	2,020	590	2,610
8' high				
Single gate				
3' wide	EA.	420	240	660
4' wide	"	460	240	700
Double gate				
10' wide	EA.	1,110	320	1,430
12' wide	"	1,180	320	1,500
14' wide	"	1,270	320	1,590
16' wide	"	1,390	320	1,710
18' wide	"	1,440	380	1,820
20' wide	"	1,500	380	1,880
22' wide	"	1,580	380	1,960
24' wide	"	1,750	470	2,220
26' wide	"	1,810	470	2,280
28' wide	"	1,930	590	2,520
30' wide	"	2,120	590	2,710
Vinyl coated swing gates, 4' high				
Single gate				
3' wide	EA.	350	120	470
4' wide	"	390	120	510
Double gate				
10' wide	EA.	910	190	1,100
12' wide	"	990	190	1,180
14' wide	"	1,010	190	1,200

02 SITE CONSTRUCTION

Site Improvements

02820.10 Chain Link Fence (Cont.)

	UNIT	MAT.	INST.	TOTAL
16' wide	EA.	1,130	190	1,320
18' wide	"	1,220	270	1,490
20' wide	"	1,280	270	1,550
22' wide	"	1,430	270	1,700
24' wide	"	1,470	320	1,790
26' wide	"	1,530	320	1,850
28' wide	"	1,630	380	2,010
30' wide	"	1,730	380	2,110
5' high				
Single gate				
3' wide	EA.	370	160	530
4' wide	"	420	160	580
Double gate				
10' wide	EA.	990	240	1,230
12' wide	"	1,050	240	1,290
14' wide	"	1,810	240	2,050
16' wide	"	1,180	240	1,420
18' wide	"	1,220	270	1,490
20' wide	"	1,380	270	1,650
22' wide	"	1,450	270	1,720
24' wide	"	1,510	320	1,830
26' wide	"	1,580	320	1,900
28' wide	"	1,760	380	2,140
30' wide	"	1,830	380	2,210
6' high				
Single gate				
3' wide	EA.	460	160	620
4' wide	"	500	160	660
Double gate				
10' wide	EA.	1,120	240	1,360
12' wide	"	1,270	240	1,510
14' wide	"	1,350	240	1,590
16' wide	"	1,450	240	1,690
18' wide	"	1,550	270	1,820
20' wide	"	1,600	270	1,870
22' wide	"	1,720	270	1,990
24' wide	"	1,850	320	2,170
26' wide	"	1,900	320	2,220
28' wide	"	2,070	380	2,450
30' wide	"	2,180	380	2,560
7' high				
Single gate				
3' wide	EA.	570	240	810
4' wide	"	630	240	870
Double gate				
10' wide	EA.	1,460	320	1,780
12' wide	"	1,600	320	1,920
14' wide	"	1,720	320	2,040
16' wide	"	1,840	320	2,160
18' wide	"	1,960	380	2,340
20' wide	"	2,080	380	2,460
22' wide	"	2,190	380	2,570
24' wide	"	2,350	470	2,820

02 SITE CONSTRUCTION

Site Improvements

02820.10 Chain Link Fence *(Cont.)*

	UNIT	MAT.	INST.	TOTAL
26' wide	EA.	2,510	470	2,980
28' wide	"	2,640	590	3,230
30' wide	"	3,020	590	3,610
8' high				
Single gate				
3' wide	EA.	640	240	880
4' wide	"	680	240	920
Double gate				
10' wide	EA.	1,670	320	1,990
12' wide	"	1,770	320	2,090
14' wide	"	1,900	320	2,220
16' wide	"	2,080	320	2,400
18' wide	"	2,160	380	2,540
20' wide	"	2,250	380	2,630
22' wide	"	2,370	380	2,750
24' wide	"	2,630	470	3,100
28' wide	"	2,720	470	3,190
30' wide	"	2,890	590	3,480
Motor operator for gates, no wiring	"			6,690
Drilling fence post holes				
In soil				
By hand	EA.		23.75	23.75
By machine auger	"		15.00	15.00
In rock				
By jackhammer	EA.		200	200
By rock drill	"		60.00	60.00
Aluminum privacy slats, installed vertically	S.F.	1.14	1.18	2.32
Post hole, dig by hand	EA.		31.50	31.50
Set fence post in concrete	"	11.25	23.75	35.00

02840.30 Guardrails

	UNIT	MAT.	INST.	TOTAL
Pipe bollard, steel pipe, concrete filled, painted				
6" dia.	EA.	250	39.50	290
8" dia.	"	330	59.00	389
12" dia.	"	520	160	680
Corrugated steel, guardrail, galvanized	L.F.	31.25	4.10	35.35
End section, wrap around or flared	EA.	83.00	47.50	131
Timber guardrail, 4" x 8"	L.F.	36.50	3.07	39.57
Guard rail, 3 cables, 3/4" dia.				
Steel posts	L.F.	16.25	12.25	28.50
Wood posts	"	16.50	9.84	26.34
Steel box beam				
6" x 6"	L.F.	67.00	13.75	80.75
6" x 8"	"	71.00	15.25	86.25
Concrete posts	EA.	43.75	23.75	67.50
Barrel type impact barrier	"	550	47.50	598
Light shield, 6' high	L.F.	36.25	9.48	45.73

02 SITE CONSTRUCTION

Site Improvements	UNIT	MAT.	INST.	TOTAL
02840.40 — Parking Barriers				
Timber, treated, 4' long				
4" x 4"	EA.	13.75	39.50	53.25
6" x 6"	"	25.50	47.50	73.00
Precast concrete, 6' long, with dowels				
12" x 6"	EA.	63.00	23.75	86.75
12" x 8"	"	75.00	26.25	101
02840.60 — Signage				
Traffic signs				
Reflectorized per OSHA stds., incl. post				
Stop, 24"x24"	EA.	87.00	31.50	119
Yield, 30" triangle	"	47.50	31.50	79.00
Speed limit, 12"x18"	"	54.00	31.50	85.50
Directional, 12"x18"	"	66.00	31.50	97.50
Exit, 12"x18"	"	66.00	31.50	97.50
Entry, 12"x18"	"	66.00	31.50	97.50
Warning, 24"x24"	"	86.00	31.50	118
Informational, 12"x18"	"	44.00	31.50	75.50
Handicap parking, 12"x18"	"	45.00	31.50	76.50
02880.40 — Recreational Facilities				
Bleachers, outdoor, portable, per seat				
10 tiers				
Minimum	EA.	77.00	15.25	92.25
Maximum	"	98.00	20.50	119
20 tiers				
Minimum	EA.	82.00	14.50	96.50
Maximum	"	110	19.00	129
Grandstands, fixed, wood seat, steel frame				
Per seat, 15 tiers				
Minimum	EA.	74.00	24.50	98.50
Maximum	"	130	41.00	171
30 tiers				
Minimum	EA.	76.00	22.25	98.25
Maximum	"	160	35.25	195
Seats				
Seat backs only				
Fiberglass	EA.	42.00	4.74	46.74
Steel and wood seat	"	61.00	4.74	65.74
Seat restoration, fiberglass on wood				
Seats	EA.	30.25	9.48	39.73
Plain bench, no backs	"	18.75	3.95	22.70
Benches				
Park, precast concrete with backs				
4' long	EA.	1,160	160	1,320
8' long	"	2,570	240	2,810
Fiberglass, with backs				
4' long	EA.	930	120	1,050
8' long	"	1,780	160	1,940
Wood, with backs and fiberglass supports				
4' long	EA.	520	120	640
8' long	"	540	160	700
Steel frame, 6' long				

02 SITE CONSTRUCTION

02880.40 Site Improvements — Recreational Facilities (Cont.)

	UNIT	MAT.	INST.	TOTAL
All steel	EA.	440	120	560
Hardwood boards	"	310	120	430
Players bench, steel frame, fir seat, 10' long	"	340	160	500
Backstops				
Handball or squash court, outdoor				
Wood	EA.			51,740
Masonry	"			38,610
Soccer goal posts	PAIR			3,450
Running track				
Gravel and cinders over stone base	S.Y.	10.50	6.15	16.65
Rubber-cork base resilient pavement	"	15.00	49.25	64.25
For colored surfaces, add	"	9.42	4.92	14.34
Colored rubberized asphalt	"	20.00	62.00	82.00
Artificial resilient mat over asphalt	"	47.25	120	167
Tennis courts				
Bituminous pavement, 2-1/2" thick	S.Y.	32.50	15.25	47.75
Colored sealer, acrylic emulsion				
3 coats	S.Y.	7.93	3.16	11.09
For 2 color seal coating, add	"	1.44	0.47	1.91
For preparing old courts, add	"	3.35	0.31	3.66
Net, nylon, 42' long	EA.	460	59.00	519
Paint markings on asphalt, 2 coats	"	180	470	650
Complete court with fence, etc., bituminous				
Minimum	EA.			24,660
Average	"			42,440
Maximum	"			60,220
Clay court				
Minimum	EA.			25,660
Average	"			34,980
Maximum	"			53,050
Playground equipment				
Basketball backboard				
Minimum	EA.	1,000	120	1,120
Maximum	"	1,890	140	2,030
Bike rack, 10' long	"	630	95.00	725
Golf shelter, fiberglass	"	3,170	120	3,290
Ground socket for movable posts				
Minimum	EA.	150	29.50	180
Maximum	"	290	29.50	320
Horizontal monkey ladder, 14' long	"	970	79.00	1,049
Posts, tether ball	"	470	23.75	494
Multiple purpose, 10' long	"	480	47.50	528
See-saw, steel				
Minimum	EA.	1,140	190	1,330
Average	"	2,210	240	2,450
Maximum	"	3,250	320	3,570
Slide				
Minimum	EA.	2,210	380	2,590
Maximum	"	5,850	430	6,280
Swings, plain seats				
8' high				
Minimum	EA.	1,050	320	1,370
Maximum	"	2,000	360	2,360

02 SITE CONSTRUCTION

Site Improvements	UNIT	MAT.	INST.	TOTAL
02880.40 Recreational Facilities (Cont.)				
12' high				
Minimum	EA.	1,620	360	1,980
Maximum	"	2,920	530	3,450
02880.70 Recreational Courts				
Walls, galvanized steel				
8' high	L.F.	17.00	9.48	26.48
10' high	"	20.25	10.50	30.75
12' high	"	23.25	12.50	35.75
Vinyl coated				
8' high	L.F.	16.50	9.48	25.98
10' high	"	20.00	10.50	30.50
12' high	"	22.25	12.50	34.75
Gates, galvanized steel				
Single, 3' transom				
3'x7'	EA.	400	240	640
4'x7'	"	420	270	690
5'x7'	"	580	320	900
6'x7'	"	630	380	1,010
Double, 3' transom				
10'x7'	EA.	970	950	1,920
12'x7'	"	1,250	1,050	2,300
14'x7'	"	1,490	1,180	2,670
Double, no transom				
10'x10'	EA.	1,050	790	1,840
12'x10'	"	1,260	950	2,210
14'x10'	"	1,440	1,050	2,490
Vinyl coated				
Single, 3' transom				
3'x7'	EA.	780	240	1,020
4'x7'	"	850	270	1,120
5'x7'	"	850	320	1,170
6'x7'	"	880	380	1,260
Double, 3'				
10'x7'	EA.	2,310	950	3,260
12'x7'	"	2,370	1,050	3,420
14'x7'	"	2,560	1,180	3,740
Double, no transom				
10'x10'	EA.	2,300	790	3,090
12'x10'	"	2,340	950	3,290
14'x10'	"	2,560	1,050	3,610
Baseball backstop, regulation				
Galvanized	EA.			9,700
Vinyl coated	"			13,510
Softball backstop, regulation				
14' high				
Galvanized	EA.			11,050
Vinyl coated	"			17,550
18' high				
Galvanized	EA.			12,210
Vinyl coated	"			17,660
20' high				
Galvanized	EA.			14,470

02 SITE CONSTRUCTION

Site Improvements	UNIT	MAT.	INST.	TOTAL
02880.70 Recreational Courts *(Cont.)*				
Vinyl coated	EA.			20,850
22' high				
Galvanized	EA.			16,730
Vinyl coated	"			24,440
24' high				
Galvanized	EA.			17,690
Vinyl coated	"			29,100
Wire and miscellaneous metal fences				
Chicken wire, post 4' o.c.				
2" mesh				
4' high	L.F.	2.31	2.37	4.68
6' high	"	2.59	3.16	5.75
Galvanized steel				
12 gauge, 2" by 4" mesh, posts 5' o.c.				
3' high	L.F.	3.69	2.37	6.06
5' high	"	5.26	2.96	8.22
14 gauge, 1" by 2" mesh, posts 5' o.c.				
3' high	L.F.	3.09	2.37	5.46
5' high	"	5.03	2.96	7.99

Planting	UNIT	MAT.	INST.	TOTAL
02910.10 Topsoil				
Spread topsoil, with equipment				
Minimum	C.Y.		13.25	13.25
Maximum	"		16.50	16.50
By hand				
Minimum	C.Y.		47.50	47.50
Maximum	"		59.00	59.00
Area prep. seeding (grade, rake and clean)				
Square yard	S.Y.		0.37	0.37
By acre	ACRE		1,900	1,900
Remove topsoil and stockpile on site				
4" deep	C.Y.		11.00	11.00
6" deep	"		10.25	10.25
Spreading topsoil from stock pile				
By loader	C.Y.		12.00	12.00
By hand	"		130	130
Top dress by hand	S.Y.		1.32	1.32
Place imported top soil				
By loader				
4" deep	S.Y.		1.32	1.32
6" deep	"		1.47	1.47
By hand				
4" deep	S.Y.		5.26	5.26
6" deep	"		5.92	5.92

02 SITE CONSTRUCTION

Planting	UNIT	MAT.	INST.	TOTAL
02910.10 Topsoil *(Cont.)*				
Plant bed preparation, 18" deep				
With backhoe/loader	S.Y.		3.31	3.31
By hand	"		7.90	7.90
02920.10 Fertilizing				
Fertilizing (23#/1000 sf)				
By square yard	S.Y.	0.03	0.15	0.18
By acre	ACRE	190	750	940
Liming (70#/1000 sf)				
By square yard	S.Y.	0.03	0.20	0.23
By acre	ACRE	190	1,010	1,200
02920.30 Seeding				
Mechanical seeding, 175 lb/acre				
By square yard	S.Y.	0.23	0.12	0.35
By acre	ACRE	910	600	1,510
450 lb/acre				
By square yard	S.Y.	0.58	0.15	0.73
By acre	ACRE	2,260	750	3,010
Seeding by hand, 10 lb per 100 s.y.				
By square yard	S.Y.	0.65	0.15	0.80
By acre	ACRE	2,520	790	3,310
Reseed disturbed areas	S.F.	0.06	0.23	0.29
02930.10 Plants				
Euonymus coloratus, 18" (Purple Wintercreeper)	EA.	2.79	7.90	10.69
Hedera Helix, 2-1/4" pot (English ivy)	"	1.17	7.90	9.07
Liriope muscari, 2" clumps	"	4.87	4.74	9.61
Santolina, 12"	"	5.59	4.74	10.33
Vinca major or minor, 3" pot	"	0.91	4.74	5.65
Cortaderia argentia, 2 gallon (Pampas Grass)	"	17.75	4.74	22.49
Ophiopogan japonicus, 1 quart (4" pot)	"	4.87	4.74	9.61
Ajuga reptans, 2-3/4" pot (carpet bugle)	"	0.91	4.74	5.65
Pachysandra terminalis, 2-3/4" pot (Japanese Spurge)	"	1.23	4.74	5.97
02930.30 Shrubs				
Juniperus conferia litoralis, 18"-24" (Shore Juniper)	EA.	41.00	19.00	60.00
Horizontalis plumosa, 18"-24" (Andorra Juniper)	"	43.50	19.00	62.50
Sabina tamar-iscfolia-tamarix juniper, 18"-24"	"	43.50	19.00	62.50
Chin San Jose, 18"-24" (San Jose Juniper)	"	43.50	19.00	62.50
Sargenti, 18"-24" (Sargent's Juniper)	"	41.00	19.00	60.00
Nandina domestica, 18"-24" (Heavenly Bamboo)	"	27.50	19.00	46.50
Raphiolepis Indica Springtime, 18"-24"	"	29.50	19.00	48.50
Osmanthus Heterophyllus Gulftide, 18"-24"	"	31.50	19.00	50.50
Ilex Cornuta Burfordi Nana, 18"-24"	"	36.00	19.00	55.00
Glabra, 18"-24" (Inkberry Holly)	"	33.75	19.00	52.75
Azalea, Indica types, 18"-24"	"	38.25	19.00	57.25
Kurume types, 18"-24"	"	42.75	19.00	61.75
Berberis Julianae, 18"-24" (Wintergreen Barberry)	"	25.00	19.00	44.00
Pieris Japonica Japanese, 18"-24"	"	25.00	19.00	44.00
Ilex Cornuta Rotunda, 18"-24"	"	29.75	19.00	48.75
Juniperus Horiz. Plumosa, 24"-30"	"	27.25	23.75	51.00
Rhodopendrow Hybrids, 24"-30"	"	73.00	23.75	96.75

02 SITE CONSTRUCTION

Planting	UNIT	MAT.	INST.	TOTAL
02930.30 **Shrubs** *(Cont.)*				
Aucuba Japonica Varigata, 24"-30"	EA.	25.00	23.75	48.75
Ilex Crenata Willow Leaf, 24"-30"	"	27.25	23.75	51.00
Cleyera Japonica, 30"-36"	"	32.00	29.50	61.50
Pittosporum Tobira, 30"-36"	"	36.50	29.50	66.00
Prumus Laurocerasus, 30"-36"	"	69.00	29.50	98.50
Ilex Cornuta Burfordi, 30"-36" (Burford Holly)	"	36.50	29.50	66.00
Abelia Grandiflora, 24"-36" (Yew Podocarpus)	"	25.00	23.75	48.75
Podocarpos Macrophylla, 24"-36"	"	41.00	23.75	64.75
Pyracantha Coccinea Lalandi, 3'-4' (Firethorn)	"	23.25	29.50	52.75
Photinia Frazieri, 3'-4' (Red Photinia)	"	37.00	29.50	66.50
Forsythia Suspensa, 3'-4' (Weeping Forsythia)	"	23.25	29.50	52.75
Camellia Japonica, 3'-4' (Common Camellia)	"	41.25	29.50	70.75
Juniperus Chin Torulosa, 3'-4' (Hollywood Juniper)	"	43.75	29.50	73.25
Cupressocyparis Leylandi, 3'-4'	"	36.75	29.50	66.25
Ilex Opaca Fosteri, 5'-6' (Foster's Holly)	"	150	39.50	190
Opaca, 5'-6' (American Holly)	"	210	39.50	250
Nyrica Cerifera, 4'-5' (Southern Wax Myrtles)	"	46.50	33.75	80.25
Ligustrum Japonicum, 4'-5' (Japanese Privet)	"	36.50	33.75	70.25
02930.60 **Trees**				
Cornus Florida, 5'-6' (White flowering Dogwood)	EA.	110	39.50	150
Prunus Serrulata Kwanzan, 6'-8' (Kwanzan Cherry)	"	120	47.50	168
Caroliniana, 6'-8' (Carolina Cherry Laurel)	"	140	47.50	188
Cercis Canadensis, 6'-8' (Eastern Redbud)	"	97.00	47.50	145
Koelreuteria Paniculata, 8'-10' (Goldenrain Tree)	"	170	59.00	229
Acer Platanoides, 1-3/4"-2" (11'-13')	"	220	79.00	299
Rubrum, 1-3/4"-2" (11'-13') (Red Maple)	"	170	79.00	249
Saccharum, 1-3/4"-2" (Sugar Maple)	"	300	79.00	379
Fraxinus Pennsylvanica, 1-3/4"-2"	"	140	79.00	219
Celtis Occidentalis, 1-3/4"-2"	"	210	79.00	289
Glenditsia Triacantos Inermis, 2"	"	200	79.00	279
Prunus Cerasifera 'Thundercloud', 6'-8'	"	110	47.50	158
Yeodensis, 6'-8' (Yoshino Cherry)	"	120	47.50	168
Lagerstroemia Indica, 8'-10' (Crapemyrtle)	"	190	59.00	249
Crataegus Phaenopyrum, 8'-10'	"	300	59.00	359
Quercus Borealis, 1-3/4"-2" (Northern Red Oak)	"	180	79.00	259
Quercus Acutissima, 1-3/4"-2" (8'-10')	"	170	79.00	249
Saliz Babylonica, 1-3/4"-2" (Weeping Willow)	"	83.00	79.00	162
Tilia Cordata Greenspire, 1-3/4"-2" (10'-12')	"	370	79.00	449
Malus, 2"-2-1/2" (8'-10') (Flowering Crabapple)	"	180	79.00	259
Platanus Occidentalis, (12'-14')	"	280	95.00	375
Pyrus Calleryana Bradford, 2"-2-1/2"	"	210	79.00	289
Quercus Palustris, 2"-2-1/2" (12'-14') (Pin Oak)	"	240	79.00	319
Phellos, 2-1/2"-3" (Willow Oak)	"	260	95.00	355
Nigra, 2"-2-1/2" (Water Oak)	"	220	79.00	299
Magnolia Soulangeana, 4'-5' (Saucer Magnolia)	"	130	39.50	170
Grandiflora, 6'-8' (Southern Magnolia)	"	180	47.50	228
Cedrus Deodara, 10'-12' (Deodare Cedar)	"	290	79.00	369
Gingko Biloba, 10'-12' (2"-2-1/2")	"	280	79.00	359
Pinus Thunbergi, 5'-6' (Japanese Black Pine)	"	110	39.50	150
Strobus, 6'-8' (White Pine)	"	120	47.50	168
Taeda, 6'-8' (Loblolly Pine)	"	100	47.50	148
Quercus Virginiana, 2"-2-1/2" (Live Oak)	"	260	95.00	355

02 SITE CONSTRUCTION

Planting

02935.10 Shrub & Tree Maintenance

	UNIT	MAT.	INST.	TOTAL
Moving shrubs on site				
12" ball	EA.		59.00	59.00
24" ball	"		79.00	79.00
3' high	"		47.50	47.50
4' high	"		53.00	53.00
5' high	"		59.00	59.00
18" spread	"		68.00	68.00
30" spread	"		79.00	79.00
Moving trees on site				
24" ball	EA.		120	120
48" ball	"		160	160
Trees				
3' high	EA.		49.25	49.25
6' high	"		55.00	55.00
8' high	"		62.00	62.00
10' high	"		82.00	82.00
Palm trees				
7' high	EA.		62.00	62.00
10' high	"		82.00	82.00
20' high	"		250	250
40' high	"		490	490
Guying trees				
4" dia.	EA.	9.80	23.75	33.55
8" dia.	"	9.80	29.50	39.30

02935.30 Weed Control

	UNIT	MAT.	INST.	TOTAL
Weed control, bromicil, 15 lb./acre, wettable powder	ACRE	340	240	580
Vegetation control, by application of plant killer	S.Y.	0.02	0.18	0.20
Weed killer, lawns and fields	"	0.28	0.09	0.37

02945.10 Prefabricated Planters

	UNIT	MAT.	INST.	TOTAL
Concrete precast, circular				
24" dia., 18" high	EA.	420	47.50	468
42" dia., 30" high	"	550	59.00	609
Fiberglass, circular				
36" dia., 27" high	EA.	700	23.75	724
60" dia., 39" high	"	1,610	26.25	1,636
Tapered, circular				
24" dia., 36" high	EA.	550	21.50	572
40" dia., 36" high	"	920	23.75	944
Square				
2' by 2', 17" high	EA.	470	21.50	492
4' by 4', 39" high	"	1,610	26.25	1,636
Rectangular				
4' by 1', 18" high	EA.	520	23.75	544

02945.20 Landscape Accessories

	UNIT	MAT.	INST.	TOTAL
Steel edging, 3/16" x 4"	L.F.	0.76	0.59	1.35
Landscaping stepping stones, 15"x15", white	EA.	6.89	2.37	9.26
Wood chip mulch	C.Y.	48.00	31.50	79.50
2" thick	S.Y.	2.95	0.94	3.89
4" thick	"	5.56	1.35	6.91
6" thick	"	8.31	1.72	10.03

02 SITE CONSTRUCTION

Planting

	UNIT	MAT.	INST.	TOTAL
02945.20 Landscape Accessories *(Cont.)*				
Gravel mulch, 3/4" stone	C.Y.	38.00	47.50	85.50
White marble chips, 1" deep	S.F.	0.76	0.47	1.23
Peat moss				
2" thick	S.Y.	4.09	1.05	5.14
4" thick	"	7.87	1.58	9.45
6" thick	"	12.00	1.97	13.97
Landscaping timbers, treated lumber				
4" x 4"	L.F.	1.59	1.58	3.17
6" x 6"	"	3.18	1.69	4.87
8" x 8"	"	5.20	1.97	7.17

Site Restoration

	UNIT	MAT.	INST.	TOTAL
02955.10 Pipeline Restoration				
Relining existing water main				
6" dia.	L.F.	9.24	36.50	45.74
8" dia.	"	10.50	38.25	48.75
10" dia.	"	11.75	40.50	52.25
12" dia.	"	12.75	42.75	55.50
14" dia.	"	13.75	45.50	59.25
16" dia.	"	14.75	48.50	63.25
18" dia.	"	16.00	52.00	68.00
20" dia.	"	17.50	56.00	73.50
24" dia.	"	18.75	61.00	79.75
36" dia.	"	20.25	73.00	93.25
48" dia.	"	22.75	81.00	104
72" dia.	"	29.00	91.00	120
Replacing in line gate valves				
6" valve	EA.	990	490	1,480
8" valve	"	1,550	610	2,160
10" valve	"	2,340	730	3,070
12" valve	"	4,060	910	4,970
16" valve	"	9,220	1,040	10,260
18" valve	"	13,950	1,210	15,160
20" valve	"	19,180	1,460	20,640
24" valve	"	27,400	1,820	29,220
36" valve	"	82,550	2,430	84,980

03 CONCRETE

Formwork

Formwork	UNIT	MAT.	INST.	TOTAL
03110.05 — Beam Formwork				
Beam forms, job built				
Beam bottoms				
1 use	S.F.	5.09	10.00	15.09
2 uses	"	3.00	9.59	12.59
3 uses	"	2.30	9.29	11.59
4 uses	"	1.89	8.88	10.77
5 uses	"	1.74	8.63	10.37
Beam sides				
1 use	S.F.	3.64	6.71	10.35
2 uses	"	2.15	6.36	8.51
3 uses	"	1.89	6.04	7.93
4 uses	"	1.75	5.75	7.50
5 uses	"	1.54	5.49	7.03
03110.10 — Box Culvert Formwork				
Box culverts, job built				
6' x 6'				
1 use	S.F.	3.48	6.04	9.52
2 uses	"	1.89	5.75	7.64
3 uses	"	1.57	5.49	7.06
4 uses	"	1.32	5.25	6.57
5 uses	"	1.17	5.03	6.20
8' x 12'				
1 use	S.F.	3.48	5.03	8.51
2 uses	"	1.89	4.83	6.72
3 uses	"	1.57	4.64	6.21
4 uses	"	1.32	4.47	5.79
5 uses	"	1.17	4.31	5.48
03110.15 — Column Formwork				
Column, square forms, job built				
8" x 8" columns				
1 use	S.F.	4.00	12.00	16.00
2 uses	"	2.15	11.50	13.65
3 uses	"	1.82	11.25	13.07
4 uses	"	1.66	10.75	12.41
5 uses	"	1.41	10.50	11.91
12" x 12" columns				
1 use	S.F.	3.65	11.00	14.65
2 uses	"	2.02	10.50	12.52
3 uses	"	1.62	10.25	11.87
4 uses	"	1.41	9.90	11.31
5 uses	"	1.19	9.59	10.78
16" x 16" columns				
1 use	S.F.	3.48	10.00	13.48
2 uses	"	1.83	9.74	11.57
3 uses	"	1.46	9.44	10.90
4 uses	"	1.33	9.15	10.48
5 uses	"	1.10	8.88	9.98
24" x 24" columns				
1 use	S.F.	3.48	9.29	12.77
2 uses	"	1.62	9.02	10.64
3 uses	"	1.35	8.75	10.10

03 CONCRETE

Formwork	UNIT	MAT.	INST.	TOTAL
03110.15 Column Formwork (Cont.)				
4 uses	S.F.	1.10	8.51	9.61
5 uses	"	1.01	8.27	9.28
36" x 36" columns				
1 use	S.F.	3.51	8.63	12.14
2 uses	"	1.65	8.39	10.04
3 uses	"	1.35	8.16	9.51
4 uses	"	1.17	7.95	9.12
5 uses	"	1.09	7.74	8.83
Round fiber forms, 1 use				
10" dia.	L.F.	5.22	12.00	17.22
12" dia.	"	6.42	12.25	18.67
14" dia.	"	8.43	12.75	21.18
16" dia.	"	11.00	13.50	24.50
18" dia.	"	18.00	14.50	32.50
24" dia.	"	22.00	15.50	37.50
30" dia.	"	33.25	16.75	50.00
36" dia.	"	41.25	18.25	59.50
42" dia.	"	75.00	20.25	95.25
03110.18 Curb Formwork				
Curb forms				
Straight, 6" high				
1 use	L.F.	2.37	6.04	8.41
2 uses	"	1.42	5.75	7.17
3 uses	"	1.07	5.49	6.56
4 uses	"	0.96	5.25	6.21
5 uses	"	0.86	5.03	5.89
Curved, 6" high				
1 use	L.F.	2.56	7.55	10.11
2 uses	"	1.61	7.11	8.72
3 uses	"	1.23	6.71	7.94
4 uses	"	1.12	6.42	7.54
5 uses	"	1.04	6.16	7.20
03110.20 Elevated Slab Formwork				
Elevated slab formwork				
Slab, with drop panels				
1 use	S.F.	4.26	4.83	9.09
2 uses	"	2.46	4.64	7.10
3 uses	"	1.92	4.47	6.39
4 uses	"	1.70	4.31	6.01
5 uses	"	1.52	4.16	5.68
Floor slab, hung from steel beams				
1 use	S.F.	3.42	4.64	8.06
2 uses	"	1.87	4.47	6.34
3 uses	"	1.71	4.31	6.02
4 uses	"	1.48	4.16	5.64
5 uses	"	1.26	4.02	5.28
Floor slab, with pans or domes				
1 use	S.F.	6.11	5.49	11.60
2 uses	"	3.97	5.25	9.22
3 uses	"	3.70	5.03	8.73
4 uses	"	3.45	4.83	8.28

03 CONCRETE

Formwork

03110.20 Elevated Slab Formwork (Cont.)

	UNIT	MAT.	INST.	TOTAL
5 uses	S.F.	3.05	4.64	7.69
Equipment curbs, 12" high				
1 use	L.F.	3.15	6.04	9.19
2 uses	"	2.02	5.75	7.77
3 uses	"	1.76	5.49	7.25
4 uses	"	1.59	5.25	6.84
5 uses	"	1.37	5.03	6.40

03110.25 Equipment Pad Formwork

	UNIT	MAT.	INST.	TOTAL
Equipment pad, job built				
1 use	S.F.	4.11	7.55	11.66
2 uses	"	2.46	7.11	9.57
3 uses	"	1.98	6.71	8.69
4 uses	"	1.54	6.36	7.90
5 uses	"	1.23	6.04	7.27

03110.35 Footing Formwork

	UNIT	MAT.	INST.	TOTAL
Wall footings, job built, continuous				
1 use	S.F.	1.94	6.04	7.98
2 uses	"	1.37	5.75	7.12
3 uses	"	1.12	5.49	6.61
4 uses	"	1.00	5.25	6.25
5 uses	"	0.86	5.03	5.89
Column footings, spread				
1 use	S.F.	2.05	7.55	9.60
2 uses	"	1.53	7.11	8.64
3 uses	"	1.09	6.71	7.80
4 uses	"	0.91	6.36	7.27
5 uses	"	0.83	6.04	6.87

03110.50 Grade Beam Formwork

	UNIT	MAT.	INST.	TOTAL
Grade beams, job built				
1 use	S.F.	3.06	6.04	9.10
2 uses	"	1.71	5.75	7.46
3 uses	"	1.34	5.49	6.83
4 uses	"	1.11	5.25	6.36
5 uses	"	0.93	5.03	5.96

03110.53 Pile Cap Formwork

	UNIT	MAT.	INST.	TOTAL
Pile cap forms, job built				
Square				
1 use	S.F.	3.48	7.55	11.03
2 uses	"	2.01	7.11	9.12
3 uses	"	1.60	6.71	8.31
4 uses	"	1.41	6.36	7.77
5 uses	"	1.16	6.04	7.20
Triangular				
1 use	S.F.	3.70	8.63	12.33
2 uses	"	2.44	8.05	10.49
3 uses	"	1.94	7.55	9.49
4 uses	"	1.60	7.11	8.71
5 uses	"	1.27	6.71	7.98

03 CONCRETE

Formwork

Formwork	UNIT	MAT.	INST.	TOTAL
03110.55 — Slab / Mat Formwork				
Mat foundations, job built				
1 use	S.F.	3.04	7.55	10.59
2 uses	"	1.75	7.11	8.86
3 uses	"	1.28	6.71	7.99
4 uses	"	1.09	6.36	7.45
5 uses	"	0.87	6.04	6.91
Edge forms				
6" high				
1 use	L.F.	3.05	5.49	8.54
2 uses	"	1.75	5.25	7.00
3 uses	"	1.28	5.03	6.31
4 uses	"	1.09	4.83	5.92
5 uses	"	0.87	4.64	5.51
12" high				
1 use	L.F.	2.88	6.04	8.92
2 uses	"	1.63	5.75	7.38
3 uses	"	1.19	5.49	6.68
4 uses	"	0.98	5.25	6.23
5 uses	"	0.80	5.03	5.83
Formwork for openings				
1 use	S.F.	4.11	12.00	16.11
2 uses	"	2.37	11.00	13.37
3 uses	"	1.98	10.00	11.98
4 uses	"	1.53	9.29	10.82
5 uses	"	1.27	8.63	9.90
03110.60 — Stair Formwork				
Stairway forms, job built				
1 use	S.F.	4.92	12.00	16.92
2 uses	"	2.75	11.00	13.75
3 uses	"	2.14	10.00	12.14
4 uses	"	1.96	9.29	11.25
5 uses	"	1.65	8.63	10.28
Stairs, elevated				
1 use	S.F.	5.94	12.00	17.94
2 uses	"	3.15	10.00	13.15
3 uses	"	2.75	8.63	11.38
4 uses	"	2.37	8.05	10.42
5 uses	"	1.96	7.55	9.51
03110.65 — Wall Formwork				
Wall forms, exterior, job built				
Up to 8' high wall				
1 use	S.F.	3.19	6.04	9.23
2 uses	"	1.76	5.75	7.51
3 uses	"	1.56	5.49	7.05
4 uses	"	1.33	5.25	6.58
5 uses	"	1.17	5.03	6.20
Over 8' high wall				
1 use	S.F.	3.51	7.55	11.06
2 uses	"	2.01	7.11	9.12
3 uses	"	1.83	6.71	8.54
4 uses	"	1.66	6.36	8.02

03 CONCRETE

03110.65 Wall Formwork (Cont.)

Formwork	UNIT	MAT.	INST.	TOTAL
5 uses	S.F.	1.44	6.04	7.48
Over 16' high wall				
1 use	S.F.	3.69	8.63	12.32
2 uses	"	2.21	8.05	10.26
3 uses	"	2.01	7.55	9.56
4 uses	"	1.83	7.11	8.94
5 uses	"	1.66	6.71	8.37
Radial wall forms				
1 use	S.F.	3.44	9.29	12.73
2 uses	"	2.06	8.63	10.69
3 uses	"	1.91	8.05	9.96
4 uses	"	1.72	7.55	9.27
5 uses	"	1.56	7.11	8.67
Retaining wall forms				
1 use	S.F.	2.97	6.71	9.68
2 uses	"	1.58	6.36	7.94
3 uses	"	1.36	6.04	7.40
4 uses	"	1.18	5.75	6.93
5 uses	"	1.01	5.49	6.50
Radial retaining wall forms				
1 use	S.F.	3.14	10.00	13.14
2 uses	"	1.93	9.29	11.22
3 uses	"	1.65	8.63	10.28
4 uses	"	1.57	8.05	9.62
5 uses	"	1.35	7.55	8.90
Column pier and pilaster				
1 use	S.F.	3.51	12.00	15.51
2 uses	"	2.08	11.00	13.08
3 uses	"	1.93	10.00	11.93
4 uses	"	1.76	9.29	11.05
5 uses	"	1.58	8.63	10.21
Interior wall forms				
Up to 8' high				
1 use	S.F.	3.19	5.49	8.68
2 uses	"	1.78	5.25	7.03
3 uses	"	1.58	5.03	6.61
4 uses	"	1.35	4.83	6.18
5 uses	"	1.14	4.64	5.78
Over 8' high				
1 use	S.F.	3.51	6.71	10.22
2 uses	"	2.01	6.36	8.37
3 uses	"	1.83	6.04	7.87
4 uses	"	1.66	5.75	7.41
5 uses	"	1.45	5.49	6.94
Over 16' high				
1 use	S.F.	3.67	7.55	11.22
2 uses	"	2.21	7.11	9.32
3 uses	"	2.01	6.71	8.72
4 uses	"	1.83	6.36	8.19
5 uses	"	1.66	6.04	7.70
Radial wall forms				
1 use	S.F.	3.44	8.05	11.49
2 uses	"	2.06	7.55	9.61

03 CONCRETE

Formwork	UNIT	MAT.	INST.	TOTAL

03110.65 — Wall Formwork *(Cont.)*

	UNIT	MAT.	INST.	TOTAL
3 uses	S.F.	1.91	7.11	9.02
4 uses	"	1.72	6.71	8.43
5 uses	"	1.56	6.36	7.92
Curved wall forms, 24" sections				
1 use	S.F.	3.28	12.00	15.28
2 uses	"	1.96	11.00	12.96
3 uses	"	1.82	10.00	11.82
4 uses	"	1.65	9.29	10.94
5 uses	"	1.48	8.63	10.11
PVC form liner, per side, smooth finish				
1 use	S.F.	8.21	5.03	13.24
2 uses	"	4.51	4.83	9.34
3 uses	"	3.82	4.64	8.46
4 uses	"	2.95	4.31	7.26
5 uses	"	2.35	4.02	6.37

03110.90 — Miscellaneous Formwork

	UNIT	MAT.	INST.	TOTAL
Keyway forms (5 uses)				
2 x 4	L.F.	0.27	3.02	3.29
2 x 6	"	0.40	3.35	3.75
Bulkheads				
Walls, with keyways				
2 piece	L.F.	4.57	5.49	10.06
3 piece	"	5.78	6.04	11.82
Elevated slab, with keyway				
2 piece	L.F.	5.24	5.03	10.27
3 piece	"	7.71	5.49	13.20
Ground slab, with keyway				
2 piece	L.F.	5.41	4.31	9.72
3 piece	"	6.61	4.64	11.25
Chamfer strips				
Wood				
1/2" wide	L.F.	0.25	1.34	1.59
3/4" wide	"	0.34	1.34	1.68
1" wide	"	0.44	1.34	1.78
PVC				
1/2" wide	L.F.	1.15	1.34	2.49
3/4" wide	"	1.26	1.34	2.60
1" wide	"	1.82	1.34	3.16
Radius				
1"	L.F.	1.36	1.43	2.79
1-1/2"	"	2.45	1.43	3.88
Reglets				
Galvanized steel, 24 ga.	L.F.	1.76	2.41	4.17
Metal formwork				
Straight edge forms				
4" high	L.F.	21.50	3.77	25.27
6" high	"	23.75	4.02	27.77
8" high	"	32.50	4.31	36.81
12" high	"	37.75	4.64	42.39
16" high	"	44.50	5.03	49.53

03 CONCRETE

Formwork

	UNIT	MAT.	INST.	TOTAL
03110.90 Miscellaneous Formwork *(Cont.)*				
Curb form, S-shape				
12" x				
1'-6"	L.F.	47.50	8.63	56.13
2'	"	52.00	8.05	60.05
2'-6"	"	57.00	7.55	64.55
3'	"	62.00	6.71	68.71

Reinforcement

	UNIT	MAT.	INST.	TOTAL
03210.05 Beam Reinforcing				
Beam-girders				
#3 - #4	TON	1,560	1,550	3,110
#5 - #6	"	1,370	1,240	2,610
#7 - #8	"	1,300	1,030	2,330
#9 - #10	"	1,300	880	2,180
#11	"	1,300	820	2,120
#14	"	1,300	770	2,070
Galvanized				
#3 - #4	TON	2,660	1,550	4,210
#5 - #6	"	2,510	1,240	3,750
#7 - #8	"	2,420	1,030	3,450
#9 - #10	"	2,420	880	3,300
#11	"	2,420	820	3,240
#14	"	2,420	770	3,190
Bond Beams				
#3 - #4	TON	1,560	2,060	3,620
#5 - #6	"	1,370	1,550	2,920
#7 - #8	"	1,300	1,370	2,670
Galvanized				
#3 - #4	TON	2,540	2,060	4,600
#5 - #6	"	2,510	1,550	4,060
#7 - #8	"	2,420	1,370	3,790
03210.10 Box Culvert Reinforcing				
Box culverts				
#3 - #4	TON	1,560	770	2,330
#5 - #6	"	1,370	690	2,060
#7 - #8	"	1,300	620	1,920
#9 - #10	"	1,300	560	1,860
#11	"	1,300	520	1,820
Galvanized				
#3 - #4	TON	2,540	770	3,310
#5 - #6	"	2,510	690	3,200
#7 - #8	"	2,420	620	3,040
#9 - #10	"	2,420	560	2,980
#11	"	2,420	520	2,940

03 CONCRETE

Reinforcement	UNIT	MAT.	INST.	TOTAL
03210.15 — Column Reinforcing				
Columns				
#3 - #4	TON	1,560	1,770	3,330
#5 - #6	"	1,370	1,370	2,740
#7 - #8	"	1,300	1,240	2,540
#9 - #10	"	1,300	1,120	2,420
#11	"	1,300	1,030	2,330
#14	"	1,300	950	2,250
#18	"	1,300	880	2,180
Galvanized				
#3 - #4	TON	2,660	1,770	4,430
#5 - #6	"	2,510	1,370	3,880
#7 - #8	"	2,420	1,240	3,660
#9 - #10	"	2,420	1,120	3,540
#11	"	2,420	1,030	3,450
#14	"	2,420	950	3,370
#18	"	2,420	880	3,300
Spirals				
8" to 24" dia.	TON	2,410	1,550	3,960
24" to 48" dia.	"	2,410	1,370	3,780
48" to 84" dia.	"	2,660	1,240	3,900
03210.20 — Elevated Slab Reinforcing				
Elevated slab				
#3 - #4	TON	1,560	770	2,330
#5 - #6	"	1,370	690	2,060
#7 - #8	"	1,300	620	1,920
#9 - #10	"	1,300	560	1,860
#11	"	1,300	520	1,820
Galvanized				
#3 - #4	TON	2,540	770	3,310
#5 - #6	"	2,510	690	3,200
#7 - #8	"	2,420	620	3,040
#9 - #10	"	2,420	560	2,980
#11	"	2,420	520	2,940
03210.25 — Equip. Pad Reinforcing				
Equipment pad				
#3 - #4	TON	1,560	1,240	2,800
#5 - #6	"	1,370	1,120	2,490
#7 - #8	"	1,300	1,030	2,330
#9 - #10	"	1,300	950	2,250
#11	"	1,300	880	2,180
03210.35 — Footing Reinforcing				
Footings				
Grade 50				
#3 - #4	TON	1,560	1,030	2,590
#5 - #6	"	1,370	880	2,250
#7 - #8	"	1,300	770	2,070
#9 - #10	"	1,300	690	1,990
Grade 60				
#3 - #4	TON	1,560	1,030	2,590
#5 - #6	"	1,370	880	2,250

03 CONCRETE

Reinforcement	UNIT	MAT.	INST.	TOTAL
03210.35 **Footing Reinforcing** *(Cont.)*				
#7 - #8	TON	1,300	770	2,070
#9 - #10	"	1,300	690	1,990
Grade 70				
#3 - #4	TON	1,560	1,030	2,590
#5 - #6	"	1,370	880	2,250
#7 - #8	"	1,300	770	2,070
#9 - #10	"	1,300	690	1,990
#11	"	1,300	620	1,920
Straight dowels, 24" long				
1" dia. (#8)	EA.	4.85	6.18	11.03
3/4" dia. (#6)	"	4.36	6.18	10.54
5/8" dia. (#5)	"	3.77	5.15	8.92
1/2" dia. (#4)	"	2.84	4.41	7.25
03210.45 **Foundation Reinforcing**				
Foundations				
#3 - #4	TON	1,560	1,030	2,590
#5 - #6	"	1,370	880	2,250
#7 - #8	"	1,300	770	2,070
#9 - #10	"	1,300	690	1,990
#11	"	1,300	620	1,920
Galvanized				
#3 - #4	TON	2,660	1,030	3,690
#5 - #6	"	2,510	880	3,390
#7 - #8	"	2,420	770	3,190
#9 - #10	"	2,420	690	3,110
#11	"	2,420	620	3,040
03210.50 **Grade Beam Reinforcing**				
Grade beams				
#3 - #4	TON	1,560	950	2,510
#5 - #6	"	1,370	820	2,190
#7 - #8	"	1,300	730	2,030
#9 - #10	"	1,300	650	1,950
#11	"	1,300	590	1,890
Galvanized				
#3 - #4	TON	2,660	950	3,610
#5 - #6	"	2,510	820	3,330
#7 - #8	"	2,420	730	3,150
#9 - #10	"	2,420	650	3,070
#11	"	2,420	590	3,010
03210.53 **Pile Cap Reinforcing**				
Pile caps				
#3 - #4	TON	1,560	1,550	3,110
#5 - #6	"	1,370	1,370	2,740
#7 - #8	"	1,300	1,240	2,540
#9 - #10	"	1,300	1,120	2,420
#11	"	1,300	1,030	2,330
Galvanized				
#3 - #4	TON	2,660	1,550	4,210
#5 - #6	"	2,510	1,370	3,880
#7 - #8	"	2,420	1,240	3,660

03 CONCRETE

Reinforcement	UNIT	MAT.	INST.	TOTAL
03210.53 Pile Cap Reinforcing (Cont.)				
#9 - #10	TON	2,420	1,120	3,540
#11	"	2,420	1,030	3,450
03210.55 Slab / Mat Reinforcing				
Bars, slabs				
#3 - #4	TON	1,560	1,030	2,590
#5 - #6	"	1,370	880	2,250
#7 - #8	"	1,300	770	2,070
#9 - #10	"	1,300	690	1,990
#11	"	1,300	620	1,920
Galvanized				
#3 - #4	TON	2,660	1,030	3,690
#5 - #6	"	2,510	880	3,390
#7 - #8	"	2,420	770	3,190
#9 - #10	"	2,420	690	3,110
#11	"	2,420	620	3,040
Wire mesh, slabs				
Galvanized				
4x4				
W1.4xW1.4	S.F.	0.39	0.41	0.80
W2.0xW2.0	"	0.50	0.44	0.94
W2.9xW2.9	"	0.70	0.47	1.17
W4.0xW4.0	"	1.04	0.51	1.55
6x6				
W1.4xW1.4	S.F.	0.36	0.30	0.66
W2.0xW2.0	"	0.50	0.34	0.84
W2.9xW2.9	"	0.67	0.36	1.03
W4.0xW4.0	"	0.73	0.41	1.14
Standard				
2x2				
W.9xW.9	S.F.	0.39	0.41	0.80
4x4				
W1.4xW1.4	S.F.	0.25	0.41	0.66
W2.0xW2.0	"	0.33	0.44	0.77
W2.9xW2.9	"	0.45	0.47	0.92
W4.0xW4.0	"	0.70	0.51	1.21
6x6				
W1.4xW1.4	S.F.	0.16	0.30	0.46
W2.0xW2.0	"	0.22	0.34	0.56
W2.9xW2.9	"	0.33	0.36	0.69
W4.0xW4.0	"	0.47	0.41	0.88
03210.60 Stair Reinforcing				
Stairs				
#3 - #4	TON	1,560	1,240	2,800
#5 - #6	"	1,370	1,030	2,400
#7 - #8	"	1,300	880	2,180
#9 - #10	"	1,300	770	2,070
Galvanized				
#3 - #4	TON	2,660	1,240	3,900
#5 - #6	"	2,510	1,030	3,540
#7 - #8	"	2,420	880	3,300
#9 - #10	"	2,420	770	3,190

03 CONCRETE

Reinforcement

03210.65 Wall Reinforcing

	UNIT	MAT.	INST.	TOTAL
Walls				
#3 - #4	TON	1,560	880	2,440
#5 - #6	"	1,370	770	2,140
#7 - #8	"	1,300	690	1,990
#9 - #10	"	1,300	620	1,920
Galvanized				
#3 - #4	TON	2,660	880	3,540
#5 - #6	"	2,510	770	3,280
#7 - #8	"	2,420	690	3,110
#9 - #10	"	2,420	620	3,040
Masonry wall (horizontal)				
#3 - #4	TON	1,560	2,470	4,030
#5 - #6	"	1,370	2,060	3,430
Galvanized				
#3 - #4	TON	2,660	2,470	5,130
#5 - #6	"	2,510	2,060	4,570
Masonry wall (vertical)				
#3 - #4	TON	1,560	3,090	4,650
#5 - #6	"	1,370	2,470	3,840
Galvanized				
#3 - #4	TON	2,660	3,090	5,750
#5 - #6	"	2,510	2,470	4,980

Accessories

03250.40 Concrete Accessories

	UNIT	MAT.	INST.	TOTAL
Expansion joint, poured				
Asphalt				
1/2" x 1"	L.F.	0.97	0.94	1.91
1" x 2"	"	3.05	1.03	4.08
Liquid neoprene, cold applied				
1/2" x 1"	L.F.	3.83	0.96	4.79
1" x 2"	"	15.75	1.05	16.80
Polyurethane, 2 parts				
1/2" x 1"	L.F.	3.67	1.58	5.25
1" x 2"	"	14.50	1.72	16.22
Rubberized asphalt, cold				
1/2" x 1"	L.F.	0.94	0.94	1.88
1" x 2"	"	2.82	1.03	3.85
Hot, fuel resistant				
1/2" x 1"	L.F.	1.70	0.94	2.64
1" x 2"	"	8.17	1.03	9.20
Expansion joint, premolded, in slabs				
Asphalt				
1/2" x 6"	L.F.	1.21	1.18	2.39
1" x 12"	"	2.02	1.58	3.60

03 CONCRETE

Accessories	UNIT	MAT.	INST.	TOTAL
03250.40 Concrete Accessories *(Cont.)*				
Cork				
1/2" x 6"	L.F.	2.40	1.18	3.58
1" x 12"	"	9.12	1.58	10.70
Neoprene sponge				
1/2" x 6"	L.F.	3.54	1.18	4.72
1" x 12"	"	13.00	1.58	14.58
Polyethylene foam				
1/2" x 6"	L.F.	1.37	1.18	2.55
1" x 12"	"	6.29	1.58	7.87
Polyurethane foam				
1/2" x 6"	L.F.	1.80	1.18	2.98
1" x 12"	"	3.97	1.58	5.55
Polyvinyl chloride foam				
1/2" x 6"	L.F.	3.86	1.18	5.04
1" x 12"	"	8.31	1.58	9.89
Rubber, gray sponge				
1/2" x 6"	L.F.	6.00	1.18	7.18
1" x 12"	"	26.00	1.58	27.58
Asphalt felt control joints or bond breaker, screed joints				
4" slab	L.F.	1.63	0.94	2.57
6" slab	"	2.03	1.05	3.08
8" slab	"	2.65	1.18	3.83
10" slab	"	3.73	1.35	5.08
Keyed cold expansion and control joints, 24 ga.				
4" slab	L.F.	1.33	2.96	4.29
5" slab	"	1.76	2.96	4.72
6" slab	"	2.03	3.16	5.19
8" slab	"	2.47	3.38	5.85
10" slab	"	2.71	3.64	6.35
Waterstops				
Polyvinyl chloride				
Ribbed				
3/16" thick x				
4" wide	L.F.	1.93	2.37	4.30
6" wide	"	2.92	2.63	5.55
1/2" thick x				
9" wide	L.F.	7.78	2.96	10.74
Ribbed with center bulb				
3/16" thick x 9" wide	L.F.	6.54	2.96	9.50
3/8" thick x 9" wide	"	7.68	2.96	10.64
Dumbbell type, 3/8" thick x 6" wide	"	7.78	2.63	10.41
Plain, 3/8" thick x 9" wide	"	10.25	2.96	13.21
Center bulb, 3/8" thick x 9" wide	"	12.50	2.96	15.46
Rubber				
Flat dumbbell				
3/8" thick x				
6" wide	L.F.	9.57	2.63	12.20
9" wide	"	10.25	2.96	13.21
Center bulb				
3/8" thick x				
6" wide	L.F.	10.75	2.63	13.38
9" wide	"	13.50	2.96	16.46
Vapor barrier				

03 CONCRETE

Accessories

03250.40 Concrete Accessories *(Cont.)*

	UNIT	MAT.	INST.	TOTAL
4 mil polyethylene	S.F.	0.07	0.15	0.22
6 mil polyethylene	"	0.10	0.15	0.25
Gravel porous fill, under floor slabs, 3/4" stone	C.Y.	27.00	79.00	106
Reinforcing accessories				
Beam bolsters				
1-1/2" high, plain	L.F.	0.72	0.61	1.33
Galvanized	"	1.57	0.61	2.18
3" high				
Plain	L.F.	1.02	0.77	1.79
Galvanized	"	2.52	0.77	3.29
Slab bolsters				
1" high				
Plain	L.F.	0.76	0.30	1.06
Galvanized	"	1.54	0.30	1.84
2" high				
Plain	L.F.	0.85	0.34	1.19
Galvanized	"	1.79	0.34	2.13
Chairs, high chairs				
3" high				
Plain	EA.	2.06	1.54	3.60
Galvanized	"	2.27	1.54	3.81
5" high				
Plain	EA.	2.12	1.62	3.74
Galvanized	"	4.26	1.62	5.88
8" high				
Plain	EA.	3.47	1.76	5.23
Galvanized	"	5.90	1.76	7.66
12" high				
Plain	EA.	6.30	2.06	8.36
Galvanized	"	12.00	2.06	14.06
Continuous, high chair				
3" high				
Plain	L.F.	2.86	0.41	3.27
Galvanized	"	3.54	0.41	3.95
5" high				
Plain	L.F.	3.07	0.44	3.51
Galvanized	"	4.33	0.44	4.77
8" high				
Plain	L.F.	3.53	0.47	4.00
Galvanized	"	4.59	0.47	5.06
12" high				
Plain	L.F.	4.54	0.51	5.05
Galvanized	"	5.90	0.51	6.41

03 CONCRETE

Cast-in-place Concrete

03300.10 Concrete Admixtures

	UNIT	MAT.	INST.	TOTAL
Concrete admixtures				
Water reducing admixture	GAL			12.25
Set retarder	"			26.50
Air entraining agent	"			11.50

03350.10 Concrete Finishes

	UNIT	MAT.	INST.	TOTAL
Floor finishes				
Broom	S.F.		0.67	0.67
Screed	"		0.59	0.59
Darby	"		0.59	0.59
Steel float	"		0.79	0.79
Granolithic topping				
1/2" thick	S.F.	0.47	2.15	2.62
1" thick	"	0.86	2.37	3.23
2" thick	"	1.03	2.63	3.66
Wall finishes				
Burlap rub, with cement paste	S.F.	0.13	0.79	0.92
Float finish	"	0.13	1.18	1.31
Etch with acid	"	0.40	0.79	1.19
Sandblast				
Minimum	S.F.	0.14	1.20	1.34
Maximum	"	0.51	1.20	1.71
Bush hammer				
Green concrete	S.F.		2.37	2.37
Cured concrete	"		3.64	3.64
Break ties and patch holes	"		0.94	0.94
Carborundum				
Dry rub	S.F.		1.58	1.58
Wet rub	"		2.37	2.37
Floor hardeners				
Metallic				
Light service	S.F.	0.43	0.59	1.02
Heavy service	"	1.31	0.79	2.10
Non-metallic				
Light service	S.F.	0.22	0.59	0.81
Heavy service	"	0.91	0.79	1.70
Rusticated concrete finish				
Beveled edge	L.F.	0.42	2.63	3.05
Square edge	"	0.60	3.38	3.98
Solid board concrete finish				
Standard	S.F.	1.07	3.95	5.02
Rustic	"	0.99	4.74	5.73

03360.10 Pneumatic Concrete

	UNIT	MAT.	INST.	TOTAL
Pneumatic applied concrete (gunite)				
2" thick	S.F.	6.32	3.07	9.39
3" thick	"	7.77	4.10	11.87
4" thick	"	9.46	4.92	14.38
Finish surface				
Minimum	S.F.		3.02	3.02
Maximum	"		6.04	6.04

03 CONCRETE

Cast-in-place Concrete

03370.10 Curing Concrete

	UNIT	MAT.	INST.	TOTAL
Sprayed membrane				
Slabs	S.F.	0.06	0.09	0.15
Walls	"	0.09	0.11	0.20
Curing paper				
Slabs	S.F.	0.09	0.12	0.21
Walls	"	0.09	0.13	0.22
Burlap				
7.5 oz.	S.F.	0.07	0.15	0.22
12 oz.	"	0.10	0.16	0.26

Placing Concrete

03380.05 Beam Concrete

	UNIT	MAT.	INST.	TOTAL
Beams and girders				
2500# or 3000# concrete				
By crane	C.Y.	130	90.00	220
By pump	"	130	82.00	212
By hand buggy	"	130	47.50	178
3500# or 4000# concrete				
By crane	C.Y.	140	90.00	230
By pump	"	140	82.00	222
By hand buggy	"	140	47.50	188
5000# concrete				
By crane	C.Y.	150	90.00	240
By pump	"	150	82.00	232
By hand buggy	"	150	47.50	198
Bond beam, 3000# concrete				
By pump				
8" high				
4" wide	L.F.	0.35	1.79	2.14
6" wide	"	0.86	2.04	2.90
8" wide	"	1.10	2.24	3.34
10" wide	"	1.45	2.49	3.94
12" wide	"	1.96	2.80	4.76
16" high				
8" wide	L.F.	2.70	2.80	5.50
10" wide	"	3.62	3.21	6.83
12" wide	"	4.80	3.74	8.54
By crane				
8" high				
4" wide	L.F.	0.45	1.95	2.40
6" wide	"	0.80	2.14	2.94
8" wide	"	1.04	2.24	3.28
10" wide	"	1.36	2.49	3.85
12" wide	"	1.84	2.80	4.64
16" high				

03 CONCRETE

Placing Concrete	UNIT	MAT.	INST.	TOTAL
03380.05 Beam Concrete *(Cont.)*				
8" wide	L.F.	2.55	2.80	5.35
10" wide	"	3.40	2.99	6.39
12" wide	"	4.53	3.45	7.98
03380.15 Column Concrete				
Columns				
2500# or 3000# concrete				
By crane	C.Y.	130	82.00	212
By pump	"	130	75.00	205
3500# or 4000# concrete				
By crane	C.Y.	140	82.00	222
By pump	"	140	75.00	215
5000# concrete				
By crane	C.Y.	150	82.00	232
By pump	"	150	75.00	225
03380.20 Elevated Slab Concrete				
Elevated slab				
2500# or 3000# concrete				
By crane	C.Y.	130	45.00	175
By pump	"	130	34.50	165
By hand buggy	"	130	47.50	178
3500# or 4000# concrete				
By crane	C.Y.	140	45.00	185
By pump	"	140	34.50	175
By hand buggy	"	140	47.50	188
5000# concrete				
By crane	C.Y.	150	45.00	195
By pump	"	150	34.50	185
By hand buggy	"	150	47.50	198
Topping				
2500# or 3000# concrete				
By crane	C.Y.	130	45.00	175
By pump	"	130	34.50	165
By hand buggy	"	130	47.50	178
3500# or 4000# concrete				
By crane	C.Y.	140	45.00	185
By pump	"	140	34.50	175
By hand buggy	"	140	47.50	188
5000# concrete				
By crane	C.Y.	150	45.00	195
By pump	"	150	34.50	185
By hand buggy	"	150	47.50	198
03380.25 Equipment Pad Concrete				
Equipment pad				
2500# or 3000# concrete				
By chute	C.Y.	130	15.75	146
By pump	"	130	64.00	194
By crane	"	130	75.00	205
3500# or 4000# concrete				
By chute	C.Y.	130	15.75	146

03 CONCRETE

Placing Concrete	UNIT	MAT.	INST.	TOTAL
03380.25 **Equipment Pad Concrete** *(Cont.)*				
By pump	C.Y.	130	64.00	194
By crane	"	130	75.00	205
5000# concrete				
By chute	C.Y.	150	15.75	166
By pump	"	150	64.00	214
By crane	"	150	75.00	225
03380.35 **Footing Concrete**				
Continuous footing				
2500# or 3000# concrete				
By chute	C.Y.	130	15.75	146
By pump	"	130	56.00	186
By crane	"	130	64.00	194
3500# or 4000# concrete				
By chute	C.Y.	130	15.75	146
By pump	"	130	56.00	186
By crane	"	130	64.00	194
5000# concrete				
By chute	C.Y.	150	15.75	166
By pump	"	150	56.00	206
By crane	"	150	64.00	214
Spread footing				
2500# or 3000# concrete				
Under 5 cy				
By chute	C.Y.	130	15.75	146
By pump	"	130	60.00	190
By crane	"	130	69.00	199
Over 5 cy				
By chute	C.Y.	130	11.75	142
By pump	"	130	53.00	183
By crane	"	130	60.00	190
3500# or 4000# concrete				
Under 5 c.y.				
By chute	C.Y.	140	15.75	156
By pump	"	130	60.00	190
By crane	"	130	69.00	199
Over 5 c.y.				
By pump	C.Y.	130	53.00	183
By crane	"	130	60.00	190
5000# concrete				
Under 5 c.y.				
By chute	C.Y.	140	15.75	156
By pump	"	140	60.00	200
By crane	"	140	69.00	209
Over 5 c.y.				
By chute	C.Y.	140	11.75	152
By pump	"	140	53.00	193
By crane	"	140	60.00	200

03 CONCRETE

Placing Concrete

03380.50 Grade Beam Concrete

	UNIT	MAT.	INST.	TOTAL
Grade beam				
2500# or 3000# concrete				
By chute	C.Y.	130	15.75	146
By crane	"	130	64.00	194
By pump	"	130	56.00	186
By hand buggy	"	130	47.50	178
3500# or 4000# concrete				
By chute	C.Y.	140	15.75	156
By crane	"	140	64.00	204
By pump	"	140	56.00	196
By hand buggy	"	140	47.50	188
5000# concrete				
By chute	C.Y.	150	15.75	166
By crane	"	150	64.00	214
By pump	"	150	56.00	206
By hand buggy	"	150	47.50	198

03380.53 Pile Cap Concrete

	UNIT	MAT.	INST.	TOTAL
Pile cap				
2500# or 3000 concrete				
By chute	C.Y.	130	15.75	146
By crane	"	130	75.00	205
By pump	"	130	64.00	194
By hand buggy	"	130	47.50	178
3500# or 4000# concrete				
By chute	C.Y.	140	15.75	156
By crane	"	140	75.00	215
By pump	"	140	64.00	204
By hand buggy	"	140	47.50	188
5000# concrete				
By chute	C.Y.	150	15.75	166
By crane	"	150	75.00	225
By pump	"	150	64.00	214
By hand buggy	"	150	47.50	198

03380.55 Slab / Mat Concrete

	UNIT	MAT.	INST.	TOTAL
Slab on grade				
2500# or 3000# concrete				
By chute	C.Y.	130	11.75	142
By crane	"	130	37.50	168
By pump	"	130	32.00	162
By hand buggy	"	130	31.50	162
3500# or 4000# concrete				
By chute	C.Y.	140	11.75	152
By crane	"	140	37.50	178
By pump	"	140	32.00	172
By hand buggy	"	140	31.50	172
5000# concrete				
By chute	C.Y.	150	11.75	162
By crane	"	150	37.50	188
By pump	"	150	32.00	182
By hand buggy	"	150	31.50	182

03 CONCRETE

Placing Concrete	UNIT	MAT.	INST.	TOTAL
03380.55 — Slab / Mat Concrete (Cont.)				
Foundation mat				
2500# or 3000# concrete, over 20 cy				
By chute	C.Y.	130	9.48	139
By crane	"	130	32.00	162
By pump	"	130	28.00	158
By hand buggy	"	130	23.75	154
03380.58 — Sidewalks				
Walks, cast in place with wire mesh, base not incl.				
4" thick	S.F.	1.84	1.58	3.42
5" thick	"	2.49	1.89	4.38
6" thick	"	3.06	2.37	5.43
03380.60 — Stair Concrete				
Stairs				
2500# or 3000# concrete				
By chute	C.Y.	130	15.75	146
By crane	"	130	75.00	205
By pump	"	130	64.00	194
By hand buggy	"	130	47.50	178
3500# or 4000# concrete				
By chute	C.Y.	140	15.75	156
By crane	"	140	75.00	215
By pump	"	140	64.00	204
By hand buggy	"	140	47.50	188
5000# concrete				
By chute	C.Y.	150	15.75	166
By crane	"	150	75.00	225
By pump	"	150	64.00	214
By hand buggy	"	150	47.50	198
03380.65 — Wall Concrete				
Walls				
2500# or 3000# concrete				
To 4'				
By chute	C.Y.	130	13.50	144
By crane	"	130	75.00	205
By pump	"	130	69.00	199
To 8'				
By crane	C.Y.	130	82.00	212
By pump	"	130	75.00	205
To 16'				
By crane	C.Y.	130	90.00	220
By pump	"	130	82.00	212
Over 16'				
By crane	C.Y.	130	100	230
By pump	"	130	90.00	220
3500# or 4000# concrete				
To 4'				
By chute	C.Y.	140	13.50	154
By crane	"	140	75.00	215
By pump	"	140	69.00	209
To 8'				

03 CONCRETE

Placing Concrete	UNIT	MAT.	INST.	TOTAL
03380.65 **Wall Concrete** (Cont.)				
By crane	C.Y.	140	82.00	222
By pump	"	140	75.00	215
To 16'				
By crane	C.Y.	140	90.00	230
By pump	"	140	82.00	222
Over 16'				
By crane	C.Y.	140	100	240
By pump	"	140	90.00	230
5000# concrete				
To 4'				
By chute	C.Y.	150	13.50	164
By crane	"	150	75.00	225
By pump	"	150	69.00	219
To 8'				
By crane	C.Y.	150	82.00	232
By pump	"	150	75.00	225
To 16'				
By crane	C.Y.	150	90.00	240
By pump	"	150	82.00	232
Filled block (CMU)				
3000# concrete, by pump				
4" wide	S.F.	0.49	3.21	3.70
6" wide	"	1.11	3.74	4.85
8" wide	"	1.73	4.49	6.22
10" wide	"	2.32	5.28	7.60
12" wide	"	2.98	6.42	9.40
Pilasters, 3000# concrete	C.F.	6.76	90.00	96.76
Wall cavity, 2" thick, 3000# concrete	S.F.	1.25	2.99	4.24

Precast Concrete	UNIT	MAT.	INST.	TOTAL
03400.10 **Precast Beams**				
Prestressed, double tee, 24" deep, 8' wide				
35' span				
115 psf	S.F.	13.00	1.21	14.21
140 psf	"	13.75	1.21	14.96
40' span				
80 psf	S.F.	12.50	1.29	13.79
143 psf	"	13.00	1.29	14.29
45' span				
50 psf	S.F.	11.50	1.11	12.61
70 psf	"	12.50	1.11	13.61
100 psf	"	12.50	1.11	13.61
130 psf	"	14.25	1.11	15.36
50' span				
75 psf	S.F.	11.50	1.01	12.51

03 CONCRETE

Precast Concrete

	UNIT	MAT.	INST.	TOTAL
03400.10 — Precast Beams *(Cont.)*				
100 psf	S.F.	12.50	1.01	13.51
Precast beams, girders and joists				
1000 lb/lf live load				
10' span	L.F.	110	24.25	134
20' span	"	110	14.50	125
30' span	"	140	12.00	152
3000 lb/lf live load				
10' span	L.F.	110	24.25	134
20' span	"	130	14.50	145
30' span	"	170	12.00	182
5000 lb/lf live load				
10' span	L.F.	110	24.25	134
20' span	"	150	14.50	165
30' span	"	190	12.00	202
03400.20 — Precast Columns				
Prestressed concrete columns				
10" x 10"				
10' long	EA.	310	150	460
15' long	"	460	150	610
20' long	"	620	160	780
25' long	"	800	170	970
30' long	"	950	180	1,130
12" x 12"				
20' long	EA.	850	180	1,030
25' long	"	1,030	200	1,230
30' long	"	1,310	210	1,520
16" x 16"				
20' long	EA.	1,310	180	1,490
25' long	"	1,930	200	2,130
30' long	"	2,290	210	2,500
20" x 20"				
20' long	EA.	2,360	190	2,550
25' long	"	3,220	200	3,420
30' long	"	3,690	210	3,900
24" x 24"				
20' long	EA.	3,610	200	3,810
25' long	"	4,390	210	4,600
30' long	"	5,420	230	5,650
28" x 28"				
20' long	EA.	4,900	230	5,130
25' long	"	5,930	240	6,170
30' long	"	7,350	260	7,610
32" x 32"				
20' long	EA.	6,190	240	6,430
25' long	"	7,990	260	8,250
30' long	"	9,150	280	9,430
36" x 36"				
20' long	EA.	7,730	260	7,990
25' long	"	9,670	280	9,950
30' long	"	11,600	300	11,900

03 CONCRETE

Precast Concrete

03400.30 Precast Slabs

	UNIT	MAT.	INST.	TOTAL
Prestressed flat slab				
6" thick, 4' wide				
20' span				
80 psf	S.F.	18.75	3.03	21.78
110 psf	"	18.75	3.03	21.78
25' span				
80 psf	S.F.	19.25	2.91	22.16
Cored slab				
6" thick, 4' wide				
20' span				
80 psf	S.F.	9.90	3.03	12.93
100 psf	"	10.00	3.03	13.03
130 psf	"	10.25	3.03	13.28
8" thick, 4' wide				
25' span				
70 psf	S.F.	9.81	2.91	12.72
125 psf	"	10.50	2.91	13.41
170 psf	"	10.50	2.91	13.41
30' span				
70 psf	S.F.	9.70	2.42	12.12
90 psf	"	10.00	2.42	12.42
35' span				
70 psf	S.F.	10.25	2.27	12.52
10" thick, 4' wide				
30' span				
75 psf	S.F.	10.25	2.42	12.67
100 psf	"	10.50	2.42	12.92
130 psf	"	10.75	2.42	13.17
35' span				
60 psf	S.F.	10.50	2.27	12.77
80 psf	"	10.75	2.27	13.02
120 psf	"	11.25	2.27	13.52
40' span				
65 psf	S.F.	11.25	1.81	13.06
Slabs, roof and floor members, 4' wide				
6" thick, 25' span	S.F.	9.01	2.91	11.92
8" thick, 30' span	"	10.25	2.20	12.45
10" thick, 40' span	"	12.75	1.96	14.71
Tee members				
Multiple tee, roof and floor				
Minimum	S.F.	11.75	1.81	13.56
Maximum	"	14.75	3.63	18.38
Double tee wall member				
Minimum	S.F.	10.75	2.07	12.82
Maximum	"	13.50	4.04	17.54
Single tee				
Short span, roof members				
Minimum	S.F.	12.25	2.20	14.45
Maximum	"	14.75	4.54	19.29
Long span, roof members				
Minimum	S.F.	15.25	1.81	17.06
Maximum	"	18.25	3.63	21.88

03 CONCRETE

Precast Concrete

	UNIT	MAT.	INST.	TOTAL

03400.40 Precast Walls

	UNIT	MAT.	INST.	TOTAL
Wall panel, 8' x 20'				
Gray cement				
Liner finish				
4" wall	S.F.	14.25	2.07	16.32
5" wall	"	15.50	2.14	17.64
6" wall	"	17.75	2.20	19.95
8" wall	"	18.00	2.27	20.27
Sandblast finish				
4" wall	S.F.	16.75	2.07	18.82
5" wall	"	18.25	2.14	20.39
6" wall	"	20.00	2.20	22.20
8" wall	"	20.75	2.27	23.02
White cement				
Liner finish				
4" wall	S.F.	17.50	2.07	19.57
5" wall	"	18.75	2.14	20.89
6" wall	"	20.50	2.20	22.70
8" wall	"	21.50	2.27	23.77
Sandblast finish				
4" wall	S.F.	18.75	2.07	20.82
5" wall	"	20.00	2.14	22.14
6" wall	"	20.50	2.20	22.70
8" wall	"	22.75	2.27	25.02
Double tee wall panel, 24" deep				
Gray cement				
Liner finish	S.F.	10.25	2.42	12.67
Sandblast finish	"	12.50	2.42	14.92
White cement				
Form liner finish	S.F.	14.25	2.42	16.67
Sandblast finish	"	18.25	2.42	20.67
Partition panels				
4" wall	S.F.	15.25	2.42	17.67
5" wall	"	16.50	2.42	18.92
6" wall	"	18.00	2.42	20.42
8" wall	"	19.50	2.42	21.92
Cladding panels				
4" wall	S.F.	15.00	2.59	17.59
5" wall	"	16.50	2.59	19.09
6" wall	"	18.25	2.59	20.84
8" wall	"	19.50	2.59	22.09
Sandwich panel, 2.5" cladding panel, 2" insulation				
5" wall	S.F.	22.25	2.59	24.84
6" wall	"	23.25	2.59	25.84
8" wall	"	24.50	2.59	27.09
Adjustable tilt-up brace	EA.		11.75	11.75

03400.90 Precast Specialties

	UNIT	MAT.	INST.	TOTAL
Precast concrete, coping, 4' to 8' long				
12" wide	L.F.	10.25	6.15	16.40
10" wide	"	9.04	7.02	16.06
Splash block, 30"x12"x4"	EA.	15.50	41.00	56.50
Stair unit, per riser	"	99.00	41.00	140
Sun screen and trellis, 8' long, 12" high				

03 CONCRETE

Precast Concrete

	UNIT	MAT.	INST.	TOTAL
03400.90 Precast Specialties *(Cont.)*				
4" thick blades	EA.	110	30.75	141
5" thick blades	"	130	30.75	161
6" thick blades	"	160	32.75	193
8" thick blades	"	210	32.75	243
Bearing pads for precast members, 2" wide strips				
1/8" thick	L.F.	0.36	0.18	0.54
1/4" thick	"	0.48	0.18	0.66
1/2" thick	"	0.53	0.18	0.71
3/4" thick	"	1.06	0.21	1.27
1" thick	"	1.10	0.23	1.33
1-1/2" thick	"	1.40	0.23	1.63

Cementitous Toppings

	UNIT	MAT.	INST.	TOTAL
03550.10 Concrete Toppings				
Gypsum fill				
2" thick	S.F.	1.97	0.46	2.43
2-1/2" thick	"	2.25	0.47	2.72
3" thick	"	2.77	0.48	3.25
3-1/2" thick	"	3.16	0.49	3.65
4" thick	"	3.69	0.56	4.25
Formboard				
Mineral fiber board				
1" thick	S.F.	1.76	1.18	2.94
1-1/2" thick	"	4.64	1.35	5.99
Cement fiber board				
1" thick	S.F.	1.37	1.58	2.95
1-1/2" thick	"	1.76	1.82	3.58
Glass fiber board				
1" thick	S.F.	2.17	1.18	3.35
1-1/2" thick	"	2.94	1.35	4.29
Poured deck				
Vermiculite or perlite				
1 to 4 mix	C.Y.	190	75.00	265
1 to 6 mix	"	170	69.00	239
Vermiculite or perlite				
2" thick				
1 to 4 mix	S.F.	1.76	0.47	2.23
1 to 6 mix	"	1.28	0.42	1.70
3" thick				
1 to 4 mix	S.F.	2.41	0.69	3.10
1 to 6 mix	"	1.90	0.64	2.54
Concrete plank, lightweight				
2" thick	S.F.	9.58	3.63	13.21
2-1/2" thick	"	9.83	3.63	13.46
3-1/2" thick	"	10.25	4.04	14.29

03 CONCRETE

Cementitous Toppings	UNIT	MAT.	INST.	TOTAL
03550.10 Concrete Toppings *(Cont.)*				
4" thick	S.F.	10.50	4.04	14.54
Channel slab, lightweight, straight				
2-3/4" thick	S.F.	7.73	3.63	11.36
3-1/2" thick	"	7.95	3.63	11.58
3-3/4" thick	"	8.59	3.63	12.22
4-3/4" thick	"	10.75	4.04	14.79
Gypsum plank				
2" thick	S.F.	3.58	3.63	7.21
3" thick	"	3.75	3.63	7.38
Cement fiber, T and G planks				
1" thick	S.F.	1.97	3.30	5.27
1-1/2" thick	"	2.09	3.30	5.39
2" thick	"	2.50	3.63	6.13
2-1/2" thick	"	2.65	3.63	6.28
3" thick	"	3.44	3.63	7.07
3-1/2" thick	"	3.97	4.04	8.01
4" thick	"	4.37	4.04	8.41

Grout	UNIT	MAT.	INST.	TOTAL
03600.10 Grouting				
Grouting for bases				
Nonshrink				
Metallic grout				
1" deep	S.F.	8.35	12.00	20.35
2" deep	"	15.75	13.50	29.25
Non-metallic grout				
1" deep	S.F.	6.28	12.00	18.28
2" deep	"	12.00	13.50	25.50
Fluid type				
Non-metallic				
1" deep	S.F.	6.43	12.00	18.43
2" deep	"	11.50	13.50	25.00
Grouting for joints				
Portland cement grout (1 cement to 3 sand)				
1/2" joint thickness				
6" wide joints	L.F.	0.21	2.01	2.22
8" wide joints	"	0.23	2.41	2.64
1" joint thickness				
4" wide joints	L.F.	0.23	1.88	2.11
6" wide joints	"	0.39	2.08	2.47
8" wide joints	"	0.47	2.51	2.98
Nonshrink, nonmetallic grout				
1/2" joint thickness				
4" wide joint	L.F.	1.14	1.72	2.86

03 CONCRETE

Grout	UNIT	MAT.	INST.	TOTAL
03600.10 Grouting *(Cont.)*				
6" wide joint	L.F.	1.48	2.01	3.49
8" wide joint	"	1.94	2.41	4.35
1" joint thickness				
4" wide joint	L.F.	1.94	1.88	3.82
6" wide joint	"	2.94	2.08	5.02
8" wide joint	"	3.97	2.51	6.48

Concrete Restoration	UNIT	MAT.	INST.	TOTAL
03730.10 Concrete Repair				
Epoxy grout floor patch, 1/4" thick	S.F.	7.67	4.74	12.41
Grout, epoxy, 2 component system	C.F.			370
Epoxy sand	BAG			25.00
Epoxy modifier	GAL			160
Epoxy gel grout	S.F.	3.74	47.50	51.24
Injection valve, 1 way, threaded plastic	EA.	10.25	9.48	19.73
Grout crack seal, 2 component	C.F.	860	47.50	908
Grout, non shrink	"	88.00	47.50	136
Concrete, epoxy modified				
Sand mix	C.F.	140	19.00	159
Gravel mix	"	110	17.50	128
Concrete repair				
Soffit repair				
16" wide	L.F.	4.38	9.48	13.86
18" wide	"	4.66	9.87	14.53
24" wide	"	5.56	10.50	16.06
30" wide	"	6.26	11.25	17.51
32" wide	"	6.68	11.75	18.43
Edge repair				
2" spall	L.F.	2.08	11.75	13.83
3" spall	"	2.08	12.50	14.58
4" spall	"	2.22	12.75	14.97
6" spall	"	2.29	13.25	15.54
8" spall	"	2.43	14.00	16.43
9" spall	"	2.50	15.75	18.25
Crack repair, 1/8" crack	"	4.10	4.74	8.84
Reinforcing steel repair				
1 bar, 4 ft				
#4 bar	L.F.	0.64	7.73	8.37
#5 bar	"	0.87	7.73	8.60
#6 bar	"	1.05	8.24	9.29
#8 bar	"	1.92	8.24	10.16
#9 bar	"	2.45	8.83	11.28
#11 bar	"	3.82	8.83	12.65
Form fabric, nylon				
18" diameter	L.F.			16.50

03 CONCRETE

Concrete Restoration

03730.10 Concrete Repair (Cont.)

	UNIT	MAT.	INST.	TOTAL
20" diameter	L.F.			16.75
24" diameter	"			27.75
30" diameter	"			28.50
36" diameter	"			32.50
Pile repairs				
Polyethylene wrap				
30 mil thick				
60" wide	S.F.	18.00	15.75	33.75
72" wide	"	19.75	19.00	38.75
60 mil thick				
60" wide	S.F.	21.25	15.75	37.00
80" wide	"	24.75	21.50	46.25
Pile spall, average repair 3'				
18" x 18"	EA.	56.00	39.50	95.50
20" x 20"	"	74.00	47.50	122

04 MASONRY

Mortar And Grout	UNIT	MAT.	INST.	TOTAL
04100.10 Masonry Grout				
Grout, non shrink, non-metallic, trowelable	C.F.	6.02	1.64	7.66
Grout door frame, hollow metal				
Single	EA.	15.00	62.00	77.00
Double	"	21.00	65.00	86.00
Grout-filled concrete block (CMU)				
4" wide	S.F.	0.40	2.05	2.45
6" wide	"	1.05	2.23	3.28
8" wide	"	1.55	2.46	4.01
12" wide	"	2.54	2.58	5.12
Grout-filled individual CMU cells				
4" wide	L.F.	0.33	1.23	1.56
6" wide	"	0.45	1.23	1.68
8" wide	"	0.60	1.23	1.83
10" wide	"	0.74	1.40	2.14
12" wide	"	0.91	1.40	2.31
Bond beams or lintels, 8" deep				
6" thick	L.F.	0.91	2.04	2.95
8" thick	"	1.20	2.24	3.44
10" thick	"	1.51	2.49	4.00
12" thick	"	1.80	2.80	4.60
Cavity walls				
2" thick	S.F.	1.00	2.99	3.99
3" thick	"	1.51	2.99	4.50
4" thick	"	2.00	3.21	5.21
6" thick	"	3.01	3.74	6.75
04150.10 Masonry Accessories				
Foundation vents	EA.	38.25	23.00	61.25
Bar reinforcing				
Horizontal				
#3 - #4	Lb.	0.75	2.30	3.05
#5 - #6	"	0.75	1.92	2.67
Vertical				
#3 - #4	Lb.	0.75	2.88	3.63
#5 - #6	"	0.75	2.30	3.05
Horizontal joint reinforcing				
Truss type				
4" wide, 6" wall	L.F.	0.24	0.23	0.47
6" wide, 8" wall	"	0.24	0.24	0.48
8" wide, 10" wall	"	0.30	0.25	0.55
10" wide, 12" wall	"	0.30	0.26	0.56
12" wide, 14" wall	"	0.36	0.27	0.63
Ladder type				
4" wide, 6" wall	L.F.	0.17	0.23	0.40
6" wide, 8" wall	"	0.20	0.24	0.44
8" wide, 10" wall	"	0.21	0.25	0.46
10" wide, 12" wall	"	0.25	0.25	0.50
Rectangular wall ties				
3/16" dia., galvanized				
2" x 6"	EA.	0.45	0.96	1.41
2" x 8"	"	0.48	0.96	1.44
2" x 10"	"	0.55	0.96	1.51
2" x 12"	"	0.63	0.96	1.59

04 MASONRY

Mortar And Grout

04150.10 Masonry Accessories (Cont.)

	UNIT	MAT.	INST.	TOTAL
4" x 6"	EA.	0.51	1.15	1.66
4" x 8"	"	0.58	1.15	1.73
4" x 10"	"	0.75	1.15	1.90
4" x 12"	"	0.87	1.15	2.02
1/4" dia., galvanized				
2" x 6"	EA.	0.84	0.96	1.80
2" x 8"	"	0.94	0.96	1.90
2" x 10"	"	1.07	0.96	2.03
2" x 12"	"	1.23	0.96	2.19
4" x 6"	"	0.97	1.15	2.12
4" x 8"	"	1.07	1.15	2.22
4" x 10"	"	1.23	1.15	2.38
4" x 12"	"	1.28	1.15	2.43
"Z" type wall ties, galvanized				
6" long				
1/8" dia.	EA.	0.40	0.96	1.36
3/16" dia.	"	0.43	0.96	1.39
1/4" dia.	"	0.45	0.96	1.41
8" long				
1/8" dia.	EA.	0.43	0.96	1.39
3/16" dia.	"	0.45	0.96	1.41
1/4" dia.	"	0.48	0.96	1.44
10" long				
1/8" dia.	EA.	0.45	0.96	1.41
3/16" dia.	"	0.51	0.96	1.47
1/4" dia.	"	0.58	0.96	1.54
Dovetail anchor slots				
Galvanized steel, filled				
24 ga.	L.F.	0.97	1.44	2.41
20 ga.	"	2.04	1.44	3.48
16 oz. copper, foam filled	"	2.40	1.44	3.84
Dovetail anchors				
16 ga.				
3-1/2" long	EA.	0.29	0.96	1.25
5-1/2" long	"	0.36	0.96	1.32
12 ga.				
3-1/2" long	EA.	0.39	0.96	1.35
5-1/2" long	"	0.65	0.96	1.61
Dovetail, triangular galvanized ties, 12 ga.				
3" x 3"	EA.	0.66	0.96	1.62
5" x 5"	"	0.71	0.96	1.67
7" x 7"	"	0.80	0.96	1.76
7" x 9"	"	0.85	0.96	1.81
Brick anchors				
Corrugated, 3-1/2" long				
16 ga.	EA.	0.27	0.96	1.23
12 ga.	"	0.47	0.96	1.43
Non-corrugated, 3-1/2" long				
16 ga.	EA.	0.38	0.96	1.34
12 ga.	"	0.68	0.96	1.64
Cavity wall anchors, corrugated, galvanized				
5" long				
16 ga.	EA.	0.84	0.96	1.80

04 MASONRY

Mortar And Grout	UNIT	MAT.	INST.	TOTAL
04150.10 Masonry Accessories *(Cont.)*				
12 ga.	EA.	1.25	0.96	2.21
7" long				
28 ga.	EA.	0.92	0.96	1.88
24 ga.	"	1.17	0.96	2.13
22 ga.	"	1.20	0.96	2.16
16 ga.	"	1.35	0.96	2.31
Mesh ties, 16 ga., 3" wide				
8" long	EA.	1.13	0.96	2.09
12" long	"	1.25	0.96	2.21
20" long	"	1.73	0.96	2.69
24" long	"	1.91	0.96	2.87
04150.20 Masonry Control Joints				
Control joint, cross shaped PVC	L.F.	2.57	1.44	4.01
Closed cell joint filler				
1/2"	L.F.	0.45	1.44	1.89
3/4"	"	0.91	1.44	2.35
Rubber, for				
4" wall	L.F.	2.96	1.44	4.40
6" wall	"	3.66	1.51	5.17
8" wall	"	4.42	1.60	6.02
PVC, for				
4" wall	L.F.	1.54	1.44	2.98
6" wall	"	2.60	1.51	4.11
8" wall	"	3.93	1.60	5.53
04150.50 Masonry Flashing				
Through-wall flashing				
5 oz. coated copper	S.F.	3.96	4.80	8.76
0.030" elastomeric	"	1.30	3.84	5.14

Unit Masonry	UNIT	MAT.	INST.	TOTAL
04210.10 Brick Masonry				
Standard size brick, running bond				
Face brick, red (6.4/sf)				
Veneer	S.F.	6.37	9.61	15.98
Cavity wall	"	6.37	8.24	14.61
9" solid wall	"	12.75	16.50	29.25
Common brick (6.4/sf)				
Select common for veneers	S.F.	4.16	9.61	13.77
Back-up				
4" thick	S.F.	3.74	7.21	10.95
8" thick	"	7.48	11.50	18.98
Firewall				
12" thick	S.F.	12.00	19.25	31.25

04 MASONRY

Unit Masonry

04210.10 Brick Masonry *(Cont.)*

	UNIT	MAT.	INST.	TOTAL
16" thick	S.F.	16.00	26.25	42.25
Glazed brick (7.4/sf)				
Veneer	S.F.	17.25	10.50	27.75
Buff or gray face brick (6.4/sf)				
Veneer	S.F.	7.41	9.61	17.02
Cavity wall	"	7.41	8.24	15.65
Jumbo or oversize brick (3/sf)				
4" veneer	S.F.	4.87	5.76	10.63
4" back-up	"	4.87	4.80	9.67
8" back-up	"	5.65	8.24	13.89
12" firewall	"	7.60	14.50	22.10
16" firewall	"	10.75	19.25	30.00
Norman brick, red face, (4.5/sf)				
4" veneer	S.F.	7.02	7.21	14.23
Cavity wall	"	7.02	6.40	13.42
Chimney, standard brick, including flue				
16" x 16"	L.F.	30.00	58.00	88.00
16" x 20"	"	51.00	58.00	109
16" x 24"	"	54.00	58.00	112
20" x 20"	"	42.25	72.00	114
20" x 24"	"	57.00	72.00	129
20" x 32"	"	64.00	82.00	146
Window sill, face brick on edge	"	3.51	14.50	18.01

04210.20 Structural Tile

	UNIT	MAT.	INST.	TOTAL
Structural glazed tile				
6T series, 5-1/2" x 12"				
Glazed on one side				
2" thick	S.F.	12.75	5.76	18.51
4" thick	"	15.75	5.76	21.51
6" thick	"	24.50	6.40	30.90
8" thick	"	30.25	7.21	37.46
Glazed on two sides				
4" thick	S.F.	23.00	7.21	30.21
6" thick	"	31.25	8.24	39.49
Special shapes				
Group 1	S.F.	13.25	11.50	24.75
Group 2	"	16.75	11.50	28.25
Group 3	"	22.25	11.50	33.75
Group 4	"	44.50	11.50	56.00
Group 5	"	54.00	11.50	65.50
Fire rated				
4" thick, 1 hr rating	S.F.	21.25	5.76	27.01
6" thick, 2 hr rating	"	26.00	6.40	32.40
8W series, 8" x 16"				
Glazed on one side				
2" thick	S.F.	13.25	3.84	17.09
4" thick	"	14.25	3.84	18.09
6" thick	"	23.50	4.43	27.93
8" thick	"	25.75	4.43	30.18
Glazed on two sides				
4" thick	S.F.	22.50	4.80	27.30
6" thick	"	31.50	5.76	37.26

04 MASONRY

Unit Masonry	UNIT	MAT.	INST.	TOTAL
04210.20 Structural Tile (Cont.)				
8" thick	S.F.	38.25	5.76	44.01
Special shapes				
Group 1	S.F.	20.25	8.24	28.49
Group 2	"	24.75	8.24	32.99
Group 3	"	27.00	8.24	35.24
Group 4	"	45.00	8.24	53.24
Group 5	"	57.00	8.24	65.24
Fire rated				
4" thick, 1 hr rating	S.F.	33.75	8.24	41.99
6" thick, 2 hr rating	"	40.50	8.24	48.74
04210.60 Pavers, Masonry				
Brick walk laid on sand, sand joints				
Laid flat, (4.5 per sf)	S.F.	4.47	6.40	10.87
Laid on edge, (7.2 per sf)	"	7.15	9.61	16.76
Precast concrete patio blocks				
2" thick				
Natural	S.F.	3.17	1.92	5.09
Colors	"	4.42	1.92	6.34
Exposed aggregates, local aggregate				
Natural	S.F.	9.55	1.92	11.47
Colors	"	9.55	1.92	11.47
Granite or limestone aggregate	"	9.98	1.92	11.90
White tumblestone aggregate	"	7.15	1.92	9.07
Stone pavers, set in mortar				
Bluestone				
1" thick				
Irregular	S.F.	7.26	14.50	21.76
Snapped rectangular	"	11.00	11.50	22.50
1-1/2" thick, random rectangular	"	12.75	14.50	27.25
2" thick, random rectangular	"	15.00	16.50	31.50
Slate				
Natural cleft				
Irregular, 3/4" thick	S.F.	9.11	16.50	25.61
Random rectangular				
1-1/4" thick	S.F.	19.75	14.50	34.25
1-1/2" thick	"	22.25	16.00	38.25
Granite blocks				
3" thick, 3" to 6" wide				
4" to 12" long	S.F.	13.00	19.25	32.25
6" to 15" long	"	8.45	16.50	24.95
Crushed stone, white marble, 3" thick	"	2.01	0.94	2.95
04220.10 Concrete Masonry Units				
Hollow, load bearing				
4"	S.F.	1.83	4.27	6.10
6"	"	2.68	4.43	7.11
8"	"	3.08	4.80	7.88
10"	"	4.25	5.24	9.49
12"	"	4.88	5.76	10.64
Solid, load bearing				
4"	S.F.	2.87	4.27	7.14
6"	"	3.22	4.43	7.65

04 MASONRY

Unit Masonry

04220.10 Concrete Masonry Units *(Cont.)*

	UNIT	MAT.	INST.	TOTAL
8"	S.F.	4.40	4.80	9.20
10"	"	4.69	5.24	9.93
12"	"	6.97	5.76	12.73
Back-up block, 8" x 16"				
2"	S.F.	1.92	3.29	5.21
4"	"	2.00	3.39	5.39
6"	"	2.93	3.60	6.53
8"	"	3.37	3.84	7.21
10"	"	4.65	4.12	8.77
12"	"	5.34	4.43	9.77
Foundation wall, 8" x 16"				
6"	S.F.	2.93	4.12	7.05
8"	"	3.37	4.43	7.80
10"	"	4.65	4.80	9.45
12"	"	5.35	5.24	10.59
Solid				
6"	S.F.	3.54	4.43	7.97
8"	"	4.83	4.80	9.63
10"	"	5.15	5.24	10.39
12"	"	7.65	5.76	13.41
Exterior, styrofoam inserts, std weight, 8" x 16"				
6"	S.F.	5.14	4.43	9.57
8"	"	5.56	4.80	10.36
10"	"	7.20	5.24	12.44
12"	"	9.88	5.76	15.64
Lightweight				
6"	S.F.	5.73	4.43	10.16
8"	"	6.44	4.80	11.24
10"	"	6.86	5.24	12.10
12"	"	9.05	5.76	14.81
Acoustical slotted block				
4"	S.F.	5.97	5.24	11.21
6"	"	6.26	5.24	11.50
8"	"	7.81	5.76	13.57
Filled cavities				
4"	S.F.	6.40	6.40	12.80
6"	"	7.38	6.78	14.16
8"	"	9.45	7.21	16.66
Hollow, split face				
4"	S.F.	4.09	4.27	8.36
6"	"	4.74	4.43	9.17
8"	"	4.97	4.80	9.77
10"	"	5.57	5.24	10.81
12"	"	5.94	5.76	11.70
Split rib profile				
4"	S.F.	4.97	5.24	10.21
6"	"	5.78	5.24	11.02
8"	"	6.29	5.76	12.05
10"	"	6.88	5.76	12.64
12"	"	7.46	5.76	13.22
High strength block, 3500 psi				
2"	S.F.	1.93	4.27	6.20
4"	"	2.42	4.43	6.85

04 MASONRY

Unit Masonry	UNIT	MAT.	INST.	TOTAL
04220.10 — Concrete Masonry Units (Cont.)				
6"	S.F.	2.89	4.43	7.32
8"	"	3.26	4.80	8.06
10"	"	3.80	5.24	9.04
12"	"	4.50	5.76	10.26
Solar screen concrete block				
4" thick				
6" x 6"	S.F.	4.60	12.75	17.35
8" x 8"	"	5.50	11.50	17.00
12" x 12"	"	5.63	8.87	14.50
8" thick				
8" x 16"	S.F.	5.63	8.24	13.87
Glazed block				
Cove base, glazed 1 side, 2"	L.F.	12.25	6.40	18.65
4"	"	12.75	6.40	19.15
6"	"	13.00	7.21	20.21
8"	"	14.00	7.21	21.21
Single face				
2"	S.F.	13.00	4.80	17.80
4"	"	16.00	4.80	20.80
6"	"	17.25	5.24	22.49
8"	"	18.00	5.76	23.76
10"	"	20.25	6.40	26.65
12"	"	21.75	6.78	28.53
Double face				
4"	S.F.	19.25	6.07	25.32
6"	"	22.75	6.40	29.15
8"	"	23.75	7.21	30.96
Corner or bullnose				
2"	EA.	13.00	7.21	20.21
4"	"	16.75	8.24	24.99
6"	"	20.25	8.24	28.49
8"	"	22.25	9.61	31.86
10"	"	24.00	10.50	34.50
12"	"	26.00	11.50	37.50
Gypsum unit masonry				
Partition blocks (12"x30")				
Solid				
2"	S.F.	1.31	2.30	3.61
Hollow				
3"	S.F.	1.32	2.30	3.62
4"	"	1.52	2.40	3.92
6"	"	1.62	2.62	4.24
Vertical reinforcing				
4' o.c., add 5% to labor				
2'8" o.c., add 15% to labor				
Interior partitions, add 10% to labor				
04220.90 — Bond Beams & Lintels				
Bond beam, no grout or reinforcement				
8" x 16" x				
4" thick	L.F.	1.73	4.43	6.16
6" thick	"	2.64	4.61	7.25
8" thick	"	3.03	4.80	7.83

04 MASONRY

Unit Masonry

04220.90 Bond Beams & Lintels (Cont.)

	UNIT	MAT.	INST.	TOTAL
10" thick	L.F.	3.75	5.01	8.76
12" thick	"	4.26	5.24	9.50
Beam lintel, no grout or reinforcement				
8" x 16" x				
10" thick	L.F.	6.93	5.76	12.69
12" thick	"	8.80	6.40	15.20
Precast masonry lintel				
6 lf, 8" high x				
4" thick	L.F.	6.34	9.61	15.95
6" thick	"	8.09	9.61	17.70
8" thick	"	9.16	10.50	19.66
10" thick	"	11.00	10.50	21.50
10 lf, 8" high x				
4" thick	L.F.	7.96	5.76	13.72
6" thick	"	9.82	5.76	15.58
8" thick	"	11.00	6.40	17.40
10" thick	"	14.75	6.40	21.15
Steel angles and plates				
Minimum	Lb.	1.29	0.82	2.11
Maximum	"	1.91	1.44	3.35
Various size angle lintels				
1/4" stock				
3" x 3"	L.F.	6.68	3.60	10.28
3" x 3-1/2"	"	7.37	3.60	10.97
3/8" stock				
3" x 4"	L.F.	11.50	3.60	15.10
3-1/2" x 4"	"	12.25	3.60	15.85
4" x 4"	"	13.25	3.60	16.85
5" x 3-1/2"	"	14.25	3.60	17.85
6" x 3-1/2"	"	16.00	3.60	19.60
1/2" stock				
6" x 4"	L.F.	17.50	3.60	21.10

04240.10 Clay Tile

	UNIT	MAT.	INST.	TOTAL
Hollow clay tile, for back-up, 12" x 12"				
Scored face				
Load bearing				
4" thick	S.F.	7.76	4.12	11.88
6" thick	"	9.04	4.27	13.31
8" thick	"	11.25	4.43	15.68
10" thick	"	14.00	4.61	18.61
12" thick	"	23.50	4.80	28.30
Non-load bearing				
3" thick	S.F.	15.25	3.97	19.22
4" thick	"	7.28	4.12	11.40
6" thick	"	8.42	4.27	12.69
8" thick	"	10.75	4.43	15.18
12" thick	"	15.25	4.80	20.05
Partition, 12" x 12"				
In walls				
3" thick	S.F.	6.29	4.80	11.09
4" thick	"	7.29	4.80	12.09
6" thick	"	8.03	5.01	13.04

04 MASONRY

Unit Masonry	UNIT	MAT.	INST.	TOTAL
04240.10 — **Clay Tile** *(Cont.)*				
8" thick	S.F.	10.50	5.24	15.74
10" thick	"	12.50	5.49	17.99
12" thick	"	18.25	5.76	24.01
Clay tile floors				
4" thick	S.F.	7.28	3.20	10.48
6" thick	"	9.04	3.39	12.43
8" thick	"	11.25	3.60	14.85
10" thick	"	14.00	3.84	17.84
12" thick	"	20.75	4.12	24.87
Terra cotta				
Coping, 10" or 12" wide, 3" thick	L.F.	17.25	11.50	28.75
04270.10 — **Glass Block**				
Glass block, 4" thick				
6" x 6"	S.F.	35.75	19.25	55.00
8" x 8"	"	22.50	14.50	37.00
12" x 12"	"	28.75	11.50	40.25
Replacement glass blocks, 4" x 8" x 8"				
Minimum	S.F.	24.75	58.00	82.75
Maximum	"	31.50	120	152
04295.10 — **Parging / Masonry Plaster**				
Parging				
1/2" thick	S.F.	0.35	3.84	4.19
3/4" thick	"	0.44	4.80	5.24
1" thick	"	0.59	5.76	6.35

Stone	UNIT	MAT.	INST.	TOTAL
04400.10 — **Stone**				
Rubble stone				
Walls set in mortar				
8" thick	S.F.	17.75	14.50	32.25
12" thick	"	21.50	23.00	44.50
18" thick	"	28.50	28.75	57.25
24" thick	"	35.75	38.50	74.25
Dry set wall				
8" thick	S.F.	20.00	9.61	29.61
12" thick	"	22.50	14.50	37.00
18" thick	"	31.25	19.25	50.50
24" thick	"	38.00	23.00	61.00
Cut stone				
Imported marble				
Facing panels				
3/4" thick	S.F.	42.00	23.00	65.00
1-1/2" thick	"	66.00	26.25	92.25

04 MASONRY

Stone	UNIT	MAT.	INST.	TOTAL
04400.10 Stone (Cont.)				
2-1/4" thick	S.F.	79.00	32.00	111
Base				
1" thick				
4" high	L.F.	20.75	28.75	49.50
6" high	"	25.25	28.75	54.00
Columns, solid				
Plain faced	C.F.	150	380	530
Fluted	"	400	380	780
Flooring, travertine, minimum	S.F.	20.25	8.87	29.12
Average	"	27.00	11.50	38.50
Maximum	"	49.75	12.75	62.50
Domestic marble				
Facing panels				
7/8" thick	S.F.	37.50	23.00	60.50
1-1/2" thick	"	56.00	26.25	82.25
2-1/4" thick	"	68.00	32.00	100
Stairs				
12" treads	L.F.	33.75	28.75	62.50
6" risers	"	25.00	19.25	44.25
Thresholds, 7/8" thick, 3' long, 4" to 6" wide				
Plain	EA.	32.00	48.00	80.00
Beveled	"	35.50	48.00	83.50
Window sill				
6" wide, 2" thick	L.F.	18.00	23.00	41.00
Stools				
5" wide, 7/8" thick	L.F.	26.25	23.00	49.25
Limestone panels up to 12' x 5', smooth finish				
2" thick	S.F.	35.00	9.84	44.84
3" thick	"	41.00	9.84	50.84
4" thick	"	59.00	9.84	68.84
Miscellaneous limestone items				
Steps, 14" wide, 6" deep	L.F.	75.00	38.50	114
Coping, smooth finish	C.F.	100	19.25	119
Sills, lintels, jambs, smooth finish	"	100	23.00	123
Granite veneer facing panels, polished				
7/8" thick				
Black	S.F.	60.00	23.00	83.00
Gray	"	47.25	23.00	70.25
Base				
4" high	L.F.	25.25	11.50	36.75
6" high	"	30.50	12.75	43.25
Curbing, straight, 6" x 16"	"	28.00	41.00	69.00
Radius curbs, radius over 5'	"	34.25	55.00	89.25
Ashlar veneer				
4" thick, random	S.F.	42.25	23.00	65.25
Pavers, 4" x 4" split				
Gray	S.F.	41.50	11.50	53.00
Pink	"	40.75	11.50	52.25
Black	"	40.25	11.50	51.75
Slate, panels				
1" thick	S.F.	29.50	23.00	52.50
2" thick	"	40.00	26.25	66.25
Sills or stools				

04 MASONRY

Stone

	UNIT	MAT.	INST.	TOTAL
04400.10 Stone (Cont.)				
1" thick				
6" wide	L.F.	13.75	23.00	36.75
10" wide	"	22.50	25.00	47.50
2" thick				
6" wide	L.F.	22.50	26.25	48.75
10" wide	"	37.25	28.75	66.00

Masonry Restoration

	UNIT	MAT.	INST.	TOTAL
04520.10 Restoration And Cleaning				
Masonry cleaning				
Washing brick				
Smooth surface	S.F.	0.25	0.96	1.21
Rough surface	"	0.35	1.28	1.63
Steam clean masonry				
Smooth face				
Minimum	S.F.		0.75	0.75
Maximum	"		1.09	1.09
Rough face				
Minimum	S.F.		1.00	1.00
Maximum	"		1.51	1.51
Sandblast masonry				
Minimum	S.F.	0.48	1.20	1.68
Maximum	"	0.71	2.01	2.72
Pointing masonry				
Brick	S.F.	1.40	2.30	3.70
Concrete block	"	0.62	1.64	2.26
Cut and repoint				
Brick				
Minimum	S.F.	0.44	2.88	3.32
Maximum	"	0.75	5.76	6.51
Stone work	L.F.	1.16	4.43	5.59
Cut and recaulk				
Oil base caulks	L.F.	1.56	3.84	5.40
Butyl caulks	"	1.36	3.84	5.20
Polysulfides and acrylics	"	2.65	3.84	6.49
Silicones	"	3.09	3.84	6.93
Cement and sand grout on walls, to 1/8" thick				
Minimum	S.F.	0.76	2.30	3.06
Maximum	"	1.93	2.88	4.81
Brick removal and replacement				
Minimum	EA.	0.80	7.21	8.01
Average	"	1.05	9.61	10.66
Maximum	"	2.13	28.75	30.88

04 MASONRY

Masonry Restoration	UNIT	MAT.	INST.	TOTAL
04550.10 Refractories				
Flue liners				
Rectangular				
8" x 12"	L.F.	9.35	9.61	18.96
12" x 12"	"	11.75	10.50	22.25
12" x 18"	"	20.50	11.50	32.00
16" x 16"	"	22.25	12.75	35.00
18" x 18"	"	27.50	13.75	41.25
20" x 20"	"	45.75	14.50	60.25
24" x 24"	"	55.00	16.50	71.50
Round				
18" dia.	L.F.	42.25	13.75	56.00
24" dia.	"	83.00	16.50	99.50

05 METALS

Metal Fastening	UNIT	MAT.	INST.	TOTAL
05050.10 **Structural Welding**				
Welding				
Single pass				
1/8"	L.F.	0.39	3.34	3.73
3/16"	"	0.65	4.45	5.10
1/4"	"	0.91	5.57	6.48
Miscellaneous steel shapes				
Plain	Lb.	1.70	0.13	1.83
Galvanized	"	2.13	0.22	2.35
Plates				
Plain	Lb.	1.53	0.16	1.69
Galvanized	"	1.96	0.26	2.22
05050.90 **Metal Anchors**				
Anchor bolts				
3/8" x				
8" long	EA.			1.19
10" long	"			1.29
12" long	"			1.41
1/2" x				
8" long	EA.			1.78
10" long	"			1.90
12" long	"			2.07
18" long	"			2.26
5/8" x				
8" long	EA.			1.66
10" long	"			1.83
12" long	"			1.95
18" long	"			2.07
24" long	"			2.26
3/4" x				
8" long	EA.			2.37
12" long	"			2.66
18" long	"			3.66
24" long	"			4.86
7/8" x				
8" long	EA.			2.37
12" long	"			2.66
18" long	"			3.66
24" long	"			4.86
1" x				
12" long	EA.			4.74
18" long	"			5.93
24" long	"			7.11
36" long	"			10.75
Expansion shield				
1/4"	EA.			0.74
3/8"	"			1.23
1/2"	"			2.41
5/8"	"			3.51
3/4"	"			4.29
1"	"			5.81
Non-drilling anchor				
1/4"	EA.			0.76

05 METALS

Metal Fastening

	UNIT	MAT.	INST.	TOTAL
05050.90 — Metal Anchors (Cont.)				
3/8"	EA.			0.94
1/2"	"			1.45
5/8"	"			2.39
3/4"	"			4.09
Self-drilling anchor				
1/4"	EA.			1.92
5/16"	"			2.40
3/8"	"			2.88
1/2"	"			3.84
5/8"	"			7.30
3/4"	"			9.59
7/8"	"			13.50
Add 25% for galvanized anchor bolts				
Channel door frame, with anchors	Lb.	2.10	0.74	2.84
Corner guard angle, with anchors	"	1.88	1.11	2.99
05050.95 — Metal Lintels				
Lintels, steel				
Plain	Lb.	1.57	1.67	3.24
Galvanized	"	2.36	1.67	4.03
05120.10 — Structural Steel				
Beams and girders, A-36				
Welded	TON	3,290	730	4,020
Bolted	"	3,210	660	3,870
Columns				
Pipe				
6" dia.	Lb.	1.85	0.72	2.57
12" dia.	"	1.58	0.60	2.18
Purlins and girts				
Welded	TON	3,450	1,210	4,660
Bolted	"	3,400	1,040	4,440
Column base plates				
Up to 150 lb each	Lb.	1.99	0.44	2.43
Over 150 lb each	"	1.62	0.55	2.17
Structural pipe				
3" to 5" o.d.	TON	3,720	1,460	5,180
6" to 12" o.d.	"	3,450	1,040	4,490
Structural tube				
6" square				
Light sections	TON	4,320	1,460	5,780
Heavy sections	"	4,050	1,040	5,090
6" wide rectangular				
Light sections	TON	4,320	1,210	5,530
Heavy sections	"	4,050	910	4,960
Greater than 6" wide rectangular				
Light sections	TON	4,750	1,210	5,960
Heavy sections	"	4,470	910	5,380
Miscellaneous structural shapes				
Steel angle	TON	3,070	1,820	4,890
Steel plate	"	3,490	1,210	4,700
Trusses, field welded				
60 lb/lf	TON	4,470	910	5,380

05 METALS

Metal Fastening	UNIT	MAT.	INST.	TOTAL
05120.10 — **Structural Steel** *(Cont.)*				
100 lb/lf	TON	3,910	730	4,640
150 lb/lf	"	3,690	610	4,300
Bolted				
60 lb/lf	TON	4,410	810	5,220
100 lb/lf	"	3,860	660	4,520
150 lb/lf	"	3,660	560	4,220
Add for galvanizing	"			960
05200.10 — **Metal Joists**				
Joist				
DLH series	TON	2,050	490	2,540
K series	"	2,050	490	2,540
LH series	"	2,050	490	2,540
05300.10 — **Metal Decking**				
Roof, 1-1/2" deep, non-composite				
16 ga.				
Primed	S.F.	3.76	1.21	4.97
Galvanized	"	4.01	1.21	5.22
18 ga.				
Primed	S.F.	3.14	1.21	4.35
Galvanized	"	3.45	1.21	4.66
20 ga.				
Primed	S.F.	2.31	1.21	3.52
Galvanized	"	2.74	1.21	3.95
22 ga.				
Primed	S.F.	1.88	1.21	3.09
Galvanized	"	2.10	1.21	3.31
Open type decking, galvanized				
1-1/2" deep				
18 ga.	S.F.	3.02	1.21	4.23
20 ga.	"	2.31	1.21	3.52
22 ga.	"	1.99	1.21	3.20
3" deep				
16 ga.	S.F.	4.68	1.32	6.00
18 ga.	"	4.41	1.32	5.73
20 ga.	"	3.34	1.32	4.66
22 ga.	"	3.24	1.32	4.56
4-1/2" deep				
16 ga.	S.F.	6.96	1.45	8.41
18 ga.	"	5.75	1.45	7.20
6" deep				
16 ga.	S.F.	9.61	1.61	11.22
18 ga.	"	8.36	1.61	9.97
7-1/2" deep				
16 ga.	S.F.	10.50	1.69	12.19
18 ga.	"	9.84	1.69	11.53
Cellular type				
1-1/2" deep, galvanized				
18-18 ga.	S.F.	7.55	1.45	9.00
22-18 ga.	"	6.86	1.45	8.31
3" deep, galvanized				
16-16 ga.	S.F.	10.75	1.61	12.36

05 METALS

Metal Fastening	UNIT	MAT.	INST.	TOTAL
05300.10 **Metal Decking** (Cont.)				
18-16 ga.	S.F.	10.00	1.61	11.61
18-18 ga.	"	9.25	1.61	10.86
20-18 ga.	"	8.20	1.61	9.81
4-1/2" deep, galvanized				
16-16 ga.	S.F.	16.25	1.73	17.98
18-16 ga.	"	14.75	1.73	16.48
18-18 ga.	"	14.25	1.73	15.98
20-18 ga.	"	13.00	1.73	14.73
Composite deck, non-cellular, galvanized				
1-1/2" deep				
18 ga.	S.F.	2.54	1.32	3.86
20 ga.	"	2.31	1.32	3.63
22 ga.	"	1.88	1.32	3.20
3" deep				
18 ga.	S.F.	3.57	1.39	4.96
20 ga.	"	2.31	1.39	3.70
22 ga.	"	1.88	1.39	3.27
Slab form floor deck				
9/16" deep				
28 ga.	S.F.	1.13	1.32	2.45
1-5/16" deep				
24 ga.	S.F.	1.74	1.37	3.11
22 ga.	"	2.00	1.37	3.37

Cold Formed Framing	UNIT	MAT.	INST.	TOTAL
05410.10 **Metal Framing**				
Furring channel, galvanized				
Beams and columns, 3/4"				
12" o.c.	S.F.	0.46	6.68	7.14
16" o.c.	"	0.36	6.07	6.43
Walls, 3/4"				
12" o.c.	S.F.	0.46	3.34	3.80
16" o.c.	"	0.36	2.78	3.14
24" o.c.	"	0.26	2.22	2.48
1-1/2"				
12" o.c.	S.F.	0.78	3.34	4.12
16" o.c.	"	0.58	2.78	3.36
24" o.c.	"	0.40	2.22	2.62
Stud, load bearing				
16" o.c.				
16 ga.				
2-1/2"	S.F.	1.43	2.97	4.40
3-5/8"	"	1.69	2.97	4.66
4"	"	1.75	2.97	4.72
6"	"	2.21	3.34	5.55

05 METALS

Cold Formed Framing	UNIT	MAT.	INST.	TOTAL
05410.10 **Metal Framing** (Cont.)				
18 ga.				
2-1/2"	S.F.	1.17	2.97	4.14
3-5/8"	"	1.43	2.97	4.40
4"	"	1.49	2.97	4.46
6"	"	1.88	3.34	5.22
8"	"	2.27	3.34	5.61
20 ga.				
2-1/2"	S.F.	0.65	2.97	3.62
3-5/8"	"	0.78	2.97	3.75
4"	"	0.84	2.97	3.81
6"	"	1.04	3.34	4.38
8"	"	1.23	3.34	4.57
24" o.c.				
16 ga.				
2-1/2"	S.F.	0.97	2.57	3.54
3-5/8"	"	1.17	2.57	3.74
4"	"	1.23	2.57	3.80
6"	"	1.49	2.78	4.27
8"	"	1.88	2.78	4.66
18 ga.				
2-1/2"	S.F.	0.78	2.57	3.35
3-5/8"	"	0.91	2.57	3.48
4"	"	0.97	2.57	3.54
6"	"	1.23	2.78	4.01
8"	"	1.49	2.78	4.27
20 ga.				
2-1/2"	S.F.	0.52	2.57	3.09
3-5/8"	"	0.58	2.57	3.15
4"	"	0.65	2.57	3.22
6"	"	0.84	2.78	3.62
8"	"	1.04	2.78	3.82

Metal Fabrications	UNIT	MAT.	INST.	TOTAL
05510.10 **Stairs**				
Stock unit, steel, complete, per riser				
Tread				
3'-6" wide	EA.	160	84.00	244
4' wide	"	190	95.00	285
5' wide	"	220	110	330
Metal pan stair, cement filled, per riser				
3'-6" wide	EA.	180	67.00	247
4' wide	"	200	74.00	274
5' wide	"	230	84.00	314
Landing, steel pan	S.F.	70.00	16.75	86.75
Cast iron tread, steel stringers, stock units, per riser				

05 METALS

Metal Fabrications

05510.10 Stairs (Cont.)

	UNIT	MAT.	INST.	TOTAL
Tread				
3'-6" wide	EA.	310	84.00	394
4' wide	"	360	95.00	455
5' wide	"	430	110	540
Stair treads, abrasive, 12" x 3'-6"				
Cast iron				
3/8"	EA.	170	33.50	204
1/2"	"	220	33.50	254
Cast aluminum				
5/16"	EA.	200	33.50	234
3/8"	"	220	33.50	254
1/2"	"	260	33.50	294

05515.10 Ladders

	UNIT	MAT.	INST.	TOTAL
Ladder, 18" wide				
With cage	L.F.	110	44.50	155
Without cage	"	72.00	33.50	106

05520.10 Railings

	UNIT	MAT.	INST.	TOTAL
Railing, pipe				
1-1/4" diameter, welded steel				
2-rail				
Primed	L.F.	35.75	13.25	49.00
Galvanized	"	45.75	13.25	59.00
3-rail				
Primed	L.F.	45.75	16.75	62.50
Galvanized	"	59.00	16.75	75.75
Wall mounted, single rail, welded steel				
Primed	L.F.	23.75	10.25	34.00
Galvanized	"	31.00	10.25	41.25
1-1/2" diameter, welded steel				
2-rail				
Primed	L.F.	38.75	13.25	52.00
Galvanized	"	50.00	13.25	63.25
3-rail				
Primed	L.F.	48.75	16.75	65.50
Galvanized	"	63.00	16.75	79.75
Wall mounted, single rail, welded steel				
Primed	L.F.	25.75	10.25	36.00
Galvanized	"	33.50	10.25	43.75
2" diameter, welded steel				
2-rail				
Primed	L.F.	48.75	14.75	63.50
Galvanized	"	63.00	14.75	77.75
3-rail				
Primed	L.F.	62.00	19.00	81.00
Galvanized	"	80.00	19.00	99.00
Wall mounted, single rail, welded steel				
Primed	L.F.	28.00	11.25	39.25
Galvanized	"	36.50	11.25	47.75

05 METALS

Metal Fabrications

05530.10 Metal Grating

	UNIT	MAT.	INST.	TOTAL
Floor plate, checkered, steel				
1/4"				
Primed	S.F.	13.75	0.95	14.70
Galvanized	"	18.25	0.95	19.20
3/8"				
Primed	S.F.	18.25	1.02	19.27
Galvanized	"	24.25	1.02	25.27
Aluminum grating, pressure-locked bearing bars				
3/4" x 1/8"	S.F.	32.75	1.67	34.42
1" x 1/8"	"	36.25	1.67	37.92
1-1/4" x 1/8"	"	46.00	1.67	47.67
1-1/4" x 3/16"	"	49.25	1.67	50.92
1-1/2" x 1/8"	"	52.00	1.67	53.67
1-3/4" x 3/16"	"	59.00	1.67	60.67
Miscellaneous expenses				
Cutting				
Minimum	L.F.		4.45	4.45
Maximum	"		6.68	6.68
Banding				
Minimum	L.F.		11.25	11.25
Maximum	"		13.25	13.25
Toe plates				
Minimum	L.F.		13.25	13.25
Maximum	"		16.75	16.75
Steel grating, primed				
3/4" x 1/8"	S.F.	10.75	2.22	12.97
1" x 1/8"	"	11.00	2.22	13.22
1-1/4" x 1/8"	"	12.25	2.22	14.47
1-1/4" x 3/16"	"	16.50	2.22	18.72
1-1/2" x 1/8"	"	15.00	2.22	17.22
1-3/4" x 3/16"	"	21.25	2.22	23.47
Galvanized				
3/4" x 1/8"	S.F.	13.50	2.22	15.72
1" x 1/8"	"	13.75	2.22	15.97
1-1/4" x 1/8"	"	15.50	2.22	17.72
1-1/4" x 3/16"	"	20.50	2.22	22.72
1-1/2" x 1/8"	"	18.75	2.22	20.97
1-3/4" x 3/16"	"	26.75	2.22	28.97
Miscellaneous expenses				
Cutting				
Minimum	L.F.		4.77	4.77
Maximum	"		7.42	7.42
Banding				
Minimum	L.F.		12.25	12.25
Maximum	"		14.75	14.75
Toe plates				
Minimum	L.F.		14.75	14.75
Maximum	"		19.00	19.00

05 METALS

Metal Fabrications

	UNIT	MAT.	INST.	TOTAL
05540.10 Castings				
Miscellaneous castings				
Light sections	Lb.	2.70	1.33	4.03
Heavy sections	"	1.65	0.95	2.60
Manhole covers and frames				
Regular, city type				
18" dia.				
100 lb	EA.	400	130	530
24" dia.				
200 lb	EA.	410	130	540
300 lb	"	420	150	570
400 lb	"	440	150	590
26" dia., 475 lb	"	540	170	710
30" dia., 600 lb	"	630	190	820
8" square, 75 lb	"	180	26.75	207
24" square				
126 lb	EA.	390	130	520
500 lb	"	630	170	800
Watertight type				
20" dia., 200 lb	EA.	360	170	530
24" dia., 350 lb	"	610	220	830
Steps, cast iron				
7" x 9"	EA.	18.25	13.25	31.50
8" x 9"	"	26.00	14.75	40.75
Manhole covers and frames, aluminum				
12" x 12"	EA.	89.00	26.75	116
18" x 18"	"	92.00	26.75	119
24" x 24"	"	100	33.50	134
Corner protection				
Steel angle guard with anchors				
2" x 2" x 3/16"	L.F.	18.75	9.54	28.29
2" x 3" x 1/4"	"	21.25	9.54	30.79
3" x 3" x 5/16"	"	24.75	9.54	34.29
3" x 4" x 5/16"	"	29.75	10.25	40.00
4" x 4" x 5/16"	"	30.75	10.25	41.00

Misc. Fabrications

	UNIT	MAT.	INST.	TOTAL
05580.10 Metal Specialties				
Kick plate				
4" high x 1/4" thick				
Primed	L.F.	9.10	13.25	22.35
Galvanized	"	10.25	13.25	23.50
6" high x 1/4" thick				
Primed	L.F.	9.36	14.75	24.11
Galvanized	"	11.00	14.75	25.75

05 METALS

Misc. Fabrications

	UNIT	MAT.	INST.	TOTAL
05700.10 Ornamental Metal				
Railings, square bars, 6" o.c., shaped top rails				
Steel	L.F.	110	33.50	144
Aluminum	"	130	33.50	164
Bronze	"	220	44.50	265
Stainless steel	"	230	44.50	275
Laminated metal or wood handrails				
2-1/2" round or oval shape	L.F.	330	33.50	364
Grilles and louvers				
Fixed type louvers				
4 through 10 sf	S.F.	33.25	11.25	44.50
Over 10 sf	"	39.50	8.35	47.85
Movable type louvers				
4 through 10 sf	S.F.	39.50	11.25	50.75
Over 10 sf	"	43.50	8.35	51.85
Aluminum louvers				
Residential use, fixed type, with screen				
8" x 8"	EA.	21.00	33.50	54.50
12" x 12"	"	23.00	33.50	56.50
12" x 18"	"	27.75	33.50	61.25
14" x 24"	"	39.75	33.50	73.25
18" x 24"	"	44.75	33.50	78.25
30" x 24"	"	61.00	37.25	98.25
05800.10 Expansion Control				
Expansion joints with covers, floor assembly type				
With 1" space				
Aluminum	L.F.	32.25	11.25	43.50
Bronze	"	67.00	11.25	78.25
Stainless steel	"	52.00	11.25	63.25
With 2" space				
Aluminum	L.F.	40.00	11.25	51.25
Bronze	"	71.00	11.25	82.25
Stainless steel	"	62.00	11.25	73.25
Ceiling and wall assembly type				
With 1" space				
Aluminum	L.F.	20.75	13.25	34.00
Bronze	"	71.00	13.25	84.25
Stainless steel	"	64.00	13.25	77.25
With 2" space				
Aluminum	L.F.	23.00	13.25	36.25
Bronze	"	77.00	13.25	90.25
Stainless steel	"	67.00	13.25	80.25
Exterior roof and wall, aluminum				
Roof to roof				
With 1" space	L.F.	60.00	11.25	71.25
With 2" space	"	65.00	11.25	76.25
Roof to wall				
With 1" space	L.F.	46.75	12.25	59.00
With 2" space	"	56.00	12.25	68.25
Flat wall to wall				
With 1" space	L.F.	22.75	11.25	34.00
With 2" space	"	24.75	11.25	36.00
Corner to flat wall				

05 METALS

Misc. Fabrications	UNIT	MAT.	INST.	TOTAL
05800.10 **Expansion Control** *(Cont.)*				
With 1" space	L.F.	27.25	13.25	40.50
With 2" in space	"	27.75	13.25	41.00

06 WOOD AND PLASTICS

Fasteners And Adhesives	UNIT	MAT.	INST.	TOTAL
06050.10 **Accessories**				
Column/post base, cast aluminum				
4" x 4"	EA.	18.50	15.00	33.50
6" x 6"	"	25.75	15.00	40.75
Bridging, metal, per pair				
12" o.c.	EA.	2.52	6.04	8.56
16" o.c.	"	2.32	5.49	7.81
Anchors				
Bolts, threaded two ends, with nuts and washers				
1/2" dia.				
4" long	EA.	2.93	3.77	6.70
7-1/2" long	"	3.41	3.77	7.18
3/4" dia.				
7-1/2" long	EA.	6.48	3.77	10.25
15" long	"	9.75	3.77	13.52
Framing anchors				
10 gauge	EA.	1.16	5.03	6.19
Bolts, carriage				
1/4 x 4	EA.	0.76	6.04	6.80
5/16 x 6	"	1.72	6.36	8.08
3/8 x 6	"	3.49	6.36	9.85
1/2 x 6	"	4.87	6.36	11.23
Joist and beam hangers				
18 ga.				
2 x 4	EA.	1.37	6.04	7.41
2 x 6	"	1.64	6.04	7.68
2 x 8	"	1.92	6.04	7.96
2 x 10	"	2.06	6.71	8.77
2 x 12	"	2.67	7.55	10.22
16 ga.				
3 x 6	EA.	4.80	6.71	11.51
3 x 8	"	5.83	6.71	12.54
3 x 10	"	6.59	7.11	13.70
3 x 12	"	7.42	8.05	15.47
3 x 14	"	8.03	8.63	16.66
4 x 6	"	8.24	6.71	14.95
4 x 8	"	9.61	6.71	16.32
4 x 10	"	11.00	7.11	18.11
4 x 12	"	14.25	8.05	22.30
4 x 14	"	15.00	8.63	23.63
Rafter anchors, 18 ga., 1-1/2" wide				
5-1/4" long	EA.	1.04	5.03	6.07
10-3/4" long	"	1.52	5.03	6.55
Shear plates				
2-5/8" dia.	EA.	3.68	4.64	8.32
4" dia.	"	7.64	5.03	12.67
Sill anchors				
Embedded in concrete	EA.	2.70	6.04	8.74
Split rings				
2-1/2" dia.	EA.	2.22	6.71	8.93
4" dia.	"	4.09	7.55	11.64
Strap ties, 14 ga., 1-3/8" wide				
12" long	EA.	2.77	5.03	7.80

06 WOOD AND PLASTICS

Fasteners And Adhesives

	UNIT	MAT.	INST.	TOTAL
06050.10 Accessories (Cont.)				
18" long	EA.	2.98	5.49	8.47
24" long	"	4.44	6.04	10.48
36" long	"	6.11	6.71	12.82
Toothed rings				
2-5/8" dia.	EA.	2.57	10.00	12.57
4" dia.	"	2.98	12.00	14.98

Rough Carpentry

	UNIT	MAT.	INST.	TOTAL
06110.10 Blocking				
Steel construction				
Walls				
2x4	L.F.	0.46	4.02	4.48
2x6	"	0.69	4.64	5.33
2x8	"	0.92	5.03	5.95
2x10	"	1.22	5.49	6.71
2x12	"	1.58	6.04	7.62
Ceilings				
2x4	L.F.	0.46	4.64	5.10
2x6	"	0.69	5.49	6.18
2x8	"	0.92	6.04	6.96
2x10	"	1.22	6.71	7.93
2x12	"	1.58	7.55	9.13
Wood construction				
Walls				
2x4	L.F.	0.46	3.35	3.81
2x6	"	0.69	3.77	4.46
2x8	"	0.92	4.02	4.94
2x10	"	1.22	4.31	5.53
2x12	"	1.58	4.64	6.22
Ceilings				
2x4	L.F.	0.46	3.77	4.23
2x6	"	0.69	4.31	5.00
2x8	"	0.92	4.64	5.56
2x10	"	1.22	5.03	6.25
2x12	"	1.58	5.49	7.07
06110.20 Ceiling Framing				
Ceiling joists				
12" o.c.				
2x4	S.F.	0.68	1.43	2.11
2x6	"	0.98	1.51	2.49
2x8	"	1.44	1.59	3.03
2x10	"	1.64	1.67	3.31
2x12	"	3.02	1.77	4.79
16" o.c.				

06 WOOD AND PLASTICS

Rough Carpentry

06110.20 Ceiling Framing (Cont.)

	UNIT	MAT.	INST.	TOTAL
2x4	S.F.	0.55	1.16	1.71
2x6	"	0.82	1.20	2.02
2x8	"	1.17	1.25	2.42
2x10	"	1.31	1.31	2.62
2x12	"	2.47	1.37	3.84
24" o.c.				
2x4	S.F.	0.39	0.95	1.34
2x6	"	0.65	1.00	1.65
2x8	"	0.98	1.06	2.04
2x10	"	1.17	1.11	2.28
2x12	"	2.96	1.18	4.14
Headers and nailers				
2x4	L.F.	0.46	1.94	2.40
2x6	"	0.69	2.01	2.70
2x8	"	0.92	2.15	3.07
2x10	"	1.22	2.32	3.54
2x12	"	1.51	2.51	4.02
Sister joists for ceilings				
2x4	L.F.	0.46	4.31	4.77
2x6	"	0.69	5.03	5.72
2x8	"	0.92	6.04	6.96
2x10	"	1.22	7.55	8.77
2x12	"	1.51	10.00	11.51

06110.30 Floor Framing

	UNIT	MAT.	INST.	TOTAL
Floor joists				
12" o.c.				
2x6	S.F.	0.84	1.20	2.04
2x8	"	1.23	1.23	2.46
2x10	"	1.71	1.25	2.96
2x12	"	2.50	1.31	3.81
2x14	"	3.81	1.25	5.06
3x6	"	2.83	1.28	4.11
3x8	"	3.68	1.31	4.99
3x10	"	4.60	1.37	5.97
3x12	"	5.53	1.43	6.96
3x14	"	6.32	1.51	7.83
4x6	"	3.68	1.25	4.93
4x8	"	4.74	1.31	6.05
4x10	"	6.05	1.37	7.42
4x12	"	7.37	1.43	8.80
4x14	"	8.55	1.51	10.06
16" o.c.				
2x6	S.F.	0.72	1.00	1.72
2x8	"	1.01	1.02	2.03
2x10	"	1.23	1.04	2.27
2x12	"	1.54	1.07	2.61
2x14	"	3.42	1.11	4.53
3x6	"	2.37	1.04	3.41
3x8	"	3.02	1.07	4.09
3x10	"	3.81	1.11	4.92
3x12	"	4.60	1.16	5.76
3x14	"	5.46	1.20	6.66

06 WOOD AND PLASTICS

Rough Carpentry	UNIT	MAT.	INST.	TOTAL
06110.30 Floor Framing (Cont.)				
4x6	S.F.	3.02	1.04	4.06
4x8	"	4.14	1.07	5.21
4x10	"	5.13	1.11	6.24
4x12	"	6.05	1.16	7.21
4x14	"	7.24	1.20	8.44
Sister joists for floors				
2x4	L.F.	0.46	3.77	4.23
2x6	"	0.69	4.31	5.00
2x8	"	0.92	5.03	5.95
2x10	"	1.22	6.04	7.26
2x12	"	1.58	7.55	9.13
3x6	"	2.30	6.04	8.34
3x8	"	2.83	6.71	9.54
3x10	"	3.75	7.55	11.30
3x12	"	4.54	8.63	13.17
4x6	"	2.96	6.04	9.00
4x8	"	3.95	6.71	10.66
4x10	"	5.13	7.55	12.68
4x12	"	5.72	8.63	14.35
06110.40 Furring				
Furring, wood strips				
Walls				
On masonry or concrete walls				
1x2 furring				
12" o.c.	S.F.	0.36	1.88	2.24
16" o.c.	"	0.31	1.72	2.03
24" o.c.	"	0.30	1.59	1.89
1x3 furring				
12" o.c.	S.F.	0.46	1.88	2.34
16" o.c.	"	0.42	1.72	2.14
24" o.c.	"	0.32	1.59	1.91
On wood walls				
1x2 furring				
12" o.c.	S.F.	0.36	1.34	1.70
16" o.c.	"	0.31	1.20	1.51
24" o.c.	"	0.29	1.09	1.38
1x3 furring				
12" o.c.	S.F.	0.47	1.34	1.81
16" o.c.	"	0.39	1.20	1.59
24" o.c.	"	0.32	1.09	1.41
Ceilings				
On masonry or concrete ceilings				
1x2 furring				
12" o.c.	S.F.	0.36	3.35	3.71
16" o.c.	"	0.31	3.02	3.33
24" o.c.	"	0.29	2.74	3.03
1x3 furring				
12" o.c.	S.F.	0.46	3.35	3.81
16" o.c.	"	0.39	3.02	3.41
24" o.c.	"	0.32	2.74	3.06
On wood ceilings				
1x2 furring				

06 WOOD AND PLASTICS

Rough Carpentry	UNIT	MAT.	INST.	TOTAL
06110.40 **Furring** *(Cont.)*				
12" o.c.	S.F.	0.36	2.23	2.59
16" o.c.	"	0.31	2.01	2.32
24" o.c.	"	0.29	1.83	2.12
1x3				
12" o.c.	S.F.	0.46	2.23	2.69
16" o.c.	"	0.39	2.01	2.40
24" o.c.	"	0.32	1.83	2.15
06110.50 **Roof Framing**				
Roof framing				
Rafters, gable end				
0-2 pitch (flat to 2-in-12)				
12" o.c.				
2x4	S.F.	0.65	1.25	1.90
2x6	"	0.92	1.31	2.23
2x8	"	1.31	1.37	2.68
2x10	"	1.64	1.43	3.07
2x12	"	3.02	1.51	4.53
16" o.c.				
2x6	S.F.	0.82	1.07	1.89
2x8	"	1.15	1.11	2.26
2x10	"	1.31	1.16	2.47
2x12	"	2.43	1.20	3.63
24" o.c.				
2x6	S.F.	0.46	0.91	1.37
2x8	"	0.96	0.94	1.90
2x10	"	1.11	0.97	2.08
2x12	"	1.97	1.00	2.97
4-6 pitch (4-in-12 to 6-in-12)				
12" o.c.				
2x4	S.F.	0.65	1.31	1.96
2x6	"	0.98	1.37	2.35
2x8	"	1.51	1.43	2.94
2x10	"	1.71	1.51	3.22
2x12	"	2.63	1.59	4.22
16" o.c.				
2x6	S.F.	0.82	1.11	1.93
2x8	"	1.31	1.16	2.47
2x10	"	1.51	1.20	2.71
2x12	"	2.23	1.25	3.48
24" o.c.				
2x6	S.F.	0.65	0.94	1.59
2x8	"	1.11	0.97	2.08
2x10	"	1.18	1.04	2.22
2x12	"	1.84	1.16	3.00
8-12 pitch (8-in-12 to 12-in-12)				
12" o.c.				
2x4	S.F.	0.72	1.37	2.09
2x6	"	1.11	1.43	2.54
2x8	"	1.58	1.51	3.09
2x10	"	1.84	1.59	3.43
2x12	"	2.83	1.67	4.50
16" o.c.				

06 WOOD AND PLASTICS

Rough Carpentry

06110.50 Roof Framing (Cont.)

	UNIT	MAT.	INST.	TOTAL
2x6	S.F.	0.92	1.16	2.08
2x8	"	1.47	1.20	2.67
2x10	"	1.64	1.25	2.89
2x12	"	2.37	1.31	3.68
24" o.c.				
2x6	S.F.	0.72	0.97	1.69
2x8	"	1.17	1.00	2.17
2x10	"	1.31	1.04	2.35
2x12	"	2.10	1.07	3.17
Ridge boards				
2x6	L.F.	0.69	3.02	3.71
2x8	"	0.92	3.35	4.27
2x10	"	1.22	3.77	4.99
2x12	"	1.58	4.31	5.89
Hip rafters				
2x6	L.F.	0.69	2.15	2.84
2x8	"	0.92	2.23	3.15
2x10	"	1.22	2.32	3.54
2x12	"	1.58	2.41	3.99
Jack rafters				
4-6 pitch (4-in-12 to 6-in-12)				
16" o.c.				
2x6	S.F.	0.85	1.77	2.62
2x8	"	1.31	1.83	3.14
2x10	"	1.51	1.94	3.45
2x12	"	2.23	2.01	4.24
24" o.c.				
2x6	S.F.	0.65	1.37	2.02
2x8	"	1.11	1.40	2.51
2x10	"	1.31	1.47	2.78
2x12	"	1.90	1.51	3.41
8-12 pitch (8-in-12 to 12-in-12)				
16" o.c.				
2x6	S.F.	1.31	1.88	3.19
2x8	"	1.64	1.94	3.58
2x10	"	2.37	2.01	4.38
2x12	"	3.29	2.08	5.37
24" o.c.				
2x6	S.F.	1.05	1.43	2.48
2x8	"	1.31	1.47	2.78
2x10	"	2.10	1.51	3.61
2x12	"	3.02	1.54	4.56
Sister rafters				
2x4	L.F.	0.46	4.31	4.77
2x6	"	0.69	5.03	5.72
2x8	"	0.92	6.04	6.96
2x10	"	1.22	7.55	8.77
2x12	"	1.58	10.00	11.58
Fascia boards				
2x4	L.F.	0.46	3.02	3.48
2x6	"	0.69	3.02	3.71
2x8	"	0.92	3.35	4.27
2x10	"	1.22	3.35	4.57

06 WOOD AND PLASTICS

Rough Carpentry	UNIT	MAT.	INST.	TOTAL
06110.50 Roof Framing *(Cont.)*				
2x12	L.F.	1.58	3.77	5.35
Cant strips				
Fiber				
3x3	L.F.	0.36	1.72	2.08
4x4	"	0.51	1.83	2.34
Wood				
3x3	L.F.	1.90	1.83	3.73
06110.60 Sleepers				
Sleepers, over concrete				
12" o.c.				
1x2	S.F.	0.26	1.37	1.63
1x3	"	0.39	1.43	1.82
2x4	"	0.85	1.67	2.52
2x6	"	1.24	1.77	3.01
16" o.c.				
1x2	S.F.	0.23	1.20	1.43
1x3	"	0.34	1.20	1.54
2x4	"	0.70	1.43	2.13
2x6	"	1.05	1.51	2.56
06110.65 Soffits				
Soffit framing				
2x3	L.F.	0.36	4.31	4.67
2x4	"	0.45	4.64	5.09
2x6	"	0.68	5.03	5.71
2x8	"	0.94	5.49	6.43
06110.70 Wall Framing				
Framing wall, studs				
12" o.c.				
2x3	S.F.	0.48	1.11	1.59
2x4	"	0.68	1.11	1.79
2x6	"	0.98	1.20	2.18
2x8	"	1.31	1.25	2.56
16" o.c.				
2x3	S.F.	0.39	0.94	1.33
2x4	"	0.55	0.94	1.49
2x6	"	0.78	1.00	1.78
2x8	"	1.23	1.04	2.27
24" o.c.				
2x3	S.F.	0.30	0.81	1.11
2x4	"	0.42	0.81	1.23
2x6	"	0.65	0.86	1.51
2x8	"	0.85	0.88	1.73
Plates, top or bottom				
2x3	L.F.	0.36	1.77	2.13
2x4	"	0.45	1.88	2.33
2x6	"	0.68	2.01	2.69
2x8	"	0.94	2.15	3.09
Headers, door or window				
2x6				
Single				

06 WOOD AND PLASTICS

Rough Carpentry	UNIT	MAT.	INST.	TOTAL
06110.70 **Wall Framing** *(Cont.)*				
3' long	EA.	2.20	30.25	32.45
6' long	"	4.41	37.75	42.16
Double				
3' long	EA.	4.42	33.50	37.92
6' long	"	8.86	43.25	52.11
2x8				
Single				
4' long	EA.	4.04	37.75	41.79
8' long	"	8.07	46.50	54.57
Double				
4' long	EA.	8.07	43.25	51.32
8' long	"	16.25	55.00	71.25
2x10				
Single				
5' long	EA.	6.10	46.50	52.60
10' long	"	12.25	60.00	72.25
Double				
5' long	EA.	12.25	50.00	62.25
10' long	"	24.50	60.00	84.50
2x12				
Single				
6' long	EA.	8.86	46.50	55.36
12' long	"	17.50	60.00	77.50
Double				
6' long	EA.	17.50	55.00	72.50
12' long	"	34.75	67.00	102
06115.10 **Floor Sheathing**				
Sub-flooring, plywood, CDX				
1/2" thick	S.F.	0.49	0.75	1.24
5/8" thick	"	0.70	0.86	1.56
3/4" thick	"	1.31	1.00	2.31
Structural plywood				
1/2" thick	S.F.	0.78	0.75	1.53
5/8" thick	"	1.24	0.86	2.10
3/4" thick	"	1.31	0.92	2.23
Board type subflooring				
1x6				
Minimum	S.F.	1.18	1.34	2.52
Maximum	"	1.51	1.51	3.02
1x8				
Minimum	S.F.	1.31	1.27	2.58
Maximum	"	1.53	1.42	2.95
1x10				
Minimum	S.F.	1.83	1.20	3.03
Maximum	"	1.96	1.34	3.30
Underlayment				
Hardboard, 1/4" tempered	S.F.	0.73	0.75	1.48
Plywood, CDX				
3/8" thick	S.F.	0.76	0.75	1.51
1/2" thick	"	0.90	0.80	1.70
5/8" thick	"	1.05	0.86	1.91
3/4" thick	"	1.31	0.92	2.23

06 WOOD AND PLASTICS

Rough Carpentry	UNIT	MAT.	INST.	TOTAL
06115.20 — Roof Sheathing				
Sheathing				
Plywood, CDX				
3/8" thick	S.F.	0.76	0.77	1.53
1/2" thick	"	0.90	0.80	1.70
5/8" thick	"	1.05	0.86	1.91
3/4" thick	"	1.31	0.92	2.23
Structural plywood				
3/8" thick	S.F.	0.48	0.77	1.25
1/2" thick	"	0.63	0.80	1.43
5/8" thick	"	0.76	0.86	1.62
3/4" thick	"	0.91	0.92	1.83
06115.30 — Wall Sheathing				
Sheathing				
Plywood, CDX				
3/8" thick	S.F.	0.76	0.89	1.65
1/2" thick	"	0.90	0.92	1.82
5/8" thick	"	1.05	1.00	2.05
3/4" thick	"	1.31	1.09	2.40
Waferboard				
3/8" thick	S.F.	0.48	0.89	1.37
1/2" thick	"	0.63	0.92	1.55
5/8" thick	"	0.76	1.00	1.76
3/4" thick	"	0.84	1.09	1.93
Structural plywood				
3/8" thick	S.F.	0.76	0.89	1.65
1/2" thick	"	0.90	0.92	1.82
5/8" thick	"	1.05	1.00	2.05
3/4" thick	"	0.90	1.09	1.99
Gypsum, 1/2" thick	"	0.48	0.92	1.40
Asphalt impregnated fiberboard, 1/2" thick	"	1.16	0.92	2.08
06125.10 — Wood Decking				
Decking, T&G solid				
Cedar				
3" thick	S.F.	10.50	1.51	12.01
4" thick	"	13.25	1.61	14.86
Fir				
3" thick	S.F.	4.59	1.51	6.10
4" thick	"	5.58	1.61	7.19
Southern yellow pine				
3" thick	S.F.	4.59	1.72	6.31
4" thick	"	4.85	1.85	6.70
White pine				
3" thick	S.F.	5.58	1.51	7.09
4" thick	"	7.54	1.61	9.15
06130.10 — Heavy Timber				
Mill framed structures				
Beams to 20' long				
Douglas fir				
6x8	L.F.	7.31	7.49	14.80
6x10	"	8.62	7.74	16.36

06 WOOD AND PLASTICS

Rough Carpentry

06130.10 Heavy Timber (Cont.)

	UNIT	MAT.	INST.	TOTAL
6x12	L.F.	10.25	8.32	18.57
6x14	"	12.50	8.64	21.14
6x16	"	13.50	8.98	22.48
8x10	"	11.50	7.74	19.24
8x12	"	13.50	8.32	21.82
8x14	"	15.50	8.64	24.14
8x16	"	17.75	8.98	26.73
Southern yellow pine				
6x8	L.F.	5.80	7.49	13.29
6x10	"	7.04	7.74	14.78
6x12	"	9.01	8.32	17.33
6x14	"	10.25	8.64	18.89
6x16	"	11.50	8.98	20.48
8x10	"	9.55	7.74	17.29
8x12	"	11.50	8.32	19.82
8x14	"	13.25	8.64	21.89
8x16	"	15.25	8.98	24.23
Columns to 12' high				
Douglas fir				
6x6	L.F.	5.25	11.25	16.50
8x8	"	9.01	11.25	20.26
10x10	"	15.75	12.50	28.25
12x12	"	19.50	12.50	32.00
Southern yellow pine				
6x6	L.F.	4.52	11.25	15.77
8x8	"	7.60	11.25	18.85
10x10	"	11.75	12.50	24.25
12x12	"	16.50	12.50	29.00
Posts, treated				
4x4	L.F.	1.81	2.41	4.22
6x6	"	5.25	3.02	8.27

06190.20 Wood Trusses

	UNIT	MAT.	INST.	TOTAL
Truss, fink, 2x4 members				
3-in-12 slope				
24' span	EA.	110	64.00	174
26' span	"	110	64.00	174
28' span	"	120	68.00	188
30' span	"	130	68.00	198
34' span	"	130	72.00	202
38' span	"	130	72.00	202
5-in-12 slope				
24' span	EA.	110	66.00	176
28' span	"	120	68.00	188
30' span	"	130	70.00	200
32' span	"	140	70.00	210
40' span	"	190	75.00	265
Gable, 2x4 members				
5-in-12 slope				
24' span	EA.	130	66.00	196
26' span	"	140	66.00	206
28' span	"	160	68.00	228
30' span	"	170	70.00	240

06 WOOD AND PLASTICS

Rough Carpentry

	UNIT	MAT.	INST.	TOTAL
06190.20 Wood Trusses (Cont.)				
32' span	EA.	170	70.00	240
36' span	"	180	72.00	252
40' span	"	200	75.00	275
King post type, 2x4 members				
4-in-12 slope				
16' span	EA.	79.00	61.00	140
18' span	"	86.00	62.00	148
24' span	"	92.00	66.00	158
26' span	"	99.00	66.00	165
30' span	"	130	70.00	200
34' span	"	130	70.00	200
38' span	"	160	72.00	232
42' span	"	190	77.00	267

Finish Carpentry

	UNIT	MAT.	INST.	TOTAL
06200.10 Finish Carpentry				
Mouldings and trim				
Apron, flat				
9/16 x 2	L.F.	1.71	3.02	4.73
9/16 x 3-1/2	"	3.95	3.18	7.13
Base				
Colonial				
7/16 x 2-1/4	L.F.	2.04	3.02	5.06
7/16 x 3	"	2.63	3.02	5.65
7/16 x 3-1/4	"	2.69	3.02	5.71
9/16 x 3	"	2.63	3.18	5.81
9/16 x 3-1/4	"	2.76	3.18	5.94
11/16 x 2-1/4	"	2.89	3.35	6.24
Ranch				
7/16 x 2-1/4	L.F.	2.23	3.02	5.25
7/16 x 3-1/4	"	2.63	3.02	5.65
9/16 x 2-1/4	"	2.43	3.18	5.61
9/16 x 3	"	2.63	3.18	5.81
9/16 x 3-1/4	"	2.69	3.18	5.87
Casing				
11/16 x 2-1/2	L.F.	2.10	2.74	4.84
11/16 x 3-1/2	"	2.37	2.87	5.24
Chair rail				
9/16 x 2-1/2	L.F.	2.23	3.02	5.25
9/16 x 3-1/2	"	3.09	3.02	6.11
Closet pole				
1-1/8" dia.	L.F.	1.51	4.02	5.53
1-5/8" dia.	"	2.23	4.02	6.25
Cove				
9/16 x 1-3/4	L.F.	1.71	3.02	4.73

06 WOOD AND PLASTICS

Finish Carpentry	UNIT	MAT.	INST.	TOTAL
06200.10 Finish Carpentry *(Cont.)*				
11/16 x 2-3/4	L.F.	2.63	3.02	5.65
Crown				
9/16 x 1-5/8	L.F.	2.23	4.02	6.25
9/16 x 2-5/8	"	2.43	4.64	7.07
11/16 x 3-5/8	"	2.63	5.03	7.66
11/16 x 4-1/4	"	3.95	5.49	9.44
11/16 x 5-1/4	"	4.41	6.04	10.45
Drip cap				
1-1/16 x 1-5/8	L.F.	2.37	3.02	5.39
Glass bead				
3/8 x 3/8	L.F.	0.85	3.77	4.62
1/2 x 9/16	"	1.05	3.77	4.82
5/8 x 5/8	"	1.11	3.77	4.88
3/4 x 3/4	"	1.31	3.77	5.08
Half round				
1/2	L.F.	0.98	2.41	3.39
5/8	"	1.31	2.41	3.72
3/4	"	1.77	2.41	4.18
Lattice				
1/4 x 7/8	L.F.	0.79	2.41	3.20
1/4 x 1-1/8	"	0.85	2.41	3.26
1/4 x 1-3/8	"	0.92	2.41	3.33
1/4 x 1-3/4	"	1.02	2.41	3.43
1/4 x 2	"	1.18	2.41	3.59
Ogee molding				
5/8 x 3/4	L.F.	1.58	3.02	4.60
11/16 x 1-1/8	"	3.68	3.02	6.70
11/16 x 1-3/8	"	2.89	3.02	5.91
Parting bead				
3/8 x 7/8	L.F.	1.31	3.77	5.08
Quarter round				
1/4 x 1/4	L.F.	0.46	2.41	2.87
3/8 x 3/8	"	0.65	2.41	3.06
1/2 x 1/2	"	0.85	2.41	3.26
11/16 x 11/16	"	0.85	2.62	3.47
3/4 x 3/4	"	1.58	2.62	4.20
1-1/16 x 1-1/16	"	1.25	2.74	3.99
Railings, balusters				
1-1/8 x 1-1/8	L.F.	4.21	6.04	10.25
1-1/2 x 1-1/2	"	4.93	5.49	10.42
Screen moldings				
1/4 x 3/4	L.F.	1.05	5.03	6.08
5/8 x 5/16	"	1.31	5.03	6.34
Shoe				
7/16 x 11/16	L.F.	1.31	2.41	3.72
Sash beads				
1/2 x 3/4	L.F.	1.51	5.03	6.54
1/2 x 7/8	"	1.71	5.03	6.74
1/2 x 1-1/8	"	1.84	5.49	7.33
5/8 x 7/8	"	1.84	5.49	7.33
Stop				
5/8 x 1-5/8				
Colonial	L.F.	0.92	3.77	4.69

06 WOOD AND PLASTICS

Finish Carpentry

	UNIT	MAT.	INST.	TOTAL
06200.10 Finish Carpentry (Cont.)				
Ranch	L.F.	0.92	3.77	4.69
Stools				
11/16 x 2-1/4	L.F.	4.01	6.71	10.72
11/16 x 2-1/2	"	4.21	6.71	10.92
11/16 x 5-1/4	"	4.34	7.55	11.89
Exterior trim, casing, select pine, 1x3	"	2.89	3.02	5.91
Douglas fir				
1x3	L.F.	1.38	3.02	4.40
1x4	"	1.71	3.02	4.73
1x6	"	2.23	3.35	5.58
1x8	"	3.09	3.77	6.86
Cornices, white pine, #2 or better				
1x2	L.F.	0.85	3.02	3.87
1x4	"	1.05	3.02	4.07
1x6	"	1.71	3.35	5.06
1x8	"	2.10	3.55	5.65
1x10	"	2.69	3.77	6.46
1x12	"	3.35	4.02	7.37
Shelving, pine				
1x8	L.F.	1.51	4.64	6.15
1x10	"	1.97	4.83	6.80
1x12	"	2.50	5.03	7.53
Plywood shelf, 3/4", with edge band, 12" wide	"	2.69	6.04	8.73
Adjustable shelf, and rod, 12" wide				
3' to 4' long	EA.	21.50	15.00	36.50
5' to 8' long	"	40.25	20.25	60.50
Prefinished wood shelves with brackets and supports				
8" wide				
3' long	EA.	63.00	15.00	78.00
4' long	"	73.00	15.00	88.00
6' long	"	110	15.00	125
10" wide				
3' long	EA.	70.00	15.00	85.00
4' long	"	100	15.00	115
6' long	"	110	15.00	125
06220.10 Millwork				
Countertop, laminated plastic				
25" x 7/8" thick				
Minimum	L.F.	16.75	15.00	31.75
Average	"	32.00	20.25	52.25
Maximum	"	47.00	24.25	71.25
25" x 1-1/4" thick				
Minimum	L.F.	20.50	20.25	40.75
Average	"	40.75	24.25	65.00
Maximum	"	61.00	30.25	91.25
Add for cutouts	EA.		37.75	37.75
Backsplash, 4" high, 7/8" thick	L.F.	22.50	12.00	34.50
Plywood, sanded, A-C				
1/4" thick	S.F.	1.44	2.01	3.45
3/8" thick	"	1.58	2.15	3.73
1/2" thick	"	1.77	2.32	4.09
A-D				

06 WOOD AND PLASTICS

Finish Carpentry	UNIT	MAT.	INST.	TOTAL
06220.10 Millwork (Cont.)				
1/4" thick	S.F.	1.38	2.01	3.39
3/8" thick	"	1.58	2.15	3.73
1/2" thick	"	1.71	2.32	4.03
Base cab., 34-1/2" high, 24" deep, hardwood				
Minimum	L.F.	220	24.25	244
Average	"	250	30.25	280
Maximum	"	280	40.25	320
Wall cabinets				
Minimum	L.F.	67.00	20.25	87.25
Average	"	91.00	24.25	115
Maximum	"	110	30.25	140

Wood Treatment	UNIT	MAT.	INST.	TOTAL
06300.10 Wood Treatment				
Creosote preservative treatment				
8 lb/cf	B.F.			0.72
10 lb/cf	"			0.85
Salt preservative treatment				
Oil borne				
Minimum	B.F.			0.65
Maximum	"			0.92
Water borne				
Minimum	B.F.			0.46
Maximum	"			0.72
Fire retardant treatment				
Minimum	B.F.			0.92
Maximum	"			1.11
Kiln dried, softwood, add to framing costs				
1" thick	B.F.			0.32
2" thick	"			0.46
3" thick	"			0.59
4" thick	"			0.72

06 WOOD AND PLASTICS

Architectural Woodwork

06420.10 Panel Work

	UNIT	MAT.	INST.	TOTAL
Hardboard, tempered, 1/4" thick				
Natural faced	S.F.	1.05	1.51	2.56
Plastic faced	"	1.58	1.72	3.30
Pegboard, natural	"	1.31	1.51	2.82
Plastic faced	"	1.58	1.72	3.30
Untempered, 1/4" thick				
Natural faced	S.F.	0.98	1.51	2.49
Plastic faced	"	1.71	1.72	3.43
Pegboard, natural	"	1.05	1.51	2.56
Plastic faced	"	1.51	1.72	3.23
Plywood unfinished, 1/4" thick				
Birch				
Natural	S.F.	1.11	2.01	3.12
Select	"	1.64	2.01	3.65
Knotty pine	"	2.17	2.01	4.18
Cedar (closet lining)				
Standard boards T&G	S.F.	2.69	2.01	4.70
Particle board	"	1.64	2.01	3.65
Plywood, prefinished, 1/4" thick, premium grade				
Birch veneer	S.F.	3.95	2.41	6.36
Cherry veneer	"	4.60	2.41	7.01
Chestnut veneer	"	8.88	2.41	11.29
Lauan veneer	"	1.71	2.41	4.12
Mahogany veneer	"	4.54	2.41	6.95
Oak veneer (red)	"	4.54	2.41	6.95
Pecan veneer	"	5.72	2.41	8.13
Rosewood veneer	"	8.88	2.41	11.29
Teak veneer	"	5.86	2.41	8.27
Walnut veneer	"	5.07	2.41	7.48

06430.10 Stairwork

	UNIT	MAT.	INST.	TOTAL
Risers, 1x8, 42" wide				
White oak	EA.	47.50	30.25	77.75
Pine	"	42.25	30.25	72.50
Treads, 1-1/16" x 9-1/2" x 42"				
White oak	EA.	57.00	37.75	94.75

06440.10 Columns

	UNIT	MAT.	INST.	TOTAL
Column, hollow, round wood				
12" diameter				
10' high	EA.	830	82.00	912
12' high	"	1,020	88.00	1,108
14' high	"	1,220	98.00	1,318
16' high	"	1,510	120	1,630
24" diameter				
16' high	EA.	3,460	120	3,580
18' high	"	3,940	130	4,070
20' high	"	4,840	130	4,970
22' high	"	5,090	140	5,230
24' high	"	5,560	140	5,700

07 THERMAL AND MOISTURE

Moisture Protection

07100.10 Waterproofing

	UNIT	MAT.	INST.	TOTAL
Membrane waterproofing, elastomeric				
Butyl				
1/32" thick	S.F.	1.43	1.89	3.32
1/16" thick	"	1.86	1.97	3.83
Butyl with nylon				
1/32" thick	S.F.	1.67	1.89	3.56
1/16" thick	"	2.01	1.97	3.98
Neoprene				
1/32" thick	S.F.	2.45	1.89	4.34
1/16" thick	"	3.51	1.97	5.48
Neoprene with nylon				
1/32" thick	S.F.	2.55	1.89	4.44
1/16" thick	"	4.11	1.97	6.08
Plastic vapor barrier (polyethylene)				
4 mil	S.F.	0.05	0.18	0.23
6 mil	"	0.07	0.18	0.25
10 mil	"	0.11	0.23	0.34
Bituminous membrane, asphalt felt, 15 lb.				
One ply	S.F.	0.74	1.18	1.92
Two ply	"	0.87	1.43	2.30
Three ply	"	1.07	1.69	2.76
Four ply	"	1.22	1.97	3.19
Five ply	"	1.32	2.49	3.81
Modified asphalt membrane, fibrous asphalt				
One ply	S.F.	0.52	1.97	2.49
Two ply	"	1.04	2.37	3.41
Three ply	"	1.62	2.63	4.25
Four ply	"	2.14	3.16	5.30
Five ply	"	2.60	3.79	6.39
Asphalt coated protective board				
1/8" thick	S.F.	0.58	1.18	1.76
1/4" thick	"	0.79	1.18	1.97
3/8" thick	"	0.88	1.18	2.06
1/2" thick	"	1.06	1.24	2.30
Cement protective board				
3/8" thick	S.F.	1.63	1.58	3.21
1/2" thick	"	2.28	1.58	3.86
Fluid applied, neoprene				
50 mil	S.F.	2.28	1.58	3.86
90 mil	"	3.78	1.58	5.36
Tab extended polyurethane				
.050" thick	S.F.	2.11	1.18	3.29
Fluid applied rubber based polyurethane				
6 mil	S.F.	1.11	1.48	2.59
15 mil	"	2.11	1.18	3.29
Bentonite waterproofing, panels				
3/16" thick	S.F.	1.89	1.18	3.07
1/4" thick	"	2.15	1.18	3.33
5/8" thick	"	3.21	1.24	4.45
Granular admixtures, trowel on, 3/8" thick	"	1.95	1.18	3.13
Metallic oxide, iron compound, troweled				
5/8" thick	S.F.	1.95	1.18	3.13
3/4" thick	"	2.36	1.35	3.71

07 THERMAL AND MOISTURE

Moisture Protection

	UNIT	MAT.	INST.	TOTAL
07150.10 Dampproofing				
Silicone dampproofing, sprayed on				
Concrete surface				
1 coat	S.F.	0.76	0.26	1.02
2 coats	"	1.26	0.36	1.62
Concrete block				
1 coat	S.F.	0.76	0.31	1.07
2 coats	"	1.26	0.43	1.69
Brick				
1 coat	S.F.	0.88	0.36	1.24
2 coats	"	1.36	0.47	1.83
07160.10 Bituminous Dampproofing				
Building paper, asphalt felt				
15 lb	S.F.	0.23	1.89	2.12
30 lb	"	0.43	1.97	2.40
Asphalt, troweled, cold, primer plus				
1 coat	S.F.	0.80	1.58	2.38
2 coats	"	1.69	2.37	4.06
3 coats	"	2.41	2.96	5.37
Fibrous asphalt, hot troweled, primer plus				
1 coat	S.F.	0.80	1.89	2.69
2 coats	"	1.69	2.63	4.32
3 coats	"	2.41	3.38	5.79
Asphaltic paint dampproofing, per coat				
Brush on	S.F.	0.41	0.67	1.08
Spray on	"	0.58	0.52	1.10
07190.10 Vapor Barriers				
Vapor barrier, polyethylene				
2 mil	S.F.	0.01	0.23	0.24
6 mil	"	0.07	0.23	0.30
8 mil	"	0.09	0.26	0.35
10 mil	"	0.10	0.26	0.36

Insulation

	UNIT	MAT.	INST.	TOTAL
07210.10 Batt Insulation				
Ceiling, fiberglass, unfaced				
3-1/2" thick, R11	S.F.	0.45	0.55	1.00
6" thick, R19	"	0.59	0.63	1.22
9" thick, R30	"	1.18	0.72	1.90
Suspended ceiling, unfaced				
3-1/2" thick, R11	S.F.	0.45	0.52	0.97
6" thick, R19	"	0.59	0.59	1.18
9" thick, R30	"	1.18	0.67	1.85
Crawl space, unfaced				

07 THERMAL AND MOISTURE

Insulation	UNIT	MAT.	INST.	TOTAL
07210.10 Batt Insulation (Cont.)				
3-1/2" thick, R11	S.F.	0.45	0.72	1.17
6" thick, R19	"	0.59	0.79	1.38
9" thick, R30	"	1.18	0.86	2.04
Wall, fiberglass				
Paper backed				
2" thick, R7	S.F.	0.29	0.49	0.78
3" thick, R8	"	0.32	0.52	0.84
4" thick, R11	"	0.53	0.55	1.08
6" thick, R19	"	0.79	0.59	1.38
Foil backed, 1 side				
2" thick, R7	S.F.	0.68	0.49	1.17
3" thick, R11	"	0.72	0.52	1.24
4" thick, R14	"	0.76	0.55	1.31
6" thick, R21	"	1.00	0.59	1.59
Foil backed, 2 sides				
2" thick, R7	S.F.	0.78	0.55	1.33
3" thick, R11	"	0.98	0.59	1.57
4" thick, R14	"	1.17	0.63	1.80
6" thick, R21	"	1.26	0.67	1.93
Unfaced				
2" thick, R7	S.F.	0.44	0.49	0.93
3" thick, R9	"	0.49	0.52	1.01
4" thick, R11	"	0.53	0.55	1.08
6" thick, R19	"	0.68	0.59	1.27
Mineral wool batts				
Paper backed				
2" thick, R6	S.F.	0.28	0.49	0.77
4" thick, R12	"	0.62	0.52	1.14
6" thick, R19	"	0.79	0.59	1.38
Fasteners, self adhering, attached to ceiling deck				
2-1/2" long	EA.	0.23	0.79	1.02
4-1/2" long	"	0.26	0.86	1.12
Capped, self-locking washers	"	0.23	0.47	0.70
07210.20 Board Insulation				
Insulation, rigid				
Fiberglass, roof				
0.75" thick, R2.78	S.F.	0.65	0.43	1.08
1.06" thick, R4.17	"	1.00	0.45	1.45
1.31" thick, R5.26	"	1.33	0.47	1.80
1.63" thick, R6.67	"	1.65	0.49	2.14
2.25" thick, R8.33	"	1.82	0.52	2.34
Composite board, roof				
1-1/2" thick, R6.67	S.F.	1.41	0.47	1.88
1-5/8" thick, R7.69	"	1.50	0.49	1.99
2" thick, R10.0	"	2.71	0.52	3.23
2-1/4" thick, R12.50	"	3.01	0.55	3.56
2-1/2" thick, R14.29	"	3.27	0.59	3.86
2-3/4" thick, R16.67	"	3.60	0.63	4.23
3-1/4" thick, R20.00	"	4.62	0.67	5.29
Perlite board, roof				
1.00" thick, R2.78	S.F.	0.68	0.39	1.07
1.50" thick, R4.17	"	1.06	0.41	1.47

07 THERMAL AND MOISTURE

Insulation

07210.20 Board Insulation (Cont.)

	UNIT	MAT.	INST.	TOTAL
2.00" thick, R5.92	S.F.	1.31	0.43	1.74
2.50" thick, R6.67	"	1.59	0.45	2.04
3.00" thick, R8.33	"	2.01	0.47	2.48
4.00" thick, R10.00	"	2.23	0.49	2.72
5.25" thick, R14.29	"	2.45	0.52	2.97
Rigid urethane				
Roof				
1" thick, R6.67	S.F.	1.22	0.39	1.61
1.20" thick, R8.33	"	1.40	0.40	1.80
1.50" thick, R11.11	"	1.66	0.41	2.07
2" thick, R14.29	"	2.15	0.43	2.58
2.25" thick, R16.67	"	2.82	0.45	3.27
Wall				
1" thick, R6.67	S.F.	1.22	0.49	1.71
1.5" thick, R11.11	"	1.66	0.52	2.18
2" thick, R14.29	"	2.21	0.55	2.76
Polystyrene				
Roof				
1.0" thick, R4.17	S.F.	0.49	0.39	0.88
1.5" thick, R6.26	"	0.75	0.41	1.16
2.0" thick, R8.33	"	0.91	0.43	1.34
Wall				
1.0" thick, R4.17	S.F.	0.49	0.49	0.98
1.5" thick, R6.26	"	0.75	0.52	1.27
2.0" thick, R8.33	"	0.91	0.55	1.46
Rigid board insulation, deck				
Mineral fiberboard				
1" thick, R3.0	S.F.	0.71	0.39	1.10
2" thick, R5.26	"	1.56	0.43	1.99
Fiberglass				
1" thick, R4.3	S.F.	1.20	0.39	1.59
2" thick, R8.5	"	1.79	0.43	2.22
Polystyrene				
1" thick, R5.4	S.F.	0.49	0.39	0.88
2" thick, R10.8	"	1.23	0.43	1.66
Urethane				
.75" thick, R5.4	S.F.	1.11	0.39	1.50
1" thick, R6.4	"	1.32	0.39	1.71
1.5" thick, R10.7	"	1.59	0.41	2.00
2" thick, R14.3	"	1.82	0.43	2.25
Foamglass				
1" thick, R1.8	S.F.	1.80	0.39	2.19
2" thick, R5.26	"	2.31	0.43	2.74
Wood fiber				
1" thick, R3.85	S.F.	2.05	0.39	2.44
2" thick, R7.7	"	2.47	0.43	2.90
Particle board				
3/4" thick, R2.08	S.F.	1.07	0.39	1.46
1" thick, R2.77	"	1.13	0.39	1.52
2" thick, R5.50	"	1.49	0.43	1.92

07 THERMAL AND MOISTURE

Insulation

	UNIT	MAT.	INST.	TOTAL
07210.60 — Loose Fill Insulation				
Blown-in type				
Fiberglass				
5" thick, R11	S.F.	0.42	0.39	0.81
6" thick, R13	"	0.49	0.47	0.96
9" thick, R19	"	0.59	0.67	1.26
Rockwool, attic application				
6" thick, R13	S.F.	0.39	0.47	0.86
8" thick, R19	"	0.46	0.59	1.05
10" thick, R22	"	0.55	0.72	1.27
12" thick, R26	"	0.70	0.79	1.49
15" thick, R30	"	0.84	0.94	1.78
Poured type				
Fiberglass				
1" thick, R4	S.F.	0.46	0.29	0.75
2" thick, R8	"	0.87	0.33	1.20
3" thick, R12	"	1.27	0.39	1.66
4" thick, R16	"	1.68	0.47	2.15
Mineral wool				
1" thick, R3	S.F.	0.51	0.29	0.80
2" thick, R6	"	0.95	0.33	1.28
3" thick, R9	"	1.45	0.39	1.84
4" thick, R12	"	1.68	0.47	2.15
Vermiculite or perlite				
2" thick, R4.8	S.F.	0.91	0.33	1.24
3" thick, R7.2	"	1.30	0.39	1.69
4" thick, R9.6	"	1.69	0.47	2.16
Masonry, poured vermiculite or perlite				
4" block	S.F.	0.39	0.23	0.62
6" block	"	0.58	0.29	0.87
8" block	"	0.84	0.33	1.17
10" block	"	1.11	0.36	1.47
12" block	"	1.39	0.39	1.78
07210.70 — Sprayed Insulation				
Foam, sprayed on				
Polystyrene				
1" thick, R4	S.F.	0.72	0.47	1.19
2" thick, R8	"	1.41	0.63	2.04
Urethane				
1" thick, R4	S.F.	0.68	0.47	1.15
2" thick, R8	"	1.31	0.63	1.94
07250.10 — Fireproofing				
Sprayed on				
1" thick				
On beams	S.F.	0.84	1.05	1.89
On columns	"	0.85	0.94	1.79
On decks				
Flat surface	S.F.	0.85	0.47	1.32
Fluted surface	"	1.09	0.59	1.68
1-1/2" thick				
On beams	S.F.	1.52	1.35	2.87
On columns	"	1.71	1.18	2.89

07 THERMAL AND MOISTURE

Insulation

07250.10 Fireproofing *(Cont.)*

	UNIT	MAT.	INST.	TOTAL
On decks				
Flat surface	S.F.	1.28	0.59	1.87
Fluted surface	"	1.52	0.79	2.31

Shingles And Tiles

07310.10 Asphalt Shingles

	UNIT	MAT.	INST.	TOTAL
Standard asphalt shingles, strip shingles				
210 lb/square	SQ.	93.00	58.00	151
235 lb/square	"	97.00	64.00	161
240 lb/square	"	100	72.00	172
260 lb/square	"	140	82.00	222
300 lb/square	"	160	96.00	256
385 lb/square	"	220	120	340
Roll roofing, mineral surface				
90 lb	SQ.	57.00	41.25	98.25
110 lb	"	94.00	48.00	142
140 lb	"	97.00	58.00	155

07310.50 Metal Shingles

	UNIT	MAT.	INST.	TOTAL
Aluminum, .020" thick				
Plain	SQ.	300	120	420
Colors	"	330	120	450
Steel, galvanized				
26 ga.				
Plain	SQ.	370	120	490
Colors	"	480	120	600
24 ga.				
Plain	SQ.	400	120	520
Colors	"	500	120	620
Porcelain enamel, 22 ga.				
Minimum	SQ.	940	140	1,080
Average	"	1,090	140	1,230
Maximum	"	1,220	140	1,360

07310.60 Slate Shingles

	UNIT	MAT.	INST.	TOTAL
Slate shingles				
Pennsylvania				
Ribbon	SQ.	740	290	1,030
Clear	"	960	290	1,250
Vermont				
Black	SQ.	800	290	1,090
Gray	"	880	290	1,170
Green	"	900	290	1,190
Red	"	1,630	290	1,920
Replacement shingles				

07 THERMAL AND MOISTURE

Shingles And Tiles	UNIT	MAT.	INST.	TOTAL
07310.60 — **Slate Shingles** *(Cont.)*				
Small jobs	EA.	14.25	19.25	33.50
Large jobs	S.F.	11.25	9.62	20.87
07310.70 — **Wood Shingles**				
Wood shingles, on roofs				
White cedar, #1 shingles				
4" exposure	SQ.	270	190	460
5" exposure	"	240	140	380
#2 shingles				
4" exposure	SQ.	200	190	390
5" exposure	"	170	140	310
Resquared and rebutted				
4" exposure	SQ.	240	190	430
5" exposure	"	200	140	340
On walls				
White cedar, #1 shingles				
4" exposure	SQ.	270	290	560
5" exposure	"	240	230	470
6" exposure	"	200	190	390
#2 shingles				
4" exposure	SQ.	200	290	490
5" exposure	"	170	230	400
6" exposure	"	140	190	330
Add for fire retarding	"			130
07310.80 — **Wood Shakes**				
Shakes, hand split, 24" red cedar, on roofs				
5" exposure	SQ.	310	290	600
7" exposure	"	290	230	520
9" exposure	"	270	190	460
On walls				
6" exposure	SQ.	290	290	580
8" exposure	"	280	230	510
10" exposure	"	270	190	460
Add for fire retarding	"			88.00

Roofing And Siding	UNIT	MAT.	INST.	TOTAL
07410.10 — **Manufactured Roofs**				
Aluminum roof panels, for steel framing				
Corrugated				
Unpainted finish				
.024"	S.F.	2.33	1.44	3.77
.030"	"	2.71	1.44	4.15
Painted finish				
.024"	S.F.	2.94	1.44	4.38

07 THERMAL AND MOISTURE

Roofing And Siding	UNIT	MAT.	INST.	TOTAL
07410.10 Manufactured Roofs (Cont.)				
.030"	S.F.	3.60	1.44	5.04
V-beam				
Unpainted finish				
.032"	S.F.	3.03	1.44	4.47
.040"	"	3.64	1.44	5.08
.050"	"	4.59	1.44	6.03
Painted finish				
.032"	S.F.	3.94	1.44	5.38
.040"	"	4.71	1.44	6.15
.050"	"	5.64	1.44	7.08
Steel roof panels, for structural steel framing				
Corrugated, painted				
18 ga.	S.F.	5.30	1.44	6.74
20 ga.	"	4.94	1.44	6.38
22 ga.	"	4.36	1.44	5.80
Box rib, painted				
18 ga.	S.F.	6.18	1.51	7.69
20 ga.	"	5.09	1.51	6.60
22 ga.	"	4.52	1.51	6.03
4" rib, painted				
18 ga.	S.F.	7.12	1.60	8.72
20 ga.	"	6.03	1.60	7.63
22 ga.	"	5.40	1.60	7.00
Standing seam roof				
2" high seam, painted				
22 ga.	S.F.	6.91	2.30	9.21
24 ga.	"	6.76	2.30	9.06
26 ga.	"	6.55	2.30	8.85
07410.30 Manufactured Walls				
Sandwich panels with 1-1/2" fiberglass insulation				
Galvanized 18 ga. steel interior panels				
Exterior panels				
16 ga. aluminum	S.F.	8.17	8.91	17.08
18 ga. galvanized steel	"	11.50	8.91	20.41
20 ga. painted steel	"	11.75	8.91	20.66
20 ga. stainless steel	"	12.00	8.91	20.91
Metal liner panels, 1-3/8" thick, 24" wide				
Galvanized				
22 ga.	S.F.	4.30	2.22	6.52
20 ga.	"	4.74	2.22	6.96
18 ga.	"	5.83	2.22	8.05
Primed				
22 ga.	S.F.	3.35	2.22	5.57
20 ga.	"	3.86	2.22	6.08
18 ga.	"	4.74	2.22	6.96
07440.10 Aggregate Coated Panels				
Dryvit type system				
1" thick	S.F.	3.13	2.22	5.35
1-1/2" thick	"	3.26	2.38	5.64
2" thick	"	3.70	2.78	6.48

07 THERMAL AND MOISTURE

Roofing And Siding

	UNIT	MAT.	INST.	TOTAL
07460.10 — Metal Siding Panels				
Aluminum siding panels				
Corrugated				
Plain finish				
.024"	S.F.	2.65	2.67	5.32
.032"	"	3.12	2.67	5.79
Painted finish				
.024"	S.F.	3.30	2.67	5.97
.032"	"	3.78	2.67	6.45
V. beam				
Plain finish				
.032"	S.F.	3.80	2.67	6.47
.040"	"	4.39	2.67	7.06
.050"	"	5.56	2.67	8.23
Painted finish				
.032"	S.F.	4.69	2.67	7.36
.040"	"	5.53	2.67	8.20
.050"	"	7.15	2.67	9.82
4" rib				
Plain finish				
.032"	S.F.	3.49	3.03	6.52
.040"	"	3.88	3.03	6.91
.050"	"	4.73	3.03	7.76
Painted finish				
.032"	S.F.	4.53	3.03	7.56
.040"	"	4.71	3.03	7.74
.050"	"	5.39	3.03	8.42
Steel siding panels				
Corrugated				
22 ga.	S.F.	2.66	4.45	7.11
24 ga.	"	2.43	4.45	6.88
26 ga.	"	2.21	4.45	6.66
Box rib				
20 ga.	S.F.	3.99	4.45	8.44
22 ga.	"	3.34	4.45	7.79
24 ga.	"	2.93	4.45	7.38
26 ga.	"	2.40	4.45	6.85
07460.50 — Plastic Siding				
Horizontal vinyl siding, solid				
8" wide				
Standard	S.F.	1.45	2.32	3.77
Insulated	"	1.76	2.32	4.08
10" wide				
Standard	S.F.	1.50	2.15	3.65
Insulated	"	1.80	2.15	3.95
Vinyl moldings for doors and windows	L.F.	0.93	2.41	3.34
07460.60 — Plywood Siding				
Rough sawn cedar, 3/8" thick	S.F.	2.11	2.01	4.12
Fir, 3/8" thick	"	1.17	2.01	3.18
Texture 1-11, 5/8" thick				
Cedar	S.F.	2.86	2.15	5.01
Fir	"	2.00	2.15	4.15

07 THERMAL AND MOISTURE

Roofing And Siding	UNIT	MAT.	INST.	TOTAL
07460.60 Plywood Siding *(Cont.)*				
Redwood	S.F.	3.06	2.06	5.12
Southern Yellow Pine	"	1.62	2.15	3.77
07460.70 Steel Siding				
Ribbed, sheets, galvanized				
22 ga.	S.F.	2.96	2.67	5.63
24 ga.	"	2.67	2.67	5.34
26 ga.	"	2.07	2.67	4.74
28 ga.	"	1.77	2.67	4.44
Primed				
24 ga.	S.F.	3.48	2.67	6.15
26 ga.	"	2.47	2.67	5.14
28 ga.	"	2.07	2.67	4.74
07460.80 Wood Siding				
Beveled siding, cedar				
A grade				
1/2 x 6	S.F.	5.37	3.02	8.39
1/2 x 8	"	5.48	2.41	7.89
3/4 x 10	"	7.05	2.01	9.06
Clear				
1/2 x 6	S.F.	5.98	3.02	9.00
1/2 x 8	"	6.11	2.41	8.52
3/4 x 10	"	8.19	2.01	10.20
B grade				
1/2 x 6	S.F.	5.78	3.02	8.80
1/2 x 8	"	6.52	24.25	30.77
3/4 x 10	"	6.14	2.01	8.15
Board and batten				
Cedar				
1x6	S.F.	6.35	3.02	9.37
1x8	"	5.78	2.41	8.19
1x10	"	5.22	2.15	7.37
1x12	"	4.68	1.94	6.62
Pine				
1x6	S.F.	1.61	3.02	4.63
1x8	"	1.57	2.41	3.98
1x10	"	1.50	2.15	3.65
1x12	"	1.39	1.94	3.33
Redwood				
1x6	S.F.	6.90	3.02	9.92
1x8	"	6.43	2.41	8.84
1x10	"	5.96	2.15	8.11
1x12	"	5.51	1.94	7.45
Tongue and groove				
Cedar				
1x4	S.F.	5.96	3.35	9.31
1x6	"	5.74	3.18	8.92
1x8	"	5.38	3.02	8.40
1x10	"	5.29	2.87	8.16
Pine				
1x4	S.F.	1.79	3.35	5.14
1x6	"	1.69	3.18	4.87

07 THERMAL AND MOISTURE

Roofing And Siding	UNIT	MAT.	INST.	TOTAL
07460.80 Wood Siding *(Cont.)*				
1x8	S.F.	1.58	3.02	4.60
1x10	"	1.50	2.87	4.37
Redwood				
1x4	S.F.	6.31	3.35	9.66
1x6	"	6.08	3.18	9.26
1x8	"	5.87	3.02	8.89
1x10	"	5.60	2.87	8.47

Membrane Roofing	UNIT	MAT.	INST.	TOTAL
07510.10 Built-up Asphalt Roofing				
Built-up roofing, asphalt felt, including gravel				
2 ply	SQ.	100	140	240
3 ply	"	140	190	330
4 ply	"	200	230	430
Walkway, for built-up roofs				
3' x 3' x				
1/2" thick	S.F.	2.77	1.92	4.69
3/4" thick	"	4.29	1.92	6.21
1" thick	"	4.64	1.92	6.56
Roof bonds				
Asphalt felt				
10 yrs	SQ.			44.00
20 yrs	"			50.00
Cant strip, 4" x 4"				
Treated wood	L.F.	2.75	1.64	4.39
Foamglass	"	2.35	1.44	3.79
Mineral fiber	"	0.47	1.44	1.91
New gravel for built-up roofing, 400 lb/sq	SQ.	41.50	120	162
Roof gravel (ballast)	C.Y.	27.75	290	318
Aluminum coating, top surfacing, for built-up roofing	SQ.	53.00	96.00	149
Remove 4-ply built-up roof (includes gravel)	"		290	290
Remove & replace gravel, includes flood coat	"	63.00	190	253
07530.10 Single-ply Roofing				
Elastic sheet roofing				
Neoprene, 1/16" thick	S.F.	3.35	0.72	4.07
EPDM rubber				
45 mil	S.F.	1.74	0.72	2.46
60 mil	"	2.39	0.72	3.11
PVC				
45 mil	S.F.	2.40	0.72	3.12
60 mil	"	2.86	0.72	3.58
Flashing				
Pipe flashing, 90 mil thick				
1" pipe	EA.	38.75	14.50	53.25

07 THERMAL AND MOISTURE

Membrane Roofing

07530.10 Single-ply Roofing (Cont.)

Description	UNIT	MAT.	INST.	TOTAL
2" pipe	EA.	41.75	14.50	56.25
3" pipe	"	42.00	15.25	57.25
4" pipe	"	45.50	15.25	60.75
5" pipe	"	48.75	16.00	64.75
6" pipe	"	53.00	16.00	69.00
8" pipe	"	60.00	17.00	77.00
10" pipe	"	69.00	19.25	88.25
12" pipe	"	84.00	19.25	103
Neoprene flashing, 60 mil thick strip				
6" wide	L.F.	2.25	4.81	7.06
12" wide	"	4.44	7.21	11.65
18" wide	"	6.53	9.62	16.15
24" wide	"	8.59	14.50	23.09
Adhesives				
Mastic sealer, applied at joints only				
1/4" bead	L.F.	0.13	0.28	0.41
Fluid applied roofing				
Urethane, 2 part, elastomeric membrane				
1" thick	S.F.	3.51	0.96	4.47
Vinyl liquid roofing, 2 coats, 2 mils per coat	"	5.49	0.82	6.31
Silicone roofing, 2 coats sprayed, 16 mil per coat	"	4.10	0.96	5.06
Inverted roof system				
Insulated membrane with coarse gravel ballast				
3 ply with 2" polystyrene	S.F.	7.91	0.96	8.87
Ballast, 3/4" through 1-1/2" gravel, 100lb/sf	"	0.48	58.00	58.48
Walkway for membrane roofs, 1/2" thick	"	2.52	1.92	4.44

Flashing And Sheet Metal

07610.10 Metal Roofing

Description	UNIT	MAT.	INST.	TOTAL
Sheet metal roofing, copper, 16 oz, batten seam	SQ.	2,130	380	2,510
Standing seam	"	2,080	360	2,440
Aluminum roofing, natural finish				
Corrugated, on steel frame				
.0175" thick	SQ.	140	160	300
.0215" thick	"	190	160	350
.024" thick	"	230	160	390
.032" thick	"	280	160	440
V-beam, on steel frame				
.032" thick	SQ.	290	160	450
.040" thick	"	320	160	480
.050" thick	"	400	160	560
Ridge cap				
.019" thick	L.F.	4.55	1.92	6.47
Corrugated galvanized steel roofing, on steel frame				
28 ga.	SQ.	260	160	420

07 THERMAL AND MOISTURE

Flashing And Sheet Metal	UNIT	MAT.	INST.	TOTAL
07610.10 **Metal Roofing** *(Cont.)*				
26 ga.	SQ.	300	160	460
24 ga.	"	340	160	500
22 ga.	"	370	160	530
26 ga., factory insulated with 1" polystyrene	"	570	230	800
Ridge roll				
10" wide	L.F.	2.60	1.92	4.52
20" wide	"	5.29	2.30	7.59
07620.10 **Flashing And Trim**				
Counter flashing				
Aluminum, .032"	S.F.	2.15	5.77	7.92
Stainless steel, .015"	"	6.87	5.77	12.64
Copper				
16 oz.	S.F.	11.00	5.77	16.77
20 oz.	"	13.25	5.77	19.02
24 oz.	"	15.75	5.77	21.52
32 oz.	"	19.50	5.77	25.27
Valley flashing				
Aluminum, .032"	S.F.	1.87	3.60	5.47
Stainless steel, .015	"	5.98	3.60	9.58
Copper				
16 oz.	S.F.	11.00	3.60	14.60
20 oz.	"	13.25	4.81	18.06
24 oz.	"	15.75	3.60	19.35
32 oz.	"	19.50	3.60	23.10
Base flashing				
Aluminum, .040"	S.F.	3.08	4.81	7.89
Stainless steel, .018"	"	7.15	4.81	11.96
Copper				
16 oz.	S.F.	11.00	4.81	15.81
20 oz.	"	13.25	3.60	16.85
24 oz.	"	15.75	4.81	20.56
32 oz.	"	19.50	4.81	24.31
Waterstop, "T" section, 22 ga.				
1-1/2" x 3"	L.F.	3.84	2.88	6.72
2" x 2"	"	4.26	2.88	7.14
4" x 3"	"	4.74	2.88	7.62
6" x 4"	"	5.01	2.88	7.89
8" x 4"	"	6.22	2.88	9.10
Scupper outlets				
10" x 10" x 4"	EA.	40.25	14.50	54.75
22" x 4" x 4"	"	49.50	14.50	64.00
8" x 8" x 5"	"	40.25	14.50	54.75
Flashing and trim, aluminum				
.019" thick	S.F.	1.52	4.12	5.64
.032" thick	"	1.85	4.12	5.97
.040" thick	"	3.18	4.44	7.62
Neoprene sheet flashing, .060" thick	"	2.53	3.60	6.13
Copper, paper backed				
2 oz.	S.F.	3.25	5.77	9.02
5 oz.	"	4.19	5.77	9.96
Drainage boots, roof, cast iron				
2 x 3	L.F.	110	7.21	117

07 THERMAL AND MOISTURE

Flashing And Sheet Metal	UNIT	MAT.	INST.	TOTAL
07620.10 Flashing And Trim *(Cont.)*				
3 x 4	L.F.	140	7.21	147
4 x 5	"	190	7.69	198
4 x 6	"	190	7.69	198
5 x 7	"	220	8.24	228
Pitch pocket, copper, 16 oz.				
4 x 4	EA.	200	14.50	215
6 x 6	"	210	14.50	225
8 x 8	"	230	14.50	245
8 x 10	"	250	14.50	265
8 x 12	"	300	14.50	315
Reglets, copper 10 oz.	L.F.	7.93	3.84	11.77
Stainless steel, .020"	"	3.60	3.84	7.44
Gravel stop				
Aluminum, .032"				
4"	L.F.	1.10	1.92	3.02
6"	"	1.62	1.92	3.54
8"	"	2.18	2.22	4.40
10"	"	2.73	2.22	4.95
Copper, 16 oz.				
4"	L.F.	4.63	1.92	6.55
6"	"	6.90	1.92	8.82
8"	"	9.26	2.22	11.48
10"	"	11.50	2.22	13.72
07620.20 Gutters And Downspouts				
Copper gutter and downspout				
Downspouts, 16 oz. copper				
Round				
3" dia.	L.F.	15.50	3.84	19.34
4" dia.	"	19.25	3.84	23.09
Rectangular, corrugated				
2" x 3"	L.F.	15.00	3.60	18.60
3" x 4"	"	18.50	3.60	22.10
Rectangular, flat surface				
2" x 3"	L.F.	17.00	3.84	20.84
3" x 4"	"	24.25	3.84	28.09
Lead-coated copper downspouts				
Round				
3" dia.	L.F.	20.25	3.60	23.85
4" dia.	"	24.50	4.12	28.62
Rectangular, corrugated				
2" x 3"	L.F.	20.25	3.84	24.09
3" x 4"	"	24.25	3.84	28.09
Rectangular, plain				
2" x 3"	L.F.	14.25	3.84	18.09
3" x 4"	"	16.50	3.84	20.34
Gutters, 16 oz. copper				
Half round				
4" wide	L.F.	13.75	5.77	19.52
5" wide	"	17.00	6.41	23.41
Type K				
4" wide	L.F.	15.50	5.77	21.27
5" wide	"	16.25	6.41	22.66

07 THERMAL AND MOISTURE

Flashing And Sheet Metal

07620.20 Gutters And Downspouts (Cont.)

	UNIT	MAT.	INST.	TOTAL
Lead-coated copper gutters				
Half round				
4" wide	L.F.	17.00	5.77	22.77
6" wide	"	23.25	6.41	29.66
Type K				
4" wide	L.F.	18.50	5.77	24.27
5" wide	"	24.00	6.41	30.41
Aluminum gutter and downspout				
Downspouts				
2" x 3"	L.F.	1.56	3.84	5.40
3" x 4"	"	2.15	4.12	6.27
4" x 5"	"	2.38	4.44	6.82
Round				
3" dia.	L.F.	2.63	3.84	6.47
4" dia.	"	3.37	4.12	7.49
Gutters, stock units				
4" wide	L.F.	2.55	6.07	8.62
5" wide	"	3.04	6.41	9.45
Galvanized steel gutter and downspout				
Downspouts, round corrugated				
3" dia.	L.F.	2.29	3.84	6.13
4" dia.	"	3.08	3.84	6.92
5" dia.	"	4.58	4.12	8.70
6" dia.	"	6.08	4.12	10.20
Rectangular				
2" x 3"	L.F.	2.07	3.84	5.91
3" x 4"	"	2.97	3.60	6.57
4" x 4"	"	3.72	3.60	7.32
Gutters, stock units				
5" wide				
Plain	L.F.	2.00	6.41	8.41
Painted	"	2.18	6.41	8.59
6" wide				
Plain	L.F.	2.79	6.79	9.58
Painted	"	3.13	6.79	9.92

Roofing Specialties

07700.10 Manufactured Specialties

	UNIT	MAT.	INST.	TOTAL
Moisture relief vent				
Aluminum	EA.	22.50	8.24	30.74
Copper	"	62.00	8.24	70.24
Expansion joint				
Aluminum				
Opening to 2.5"	L.F.	18.25	4.12	22.37
Opening to 3.5"	"	18.00	4.44	22.44

07 THERMAL AND MOISTURE

Roofing Specialties

07700.10 Manufactured Specialties (Cont.)

	UNIT	MAT.	INST.	TOTAL
Copper, 16 oz.				
Opening to 2.5"	L.F.	36.25	4.12	40.37
Opening to 3.5"	"	44.75	4.44	49.19
Butyl or neoprene				
4" wide				
16 oz. copper bellows	L.F.	34.50	4.81	39.31
28 ga. stainless steel bellows	"	21.50	4.81	26.31
6" wide				
Copper bellows	L.F.	37.75	5.24	42.99
Stainless steel				
Opening to 2.5"	L.F.	22.00	4.12	26.12
Opening to 3.5"	"	28.00	4.44	32.44
Smoke vent, 48" x 48"				
Aluminum	EA.	2,370	140	2,510
Galvanized steel	"	2,080	140	2,220
Heat/smoke vent, 48" x 96"				
Aluminum	EA.	3,310	190	3,500
Galvanized steel	"	2,830	190	3,020
Ridge vent strips				
Mill finish	L.F.	4.61	3.84	8.45
Connectors	EA.	4.55	14.50	19.05
End cap	"	2.27	16.50	18.77
Soffit vents				
Mill finish				
2-1/2" wide	L.F.	0.58	2.30	2.88
3" wide	"	0.71	2.30	3.01
6" wide	"	1.23	2.30	3.53
Roof hatches				
Steel, plain, primed				
2'6" x 3'0"	EA.	750	140	890
2'6" x 4'6"	"	1,100	190	1,290
2'6" x 8'0"	"	1,700	290	1,990
Galvanized steel				
2'6" x 3'0"	EA.	770	140	910
2'6" x 4'6"	"	1,160	190	1,350
2'6" x 8'0"	"	1,810	290	2,100
Aluminum				
2'6" x 3'0"	EA.	930	140	1,070
2'6" x 4'6"	"	1,400	190	1,590
2'6" x 8'0"	"	2,590	290	2,880
Ceiling access doors				
Swing up model, metal frame				
Steel door				
2'6" x 2'6"	EA.	640	58.00	698
2'6" x 3'0"	"	690	58.00	748
Aluminum door				
2'6" x 2'6"	EA.	790	58.00	848
2'6" x 3'0"	"	860	58.00	918
Swing down model, metal frame				
Steel door				
2'6" x 2'6"	EA.	620	58.00	678
2'6" x 3'0"	"	670	58.00	728
Aluminum door				

07 THERMAL AND MOISTURE

Roofing Specialties

07700.10 Manufactured Specialties (Cont.)

	UNIT	MAT.	INST.	TOTAL
2'6" x 2'6"	EA.	740	58.00	798
2'6" x 3'0"	"	790	58.00	848
Gravity ventilators, with curb, base, damper and screen				
Stationary siphon				
6" dia.	EA.	58.00	38.50	96.50
12" dia.	"	100	38.50	139
24" dia.	"	370	58.00	428
36" dia.	"	770	58.00	828
Wind driven spinner				
6" dia.	EA.	88.00	38.50	127
12" dia.	"	120	38.50	159
24" dia.	"	440	58.00	498
36" dia.	"	900	58.00	958
Stationary mushroom				
16" dia.	EA.	680	58.00	738
30" dia.	"	1,540	72.00	1,612
36" dia.	"	1,980	96.00	2,076
42" dia.	"	2,960	120	3,080

Skylights

07810.10 Plastic Skylights

	UNIT	MAT.	INST.	TOTAL
Single thickness, not including mounting curb				
2' x 4'	EA.	440	72.00	512
4' x 4'	"	590	96.00	686
5' x 5'	"	790	140	930
6' x 8'	"	1,670	190	1,860
Double thickness, not including mounting curb				
2' x 4'	EA.	580	72.00	652
4' x 4'	"	720	96.00	816
5' x 5'	"	1,060	140	1,200
6' x 8'	"	1,860	190	2,050
Metal framed skylights				
Translucent panels, 2-1/2" thick	S.F.	52.00	5.77	57.77
Continuous vaults, 8' wide				
Single glazed	S.F.	70.00	7.21	77.21
Double glazed	"	110	8.24	118

07 THERMAL AND MOISTURE

Joint Sealers	UNIT	MAT.	INST.	TOTAL
07920.10 **Caulking**				
Caulk exterior, two component				
1/4 x 1/2	L.F.	0.46	3.02	3.48
3/8 x 1/2	"	0.71	3.35	4.06
1/2 x 1/2	"	0.97	3.77	4.74
Caulk interior, single component				
1/4 x 1/2	L.F.	0.31	2.87	3.18
3/8 x 1/2	"	0.44	3.18	3.62
1/2 x 1/2	"	0.58	3.55	4.13
Butyl rubber fillers				
1/4" x 1/4"	L.F.	0.89	1.20	2.09
1/2" x 1/2"	"	1.30	2.01	3.31
1/2" x 3/4"	"	1.98	2.41	4.39
3/4" x 3/4"	"	2.62	2.41	5.03
1" x 1"	"	3.08	2.68	5.76
Seals, "O" ring type cord				
1/4" dia.	L.F.	0.98	1.51	2.49
1/2" dia.	"	2.93	1.59	4.52
1" dia.	"	10.50	1.67	12.17
1-1/4" dia.	"	13.25	1.77	15.02
1-1/2" dia.	"	17.50	1.88	19.38
1-3/4" dia.	"	25.75	1.94	27.69
2" dia.	"	33.50	2.01	35.51
Polyvinyl chloride, closed cell				
1/4" x 2"	L.F.	0.67	2.15	2.82
1/4" x 6"	"	2.10	2.74	4.84
Silicon foam penetration seal				
1/4" x 1/2"	L.F.	0.24	0.75	0.99
1/2" x 1/2"	"	0.42	1.00	1.42
1/2" x 3/4"	"	0.66	1.20	1.86
3/4" x 3/4"	"	0.98	1.51	2.49
1/8" x 1"	"	0.24	0.75	0.99
1/8" x 3"	"	0.66	1.20	1.86
1/4" x 3"	"	1.30	1.51	2.81
1/4" x 6"	"	2.69	2.01	4.70
1/2" x 6"	"	5.27	4.64	9.91
1/2" x 9"	"	7.93	7.55	15.48
1/2" x 12"	"	10.50	11.00	21.50
Oil base sealants and caulking				
1/4" x 1/4"	L.F.	0.05	1.51	1.56
1/4" x 3/8"	"	0.11	1.56	1.67
1/4" x 1/2"	"	0.13	1.63	1.76
3/8" x 3/8"	"	0.15	1.72	1.87
3/8" x 1/2"	"	0.18	1.83	2.01
3/8" x 5/8"	"	0.32	1.94	2.26
3/8" x 3/4"	"	0.36	2.08	2.44
1/2" x 1/2"	"	0.32	2.32	2.64
1/2" x 5/8"	"	0.40	2.62	3.02
1/2" x 3/4"	"	0.46	3.02	3.48
1/2" x 7/8"	"	0.54	3.09	3.63
1/2" x 1"	"	0.59	3.18	3.77
3/4" x 3/4"	"	0.70	3.26	3.96
1" x 1"	"	1.22	3.35	4.57
Polyurethane compounds				

07 THERMAL AND MOISTURE

Joint Sealers

07920.10 Caulking (Cont.)

	UNIT	MAT.	INST.	TOTAL
1/4" x 1/4"	L.F.	0.24	1.51	1.75
1/4" x 3/8"	"	0.44	1.56	2.00
1/4" x 1/2"	"	0.55	1.63	2.18
3/8" x 3/8"	"	0.62	1.72	2.34
3/8" x 1/2"	"	0.79	1.83	2.62
3/8" x 5/8"	"	0.98	1.94	2.92
3/8" x 3/4"	"	1.19	2.08	3.27
1/2" x 1/2"	"	1.14	2.32	3.46
1/2" x 5/8"	"	1.24	2.62	3.86
1/2" x 3/4"	"	1.43	3.02	4.45
1/2" x 7/8"	"	1.82	3.09	4.91
1/2" x 1"	"	2.30	3.35	5.65
3/4" x 3/4"	"	2.34	3.26	5.60
3/4" x 1"	"	2.54	3.35	5.89
Backer rod, polyethylene				
1/4"	L.F.	0.06	1.51	1.57
1/2"	"	0.11	1.59	1.70
3/4"	"	0.14	1.67	1.81
1"	"	0.19	1.77	1.96

08 DOORS AND WINDOWS

Metal	UNIT	MAT.	INST.	TOTAL
08110.10 — Metal Doors				
Flush hollow metal, std. duty, 20 ga., 1-3/8" thick				
2-6 x 6-8	EA.	360	67.00	427
2-8 x 6-8	"	400	67.00	467
3-0 x 6-8	"	430	67.00	497
1-3/4" thick				
2-6 x 6-8	EA.	420	67.00	487
2-8 x 6-8	"	450	67.00	517
3-0 x 6-8	"	480	67.00	547
2-6 x 7-0	"	470	67.00	537
2-8 x 7-0	"	480	67.00	547
3-0 x 7-0	"	510	67.00	577
Heavy duty, 20 ga., unrated, 1-3/4"				
2-8 x 6-8	EA.	470	67.00	537
3-0 x 6-8	"	500	67.00	567
2-8 x 7-0	"	530	67.00	597
3-0 x 7-0	"	510	67.00	577
3-4 x 7-0	"	530	67.00	597
18 ga., 1-3/4", unrated door				
2-0 x 7-0	EA.	490	67.00	557
2-4 x 7-0	"	490	67.00	557
2-6 x 7-0	"	490	67.00	557
2-8 x 7-0	"	540	67.00	607
3-0 x 7-0	"	560	67.00	627
3-4 x 7-0	"	570	67.00	637
2", unrated door				
2-0 x 7-0	EA.	540	76.00	616
2-4 x 7-0	"	540	76.00	616
2-6 x 7-0	"	540	76.00	616
2-8 x 7-0	"	590	76.00	666
3-0 x 7-0	"	610	76.00	686
3-4 x 7-0	"	630	76.00	706
Galvanized metal door				
3-0 x 7-0	EA.	630	76.00	706
For lead lining in doors	"			1,150
For sound attenuation	"			100
Vision glass				
8" x 8"	EA.	140	76.00	216
8" x 48"	"	200	76.00	276
Fixed metal louver	"	300	60.00	360
For fire rating, add				
3 hr door	EA.			480
1-1/2 hr door	"			210
3/4 hr door	"			110
1' extra height, add to material, 20%				
1'6" extra height, add to material, 60%				
For dutch doors with shelf, add to material, 100%				
08110.40 — Metal Door Frames				
Hollow metal, stock, 18 ga., 4-3/4" x 1-3/4"				
2-0 x 7-0	EA.	170	76.00	246
2-4 x 7-0	"	190	76.00	266
2-6 x 7-0	"	190	76.00	266
2-8 x 7-0	"	190	76.00	266

08 DOORS AND WINDOWS

Metal	UNIT	MAT.	INST.	TOTAL
08110.40 **Metal Door Frames** *(Cont.)*				
3-0 x 7-0	EA.	190	76.00	266
4-0 x 7-0	"	210	100	310
5-0 x 7-0	"	220	100	320
6-0 x 7-0	"	270	100	370
16 ga., 6-3/4" x 1-3/4"				
2-0 x 7-0	EA.	190	84.00	274
2-4 x 7-0	"	180	84.00	264
2-6 x 7-0	"	180	84.00	264
2-8 x 7-0	"	190	84.00	274
3-0 x 7-0	"	200	84.00	284
4-0 x 7-0	"	240	110	350
6-0 x 7-0	"	270	110	380
Transom frame				
3-4 x 1-6	EA.	110	84.00	194
3-8 x 1-6	"	110	84.00	194
6-4 x 1-6	"	160	84.00	244
Transom sash				
3-0 x 1-4	EA.	98.00	84.00	182
3-4 x 1-4	"	110	84.00	194
6-0 x 1-4	"	160	84.00	244
1' extension of frame, add	"			24.00
14 ga. frame, add	"			24.00
For fire rating, add				
3 hour	EA.			65.00
1-1/2 hour	"			52.00
3/4 hour	"			45.75
Lead lining in frame, add	"			140
Sidelights, complete				
1-0 x 7-2	EA.	450	84.00	534
1-4 x 7-2	"	500	84.00	584
1-0 x 8-8	"	520	84.00	604
1-6 x 8-8	"	540	84.00	624
16 ga., 4-3/4" x 1-3/4"				
2-0 x 7-0	EA.	160	84.00	244
2-4 x 7-0	"	160	84.00	244
2-6 x 7-0	"	170	84.00	254
2-8 x 7-0	"	170	84.00	254
3-0 x 7-0	"	170	84.00	254
4-0 x 7-0	"	180	110	290
6-0 x 7-0	"	190	110	300
Transom frame				
3-4 x 1-6	EA.	110	84.00	194
3-8 x 1-6	"	110	84.00	194
6-4 x 1-6	"	160	84.00	244
Transom sash				
3-0 x 1-4	EA.	98.00	84.00	182
3-4 x 1-4	"	110	84.00	194
6-0 x 1-4	"	160	84.00	244
1' extension of door frame, add	"			24.00
14 ga., metal frame, add	"			24.00
For fire rating, add				
3 hour	EA.			62.00
1-1/2 hour	"			50.00

08 DOORS AND WINDOWS

Metal	UNIT	MAT.	INST.	TOTAL
08110.40 **Metal Door Frames** *(Cont.)*				
3/4 hour	EA.			43.75
Lead lining in frame, add	"			130
Sidelights, complete				
1-0 x 7-2	EA.	430	84.00	514
1-4 x 7-2	"	450	84.00	534
1-0 x 8-8	"	490	84.00	574
1-4 x 8-8	"	500	84.00	584
16 ga., 5-3/4" x 1-3/4"				
2-0 x 7-0	EA.	170	76.00	246
2-4 x 7-0	"	170	76.00	246
2-6 x 7-0	"	180	76.00	256
2-8 x 7-0	"	180	76.00	256
3-0 x 7-0	"	190	76.00	266
4-0 x 7-0	"	200	100	300
5-0 x 7-0	"	210	100	310
6-0 x 7-0	"	220	100	320
Mullions, vertical				
5-1/4" x 1-3/4"	L.F.	18.75	7.55	26.30
5-1/4" x 2"	"	23.50	7.55	31.05
Horizontal				
5-1/4" x 1-3/4"	L.F.	18.75	7.55	26.30
5-1/4" x 2"	"	23.50	7.55	31.05
08120.10 **Aluminum Doors**				
Aluminum doors, commercial				
Narrow stile				
2-6 x 7-0	EA.	830	330	1,160
3-0 x 7-0	"	870	330	1,200
3-6 x 7-0	"	890	330	1,220
Pair				
5-0 x 7-0	EA.	1,380	670	2,050
6-0 x 7-0	"	1,400	670	2,070
7-0 x 7-0	"	1,460	670	2,130
Wide stile				
2-6 x 7-0	EA.	1,270	330	1,600
3-0 x 7-0	"	1,320	330	1,650
3-6 x 7-0	"	1,360	330	1,690
Pair				
5-0 x 7-0	EA.	2,340	670	3,010
6-0 x 7-0	"	2,440	670	3,110
7-0 x 7-0	"	2,490	670	3,160

08 DOORS AND WINDOWS

Wood And Plastic

08210.10 Wood Doors

	UNIT	MAT.	INST.	TOTAL
Solid core, 1-3/8" thick				
Birch faced				
2-4 x 7-0	EA.	180	76.00	256
2-8 x 7-0	"	190	76.00	266
3-0 x 7-0	"	190	76.00	266
3-4 x 7-0	"	370	76.00	446
2-4 x 6-8	"	180	76.00	256
2-6 x 6-8	"	180	76.00	256
2-8 x 6-8	"	190	76.00	266
3-0 x 6-8	"	190	76.00	266
Lauan faced				
2-4 x 6-8	EA.	160	76.00	236
2-8 x 6-8	"	170	76.00	246
3-0 x 6-8	"	180	76.00	256
3-4 x 6-8	"	190	76.00	266
Tempered hardboard faced				
2-4 x 7-0	EA.	200	76.00	276
2-8 x 7-0	"	220	76.00	296
3-0 x 7-0	"	250	76.00	326
3-4 x 7-0	"	250	76.00	326
Hollow core, 1-3/8" thick				
Birch faced				
2-4 x 7-0	EA.	160	76.00	236
2-8 x 7-0	"	160	76.00	236
3-0 x 7-0	"	170	76.00	246
3-4 x 7-0	"	180	76.00	256
Lauan faced				
2-4 x 6-8	EA.	69.00	76.00	145
2-6 x 6-8	"	74.00	76.00	150
2-8 x 6-8	"	93.00	76.00	169
3-0 x 6-8	"	97.00	76.00	173
3-4 x 6-8	"	110	76.00	186
Tempered hardboard faced				
2-4 x 7-0	EA.	84.00	76.00	160
2-6 x 7-0	"	90.00	76.00	166
2-8 x 7-0	"	100	76.00	176
3-0 x 7-0	"	110	76.00	186
3-4 x 7-0	"	120	76.00	196
Solid core, 1-3/4" thick				
Birch faced				
2-4 x 7-0	EA.	280	76.00	356
2-6 x 7-0	"	280	76.00	356
2-8 x 7-0	"	290	76.00	366
3-0 x 7-0	"	270	76.00	346
3-4 x 7-0	"	280	76.00	356
Lauan faced				
2-4 x 7-0	EA.	190	76.00	266
2-6 x 7-0	"	220	76.00	296
2-8 x 7-0	"	230	76.00	306
3-4 x 7-0	"	240	76.00	316
3-0 x 7-0	"	260	76.00	336
Tempered hardboard faced				
2-4 x 7-0	EA.	250	76.00	326

08 DOORS AND WINDOWS

Wood And Plastic	UNIT	MAT.	INST.	TOTAL
08210.10 **Wood Doors** *(Cont.)*				
2-6 x 7-0	EA.	280	76.00	356
2-8 x 7-0	"	300	76.00	376
3-0 x 7-0	"	320	76.00	396
3-4 x 7-0	"	340	76.00	416
Hollow core, 1-3/4" thick				
Birch faced				
2-4 x 7-0	EA.	190	76.00	266
2-6 x 7-0	"	190	76.00	266
2-8 x 7-0	"	200	76.00	276
3-0 x 7-0	"	200	76.00	276
3-4 x 7-0	"	220	76.00	296
Lauan faced				
2-4 x 6-8	EA.	110	76.00	186
2-6 x 6-8	"	130	76.00	206
2-8 x 6-8	"	110	76.00	186
3-0 x 6-8	"	120	76.00	196
3-4 x 6-8	"	130	76.00	206
Tempered hardboard				
2-4 x 7-0	EA.	100	76.00	176
2-6 x 7-0	"	110	76.00	186
2-8 x 7-0	"	110	76.00	186
3-0 x 7-0	"	120	76.00	196
3-4 x 7-0	"	130	76.00	206
Add-on, louver	"	41.50	60.00	102
Glass	"	130	60.00	190
Exterior doors, 3-0 x 7-0 x 2-1/2", solid core				
Carved				
One face	EA.	1,730	150	1,880
Two faces	"	2,380	150	2,530
Closet doors, 1-3/4" thick				
Bi-fold or bi-passing, includes frame and trim				
Paneled				
4-0 x 6-8	EA.	600	100	700
6-0 x 6-8	"	680	100	780
Louvered				
4-0 x 6-8	EA.	420	100	520
6-0 x 6-8	"	500	100	600
Flush				
4-0 x 6-8	EA.	300	100	400
6-0 x 6-8	"	390	100	490
Primed				
4-0 x 6-8	EA.	330	100	430
6-0 x 6-8	"	370	100	470
08210.90 **Wood Frames**				
Frame, interior, pine				
2-6 x 6-8	EA.	92.00	86.00	178
2-8 x 6-8	"	99.00	86.00	185
3-0 x 6-8	"	100	86.00	186
5-0 x 6-8	"	110	86.00	196
6-0 x 6-8	"	110	86.00	196
2-6 x 7-0	"	110	86.00	196
2-8 x 7-0	"	120	86.00	206

08 DOORS AND WINDOWS

Wood And Plastic	UNIT	MAT.	INST.	TOTAL
08210.90 **Wood Frames** *(Cont.)*				
3-0 x 7-0	EA.	120	86.00	206
5-0 x 7-0	"	140	120	260
6-0 x 7-0	"	140	120	260
Exterior, custom, with threshold, including trim				
Walnut				
3-0 x 7-0	EA.	410	150	560
6-0 x 7-0	"	470	150	620
Oak				
3-0 x 7-0	EA.	370	150	520
6-0 x 7-0	"	420	150	570
Pine				
2-4 x 7-0	EA.	150	120	270
2-6 x 7-0	"	160	120	280
2-8 x 7-0	"	160	120	280
3-0 x 7-0	"	170	120	290
3-4 x 7-0	"	190	120	310
6-0 x 7-0	"	200	200	400
08300.10 **Special Doors**				
Vault door and frame, class 5, steel	EA.	8,120	600	8,720
Overhead door, coiling insulated				
Chain gear, no frame, 12' x 12'	EA.	4,030	760	4,790
Aluminum, bronze glass panels, 12-9 x 13-0	"	4,550	600	5,150
Garage, flush, ins. metal, primed, 9-0 x 7-0	"	1,230	200	1,430
Sliding fire doors, motorized, fusible link, 3 hr.				
3-0 x 6-8	EA.	4,610	1,210	5,820
3-8 x 6-8	"	4,610	1,210	5,820
4-0 x 8-0	"	5,000	1,210	6,210
5-0 x 8-0	"	5,060	1,210	6,270
Metal clad doors, including electric motor				
Light duty				
Minimum	S.F.	51.00	10.00	61.00
Maximum	"	82.00	24.25	106
Heavy duty				
Minimum	S.F.	79.00	30.25	109
Maximum	"	130	37.75	168
Hangar doors, based on 150' openings				
To 20' high	S.F.	72.00	8.98	80.98
20' to 40' high	"	79.00	5.61	84.61
40' to 60' high	"	83.00	3.74	86.74
60' to 80' high	"	86.00	2.24	88.24
Over 80' high	"	120	1.79	122
Counter doors, (roll-up shutters), std, manual				
Opening, 4' high				
4' wide	EA.	1,530	500	2,030
6' wide	"	2,080	500	2,580
8' wide	"	2,340	550	2,890
10' wide	"	2,600	760	3,360
14' wide	"	3,250	760	4,010
6' high				
4' wide	EA.	1,820	500	2,320
6' wide	"	2,370	550	2,920
8' wide	"	2,600	600	3,200

08 DOORS AND WINDOWS

Wood And Plastic

08300.10 Special Doors (Cont.)

	UNIT	MAT.	INST.	TOTAL
10' wide	EA.	2,930	760	3,690
14' wide	"	3,310	860	4,170
For stainless steel, add to material, 40%				
For motor operator, add	EA.			1,790
Service doors, (roll up shutters), std, manual				
Opening				
8' high x 8' wide	EA.	1,950	340	2,290
10' high x 10' wide	"	2,440	500	2,940
12' high x 12' wide	"	2,730	760	3,490
14' high x 14' wide	"	3,580	1,010	4,590
16' high x 14' wide	"	5,000	1,010	6,010
20' high x 14' wide	"	5,650	1,510	7,160
24' high x 16' wide	"	9,230	1,340	10,570
For motor operator				
Up to 12-0 x 12-0, add	EA.			1,830
Over 12-0 x 12-0, add	"			2,340
Roll-up doors				
13-0 high x 14-0 wide	EA.	1,660	860	2,520
12-0 high x 14-0 wide	"	2,110	860	2,970
Top coiling grilles, manual, steel or aluminum				
Opening, 4' high x				
4' wide	EA.	1,950	240	2,190
6' wide	"	2,010	240	2,250
8' wide	"	2,330	340	2,670
12' wide	"	2,700	340	3,040
16' wide	"	3,090	500	3,590
6' high x				
4' wide	EA.	2,050	500	2,550
6' wide	"	2,280	550	2,830
8' wide	"	2,370	600	2,970
12' wide	"	2,860	670	3,530
16' wide	"	3,640	860	4,500
Side coiling grilles, manual, aluminum				
Opening, 8' high x				
18' wide	EA.	4,230	5,620	9,850
24' wide	"	5,460	6,420	11,880
12' high x				
12' wide	EA.	4,290	5,620	9,910
18' wide	"	5,460	6,420	11,880
24' wide	"	7,800	7,490	15,290
Accordion folding, tracks and fittings included				
Vinyl covered, 2 layers	S.F.	16.25	24.25	40.50
Woven mahogany and vinyl	"	20.50	24.25	44.75
Economy vinyl	"	13.75	24.25	38.00
Rigid polyvinyl chloride	"	22.50	24.25	46.75
Sectional wood overhead, frames not incl.				
Commercial grade, HD, 1-3/4" thick, manual				
8' x 8'	EA.	1,280	500	1,780
10' x 10'	"	1,840	550	2,390
12' x 12'	"	2,400	600	3,000
Chain hoist				
12' x 16' high	EA.	3,550	1,010	4,560
14' x 14' high	"	3,890	760	4,650

08 DOORS AND WINDOWS

Wood And Plastic

08300.10 Special Doors (Cont.)

	UNIT	MAT.	INST.	TOTAL
20' x 8' high	EA.	3,340	1,210	4,550
16' high	"	7,510	1,510	9,020
Sectional metal overhead doors, complete				
Residential grade, manual				
9' x 7'	EA.	880	240	1,120
16' x 7'	"	1,570	300	1,870
Commercial grade				
8' x 8'	EA.	1,020	500	1,520
10' x 10'	"	1,360	550	1,910
12' x 12'	"	2,250	600	2,850
20' x 14', with chain hoist	"	5,390	1,210	6,600
Sliding glass doors				
Tempered plate glass, 1/4" thick				
6' wide				
Economy grade	EA.	1,260	200	1,460
Premium grade	"	1,450	200	1,650
12' wide				
Economy grade	EA.	1,770	300	2,070
Premium grade	"	2,660	300	2,960
Insulating glass, 5/8" thick				
6' wide				
Economy grade	EA.	1,560	200	1,760
Premium grade	"	2,000	200	2,200
12' wide				
Economy grade	EA.	1,940	300	2,240
Premium grade	"	3,110	300	3,410
1" thick				
6' wide				
Economy grade	EA.	1,960	200	2,160
Premium grade	"	2,260	200	2,460
12' wide				
Economy grade	EA.	3,040	300	3,340
Premium grade	"	4,440	300	4,740
Added costs				
Custom quality, add to material, 30%				
Tempered glass, 6' wide, add	S.F.			5.46
Vertical lift doors, channel frame construction				
20' high x				
10' wide	EA.	42,790	980	43,770
15' wide	"	53,490	980	54,470
20' wide	"	58,840	1,760	60,600
25' wide	"	64,190	1,760	65,950
25' high x				
20' wide	EA.	66,860	1,760	68,620
25' wide	"	74,890	2,050	76,940
30' high x				
25' wide	EA.	80,240	2,050	82,290
30' wide	"	88,260	2,050	90,310
35' wide	"	104,300	2,050	106,350
Residential storm door				
Minimum	EA.	200	100	300
Average	"	260	100	360
Maximum	"	570	150	720

08 DOORS AND WINDOWS

Storefronts

Storefronts	UNIT	MAT.	INST.	TOTAL
08410.10 — Storefronts				
Storefront, aluminum and glass				
Minimum	S.F.	28.50	8.35	36.85
Average	"	42.25	9.54	51.79
Maximum	"	85.00	11.25	96.25
Entrance doors, premium, closers, panic dev.,etc.				
1/2" thick glass				
3' x 7'	EA.	3,770	560	4,330
6' x 7'	"	6,430	840	7,270
3/4" thick glass				
3' x 7'	EA.	3,900	560	4,460
6' x 7'	"	6,500	840	7,340
1" thick glass				
3' x 7'	EA.	4,230	560	4,790
6' x 7'	"	7,480	840	8,320
Revolving doors				
7' diameter, 7' high				
Minimum	EA.	29,440	6,150	35,590
Average	"	37,050	9,840	46,890
Maximum	"	47,810	12,300	60,110

Metal Windows

Metal Windows	UNIT	MAT.	INST.	TOTAL
08510.10 — Steel Windows				
Steel windows, primed				
Casements				
Operable				
Minimum	S.F.	45.75	3.93	49.68
Maximum	"	69.00	4.45	73.45
Fixed sash	"	36.75	3.34	40.09
Double hung	"	69.00	3.71	72.71
Industrial windows				
Horizontally pivoted sash	S.F.	64.00	4.45	68.45
Fixed sash	"	50.00	3.71	53.71
Security sash				
Operable	S.F.	80.00	4.45	84.45
Fixed	"	71.00	3.71	74.71
Picture window	"	34.50	3.71	38.21
Projecting sash				
Minimum	S.F.	60.00	4.17	64.17
Maximum	"	73.00	4.17	77.17
Mullions	L.F.	15.75	3.34	19.09

08 DOORS AND WINDOWS

Metal Windows	UNIT	MAT.	INST.	TOTAL
08520.10 **Aluminum Windows**				
Jalousie				
3-0 x 4-0	EA.	380	84.00	464
3-0 x 5-0	"	440	84.00	524
Fixed window				
6 sf to 8 sf	S.F.	18.75	9.54	28.29
12 sf to 16 sf	"	16.75	7.42	24.17
Projecting window				
6 sf to 8 sf	S.F.	41.50	16.75	58.25
12 sf to 16 sf	"	37.25	11.25	48.50
Horizontal sliding				
6 sf to 8 sf	S.F.	27.00	8.35	35.35
12 sf to 16 sf	"	25.00	6.68	31.68
Double hung				
6 sf to 8 sf	S.F.	37.25	13.25	50.50
10 sf to 12 sf	"	33.25	11.25	44.50
Storm window, 0.5 cfm, up to				
60 u.i. (united inches)	EA.	87.00	33.50	121
70 u.i.	"	89.00	33.50	123
80 u.i.	"	100	33.50	134
90 u.i.	"	100	37.25	137
100 u.i.	"	100	37.25	137
2.0 cfm, up to				
60 u.i.	EA.	110	33.50	144
70 u.i.	"	110	33.50	144
80 u.i.	"	120	33.50	154
90 u.i.	"	120	37.25	157
100 u.i.	"	130	37.25	167

Wood And Plastic	UNIT	MAT.	INST.	TOTAL
08600.10 **Wood Windows**				
Double hung				
24" x 36"				
Minimum	EA.	260	60.00	320
Average	"	380	76.00	456
Maximum	"	510	100	610
24" x 48"				
Minimum	EA.	300	60.00	360
Average	"	440	76.00	516
Maximum	"	610	100	710
30" x 48"				
Minimum	EA.	320	67.00	387
Average	"	440	86.00	526
Maximum	"	640	120	760
30" x 60"				
Minimum	EA.	340	67.00	407

08 DOORS AND WINDOWS

Wood And Plastic

08600.10 Wood Windows *(Cont.)*

	UNIT	MAT.	INST.	TOTAL
Average	EA.	550	86.00	636
Maximum	"	670	120	790
Casement				
1 leaf, 22" x 38" high				
Minimum	EA.	380	60.00	440
Average	"	460	76.00	536
Maximum	"	540	100	640
2 leaf, 50" x 50" high				
Minimum	EA.	1,010	76.00	1,086
Average	"	1,320	100	1,420
Maximum	"	1,510	150	1,660
3 leaf, 71" x 62" high				
Minimum	EA.	1,670	76.00	1,746
Average	"	1,690	100	1,790
Maximum	"	2,030	150	2,180
4 leaf, 95" x 75" high				
Minimum	EA.	2,210	86.00	2,296
Average	"	2,520	120	2,640
Maximum	"	3,220	200	3,420
5 leaf, 119" x 75" high				
Minimum	EA.	2,860	86.00	2,946
Average	"	3,080	120	3,200
Maximum	"	3,950	200	4,150
Picture window, fixed glass, 54" x 54" high				
Minimum	EA.	590	76.00	666
Average	"	660	86.00	746
Maximum	"	1,180	100	1,280
68" x 55" high				
Minimum	EA.	1,070	76.00	1,146
Average	"	1,220	86.00	1,306
Maximum	"	1,600	100	1,700
Sliding, 40" x 31" high				
Minimum	EA.	350	60.00	410
Average	"	540	76.00	616
Maximum	"	640	100	740
52" x 39" high				
Minimum	EA.	440	76.00	516
Average	"	650	86.00	736
Maximum	"	730	100	830
64" x 72" high				
Minimum	EA.	680	76.00	756
Average	"	1,080	100	1,180
Maximum	"	1,190	120	1,310
Awning windows				
34" x 21" high				
Minimum	EA.	370	60.00	430
Average	"	430	76.00	506
Maximum	"	500	100	600
40" x 21" high				
Minimum	EA.	440	67.00	507
Average	"	490	86.00	576
Maximum	"	540	120	660
48" x 27" high				

08 DOORS AND WINDOWS

Wood And Plastic

08600.10 Wood Windows (Cont.)

	UNIT	MAT.	INST.	TOTAL
Minimum	EA.	460	67.00	527
Average	"	550	86.00	636
Maximum	"	640	120	760
60" x 36" high				
Minimum	EA.	480	76.00	556
Average	"	860	100	960
Maximum	"	970	120	1,090
Window frame, milled				
Minimum	L.F.	6.86	12.00	18.86
Average	"	7.65	15.00	22.65
Maximum	"	11.50	20.25	31.75

Hardware

08710.10 Hinges

	UNIT	MAT.	INST.	TOTAL
Hinges				
3 x 3 butts, steel, interior, plain bearing	PAIR			24.50
4 x 4 butts, steel, standard	"			36.25
5 x 4-1/2 butts, bronze/s. steel, heavy duty	"			94.00
Pivot hinges				
Top pivot	EA.			100
Intermediate pivot	"			110
Bottom pivot	"			220
BHMA specifications				
3-1/2 x 3-1/2, full mortise butts				
Plain bearing	PAIR			29.25
Ball bearing	"			35.00
Half surface butts	"			50.00
4 x 4				
Full mortise butts, plain bearing, standard duty	PAIR			31.75
Full mortise butts, ball bearing	"			38.25
Half surface butts				
Standard duty	PAIR			50.00
Ball bearing	"			50.00
4-1/2 x 4-1/2				
Full mortise butts, plain bearing	PAIR			42.25
Ball bearing, heavy duty	"			86.00
Half mortise and half surface butts				
Plain bearing	PAIR			50.00
Full surface and half surface butts				
Standard duty	PAIR			85.00
Heavy duty	"			160
Full mortise and full slide-in butts, ball bearing	"			39.00
Half mortise butts, ball bearing				
Standard duty	PAIR			180
Heavy duty	"			200

08 DOORS AND WINDOWS

Hardware	UNIT	MAT.	INST.	TOTAL
08710.10 — **Hinges** *(Cont.)*				
5 x 5, ball bearing				
Full mortise butts	PAIR			82.00
Half mortise, full & half surface butts	"			170
Full mortise, full surface and half surface butts	"			230
4 x 4				
Full mortise butts, plain bearing, standard duty	PAIR			19.50
5 x 4-1/2				
Full mortise butts, ball bearing, heavy duty	PAIR			91.00
08710.20 — **Locksets**				
Latchset, heavy duty				
Cylindrical	EA.	230	37.75	268
Mortise	"	230	60.00	290
Lockset, heavy duty				
Cylindrical	EA.	360	37.75	398
Mortise	"	410	60.00	470
Mortise locks and latchsets, chrome				
Latchset passage or closet latch	EA.	290	50.00	340
Privacy (bath or bedroom)	"	300	50.00	350
Entry lockset	"	370	50.00	420
Classroom lockset (outside key operated)	"	370	50.00	420
Storeroom lock	"	370	50.00	420
Front door lock	"	370	50.00	420
Dormitory or exit lock	"	370	50.00	420
Preassembled locks and latches, brass				
Latchset, passage or closet latch	EA.	330	50.00	380
Lockset				
Privacy (bath or bathroom)	EA.	400	50.00	450
Entry lock	"	570	50.00	620
Classroom lock (outside key, operated)	"	570	50.00	620
Storeroom lock	"	630	50.00	680
Bored locks and latches, satin chrome plated				
Latchset passage or closet latch	EA.	190	50.00	240
Lockset				
Privacy (bath or bedroom)	EA.	260	50.00	310
Entry lock	"	290	50.00	340
Classroom lock	"	290	50.00	340
Corridor lock	"	290	50.00	340
Miscellaneous locks				
Exit lock with alarm, single door	EA.	920	240	1,160
Electric strike				
Rim mounted wrought steel	EA.	680	150	830
Mortised, wrought steel with bronze plating	"	270	240	510
Dead bolt				
Bored, wrought brass, keyed both sides	EA.	150	100	250
Mortised, cast brass	"	390	100	490
Lockset, cipher, mechanical	"	2,390	60.00	2,450

08 DOORS AND WINDOWS

08710.30 Hardware — Closers

Hardware	UNIT	MAT.	INST.	TOTAL
Door closers				
Surface mounted, traditional type, parallel arm				
Standard	EA.	290	76.00	366
Heavy duty	"	340	76.00	416
Modern type, parallel arm, standard duty	"	340	76.00	416
Overhead, concealed, pivot hung, single acting				
Interior	EA.	540	76.00	616
Exterior	"	790	76.00	866
Floor concealed, single acting, offset, pivoted				
Interior	EA.	870	200	1,070
Exterior	"	1,100	200	1,300

08710.40 Door Trim

Hardware	UNIT	MAT.	INST.	TOTAL
Door bumper, bronze, wall type	EA.	7.15	12.00	19.15
Wall type, 4" dia. with convex rubber pad, aluminum	"	13.25	12.00	25.25
Floor type				
Aluminum	EA.	6.04	12.00	18.04
Brass	"	7.28	12.00	19.28
Door holders				
Wall type, bronze	EA.	36.00	12.00	48.00
Overhead	"	30.25	30.25	60.50
Floor type	"	30.25	30.25	60.50
Plunger type	"	29.25	30.25	59.50
Wall type, aluminum	"	28.50	30.25	58.75
Surface bolt	"	25.25	12.00	37.25
Panic device				
Rim type with thumb piece	EA.	730	150	880
Mortise	"	920	150	1,070
Vertical rod	"	1,380	150	1,530
Labeled, rim type	"	960	150	1,110
Mortise	"	1,250	150	1,400
Vertical rod	"	1,330	150	1,480
Silencers, rubber type	"	3.44	1.20	4.64
Dust proof strike with plate, brass	"	21.00	20.25	41.25
Flush bolt, lever extension, brass, rated	"	38.75	12.00	50.75
Surface bolt with strike, brass, 6" long	"	29.25	12.00	41.25
Door coordinator, labeled, brass, satin chrome	"	140	43.25	183
Door plates				
Kick plate, aluminum, 3 beveled edges				
10" x 28"	EA.	34.25	30.25	64.50
10" x 30"	"	37.75	30.25	68.00
10" x 34"	"	41.25	30.25	71.50
10" x 38"	"	44.50	30.25	74.75
Push plate, 4" x 16"				
Aluminum	EA.	31.25	12.00	43.25
Bronze	"	99.00	12.00	111
Stainless steel	"	79.00	12.00	91.00
Armor plate, 40" x 34"	"	91.00	24.25	115
Pull handle, 4" x 16"				
Aluminum	EA.	110	12.00	122
Bronze	"	210	12.00	222
Stainless steel	"	160	12.00	172
Hasp assembly				

08 DOORS AND WINDOWS

Hardware	UNIT	MAT.	INST.	TOTAL
08710.40 Door Trim *(Cont.)*				
3"	EA.	5.20	10.00	15.20
4-1/2"	"	6.50	13.50	20.00
6"	"	10.25	17.25	27.50
Electro-magnetic door holder				
Wall mounted	EA.	220	200	420
Floor mounted	"	390	200	590
Smoke detector door holder				
Photo electric type	EA.	340	200	540
Ionization type	"	340	200	540
Pneumatic operators, activated by rubber mats				
Swing				
Single	EA.	4,880	500	5,380
Double	"	8,060	760	8,820
Sliding				
Single	EA.	5,400	500	5,900
Double	"	9,360	760	10,120
08710.60 Weatherstripping				
Weatherstrip, head and jamb, metal strip, neoprene bulb				
Standard duty	L.F.	5.85	3.35	9.20
Heavy duty	"	6.50	3.77	10.27
Spring type				
Metal doors	EA.	64.00	150	214
Wood doors	"	64.00	200	264
Sponge type with adhesive backing	"	60.00	60.00	120
Astragal				
1-3/4" x 13 ga., aluminum	L.F.	7.99	5.03	13.02
1-3/8" x 5/8", oak	"	6.50	4.02	10.52
Thresholds				
Bronze	L.F.	63.00	15.00	78.00
Aluminum				
Plain	L.F.	43.25	15.00	58.25
Vinyl insert	"	46.50	15.00	61.50
Aluminum with grit	"	44.25	15.00	59.25
Steel				
Plain	L.F.	35.00	15.00	50.00
Interlocking	"	46.75	50.00	96.75

Glazing	UNIT	MAT.	INST.	TOTAL
08810.10 Glazing				
Sheet glass, 1/8" thick	S.F.	9.10	3.71	12.81
Plate glass, bronze or grey, 1/4" thick	"	13.25	6.07	19.32
Clear	"	10.50	6.07	16.57
Polished	"	12.25	6.07	18.32
Plexiglass				

08 DOORS AND WINDOWS

Glazing	UNIT	MAT.	INST.	TOTAL
08810.10 **Glazing** *(Cont.)*				
1/8" thick	S.F.	5.85	6.07	11.92
1/4" thick	"	10.50	3.71	14.21
Float glass, clear				
3/16" thick	S.F.	7.08	5.57	12.65
1/4" thick	"	7.21	6.07	13.28
5/16" thick	"	13.50	6.68	20.18
3/8" thick	"	14.50	8.35	22.85
1/2" thick	"	24.50	11.25	35.75
5/8" thick	"	32.25	13.25	45.50
3/4" thick	"	35.25	16.75	52.00
1" thick	"	62.00	22.25	84.25
Tinted glass, polished plate, twin ground				
3/16" thick	S.F.	9.75	5.57	15.32
1/4" thick	"	9.75	6.07	15.82
3/8" thick	"	15.75	8.35	24.10
1/2" thick	"	25.25	11.25	36.50
Total, full vision, all glass window system				
To 10' high				
Minimum	S.F.	65.00	16.75	81.75
Average	"	86.00	16.75	103
Maximum	"	100	16.75	117
10' to 20' high				
Minimum	S.F.	79.00	16.75	95.75
Average	"	96.00	16.75	113
Maximum	"	120	16.75	137
Insulated glass, bronze or gray				
1/2" thick	S.F.	20.00	11.25	31.25
1" thick	"	23.75	16.75	40.50
Spandrel, polished, 1 side, 1/4" thick	"	16.00	6.07	22.07
Tempered glass (safety)				
Clear sheet glass				
1/8" thick	S.F.	11.00	3.71	14.71
3/16" thick	"	13.25	5.14	18.39
Clear float glass				
1/4" thick	S.F.	11.50	5.57	17.07
5/16" thick	"	20.50	6.68	27.18
3/8" thick	"	25.00	8.35	33.35
1/2" thick	"	34.25	11.25	45.50
5/8" thick	"	38.75	13.25	52.00
3/4" thick	"	48.00	22.25	70.25
Tinted float glass				
3/16" thick	S.F.	13.75	5.14	18.89
1/4" thick	"	15.00	5.57	20.57
3/8" thick	"	27.50	8.35	35.85
1/2" thick	"	36.50	11.25	47.75
Laminated glass				
Float safety glass with polyvinyl plastic layer				
1/4", sheet or float				
Two lites, 1/8" thick, clear glass	S.F.	14.50	5.57	20.07
1/2" thick, float glass				
Two lites, 1/4" thick, clear glass	S.F.	22.00	11.25	33.25
Tinted glass	"	26.00	11.25	37.25
Insulating glass, two lites, clear float glass				

08 DOORS AND WINDOWS

Glazing	UNIT	MAT.	INST.	TOTAL
08810.10 Glazing (Cont.)				
1/2" thick	S.F.	14.00	11.25	25.25
5/8" thick	"	16.25	13.25	29.50
3/4" thick	"	18.00	16.75	34.75
7/8" thick	"	19.00	19.00	38.00
1" thick	"	25.25	22.25	47.50
Glass seal edge				
3/8" thick	S.F.	12.00	11.25	23.25
Tinted glass				
1/2" thick	S.F.	24.25	11.25	35.50
1" thick	"	26.00	22.25	48.25
Tempered, clear				
1" thick	S.F.	47.50	22.25	69.75
Wire reinforced	"	60.00	22.25	82.25
Plate mirror glass				
1/4" thick				
15 sf	S.F.	12.50	6.68	19.18
Over 15 sf	"	11.50	6.07	17.57
Door type, 1/4" thick	"	13.00	6.68	19.68
Transparent, one way vision, 1/4" thick	"	27.50	6.68	34.18
Sheet mirror glass				
3/16" thick	S.F.	12.00	6.68	18.68
1/4" thick	"	12.75	5.57	18.32
Wall tiles, 12" x 12"				
Clear glass	S.F.	4.09	3.71	7.80
Veined glass	"	5.20	3.71	8.91
Wire glass, 1/4" thick				
Clear	S.F.	24.25	22.25	46.50
Hammered	"	24.50	22.25	46.75
Obscure	"	28.50	22.25	50.75
Bullet resistant, plate, with inter-leaved vinyl				
1-3/16" thick				
To 15 sf	S.F.	110	33.50	144
Over 15 sf	"	120	33.50	154
2" thick				
To 15 sf	S.F.	150	56.00	206
Over 15 sf	"	150	56.00	206
Glazing accessories				
Neoprene glazing gaskets				
1/4" glass	L.F.	2.57	2.67	5.24
3/8" glass	"	2.86	2.78	5.64
1/2" glass	"	3.00	2.90	5.90
3/4" glass	"	4.29	3.03	7.32
1" glass	"	5.00	3.34	8.34
Mullion section				
1/4" glass	L.F.	0.79	1.33	2.12
3/8" glass	"	1.00	1.67	2.67
1/2" glass	"	1.43	1.90	3.33
3/4" glass	"	2.14	2.22	4.36
1" glass	"	2.86	2.67	5.53
Molded corners	EA.	3.08	44.50	47.58

08 DOORS AND WINDOWS

Glazed Curtain Walls

08910.10 Glazed Curtain Walls

	UNIT	MAT.	INST.	TOTAL
Curtain wall, aluminum system, framing sections				
2" x 3"				
Jamb	L.F.	13.25	5.57	18.82
Horizontal	"	13.50	5.57	19.07
Mullion	"	18.00	5.57	23.57
2" x 4"				
Jamb	L.F.	18.00	8.35	26.35
Horizontal	"	18.50	8.35	26.85
Mullion	"	18.00	8.35	26.35
3" x 5-1/2"				
Jamb	L.F.	23.75	8.35	32.10
Horizontal	"	26.50	8.35	34.85
Mullion	"	24.00	8.35	32.35
4" corner mullion	"	31.75	11.25	43.00
Coping sections				
1/8" x 8"	L.F.	33.25	11.25	44.50
1/8" x 9"	"	33.50	11.25	44.75
1/8" x 12-1/2"	"	34.25	13.25	47.50
Sill section				
1/8" x 6"	L.F.	32.75	6.68	39.43
1/8" x 7"	"	33.25	6.68	39.93
1/8" x 8-1/2"	"	33.75	6.68	40.43
Column covers, aluminum				
1/8" x 26"	L.F.	32.75	16.75	49.50
1/8" x 34"	"	33.25	17.50	50.75
1/8" x 38"	"	33.50	17.50	51.00
Doors				
Aluminum framed, standard hardware				
Narrow stile				
2-6 x 7-0	EA.	740	330	1,070
3-0 x 7-0	"	750	330	1,080
3-6 x 7-0	"	780	330	1,110
Wide stile				
2-6 x 7-0	EA.	1,270	330	1,600
3-0 x 7-0	"	1,370	330	1,700
3-6 x 7-0	"	1,470	330	1,800
Flush panel doors, to match adjacent wall panels				
2-6 x 7-0	EA.	1,070	420	1,490
3-0 x 7-0	"	1,130	420	1,550
3-6 x 7-0	"	1,170	420	1,590
Wall panel, insulated				
"U"=.08	S.F.	13.50	5.57	19.07
"U"=.10	"	12.75	5.57	18.32
"U"=.15	"	11.50	5.57	17.07
Window wall system, complete				
Minimum	S.F.	35.75	6.68	42.43
Average	"	57.00	7.42	64.42
Maximum	"	130	9.54	140
Added costs				
For bronze, add 20% to material				
For stainless steel, add 50% to material				

09 FINISHES

Support Systems

09110.10 Metal Studs

	UNIT	MAT.	INST.	TOTAL
Studs, non load bearing, galvanized				
2-1/2", 20 ga.				
12" o.c.	S.F.	0.84	1.25	2.09
16" o.c.	"	0.65	1.00	1.65
25 ga.				
12" o.c.	S.F.	0.57	1.25	1.82
16" o.c.	"	0.45	1.00	1.45
24" o.c.	"	0.35	0.83	1.18
3-5/8", 20 ga.				
12" o.c.	S.F.	1.01	1.51	2.52
16" o.c.	"	0.78	1.20	1.98
24" o.c.	"	0.58	1.00	1.58
25 ga.				
12" o.c.	S.F.	0.67	1.51	2.18
16" o.c.	"	0.54	1.20	1.74
24" o.c.	"	0.41	1.00	1.41
4", 20 ga.				
12" o.c.	S.F.	1.10	1.51	2.61
16" o.c.	"	0.84	1.20	2.04
24" o.c.	"	0.65	1.00	1.65
25 ga.				
12" o.c.	S.F.	0.75	1.51	2.26
16" o.c.	"	0.58	1.20	1.78
24" o.c.	"	0.44	1.00	1.44
6", 20 ga.				
12" o.c.	S.F.	1.41	1.88	3.29
16" o.c.	"	1.04	1.51	2.55
24" o.c.	"	0.84	1.25	2.09
25 ga.				
12" o.c.	S.F.	0.92	1.88	2.80
16" o.c.	"	0.72	1.51	2.23
24" o.c.	"	0.54	1.25	1.79
Load bearing studs, galvanized				
3-5/8", 16 ga.				
12" o.c.	S.F.	1.83	1.51	3.34
16" o.c.	"	1.69	1.20	2.89
18 ga.				
12" o.c.	S.F.	1.31	1.00	2.31
16" o.c.	"	1.43	1.20	2.63
4", 16 ga.				
12" o.c.	S.F.	1.92	1.51	3.43
16" o.c.	"	1.75	1.20	2.95
6", 16 ga.				
12" o.c.	S.F.	2.45	1.88	4.33
16" o.c.	"	2.21	1.51	3.72
Furring				
On beams and columns				
7/8" channel	L.F.	0.65	4.02	4.67
1-1/2" channel	"	0.78	4.64	5.42
On ceilings				
3/4" furring channels				
12" o.c.	S.F.	0.46	2.51	2.97
16" o.c.	"	0.36	2.41	2.77

09 FINISHES

Support Systems	UNIT	MAT.	INST.	TOTAL
09110.10 Metal Studs (Cont.)				
24" o.c.	S.F.	0.26	2.15	2.41
1-1/2" furring channels				
12" o.c.	S.F.	0.78	2.74	3.52
16" o.c.	"	0.58	2.51	3.09
24" o.c.	"	0.40	2.32	2.72
On walls				
3/4" furring channels				
12" o.c.	S.F.	0.46	2.01	2.47
16" o.c.	"	0.36	1.88	2.24
24" o.c.	"	0.26	1.77	2.03
1-1/2" furring channels				
12" o.c.	S.F.	0.78	2.15	2.93
16" o.c.	"	0.58	2.01	2.59
24" o.c.	"	0.40	1.88	2.28

Lath And Plaster	UNIT	MAT.	INST.	TOTAL
09205.10 Gypsum Lath				
Gypsum lath, 1/2" thick				
Clipped	S.Y.	8.51	3.35	11.86
Nailed	"	8.51	3.77	12.28
09205.20 Metal Lath				
Diamond expanded, galvanized				
2.5 lb., on walls				
Nailed	S.Y.	4.99	7.55	12.54
Wired	"	4.99	8.63	13.62
On ceilings				
Nailed	S.Y.	4.99	8.63	13.62
Wired	"	4.99	10.00	14.99
3.4 lb., on walls				
Nailed	S.Y.	6.77	7.55	14.32
Wired	"	6.77	8.63	15.40
On ceilings				
Nailed	S.Y.	6.77	8.63	15.40
Wired	"	6.77	10.00	16.77
Flat rib				
2.75 lb., on walls				
Nailed	S.Y.	4.71	7.55	12.26
Wired	"	4.71	8.63	13.34
On ceilings				
Nailed	S.Y.	4.71	8.63	13.34
Wired	"	4.71	10.00	14.71
3.4 lb., on walls				
Nailed	S.Y.	5.66	7.55	13.21
Wired	"	5.66	8.63	14.29

09 FINISHES

Lath And Plaster

	UNIT	MAT.	INST.	TOTAL
09205.20 Metal Lath (Cont.)				
On ceilings				
Nailed	S.Y.	5.66	8.63	14.29
Wired	"	5.66	10.00	15.66
Stucco lath				
1.8 lb.	S.Y.	5.86	7.55	13.41
3.6 lb.	"	6.57	7.55	14.12
Paper backed				
Minimum	S.Y.	4.55	6.04	10.59
Maximum	"	7.34	8.63	15.97
09205.60 Plaster Accessories				
Expansion joint, 3/4", 26 ga., galv.	L.F.	1.75	1.51	3.26
Plaster corner beads, 3/4", galvanized	"	0.49	1.72	2.21
Casing bead, expanded flange, galvanized	"	0.66	1.51	2.17
Expanded wing, 1-1/4" wide, galvanized	"	0.78	1.51	2.29
Joint clips for lath	EA.	0.20	0.30	0.50
Metal base, galvanized, 2-1/2" high	L.F.	0.89	2.01	2.90
Stud clips for gypsum lath	EA.	0.20	0.30	0.50
Tie wire galvanized, 18 ga., 25 lb. hank	"			55.00
Sound deadening board, 1/4"	S.F.	0.37	1.00	1.37
09210.10 Plaster				
Gypsum plaster, trowel finish, 2 coats				
Ceilings	S.Y.	7.33	17.50	24.83
Walls	"	7.33	16.50	23.83
3 coats				
Ceilings	S.Y.	10.25	24.50	34.75
Walls	"	10.25	21.75	32.00
Vermiculite plaster				
2 coats				
Ceilings	S.Y.	8.34	26.75	35.09
Walls	"	8.34	24.50	32.84
3 coats				
Ceilings	S.Y.	13.00	33.25	46.25
Walls	"	13.00	29.75	42.75
Keenes cement plaster				
2 coats				
Ceilings	S.Y.	2.92	21.75	24.67
Walls	"	2.92	18.75	21.67
3 coats				
Ceilings	S.Y.	2.92	24.50	27.42
Walls	"	2.92	21.75	24.67
On columns, add to installation, 50%	"			
Chases, fascia, and soffits, add to installation, 50%	"			
Beams, add to installation, 50%	"			
Patch holes, average size holes				
1 sf to 5 sf				
Minimum	S.F.	2.92	9.40	12.32
Average	"	2.92	11.25	14.17
Maximum	"	2.92	14.00	16.92
Over 5 sf				
Minimum	S.F.	2.92	5.64	8.56

09 FINISHES

Lath And Plaster	UNIT	MAT.	INST.	TOTAL
09210.10 Plaster (Cont.)				
Average	S.F.	2.92	8.05	10.97
Maximum	"	2.92	9.40	12.32
Patch cracks				
Minimum	S.F.	2.92	1.88	4.80
Average	"	2.92	2.82	5.74
Maximum	"	2.92	5.64	8.56
09220.10 Portland Cement Plaster				
Stucco, portland, gray, 3 coat, 1" thick				
Sand finish	S.Y.	9.16	24.50	33.66
Trowel finish	"	9.16	25.75	34.91
White cement				
Sand finish	S.Y.	10.50	25.75	36.25
Trowel finish	"	10.50	28.25	38.75
Scratch coat				
For ceramic tile	S.Y.	3.32	5.64	8.96
For quarry tile	"	3.32	5.64	8.96
Portland cement plaster				
2 coats, 1/2"	S.Y.	6.60	11.25	17.85
3 coats, 7/8"	"	7.87	14.00	21.87
09250.10 Gypsum Board				
Drywall, plasterboard, 3/8" clipped to				
Metal furred ceiling	S.F.	0.41	0.67	1.08
Columns and beams	"	0.41	1.51	1.92
Walls	"	0.41	0.60	1.01
Nailed or screwed to				
Wood framed ceiling	S.F.	0.41	0.60	1.01
Columns and beams	"	0.41	1.34	1.75
Walls	"	0.41	0.54	0.95
1/2", clipped to				
Metal furred ceiling	S.F.	0.42	0.67	1.09
Columns and beams	"	0.39	1.51	1.90
Walls	"	0.39	0.60	0.99
Nailed or screwed to				
Wood framed ceiling	S.F.	0.39	0.60	0.99
Columns and beams	"	0.39	1.34	1.73
Walls	"	0.39	0.54	0.93
5/8", clipped to				
Metal furred ceiling	S.F.	0.42	0.75	1.17
Columns and beams	"	0.42	1.67	2.09
Walls	"	0.42	0.67	1.09
Nailed or screwed to				
Wood framed ceiling	S.F.	0.42	0.75	1.17
Columns and beams	"	0.42	1.67	2.09
Walls	"	0.42	0.67	1.09
Vinyl faced, clipped to metal studs				
1/2"	S.F.	1.04	0.75	1.79
5/8"	"	1.10	0.75	1.85
Add for				
Fire resistant	S.F.			0.13
Water resistant	"			0.20
Water and fire resistant	"			0.26

09 FINISHES

Lath And Plaster

09250.10 Gypsum Board *(Cont.)*

	UNIT	MAT.	INST.	TOTAL
Taping and finishing joints				
Minimum	S.F.	0.05	0.40	0.45
Average	"	0.07	0.50	0.57
Maximum	"	0.11	0.60	0.71
Casing bead				
Minimum	L.F.	0.18	1.72	1.90
Average	"	0.19	2.01	2.20
Maximum	"	0.24	3.02	3.26
Corner bead				
Minimum	L.F.	0.19	1.72	1.91
Average	"	0.24	2.01	2.25
Maximum	"	0.29	3.02	3.31

Tile

09310.10 Ceramic Tile

	UNIT	MAT.	INST.	TOTAL
Glazed wall tile, 4-1/4" x 4-1/4"				
Minimum	S.F.	2.74	4.12	6.86
Average	"	4.35	4.80	9.15
Maximum	"	15.50	5.76	21.26
6" x 6"				
Minimum	S.F.	1.67	3.60	5.27
Average	"	2.24	4.12	6.36
Maximum	"	2.80	4.80	7.60
Base, 4-1/4" high				
Minimum	L.F.	4.53	7.21	11.74
Average	"	5.27	7.21	12.48
Maximum	"	6.96	7.21	14.17
Glazed modlings and trim, 12" x 12"				
Minimum	L.F.	2.49	5.76	8.25
Average	"	3.80	5.76	9.56
Maximum	"	5.12	5.76	10.88
Unglazed floor tile				
Portland cem., cushion edge, face mtd				
1" x 1"	S.F.	7.93	5.24	13.17
2" x 2"	"	8.38	4.80	13.18
4" x 4"	"	7.80	4.80	12.60
6" x 6"	"	2.79	4.12	6.91
12" x 12"	"	2.45	3.60	6.05
16" x 16"	"	2.13	3.20	5.33
18" x 18"	"	2.06	2.88	4.94
Adhesive bed, with white grout				
1" x 1"	S.F.	7.93	5.24	13.17
2" x 2"	"	8.38	4.80	13.18
4" x 4"	"	8.38	4.80	13.18
6" x 6"	"	2.79	4.12	6.91

09 FINISHES

Tile

09310.10 Ceramic Tile *(Cont.)*

	UNIT	MAT.	INST.	TOTAL
12" x 12"	S.F.	2.45	3.60	6.05
16" x 16"	"	2.13	3.20	5.33
18" x 18"	"	2.06	2.88	4.94
Organic adhesive bed, thin set, back mounted				
1" x 1"	S.F.	7.93	5.24	13.17
2" x 2"	"	8.38	4.80	13.18
For group 2 colors, add to material, 10%				
For group 3 colors, add to material, 20%				
For abrasive surface, add to material, 25%				
Porcelain floor tile				
1" x 1"	S.F.	10.75	5.24	15.99
2" x 2"	"	9.72	5.01	14.73
4" x 4"	"	9.03	4.80	13.83
6" x 6"	"	3.25	4.12	7.37
12" x 12"	"	2.92	3.60	6.52
16" x 16"	"	2.32	3.20	5.52
18" x 18"	"	2.19	2.88	5.07
Unglazed wall tile				
Organic adhesive, face mounted cushion edge				
1" x 1"				
Minimum	S.F.	4.22	4.80	9.02
Average	"	5.52	5.24	10.76
Maximum	"	8.25	5.76	14.01
2" x 2"				
Minimum	S.F.	4.87	4.43	9.30
Average	"	5.52	4.80	10.32
Maximum	"	9.03	5.24	14.27
Back mounted				
1" x 1"				
Minimum	S.F.	4.22	4.80	9.02
Average	"	5.52	5.24	10.76
Maximum	"	8.25	5.76	14.01
2" x 2"				
Minimum	S.F.	4.87	4.43	9.30
Average	"	5.52	4.80	10.32
Maximum	"	9.03	5.24	14.27
For glazed finish, add to material, 25%				
For glazed mosaic, add to material, 100%				
For metallic colors, add to material, 125%				
For exterior wall use, add to total, 25%				
For exterior soffit, add to total, 25%				
For portland cement bed, add to total, 25%				
For dry set portland cement bed, add to total, 10%				
Conductive floor tile, unglazed square edged				
Portland cement bed				
1 x 1	S.F.	8.19	7.21	15.40
1-9/16 x 1-9/16	"	7.54	7.21	14.75
Dry set				
1 x 1	S.F.	8.19	7.21	15.40
1-9/16 x 1-9/16	"	7.54	7.21	14.75
Epoxy bed with epoxy joints				
1 x 1	S.F.	8.19	7.21	15.40
1-9/16 x 1-9/16	"	7.54	7.21	14.75

09 FINISHES

Tile

	UNIT	MAT.	INST.	TOTAL
09310.10 — Ceramic Tile (Cont.)				
For WWF in bed add to total, 15%				
For abrasive surface, add to material, 40%				
Ceramic accessories				
Towel bar, 24" long				
Minimum	EA.	19.25	23.00	42.25
Average	"	23.75	28.75	52.50
Maximum	"	64.00	38.50	103
Soap dish				
Minimum	EA.	9.10	38.50	47.60
Average	"	12.25	48.00	60.25
Maximum	"	32.50	58.00	90.50
09330.10 — Quarry Tile				
Floor				
4 x 4 x 1/2"	S.F.	7.40	7.69	15.09
6 x 6 x 1/2"	"	7.25	7.21	14.46
6 x 6 x 3/4"	"	9.02	7.21	16.23
12 x 12 x 3/4"	"	12.50	6.40	18.90
16 x 16 x 3/4"	"	8.83	5.76	14.59
18 x 18 x 3/4"	"	6.22	4.80	11.02
Medallion				
36" dia.	EA.	380	140	520
48" dia.	"	440	140	580
Wall, applied to 3/4" portland cement bed				
4 x 4 x 1/2"	S.F.	5.46	11.50	16.96
6 x 6 x 3/4"	"	6.10	9.61	15.71
Cove base				
5 x 6 x 1/2" straight top	L.F.	7.15	9.61	16.76
6 x 6 x 3/4" round top	"	6.64	9.61	16.25
Moldings				
2 x 12	L.F.	11.50	5.76	17.26
4 x 12	"	17.75	5.76	23.51
Stair treads 6 x 6 x 3/4"	"	9.79	14.50	24.29
Window sill 6 x 8 x 3/4"	"	8.93	11.50	20.43
For abrasive surface, add to material, 25%				
09410.10 — Terrazzo				
Floors on concrete, 1-3/4" thick, 5/8" topping				
Gray cement	S.F.	5.00	8.05	13.05
White cement	"	5.46	8.05	13.51
Sand cushion, 3" thick, 5/8" top, 1/4"				
Gray cement	S.F.	5.91	9.40	15.31
White cement	"	6.56	9.40	15.96
Monolithic terrazzo, 3-1/2" base slab, 5/8" topping	"	4.69	7.05	11.74
Terrazzo wainscot, cast-in-place, 1/2" thick	"	8.84	14.00	22.84
Base, cast in place, terrazzo cove type, 6" high	L.F.	10.50	8.05	18.55
Curb, cast in place, 6" wide x 6" high, polished top	"	11.75	28.25	40.00
For venetian type terrazzo, add to material, 10%				
For abrasive heavy duty terrazzo, add to material, 15%				
Divider strips				
Zinc	L.F.			1.78
Brass	"			3.31
Stairs, cast-in-place, topping on concrete or metal				

09 FINISHES

Tile

09410.10 Terrazzo (Cont.)

	UNIT	MAT.	INST.	TOTAL
1-1/2" thick treads, 12" wide	L.F.	6.98	28.25	35.23
Combined tread and riser	"	10.50	71.00	81.50
Precast terrazzo, thin set				
Terrazzo tiles, non-slip surface				
9" x 9" x 1" thick	S.F.	22.25	8.05	30.30
12" x 12"				
1" thick	S.F.	24.00	7.52	31.52
1-1/2" thick	"	25.00	8.05	33.05
18" x 18" x 1-1/2" thick	"	32.50	8.05	40.55
24" x 24" x 1-1/2" thick	"	42.00	6.63	48.63
For white cement, add to material, 10%				
For venetian type terrazzo, add to material, 25%				
Terrazzo wainscot				
12" x 12" x 1" thick	S.F.	11.00	14.00	25.00
18" x 18" x 1-1/2" thick	"	18.00	16.00	34.00
Base				
6" high				
Straight	L.F.	15.75	4.33	20.08
Coved	"	18.75	4.33	23.08
8" high				
Straight	L.F.	17.75	4.70	22.45
Coved	"	20.75	4.70	25.45
Terrazzo curbs				
8" wide x 8" high	L.F.	41.00	22.50	63.50
6" wide x 6" high	"	37.00	18.75	55.75
Precast terrazzo stair treads, 12" wide				
1-1/2" thick				
Diamond pattern	L.F.	49.50	10.25	59.75
Non-slip surface	"	52.00	10.25	62.25
2" thick				
Diamond pattern	L.F.	52.00	10.25	62.25
Non-slip surface	"	55.00	11.25	66.25
Stair risers, 1" thick to 6" high				
Straight sections	L.F.	16.75	5.64	22.39
Cove sections	"	19.50	5.64	25.14
Combined tread and riser				
Straight sections				
1-1/2" tread, 3/4" riser	L.F.	72.00	16.00	88.00
3" tread, 1" riser	"	85.00	16.00	101
Curved sections				
2" tread, 1" riser	L.F.	91.00	18.75	110
3" tread, 1" riser	"	94.00	18.75	113
Stair stringers, notched for treads and risers				
1" thick	L.F.	43.00	14.00	57.00
2" thick	"	44.75	18.75	63.50
Landings, structural, nonslip				
1-1/2" thick	S.F.	40.50	9.40	49.90
3" thick	"	57.00	11.25	68.25
Conductive terrazzo, spark proof industrial floor				
Epoxy terrazzo				
Floor	S.F.	7.80	3.52	11.32
Base	"	7.80	4.70	12.50
Polyacrylate				

09 FINISHES

Tile	UNIT	MAT.	INST.	TOTAL
09410.10 Terrazzo *(Cont.)*				
Floor	S.F.	12.50	3.52	16.02
Base	"	12.50	4.70	17.20
Polyester				
Floor	S.F.	4.48	2.25	6.73
Base	"	4.48	2.82	7.30
Synthetic latex mastic				
Floor	S.F.	7.28	3.52	10.80
Base	"	7.28	4.70	11.98

Acoustical Treatment	UNIT	MAT.	INST.	TOTAL
09510.10 Ceilings And Walls				
Acoustical panels, suspension system not included				
Fiberglass panels				
5/8" thick				
2' x 2'	S.F.	1.80	0.86	2.66
2' x 4'	"	1.70	0.67	2.37
3/4" thick				
2' x 2'	S.F.	3.21	0.86	4.07
2' x 4'	"	2.84	0.67	3.51
Glass cloth faced fiberglass panels				
3/4" thick	S.F.	2.96	1.00	3.96
1" thick	"	3.30	1.00	4.30
Mineral fiber panels				
5/8" thick				
2' x 2'	S.F.	1.32	0.86	2.18
2' x 4'	"	1.32	0.67	1.99
3/4" thick				
2' x 2'	S.F.	1.79	0.86	2.65
2' x 4'	"	1.79	0.67	2.46
For aluminum faced panels, add to material, 80%				
For vinyl faced panels, add to total, 125%				
For fire rated panels, add to material, 75%				
Wood fiber panels				
1/2" thick				
2' x 2'	S.F.	1.98	0.86	2.84
2' x 4'	"	1.98	0.67	2.65
5/8" thick				
2' x 2'	S.F.	2.18	0.86	3.04
2' x 4'	"	2.18	0.67	2.85
3/4" thick				
2' x 2'	S.F.	2.67	0.86	3.53
2' x 4'	"	2.67	0.67	3.34
2" thick				
2' x 2'	S.F.	3.12	1.00	4.12
2' x 4'	"	3.12	0.75	3.87

09 FINISHES

Acoustical Treatment

09510.10 Ceilings And Walls *(Cont.)*

	UNIT	MAT.	INST.	TOTAL
For flameproofing, add to material, 10%				
For sculptured finish, add to material, 15%				
Air distributing panels				
3/4" thick	S.F.	3.12	1.51	4.63
5/8" thick	"	2.67	1.20	3.87
Acoustical tiles, suspension system not included				
Fiberglass tile, 12" x 12"				
5/8" thick	S.F.	1.65	1.09	2.74
3/4" thick	"	1.91	1.34	3.25
Glass cloth faced fiberglass tile				
3/4" thick	S.F.	3.83	1.34	5.17
3" thick	"	4.29	1.51	5.80
Mineral fiber tile, 12" x 12"				
5/8" thick				
Standard	S.F.	0.97	1.20	2.17
Vinyl faced	"	1.95	1.20	3.15
3/4" thick				
Standard	S.F.	1.43	1.20	2.63
Vinyl faced	"	2.49	1.20	3.69
Fire rated	"	3.17	1.20	4.37
Aluminum or mylar faced	"	6.04	1.20	7.24
Wood fiber tile, 12" x 12"				
1/2" thick	S.F.	1.57	1.20	2.77
3/4" thick	"	2.27	1.20	3.47
For flameproofing, add to material, 10%				
For sculptured 3 dimensional, add to material, 50%				
Metal pan units, 24 ga. steel				
12" x 12"	S.F.	6.17	2.41	8.58
12" x 24"	"	6.38	2.01	8.39
Aluminum, .025" thick				
12" x 12"	S.F.	6.56	2.41	8.97
12" x 24"	"	6.77	2.01	8.78
Anodized aluminum, 0.25" thick				
12" x 12"	S.F.	7.56	2.41	9.97
12" x 24"	"	8.97	2.01	10.98
Stainless steel, 24 ga.				
12" x 12"	S.F.	15.25	2.41	17.66
12" x 24"	"	15.50	2.01	17.51
For flameproof sound absorbing pads, add to material	"			2.35
Metal ceiling systems				
.020" thick panels				
10', 12', and 16' lengths	S.F.	5.90	1.72	7.62
Custom lengths, 3' to 20'	"	5.96	1.72	7.68
.025" thick panels				
32 sf, 38 sf, and 52 sf pieces	S.F.	5.92	2.01	7.93
Custom lengths, 10 sf to 65 sf	"	6.91	2.01	8.92
Carriers, black, add	"			3.56
Recess filler strip, add	"			1.19
Custom lengths, add	"			1.78
Sound absorption walls, with fabric cover				
2-6" x 9' x 3/4"	S.F.	11.50	2.01	13.51
2' x 9' x 1"	"	12.75	2.01	14.76
Starter spline	L.F.	1.85	1.51	3.36

09 FINISHES

Acoustical Treatment

09510.10 Ceilings And Walls (Cont.)

	UNIT	MAT.	INST.	TOTAL
Internal spline	L.F.	1.62	1.51	3.13
Acoustical treatment				
Barriers for plenums				
Leaded vinyl				
0.48 lb per sf	S.F.	3.52	2.87	6.39
0.87 lb per sf	"	3.52	3.02	6.54
Aluminum foil, fiberglass reinforcement				
Minimum	S.F.	1.31	2.01	3.32
Maximum	"	1.49	3.02	4.51
Aluminum mesh, paper backed	"	1.23	2.01	3.24
Fibered cement sheet, 3/16" thick	"	2.53	2.15	4.68
Sheet lead, 1/64" thick	"	3.62	1.51	5.13
Sound attenuation blanket				
1" thick	S.F.	0.46	6.04	6.50
1-1/2" thick	"	0.65	6.04	6.69
2" thick	"	0.80	6.04	6.84
3" thick	"	0.97	6.71	7.68
Ceiling suspension systems				
T bar system				
2' x 4'	S.F.	1.36	0.60	1.96
2' x 2'	"	1.48	0.67	2.15
Concealed Z bar suspension system, 12" module	"	1.27	1.00	2.27
For 1-1/2" carrier channels, 4' o.c., add	"			0.45
Carrier channel for recessed light fixtures	"			0.81

Flooring

09550.10 Wood Flooring

	UNIT	MAT.	INST.	TOTAL
Wood strip flooring, unfinished				
Fir floor				
C and better				
Vertical grain	S.F.	4.16	2.01	6.17
Flat grain	"	3.92	2.01	5.93
Oak floor				
Minimum	S.F.	4.39	2.87	7.26
Average	"	6.05	2.87	8.92
Maximum	"	8.77	2.87	11.64
Maple floor				
25/32" x 2-1/4"				
Minimum	S.F.	4.10	2.87	6.97
Maximum	"	5.99	2.87	8.86
33/32" x 3-1/4"				
Minimum	S.F.	6.57	2.87	9.44
Maximum	"	7.43	2.87	10.30
Added costs				
For factory finish, add to material, 10%				

09 FINISHES

Flooring

09550.10 Wood Flooring (Cont.)

	UNIT	MAT.	INST.	TOTAL
For random width floor, add to total, 20%				
For simulated pegs, add to total, 10%				
Wood block industrial flooring				
Creosoted				
2" thick	S.F.	4.94	1.59	6.53
2-1/2" thick	"	5.13	1.88	7.01
3" thick	"	5.33	2.01	7.34
Parquet, 5/16", white oak				
Finished	S.F.	10.75	3.02	13.77
Unfinished	"	5.20	3.02	8.22
Gym floor, 2 ply felt, 25/32" maple, finished, in mastic	"	9.16	3.35	12.51
Over wood sleepers	"	10.25	3.77	14.02
Finishing, sand, fill, finish, and wax	"	0.78	1.51	2.29
Refinish sand, seal, and 2 coats of polyurethane	"	1.36	2.01	3.37
Clean and wax floors	"	0.23	0.30	0.53

09630.10 Unit Masonry Flooring

	UNIT	MAT.	INST.	TOTAL
Clay brick				
9 x 4-1/2 x 3" thick				
Glazed	S.F.	9.49	5.03	14.52
Unglazed	"	9.10	5.03	14.13
8 x 4 x 3/4" thick				
Glazed	S.F.	8.58	5.25	13.83
Unglazed	"	8.45	5.25	13.70
For herringbone pattern, add to labor, 15%				

09660.10 Resilient Tile Flooring

	UNIT	MAT.	INST.	TOTAL
Solid vinyl tile, 1/8" thick, 12" x 12"				
Marble patterns	S.F.	5.26	1.51	6.77
Solid colors	"	6.82	1.51	8.33
Travertine patterns	"	7.67	1.51	9.18
Conductive resilient flooring, vinyl tile				
1/8" thick, 12" x 12"	S.F.	7.93	1.72	9.65

09665.10 Resilient Sheet Flooring

	UNIT	MAT.	INST.	TOTAL
Vinyl sheet flooring				
Minimum	S.F.	4.53	0.60	5.13
Average	"	7.31	0.72	8.03
Maximum	"	12.25	1.00	13.25
Cove, to 6"	L.F.	2.70	1.20	3.90
Fluid applied resilient flooring				
Polyurethane, poured in place, 3/8" thick	S.F.	12.50	5.03	17.53
Vinyl sheet goods, backed				
0.070" thick	S.F.	4.61	0.75	5.36
0.093" thick	"	7.15	0.75	7.90
0.125" thick	"	8.25	0.75	9.00
0.250" thick	"	9.49	0.75	10.24

09 FINISHES

Flooring

	UNIT	MAT.	INST.	TOTAL
09678.10 — Resilient Base And Accessories				
Wall base, vinyl				
Group 1				
4" high	L.F.	1.19	2.01	3.20
6" high	"	1.62	2.01	3.63
Group 2				
4" high	L.F.	1.05	2.01	3.06
6" high	"	1.66	2.01	3.67
Group 3				
4" high	L.F.	2.37	2.01	4.38
6" high	"	2.66	2.01	4.67
Stair accessories				
Treads, 1/4" x 12", rubber diamond surface				
Marbled	L.F.	15.00	5.03	20.03
Plain	"	15.50	5.03	20.53
Grit strip safety tread, 12" wide, colors				
3/16" thick	L.F.	15.00	5.03	20.03
5/16" thick	"	19.75	5.03	24.78
Risers, 7" high, 1/8" thick, colors				
Flat	L.F.	6.11	3.02	9.13
Coved	"	4.22	3.02	7.24
Nosing, rubber				
3/16" thick, 3" wide				
Black	L.F.	4.74	3.02	7.76
Colors	"	4.94	3.02	7.96
6" wide				
Black	L.F.	5.91	5.03	10.94
Colors	"	6.17	5.03	11.20

Carpet

	UNIT	MAT.	INST.	TOTAL
09680.10 — Floor Leveling				
Repair and level floors to receive new flooring				
Minimum	S.Y.	1.04	2.01	3.05
Average	"	4.03	5.03	9.06
Maximum	"	5.98	6.04	12.02
09682.10 — Carpet Padding				
Carpet padding				
Foam rubber, waffle type, 0.3" thick	S.Y.	7.28	3.02	10.30
Jute padding				
Minimum	S.Y.	4.94	2.74	7.68
Average	"	6.43	3.02	9.45
Maximum	"	9.68	3.35	13.03
Sponge rubber cushion				
Minimum	S.Y.	5.85	2.74	8.59

09 FINISHES

Carpet	UNIT	MAT.	INST.	TOTAL
09682.10 Carpet Padding *(Cont.)*				
Average	S.Y.	7.80	3.02	10.82
Maximum	"	11.00	3.35	14.35
Urethane cushion, 3/8" thick				
Minimum	S.Y.	5.85	2.74	8.59
Average	"	6.82	3.02	9.84
Maximum	"	8.90	3.35	12.25
09685.10 Carpet				
Carpet, acrylic				
24 oz., light traffic	S.Y.	19.00	6.71	25.71
28 oz., medium traffic	"	22.75	6.71	29.46
Residential				
Nylon				
15 oz., light traffic	S.Y.	26.50	6.71	33.21
28 oz., medium traffic	"	34.25	6.71	40.96
Commercial				
Nylon				
28 oz., medium traffic	S.Y.	33.00	6.71	39.71
35 oz., heavy traffic	"	40.00	6.71	46.71
Wool				
30 oz., medium traffic	S.Y.	54.00	6.71	60.71
36 oz., medium traffic	"	57.00	6.71	63.71
42 oz., heavy traffic	"	76.00	6.71	82.71
Carpet tile				
Foam backed				
Minimum	S.F.	4.17	1.20	5.37
Average	"	4.83	1.34	6.17
Maximum	"	7.65	1.51	9.16
Tufted loop or shag				
Minimum	S.F.	4.51	1.20	5.71
Average	"	5.44	1.34	6.78
Maximum	"	8.76	1.51	10.27
Clean and vacuum carpet				
Minimum	S.Y.	0.38	0.23	0.61
Average	"	0.60	0.40	1.00
Maximum	"	0.82	0.60	1.42
09700.10 Special Flooring				
Epoxy flooring, marble chips				
Epoxy with colored quartz chips in 1/4" base	S.F.	7.15	3.35	10.50
Heavy duty epoxy topping, 3/16" thick	"	5.85	3.35	9.20
Epoxy terrazzo				
1/4" thick chemical resistant	S.F.	8.90	3.77	12.67

09 FINISHES

Painting

09905.10 Painting Preparation

	UNIT	MAT.	INST.	TOTAL
Dropcloths				
Minimum	S.F.	0.03	0.03	0.06
Average	"	0.05	0.04	0.09
Maximum	"	0.06	0.05	0.11
Masking				
Paper and tape				
Minimum	L.F.	0.02	0.50	0.52
Average	"	0.03	0.63	0.66
Maximum	"	0.05	0.84	0.89
Doors				
Minimum	EA.	0.05	6.31	6.36
Average	"	0.06	8.42	8.48
Maximum	"	0.07	11.25	11.32
Windows				
Minimum	EA.	0.05	6.31	6.36
Average	"	0.06	8.42	8.48
Maximum	"	0.07	11.25	11.32
Sanding				
Walls and flat surfaces				
Minimum	S.F.		0.33	0.33
Average	"		0.42	0.42
Maximum	"		0.50	0.50
Doors and windows				
Minimum	EA.		8.42	8.42
Average	"		12.50	12.50
Maximum	"		16.75	16.75
Trim				
Minimum	L.F.		0.63	0.63
Average	"		0.84	0.84
Maximum	"		1.12	1.12
Puttying				
Minimum	S.F.	0.01	0.77	0.78
Average	"	0.02	1.01	1.03
Maximum	"	0.03	1.26	1.29
Chemical Preparation				
Concrete floors				
Acid Etch				
Minimum	S.F.	0.06	0.12	0.18
Average	"	0.07	0.21	0.28
Maximum	"	0.10	0.28	0.38
Chemical Stripping				
Minimum	S.F.	0.06	0.84	0.90
Average	"	0.07	1.26	1.33
Maximum	"	0.10	2.02	2.12
Stone				
Chemical cleaning				
Minimum	S.F.	0.03	2.52	2.55
Average	"	0.06	3.36	3.42
Maximum	"	0.07	4.21	4.28
Wood				
Bleaching				
Minimum	S.F.	0.03	0.56	0.59
Average	"	0.06	0.72	0.78

09 FINISHES

Painting	UNIT	MAT.	INST.	TOTAL
09905.10 **Painting Preparation** *(Cont.)*				
Maximum	S.F.	0.07	0.91	0.98
Chemical stripping				
Minimum	S.F.	0.03	2.02	2.05
Average	"	0.06	5.05	5.11
Maximum	"	0.07	8.42	8.49
Water cleaning/preparation				
Washing (General)				
Minimum	S.F.		0.03	0.03
Average	"		0.04	0.04
Maximum	"		0.06	0.06
Mildew eradication				
Minimum	S.F.	0.03	0.06	0.09
Average	"	0.05	0.10	0.15
Maximum	"	0.06	0.16	0.22
Remove loose paint				
Minimum	S.F.		0.10	0.10
Average	"		0.16	0.16
Maximum	"		0.25	0.25
Steam clean				
Minimum	S.F.		0.12	0.12
Average	"		0.16	0.16
Maximum	"		0.25	0.25
09910.05 **Ext. Painting, Sitework**				
Benches				
Brush				
First Coat				
Minimum	S.F.	0.19	0.50	0.69
Average	"	0.19	0.63	0.82
Maximum	"	0.19	0.84	1.03
Second Coat				
Minimum	S.F.	0.18	0.31	0.49
Average	"	0.18	0.36	0.54
Maximum	"	0.18	0.42	0.60
Roller				
First Coat				
Minimum	S.F.	0.19	0.25	0.44
Average	"	0.19	0.28	0.47
Maximum	"	0.19	0.31	0.50
Second Coat				
Minimum	S.F.	0.18	0.18	0.36
Average	"	0.18	0.21	0.39
Maximum	"	0.18	0.22	0.40
Brickwork				
Brush				
First Coat				
Minimum	S.F.	0.19	0.31	0.50
Average	"	0.19	0.42	0.61
Maximum	"	0.19	0.63	0.82
Second Coat				
Minimum	S.F.	0.19	0.28	0.47
Average	"	0.19	0.33	0.52
Maximum	"	0.19	0.42	0.61

09 FINISHES

Painting

09910.05 Ext. Painting, Sitework *(Cont.)*

	UNIT	MAT.	INST.	TOTAL
Roller				
First Coat				
Minimum	S.F.	0.19	0.25	0.44
Average	"	0.19	0.31	0.50
Maximum	"	0.19	0.42	0.61
Second Coat				
Minimum	S.F.	0.19	0.21	0.40
Average	"	0.19	0.25	0.44
Maximum	"	0.19	0.31	0.50
Spray				
First Coat				
Minimum	S.F.	0.15	0.14	0.29
Average	"	0.15	0.18	0.33
Maximum	"	0.15	0.22	0.37
Second Coat				
Minimum	S.F.	0.15	0.13	0.28
Average	"	0.15	0.16	0.31
Maximum	"	0.15	0.21	0.36
Concrete Block				
Roller				
First Coat				
Minimum	S.F.	0.19	0.25	0.44
Average	"	0.19	0.33	0.52
Maximum	"	0.19	0.50	0.69
Second Coat				
Minimum	S.F.	0.19	0.21	0.40
Average	"	0.19	0.28	0.47
Maximum	"	0.19	0.42	0.61
Spray				
First Coat				
Minimum	S.F.	0.15	0.14	0.29
Average	"	0.15	0.16	0.31
Maximum	"	0.15	0.19	0.34
Second Coat				
Minimum	S.F.	0.15	0.09	0.24
Average	"	0.15	0.11	0.26
Maximum	"	0.15	0.15	0.30
Fences, Chain Link				
Brush				
First Coat				
Minimum	S.F.	0.13	0.50	0.63
Average	"	0.13	0.56	0.69
Maximum	"	0.13	0.63	0.76
Second Coat				
Minimum	S.F.	0.13	0.33	0.46
Average	"	0.13	0.38	0.51
Maximum	"	0.13	0.45	0.58
Roller				
First Coat				
Minimum	S.F.	0.13	0.36	0.49
Average	"	0.13	0.42	0.55
Maximum	"	0.13	0.48	0.61
Second Coat				

09 FINISHES

Painting

09910.05 Ext. Painting, Sitework (Cont.)

	UNIT	MAT.	INST.	TOTAL
Minimum	S.F.	0.13	0.21	0.34
Average	"	0.13	0.25	0.38
Maximum	"	0.13	0.31	0.44
Spray				
First Coat				
Minimum	S.F.	0.10	0.15	0.25
Average	"	0.10	0.18	0.28
Maximum	"	0.10	0.21	0.31
Second Coat				
Minimum	S.F.	0.10	0.12	0.22
Average	"	0.10	0.14	0.24
Maximum	"	0.10	0.15	0.25
Fences, Wood or Masonry				
Brush				
First Coat				
Minimum	S.F.	0.19	0.53	0.72
Average	"	0.19	0.63	0.82
Maximum	"	0.19	0.84	1.03
Second Coat				
Minimum	S.F.	0.19	0.31	0.50
Average	"	0.19	0.38	0.57
Maximum	"	0.19	0.50	0.69
Roller				
First Coat				
Minimum	S.F.	0.19	0.28	0.47
Average	"	0.19	0.33	0.52
Maximum	"	0.19	0.38	0.57
Second Coat				
Minimum	S.F.	0.19	0.19	0.38
Average	"	0.19	0.24	0.43
Maximum	"	0.19	0.31	0.50
Spray				
First Coat				
Minimum	S.F.	0.15	0.18	0.33
Average	"	0.15	0.22	0.37
Maximum	"	0.15	0.31	0.46
Second Coat				
Minimum	S.F.	0.15	0.12	0.27
Average	"	0.15	0.15	0.30
Maximum	"	0.15	0.21	0.36
Storage Tanks				
Roller				
First Coat				
Minimum	S.F.	0.15	0.21	0.36
Average	"	0.15	0.25	0.40
Maximum	"	0.15	0.31	0.46
Second Coat				
Minimum	S.F.	0.15	0.16	0.31
Average	"	0.15	0.20	0.35
Maximum	"	0.15	0.25	0.40
Spray				
First Coat				
Minimum	S.F.	0.13	0.12	0.25

09 FINISHES

Painting

	UNIT	MAT.	INST.	TOTAL
09910.05 Ext. Painting, Sitework *(Cont.)*				
Average	S.F.	0.13	0.14	0.27
Maximum	"	0.13	0.18	0.31
Second Coat				
Minimum	S.F.	0.13	0.10	0.23
Average	"	0.13	0.11	0.24
Maximum	"	0.13	0.12	0.25
09910.15 Ext. Painting, Buildings				
Decks, Metal				
Spray				
First Coat				
Minimum	S.F.	0.13	0.22	0.35
Average	"	0.13	0.25	0.38
Maximum	"	0.13	0.28	0.41
Second Coat				
Minimum	S.F.	0.11	0.15	0.26
Average	"	0.11	0.18	0.29
Maximum	"	0.11	0.21	0.32
Decks, Wood, Stained				
Brush				
First Coat				
Minimum	S.F.	0.15	0.25	0.40
Average	"	0.15	0.28	0.43
Maximum	"	0.15	0.31	0.46
Second Coat				
Minimum	S.F.	0.15	0.18	0.33
Average	"	0.15	0.19	0.34
Maximum	"	0.15	0.21	0.36
Roller				
First Coat				
Minimum	S.F.	0.15	0.18	0.33
Average	"	0.15	0.19	0.34
Maximum	"	0.15	0.21	0.36
Second Coat				
Minimum	S.F.	0.15	0.15	0.30
Average	"	0.15	0.16	0.31
Maximum	"	0.15	0.19	0.34
Spray				
First Coat				
Minimum	S.F.	0.13	0.15	0.28
Average	"	0.13	0.16	0.29
Maximum	"	0.13	0.19	0.32
Second Coat				
Minimum	S.F.	0.13	0.14	0.27
Average	"	0.13	0.15	0.28
Maximum	"	0.13	0.16	0.29
Doors, Metal				
Roller				
First Coat				
Minimum	S.F.	0.15	0.36	0.51
Average	"	0.15	0.42	0.57
Maximum	"	0.15	0.50	0.65
Second Coat				

09 FINISHES

Painting

09910.15 Ext. Painting, Buildings *(Cont.)*

	UNIT	MAT.	INST.	TOTAL
Minimum	S.F.	0.15	0.25	0.40
Average	"	0.15	0.28	0.43
Maximum	"	0.15	0.31	0.46
Spray				
First Coat				
Minimum	S.F.	0.13	0.31	0.44
Average	"	0.13	0.36	0.49
Maximum	"	0.13	0.42	0.55
Second Coat				
Minimum	S.F.	0.13	0.22	0.35
Average	"	0.13	0.25	0.38
Maximum	"	0.13	0.28	0.41
Door Frames, Metal				
Brush				
First Coat				
Minimum	L.F.	0.19	0.63	0.82
Average	"	0.19	0.78	0.97
Maximum	"	0.19	0.91	1.10
Second Coat				
Minimum	L.F.	0.19	0.36	0.55
Average	"	0.19	0.42	0.61
Maximum	"	0.19	0.50	0.69
Spray				
First Coat				
Minimum	L.F.	0.13	0.28	0.41
Average	"	0.13	0.36	0.49
Maximum	"	0.13	0.50	0.63
Second Coat				
Minimum	L.F.	0.13	0.22	0.35
Average	"	0.13	0.25	0.38
Maximum	"	0.13	0.28	0.41
Doors, Wood				
Brush				
First Coat				
Minimum	S.F.	0.15	0.77	0.92
Average	"	0.15	1.01	1.16
Maximum	"	0.15	1.26	1.41
Second Coat				
Minimum	S.F.	0.15	0.63	0.78
Average	"	0.15	0.72	0.87
Maximum	"	0.15	0.84	0.99
Roller				
First Coat				
Minimum	S.F.	0.15	0.33	0.48
Average	"	0.15	0.42	0.57
Maximum	"	0.15	0.63	0.78
Second Coat				
Minimum	S.F.	0.15	0.25	0.40
Average	"	0.15	0.28	0.43
Maximum	"	0.15	0.42	0.57
Spray				
First Coat				
Minimum	S.F.	0.13	0.15	0.28

09 FINISHES

Painting

09910.15 Ext. Painting, Buildings *(Cont.)*

	UNIT	MAT.	INST.	TOTAL
Average	S.F.	0.13	0.19	0.32
Maximum	"	0.13	0.25	0.38
Second Coat				
Minimum	S.F.	0.13	0.12	0.25
Average	"	0.13	0.14	0.27
Maximum	"	0.13	0.16	0.29
Gutters and Downspouts				
Brush				
First Coat				
Minimum	L.F.	0.19	0.63	0.82
Average	"	0.19	0.72	0.91
Maximum	"	0.19	0.84	1.03
Second Coat				
Minimum	L.F.	0.19	0.42	0.61
Average	"	0.19	0.50	0.69
Maximum	"	0.19	0.63	0.82
Siding, Metal				
Roller				
First Coat				
Minimum	S.F.	0.15	0.21	0.36
Average	"	0.15	0.22	0.37
Maximum	"	0.15	0.25	0.40
Second Coat				
Minimum	S.F.	0.15	0.19	0.34
Average	"	0.15	0.21	0.36
Maximum	"	0.15	0.22	0.37
Spray				
First Coat				
Minimum	S.F.	0.13	0.15	0.28
Average	"	0.13	0.18	0.31
Maximum	"	0.13	0.21	0.34
Second Coat				
Minimum	S.F.	0.13	0.10	0.23
Average	"	0.13	0.12	0.25
Maximum	"	0.13	0.16	0.29
Siding, Wood				
Roller				
First Coat				
Minimum	S.F.	0.13	0.18	0.31
Average	"	0.13	0.21	0.34
Maximum	"	0.13	0.22	0.35
Second Coat				
Minimum	S.F.	0.13	0.21	0.34
Average	"	0.13	0.22	0.35
Maximum	"	0.13	0.25	0.38
Spray				
First Coat				
Minimum	S.F.	0.13	0.16	0.29
Average	"	0.13	0.18	0.31
Maximum	"	0.13	0.19	0.32
Second Coat				
Minimum	S.F.	0.13	0.12	0.25
Average	"	0.13	0.16	0.29

09 FINISHES

Painting

09910.15 Ext. Painting, Buildings *(Cont.)*

	UNIT	MAT.	INST.	TOTAL
Maximum	S.F.	0.13	0.25	0.38
Stucco				
Roller				
First Coat				
Minimum	S.F.	0.19	0.22	0.41
Average	"	0.19	0.26	0.45
Maximum	"	0.19	0.31	0.50
Second Coat				
Minimum	S.F.	0.19	0.18	0.37
Average	"	0.19	0.21	0.40
Maximum	"	0.19	0.25	0.44
Spray				
First Coat				
Minimum	S.F.	0.15	0.15	0.30
Average	"	0.15	0.18	0.33
Maximum	"	0.15	0.21	0.36
Second Coat				
Minimum	S.F.	0.15	0.12	0.27
Average	"	0.15	0.14	0.29
Maximum	"	0.15	0.16	0.31
Trim				
Brush				
First Coat				
Minimum	L.F.	0.19	0.21	0.40
Average	"	0.19	0.25	0.44
Maximum	"	0.19	0.31	0.50
Second Coat				
Minimum	L.F.	0.19	0.15	0.34
Average	"	0.19	0.21	0.40
Maximum	"	0.19	0.31	0.50
Walls				
Roller				
First Coat				
Minimum	S.F.	0.15	0.18	0.33
Average	"	0.15	0.18	0.33
Maximum	"	0.15	0.20	0.35
Second Coat				
Minimum	S.F.	0.15	0.15	0.30
Average	"	0.15	0.16	0.31
Maximum	"	0.15	0.19	0.34
Spray				
First Coat				
Minimum	S.F.	0.11	0.07	0.18
Average	"	0.11	0.10	0.21
Maximum	"	0.11	0.12	0.23
Second Coat				
Minimum	S.F.	0.11	0.06	0.17
Average	"	0.11	0.08	0.19
Maximum	"	0.11	0.11	0.22
Windows				
Brush				
First Coat				
Minimum	S.F.	0.13	0.84	0.97

09 FINISHES

Painting

	UNIT	MAT.	INST.	TOTAL
09910.15 Ext. Painting, Buildings (Cont.)				
Average	S.F.	0.13	1.01	1.14
Maximum	"	0.13	1.26	1.39
Second Coat				
Minimum	S.F.	0.13	0.72	0.85
Average	"	0.13	0.84	0.97
Maximum	"	0.13	1.01	1.14
09910.25 Ext. Painting, Misc.				
Gratings, Metal				
Roller				
First Coat				
Minimum	S.F.	0.15	1.44	1.59
Average	"	0.15	1.68	1.83
Maximum	"	0.15	2.02	2.17
Second Coat				
Minimum	S.F.	0.15	1.01	1.16
Average	"	0.15	1.26	1.41
Maximum	"	0.15	1.68	1.83
Spray				
First Coat				
Minimum	S.F.	0.13	0.72	0.85
Average	"	0.13	0.84	0.97
Maximum	"	0.13	1.01	1.14
Second Coat				
Minimum	S.F.	0.13	0.56	0.69
Average	"	0.13	0.63	0.76
Maximum	"	0.13	0.72	0.85
Ladders				
Brush				
First Coat				
Minimum	L.F.	0.19	1.26	1.45
Average	"	0.19	1.44	1.63
Maximum	"	0.19	1.68	1.87
Second Coat				
Minimum	L.F.	0.19	1.01	1.20
Average	"	0.19	1.12	1.31
Maximum	"	0.19	1.26	1.45
Spray				
First Coat				
Minimum	L.F.	0.13	0.84	0.97
Average	"	0.13	0.91	1.04
Maximum	"	0.13	1.01	1.14
Second Coat				
Minimum	L.F.	0.13	0.72	0.85
Average	"	0.13	0.77	0.90
Maximum	"	0.13	0.84	0.97
Shakes				
Spray				
First Coat				
Minimum	S.F.	0.14	0.21	0.35
Average	"	0.14	0.22	0.36
Maximum	"	0.14	0.25	0.39
Second Coat				

09 FINISHES

Painting

09910.25 Ext. Painting, Misc. (Cont.)

	UNIT	MAT.	INST.	TOTAL
Minimum	S.F.	0.14	0.19	0.33
Average	"	0.14	0.21	0.35
Maximum	"	0.14	0.22	0.36
Shingles, Wood				
Roller				
First Coat				
Minimum	S.F.	0.15	0.28	0.43
Average	"	0.15	0.31	0.46
Maximum	"	0.15	0.36	0.51
Second Coat				
Minimum	S.F.	0.15	0.19	0.34
Average	"	0.15	0.21	0.36
Maximum	"	0.15	0.22	0.37
Spray				
First Coat				
Minimum	L.F.	0.13	0.19	0.32
Average	"	0.13	0.21	0.34
Maximum	"	0.13	0.22	0.35
Second Coat				
Minimum	L.F.	0.13	0.14	0.27
Average	"	0.13	0.15	0.28
Maximum	"	0.13	0.16	0.29
Shutters and Louvres				
Brush				
First Coat				
Minimum	EA.	0.19	10.00	10.19
Average	"	0.19	12.50	12.69
Maximum	"	0.19	16.75	16.94
Second Coat				
Minimum	EA.	0.19	6.31	6.50
Average	"	0.19	7.77	7.96
Maximum	"	0.19	10.00	10.19
Spray				
First Coat				
Minimum	EA.	0.14	3.36	3.50
Average	"	0.14	4.04	4.18
Maximum	"	0.14	5.05	5.19
Second Coat				
Minimum	EA.	0.14	2.52	2.66
Average	"	0.14	3.36	3.50
Maximum	"	0.14	4.04	4.18
Stairs, metal				
Brush				
First Coat				
Minimum	S.F.	0.19	0.56	0.75
Average	"	0.19	0.63	0.82
Maximum	"	0.19	0.72	0.91
Second Coat				
Minimum	S.F.	0.19	0.31	0.50
Average	"	0.19	0.36	0.55
Maximum	"	0.19	0.42	0.61
Spray				
First Coat				

09 FINISHES

Painting

09910.25 — Ext. Painting, Misc. *(Cont.)*

Description	UNIT	MAT.	INST.	TOTAL
Minimum	S.F.	0.14	0.28	0.42
Average	"	0.14	0.36	0.50
Maximum	"	0.14	0.38	0.52
Second Coat				
Minimum	S.F.	0.14	0.21	0.35
Average	"	0.14	0.25	0.39
Maximum	"	0.14	0.31	0.45
Steel, Structural, Light				
Brush				
First Coat				
Minimum	S.F.	0.15	1.01	1.16
Average	"	0.15	1.26	1.41
Maximum	"	0.15	1.68	1.83
Second Coat				
Minimum	S.F.	0.14	0.72	0.86
Average	"	0.14	0.84	0.98
Maximum	"	0.14	1.01	1.15
Roller				
First Coat				
Minimum	S.F.	0.15	0.72	0.87
Average	"	0.15	0.84	0.99
Maximum	"	0.15	1.01	1.16
Second Coat				
Minimum	S.F.	0.14	0.42	0.56
Average	"	0.14	0.50	0.64
Maximum	"	0.14	0.63	0.77
Spray				
First Coat				
Minimum	S.F.	0.14	0.42	0.56
Average	"	0.14	0.50	0.64
Maximum	"	0.14	0.63	0.77
Second Coat				
Minimum	S.F.	0.10	0.36	0.46
Average	"	0.10	0.42	0.52
Maximum	"	0.10	0.50	0.60
Steel, Medium to Heavy				
Brush				
First Coat				
Minimum	S.F.	0.15	0.50	0.65
Average	"	0.15	0.56	0.71
Maximum	"	0.15	0.63	0.78
Second Coat				
Minimum	S.F.	0.14	0.42	0.56
Average	"	0.14	0.50	0.64
Maximum	"	0.14	0.59	0.73
Roller				
First Coat				
Minimum	S.F.	0.15	0.42	0.57
Average	"	0.15	0.50	0.65
Maximum	"	0.15	0.59	0.74
Second Coat				
Minimum	S.F.	0.14	0.31	0.45
Average	"	0.14	0.36	0.50

09 FINISHES

Painting

	UNIT	MAT.	INST.	TOTAL
09910.25 — Ext. Painting, Misc. (Cont.)				
Maximum	S.F.	0.14	0.42	0.56
Spray				
First Coat				
Minimum	S.F.	0.14	0.28	0.42
Average	"	0.14	0.31	0.45
Maximum	"	0.14	0.36	0.50
Second Coat				
Minimum	S.F.	0.10	0.22	0.32
Average	"	0.10	0.25	0.35
Maximum	"	0.10	0.28	0.38
09910.35 — Int. Painting, Buildings				
Acoustical Ceiling				
Roller				
First Coat				
Minimum	S.F.	0.19	0.31	0.50
Average	"	0.19	0.42	0.61
Maximum	"	0.19	0.63	0.82
Second Coat				
Minimum	S.F.	0.19	0.25	0.44
Average	"	0.19	0.31	0.50
Maximum	"	0.19	0.42	0.61
Spray				
First Coat				
Minimum	S.F.	0.15	0.14	0.29
Average	"	0.15	0.16	0.31
Maximum	"	0.15	0.21	0.36
Second Coat				
Minimum	S.F.	0.15	0.11	0.26
Average	"	0.15	0.12	0.27
Maximum	"	0.15	0.14	0.29
Cabinets and Casework				
Brush				
First Coat				
Minimum	S.F.	0.19	0.50	0.69
Average	"	0.19	0.56	0.75
Maximum	"	0.19	0.63	0.82
Second Coat				
Minimum	S.F.	0.19	0.42	0.61
Average	"	0.19	0.45	0.64
Maximum	"	0.19	0.50	0.69
Spray				
First Coat				
Minimum	S.F.	0.15	0.25	0.40
Average	"	0.15	0.29	0.44
Maximum	"	0.15	0.36	0.51
Second Coat				
Minimum	S.F.	0.15	0.20	0.35
Average	"	0.15	0.21	0.36
Maximum	"	0.15	0.28	0.43
Ceilings				
Roller				
First Coat				

09 FINISHES

Painting	UNIT	MAT.	INST.	TOTAL
09910.35 Int. Painting, Buildings *(Cont.)*				
Minimum	S.F.	0.15	0.21	0.36
Average	"	0.15	0.22	0.37
Maximum	"	0.15	0.25	0.40
Second Coat				
Minimum	S.F.	0.15	0.16	0.31
Average	"	0.15	0.19	0.34
Maximum	"	0.15	0.21	0.36
Spray				
First Coat				
Minimum	S.F.	0.13	0.12	0.25
Average	"	0.13	0.14	0.27
Maximum	"	0.13	0.15	0.28
Second Coat				
Minimum	S.F.	0.13	0.09	0.22
Average	"	0.13	0.10	0.23
Maximum	"	0.13	0.12	0.25
Doors, Metal				
Roller				
First Coat				
Minimum	L.F.	0.19	0.33	0.52
Average	"	0.19	0.38	0.57
Maximum	"	0.19	0.45	0.64
Second Coat				
Minimum	L.F.	0.19	0.24	0.43
Average	"	0.19	0.26	0.45
Maximum	"	0.19	0.29	0.48
Spray				
First Coat				
Minimum	L.F.	0.13	0.28	0.41
Average	"	0.13	0.31	0.44
Maximum	"	0.13	0.36	0.49
Second Coat				
Minimum	L.F.	0.19	0.21	0.40
Average	"	0.19	0.22	0.41
Maximum	"	0.19	0.25	0.44
Doors, Wood				
Brush				
First Coat				
Minimum	S.F.	0.19	0.72	0.91
Average	"	0.19	0.91	1.10
Maximum	"	0.19	1.12	1.31
Second Coat				
Minimum	S.F.	0.14	0.56	0.70
Average	"	0.14	0.63	0.77
Maximum	"	0.14	0.72	0.86
Spray				
First Coat				
Minimum	S.F.	0.14	0.14	0.28
Average	"	0.14	0.18	0.32
Maximum	"	0.14	0.22	0.36
Second Coat				
Minimum	S.F.	0.14	0.12	0.26
Average	"	0.14	0.13	0.27

09 FINISHES

Painting

09910.35 Int. Painting, Buildings (Cont.)

	UNIT	MAT.	INST.	TOTAL
Maximum	S.F.	0.14	0.15	0.29
Ductwork				
Brush				
Minimum	L.F.	0.15	0.63	0.78
Average	"	0.15	0.72	0.87
Maximum	"	0.15	0.84	0.99
Roller				
Minimum	S.F.	0.15	0.42	0.57
Average	"	0.15	0.45	0.60
Maximum	"	0.15	0.50	0.65
Spray				
Minimum	S.F.	0.15	0.18	0.33
Average	"	0.15	0.19	0.34
Maximum	"	0.15	0.21	0.36
Floors				
Roller				
First Coat				
Minimum	S.F.	0.15	0.15	0.30
Average	"	0.15	0.18	0.33
Maximum	"	0.15	0.21	0.36
Second Coat				
Minimum	S.F.	0.15	0.12	0.27
Average	"	0.15	0.14	0.29
Maximum	"	0.15	0.14	0.29
Spray				
First Coat				
Minimum	S.F.	0.14	0.11	0.25
Average	"	0.14	0.12	0.26
Maximum	"	0.14	0.13	0.27
Second Coat				
Minimum	S.F.	0.14	0.09	0.23
Average	"	0.14	0.10	0.24
Maximum	"	0.14	0.11	0.25
Pipes to 6" diameter				
Brush				
Minimum	L.F.	0.19	0.63	0.82
Average	"	0.19	0.72	0.91
Maximum	"	0.19	0.84	1.03
Spray				
Minimum	L.F.	0.15	0.21	0.36
Average	"	0.15	0.25	0.40
Maximum	"	0.15	0.33	0.48
Pipes to 12" diameter				
Brush				
Minimum	L.F.	0.33	1.26	1.59
Average	"	0.33	1.44	1.77
Maximum	"	0.33	1.68	2.01
Spray				
Minimum	L.F.	0.29	0.42	0.71
Average	"	0.29	0.50	0.79
Maximum	"	0.29	0.63	0.92
Trim				
Brush				

09 FINISHES

Painting	UNIT	MAT.	INST.	TOTAL
09910.35 **Int. Painting, Buildings** *(Cont.)*				
First Coat				
Minimum	L.F.	0.19	0.20	0.39
Average	"	0.19	0.22	0.41
Maximum	"	0.19	0.28	0.47
Second Coat				
Minimum	L.F.	0.19	0.14	0.33
Average	"	0.19	0.19	0.38
Maximum	"	0.19	0.28	0.47
Walls				
Roller				
First Coat				
Minimum	S.F.	0.15	0.18	0.33
Average	"	0.15	0.18	0.33
Maximum	"	0.15	0.21	0.36
Second Coat				
Minimum	S.F.	0.15	0.15	0.30
Average	"	0.15	0.16	0.31
Maximum	"	0.15	0.19	0.34
Spray				
First Coat				
Minimum	S.F.	0.13	0.07	0.20
Average	"	0.13	0.09	0.22
Maximum	"	0.13	0.12	0.25
Second Coat				
Minimum	S.F.	0.13	0.07	0.20
Average	"	0.13	0.08	0.21
Maximum	"	0.13	0.11	0.24
09955.10 **Wall Covering**				
Vinyl wall covering				
Medium duty	S.F.	0.97	0.72	1.69
Heavy duty	"	2.01	0.84	2.85
Over pipes and irregular shapes				
Lightweight, 13 oz.	S.F.	1.69	1.01	2.70
Medium weight, 25 oz.	"	2.01	1.12	3.13
Heavy weight, 34 oz.	"	2.47	1.26	3.73
Cork wall covering				
1' x 1' squares				
1/4" thick	S.F.	5.26	1.26	6.52
1/2" thick	"	6.69	1.26	7.95
3/4" thick	"	7.54	1.26	8.80
Wall fabrics				
Natural fabrics, grass cloths				
Minimum	S.F.	1.69	0.77	2.46
Average	"	1.88	0.84	2.72
Maximum	"	6.30	1.01	7.31
Flexible gypsum coated wall fabric, fire resistant	"	1.89	0.50	2.39
Vinyl corner guards				
3/4" x 3/4" x 8'	EA.	8.90	6.31	15.21
2-3/4" x 2-3/4" x 4'	"	5.26	6.31	11.57

09 FINISHES

Painting

	UNIT	MAT.	INST.	TOTAL
09980.15 Paint				
Paint, enamel				
600 sf per gal.	GAL			59.00
550 sf per gal.	"			55.00
500 sf per gal.	"			39.00
450 sf per gal.	"			36.50
350 sf per gal.	"			35.00
Filler, 60 sf per gal.	"			41.50
Latex, 400 sf per gal.	"			39.00
Aluminum				
400 sf per gal.	GAL			52.00
500 sf per gal.	"			83.00
Red lead, 350 sf per gal.	"			73.00
Primer				
400 sf per gal.	GAL			35.00
300 sf per gal.	"			35.00
Latex base, interior, white	"			39.00
Sealer and varnish				
400 sf per gal.	GAL			36.50
425 sf per gal.	"			52.00
600 sf per gal.	"			68.00

10 SPECIALTIES

Specialties	UNIT	MAT.	INST.	TOTAL
10110.10 **Chalkboards**				
Chalkboard, metal frame, 1/4" thick				
48"x60"	EA.	540	60.00	600
48"x96"	"	740	67.00	807
48"x144"	"	980	76.00	1,056
48"x192"	"	1,330	86.00	1,416
Liquid chalkboard				
48"x60"	EA.	710	60.00	770
48"x96"	"	910	67.00	977
48"x144"	"	1,360	76.00	1,436
48"x192"	"	1,560	86.00	1,646
Map rail, deluxe	L.F.	8.89	3.02	11.91
10165.10 **Toilet Partitions**				
Toilet partition, plastic laminate				
Ceiling mounted	EA.	1,210	200	1,410
Floor mounted	"	800	150	950
Metal				
Ceiling mounted	EA.	820	200	1,020
Floor mounted	"	790	150	940
Wheel chair partition, plastic laminate				
Ceiling mounted	EA.	1,800	200	2,000
Floor mounted	"	1,580	150	1,730
Painted metal				
Ceiling mounted	EA.	1,290	200	1,490
Floor mounted	"	1,170	150	1,320
Urinal screen, plastic laminate				
Wall hung	EA.	560	76.00	636
Floor mounted	"	500	76.00	576
Porcelain enameled steel, floor mounted	"	640	76.00	716
Painted metal, floor mounted	"	420	76.00	496
Stainless steel, floor mounted	"	810	76.00	886
Metal toilet partitions				
Front door and side divider, floor mounted				
Porcelain enameled steel	EA.	1,330	150	1,480
Painted steel	"	780	150	930
Stainless steel	"	1,930	150	2,080
10185.10 **Shower Stalls**				
Shower receptors				
Precast, terrazzo				
32" x 32"	EA.	720	55.00	775
32" x 48"	"	750	66.00	816
Concrete				
32" x 32"	EA.	290	55.00	345
48" x 48"	"	320	74.00	394
Shower door, trim and hardware				
Economy, 24" wide, chrome, tempered glass	EA.	330	66.00	396
Porcelain enameled steel, flush	"	590	66.00	656
Baked enameled steel, flush	"	350	66.00	416
Aluminum, tempered glass, 48" wide, sliding	"	730	83.00	813
Folding	"	700	83.00	783
Aluminum and tempered glass, molded plastic				
Complete with receptor and door				

10 SPECIALTIES

Specialties	UNIT	MAT.	INST.	TOTAL
10185.10 Shower Stalls *(Cont.)*				
32" x 32"	EA.	890	170	1,060
36" x 36"	"	1,000	170	1,170
40" x 40"	"	1,040	190	1,230
Shower compartment, precast concrete receptor				
Single entry type				
Porcelain enameled steel	EA.	2,510	660	3,170
Baked enameled steel	"	2,410	660	3,070
Stainless steel	"	2,310	660	2,970
Double entry type				
Porcelain enameled steel	EA.	4,470	830	5,300
Baked enameled steel	"	3,060	830	3,890
Stainless steel	"	4,940	830	5,770
10190.10 Cubicles				
Hospital track				
Ceiling hung	L.F.	7.93	6.71	14.64
Suspended	"	8.59	8.63	17.22
Hospital metal dividers, galvanized steel				
Baked enamel finish				
54" high				
10" glass light	L.F.	130	30.25	160
14" glass light	"	130	30.25	160
24" glass light	"	130	30.25	160
60" high				
10" glass light	L.F.	140	33.50	174
14" glass light	"	140	33.50	174
24" glass light	"	130	33.50	164
Stainless steel				
54" high				
10" glass light	L.F.	230	33.50	264
14" glass light	"	250	33.50	284
24" glass light	"	290	33.50	324
60" high				
14" glass light	L.F.	260	37.75	298
14" glass light	"	260	37.75	298
24" glass light	"	320	37.75	358
10210.10 Vents And Wall Louvers				
Block vent, 8"x16"x4" alum., w/screen, mill finish	EA.	180	22.25	202
Standard	"	100	21.00	121
Vents w/screen, 4" deep, 8" wide, 5" high				
Modular	EA.	120	21.00	141
Aluminum gable louvers	S.F.	21.25	11.25	32.50
Vent screen aluminum, 4" wide, continuous	L.F.	6.17	2.22	8.39
Louvers, aluminum, anodized, fixed blade				
Horizontal line	S.F.	62.00	16.75	78.75
Vertical line	"	62.00	16.75	78.75
Wall louvre, aluminum mill finish				
Under, 2 sf	S.F.	47.50	8.35	55.85
2 to 4 sf	"	41.50	7.42	48.92
5 to 10 sf	"	39.00	7.42	46.42
Galvanized steel				
Under 2 sf	S.F.	43.00	8.35	51.35

10 SPECIALTIES

Specialties	UNIT	MAT.	INST.	TOTAL
10210.10 Vents And Wall Louvers (Cont.)				
2 to 4 sf	S.F.	29.50	7.42	36.92
5 to 10 sf	"	27.75	7.42	35.17
10225.10 Door Louvers				
Fixed, 1" thick, enameled steel				
8"x8"	EA.	64.00	7.55	71.55
12"x8"	"	73.00	7.55	80.55
12"x12"	"	82.00	8.63	90.63
16"x12"	"	120	9.29	129
18"x12"	"	120	15.00	135
20"x8"	"	130	8.63	139
20"x12"	"	150	17.25	167
20"x16"	"	150	20.25	170
20"x20"	"	160	24.25	184
24"x12"	"	130	20.25	150
24"x16"	"	140	21.50	162
24"x18"	"	160	23.25	183
24"x20"	"	170	25.25	195
24"x24"	"	170	27.50	198
26"x26"	"	200	37.75	238
10270.40 Access & Pedestal Floor				
Panels, no covering, 2'x2'				
Plain	S.F.	11.75	0.75	12.50
Perforated	"	16.25	30.25	46.50
Pedestals				
For 6" to 12" clearance	EA.	9.42	6.04	15.46
Stringers				
2'	L.F.	2.60	2.87	5.47
6'	"	2.60	2.01	4.61
Accessories				
Ramp assembly	S.F.	55.00	2.41	57.41
Elevated floor assembly	"	87.00	2.23	89.23
Handrail	L.F.	68.00	30.25	98.25
Fascia plate	"	33.00	15.00	48.00
For carpet tiles, add	S.F.			9.10
For vinyl flooring, add	"			10.50
RF shielding components, floor liner				
Hot rolled steel sheet				
14 ga.	S.F.	16.00	1.51	17.51
11 ga.	"	22.00	4.64	26.64
10290.10 Pest Control				
Termite control				
Under slab spraying				
Minimum	S.F.	1.33	0.11	1.44
Average	"	1.33	0.23	1.56
Maximum	"	1.94	0.47	2.41

10 SPECIALTIES

Specialties	UNIT	MAT.	INST.	TOTAL
10350.10 Flagpoles				
Installed in concrete base				
Fiberglass				
25' high	EA.	1,680	400	2,080
50' high	"	4,450	1,010	5,460
Aluminum				
25' high	EA.	1,630	400	2,030
50' high	"	3,230	1,010	4,240
Bonderized steel				
25' high	EA.	1,830	460	2,290
50' high	"	3,650	1,210	4,860
Freestanding tapered, fiberglass				
30' high	EA.	2,000	430	2,430
40' high	"	2,600	550	3,150
50' high	"	6,630	600	7,230
60' high	"	7,080	710	7,790
Wall mounted, with collar, brushed aluminum finish				
15' long	EA.	1,550	300	1,850
18' long	"	1,750	300	2,050
20' long	"	1,910	320	2,230
24' long	"	2,050	360	2,410
Outrigger, wall, including base				
10' long	EA.	1,570	400	1,970
20' long	"	2,080	500	2,580
10400.10 Identifying Devices				
Directory and bulletin boards				
Open face boards				
Chrome plated steel frame	S.F.	37.00	30.25	67.25
Aluminum framed	"	64.00	30.25	94.25
Bronze framed	"	82.00	30.25	112
Stainless steel framed	"	110	30.25	140
Tack board, aluminum framed	"	26.00	30.25	56.25
Visual aid board, aluminum framed	"	26.00	30.25	56.25
Glass encased boards, hinged and keyed				
Aluminum framed	S.F.	140	76.00	216
Bronze framed	"	160	76.00	236
Stainless steel framed	"	210	76.00	286
Chrome plated steel framed	"	220	76.00	296
Metal plaque				
Cast bronze	S.F.	630	50.00	680
Aluminum	"	360	50.00	410
Metal engraved plaque				
Porcelain steel	S.F.	760	50.00	810
Stainless steel	"	610	50.00	660
Brass	"	900	50.00	950
Aluminum	"	560	50.00	610
Metal built-up plaque				
Bronze	S.F.	690	60.00	750
Copper and bronze	"	610	60.00	670
Copper and aluminum	"	670	60.00	730
Metal nameplate plaques				
Cast bronze	S.F.	680	37.75	718
Cast aluminum	"	500	37.75	538

10 SPECIALTIES

Specialties	UNIT	MAT.	INST.	TOTAL
10400.10 Identifying Devices (Cont.)				
Engraved, 1-1/2" x 6"				
Bronze	EA.	290	37.75	328
Aluminum	"	220	37.75	258
Letters, on masonry, aluminum, satin finish				
1/2" thick				
2" high	EA.	28.00	24.25	52.25
4" high	"	41.75	30.25	72.00
6" high	"	56.00	33.50	89.50
3/4" thick				
8" high	EA.	84.00	37.75	122
10" high	"	96.00	43.25	139
1" thick				
12" high	EA.	110	50.00	160
14" high	"	120	60.00	180
16" high	"	150	76.00	226
For polished aluminum add, 15%				
For clear anodized aluminum add, 15%				
For colored anodic aluminum add, 30%				
For profiled and color enameled letters add, 50%				
Cast bronze, satin finish letters				
3/8" thick				
2" high	EA.	31.75	24.25	56.00
4" high	"	47.75	30.25	78.00
1/2" thick, 6" high	"	65.00	33.50	98.50
5/8" thick, 8" high	"	97.00	37.75	135
1" thick				
10" high	EA.	120	43.25	163
12" high	"	150	50.00	200
14" high	"	180	60.00	240
16" high	"	270	76.00	346
Interior door signs, adhesive, flexible				
2" x 8"	EA.	26.00	11.75	37.75
4" x 4"	"	27.50	11.75	39.25
6" x 7"	"	34.75	11.75	46.50
6" x 9"	"	44.25	11.75	56.00
10" x 9"	"	58.00	11.75	69.75
10" x 12"	"	75.00	11.75	86.75
Hard plastic type, no frame				
3" x 8"	EA.	58.00	11.75	69.75
4" x 4"	"	57.00	11.75	68.75
4" x 12"	"	62.00	11.75	73.75
Hard plastic type, with frame				
3" x 8"	EA.	180	11.75	192
4" x 4"	"	140	11.75	152
4" x 12"	"	210	11.75	222
10450.10 Control				
Access control, 7' high, indoor or outdoor impenetrability				
Remote or card control, type B	EA.	1,920	820	2,740
Free passage, type B	"	1,560	820	2,380
Remote or card control, type AA	"	3,070	820	3,890
Free passage, type AA	"	2,780	820	3,600

10 SPECIALTIES

Specialties	UNIT	MAT.	INST.	TOTAL
10500.10 Lockers				
Locker bench, floor mounted, laminated maple				
4'	EA.	350	50.00	400
6'	"	490	50.00	540
Wardrobe locker, 12" x 60" x 15", baked on enamel				
1-tier	EA.	380	30.25	410
2-tier	"	400	30.25	430
3-tier	"	450	31.75	482
4-tier	"	490	31.75	522
12" x 72" x 15", baked on enamel				
1-tier	EA.	320	30.25	350
2-tier	"	390	30.25	420
4-tier	"	480	31.75	512
5-tier	"	480	31.75	512
15" x 60" x 15", baked on enamel				
1-tier	EA.	420	30.25	450
4-tier	"	460	31.75	492
Wardrobe locker, single tier type				
12" x 15" x 72"	EA.	290	60.00	350
18" x 15" x 72"	"	350	64.00	414
12" x 18" x 72"	"	310	67.00	377
18" x 18" x 72"	"	400	71.00	471
Double tier type				
12" x 15" x 36"	EA.	230	30.25	260
18" x 15" x 36"	"	220	30.25	250
12" x 18" x 36"	"	230	30.25	260
18" x 18" x 36"	"	250	30.25	280
Two person unit				
18" x 15" x 72"	EA.	600	100	700
18" x 18" x 72"	"	680	120	800
Duplex unit				
15" x 15" x 72"	EA.	620	60.00	680
15" x 21" x 72"	"	650	60.00	710
Basket lockers, basket sets with baskets				
24 basket set	SET	1,520	300	1,820
30 basket set	"	1,820	380	2,200
36 basket set	"	2,060	500	2,560
42 basket set	"	2,300	600	2,900
10520.10 Fire Protection				
Portable fire extinguishers				
Water pump tank type				
2.5 gal.				
Red enameled galvanized	EA.	140	31.50	172
Red enameled copper	"	210	31.50	242
Polished copper	"	280	31.50	312
Carbon dioxide type, red enamel steel				
Squeeze grip with hose and horn				
2.5 lb	EA.	230	31.50	262
5 lb	"	330	36.50	367
10 lb	"	340	47.50	388
15 lb	"	380	59.00	439
20 lb	"	460	59.00	519
Wheeled type				

10 SPECIALTIES

10520.10 Fire Protection (Cont.)

Specialties	UNIT	MAT.	INST.	TOTAL
125 lb	EA.	4,110	95.00	4,205
250 lb	"	5,200	95.00	5,295
500 lb	"	6,710	95.00	6,805
Dry chemical, pressurized type				
Red enameled steel				
2.5 lb	EA.	72.00	31.50	104
5 lb	"	98.00	36.50	135
10 lb	"	200	47.50	248
20 lb	"	260	59.00	319
30 lb	"	330	59.00	389
Chrome plated steel, 2.5 lb	"	310	31.50	342
Other type extinguishers				
2.5 gal, stainless steel, pressurized water tanks	EA.	230	31.50	262
Soda and acid type	"	190	31.50	222
Cartridge operated, water type	"	170	31.50	202
Loaded stream, water type	"	200	31.50	232
Foam type	"	270	31.50	302
40 gal, wheeled foam type	"	6,140	95.00	6,235
Fire extinguisher cabinets				
Enameled steel				
8" x 12" x 27"	EA.	160	95.00	255
8" x 16" x 38"	"	190	95.00	285
Aluminum				
8" x 12" x 27"	EA.	240	95.00	335
8" x 16" x 38"	"	290	95.00	385
8" x 12" x 27"	"	260	95.00	355
Stainless steel				
8" x 16" x 38"	EA.	270	95.00	365

10550.10 Postal Specialties

	UNIT	MAT.	INST.	TOTAL
Mail chutes				
Single mail chute				
Finished aluminum	L.F.	880	150	1,030
Bronze	"	1,220	150	1,370
Single mail chute receiving box				
Finished aluminum	EA.	1,300	300	1,600
Bronze	"	1,560	300	1,860
Twin mail chute, double parallel				
Finished aluminum	FLR	1,910	300	2,210
Bronze	"	2,500	300	2,800
Receiving box, 36" x 20" x 12"				
Finished aluminum	EA.	2,990	500	3,490
Bronze	"	4,030	500	4,530
Locked receiving mail box				
Finished aluminum	EA.	1,300	300	1,600
Bronze	"	2,540	300	2,840
Commercial postal accessories for mail chutes				
Letter slot, brass	EA.	120	100	220
Bulk mail slot, brass	"	260	100	360
Mail boxes				
Residential postal accessories				
Letter slot	EA.	98.00	30.25	128
Rural letter box	"	190	76.00	266

10 SPECIALTIES

Specialties	UNIT	MAT.	INST.	TOTAL
10550.10 Postal Specialties *(Cont.)*				
Apartment house, keyed, 3.5" x 4.5" x 16"	EA.	180	20.25	200
Ranch style	"	190	30.25	220
Commercial postal accessories				
Letter box, with combination lock	EA.	140	21.50	162
Key lock	"	150	21.50	172
Mail box, aluminum w/glass front, 4x5				
Horizontal rear load	EA.	230	17.25	247
Vertical front load	"	120	17.25	137
10600.10 Movable Partitions				
Partition, movable, 2-1/2" thick, vinyl-gypsum	S.F.	25.50	3.02	28.52
Enameled steel frame, with 1/4" thick clear glass	"	31.25	3.02	34.27
Door frame and hardware for movable partitions	EA.	870	200	1,070
Cased opening for movable partitions	"	290	100	390
Add for acoustic movable partition	S.F.			1.90
Accordion partition, 12' high				
Vinyl	S.F.	14.00	10.00	24.00
Acoustical	"	16.50	10.00	26.50
Standard office cubicles, 8' high, steel framed				
Baked enamel finish				
100% flush	L.F.	210	15.00	225
75% flush and 25% glass	"	260	16.75	277
50% flush and 50% glass	"	370	16.75	387
100% glass	"	430	20.25	450
Natural hardwood panels				
100% flush	L.F.	230	20.25	250
50% flush and 50% glass	"	300	21.50	322
Plastic laminated panels				
100% flush	L.F.	260	20.25	280
75% flush and 25% glass	"	340	21.50	362
50% flush and 50% glass	"	380	21.50	402
Vinyl covered panels				
100% flush	L.F.	230	20.75	251
75% flush and 25% glass	"	300	22.50	323
50% and 50% glass	"	370	22.50	393
Aluminum framed				
Enameled or anodized aluminum panels				
100% flush	L.F.	180	15.00	195
75% flush and 25% glass	"	170	16.75	187
50% flush and 50% glass	"	160	16.75	177
Vinyl covered panels				
100% flush	L.F.	190	20.75	211
75% flush and 25% glass	"	210	22.50	233
50% flush and 50% glass	"	240	22.50	263
60" high partitions, steel framed				
Enameled panels	L.F.	92.00	13.50	106
Natural hardwood panels, two sides	"	370	14.00	384
Plastic laminated panels	"	200	14.00	214
Vinyl covered panels	"	250	13.50	264
Aluminum framed				
Anodized or baked enamel panels	L.F.	140	13.50	154
Natural hardwood panels	"	400	14.00	414
Plastic laminated panels	"	200	14.00	214

10 SPECIALTIES

10600.10 Movable Partitions (Cont.)

Specialties	UNIT	MAT.	INST.	TOTAL
Vinyl covered panels	L.F.	250	13.50	264
Wire mesh partitions				
Wall panels				
4' x 7'	EA.	190	37.75	228
4' x 8'	"	200	40.25	240
4' x 10'	"	230	46.50	277
Wall filler panels				
1' x 7'	EA.	100	37.75	138
1' x 8'	"	110	40.25	150
1' x 10'	"	140	43.25	183
2' x 7'	"	110	37.75	148
2' x 8'	"	120	40.25	160
2' x 10'	"	150	43.25	193
3' x 7'	"	140	37.75	178
3' x 8'	"	200	40.25	240
3' x 10'	"	220	43.25	263
Ceiling panels				
10' x 2'	EA.	150	86.00	236
10' x 4'	"	220	120	340
Wall panel with service window				
5' wide				
7' high	EA.	590	37.75	628
8' high	"	620	40.25	660
10' high	"	600	43.25	643
Doors				
Sliding				
3' x 7'	EA.	400	150	550
3' x 8'	"	530	170	700
3' x 10'	"	560	240	800
4' x 7'	"	500	170	670
4' x 8'	"	560	240	800
4' x 10'	"	600	300	900
5' x 7'	"	550	240	790
5' x 8'	"	590	300	890
5' x 10'	"	670	300	970
Swing door				
3' x 7'	EA.	330	150	480
4' x 7'	"	360	170	530
Swing door, with 1' transom				
3' x 7'	EA.	480	170	650
4' x 7'	"	540	240	780
Swing door, with 3' transom				
3' x 7'	EA.	600	240	840
4' x 7'	"	630	300	930

10670.10 Shelving

	UNIT	MAT.	INST.	TOTAL
Shelving, enamel, closed side and back, 12" x 36"				
5 shelves	EA.	300	100	400
8 shelves	"	340	130	470
Open				
5 shelves	EA.	160	100	260
8 shelves	"	170	130	300
Metal storage shelving, baked enamel				

10 SPECIALTIES

Specialties	UNIT	MAT.	INST.	TOTAL
10670.10 Shelving (Cont.)				
7 shelf unit, 72" or 84" high				
10" shelf	L.F.	44.25	60.00	104
12" shelf	"	49.50	64.00	114
15" shelf	"	85.00	67.00	152
18" shelf	"	85.00	71.00	156
24" shelf	"	100	76.00	176
30" shelf	"	98.00	81.00	179
36" shelf	"	120	86.00	206
4 shelf unit, 40" high				
10" shelf	L.F.	60.00	50.00	110
12" shelf	"	65.00	55.00	120
15" shelf	"	79.00	60.00	139
18" shelf	"	89.00	64.00	153
24" shelf	"	130	67.00	197
3 shelf unit, 32" high				
10" shelf	L.F.	52.00	30.25	82.25
12" shelf	"	52.00	31.75	83.75
15" shelf	"	57.00	33.50	90.50
18" shelf	"	59.00	35.50	94.50
24" shelf	"	65.00	37.75	103
Single shelf unit, attached to masonry				
10" shelf	L.F.	18.25	10.00	28.25
12" shelf	"	20.00	11.00	31.00
15" shelf	"	20.75	11.50	32.25
18" shelf	"	23.50	12.25	35.75
24" shelf	"	28.50	13.25	41.75
For stainless steel, add to material, 120%				
For attachment to gypsum board, add to labor, 50%				
Built-in wood shelves				
Posts and trimmed plywood	L.F.	4.74	8.63	13.37
Solid clear pine	"	7.47	9.29	16.76
Closet shelf, pine with rod	"	5.07	9.29	14.36
For lumber edge band, add to material	"			2.60
For prefinished shelves, add to material, 225%				
10750.10 Telephone Enclosures				
Enclosure, wall mounted, shelf, 28" x 30" x 15"	EA.	1,950	150	2,100
Directory shelf, stainless steel, 3 binders	"	1,670	100	1,770
10800.10 Bath Accessories				
Ash receiver, wall mounted, aluminum	EA.	160	30.25	190
Grab bar, 1-1/2" dia., stainless steel, wall mounted				
24" long	EA.	54.00	30.25	84.25
36" long	"	61.00	31.75	92.75
42" long	"	68.00	33.50	102
48" long	"	75.00	35.50	111
52" long	"	82.00	37.75	120
1" dia., stainless steel				
12" long	EA.	34.00	26.25	60.25
18" long	"	40.75	27.50	68.25
24" long	"	46.25	30.25	76.50
30" long	"	54.00	31.75	85.75
36" long	"	61.00	33.50	94.50

10 SPECIALTIES

Specialties	UNIT	MAT.	INST.	TOTAL
10800.10 **Bath Accessories** *(Cont.)*				
48" long	EA.	68.00	35.50	104
Hand dryer, surface mounted, 110 volt	"	780	76.00	856
Medicine cabinet, 16 x 22, baked enamel, lighted	"	150	24.25	174
With mirror, lighted	"	220	40.25	260
Mirror, 1/4" plate glass, up to 10 sf	S.F.	12.00	6.04	18.04
Mirror, stainless steel frame				
18"x24"	EA.	91.00	20.25	111
18"x32"	"	100	24.25	124
18"x36"	"	110	30.25	140
24"x30"	"	110	30.25	140
24"x36"	"	120	33.50	154
24"x48"	"	170	50.00	220
24"x60"	"	420	60.00	480
30"x30"	"	360	60.00	420
30"x72"	"	650	76.00	726
48"x72"	"	710	100	810
With shelf, 18"x24"	"	280	24.25	304
Sanitary napkin dispenser, stainless steel	"	680	40.25	720
Shower rod, 1" diameter				
Chrome finish over brass	EA.	240	30.25	270
Stainless steel	"	170	30.25	200
Soap dish, stainless steel, wall mounted	"	150	40.25	190
Toilet tissue dispenser, stainless, wall mounted				
Single roll	EA.	76.00	15.00	91.00
Double roll	"	150	17.25	167
Towel dispenser, stainless steel				
Flush mounted	EA.	290	33.50	324
Surface mounted	"	420	30.25	450
Combination towel and waste receptacle	"	630	40.25	670
Towel bar, stainless steel				
18" long	EA.	94.00	24.25	118
24" long	"	130	27.50	158
30" long	"	130	30.25	160
36" long	"	150	33.50	184
Toothbrush and tumbler holder	"	59.00	20.25	79.25
Waste receptacle, stainless steel, wall mounted	"	480	50.00	530
10900.10 **Wardrobe Specialties**				
Hospital wardrobe units, 24" x 24" x 76", with door				
Baked enameled steel	EA.	3,090	340	3,430
Hardwood	"	1,650	340	1,990
Stainless steel	"	5,160	340	5,500
Plastic laminated	"	1,860	340	2,200
Dormitory wardrobe units, 24" x 76", with door				
Hardwood	EA.	1,740	340	2,080
Plastic laminated	"	1,570	340	1,910
Hat and coat rack				
Single tier				
Baked enameled steel	L.F.	200	15.00	215
Stainless steel	"	220	15.00	235
Aluminum	"	220	15.00	235
Double tier				
Baked enameled steel	L.F.	370	17.25	387

10 SPECIALTIES

Specialties	UNIT	MAT.	INST.	TOTAL
10900.10 Wardrobe Specialties *(Cont.)*				
Stainless steel	L.F.	510	17.25	527
Aluminum	"	440	17.25	457

11 EQUIPMENT

Architectural Equipment	UNIT	MAT.	INST.	TOTAL
11010.10 Maintenance Equipment				
Vacuum cleaning system				
3 valves				
1.5 hp	EA.	1,100	670	1,770
2.5 hp	"	1,330	860	2,190
5 valves	"	2,070	1,210	3,280
7 valves	"	2,760	1,510	4,270
11020.10 Security Equipment				
Bulletproof teller window				
4' x 4'	EA.	2,900	1,010	3,910
5' x 4'	"	3,740	1,210	4,950
Bulletproof partitions				
Up to 12' high, 2.5" thick	S.F.	250	4.02	254
Counter for banks				
Minimum	L.F.	1,020	120	1,140
Maximum	"	4,570	200	4,770
Drive-up window				
Minimum	EA.	6,360	860	7,220
Maximum	"	6,900	2,010	8,910
Night depository				
Minimum	EA.	11,930	860	12,790
Maximum	"	16,970	2,010	18,980
Office safes, 30" x 20" x 20", 1 hr rating	"	4,640	150	4,790
30" x 16" x 15", 2 hr rating	"	2,370	120	2,490
30" x 28" x 20", H&G rating	"	5,500	76.00	5,576
Service windows, pass through painted steel				
24" x 36"	EA.	4,000	600	4,600
48" x 40"	"	6,230	760	6,990
72" x 40"	"	7,830	1,210	9,040
Special doors and windows				
3' x 7' bulletproof door with frame	EA.	7,960	860	8,820
12" x 12" vision panel	"	4,610	430	5,040
Surveillance system				
Minimum	EA.	7,880	1,210	9,090
Maximum	"	14,250	6,040	20,290
Vault door, 3' wide, 6'6" high				
3-1/2" thick	EA.	5,120	7,550	12,670
7" thick	"	8,220	10,070	18,290
10" thick	"	10,270	12,090	22,360
Insulated vault door				
2 hr rating				
32" wide	EA.	5,120	600	5,720
40" wide	"	5,700	640	6,340
4 hr rating				
32" wide	EA.	5,680	670	6,350
40" wide	"	6,610	760	7,370
6 hr rating				
32" wide	EA.	6,630	670	7,300
40" wide	"	7,750	760	8,510
Insulated file room door				
1 hr rating				
32" wide	EA.	5,120	600	5,720
40" wide	"	5,700	670	6,370

11 EQUIPMENT

Architectural Equipment

	UNIT	MAT.	INST.	TOTAL
11060.10 Theater Equipment				
Roll out stage, steel frame, wood floor				
Manual	S.F.	62.00	3.77	65.77
Electric	"	60.00	6.04	66.04
Portable stages				
8" high	S.F.	24.50	3.02	27.52
18" high	"	28.50	3.35	31.85
36" high	"	33.25	3.55	36.80
48" high	"	37.00	3.77	40.77
Band risers				
Minimum	S.F.	64.00	3.02	67.02
Maximum	"	130	3.02	133
Chairs for risers				
Minimum	EA.	830	2.15	832
Maximum	"	1,330	2.15	1,332
11080.10 Police Equipment				
Firing range equipment, rifle				
3 position	EA.	21,220	2,010	23,230
4 position	"	27,180	3,020	30,200
5 position	"	33,150	3,360	36,510
6 position	"	39,780	3,560	43,340
11090.10 Checkroom Equipment				
Motorized checkroom equipment				
No shelf system, 6'4" height				
7'6" length	EA.	5,730	600	6,330
14'6" length	"	5,790	600	6,390
28' length	"	6,980	600	7,580
One shelf, 6'8" height				
7'6" length	EA.	7,020	600	7,620
14'6" length	"	7,210	600	7,810
28' length	"	8,660	600	9,260
Two shelves, 7'5" height				
7'6" length	EA.	8,620	600	9,220
14'6" length	"	10,890	600	11,490
28' length	"	11,120	600	11,720
Three shelves, 8' height				
7'6" length	EA.	8,820	1,210	10,030
14'6" length	"	11,110	1,210	12,320
28' length	"	11,200	1,210	12,410
Four shelves, 8'7" height				
7'6" length	EA.	8,910	1,210	10,120
14'6" length	"	11,370	1,210	12,580
28' length	"	11,630	1,210	12,840
11110.10 Laundry Equipment				
High capacity, heavy duty				
Washer extractors				
135 lb				
Standard	EA.	39,300	500	39,800
Pass through	"	45,160	500	45,660
200 lb				
Standard	EA.	48,270	500	48,770

11 EQUIPMENT

Architectural Equipment

	UNIT	MAT.	INST.	TOTAL
11110.10 Laundry Equipment *(Cont.)*				
Pass through	EA.	58,610	500	59,110
110 lb dryer	"	14,780	500	15,280
Hand operated presser	"	10,610	670	11,280
Mushroom press	"	6,630	670	7,300
Spreader feeders				
2 station	EA.	75,850	670	76,520
4 station	"	89,640	1,210	90,850
Delivery carts				
12 bushel	EA.	380	7.55	388
16 bushel	"	470	8.05	478
18 bushel	"	600	8.63	609
30 bushel	"	860	10.00	870
40 bushel	"	1,030	12.00	1,042
Low capacity				
Pressers				
Air operated	EA.	8,270	240	8,510
Hand operated	"	6,550	240	6,790
Extractor, low capacity	"	5,970	240	6,210
Ironer, 48"	"	4,510	120	4,630
Coin washers				
10 lb capacity	EA.	2,190	120	2,310
20 lb capacity	"	5,170	120	5,290
Coin dryer	"	1,040	76.00	1,116
Coin dry cleaner, 20 lb	"	4,510	240	4,750
11161.10 Loading Dock Equipment				
Dock leveler, 10 ton capacity				
6' x 8'	EA.	6,090	600	6,690
7' x 8'	"	6,990	600	7,590
Bumpers, laminated rubber				
4-1/2" thick				
6" x 14"	EA.	74.00	12.00	86.00
6" x 36"	"	140	13.50	154
10" x 14"	"	100	15.00	115
10" x 24"	"	150	17.25	167
10" x 36"	"	210	20.25	230
12" x 14"	"	120	16.00	136
12" x 24"	"	180	19.00	199
12" x 36"	"	260	22.50	283
6" thick				
10" x 14"	EA.	120	17.25	137
10" x 24"	"	130	20.75	151
10" x 36"	"	270	30.25	300
Extruded rubber bumpers				
T-section, 22" x 22" x 3"	EA.	180	12.00	192
Molded rubber bumpers				
24" x 12" x 3" thick	EA.	99.00	30.25	129
Door seal, 12" x 12", vinyl covered	L.F.	61.00	15.00	76.00
Dock boards, heavy duty, 5' x 5'				
5000 lb				
Minimum	EA.	1,440	500	1,940
Maximum	"	1,590	500	2,090
9000 lb				

11 EQUIPMENT

Architectural Equipment

11161.10 Loading Dock Equipment (Cont.)

	UNIT	MAT.	INST.	TOTAL
Minimum	EA.	1,660	500	2,160
Maximum	"	1,990	550	2,540
15,000 lb	"	2,260	550	2,810
Truck shelters				
Minimum	EA.	1,270	460	1,730
Maximum	"	2,120	860	2,980

11170.10 Waste Handling

	UNIT	MAT.	INST.	TOTAL
Incinerator, electric				
100 lb/hr				
Minimum	EA.	17,900	620	18,520
Maximum	"	30,760	620	31,380
400 lb/hr				
Minimum	EA.	45,080	1,230	46,310
Maximum	"	56,350	1,230	57,580
1000 lb/hr				
Minimum	EA.	106,080	1,870	107,950
Maximum	"	159,120	1,870	160,990
Incinerator, medical-waste				
25 lb/hr, 2-7 x 4-0	EA.	15,250	1,230	16,480
50 lb/hr, 2-11 x 4-11	"	29,570	1,230	30,800
75 lb/hr, 3-8 x 5-0	"	39,780	2,470	42,250
100 lb/hr, 3-8 x 6-0	"	59,670	2,470	62,140
Industrial compactor				
1 cy	EA.	15,650	680	16,330
3 cy	"	24,400	880	25,280
5 cy	"	46,410	1,230	47,640
Trash chutes steel, including sprinklers				
18" dia.	L.F.	120	300	420
24" dia.	"	150	320	470
30" dia.	"	180	340	520
36" dia.	"	220	360	580
Refuse bottom hopper	EA.	1,990	340	2,330

11400.10 Food Service Equipment

	UNIT	MAT.	INST.	TOTAL
Unit kitchens				
30" compact kitchen				
Refrigerator, with range, sink	EA.	1,540	310	1,850
Sink only	"	1,970	210	2,180
Range only	"	1,600	150	1,750
Cabinet for upper wall section	"	400	88.00	488
Stainless shield, for rear wall	"	160	24.75	185
Side wall	"	120	24.75	145
42" compact kitchen				
Refrigerator with range, sink	EA.	1,880	340	2,220
Sink only	"	1,030	310	1,340
Cabinet for upper wall section	"	800	100	900
Stainless shield, for rear wall	"	640	25.75	666
Side wall	"	180	25.75	206
54" compact kitchen				
Refrigerator, oven, range, sink	EA.	2,520	440	2,960
Cabinet for upper wall section	"	1,030	120	1,150
Stainless shield, for				

11 EQUIPMENT

Architectural Equipment

11400.10 Food Service Equipment (Cont.)

	UNIT	MAT.	INST.	TOTAL
Rear wall	EA.	640	28.00	668
Side wall	"	180	28.00	208
60" compact kitchen				
Refrigerator, oven, range, sink	EA.	3,430	440	3,870
Cabinet for upper wall section	"	230	120	350
Stainless shield, for				
Rear wall	EA.	760	28.00	788
Side wall	"	190	28.00	218
72" compact kitchen				
Refrigerator, oven, range, sink	EA.	3,620	510	4,130
Cabinet for upper wall section	"	250	120	370
Stainless shield for				
Rear wall	EA.	820	30.75	851
Side wall	"	190	30.75	221
Bake oven				
Single deck				
Minimum	EA.	3,650	77.00	3,727
Maximum	"	6,970	150	7,120
Double deck				
Minimum	EA.	6,500	100	6,600
Maximum	"	20,300	150	20,450
Triple deck				
Minimum	EA.	23,040	100	23,140
Maximum	"	41,110	210	41,320
Convection type oven, electric, 40" x 45" x 57"				
Minimum	EA.	3,590	77.00	3,667
Maximum	"	6,330	150	6,480
Broiler, without oven, 69" x 26" x 39"				
Minimum	EA.	5,730	77.00	5,807
Maximum	"	9,010	100	9,110
Coffee urns, 10 gallons				
Minimum	EA.	4,360	210	4,570
Maximum	"	4,930	310	5,240
Fryer, with submerger				
Single				
Minimum	EA.	1,790	120	1,910
Maximum	"	4,910	210	5,120
Double				
Minimum	EA.	2,960	150	3,110
Maximum	"	15,880	210	16,090
Griddle, counter				
3' long				
Minimum	EA.	2,450	100	2,550
Maximum	"	5,040	120	5,160
5' long				
Minimum	EA.	5,300	150	5,450
Maximum	"	12,230	210	12,440
Kettles, steam, jacketed				
20 gallons				
Minimum	EA.	11,670	150	11,820
Maximum	"	12,750	310	13,060
40 gallons				
Minimum	EA.	17,900	150	18,050

11 EQUIPMENT

Architectural Equipment

11400.10 Food Service Equipment (Cont.)

	UNIT	MAT.	INST.	TOTAL
Maximum	EA.	25,860	310	26,170
60 gallons				
Minimum	EA.	19,680	150	19,830
Maximum	"	27,850	310	28,160
Range				
Heavy duty, single oven, open top				
Minimum	EA.	7,520	77.00	7,597
Maximum	"	15,840	210	16,050
Fry top				
Minimum	EA.	7,690	77.00	7,767
Maximum	"	10,770	210	10,980
Steamers, electric				
27 kw				
Minimum	EA.	13,660	150	13,810
Maximum	"	24,970	210	25,180
18 kw				
Minimum	EA.	7,520	150	7,670
Maximum	"	17,640	210	17,850
Dishwasher, rack type				
Single tank, 190 racks/hr	EA.	20,020	310	20,330
Double tank				
234 racks/hr	EA.	41,350	340	41,690
265 racks/hr	"	49,420	410	49,830
Dishwasher, automatic 100 meals/hr	"	16,080	210	16,290
Disposals				
100 gal/hr	EA.	1,340	210	1,550
120 gal/hr	"	1,560	210	1,770
250 gal/hr	"	1,840	220	2,060
Exhaust hood for dishwasher, gutter 4 sides				
4'x4'x2'	EA.	3,250	230	3,480
4'x7'x2'	"	4,410	250	4,660
Food preparation machines				
Vertical cutter mixers				
25 quart	EA.	13,260	210	13,470
40 quart	"	17,120	210	17,330
80 quart	"	21,880	310	22,190
130 quart	"	29,170	510	29,680
Choppers				
5 lb	EA.	3,700	150	3,850
16 lb	"	5,830	210	6,040
40 lb	"	5,720	310	6,030
Mixers, floor models				
20 quart	EA.	4,080	77.00	4,157
60 quart	"	20,290	77.00	20,367
80 quart	"	33,150	88.00	33,238
140 quart	"	39,540	120	39,660
Ice cube maker				
50 lb per day				
Minimum	EA.	2,520	620	3,140
Maximum	"	3,710	620	4,330
500 lb per day				
Minimum	EA.	5,970	1,030	7,000
Maximum	"	7,050	1,030	8,080

11 EQUIPMENT

Architectural Equipment

11400.10 Food Service Equipment (Cont.)

	UNIT	MAT.	INST.	TOTAL
Ice flakers				
300 lb per day	EA.	4,370	620	4,990
600 lb per day	"	7,170	1,030	8,200
1000 lb per day	"	8,180	1,370	9,550
2000 lb per day	"	15,780	1,540	17,320
Refrigerated cases				
Dairy products				
Multi deck type	L.F.	1,410	41.00	1,451
For rear sliding doors, add	"			270
Delicatessen case, service deli				
Single deck	L.F.	1,010	310	1,320
Multi deck	"	1,150	390	1,540
Meat case				
Single deck	L.F.	870	360	1,230
Multi deck	"	1,010	390	1,400
Produce case				
Single deck	L.F.	1,000	360	1,360
Multi deck	"	1,080	390	1,470
Bottle coolers				
6' long				
Minimum	EA.	2,860	1,230	4,090
Maximum	"	4,240	1,230	5,470
10' long				
Minimum	EA.	3,710	2,050	5,760
Maximum	"	7,290	2,050	9,340
Frozen food cases				
Chest type	L.F.	770	360	1,130
Reach-in, glass door	"	1,070	390	1,460
Island case, single	"	960	360	1,320
Multi deck	"	1,510	390	1,900
Ice storage bins				
500 lb capacity	EA.	1,600	880	2,480
1000 lb capacity	"	2,390	1,760	4,150

11450.10 Residential Equipment

	UNIT	MAT.	INST.	TOTAL
Compactor, 4 to 1 compaction	EA.	1,600	150	1,750
Dishwasher, built-in				
2 cycles	EA.	790	310	1,100
4 or more cycles	"	2,120	310	2,430
Disposal				
Garbage disposer	EA.	210	210	420
Heaters, electric, built-in				
Ceiling type	EA.	450	210	660
Wall type				
Minimum	EA.	230	150	380
Maximum	"	780	210	990
Hood for range, 2-speed, vented				
30" wide	EA.	620	210	830
42" wide	"	1,150	210	1,360
Ice maker, automatic				
30 lb per day	EA.	2,090	88.00	2,178
50 lb per day	"	2,650	310	2,960
Folding access stairs, disappearing metal stair				

11 EQUIPMENT

Architectural Equipment	UNIT	MAT.	INST.	TOTAL
11450.10 **Residential Equipment** *(Cont.)*				
8' long	EA.	1,090	88.00	1,178
11' long	"	1,140	88.00	1,228
12' long	"	1,220	88.00	1,308
Wood frame, wood stair				
22" x 54" x 8'9" long	EA.	210	62.00	272
25" x 54" x 10' long	"	260	62.00	322
Ranges electric				
Built-in, 30", 1 oven	EA.	2,290	210	2,500
2 oven	"	2,650	210	2,860
Counter top, 4 burner, standard	"	1,330	150	1,480
With grill	"	3,310	150	3,460
Free standing, 21", 1 oven	"	1,190	210	1,400
30", 1 oven	"	2,320	120	2,440
2 oven	"	3,780	120	3,900
Water softener				
30 grains per gallon	EA.	1,300	210	1,510
70 grains per gallon	"	1,630	310	1,940
11470.10 **Darkroom Equipment**				
Dryers				
36" x 25" x 68"	EA.	11,530	330	11,860
48" x 25" x 68"	"	11,920	330	12,250
Processors, film				
Black and white	EA.	18,380	330	18,710
Color negatives	"	20,820	330	21,150
Prints	"	23,870	330	24,200
Transparencies	"	26,190	330	26,520
Sinks with cabinet and/or stand				
5" sink with stand				
24" x 48"	EA.	1,030	170	1,200
32" x 64"	"	1,580	220	1,800
38" x 52"	"	2,220	220	2,440
42" x 132"	"	3,310	330	3,640
48" x 52"	"	2,550	330	2,880
5" sink with cabinet				
24" x 48"	EA.	2,120	170	2,290
32" x 64"	"	2,650	220	2,870
38" x 52"	"	2,720	220	2,940
42" x 132"	"	4,440	330	4,770
48" x 52"	"	3,710	330	4,040
10" sink with stand				
24" x 48"	EA.	1,750	170	1,920
32" x 64"	"	1,860	220	2,080
38" x 52"	"	2,520	220	2,740
10" sink with cabinet				
24" x 48"	EA.	1,920	170	2,090
38" x 52"	"	3,560	220	3,780

11 EQUIPMENT

Architectural Equipment

	UNIT	MAT.	INST.	TOTAL
11480.10 Athletic Equipment				
Basketball backboard				
Fixed	EA.	2,530	760	3,290
Swing-up	"	4,050	1,210	5,260
Portable, hydraulic	"	19,890	300	20,190
Suspended type, standard	"	5,770	1,210	6,980
For glass backboard, add	"			1,670
For electrically operated, add	"			1,950
Bleacher, telescoping, manual				
15 tier, minimum	SEAT	150	12.00	162
Maximum	"	380	12.00	392
20 tier, minimum	"	110	13.50	124
Maximum	"	320	13.50	334
30 tier, minimum	"	86.00	20.25	106
Maximum	"	260	20.25	280
Boxing ring elevated, complete, 22' x 22'	EA.	10,610	8,630	19,240
Gym divider curtain				
Minimum	S.F.	3.44	0.80	4.24
Maximum	"	5.17	0.80	5.97
Scoreboards, single face				
Minimum	EA.	7,330	600	7,930
Maximum	"	39,780	3,020	42,800
Parallel bars				
Minimum	EA.	1,520	600	2,120
Maximum	"	8,130	1,010	9,140
11500.10 Industrial Equipment				
Vehicular paint spray booth, solid back, 14'4" x 9'6"				
24' deep	EA.	9,280	600	9,880
26'6" deep	"	10,610	600	11,210
28'6" deep	"	12,250	600	12,850
Drive through, 14'9" x 9'6"				
24' deep	EA.	9,880	600	10,480
26'6" deep	"	11,910	600	12,510
28'6" deep	"	13,590	600	14,190
Water wash, paint spray booth				
5' x 11'2" x 10'8"	EA.	6,680	600	7,280
6' x 11'2" x 10'8"	"	6,990	600	7,590
8' x 11'2" x 10'8"	"	7,550	600	8,150
10' x 11'2" x 11'2"	"	8,570	600	9,170
12' x 12'2" x 11'2"	"	9,960	600	10,560
14' x 12'2" x 11'2"	"	11,270	600	11,870
16' x 12'2" x 11'2"	"	13,720	600	14,320
20' x 12'2" x 11'2"	"	16,650	600	17,250
Dry type spray booth, with paint arrestors				
5'4" x 7'2" x 6'8"	EA.	4,040	600	4,640
6'4" x 7'2" x 6'8"	"	5,250	600	5,850
8'4" x 7'2" x 9'2"	"	5,950	600	6,550
10'4" x 7'2" x 9'2"	"	7,000	600	7,600
12'4" x 7'6" x 9'2"	"	6,960	600	7,560
14'4" x 7'6" x 9'8"	"	9,420	600	10,020
16'4" x 7'7" x 9'8"	"	10,760	600	11,360
20'4" x 7'7" x 10'8"	"	12,240	600	12,840
Air compressor, electric				

11 EQUIPMENT

Architectural Equipment	UNIT	MAT.	INST.	TOTAL
11500.10 **Industrial Equipment** *(Cont.)*				
1 hp				
115 volt	EA.	1,540	400	1,940
7.5 hp				
115 volt	EA.	4,570	600	5,170
230 volt	"	5,720	600	6,320
Hydraulic lifts				
8,000 lb capacity	EA.	3,160	1,510	4,670
11,000 lb capacity	"	5,650	2,420	8,070
24,000 lb capacity	"	9,960	4,030	13,990
Power tools				
Band saws				
10"	EA.	1,350	50.00	1,400
14"	"	2,020	60.00	2,080
Motorized shaper	"	1,080	46.50	1,127
Motorized lathe	"	1,280	50.00	1,330
Bench saws				
9" saw	EA.	3,360	40.25	3,400
10" saw	"	4,040	43.25	4,083
12" saw	"	4,970	50.00	5,020
Electric grinders				
1/3 hp	EA.	520	24.25	544
1/2 hp	"	540	26.25	566
3/4 hp	"	910	26.25	936
11600.10 **Laboratory Equipment**				
Cabinets, base				
Minimum	L.F.	490	50.00	540
Maximum	"	890	50.00	940
Full storage, 7' high				
Minimum	L.F.	470	50.00	520
Maximum	"	900	50.00	950
Wall				
Minimum	L.F.	170	60.00	230
Maximum	"	300	60.00	360
Counter tops				
Minimum	S.F.	73.00	7.55	80.55
Average	"	86.00	8.63	94.63
Maximum	"	99.00	10.00	109
Tables				
Open underneath	S.F.	160	30.25	190
Doors underneath	"	530	37.75	568
Medical laboratory equipment				
Analyzer				
Chloride	EA.	5,870	30.75	5,901
Blood	"	32,290	51.00	32,341
Bath, water, utility, countertop unit	"	1,210	62.00	1,272
Hot plate, lab, countertop	"	470	56.00	526
Stirrer	"	560	56.00	616
Incubator, anaerobic, 23x23x36"	"	9,960	310	10,270
Dry heat bath	"	1,080	100	1,180
Incinerator, for sterilizing	"	740	6.16	746
Meter, serum protein	"	1,140	7.70	1,148
Ph analog, general purpose	"	1,230	8.80	1,239

11 EQUIPMENT

Architectural Equipment	UNIT	MAT.	INST.	TOTAL
11600.10 Laboratory Equipment *(Cont.)*				
Refrigerator, blood bank, undercounter type 153 litres	EA.	9,420	100	9,520
5.4 cf, undercounter type	"	6,390	100	6,490
Refrigerator/freezer, 4.4 cf, undercounter type	"	1,210	100	1,310
Sealer, impulse, free standing, 20x12x4"	"	670	20.50	691
Timer, electric, 1-60 minutes, bench or wall mounted	"	260	34.25	294
Glassware washer - dryer, undercounter	"	11,150	770	11,920
Balance, torsion suspension, tabletop, 4.5 lb capacity	"	1,470	34.25	1,504
Binocular microscope, with in base illuminator	"	4,470	23.75	4,494
Centrifuge, table model, 19x16x13"	"	1,720	24.75	1,745
Clinical model, with four place head	"	1,850	13.75	1,864
11700.10 Medical Equipment				
Hospital equipment, lights				
Examination, portable	EA.	2,020	51.00	2,071
Meters				
Air flow meter	EA.	110	34.25	144
Oxygen flow meters	"	130	25.75	156
Racks				
40 chart, revolving open frame; mobile caddy	EA.	1,400	51.00	1,451
Scales.				
Clinical, metric with measure rod, 350 lb	EA.	750	56.00	806
Physical therapy				
Chair, hydrotherapy	EA.	780	10.00	790
Diathermy, shortwave, portable, on casters	"	3,360	24.25	3,384
Exercise bicycle, floor standing, 35" x 15"	"	3,270	20.25	3,290
Hydrocollator, 4 pack, portable, 129 x 90 x 160"	"	590	8.63	599
Lamp, infrared, mobile with variable heat control	"	790	46.50	837
Ultra violet, base mounted	"	730	46.50	777
Mirror, posture training, 27" wide and 72" high	"	800	15.00	815
Parallel bars, adjustable	"	3,490	76.00	3,566
Platform mat 10'x6', 1" thick	"	1,120	15.00	1,135
Pulley, duplex, wall mounted	"	1,910	200	2,110
Rack, crutch, wall mounted, 66 x 16 x 13"	"	460	60.00	520
Stimulator, galvanic-faradic, hand held	"	410	4.02	414
Ultrasound, stimulator, portable, 13x13x8"	"	3,500	5.03	3,505
Sandbag set, velcro straps, saddle bag type	"	180	8.63	189
Whirlpool, 85 gallon	"	6,710	300	7,010
65 gallon capacity	"	6,050	300	6,350
Radiology				
Radiographic table, motor driven tilting table	EA.	59,200	6,040	65,240
Fluoroscope image/tv system	"	98,220	12,090	110,310
Processor for washing and drying radiographs				
Water filter unit, 30" x 48-1/2" x 37-1/2"	EA.	130	1,030	1,160
Cassette transfer cabinet	"	2,830	51.00	2,881
Base storage cabinets, sectional design				
With back splash, 24" deep and 35" high	L.F.	670	51.00	721
Wall storage cabinets	"	260	77.00	337
Steam sterilizers				
For heat and moisture stable materials	EA.	5,710	62.00	5,772
For fast drying after sterilization	"	7,400	77.00	7,477
Compact unit	"	2,390	77.00	2,467
Semi-automatic	"	2,830	310	3,140
Floor loading				

11 EQUIPMENT

Architectural Equipment

11700.10 Medical Equipment (Cont.)

	UNIT	MAT.	INST.	TOTAL
Single door	EA.	83,820	510	84,330
Double door	"	91,830	620	92,450
Utensil washer, sanitizer	"	18,430	470	18,900
Automatic washer/sterilizer	"	20,180	1,230	21,410
16 x 16 x 26", including accessories	"	23,200	2,050	25,250
Steam generator, elec., 10 kw to 180 kw	"	38,350	1,230	39,580
Surgical scrub				
Minimum	EA.	2,020	210	2,230
Maximum	"	11,710	210	11,920
Gas sterilizers				
Automatic, free standing, 21x19x29"	EA.	7,200	620	7,820
Surgical tables				
Minimum	EA.	24,940	880	25,820
Maximum	"	30,340	1,230	31,570
Surgical lights, ceiling mounted				
Minimum	EA.	9,960	1,030	10,990
Maximum	"	20,320	1,230	21,550
Water stills				
4 liters/hr	EA.	4,510	210	4,720
8 liters/hr	"	7,200	210	7,410
19 liters/hr	"	14,800	510	15,310
X-ray equipment				
Mobile unit				
Minimum	EA.	13,180	310	13,490
Maximum	"	25,520	620	26,140
Film viewers				
Minimum	EA.	330	100	430
Maximum	"	1,140	210	1,350
Autopsy table				
Minimum	EA.	18,390	620	19,010
Maximum	"	25,990	620	26,610
Incubators				
15 cf	EA.	9,180	310	9,490
29 cf	"	12,440	510	12,950
Infant transport, portable	"	6,910	320	7,230
Beds				
Stretcher, with pad, 30" x 78"	EA.	5,130	150	5,280
Transfer, for patient transport	"	6,040	150	6,190
Headwall				
Aluminum, with back frame and console	EA.	5,250	310	5,560
Hospital ground detection system				
Power ground module	EA.	1,620	180	1,800
Ground slave module	"	690	130	820
Master ground module	"	610	120	730
Remote indicator	"	640	120	760
X-ray indicator	"	1,810	130	1,940
Micro ammeter	"	2,150	150	2,300
Supervisory module	"	1,810	130	1,940
Ground cords	"	170	22.75	193
Hospital isolation monitors, 5 ma				
120v	EA.	3,720	270	3,990
208v	"	3,440	270	3,710
240v	"	3,440	270	3,710

11 EQUIPMENT

Architectural Equipment

11700.10 Medical Equipment *(Cont.)*

	UNIT	MAT.	INST.	TOTAL
Digital clock-timers separate display	EA.	1,510	120	1,630
One display	"	970	120	1,090
Remote control	"	470	96.00	566
Battery pack	"	110	96.00	206
Surgical chronometer clock and 3 timers	"	2,830	190	3,020
Auxilary control	"	770	89.00	859

11700.20 Dental Equipment

	UNIT	MAT.	INST.	TOTAL
Dental care equipment				
Drill console with accessories	EA.	5,380	1,030	6,410
Amalgamator	"	580	30.75	611
Lathe	"	1,330	20.50	1,351
Finish polisher	"	1,750	41.00	1,791
Model trimmer	"	1,070	28.00	1,098
Motor, wall mounted	"	1,210	28.00	1,238
Cleaner, ultrasonic	"	3,240	62.00	3,302
Curing unit, bench mounted	"	4,970	100	5,070
Oral evacuation system, dual pump	"	6,050	77.00	6,127
Sterilizer, table top, self contained	"	2,560	34.25	2,594
Dental lights				
Light, floor or ceiling mounted	EA.	2,170	310	2,480
X-ray unit				
Portable	EA.	5,380	150	5,530
Wall mounted with remote control	"	7,530	510	8,040
Illuminator, single panel	"	770	880	1,650
X-ray film processor	"	9,400	510	9,910
Shield, portable x-ray, lead lined	"	1,750	41.00	1,791

12 FURNISHINGS

Interior	UNIT	MAT.	INST.	TOTAL
12302.10 Casework				
Kitchen base cabinet, standard, 24" deep, 35" high				
12" wide	EA.	230	60.00	290
18" wide	"	260	60.00	320
24" wide	"	340	67.00	407
27" wide	"	380	67.00	447
36" wide	"	460	76.00	536
48" wide	"	550	76.00	626
Drawer base, 24" deep, 35" high				
15" wide	EA.	290	60.00	350
18" wide	"	300	60.00	360
24" wide	"	490	67.00	557
27" wide	"	560	67.00	627
30" wide	"	650	67.00	717
Sink-ready, base cabinet				
30" wide	EA.	300	67.00	367
36" wide	"	320	67.00	387
42" wide	"	350	67.00	417
60" wide	"	410	76.00	486
Corner cabinet, 36" wide	"	570	76.00	646
Wall cabinet, 12" deep, 12" high				
30" wide	EA.	290	60.00	350
36" wide	"	300	60.00	360
15" high				
30" wide	EA.	340	67.00	407
36" wide	"	510	67.00	577
24" high				
30" wide	EA.	380	67.00	447
36" wide	"	390	67.00	457
30" high				
12" wide	EA.	220	76.00	296
18" wide	"	250	76.00	326
24" wide	"	270	76.00	346
27" wide	"	320	76.00	396
30" wide	"	360	86.00	446
36" wide	"	360	86.00	446
Corner cabinet, 30" high				
24" wide	EA.	400	100	500
30" wide	"	480	100	580
36" wide	"	530	100	630
Wardrobe	"	1,060	150	1,210
Vanity with top, laminated plastic				
24" wide	EA.	880	150	1,030
30" wide	"	980	150	1,130
36" wide	"	1,130	200	1,330
48" wide	"	1,260	240	1,500
12390.10 Counter Tops				
Stainless steel, counter top, with backsplash	S.F.	230	15.00	245
Acid-proof, kemrock surface	"	110	10.00	120

12 FURNISHINGS

Interior

	UNIT	MAT.	INST.	TOTAL
12500.10 Window Treatment				
Drapery tracks, wall or ceiling mounted				
Basic traverse rod				
50 to 90"	EA.	59.00	30.25	89.25
84 to 156"	"	78.00	33.50	112
136 to 250"	"	110	33.50	144
165 to 312"	"	170	37.75	208
Traverse rod with stationary curtain rod				
30 to 50"	EA.	88.00	30.25	118
50 to 90"	"	100	30.25	130
84 to 156"	"	140	33.50	174
136 to 250"	"	170	37.75	208
Double traverse rod				
30 to 50"	EA.	100	30.25	130
50 to 84"	"	130	30.25	160
84 to 156"	"	140	33.50	174
136 to 250"	"	180	37.75	218
12510.10 Blinds				
Venetian blinds				
2" slats	S.F.	40.75	1.51	42.26
1" slats	"	43.50	1.51	45.01
12690.40 Floor Mats				
Recessed entrance mat, 3/8" thick, aluminum link	S.F.	59.00	30.25	89.25
Steel, flexible	"	21.25	30.25	51.50

13 SPECIAL

Construction

	UNIT	MAT.	INST.	TOTAL
13056.10 **Vaults**				
Floor safes				
Class C				
1.0 cf	EA.	980	50.00	1,030
1.3 cf	"	1,080	76.00	1,156
1.9 cf	"	1,410	100	1,510
5.2 cf	"	2,880	100	2,980
13121.10 **Pre-engineered Buildings**				
Pre-engineered metal building, 40'x100'				
14' eave height	S.F.	8.42	4.85	13.27
16' eave height	"	9.54	5.59	15.13
20' eave height	"	10.75	7.27	18.02
60'x100'				
14' eave height	S.F.	10.50	4.85	15.35
16' eave height	"	11.75	5.59	17.34
20' eave height	"	13.00	7.27	20.27
80'x100'				
14' eave height	S.F.	8.15	4.85	13.00
16' eave height	"	8.42	5.59	14.01
20' eave height	"	9.41	7.27	16.68
100'x100'				
14' eave height	S.F.	7.95	4.85	12.80
16' eave height	"	8.28	5.59	13.87
20' eave height	"	9.13	7.27	16.40
100'x150'				
14' eave height	S.F.	7.09	4.85	11.94
16' eave height	"	7.35	5.59	12.94
20' eave height	"	7.88	7.27	15.15
120'x150'				
14' eave height	S.F.	7.49	4.85	12.34
16' eave height	"	7.62	5.59	13.21
20' eave height	"	7.95	7.27	15.22
140'x150'				
14' eave height	S.F.	7.09	4.85	11.94
16' eave height	"	7.26	5.59	12.85
20' eave height	"	7.88	7.27	15.15
160'x200'				
14' eave height	S.F.	5.46	4.85	10.31
16' eave height	"	5.63	5.59	11.22
20' eave height	"	5.96	7.27	13.23
200'x200'				
14' eave height	S.F.	4.70	4.85	9.55
16' eave height	"	5.17	5.59	10.76
20' eave height	"	5.50	7.27	12.77
Hollow metal door and frame, 6' x 7'	EA.			1,260
Sectional steel overhead door, manually operated				
8' x 8'	EA.			2,060
12' x 12'	"			2,740
Roll-up steel door, manually operated				
10' x 10'	EA.			1,600
12' x 12'	"			2,870
For gravity ridge ventilator with birdscreen	"			690
9" throat x 10'	"			750

13 SPECIAL

Construction	UNIT	MAT.	INST.	TOTAL
13121.10 **Pre-engineered Buildings** *(Cont.)*				
12" throat x 10'	EA.			910
For 20" rotary vent with damper	"			340
For 4' x 3' fixed louver	"			250
For 4' x 3' aluminum sliding window	"			210
For 3' x 9' fiberglass panels	"			160
Liner panel, 26 ga, painted steel	S.F.	3.02	1.67	4.69
Wall panel insulated, 26 ga. steel, foam core	"	9.48	1.67	11.15
Roof panel, 26 ga. painted steel	"	2.85	0.95	3.80
Plastic (sky light)	"	6.43	0.95	7.38
Insulation, 3-1/2" thick blanket, R11	"	1.92	0.44	2.36
13152.10 **Swimming Pool Equipment**				
Diving boards				
14' long				
Aluminum	EA.	4,910	260	5,170
Fiberglass	"	3,710	260	3,970
Ladders, heavy duty				
2 steps				
Minimum	EA.	1,160	95.00	1,255
Maximum	"	1,820	95.00	1,915
4 steps				
Minimum	EA.	1,240	120	1,360
Maximum	"	1,990	120	2,110
Lifeguard chair				
Minimum	EA.	3,300	470	3,770
Maximum	"	5,110	470	5,580
Lights, underwater				
12 volt, with transformer, 100 watt				
Incandescent	EA.	240	120	360
Halogen	"	210	120	330
LED	"	660	120	780
110 volt				
Minimum	EA.	1,110	120	1,230
Maximum	"	2,690	120	2,810
Ground fault interrupter for 110 volt, each light	"	240	39.50	280
Pool cover				
Reinforced polyethylene	S.F.	2.38	3.64	6.02
Vinyl water tube				
Minimum	S.F.	1.45	3.64	5.09
Maximum	"	2.18	3.64	5.82
Slides with water tube				
Minimum	EA.	1,160	400	1,560
Maximum	"	24,890	400	25,290
13200.10 **Storage Tanks**				
Oil storage tank, underground, single wall, no excv.				
Steel				
500 gals	EA.	4,040	310	4,350
1,000 gals	"	5,470	410	5,880
4,000 gals	"	8,380	820	9,200
5,000 gals	"	9,820	1,230	11,050
10,000 gals	"	17,490	2,460	19,950
Fiberglass, double wall				

13 SPECIAL

Construction

13200.10 Storage Tanks *(Cont.)*

	UNIT	MAT.	INST.	TOTAL
550 gals	EA.	11,370	410	11,780
1,000 gals	"	14,620	410	15,030
2,000 gals	"	15,880	620	16,500
4,000 gals	"	18,430	1,230	19,660
6,000 gals	"	20,120	1,640	21,760
8,000 gals	"	22,570	2,460	25,030
10,000 gals	"	23,140	3,080	26,220
12,000 gals	"	25,090	4,100	29,190
15,000 gals	"	26,910	5,470	32,380
20,000 gals	"	32,960	6,150	39,110
Above ground				
Steel, single wall				
275 gals	EA.	2,290	250	2,540
500 gals	"	5,720	410	6,130
1,000 gals	"	7,800	490	8,290
1,500 gals	"	9,960	620	10,580
2,000 gals	"	12,310	820	13,130
5,000 gals	"	14,460	1,230	15,690
Fill cap	"	140	66.00	206
Vent cap	"	140	66.00	206
Level indicator	"	220	66.00	286

Hazardous Waste

13280.10 Asbestos Removal

	UNIT	MAT.	INST.	TOTAL
Enclosure using wood studs & poly, install & remove	S.F.	600	1.18	601
Trailer (change room)	DAY			130
Disposal suits (4 suits per man day)	"			53.00
Type C respirator mask, includes hose & filters, per man	"			26.50
Respirator mask & filter, light contamination	"			10.50
Air monitoring test, 12 tests per day				
Off job testing	DAY			1,390
On the job testing	"			1,860
Asbestos vacuum with attachments	EA.			810
Hydraspray piston pump	"			1,080
Negative air pressure system	"			1,080
Grade D breathing air equipment	"			2,420
Glove bag, 44" x 60" x 6 mil plastic	"			7.80
40 CY asbestos dumpster				
Weekly rental	EA.			910
Pick up/delivery	"			420
Asbestos dump fee	"			260

13 SPECIAL

Hazardous Waste	UNIT	MAT.	INST.	TOTAL
13280.12 **Duct Insulation Removal**				
Remove duct insulation, duct size				
6" x 12"	L.F.	270	2.63	273
x 18"	"	200	3.64	204
x 24"	"	130	5.26	135
8" x 12"	"	180	3.95	184
x 18"	"	170	4.30	174
x 24"	"	120	5.92	126
12" x 12"	"	180	3.95	184
x 18"	"	130	5.26	135
x 24"	"	110	6.77	117
13280.15 **Pipe Insulation Removal**				
Removal, asbestos insulation				
2" thick, pipe				
1" to 3" dia.	L.F.	180	3.95	184
4" to 6" dia.	"	160	4.51	165
3" thick				
7" to 8" dia.	L.F.	150	4.74	155
9" to 10" dia.	"	150	4.98	155
11" to 12" dia.	"	130	5.26	135
13" to 14" dia.	"	130	5.57	136
15" to 18" dia.	"	120	5.92	126

14 CONVEYING

Elevators

14210.10 Elevators

	UNIT	MAT.	INST.	TOTAL
Passenger elevators, electric, geared				
Based on a shaft of 6 stops and 6 openings				
50 fpm, 2000 lb	EA.	144,190	2,460	146,650
100 fpm, 2000 lb	"	149,530	2,730	152,260
150 fpm				
2000 lb	EA.	164,970	3,080	168,050
3000 lb	"	207,790	3,510	211,300
4000 lb	"	216,220	4,100	220,320
200 fpm				
2500 lb	EA.	199,370	3,510	202,880
3000 lb	"	204,980	3,780	208,760
4000 lb	"	216,220	4,100	220,320
250 fpm				
2500 lb	EA.	206,390	3,510	209,900
3000 lb	"	217,830	3,780	221,610
4000 lb	"	222,880	4,100	226,980
300 fpm				
2500 lb	EA.	204,420	3,510	207,930
3000 lb	"	216,220	3,780	220,000
4000 lb	"	220,010	2,460	222,470
For each additional; 50 fpm, add per stop, $3000				
500 lb, add per stop, $4000				
Opening, add per stop, $4500				
Stop, add per stop, $4000				
Bonderized steel door, add per opening, $150				
Colored aluminum door, add per opening, $850				
Stainless steel door, add per opening, $600				
Cast bronze door, add per opening, $1100				
Two speed door, add per opening, $360				
Bi-parting door, add per opening, $850				
Custom cab interior add, $4800				
Based on a shaft of 8 stops and 8 openings				
300 fpm				
3000 lb	EA.	247,000	4,920	251,920
3500 lb	"	250,900	4,920	255,820
4000 lb	"	263,250	5,470	268,720
5000 lb	"	292,500	5,860	298,360
400 fpm				
3000 lb	EA.	258,770	4,920	263,690
3500 lb	"	263,250	4,920	268,170
4000 lb	"	281,450	5,470	286,920
5000 lb	"	320,450	5,860	326,310
600 fpm				
3000 lb	EA.	364,460	5,470	369,930
3500 lb	"	373,100	5,860	378,960
4000 lb	"	377,000	6,000	383,000
5000 lb	"	387,400	6,150	393,550
800 fpm				
3000 lb	EA.	432,250	5,470	437,720
3500 lb	"	436,800	5,860	442,660
4000 lb	"	442,000	6,000	448,000
5000 lb	"	444,930	6,150	451,080
For each additional; 100 fpm add per stop, $13,000				

14 CONVEYING

Elevators

14210.10 Elevators *(Cont.)*

	UNIT	MAT.	INST.	TOTAL
500 lb, add per stop, $6500				
Opening add per stop, $12,000				
Stop add per stop, $4800				
Bypass floor, add per each, $2000				
Bonderized steel door, add per opening, $150				
Colored aluminum door, add per opening, $900				
Stainless steel door, add per opening, $600				
Cast bronze door, add per opening, $600				
Two speed bi-parting door, add per opening, $1000				
Custom cab interior, add $5000				
Hydraulic, based on a shaft of 3 stops, 3 openings				
50 fpm				
2000 lb	EA.	92,560	2,050	94,610
2500 lb	"	98,980	2,050	101,030
3000 lb	"	104,450	2,140	106,590
100 fpm				
2000 lb	EA.	101,400	2,050	103,450
2500 lb	"	106,930	2,140	109,070
3000 lb	"	114,400	2,240	116,640
150 fpm				
2000 lb	EA.	109,850	2,050	111,900
2500 lb	"	120,250	2,140	122,390
3000 lb	"	128,700	2,340	131,040
For each additional; 50 fpm add per stop, $3500				
500 lb, add per stop, $3500				
Opening, add, $4200				
Stop, add per stop, $5300				
Bonderized steel door, add per opening, $400				
Colored aluminum door, add per opening, $1500				
Stainless steel door, add per opening, $650				
Cast bronze door, add per opening, $1200				
Two speed door, add per opening, $400				
Bi-parting door, add per opening, $900				
Custom cab interior, add per cab, $5000				
Small elevators, 4 to 6 passenger capacity				
Electric, push				
2 stops	EA.	33,800	2,050	35,850
3 stops	"	42,250	2,240	44,490
4 stops	"	48,100	2,460	50,560
Freight elevators, electric				
Based on a shaft of 6 stops and 6 openings				
50 fpm				
3500 lb	EA.	254,800	2,730	257,530
4000 lb	"	255,450	2,730	258,180
5000 lb	"	260,000	3,080	263,080
100 fpm				
3500 lb	EA.	266,500	3,080	269,580
4000 lb	"	270,400	3,080	273,480
5000 lb	"	262,600	3,510	266,110
200 fpm				
3500 lb	EA.	265,200	3,510	268,710
4000 lb	"	266,500	3,510	270,010
5000 lb	"	269,100	4,100	273,200

14 CONVEYING

Elevators

14210.10 Elevators (Cont.)

	UNIT	MAT.	INST.	TOTAL
For elevator with manual door, deduct 15%				
For variable voltage control, add 20%				
Based on shaft of 8 stops and 8 openings				
100 fpm				
4000 lb	EA.	269,100	3,080	272,180
6000 lb	"	258,700	3,150	261,850
8000 lb	"	263,250	3,320	266,570
150 fpm				
4000 lb	EA.	260,000	3,510	263,510
6000 lb	"	260,000	3,620	263,620
8000 lb	"	267,800	3,840	271,640
200 fpm				
4000 lb	EA.	261,300	4,100	265,400
6000 lb	"	267,800	4,240	272,040
8000 lb	"	279,500	4,470	283,970
For each additional; 50 fpm, add per stop, $2000				
500 lb, add per stop, $600				
Opening, add per stop, $7000				
Stop, add per stop, $5500				
For variable voltage, add 20%				
Hydraulic, based on 3 stops and 3 openings				
50 fpm				
3000 lb	EA.	109,200	1,760	110,960
4000 lb	"	119,600	1,820	121,420
6000 lb	"	139,100	1,890	140,990
100 fpm				
3000 lb	EA.	123,500	1,760	125,260
4000 lb	"	130,000	1,820	131,820
6000 lb	"	153,400	1,890	155,290
150 fpm				
3000 lb	EA.	135,200	1,760	136,960
4000 lb	"	145,600	1,820	147,420
6000 lb	"	166,400	1,890	168,290
For each additional; 50 fpm, add per stop, $2000				
500 lb, add per stop, $600				
Opening, add per stop, $5500				
Stop, add per stop, $5500				
For elevator with manual door deduct from total, 15%				

14300.10 Escalators

	UNIT	MAT.	INST.	TOTAL
Escalators				
32" wide, floor to floor				
12' high	EA.	181,040	4,100	185,140
15' high	"	197,800	4,920	202,720
18' high	"	213,030	6,150	219,180
22' high	"	210,810	8,200	219,010
25' high	"	239,550	9,840	249,390
48" wide				
12' high	EA.	201,700	4,240	205,940
15' high	"	220,130	5,130	225,260
18' high	"	236,470	6,470	242,940
22' high	"	264,710	8,790	273,500
25' high	"	282,650	9,840	292,490

14 CONVEYING

Lifts

	UNIT	MAT.	INST.	TOTAL
14410.10 Personnel Lifts				
Electrically operated, 1 or 2 person lift				
With attached foot platforms				
3 stops	EA.			12,090
5 stops	"			18,850
7 stops	"			21,970
For each additional stop, add $1250				
Residential stair climber, per story	EA.	5,690	510	6,200
curved	"	12,030	620	12,650
14410.20 Wheelchair Lifts				
600 lb, Residential	EA.	6,760	620	7,380
Commercial	"	15,990	620	16,610
14450.10 Vehicle Lifts				
Automotive hoist, one post, semi-hydraulic, 8,000 lb	EA.	4,100	2,460	6,560
Full hydraulic, 8,000 lb	"	4,230	2,460	6,690
2 post, semi-hydraulic, 10,000 lb	"	4,420	3,510	7,930
Full hydraulic				
10,000 lb	EA.	5,200	3,510	8,710
13,000 lb	"	6,500	6,150	12,650
18,500 lb	"	10,390	6,150	16,540
24,000 lb	"	14,630	6,150	20,780
26,000 lb	"	14,230	6,150	20,380
Pneumatic hoist, fully hydraulic				
11,000 lb	EA.	7,020	8,200	15,220
24,000 lb	"	12,680	8,200	20,880

Material Handling

	UNIT	MAT.	INST.	TOTAL
14560.10 Chutes				
Linen chutes, stainless steel, with supports				
18" dia.	L.F.	170	4.77	175
24" dia.	"	210	5.14	215
30" dia.	"	230	5.57	236
Hopper	EA.	2,730	44.50	2,775
Skylight	"	1,660	67.00	1,727
Sprinkler unit at top	"	620	74.00	694
For galvanized metal, deduct from material cost, 35%				
For aluminum, deduct from material cost, 25%				
14580.10 Pneumatic Systems				
Pneumatic message tube system				
Average, 20 station job				
3" round system	E.A.	43,990	5,600	49,590
4" round system	"	55,570	6,160	61,730
6" round system	"	95,340	6,850	102,190

14 CONVEYING

Material Handling

14580.10 Pneumatic Systems (Cont.)

	UNIT	MAT.	INST.	TOTAL
4" x 7" oval system	E.A.	100,430	12,330	112,760
Trash and linen tube system				
10 stations	EA.	29,340	12,300	41,640
15 stations	"	36,770	16,400	53,170
20 stations	"	49,010	18,920	67,930
30 stations	"	63,750	22,360	86,110

Hoists And Cranes

14600.10 Industrial Hoists

	UNIT	MAT.	INST.	TOTAL
Industrial hoists, electric, light to medium duty				
500 lb	EA.	8,870	310	9,180
1000 lb	"	9,350	320	9,670
2000 lb	"	9,760	340	10,100
3000 lb	"	10,100	360	10,460
4000 lb	"	10,650	390	11,040
5000 lb	"	12,760	410	13,170
6000 lb	"	14,470	430	14,900
7500 lb	"	16,380	440	16,820
10,000 lb	"	41,290	460	41,750
15,000 lb	"	51,800	470	52,270
20,000 lb	"	61,290	510	61,800
25,000 lb	"	63,950	560	64,510
30,000 lb	"	66,750	620	67,370
Heavy duty				
500 lb	EA.	14,390	310	14,700
1000 lb	"	20,130	320	20,450
2000 lb	"	22,250	340	22,590
3000 lb	"	23,070	360	23,430
4000 lb	"	24,090	390	24,480
5000 lb	"	24,570	410	24,980
6000 lb	"	26,620	430	27,050
7500 lb	"	30,580	440	31,020
10,000 lb	"	32,620	460	33,080
15,000 lb	"	39,590	470	40,060
20,000 lb	"	48,050	510	48,560
25,000 lb	"	53,920	560	54,480
30,000 lb	"	59,920	620	60,540
Air powered hoists				
500 lb	EA.	8,330	310	8,640
1000 lb	"	8,740	310	9,050
2000 lb	"	8,870	320	9,190
4000 lb	"	9,690	360	10,050
6000 lb	"	10,440	470	10,910
Overhead traveling bridge crane				
Single girder, 20' span				

14 CONVEYING

Hoists And Cranes

14600.10 Industrial Hoists (Cont.)

	UNIT	MAT.	INST.	TOTAL
3 ton	EA.	22,520	1,230	23,750
5 ton	"	25,870	1,230	27,100
7.5 ton	"	32,490	1,230	33,720
10 ton	"	32,760	1,540	34,300
15 ton	"	40,400	1,540	41,940
30' span				
3 ton	EA.	27,850	1,230	29,080
5 ton	"	33,720	1,230	34,950
10 ton	"	46,960	1,540	48,500
15 ton	"	53,100	1,540	54,640
Double girder, 40' span				
3 ton	EA.	41,910	2,730	44,640
5 ton	"	46,140	2,730	48,870
7.5 ton	"	45,660	2,730	48,390
10 ton	"	57,260	3,510	60,770
15 ton	"	62,930	3,510	66,440
25 ton	"	109,000	3,510	112,510
50' span				
3 ton	EA.	52,140	2,730	54,870
5 ton	"	53,240	2,730	55,970
7.5 ton	"	54,740	2,730	57,470
10 ton	"	59,920	3,510	63,430
15 ton	"	73,710	3,510	77,220
25 ton	"	97,260	3,510	100,770

14650.10 Jib Cranes

	UNIT	MAT.	INST.	TOTAL
Self supporting, swinging 8' boom, 200 deg rotation				
1000 lb	EA.	4,190	560	4,750
2000 lb	"	4,600	560	5,160
3000 lb	"	4,930	1,110	6,040
4000 lb	"	5,410	1,110	6,520
6000 lb	"	6,560	1,110	7,670
10,000 lb	"	9,260	1,110	10,370
Wall mounted, 180 deg rotation				
2000 lb	EA.	2,230	560	2,790
3000 lb	"	2,700	560	3,260
4000 lb	"	1,690	1,110	2,800
6000 lb	"	3,380	1,110	4,490
10,000 lb	"	6,420	1,110	7,530

15 MECHANICAL

Basic Materials

	UNIT	MAT.	INST.	TOTAL
15100.10 Specialties				
Wall penetration				
Concrete wall, 6" thick				
2" dia.	EA.		15.75	15.75
4" dia.	"		23.75	23.75
8" dia.	"		33.75	33.75
12" thick				
2" dia.	EA.		21.50	21.50
4" dia.	"		33.75	33.75
8" dia.	"		53.00	53.00
Non-destructive testing, piping systems				
X-ray of welds				
3" dia. pipe	EA.	15.50	66.00	81.50
4" dia. pipe	"	20.75	66.00	86.75
6" dia. pipe	"	20.75	66.00	86.75
8" dia. pipe	"	20.75	83.00	104
10" dia. pipe	"	27.25	83.00	110
Liquid penetration of welds				
2" dia. pipe	EA.	3.90	41.50	45.40
3" dia. pipe	"	3.90	41.50	45.40
4" dia. pipe	"	3.90	41.50	45.40
6" dia. pipe	"	3.90	41.50	45.40
8" dia. pipe	"	5.85	41.50	47.35
10" dia. pipe	"	5.85	41.50	47.35
15120.10 Backflow Preventers				
Backflow preventer, flanged, cast iron, with valves				
3" pipe	EA.	3,740	330	4,070
4" pipe	"	4,810	370	5,180
6" pipe	"	8,220	550	8,770
8" pipe	"	11,380	660	12,040
Threaded				
3/4" pipe	EA.	770	41.50	812
2" pipe	"	1,350	66.00	1,416
Reduced pressure assembly, bronze, threaded				
3/4"	EA.	720	41.50	762
1"	"	750	47.50	798
1-1/4"	"	1,080	55.00	1,135
1-1/2"	"	1,110	66.00	1,176
15140.10 Pipe Hangers, Heavy				
Hangers				
1/2" pipe, clevis pipe hanger				
Black steel	EA.	2.13	22.25	24.38
Galvanized	"	3.23	22.25	25.48
U bolt	"	2.00	6.64	8.64
3/4" pipe, clevis pipe hanger				
Black steel	EA.	2.20	22.25	24.45
Galvanized	"	3.30	22.25	25.55
U bolt	"	2.14	6.64	8.78
1" pipe, clevis pipe hanger				
Black steel	EA.	2.26	22.25	24.51
Galvanized	"	3.54	22.25	25.79
U bolt	"	2.35	6.64	8.99

15 MECHANICAL

Basic Materials

15140.10 Pipe Hangers, Heavy (Cont.)

Description	UNIT	MAT.	INST.	TOTAL
1-1/4" pipe, clevis pipe hanger				
Black steel	EA.	2.44	22.25	24.69
Galvanized	"	3.91	22.25	26.16
U bolt	"	2.35	6.64	8.99
1-1/2" pipe, clevis pipe hanger				
Black steel	EA.	2.56	22.25	24.81
Galvanized	"	4.21	22.25	26.46
U bolt	"	2.43	6.64	9.07
2" pipe, clevis pipe hanger				
Black steel	EA.	3.05	22.25	25.30
Galvanized	"	5.01	22.25	27.26
Adjustable pipe roll stand	"	220	83.00	303
U bolt	"	2.64	6.64	9.28
Adjustable roller hanger	"	14.75	22.25	37.00
2-1/2" pipe, clevis pipe hanger				
Black steel	EA.	4.82	22.25	27.07
Galvanized	"	8.67	22.25	30.92
Adjustable pipe roll stand	"	220	22.25	242
Adjustable roller hanger	"	16.00	22.25	38.25
3" pipe, clevis pipe hanger				
Black steel	EA.	5.98	22.25	28.23
Galvanized	"	10.50	22.25	32.75
Adjustable pipe roll stand	"	220	22.25	242
U bolt	"	4.07	6.64	10.71
Adjustable roller hanger	"	16.25	22.25	38.50
3-1/2" pipe, clevis pipe hanger				
Black steel	EA.	6.35	22.25	28.60
Galvanized	"	11.25	22.25	33.50
Adjustable pipe roll stand	"	220	22.25	242
U bolt	"	4.29	6.64	10.93
Adjustable roller hanger	"	18.25	22.25	40.50
4" pipe, clevis pipe hanger				
Black steel	EA.	7.33	22.25	29.58
Galvanized	"	13.00	22.25	35.25
Adjustable pipe roll stand	"	250	22.25	272
U bolt	"	4.36	6.64	11.00
Adjustable roller hanger	"	18.50	22.25	40.75
5" pipe, clevis pipe hanger				
Black steel	EA.	10.00	26.50	36.50
Galvanized	"	18.25	26.50	44.75
Adjustable pipe roll stand	"	250	26.50	277
U bolt	"	4.53	6.64	11.17
Adjustable roller hanger	"	26.50	26.50	53.00
6" pipe, clevis pipe hanger				
Black steel	EA.	11.75	26.50	38.25
Galvanized	"	23.75	26.50	50.25
Adjustable pipe roll stand	"	250	26.50	277
U bolt	"	9.00	8.30	17.30
Adjustable roller hanger	"	38.75	26.50	65.25
8" pipe, clevis pipe hanger				
Black steel	EA.	19.25	26.50	45.75
Galvanized	"	36.00	26.50	62.50
Adjustable pipe roll stand	"	360	26.50	387

15 MECHANICAL

Basic Materials

15140.10 Pipe Hangers, Heavy (Cont.)

	UNIT	MAT.	INST.	TOTAL
U bolt	EA.	11.00	8.30	19.30
Adjustable roller hanger	"	52.00	26.50	78.50
10" clevis pipe hanger				
Black steel	EA.	34.75	26.50	61.25
Galvanized	"	48.75	26.50	75.25
Adjustable pipe roll stand	"	360	26.50	387
Adjustable roller hanger	"	65.00	26.50	91.50
12" pipe, clevis pipe hanger				
Black steel	EA.	41.50	26.50	68.00
Galvanized	"	61.00	26.50	87.50
Adjustable pipe roll stand	"	620	26.50	647
Adjustable roller hanger	"	68.00	26.50	94.50
14" pipe, clevis pipe hanger				
Black steel	EA.	66.00	30.25	96.25
Galvanized	"	93.00	30.25	123
Threaded rod, galvanized				
3/8"	L.F.			0.71
1/2"	"			1.43
5/8"	"			2.14
3/4"	"			3.86
7/8"	"			5.72
1"	"			8.58
Hex nuts, galvanized				
3/8"	EA.			0.26
1/2"	"			0.55
5/8"	"			1.19
3/4"	"			1.58
7/8"	"			2.50
1"	"			3.49
C-clamp, steel, with lock nut				
3/8"	EA.	3.12	8.30	11.42
1/2"	"	3.49	8.30	11.79
5/8"	"	5.83	8.30	14.13
3/4"	"	7.95	8.30	16.25
7/8"	"	15.00	8.30	23.30
Angle support, medium, welded steel				
12"x18"	EA.	140	66.00	206
18"x24"	"	170	66.00	236
24"x30"	"	220	66.00	286
Heavy, welded steel				
12"x18"	EA.	200	66.00	266
18"x24"	"	280	66.00	346
24"x30"	"	320	66.00	386

15140.11 Pipe Hangers, Light

	UNIT	MAT.	INST.	TOTAL
A band, black iron				
1/2"	EA.	1.18	4.74	5.92
1"	"	1.27	4.91	6.18
1-1/4"	"	1.41	5.10	6.51
1-1/2"	"	1.48	5.53	7.01
2"	"	1.56	6.03	7.59
2-1/2"	"	2.34	6.64	8.98
3"	"	2.84	7.37	10.21

15 MECHANICAL

Basic Materials	UNIT	MAT.	INST.	TOTAL
15140.11 Pipe Hangers, Light *(Cont.)*				
4"	EA.	3.74	8.30	12.04
5"	"	3.95	8.85	12.80
6"	"	6.82	9.48	16.30
8"	"	11.00	11.00	22.00
Copper				
1/2"	EA.	1.92	4.74	6.66
3/4"	"	2.23	4.91	7.14
1"	"	2.23	4.91	7.14
1-1/4"	"	2.39	5.10	7.49
1-1/2"	"	2.57	5.53	8.10
2"	"	2.73	6.03	8.76
2-1/2"	"	5.49	6.64	12.13
3"	"	5.73	7.37	13.10
4"	"	6.31	8.30	14.61
Black riser friction hangers				
3/4"	EA.	5.66	5.53	11.19
1"	"	5.74	5.77	11.51
1-1/4"	"	7.15	6.03	13.18
1-1/2"	"	7.78	6.32	14.10
2"	"	7.94	6.64	14.58
2-1/2"	"	8.57	7.37	15.94
3"	"	8.80	8.30	17.10
4"	"	11.25	9.48	20.73
5"	"	16.50	10.25	26.75
6"	"	18.75	11.00	29.75
8"	"	32.25	12.00	44.25
10"	"	41.25	13.25	54.50
Short pattern black riser clamps				
1-1/2"	EA.	6.37	6.03	12.40
2"	"	6.68	6.32	13.00
3"	"	7.31	6.64	13.95
4"	"	8.41	7.37	15.78
Copper riser friction hanger				
1/2"	EA.	8.25	5.10	13.35
3/4"	"	8.49	5.31	13.80
1"	"	8.65	5.53	14.18
1-1/4"	"	10.75	5.77	16.52
1-1/2"	"	11.75	6.03	17.78
2"	"	12.00	6.32	18.32
2-1/2"	"	13.00	6.64	19.64
3"	"	13.25	6.64	19.89
4"	"	17.00	7.37	24.37
Auto grip hangers, galvanized				
1/2"	EA.	1.18	4.74	5.92
3/4"	"	1.23	5.10	6.33
1"	"	1.40	5.31	6.71
1-1/4"	"	1.84	5.53	7.37
1-1/2"	"	1.91	5.77	7.68
2"	"	2.34	6.03	8.37
2-1/2"	"	3.90	6.32	10.22
3"	"	4.61	6.64	11.25
4"	"	5.72	7.37	13.09
Copper				

15 MECHANICAL

Basic Materials

15140.11 Pipe Hangers, Light *(Cont.)*

	UNIT	MAT.	INST.	TOTAL
1/2"	EA.	1.95	4.74	6.69
3/4"	"	2.01	5.10	7.11
1"	"	2.27	5.31	7.58
1-1/4"	"	2.99	5.53	8.52
1-1/2"	"	3.05	5.77	8.82
2"	"	3.77	6.03	9.80
2-1/2"	"	6.24	6.32	12.56
3"	"	7.41	6.64	14.05
4"	"	9.16	7.37	16.53
Split rings (F&M), galvanized				
3/8"	EA.	3.30	4.74	8.04
1/2"	"	3.88	5.10	8.98
3/4"	"	4.29	5.31	9.60
1"	"	5.36	5.53	10.89
1-1/4"	"	5.69	5.77	11.46
1-1/2"	"	6.68	6.03	12.71
2"	"	7.68	6.32	14.00
2-1/2"	"	14.75	6.64	21.39
3"	"	18.75	6.98	25.73
4"	"	19.50	7.37	26.87
Copper				
1/4"	EA.	4.87	4.74	9.61
3/8"	"	4.95	4.74	9.69
1/2"	"	5.86	5.10	10.96
1"	"	8.09	5.53	13.62
1-1/4"	"	8.58	5.77	14.35
1-1/2"	"	10.00	6.03	16.03
2"	"	11.50	6.32	17.82
2-1/2"	"	22.25	6.64	28.89
3"	"	28.00	6.98	34.98
4"	"	29.25	7.37	36.62
F&M plates, galvanized				
3/8"	EA.	2.40	4.74	7.14
1/2"	"	3.15	4.91	8.06
Copper				
3/8"	EA.	3.60	4.74	8.34
1/2"	"	4.72	4.91	9.63
2 hole clips, galvanized				
3/4"	EA.	0.31	4.42	4.73
1"	"	0.35	4.57	4.92
1-1/4"	"	0.45	4.74	5.19
1-1/2"	"	0.55	4.91	5.46
2"	"	0.72	5.10	5.82
2-1/2"	"	1.31	5.31	6.62
3"	"	1.91	5.53	7.44
4"	"	4.09	6.03	10.12
Perforated strap				
3/4"				
Galvanized, 20 ga.	L.F.	0.48	3.32	3.80
Copper, 22 ga.	"	0.75	3.32	4.07
Threaded rod-couplings				
1/4"	EA.	1.71	4.15	5.86
3/4"	"	1.80	4.42	6.22

15 MECHANICAL

Basic Materials

15140.11 Pipe Hangers, Light *(Cont.)*

	UNIT	MAT.	INST.	TOTAL
1/2"	EA.	2.04	4.74	6.78
5/8"	"	3.13	5.10	8.23
Reducing rod coupling, 1/2" x 3/8"	"	2.84	4.74	7.58
C-clamps				
3/4"	EA.	2.45	6.64	9.09
Top beam clamp				
3/8"	EA.	3.77	5.53	9.30
1/2"	"	4.68	6.03	10.71
Side beam connector				
3/8"	EA.	1.63	5.53	7.16
1/2"	"	3.61	6.03	9.64
Hex nuts, heavy				
1"	EA.			4.16
Heavy washers				
3/8"	EA.			0.11
1/2"	"			0.28
5/8"	"			0.59
3/4"	"			1.16
Lag rod, 3/8" x				
4"	EA.			0.54
4-1/2"	"			0.54
6"	"			0.57
8"	"			1.01
10"	"			1.18
12"	"			1.45
18"	"			1.97
Drive screws				
1-1/2" x 12"	EA.			0.31
2" x 12"	"			0.49
Wood screws				
3/4"				
#8	EA.			0.06
1"				
#8	EA.			0.10
#10	"			0.10
#12	"			0.15
1-1/4"				
#8	EA.			0.10
#10	"			0.11
#12	"			0.15
1-1/2"				
#8	EA.			0.11
#10	"			0.14
2-1/2"				
#10	EA.			0.20
3"				
#12	EA.			0.36
4"				
#12	EA.			0.74
#14	"			0.92
J-Hooks				
1/2"	EA.	0.86	3.01	3.87
3/4"	"	0.93	3.01	3.94

15 MECHANICAL

Basic Materials

15140.11 Pipe Hangers, Light *(Cont.)*

	UNIT	MAT.	INST.	TOTAL
1"	EA.	0.95	3.16	4.11
1-1/4"	"	1.00	3.23	4.23
1-1/2"	"	1.02	3.32	4.34
2"	"	1.06	3.32	4.38
3"	"	1.22	3.49	4.71
4"	"	1.31	3.49	4.80
PVC coated hangers, galvanized, 28 ga.				
1-1/2" x 12"	EA.	1.49	4.42	5.91
2" x 12"	"	1.62	4.74	6.36
3" x 12"	"	1.82	5.10	6.92
4" x 12"	"	2.01	5.53	7.54
Copper, 30 ga.				
1-1/2" x 12"	EA.	2.28	4.42	6.70
2" x 12"	"	2.71	4.74	7.45
3" x 12"	"	3.00	5.10	8.10
4" x 12"	"	3.28	5.53	8.81
2" x 24"	"	5.29	5.10	10.39
3" x 24"	"	6.07	5.53	11.60
4" x 24"	"	9.50	6.03	15.53
Milford hangers				
1/2" x 6"	EA.	1.54	4.74	6.28
1/2" x 12"	"	2.12	5.10	7.22
3/4" x 6"	"	1.73	4.91	6.64
3/4" x 12"	"	2.20	5.31	7.51
1" x 6"	"	1.80	5.10	6.90
1" x 12"	"	2.20	5.31	7.51
1-1/4" x 6"	"	2.04	5.31	7.35
1-1/4" x 12"	"	2.51	5.53	8.04
1-1/2" x 6"	"	2.28	5.53	7.81
1-1/2" x 12"	"	2.67	5.77	8.44
2" x 6"	"	2.51	5.77	8.28
2" x 12"	"	2.83	6.03	8.86
Wire hook hangers				
Black wire, 1/2" x				
4"	EA.	0.50	3.32	3.82
6"	"	0.58	3.49	4.07
8"	"	0.63	3.68	4.31
10"	"	0.80	3.68	4.48
12"	"	0.96	3.90	4.86
3/4" x				
4"	EA.	0.62	3.49	4.11
6"	"	0.66	3.68	4.34
8"	"	0.67	3.90	4.57
10"	"	0.93	4.15	5.08
12"	"	0.94	4.42	5.36
1" x				
4"	EA.	0.62	3.68	4.30
6"	"	0.63	3.90	4.53
8"	"	0.67	4.15	4.82
10"	"	0.92	4.42	5.34
12"	"	1.00	4.74	5.74
1-1/4" x				
4"	EA.	0.66	3.90	4.56

15 MECHANICAL

Basic Materials

15140.11 Pipe Hangers, Light (Cont.)

	UNIT	MAT.	INST.	TOTAL
6"	EA.	0.67	4.15	4.82
8"	"	0.76	4.42	5.18
10"	"	0.88	4.74	5.62
12"	"	1.00	5.10	6.10
1-1/2" x				
6"	EA.	0.75	4.42	5.17
8"	"	0.76	4.74	5.50
10"	"	0.96	5.10	6.06
12"	"	1.04	5.53	6.57
2" x				
6"	EA.	0.76	4.74	5.50
8"	"	0.92	5.10	6.02
10"	"	0.94	5.53	6.47
12"	"	1.06	6.03	7.09
Copper wire hooks				
1/2" x				
4"	EA.	0.70	3.32	4.02
6"	"	0.79	3.49	4.28
8"	"	0.89	3.68	4.57
10"	"	1.13	3.90	5.03
12"	"	1.28	4.15	5.43
3/4" x				
4"	EA.	0.70	3.49	4.19
6"	"	0.87	3.68	4.55
8"	"	1.01	3.90	4.91
10"	"	1.14	4.15	5.29
12"	"	1.36	4.42	5.78
1" x				
4"	EA.	0.75	3.68	4.43
6"	"	0.84	3.90	4.74
8"	"	1.07	4.15	5.22
10"	"	1.28	4.42	5.70
12"	"	1.49	4.74	6.23
1-1/4" x				
6"	EA.	0.98	3.90	4.88
8"	"	1.13	4.15	5.28
10"	"	1.43	4.42	5.85
12"	"	1.49	4.74	6.23
1-1/2" x				
6"	EA.	1.17	4.42	5.59
8"	"	1.27	4.74	6.01
10"	"	1.43	5.10	6.53
12"	"	1.56	5.53	7.09
2" x				
6"	EA.	1.22	4.74	5.96
8"	"	1.36	5.10	6.46
10"	"	1.43	5.53	6.96
12"	"	1.56	6.03	7.59

15 MECHANICAL

Basic Materials

	UNIT	MAT.	INST.	TOTAL
15175.60 — Expansion Tanks				
Expansion tank, 125 psi, steel				
20 gallon	EA.	740	83.00	823
30 gallon	"	790	110	900
65 gallon	"	990	170	1,160
80 gallon	"	1,110	190	1,300
15240.10 — Vibration Control				
Vibration isolator, in-line, stainless connector				
1/2"	EA.	110	37.00	147
3/4"	"	120	39.00	159
1"	"	130	41.50	172
1-1/4"	"	180	44.25	224
1-1/2"	"	200	47.50	248
2"	"	240	51.00	291
2-1/2"	"	370	55.00	425
3"	"	430	60.00	490
4"	"	550	66.00	616
6"	"	920	74.00	994
Flanged				
8"	EA.	1,760	83.00	1,843
10"	"	2,340	95.00	2,435
12"	"	3,630	110	3,740

Insulation

	UNIT	MAT.	INST.	TOTAL
15260.10 — Fiberglass Pipe Insulation				
Fiberglass insulation on 1/2" pipe				
1" thick	L.F.	1.41	2.21	3.62
1-1/2" thick	"	2.98	2.76	5.74
3/4" pipe				
1" thick	L.F.	1.72	2.21	3.93
1-1/2" thick	"	3.13	2.76	5.89
1" pipe				
1" thick	L.F.	1.72	2.21	3.93
1-1/2" thick	"	3.29	2.76	6.05
2" thick	"	5.10	3.32	8.42
1-1/4" pipe				
1" thick	L.F.	1.96	2.76	4.72
1-1/2" thick	"	3.61	3.01	6.62
2" thick	"	5.49	3.32	8.81
1-1/2" pipe				
1" thick	L.F.	2.11	2.76	4.87
1-1/2" thick	"	3.68	3.01	6.69
2" thick	"	5.72	3.16	8.88
2-1/2" thick	"	6.82	3.32	10.14
3-1/2" thick	"	11.25	3.49	14.74

15 MECHANICAL

Insulation

15260.10 Fiberglass Pipe Insulation (Cont.)

	UNIT	MAT.	INST.	TOTAL
2" pipe				
1" thick	L.F.	2.35	2.76	5.11
1-1/2" thick	"	4.08	3.01	7.09
2" thick	"	5.96	3.32	9.28
2-1/2" thick	"	7.06	3.68	10.74
3-1/2" thick	"	11.50	4.15	15.65
2-1/2" pipe				
1" thick	L.F.	2.51	2.76	5.27
1-1/2" thick	"	4.39	3.01	7.40
2" thick	"	6.35	3.32	9.67
2-1/2" thick	"	7.61	3.68	11.29
3" thick	"	9.81	4.15	13.96
3-1/2" thick	"	11.50	4.74	16.24
3" pipe				
1" thick	L.F.	2.82	3.16	5.98
1-1/2" thick	"	4.55	3.32	7.87
2" thick	"	6.90	3.68	10.58
2-1/2" thick	"	8.71	4.15	12.86
3" thick	"	10.50	4.74	15.24
3-1/2" thick	"	12.75	5.53	18.28
4" pipe				
1" thick	L.F.	3.61	3.16	6.77
1-1/2" thick	"	5.18	3.32	8.50
2" thick	"	7.92	3.68	11.60
2-1/2" thick	"	9.65	4.15	13.80
3" thick	"	12.25	4.74	16.99
3-1/2" thick	"	14.50	5.53	20.03
5" pipe				
1" thick	L.F.	4.15	3.16	7.31
2" thick	"	8.94	3.32	12.26
3" thick	"	14.00	3.90	17.90
4" thick	"	20.75	5.10	25.85
6" pipe				
1" thick	L.F.	4.70	3.49	8.19
2" thick	"	9.73	3.68	13.41
4" thick	"	22.25	4.42	26.67
6" thick	"	42.50	6.03	48.53
8" pipe				
2" thick	L.F.	12.00	3.49	15.49
3" thick	"	19.25	3.68	22.93
4" thick	"	24.75	4.42	29.17
6" thick	"	50.00	6.03	56.03
10" pipe				
2" thick	L.F.	15.00	3.49	18.49
3" thick	"	22.25	3.68	25.93
4" thick	"	30.00	4.42	34.42
6" thick	"	55.00	6.03	61.03
12" pipe				
2" thick	L.F.	16.75	3.49	20.24
3" thick	"	25.00	3.68	28.68
4" thick	"	37.50	4.42	41.92
6" thick	"	63.00	6.03	69.03

15 MECHANICAL

Insulation	UNIT	MAT.	INST.	TOTAL
15260.20 — Calcium Silicate				
Calcium silicate insulation, 6" pipe				
2" thick	L.F.	13.50	4.74	18.24
2-1/2" thick	"	19.25	5.10	24.35
3" thick	"	25.50	5.53	31.03
4" thick	"	31.75	6.03	37.78
6" thick	"	46.50	6.64	53.14
8" pipe				
2" thick	L.F.	15.75	5.10	20.85
2-1/2" thick	"	23.25	5.53	28.78
3" thick	"	30.75	6.03	36.78
4" thick	"	37.75	6.64	44.39
6" thick	"	56.00	7.37	63.37
10" pipe				
2" thick	L.F.	20.25	5.10	25.35
2-1/2" thick	"	28.75	5.53	34.28
3" thick	"	36.25	6.03	42.28
4" thick	"	47.25	6.64	53.89
6" thick	"	64.00	7.37	71.37
12" pipe				
2" thick	L.F.	22.50	5.10	27.60
2-1/2" thick	"	31.25	5.53	36.78
3" thick	"	42.50	6.03	48.53
4" thick	"	49.25	6.64	55.89
6" thick	"	79.00	7.37	86.37
15260.60 — Exterior Pipe Insulation				
Fiberglass insulation, aluminum jacket				
1/2" pipe				
1" thick	L.F.	2.23	5.10	7.33
1-1/2" thick	"	4.19	5.53	9.72
3/4" pipe				
1" thick	L.F.	2.63	5.10	7.73
1-1/2" thick	"	4.44	5.53	9.97
1" pipe				
1" thick	L.F.	2.71	5.10	7.81
1-1/2" thick	"	4.68	5.53	10.21
3-1/2" thick	"	13.75	7.37	21.12
1-1/4" pipe				
1" thick	L.F.	3.04	6.03	9.07
1-1/2" thick	"	5.09	6.32	11.41
3-1/2" thick	"	14.00	7.37	21.37
1-1/2" pipe				
1" thick	L.F.	3.28	6.03	9.31
1-1/2" thick	"	5.26	6.32	11.58
2" thick	"	7.64	6.64	14.28
2-1/2" thick	"	9.12	6.98	16.10
3-1/2" thick	"	14.25	7.37	21.62
2" pipe				
1" thick	L.F.	3.70	6.03	9.73
1-1/2" thick	"	5.50	6.32	11.82
2" thick	"	8.05	6.64	14.69
2-1/2" thick	"	10.50	6.98	17.48
3-1/2" thick	"	14.75	7.37	22.12

15 MECHANICAL

Insulation

15260.60 Exterior Pipe Insulation (Cont.)

	UNIT	MAT.	INST.	TOTAL
2-1/2" pipe				
1" thick	L.F.	3.78	6.03	9.81
1-1/2" thick	"	5.75	6.32	12.07
2" thick	"	9.04	6.64	15.68
2-1/2" thick	"	10.75	6.98	17.73
3" thick	"	12.75	7.37	20.12
3-1/2" thick	"	15.50	7.81	23.31
3" pipe				
1" thick	L.F.	4.44	6.64	11.08
1-1/2" thick	"	6.57	6.98	13.55
2" thick	"	9.37	7.37	16.74
2-1/2" thick	"	11.50	7.81	19.31
3" thick	"	13.75	8.30	22.05
3-1/2" thick	"	16.25	8.85	25.10
4" pipe				
1" thick	L.F.	5.59	6.64	12.23
1-1/2" thick	"	7.56	6.98	14.54
2" thick	"	10.75	7.37	18.12
2-1/2" thick	"	12.75	7.81	20.56
3" thick	"	16.00	8.30	24.30
3-1/2" thick	"	18.50	8.85	27.35
5" pipe				
1" thick	L.F.	6.49	6.64	13.13
2" thick	"	12.00	6.98	18.98
2-1/2" thick	"	14.00	7.37	21.37
3" thick	"	18.00	7.81	25.81
3-1/2" thick	"	23.25	8.30	31.55
6" pipe				
1" thick	L.F.	7.40	7.37	14.77
2" thick	"	13.25	7.81	21.06
2-1/2" thick	"	15.50	8.30	23.80
3" thick	"	18.25	8.85	27.10
3-1/2" thick	"	21.00	9.48	30.48
4" thick	"	27.50	10.25	37.75
6" thick	"	49.75	11.00	60.75
8" pipe				
2" thick	L.F.	15.00	7.37	22.37
3" thick	"	22.25	7.81	30.06
3-1/2" thick	"	30.00	8.30	38.30
4" thick	"	31.50	8.85	40.35
6" thick	"	52.00	9.48	61.48
10" pipe				
2" thick	L.F.	18.75	7.37	26.12
3" thick	"	27.00	7.81	34.81
3-1/2" thick	"	31.50	8.30	39.80
4" thick	"	37.25	8.85	46.10
6" thick	"	58.00	9.48	67.48
12" pipe				
2" thick	L.F.	21.00	7.37	28.37
3" thick	"	30.00	7.81	37.81
3-1/2" thick	"	37.25	8.30	45.55
4" thick	"	40.25	8.85	49.10
6" thick	"	65.00	9.48	74.48

15 MECHANICAL

Insulation	UNIT	MAT.	INST.	TOTAL
15260.60 Exterior Pipe Insulation *(Cont.)*				
Calcium silicate with aluminum jacket, 6" pipe				
2" thick	L.F.	15.25	8.30	23.55
2-1/2" thick	"	20.75	8.85	29.60
3" thick	"	26.75	9.48	36.23
4" thick	"	33.25	11.00	44.25
8" pipe				
2" thick	L.F.	18.00	8.30	26.30
2-1/2" thick	"	25.00	8.85	33.85
3" thick	"	32.25	9.48	41.73
4" thick	"	39.25	10.25	49.50
6" thick	"	57.00	11.00	68.00
10" pipe				
2" thick	L.F.	22.75	9.48	32.23
2-1/2" thick	"	30.75	10.25	41.00
3" thick	"	39.00	11.00	50.00
4" thick	"	48.25	12.00	60.25
6" thick	"	65.00	13.25	78.25
12" pipe				
2" thick	L.F.	25.50	9.48	34.98
2-1/2" thick	"	33.50	10.25	43.75
3" thick	"	44.00	11.00	55.00
4" thick	"	51.00	12.00	63.00
6" thick	"	79.00	13.25	92.25
15260.90 Pipe Insulation Fittings				
Insulation protection saddle				
1" thick covering				
1/2" pipe	EA.	7.57	26.50	34.07
3/4" pipe	"	7.78	26.50	34.28
1" pipe	"	7.98	26.50	34.48
1-1/4" pipe	"	8.19	26.50	34.69
1-1/2" pipe	"	8.59	26.50	35.09
2" pipe	"	8.80	26.50	35.30
2-1/2" pipe	"	9.39	26.50	35.89
3" pipe	"	10.00	30.25	40.25
4" pipe	"	9.84	33.25	43.09
6" pipe	"	9.90	41.50	51.40
1-1/2" thick covering				
3/4" pipe	EA.	12.75	26.50	39.25
1" pipe	"	13.00	26.50	39.50
1-1/4" pipe	"	13.25	26.50	39.75
1-1/2" pipe	"	11.75	26.50	38.25
2" pipe	"	11.25	26.50	37.75
2-1/2" pipe	"	13.50	26.50	40.00
3" pipe	"	12.75	26.50	39.25
4" pipe	"	12.50	33.25	45.75
6" pipe	"	15.00	41.50	56.50
8" pipe	"	17.25	55.00	72.25
10" pipe	"	15.75	55.00	70.75
12" pipe	"	19.75	55.00	74.75
2" thick covering				
3/4" pipe	EA.	13.25	26.50	39.75
1" pipe	"	13.50	26.50	40.00

15 MECHANICAL

Insulation

	UNIT	MAT.	INST.	TOTAL
15260.90 Pipe Insulation Fittings (Cont.)				
1-1/4" pipe	EA.	14.00	26.50	40.50
1-1/2" pipe	"	14.50	26.50	41.00
2" pipe	"	12.50	26.50	39.00
2-1/2" pipe	"	13.25	26.50	39.75
3" pipe	"	14.75	26.50	41.25
4" pipe	"	13.75	33.25	47.00
6" pipe	"	16.50	41.50	58.00
8" pipe	"	18.25	55.00	73.25
10" pipe	"	17.25	55.00	72.25
12" pipe	"	22.75	55.00	77.75
3" thick covering				
2" pipe	EA.	16.00	26.50	42.50
2-1/2" pipe	"	17.00	26.50	43.50
3" pipe	"	17.00	26.50	43.50
4" pipe	"	16.00	33.25	49.25
6" pipe	"	23.50	41.50	65.00
8" pipe	"	21.50	55.00	76.50
10" pipe	"	22.00	55.00	77.00
12" pipe	"	26.75	55.00	81.75
15280.10 Equipment Insulation				
Equipment insulation, 2" thick, cellular glass	S.F.	3.83	4.15	7.98
Urethane, rigid, jacket, plastered finish	"	4.09	8.30	12.39
Fiberglass, rigid, with vapor barrier	"	3.83	3.68	7.51
15290.10 Ductwork Insulation				
Fiberglass duct insulation, plain blanket				
1-1/2" thick	S.F.	0.24	0.83	1.07
2" thick	"	0.32	1.10	1.42
With vapor barrier				
1-1/2" thick	S.F.	0.28	0.83	1.11
2" thick	"	0.36	1.10	1.46
Rigid with vapor barrier				
2" thick	S.F.	1.56	2.21	3.77
3" thick	"	2.14	2.65	4.79
4" thick	"	2.73	3.32	6.05
6" thick	"	4.29	4.42	8.71
Weatherproof, poly, 3" thick, w/vapor barrier	"	3.31	6.64	9.95
Urethane board with vapor barrier	"	4.68	8.30	12.98

15 MECHANICAL

Fire Protection

15330.10 Wet Sprinkler System

Description	UNIT	MAT.	INST.	TOTAL
Sprinkler head, 212 deg, brass, exposed piping	EA.	16.50	26.50	43.00
Chrome, concealed piping	"	19.75	37.00	56.75
Water motor alarm	"	380	110	490
Fire department inlet connection	"	270	130	400
Wall plate for fire dept connection	"	130	55.00	185
Swing check valve flanged iron body, 4"	"	360	220	580
Check valve, 6"	"	1,170	330	1,500
Wet pipe valve, flange to groove, 4"	"	1,020	74.00	1,094
Flange to flange				
6"	EA.	1,370	110	1,480
8"	"	2,410	220	2,630
Alarm valve, flange to flange, (wet valve)				
4"	EA.	1,560	74.00	1,634
8"	"	2,470	550	3,020
Inspector's test connection	"	72.00	55.00	127
Wall hydrant, polished brass, 2-1/2" x 2-1/2", single	"	480	47.50	528
2-way	"	1,080	47.50	1,128
3-way	"	2,210	47.50	2,258
Wet valve trim, includes retard chamber & gauges, 4"-6"	"	760	55.00	815
Retard pressure switch for wet systems	"	1,320	130	1,450
Air maintenance device	"	400	55.00	455
Wall hydrant non-freeze, 8" thick wall, vacuum breaker	"	52.00	33.25	85.25
12" thick wall	"	57.00	33.25	90.25

15330.50 Dry Sprinkler System

Description	UNIT	MAT.	INST.	TOTAL
Dry pipe valve, flange to flange				
4"	EA.	2,300	130	2,430
6"	"	2,880	170	3,050
Trim, 4" and 6", includes gauges	"	890	55.00	945
Field testing and flushing	"		550	550
Disinfection	"		550	550
Pressure switch double circuit, open/close contacts	"	430	170	600
Low air				
Supervisory unit	EA.	1,390	110	1,500
Pressure switch	"	440	55.00	495

15330.70 CO_2 System

Description	UNIT	MAT.	INST.	TOTAL
CO_2 system, high pressure, 75# cylinder with				
Valve assemblies	EA.	2,850	130	2,980
Storage rack	"	1,520	95.00	1,615
Manifold	"	1,110	470	1,580
Flexible loops	"	90.00	8.30	98.30
Beam scale for cylinders	"	750	110	860
Mechanically control head	"	640	44.25	684
Electrically control head	"	640	44.25	684
Stop valves	"	1,610	66.00	1,676
Check valves	"	730	83.00	813
Activation station	"	840	66.00	906
Nozzles	"	140	55.00	195
Hose reel with 75' of 3/4" hose	"	5,100	330	5,430
Main/reserve transfer switch	"	5,930	110	6,040
Pressure switch	"	520	66.00	586
Heat responsive device	"	830	110	940

15 MECHANICAL

Fire Protection

	UNIT	MAT.	INST.	TOTAL
15330.70 CO$_2$ System *(Cont.)*				
Battery and charger	EA.	4,750	330	5,080
Low pressure				
Battery and charger	EA.	4,750	330	5,080
Pressure switch	"	470	74.00	544
Nozzles	"	140	60.00	200
Master selector valve	"	400	110	510
Selector valve	"	5,930	110	6,040
Low pressure hose reel with 75' of 3/4" hose	"	7,360	330	7,690
Tank fill lines	"	1,660	83.00	1,743
Activation stations	"	830	55.00	885
Electro manual pilot panels	"	1,660	83.00	1,743

Plumbing

	UNIT	MAT.	INST.	TOTAL
15410.05 C.i. Pipe, Above Ground				
No hub pipe				
1-1/2" pipe	L.F.	8.88	4.74	13.62
2" pipe	"	7.86	5.53	13.39
3" pipe	"	10.75	6.64	17.39
4" pipe	"	14.25	11.00	25.25
6" pipe	"	25.00	13.25	38.25
8" pipe	"	40.00	22.25	62.25
10" pipe	"	63.00	26.50	89.50
No hub fittings, 1-1/2" pipe				
1/4 bend	EA.	10.50	22.25	32.75
1/8 bend	"	8.79	22.25	31.04
Sanitary tee	"	14.50	33.25	47.75
Sanitary cross	"	19.75	33.25	53.00
Plug	"			5.80
Coupling	"			20.75
Wye	"	18.25	33.25	51.50
Tapped tee	"	19.25	22.25	41.50
P-trap	"	16.50	22.25	38.75
Tapped cross	"	21.75	22.25	44.00
2" pipe				
1/4 bend	EA.	12.25	26.50	38.75
1/8 bend	"	9.79	26.50	36.29
Sanitary tee	"	16.50	44.25	60.75
Sanitary cross	"	28.00	44.25	72.25
Plug	"			5.80
Coupling	"			18.25
Wye	"	15.50	55.00	70.50
Double wye	"	24.00	55.00	79.00
2x1-1/2" wye & 1/8 bend	"	29.00	41.50	70.50
Double wye & 1/8 bend	"	24.00	55.00	79.00
Test tee less 2" plug	"	15.00	26.50	41.50

15 MECHANICAL

Plumbing

	UNIT	MAT.	INST.	TOTAL
15410.05 C.i. Pipe, Above Ground (Cont.)				
Tapped tee				
2"x2"	EA.	19.50	26.50	46.00
2"x1-1/2"	"	18.25	26.50	44.75
P-trap				
2"x2"	EA.	17.50	26.50	44.00
Tapped cross				
2"x1-1/2"	EA.	25.00	26.50	51.50
3" pipe				
1/4 bend	EA.	16.75	33.25	50.00
1/8 bend	"	14.00	33.25	47.25
Sanitary tee	"	20.25	41.50	61.75
3"x2" sanitary tee	"	18.25	41.50	59.75
3"x1-1/2" sanitary tee	"	19.25	41.50	60.75
Sanitary cross	"	43.25	55.00	98.25
3x2" sanitary cross	"	38.50	55.00	93.50
Plug	"			8.60
Coupling	"			21.00
Wye	"	22.00	55.00	77.00
3x2" wye	"	16.50	55.00	71.50
Double wye	"	44.25	55.00	99.25
3x2" double wye	"	37.50	55.00	92.50
3x2" wye & 1/8 bend	"	20.75	47.50	68.25
3x1-1/2" wye & 1/8 bend	"	20.75	47.50	68.25
Double wye & 1/8 bend	"	44.25	55.00	99.25
3x2" double wye & 1/8 bend	"	37.50	55.00	92.50
3x2" reducer	"	8.40	30.25	38.65
Test tee, less 3" plug	"	23.00	33.25	56.25
Plug	"			8.60
3x3" tapped tee	"	53.00	33.25	86.25
3x2" tapped tee	"	28.50	33.25	61.75
3x1-1/2" tapped tee	"	24.50	33.25	57.75
P-trap	"	38.75	33.25	72.00
3x2" tapped cross	"	36.00	33.25	69.25
3x1-1/2" tapped cross	"	33.75	33.25	67.00
Closet flange, 3-1/2" deep	"	24.50	16.50	41.00
4" pipe				
1/4 bend	EA.	24.00	33.25	57.25
1/8 bend	"	17.50	33.25	50.75
Sanitary tee	"	31.50	55.00	86.50
4x3" sanitary tee	"	29.00	55.00	84.00
4x2" sanitary tee	"	24.00	55.00	79.00
Sanitary cross	"	82.00	66.00	148
4x3" sanitary cross	"	66.00	66.00	132
4x2" sanitary cross	"	55.00	66.00	121
Plug	"			13.50
Coupling	"			20.50
Wye	"	36.00	55.00	91.00
4x3" wye	"	31.50	55.00	86.50
4x2" wye	"	23.00	55.00	78.00
Double wye	"	90.00	66.00	156
4x3" double wye	"	57.00	66.00	123
4x2" double wye	"	50.00	66.00	116
Wye & 1/8 bend	"	49.00	55.00	104

15 MECHANICAL

Plumbing

15410.05 C.i. Pipe, Above Ground *(Cont.)*

Item	UNIT	MAT.	INST.	TOTAL
4x3" wye & 1/8 bend	EA.	35.75	55.00	90.75
4x2" wye & 1/8 bend	"	27.75	55.00	82.75
Double wye & 1/8 bend	"	130	66.00	196
4x3" double wye & 1/8 bend	"	83.00	66.00	149
4x2" double wye & 1/8 bend	"	79.00	66.00	145
4x3" reducer	"	13.00	33.25	46.25
4x2" reducer	"	13.00	33.25	46.25
Test tee, less 4" plug	"	39.25	33.25	72.50
Plug	"			13.50
4x2" tapped tee	"	28.75	33.25	62.00
4x1-1/2" tapped tee	"	25.25	33.25	58.50
P-trap	"	67.00	33.25	100
4x2" tapped cross	"	52.00	33.25	85.25
4x1-1/2" tapped cross	"	40.25	33.25	73.50
Closet flange				
3" deep	EA.	26.75	33.25	60.00
8" deep	"	69.00	33.25	102
6" pipe				
1/4 bend	EA.	61.00	55.00	116
1/8 bend	"	40.75	55.00	95.75
Sanitary tee	"	91.00	66.00	157
6x4" sanitary tee	"	69.00	66.00	135
Coupling	"			52.00
Wye	"	96.00	66.00	162
6x4" wye	"	75.00	66.00	141
6x3" wye	"	73.00	66.00	139
6x2" wye	"	58.00	66.00	124
Double wye	"	160	83.00	243
6x4" double wye	"	130	83.00	213
Wye & 1/8 bend	"	150	66.00	216
6x4" wye & 1/8 bend	"	87.00	66.00	153
6x3" wye & 1/8 bend	"	84.00	66.00	150
6x2" wye & 1/8 bend	"	66.00	66.00	132
6x4" reducer	"	34.75	37.00	71.75
6x3" reducer	"	34.75	37.00	71.75
6x2" reducer	"	35.75	33.25	69.00
Test tee				
Less 6" plug	EA.	94.00	41.50	136
Plug	"			26.00
P-trap	"	160	41.50	202
8" pipe				
1/4 bend	EA.	110	55.00	165
1/8 bend	"	74.00	55.00	129
Sanitary tee	"	230	83.00	313
8x6" sanitary tee	"	130	83.00	213
Plug	"			43.75
Coupling	"			98.00
Wye	"	140	66.00	206
8x6" wye	"	100	55.00	155
8x4" wye	"	81.00	55.00	136
Double wye	"	290	55.00	345
Wye & 1/8 bend	"	220	55.00	275
8x6" wye & 1/8 bend	"	170	55.00	225

15 MECHANICAL

Plumbing	UNIT	MAT.	INST.	TOTAL
15410.05 C.i. Pipe, Above Ground *(Cont.)*				
8x4" wye & 1/8 bend	EA.	110	55.00	165
8x6" reducer	"	37.75	55.00	92.75
8x4" reducer	"	33.75	55.00	88.75
8x3" reducer	"	32.25	55.00	87.25
8x2" reducer	"	33.00	37.00	70.00
Test tee				
Less 8" plug	EA.	150	55.00	205
Plug	"			69.00
10" pipe				
1/4 bend	EA.	210	55.00	265
1/8 bend	"	140	55.00	195
Plug	"			65.00
Coupling	"			130
Wye	"	310	110	420
10x8" wye	"	270	110	380
10x6" wye	"	230	110	340
10x4" wye	"	220	110	330
10x8" reducer	"	79.00	55.00	134
10x6" reducer	"	70.00	55.00	125
10x4" reducer	"	60.00	55.00	115
15410.06 C.i. Pipe, Below Ground				
No hub pipe				
1-1/2" pipe	L.F.	8.40	3.32	11.72
2" pipe	"	8.62	3.68	12.30
3" pipe	"	12.00	4.15	16.15
4" pipe	"	15.50	5.53	21.03
6" pipe	"	26.50	6.03	32.53
8" pipe	"	41.50	7.37	48.87
10" pipe	"	69.00	8.30	77.30
Fittings, 1-1/2"				
1/4 bend	EA.	11.00	19.00	30.00
1/8 bend	"	9.20	19.00	28.20
Plug	"			5.80
Wye	"	15.50	26.50	42.00
Wye & 1/8 bend	"	16.50	19.00	35.50
P-trap	"	18.25	19.00	37.25
2"				
1/4 bend	EA.	12.00	22.25	34.25
1/8 bend	"	10.25	22.25	32.50
Plug	"			5.80
Double wye	"	24.00	41.50	65.50
Wye & 1/8 bend	"	16.75	33.25	50.00
Double wye & 1/8 bend	"	41.25	41.50	82.75
P-trap	"	17.50	22.25	39.75
3"				
1/4 bend	EA.	16.50	26.50	43.00
1/8 bend	"	14.00	26.50	40.50
Plug	"			8.60
Wye	"	22.00	41.50	63.50
3x2" wye	"	16.50	41.50	58.00
Wye & 1/8 bend	"	26.75	41.50	68.25
Double wye & 1/8 bend	"	64.00	41.50	106

15 MECHANICAL

Plumbing

15410.06 C.i. Pipe, Below Ground (Cont.)

Description	UNIT	MAT.	INST.	TOTAL
3x2" double wye & 1/8 bend	EA.	48.25	41.50	89.75
3x2" reducer	"	8.40	26.50	34.90
P-trap	"	38.75	26.50	65.25
4"				
1/4 bend	EA.	24.00	26.50	50.50
1/8 bend	"	17.50	26.50	44.00
Plug	"			13.50
Wye	"	36.00	41.50	77.50
4x3" wye	"	31.50	41.50	73.00
4x2" wye	"	23.00	41.50	64.50
Double wye	"	90.00	55.00	145
4x3" double wye	"	57.00	55.00	112
4x2" double wye	"	50.00	55.00	105
Wye & 1/8 bend	"	49.00	41.50	90.50
4x3" wye & 1/8 bend	"	35.75	41.50	77.25
4x2" wye & 1/8 bend	"	27.75	41.50	69.25
Double wye & 1/8 bend	"	130	55.00	185
4x3" double wye & 1/8 bend	"	83.00	55.00	138
4x2" double wye & 1/8 bend	"	79.00	55.00	134
4x3" reducer	"	13.00	26.50	39.50
4x2" reducer	"	13.00	26.50	39.50
6"				
1/4 bend	EA.	38.25	41.50	79.75
1/8 bend	"	25.75	41.50	67.25
Wye & 1/8 bend	"	61.00	55.00	116
6x4" wye & 1/8 bend	"	47.50	55.00	103
6x3" wye & 1/8 bend	"	46.00	55.00	101
6x2" wye & 1/8 bend	"	36.50	55.00	91.50
6x3" reducer	"	22.00	30.25	52.25
P-trap	"	100	33.25	133
8"				
1/4 bend	EA.	110	41.50	152
1/8 bend	"	74.00	41.50	116
Plug	"			43.75
Wye	"	140	55.00	195
8x6" wye	"	100	41.50	142
8x4" wye	"	81.00	41.50	123
8x6" wye & 1/8 bend	"	100	41.50	142
8x4" reducer	"	33.75	33.25	67.00
8x3" reducer	"	32.25	33.25	65.50
8x2" reducer	"	33.00	33.25	66.25
10"				
1/4 bend	EA.	210	41.50	252
1/8 bend	"	140	41.50	182
Plug	"			65.00
Wye	"	310	83.00	393
10x8" wye	"	270	83.00	353
10x6" wye	"	230	83.00	313
10x4" wye	"	220	83.00	303
10x8" reducer	"	79.00	41.50	121
10x6" reducer	"	70.00	41.50	112

15 MECHANICAL

Plumbing

15410.08 Extra Heavy Soil Pipe

	UNIT	MAT.	INST.	TOTAL
Extra heavy soil pipe, single hub				
2" x 5'	EA.	66.00	13.25	79.25
3" x 5'	"	73.00	14.00	87.00
4" x 5'	"	85.00	15.50	101
6" x 5'	"	160	16.50	177
Double hub				
2" x 5'	EA.	73.00	16.50	89.50
4" x 5'	"	93.00	17.50	111
5" x 5'	"	140	18.50	159
6" x 5'	"	170	19.00	189
Single hub				
3" x 10'	EA.	150	11.00	161
4" x 10'	"	190	11.50	202
6" x 10'	"	290	12.00	302
8" x 10'	"	500	12.75	513
"Mini", single hub				
2" x 42"	EA.	68.00	13.25	81.25
3" x 42"	"	75.00	14.00	89.00
4" x 42"	"	82.00	15.50	97.50
6" x 42"	"	110	16.50	127
8" x 42"	"	150	17.50	168
Fittings, 1/4" bend				
2"	EA.	17.00	22.25	39.25
3"	"	34.00	26.50	60.50
4"	"	51.00	26.50	77.50
1/8 bend				
2"	EA.	16.25	22.25	38.50
3"	"	28.00	26.50	54.50
4"	"	42.50	26.50	69.00
5"	"	58.00	33.25	91.25
6"	"	64.00	41.50	106
8"	"	160	41.50	202
Long sweep				
2"	EA.	28.75	22.25	51.00
3"	"	59.00	26.50	85.50
4"	"	78.00	26.50	105
Straight T				
4" x 2"	EA.	73.00	41.50	115
4" x 3"	"	58.00	41.50	99.50
4"	"	64.00	47.50	112
Sanitary T				
3" x 2"	EA.	61.00	41.50	103
3"	"	77.00	41.50	119
4" x 2"	"	78.00	47.50	126
4" x 3"	"	83.00	47.50	131
4"	"	82.00	51.00	133
Wye				
4" x 2"	EA.	77.00	47.50	125
4" x 3"	"	93.00	47.50	141
4"	"	110	51.00	161
Combination Y and 1/8 bend, 4"	"	100	51.00	151
Double wye, 4"	"	130	51.00	181
Tapped sanitary T, 4" x 2"	"	70.00	55.00	125

15 MECHANICAL

Plumbing	UNIT	MAT.	INST.	TOTAL
15410.08 Extra Heavy Soil Pipe (Cont.)				
P trap				
2"	EA.	39.25	26.50	65.75
4"	"	100	44.25	144
Dandy				
4", with 3" brass plug	EA.	71.00	37.00	108
6", 4" brass plug	"	110	41.50	152
15410.09 Service Weight Pipe				
Service weight pipe, single hub				
2" x 5'	EA.	53.00	13.25	66.25
3" x 5'	"	59.00	14.00	73.00
4" x 5'	"	68.00	14.75	82.75
5" x 5'	"	100	15.75	116
6" x 5'	"	130	16.50	147
8" x 5'	"	190	17.50	208
10" x 5'	"	290	18.50	309
12" x 5'	"	380	20.75	401
Double hub				
2" x 5'	EA.	59.00	16.50	75.50
3" x 5'	"	65.00	18.00	83.00
4" x 5'	"	74.00	19.00	93.00
5" x 5'	"	120	20.75	141
6" x 5'	"	140	22.25	162
10" x 5'	"	300	23.75	324
12' x 5'	"	380	25.50	406
Single hub				
2" x 10'	EA.	56.00	16.50	72.50
3" x 10'	"	78.00	18.00	96.00
4" x 10'	"	100	19.00	119
5" x 10'	"	140	20.75	161
6" x 10'	"	170	22.25	192
8" x 10'	"	270	23.75	294
10" x 10'	"	450	25.50	476
12" x 10'	"	650	27.75	678
Shorty				
2" x 42"	EA.	38.25	13.25	51.50
3" x 42"	"	42.75	14.00	56.75
4" x 42"	"	55.00	14.75	69.75
5" x 42"	"	84.00	15.75	99.75
6" x 42"	"	99.00	16.50	116
8" x 42"	"	150	17.50	168
10" x 42"	"	210	18.50	229
Soil plug				
2"	EA.	6.08	22.25	28.33
3"	"	8.72	23.75	32.47
4"	"	10.50	25.50	36.00
5"	"	20.50	26.50	47.00
6"	"	20.00	27.75	47.75
1/5 bend				
3"	EA.	18.50	26.50	45.00
4"	"	37.75	30.25	68.00
1/6 bend				
2"	EA.	14.50	22.25	36.75

15 MECHANICAL

Plumbing

15410.09 Service Weight Pipe (Cont.)

	UNIT	MAT.	INST.	TOTAL
3"	EA.	19.25	26.50	45.75
4"	"	24.25	30.25	54.50
1/8 bend				
2"	EA.	8.65	22.25	30.90
3"	"	13.50	26.50	40.00
4"	"	19.75	30.25	50.00
5"	"	27.50	31.50	59.00
6"	"	33.75	33.25	67.00
8"	"	100	41.50	142
10"	"	140	47.50	188
12"	"	270	55.00	325
1/16 bend				
2"	EA.	9.80	22.25	32.05
3"	"	15.00	26.50	41.50
4"	"	17.25	30.25	47.50
5"	"	28.50	31.50	60.00
6"	"	29.25	33.25	62.50
1/4 bend				
2"	EA.	12.25	22.25	34.50
3"	"	16.25	26.50	42.75
4"	"	25.50	30.25	55.75
5"	"	35.50	31.50	67.00
6"	"	44.25	33.25	77.50
8"	"	130	41.50	172
10"	"	190	47.50	238
12"	"	260	55.00	315
1/4 bend, long				
2" x 12"	EA.	30.25	23.75	54.00
4" x 18"	"	51.00	33.25	84.25
4" x 12"	"	34.25	33.25	67.50
Sweep				
2"	EA.	18.50	22.25	40.75
3"	"	26.25	26.50	52.75
4"	"	38.75	30.25	69.00
5"	"	64.00	31.50	95.50
6"	"	78.00	33.25	111
8"	"	180	41.50	222
Reducing long sweep				
3" x 2"	EA.	33.25	26.50	59.75
4" x 3"	"	52.00	33.25	85.25
Straight T				
2"	EA.	24.50	41.50	66.00
3"	"	36.50	44.25	80.75
4" x 2"	"	38.00	47.50	85.50
4" x 3"	"	42.25	51.00	93.25
4"	"	47.00	55.00	102
Sanitary T				
2"	EA.	17.00	41.50	58.50
3" x 2"	"	23.50	44.25	67.75
3"	"	27.50	47.50	75.00
4" x 2"	"	28.75	51.00	79.75
4" x 3"	"	30.75	55.00	85.75
4"	"	33.75	55.00	88.75

15 MECHANICAL

Plumbing

15410.09 Service Weight Pipe (Cont.)

	UNIT	MAT.	INST.	TOTAL
5"	EA.	67.00	60.00	127
6"	"	76.00	60.00	136
Wye				
2"	EA.	15.50	33.25	48.75
3" x 2"	"	22.25	37.00	59.25
3"	"	28.75	37.00	65.75
4" x 2"	"	29.75	39.00	68.75
4" x 3"	"	33.00	39.00	72.00
4"	"	38.50	39.00	77.50
5" x 2"	"	47.75	41.50	89.25
5" x 3"	"	51.00	41.50	92.50
5" x 4"	"	52.00	41.50	93.50
5"	"	65.00	41.50	107
6" x 2"	"	56.00	41.50	97.50
6" x 3"	"	57.00	41.50	98.50
6" x 4"	"	60.00	44.25	104
6" x 5"	"	84.00	44.25	128
6"	"	88.00	47.50	136
8" x 4"	"	110	47.50	158
8" x 6"	"	130	51.00	181
8"	"	220	55.00	275
10" x 4"	"	190	55.00	245
10" x 6"	"	210	60.00	270
10" x 8"	"	290	66.00	356
Service wye				
10"	EA.	350	83.00	433
12" x 4"	"	330	83.00	413
12" x 6"	"	340	83.00	423
12" x 8"	"	410	95.00	505
12"	"	710	110	820
Combination wye and 1/8 bend				
2"	EA.	21.25	44.25	65.50
3" x 2"	"	24.50	47.50	72.00
3"	"	32.50	47.50	80.00
4" x 2"	"	33.00	51.00	84.00
4" x 3"	"	38.75	51.00	89.75
4"	"	44.50	55.00	99.50
5" x 4"	"	81.00	55.00	136
6" x 4"	"	77.00	66.00	143
8"	"	260	83.00	343
Straight cross, 4"	"	62.00	55.00	117
Sanitary cross				
2"	EA.	36.50	47.50	84.00
3"	"	44.75	51.00	95.75
3" x 2"	"	44.25	51.00	95.25
4"	"	56.00	55.00	111
4" x 3"	"	57.00	55.00	112
4" x 2"	"	43.75	60.00	104
Double wye				
2"	EA.	35.25	47.50	82.75
3" x 2"	"	40.25	51.00	91.25
3"	"	47.00	51.00	98.00
4" x 2"	"	48.50	55.00	104

15 MECHANICAL

Plumbing		UNIT	MAT.	INST.	TOTAL
15410.09	**Service Weight Pipe** *(Cont.)*				
4" x 3"		EA.	51.00	55.00	106
4"		"	62.00	55.00	117
5"		"	110	66.00	176
6" x 4"		"	110	83.00	193
Combination double wye and 1/8 bend					
2"		EA.	35.25	47.50	82.75
3" x 2"		"	40.25	51.00	91.25
3"		"	47.00	51.00	98.00
4" x 3"		"	51.00	55.00	106
4"		"	62.00	55.00	117
Tapped sanitary T					
2" x 1-1/2"		EA.	24.00	47.50	71.50
2" x 2"		"	22.75	47.50	70.25
3" x 1-1/2"		"	27.75	51.00	78.75
3" x 2"		"	25.50	51.00	76.50
4" x 1-1/2"		"	34.75	55.00	89.75
4" x 2"		"	41.50	55.00	96.50
Tapped straight T					
3" x 1-1/2"		EA.	27.75	51.00	78.75
3" x 2"		"	27.00	51.00	78.00
4" x 2"		"	31.75	55.00	86.75
4" x 1-1/2"		"	44.75	55.00	99.75
Tapped Y					
4" x 2"		EA.	43.00	55.00	98.00
Tapped sanitary cross					
2" x 1-1/2"		EA.	26.25	47.50	73.75
2" x 2"		"	27.00	51.00	78.00
3" x 1-1/2"		"	42.00	51.00	93.00
3" x 2"		"	36.50	51.00	87.50
3" x 3"		"	47.25	51.00	98.25
4" x 1-1/2"		"	48.00	55.00	103
4" x 2"		"	47.25	55.00	102
Cleanout, dandy, with brass plug					
2", 1-1/2" plug		EA.	29.75	47.50	77.25
3", 2" plug		"	34.75	51.00	85.75
4", 3" plug		"	66.00	55.00	121
5", 4" plug		"	76.00	60.00	136
6", 4" plug		"	96.00	66.00	162
8", 6" plug		"	180	83.00	263
Reducer, 3" x 2"		"	12.00	47.50	59.50
4" x					
2"		EA.	13.75	51.00	64.75
3"		"	15.50	51.00	66.50
5" x					
2"		EA.	30.75	55.00	85.75
3"		"	32.50	55.00	87.50
4"		"	32.75	55.00	87.75
6" x					
2"		EA.	29.25	66.00	95.25
3"		"	30.25	66.00	96.25
4"		"	31.25	66.00	97.25
5"		"	33.75	66.00	99.75
8" x					

15 MECHANICAL

Plumbing	UNIT	MAT.	INST.	TOTAL
15410.09 — Service Weight Pipe (Cont.)				
4"	EA.	52.00	83.00	135
6"	"	55.00	83.00	138
10" x				
4"	EA.	79.00	95.00	174
6"	"	88.00	95.00	183
8"	"	88.00	95.00	183
12" x				
4"	EA.	130	110	240
6"	"	140	110	250
8"	"	140	110	250
10"	"	140	110	250
P trap				
2"	EA.	19.25	47.50	66.75
3"	"	29.00	51.00	80.00
4"	"	41.75	55.00	96.75
5"	"	88.00	66.00	154
6"	"	130	83.00	213
15410.10 — Copper Pipe				
Type "K" copper				
1/2"	L.F.	4.89	2.07	6.96
3/4"	"	9.12	2.21	11.33
1"	"	12.00	2.37	14.37
1-1/4"	"	14.75	2.55	17.30
1-1/2"	"	19.25	2.76	22.01
2"	"	29.75	3.01	32.76
2-1/2"	"	43.75	3.32	47.07
3"	"	61.00	3.49	64.49
4"	"	100	3.68	104
DWV, copper				
1-1/4"	L.F.	13.25	2.76	16.01
1-1/2"	"	17.00	3.01	20.01
2"	"	22.00	3.32	25.32
3"	"	37.50	3.68	41.18
4"	"	66.00	4.15	70.15
6"	"	250	4.74	255
Refrigeration tubing, copper, sealed				
1/8"	L.F.	0.95	2.65	3.60
3/16"	"	1.11	2.76	3.87
1/4"	"	1.32	2.88	4.20
5/16"	"	1.71	3.01	4.72
3/8"	"	1.95	3.16	5.11
1/2"	"	2.57	3.32	5.89
7/8"	"	6.24	3.79	10.03
1-1/8"	"	8.96	4.42	13.38
1-3/8"	"	13.75	5.10	18.85
Type "L" copper				
1/4"	L.F.	1.97	1.95	3.92
3/8"	"	3.03	1.95	4.98
1/2"	"	3.51	2.07	5.58
3/4"	"	5.61	2.21	7.82
1"	"	8.45	2.37	10.82
1-1/4"	"	12.00	2.55	14.55

15 MECHANICAL

Plumbing	UNIT	MAT.	INST.	TOTAL
15410.10 Copper Pipe *(Cont.)*				
1-1/2"	L.F.	15.50	2.76	18.26
2"	"	24.25	3.01	27.26
2-1/2"	"	36.00	3.32	39.32
3"	"	48.50	3.49	51.99
3-1/2"	"	63.00	3.58	66.58
4"	"	80.00	3.68	83.68
Type "M" copper				
1/2"	L.F.	2.48	2.07	4.55
3/4"	"	4.04	2.21	6.25
1"	"	6.57	2.37	8.94
1-1/4"	"	9.69	2.55	12.24
2"	"	21.25	3.01	24.26
2-1/2"	"	31.00	3.32	34.32
3"	"	40.75	3.49	44.24
4"	"	72.00	3.68	75.68
Type "K" tube, coil				
1/4" x 60'	EA.			150
1/2" x 60'	"			310
1/2" x 100'	"			510
3/4" x 60'	"			570
3/4" x 100'	"			940
1" x 60'	"			740
1" x 100'	"			1,230
1-1/4" x 60'	"			940
1-1/2" x 60'	"			1,230
2" x 40'	"			1,320
Type "L" tube, coil				
1/4" x 60'	EA.			160
3/8" x 60'	"			250
1/2" x 60'	"			330
1/2" x 100'	"			550
3/4" x 60'	"			520
3/4" x 100'	"			870
1" x 60'	"			760
1" x 100'	"			1,260
1-1/4" x 60'	"			1,080
1-1/2" x 60'	"			1,380
15410.11 Copper Fittings				
Coupling, with stop				
1/4"	EA.	1.19	22.25	23.44
3/8"	"	1.56	26.50	28.06
1/2"	"	1.24	28.75	29.99
5/8"	"	3.61	33.25	36.86
3/4"	"	2.48	37.00	39.48
1"	"	5.10	39.00	44.10
3"	"	62.00	66.00	128
4"	"	130	83.00	213
Reducing coupling				
1/4" x 1/8"	EA.	3.18	26.50	29.68
3/8" x 1/4"	"	3.51	28.75	32.26
1/2" x 3/8"	EA.	2.63	33.25	35.88

15 MECHANICAL

Plumbing

15410.11 Copper Fittings *(Cont.)*

	UNIT	MAT.	INST.	TOTAL
1/4"	EA.	3.19	33.25	36.44
1/8"	"	3.52	33.25	36.77
3/4" x				
3/8"	EA.	5.66	37.00	42.66
1/2"	"	4.48	37.00	41.48
1" x				
3/8"	EA.	10.25	41.50	51.75
1" x 1/2"	"	9.84	41.50	51.34
1" x 3/4"	"	8.29	41.50	49.79
1-1/4" x				
1/2"	EA.	12.25	44.25	56.50
3/4"	"	11.75	44.25	56.00
1"	"	11.75	44.25	56.00
1-1/2" x				
1/2"	EA.	20.50	47.50	68.00
3/4"	"	19.25	47.50	66.75
1"	"	19.25	47.50	66.75
1-1/4"	"	19.25	47.50	66.75
2" x				
1/2"	EA.	33.50	55.00	88.50
3/4"	"	32.25	55.00	87.25
1"	"	31.50	55.00	86.50
1-1/4"	"	30.00	55.00	85.00
1-1/2"	"	30.00	55.00	85.00
2-1/2" x				
1"	EA.	77.00	66.00	143
1-1/4"	"	76.00	66.00	142
1-1/2"	"	68.00	66.00	134
2"	"	66.00	66.00	132
3" x				
1-1/2"	EA.	93.00	83.00	176
2"	"	83.00	83.00	166
2-1/2"	"	85.00	83.00	168
4" x				
2"	EA.	180	95.00	275
2-1/2"	"	190	95.00	285
3"	"	170	95.00	265
Slip coupling				
1/4"	EA.	0.98	22.25	23.23
1/2"	"	1.66	26.50	28.16
3/4"	"	3.45	33.25	36.70
1"	"	7.30	37.00	44.30
1-1/4"	"	11.00	41.50	52.50
1-1/2"	"	15.00	44.25	59.25
2"	"	25.25	55.00	80.25
2-1/2"	"	33.00	55.00	88.00
3"	"	63.00	66.00	129
4"	"	120	83.00	203
Coupling with drain				
1/2"	EA.	12.50	33.25	45.75
3/4"	"	18.50	37.00	55.50
1"	"	23.00	41.50	64.50
Reducer				

15 MECHANICAL

Plumbing

15410.11 Copper Fittings (Cont.)

	UNIT	MAT.	INST.	TOTAL
3/8" x 1/4"	EA.	3.57	26.50	30.07
1/2" x 3/8"	"	2.88	26.50	29.38
3/4" x				
1/4"	EA.	5.85	30.25	36.10
3/8"	"	6.11	30.25	36.36
1/2"	"	6.37	30.25	36.62
1" x				
1/2"	EA.	8.77	33.25	42.02
3/4"	"	6.73	33.25	39.98
1-1/4" x				
1/2"	EA.	12.50	37.00	49.50
3/4"	"	12.50	37.00	49.50
1"	"	12.50	37.00	49.50
1-1/2" x				
1/2"	EA.	16.25	41.50	57.75
3/4"	"	16.25	41.50	57.75
1"	"	16.25	41.50	57.75
1-1/4"	"	16.25	41.50	57.75
2" x				
1/2"	EA.	32.00	47.50	79.50
3/4"	"	32.00	47.50	79.50
1"	"	32.00	47.50	79.50
1-1/4"	"	30.50	47.50	78.00
1-1/2"	"	30.50	47.50	78.00
2-1/2" x				
1"	EA.	71.00	55.00	126
1-1/4"	"	63.00	55.00	118
1-1/2"	"	62.00	55.00	117
2"	"	61.00	55.00	116
3" x				
1-1/4"	EA.	83.00	66.00	149
1-1/2"	"	85.00	66.00	151
2"	"	76.00	66.00	142
2-1/2"	"	78.00	66.00	144
4" x				
2"	EA.	180	83.00	263
3"	"	160	83.00	243
Female adapters				
1/4"	EA.	9.30	26.50	35.80
3/8"	"	9.52	30.25	39.77
1/2"	"	4.53	33.25	37.78
3/4"	"	6.21	37.00	43.21
1"	"	14.50	37.00	51.50
1-1/4"	"	21.00	41.50	62.50
1-1/2"	"	32.75	41.50	74.25
2"	"	44.75	44.25	89.00
2-1/2"	"	160	47.50	208
3"	"	250	55.00	305
4"	"	300	66.00	366
Increasing female adapters				
1/8" x				
3/8"	EA.	9.14	26.50	35.64
1/2"	"	8.51	26.50	35.01

15 MECHANICAL

Plumbing

15410.11 Copper Fittings (Cont.)

	UNIT	MAT.	INST.	TOTAL
1/4" x 1/2"	EA.	8.92	28.75	37.67
3/8" x 1/2"	"	9.57	30.25	39.82
1/2" X				
3/4"	EA.	10.25	33.25	43.50
1"	"	20.50	33.25	53.75
3/4" X				
1"	EA.	21.75	37.00	58.75
1-1/4"	"	36.50	37.00	73.50
1" x				
1-1/4"	EA.	39.00	37.00	76.00
1-1/2"	"	42.75	37.00	79.75
1-1/4" x				
1-1/2"	EA.	46.50	41.50	88.00
2"	"	59.00	41.50	101
1-1/2" x 2"	"	86.00	44.25	130
Reducing female adapters				
3/8" x 1/4"	EA.	8.24	30.25	38.49
1/2" x				
1/4"	EA.	7.09	33.25	40.34
3/8"	"	7.09	33.25	40.34
3/4" x 1/2"	"	9.88	37.00	46.88
1" x				
1/2"	EA.	26.50	37.00	63.50
3/4"	"	21.25	37.00	58.25
1-1/4" x				
1/2"	EA.	35.75	41.50	77.25
3/4"	"	44.75	41.50	86.25
1"	"	44.75	41.50	86.25
1-1/2" x				
1"	EA.	41.75	44.25	86.00
1-1/4"	"	45.25	44.25	89.50
2" x				
1"	EA.	57.00	47.50	105
1-1/4"	"	84.00	47.50	132
1-1/2"	"	73.00	47.50	121
Female fitting adapters				
1/2"	EA.	12.50	33.25	45.75
3/4"	"	16.25	33.25	49.50
3/4" x 1/2"	"	19.25	35.00	54.25
1"	"	21.50	37.00	58.50
1-1/4"	"	34.75	39.00	73.75
1-1/2"	"	45.75	41.50	87.25
2"	"	61.00	44.25	105
Male adapters				
1/4"	EA.	14.25	30.25	44.50
3/8"	"	7.09	30.25	37.34
3"	"	180	55.00	235
4"	"	250	66.00	316
Increasing male adapters				
3/8" x 1/2"	EA.	9.65	30.25	39.90
1/2" x				
3/4"	EA.	8.37	33.25	41.62
1"	"	18.75	33.25	52.00

15 MECHANICAL

Plumbing

15410.11 Copper Fittings *(Cont.)*

	UNIT	MAT.	INST.	TOTAL
3/4" x				
1"	EA.	18.50	35.00	53.50
1-1/4"	"	23.50	35.00	58.50
1" x 1-1/4"	"	23.50	37.00	60.50
1-1/2" x				
3/4"	EA.	28.50	39.00	67.50
1"	"	41.75	39.00	80.75
1-1/4"	"	45.25	39.00	84.25
2" x				
1"	EA.	110	41.50	152
1-1/4"	"	110	41.50	152
1-1/2"	"	100	41.50	142
2" x 2-1/2"	"	180	44.25	224
Reducing male adapters				
1/2" x				
1/4"	EA.	12.25	33.25	45.50
3/8"	"	10.25	33.25	43.50
3/4" x 1/2"	"	11.75	35.00	46.75
1" x				
1/2"	EA.	32.00	37.00	69.00
3/4"	"	25.50	37.00	62.50
1-1/4" x				
3/4"	EA.	44.00	39.00	83.00
1"	"	38.25	39.00	77.25
1-1/2" x				
3/4"	EA.	48.50	41.50	90.00
1"	"	71.00	41.50	113
1-1/4'	"	60.00	41.50	102
2" x				
3/4"	EA.	54.00	44.25	98.25
1"	"	54.00	44.25	98.25
1-1/4"	"	81.00	44.25	125
1-1/2"	"	90.00	44.25	134
2-1/2" x				
2"	EA.	170	51.00	221
Fitting x male adapters				
1/2"	EA.	17.50	33.25	50.75
3/4"	"	22.75	35.00	57.75
1"	"	23.00	37.00	60.00
1-1/4"	"	89.00	39.00	128
1-1/2"	"	110	41.50	152
2"	"	170	44.25	214
90 ells				
1/8"	EA.	2.66	26.50	29.16
1/4"	"	4.25	26.50	30.75
3/8"	"	4.03	30.25	34.28
1/2"	"	1.34	33.25	34.59
3/4"	"	3.01	35.00	38.01
1"	"	7.42	37.00	44.42
1-1/4"	"	11.25	39.00	50.25
1-1/2"	"	17.50	41.50	59.00
2"	"	31.75	44.25	76.00
2-1/2"	"	64.00	51.00	115

15 MECHANICAL

Plumbing

15410.11 Copper Fittings *(Cont.)*

	UNIT	MAT.	INST.	TOTAL
3"	EA.	85.00	55.00	140
4"	"	220	66.00	286
Reducing 90 ell				
3/8" x 1/4"	EA.	7.31	30.25	37.56
1/2" x				
1/4"	EA.	10.50	33.25	43.75
3/8"	"	10.50	33.25	43.75
3/4" x 1/2"	"	9.19	35.00	44.19
1" x				
1/2"	EA.	15.75	37.00	52.75
3/4"	"	15.00	37.00	52.00
1-1/4" x				
1/2"	EA.	39.75	39.00	78.75
3/4"	"	42.25	39.00	81.25
1"	"	42.25	39.00	81.25
1-1/2" x				
1/2"	EA.	41.00	41.50	82.50
3/4"	"	59.00	41.50	101
1"	"	55.00	41.50	96.50
1-1/4"	"	50.00	41.50	91.50
2" x				
3/4"	EA.	91.00	44.25	135
1"	"	79.00	44.25	123
1-1/4"	"	83.00	44.25	127
1-1/2"	"	68.00	44.25	112
2-1/2" x				
1-1/2"	EA.	190	51.00	241
2"	"	190	51.00	241
3" x				
2"	EA.	380	55.00	435
2-1/2"	"	380	55.00	435
Street ells, copper				
1/4"	EA.	7.57	26.50	34.07
3/8"	"	5.22	30.25	35.47
1/2"	"	2.10	33.25	35.35
3/4"	"	4.43	35.00	39.43
1"	"	11.50	37.00	48.50
1-1/4"	"	17.50	39.00	56.50
1-1/2"	"	22.75	41.50	64.25
2"	"	49.75	44.25	94.00
2-1/2"	"	85.00	51.00	136
3"	"	130	55.00	185
4"	"	290	66.00	356
Female, 90 ell				
1/2"	EA.	1.44	33.25	34.69
3/4"	"	3.22	35.00	38.22
1"	"	7.94	37.00	44.94
1-1/4"	"	12.00	39.00	51.00
1-1/2"	"	18.75	41.50	60.25
2"	"	34.00	44.25	78.25
Female increasing, 90 ell				
3/8" x 1/2"	EA.	16.00	30.25	46.25
1/2" x				

15 MECHANICAL

Plumbing

15410.11 Copper Fittings (Cont.)

	UNIT	MAT.	INST.	TOTAL
3/4"	EA.	11.00	33.25	44.25
1"	"	22.75	33.25	56.00
3/4" x 1"	"	20.25	35.00	55.25
1" x 1-1/4"	"	52.00	37.00	89.00
1-1/4" x 1-1/2"	"	52.00	39.00	91.00
Female reducing, 90 ell				
1/2" x 3/8"	EA.	17.75	33.25	51.00
3/4" x 1/2"	"	19.50	35.00	54.50
1" x				
1/2"	EA.	27.50	37.00	64.50
3/4"	"	29.75	37.00	66.75
1-1/4" x				
3/4"	EA.	33.00	39.00	72.00
1"	"	34.25	39.00	73.25
1-1/2"	"	53.00	39.00	92.00
1-1/2" x				
1"	EA.	53.00	41.50	94.50
1-1/4"	"	53.00	41.50	94.50
2" x 1-1/2"	"	47.75	44.25	92.00
Male, 90 ell				
1/4"	EA.	13.50	26.50	40.00
3/8"	"	14.75	30.25	45.00
1/2"	"	7.54	33.25	40.79
3/4"	"	17.00	35.00	52.00
1"	"	19.75	37.00	56.75
1-1/4"	"	25.75	39.00	64.75
1-1/2"	"	32.75	41.50	74.25
2"	"	67.00	44.25	111
Male, increasing 90 ell				
1/2" x				
3/4"	EA.	26.75	33.25	60.00
1"	"	53.00	33.25	86.25
3/4" x 1"	"	50.00	35.00	85.00
1" x 1-1/4"	"	46.75	37.00	83.75
1-1/4" x 1-1/2"	"	46.75	39.00	85.75
Male, reducing 90 ell				
1/2" x 3/8"	EA.	15.25	33.25	48.50
3/4" x 1/2"	"	26.75	35.00	61.75
1" x				
1/2"	EA.	51.00	37.00	88.00
3/4"	"	48.25	37.00	85.25
1-1/4" x 1"	"	46.75	39.00	85.75
Drop ear ells				
1/2"	EA.	9.72	33.25	42.97
Female drop ear ells				
1/2"	EA.	9.72	33.25	42.97
1/2" x 3/8"	"	17.00	33.25	50.25
3/4"	"	28.50	35.00	63.50
Female flanged sink ell				
1/2"	EA.	17.75	33.25	51.00
45 ells				
1/4"	EA.	8.11	26.50	34.61
3/8"	"	6.59	30.25	36.84

15 MECHANICAL

Plumbing

15410.11 Copper Fittings *(Cont.)*

	UNIT	MAT.	INST.	TOTAL
3"	EA.	95.00	55.00	150
4"	"	200	66.00	266
45 street ell				
1/4"	EA.	9.17	26.50	35.67
3/8"	"	9.91	30.25	40.16
1/2"	"	2.95	33.25	36.20
3/4"	"	4.43	35.00	39.43
1"	"	11.75	37.00	48.75
1-1/2"	"	18.00	41.50	59.50
2"	"	28.75	44.25	73.00
2-1/2"	"	68.00	51.00	119
3"	"	92.00	55.00	147
4"	"	290	66.00	356
Tee				
1/8"	EA.	6.89	26.50	33.39
1/4"	"	7.27	26.50	33.77
3/8"	"	5.55	30.25	35.80
3"	"	140	55.00	195
4"	"	340	66.00	406
Caps				
1/4"	EA.	1.25	26.50	27.75
3/8"	"	2.00	30.25	32.25
3"	"	70.00	55.00	125
4"	"	130	66.00	196
Test caps				
1/2"	EA.	1.17	33.25	34.42
3/4"	"	1.32	35.00	36.32
1"	"	2.36	37.00	39.36
1-1/4"	"	2.36	39.00	41.36
1-1/2"	"	2.39	41.50	43.89
2"	"	5.13	44.25	49.38
3"	"	10.50	55.00	65.50
Flush bushing				
1/4" x 1/8"	EA.	2.39	26.50	28.89
1/2" x				
1/4"	EA.	3.23	33.25	36.48
3/8"	"	2.85	33.25	36.10
3/4" x				
3/8"	EA.	6.01	35.00	41.01
1/2"	"	5.31	35.00	40.31
1" x				
1/2"	EA.	9.15	37.00	46.15
3/4"	"	8.13	37.00	45.13
1-1/4" x				
1/2"	EA.	13.00	39.00	52.00
3/4"	"	13.00	39.00	52.00
x 1"	"	10.00	39.00	49.00
1-1/2" x				
1/2"	EA.	13.25	41.50	54.75
1-1/4"	"	13.25	41.50	54.75
2" x				
1"	EA.	33.00	44.25	77.25
1-1/4"	"	33.00	44.25	77.25

15 MECHANICAL

Plumbing

15410.11 Copper Fittings (Cont.)

	UNIT	MAT.	INST.	TOTAL
1-1/2"	EA.	38.00	44.25	82.25
Female flush bushing				
1/2" x				
1/2" x 1/8"	EA.	7.22	33.25	40.47
1/4"	"	7.60	33.25	40.85
Union				
1/4"	EA.	45.00	26.50	71.50
3/8"	"	62.00	30.25	92.25
3"	"	400	55.00	455
Female				
1/2"	EA.	21.75	33.25	55.00
3/4"	"	21.75	35.00	56.75
Male				
1/2"	EA.	23.25	33.25	56.50
3/4"	"	31.00	35.00	66.00
1"	"	67.00	37.00	104
45 degree wye				
1/2"	EA.	28.75	33.25	62.00
3/4"	"	41.50	35.00	76.50
1"	"	56.00	37.00	93.00
1" x 3/4" x 3/4"	"	78.00	37.00	115
1-1/4"	"	78.00	39.00	117
1-1/4" x 1" x 1"	"	78.00	39.00	117
Twin ells				
1" x 3/4" x 3/4"	EA.	20.25	37.00	57.25
1" x 1" x 1"	"	20.25	37.00	57.25
1-1/4" x 1" x 1"	"	31.25	39.00	70.25
Companion flanges, 125#				
2" x 6"	EA.	140	44.25	184
2-1/2"	"	170	51.00	221
3" x 7-1/2"	"	200	55.00	255
4" x 9"	"	220	66.00	286
90 union ells, male				
1/2"	EA.	31.25	33.25	64.50
3/4"	"	52.00	35.00	87.00
1"	"	78.00	37.00	115
DWV fittings, coupling with stop				
1-1/4"	EA.	7.13	39.00	46.13
1-1/2"	"	8.87	41.50	50.37
1-1/2" x 1-1/4"	"	14.25	41.50	55.75
2"	"	12.25	44.25	56.50
2" x 1-1/4"	"	16.75	44.25	61.00
2" x 1-1/2"	"	16.50	44.25	60.75
3"	"	23.75	55.00	78.75
3" x 1-1/2"	"	57.00	55.00	112
3" x 2"	"	54.00	55.00	109
4"	"	76.00	66.00	142
Slip coupling				
1-1/2"	EA.	13.75	41.50	55.25
2"	"	16.50	44.25	60.75
3"	"	30.00	55.00	85.00
90 ells				
1-1/2"	EA.	16.75	41.50	58.25

15 MECHANICAL

Plumbing

15410.11 Copper Fittings (Cont.)

	UNIT	MAT.	INST.	TOTAL
1-1/2" x 1-1/4"	EA.	46.25	41.50	87.75
2"	"	30.75	44.25	75.00
2" x 1-1/2"	"	62.00	44.25	106
3"	"	82.00	55.00	137
4"	"	270	66.00	336
Street, 90 elbows				
1-1/2"	EA.	21.50	41.50	63.00
2"	"	47.00	44.25	91.25
3"	"	120	55.00	175
4"	"	300	66.00	366
Female, 90 elbows				
1-1/2"	EA.	21.25	41.50	62.75
2"	"	41.00	44.25	85.25
Male, 90 elbows				
1-1/2"	EA.	37.50	41.50	79.00
2"	"	77.00	44.25	121
90 with side inlet				
3" x 3" x 1"	EA.	110	55.00	165
3" x 3" x 1-1/2"	"	120	55.00	175
3" x 3" x 2"	"	120	55.00	175
45 ells				
1-1/4"	EA.	14.00	39.00	53.00
1-1/2"	"	11.50	41.50	53.00
2"	"	26.75	44.25	71.00
3"	"	57.00	55.00	112
4"	"	260	66.00	326
Street, 45 ell				
1-1/2"	EA.	18.50	41.50	60.00
2"	"	33.00	44.25	77.25
3"	"	95.00	55.00	150
60 ell				
1-1/2"	EA.	29.00	41.50	70.50
2"	"	54.00	44.25	98.25
3"	"	120	55.00	175
22-1/2 ell				
1-1/2"	EA.	35.25	41.50	76.75
2"	"	45.50	44.25	89.75
3"	"	79.00	55.00	134
11-1/4 ell				
1-1/2"	EA.	39.00	41.50	80.50
2"	"	55.00	44.25	99.25
3"	"	110	55.00	165
Wye				
1-1/4"	EA.	59.00	39.00	98.00
1-1/2"	"	65.00	41.50	107
2"	"	84.00	44.25	128
2" x 1-1/2" x 1-1/2"	"	94.00	44.25	138
2" x 1-1/2" x 2"	"	100	44.25	144
2" x 1-1/2" x 2"	"	100	44.25	144
3"	"	200	55.00	255
3" x 3" x 1-1/2"	"	180	55.00	235
3" x 3" x 2"	"	180	55.00	235
4"	"	410	66.00	476

15 MECHANICAL

Plumbing

15410.11 Copper Fittings *(Cont.)*

	UNIT	MAT.	INST.	TOTAL
4" x 4" x 2"	EA.	290	66.00	356
4" x 4" x 3"	"	290	66.00	356
Sanitary tee				
1-1/4"	EA.	30.00	39.00	69.00
1-1/2"	"	37.50	41.50	79.00
2"	"	43.75	44.25	88.00
2" x 1-1/2" x 1-1/2"	"	70.00	44.25	114
2" x 1-1/2" x 2"	"	71.00	44.25	115
2" x 2" x 1-1/2"	"	41.25	44.25	85.50
3"	"	160	55.00	215
3" x 3" x 1-1/2"	"	120	55.00	175
3" x 3" x 2"	"	120	55.00	175
4"	"	410	66.00	476
4" x 4" x 3"	"	340	66.00	406
Female sanitary tee				
1-1/2"	EA.	73.00	41.50	115
Long turn tee				
1-1/2"	EA.	72.00	41.50	114
2"	"	160	44.25	204
3" x 1-1/2"	"	210	55.00	265
Double wye				
1-1/2"	EA.	110	41.50	152
2"	"	190	44.25	234
2" x 2" x 1-1/2" x 1-1/2"	"	150	44.25	194
3"	"	300	55.00	355
3" x 3" x 1-1/2" x 1-1/2"	"	300	55.00	355
3" x 3" x 2" x 2"	"	300	55.00	355
4" x 4" x 1-1/2" x 1-1/2"	"	330	66.00	396
Double sanitary tee				
1-1/2"	EA.	74.00	41.50	116
2"	"	170	44.25	214
2" x 2" x 1-1/2"	"	150	44.25	194
3"	"	200	55.00	255
3" x 3" x 1-1/2" x 1-1/2"	"	260	55.00	315
3" x 3" x 2" x 2"	"	210	55.00	265
4" x 4" x 1-1/2" x 1-1/2"	"	450	66.00	516
Long				
2" x 1-1/2"	EA.	190	44.25	234
Twin elbow				
1-1/2"	EA.	96.00	41.50	138
2"	"	150	44.25	194
2" x 1-1/2" x 1-1/2"	"	130	44.25	174
Spigot adapter, manoff				
1-1/2" x 2"	EA.	63.00	41.50	105
1-1/2" x 3"	"	77.00	41.50	119
2"	"	31.25	44.25	75.50
2" x 3"	"	74.00	44.25	118
2" x 4"	"	100	44.25	144
3"	"	110	55.00	165
3" x 4"	"	200	55.00	255
4"	"	170	66.00	236
No-hub adapters				
1-1/2" x 2"	EA.	38.25	41.50	79.75

15 MECHANICAL

Plumbing

15410.11 Copper Fittings *(Cont.)*

	UNIT	MAT.	INST.	TOTAL
2"	EA.	36.00	44.25	80.25
2" x 3"	"	83.00	44.25	127
3"	"	73.00	55.00	128
3" x 4"	"	150	55.00	205
4"	"	160	66.00	226
Fitting reducers				
1-1/2" x 1-1/4"	EA.	13.50	41.50	55.00
2" x 1-1/2"	"	21.25	44.25	65.50
3" x 1-1/2"	"	60.00	55.00	115
3" x 2"	"	53.00	55.00	108
Slip joint (Desanco)				
1-1/4"	EA.	22.75	39.00	61.75
1-1/2"	"	23.75	41.50	65.25
1-1/2" x 1-1/4"	"	24.50	41.50	66.00
Street x slip joint (Desanco)				
1-1/2"	EA.	29.25	41.50	70.75
1-1/2" x 1-1/4"	"	31.25	41.50	72.75
Flush bushing				
1-1/2" x 1-1/4"	EA.	16.75	41.50	58.25
2" x 1-1/2"	"	28.75	44.25	73.00
3" x 1-1/2"	"	52.00	55.00	107
3" x 2"	"	52.00	55.00	107
Male hex trap bushing				
1-1/4" x 1-1/2"	EA.	24.50	39.00	63.50
1-1/2"	"	18.00	41.50	59.50
1-1/2" x 2"	"	27.25	41.50	68.75
2"	"	21.25	44.25	65.50
Round trap bushing				
1-1/2"	EA.	20.75	41.50	62.25
2"	"	22.25	44.25	66.50
Female adapter				
1-1/4"	EA.	24.25	39.00	63.25
1-1/2"	"	38.00	41.50	79.50
1-1/2" x 2"	"	96.00	41.50	138
2"	"	52.00	44.25	96.25
2" x 1-1/2"	"	84.00	44.25	128
3"	"	210	55.00	265
Fitting x female adapter				
1-1/2"	EA.	51.00	41.50	92.50
2"	"	68.00	44.25	112
Male adapters				
1-1/4"	EA.	21.25	39.00	60.25
1-1/4" x 1-1/2"	"	49.50	39.00	88.50
1-1/2"	"	24.50	41.50	66.00
1-1/2" x 2"	"	92.00	41.50	134
2"	"	41.00	44.25	85.25
2" x 1-1/2"	"	95.00	44.25	139
3"	"	210	55.00	265
Male x slip joint adapters				
1-1/2" x 1-1/4"	EA.	38.75	41.50	80.25
Dandy cleanout				
1-1/2"	EA.	68.00	41.50	110
2"	"	81.00	44.25	125

15 MECHANICAL

Plumbing

15410.11 Copper Fittings *(Cont.)*

	UNIT	MAT.	INST.	TOTAL
3"	EA.	290	55.00	345
End cleanout, flush pattern				
1-1/2" x 1"	EA.	41.50	41.50	83.00
2" x 1-1/2"	"	50.00	44.25	94.25
3" x 2-1/2"	"	110	55.00	165
Copper caps				
1-1/2"	EA.	14.25	41.50	55.75
2"	"	26.00	44.25	70.25
Closet flanges				
3"	EA.	62.00	55.00	117
4"	"	110	66.00	176
Drum traps, with cleanout				
1-1/2" x 3" x 6"	EA.	230	41.50	272
P-trap, swivel, with cleanout				
1-1/2"	EA.	150	41.50	192
P-trap, solder union				
1-1/2"	EA.	61.00	41.50	103
2"	"	110	44.25	154
With cleanout				
1-1/2"	EA.	67.00	41.50	109
2"	"	120	44.25	164
2" x 1-1/2"	"	120	44.25	164
Swivel joint, with cleanout				
1-1/2" x 1-1/4"	EA.	86.00	41.50	128
1-1/2"	"	110	41.50	152
2" x 1-1/2"	"	130	44.25	174
Estabrook TY, with inlets				
3", with 1-1/2" inlet	EA.	190	55.00	245
Fine thread adapters				
1/2"	EA.	4.94	33.25	38.19
1/2" x 1/2" IPS	"	5.57	33.25	38.82
1/2" x 3/4" IPS	"	9.08	33.25	42.33
1/2" x male	"	3.43	33.25	36.68
1/2" x female	"	7.00	33.25	40.25
Copper pipe fittings				
1/2"				
90 deg ell	EA.	2.17	14.75	16.92
45 deg ell	"	2.75	14.75	17.50
Tee	"	3.64	19.00	22.64
Cap	"	1.48	7.37	8.85
Coupling	"	1.60	14.75	16.35
Union	"	11.00	16.50	27.50
3/4"				
90 deg ell	EA.	4.74	16.50	21.24
45 deg ell	"	5.54	16.50	22.04
Tee	"	7.95	22.25	30.20
Cap	"	2.90	7.81	10.71
Coupling	"	3.23	16.50	19.73
Union	"	16.25	19.00	35.25
1"				
90 deg ell	EA.	11.00	22.25	33.25
45 deg ell	"	14.25	22.25	36.50
Tee	"	18.00	26.50	44.50

15 MECHANICAL

Plumbing	UNIT	MAT.	INST.	TOTAL
15410.11 **Copper Fittings** *(Cont.)*				
Cap	EA.	5.37	11.00	16.37
Coupling	"	7.95	22.25	30.20
Union	"	21.25	22.25	43.50
1-1/4"				
90 deg ell	EA.	15.00	19.00	34.00
45 deg ell	"	18.75	19.00	37.75
Tee	"	24.50	33.25	57.75
Cap	"	4.30	11.00	15.30
Union	"	35.25	23.75	59.00
1-1/2"				
90 deg ell	EA.	19.50	23.75	43.25
45 deg ell	"	23.25	23.75	47.00
Tee	"	32.00	37.00	69.00
Cap	"	4.30	11.00	15.30
Coupling	"	14.25	22.25	36.50
Union	"	54.00	30.25	84.25
2"				
90 deg ell	EA.	38.25	26.50	64.75
45 deg ell	"	35.00	41.50	76.50
Tee	"	55.00	41.50	96.50
Cap	"	8.88	13.25	22.13
Coupling	"	23.25	26.50	49.75
Union	"	58.00	33.25	91.25
2-1/2"				
90 deg ell	EA.	73.00	33.25	106
45 deg ell	"	64.00	33.25	97.25
Tee	"	73.00	47.50	121
Cap	"	18.00	16.50	34.50
Coupling	"	35.25	33.25	68.50
Union	"	110	37.00	147
15410.14 **Brass I.p.s. Fittings**				
Fittings, iron pipe size, 45 deg ell				
1/8"	EA.	11.50	26.50	38.00
1/4"	"	11.50	26.50	38.00
3/8"	"	11.50	30.25	41.75
1/2"	"	11.50	33.25	44.75
3/4"	"	16.50	35.00	51.50
1"	"	28.25	37.00	65.25
1-1/4"	"	44.75	39.00	83.75
1-1/2"	"	56.00	41.50	97.50
2"	"	91.00	44.25	135
90 deg ell				
1/8"	EA.	11.00	26.50	37.50
1/4"	"	11.00	26.50	37.50
3/8"	"	11.00	30.25	41.25
1/2"	"	11.00	33.25	44.25
3/4"	"	13.00	35.00	48.00
1"	"	23.25	37.00	60.25
1-1/4"	"	33.75	39.00	72.75
1-1/2"	"	43.50	41.50	85.00
2"	"	69.00	44.25	113
2-1/2"	"	170	51.00	221

15 MECHANICAL

Plumbing	UNIT	MAT.	INST.	TOTAL
15410.14 **Brass I.p.s. Fittings** (Cont.)				
90 deg ell, reducing				
1/4" x 1/8"	EA.	13.00	26.50	39.50
3/8" x 1/8"	"	13.00	30.25	43.25
3/8" x 1/4"	"	13.00	30.25	43.25
1/2" x 1/4"	"	13.00	33.25	46.25
1/2" x 3/8"	"	13.00	33.25	46.25
3/4" x 1/2"	"	18.75	35.00	53.75
1" x 3/8"	"	26.75	37.00	63.75
1" x 1/2"	"	26.75	37.00	63.75
1" x 3/4"	"	26.75	37.00	63.75
1-1/4" x 3/4"	"	44.25	39.00	83.25
1-1/4" x 1"	"	44.25	39.00	83.25
1-1/2" x 1/2"	"	51.00	41.50	92.50
1-1/2" x 3/4"	"	51.00	41.50	92.50
1-1/2" x 1"	"	51.00	41.50	92.50
2" x 1"	"	75.00	44.25	119
2" x 1-1/4"	"	75.00	44.25	119
2" x 1-1/2"	"	75.00	44.25	119
Street ell, 45 deg				
1/2"	EA.	13.00	33.25	46.25
3/4"	"	18.75	35.00	53.75
2"	"	91.00	44.25	135
90 deg				
1/8"	EA.	13.00	26.50	39.50
1/4"	"	13.00	26.50	39.50
3/8"	"	13.00	30.25	43.25
1/2"	"	13.00	33.25	46.25
3/4"	"	16.00	35.00	51.00
1"	"	21.00	37.00	58.00
1-1/4"	"	44.50	39.00	83.50
1-1/2"	"	46.75	41.50	88.25
2"	"	75.00	44.25	119
Tee, 1/8"	"	11.00	26.50	37.50
1/4"	"	11.00	26.50	37.50
3/8"	"	11.00	30.25	41.25
1/2"	"	11.00	33.25	44.25
3/4"	"	15.00	35.00	50.00
1"	"	22.25	37.00	59.25
1-1/4"	"	49.00	39.00	88.00
1-1/2"	"	55.00	41.50	96.50
2"	"	92.00	44.25	136
2-1/2"	"	220	51.00	271
Tee, reducing, 3/8" x				
1/4"	EA.	15.25	30.25	45.50
1/2"	"	15.25	30.25	45.50
1/2" x				
1/4"	EA.	15.25	33.25	48.50
3/8"	"	15.25	33.25	48.50
3/4"	"	18.00	33.25	51.25
3/4" x				
1/4"	EA.	18.00	35.00	53.00
1/2"	"	18.00	35.00	53.00
1"	"	36.25	35.00	71.25

15 MECHANICAL

Plumbing

15410.14 Brass I.p.s. Fittings (Cont.)

	UNIT	MAT.	INST.	TOTAL
1" x				
1/2"	EA.	36.00	37.00	73.00
3/4"	"	36.00	37.00	73.00
1-1/4" x				
1/2"	EA.	46.25	39.00	85.25
1"	"	46.25	39.00	85.25
1-1/2" x				
1/2"	EA.	67.00	41.50	109
3/4"	"	67.00	41.50	109
1"	"	67.00	41.50	109
1-1/2"	"	72.00	41.50	114
2" x				
1/2"	EA.	100	44.25	144
3/4"	"	100	44.25	144
1"	"	100	44.25	144
1-1/4"	"	100	44.25	144
1-1/2"	"	100	44.25	144
2-1/2" x 2"	"	210	51.00	261
Tee, reducing				
1/2" x 3/8" x 1/2"	EA.	14.00	33.25	47.25
3/4" x 1/2" x 1/2"	"	20.00	35.00	55.00
3/4" x 1/2" x 3/4"	"	18.75	35.00	53.75
1" x 1/2" x 1/2"	"	35.00	37.00	72.00
1" x 1/2" x 3/4"	"	35.00	37.00	72.00
1" x 3/4" x 1/2"	"	42.50	37.00	79.50
1" x 3/4" x 3/4"	"	35.00	37.00	72.00
1-1/4" x 1/2" x 1-1/4"	"	42.00	39.00	81.00
1-1/4" x 1" x 1"	"	44.75	39.00	83.75
1-1/4" x 1-1/4" x 3/4"	"	42.00	39.00	81.00
1-1/2" x 1-1/4" x 1-1/4"	"	61.00	41.50	103
Union				
1/8"	EA.	29.00	26.50	55.50
1/4"	"	29.00	26.50	55.50
3/8"	"	29.00	30.25	59.25
1/2"	"	29.00	33.25	62.25
3/4"	"	39.75	35.00	74.75
1"	"	53.00	37.00	90.00
1-1/4"	"	76.00	39.00	115
1-1/2"	"	91.00	41.50	133
2"	"	140	44.25	184
Brass face bushing				
3/8" x 1/4"	EA.	10.00	30.25	40.25
1/2" x 3/8"	"	10.00	33.25	43.25
3/4" x 1/2"	"	12.50	35.00	47.50
1" x 3/4"	"	21.25	37.00	58.25
1-1/4" x 1"	"	26.75	39.00	65.75
Hex bushing, 1/4" x 1/8"	"	7.08	26.50	33.58
1/2" x				
1/4"	EA.	6.47	33.25	39.72
3/8"	"	6.47	33.25	39.72
5/8" x				
1/8"	EA.	6.47	33.25	39.72
1/4"	"	6.47	33.25	39.72

15 MECHANICAL

Plumbing		UNIT	MAT.	INST.	TOTAL
15410.14	**Brass I.p.s. Fittings** (Cont.)				
3/4" x					
1/8"		EA.	9.39	35.00	44.39
1/4"		"	9.39	35.00	44.39
3/8"		"	8.15	35.00	43.15
1/2"		"	8.15	35.00	43.15
1" x					
1/4"		EA.	12.50	37.00	49.50
3/8"		"	12.50	37.00	49.50
1/2"		"	11.50	37.00	48.50
3/4"		"	11.50	37.00	48.50
1-1/4" x					
3/8"		EA.	19.75	39.00	58.75
1/2"		"	19.75	39.00	58.75
3/4"		"	14.50	39.00	53.50
1"		"	14.50	39.00	53.50
1-1/2" x					
1/4"		EA.	26.25	41.50	67.75
1/2"		"	26.25	41.50	67.75
3/4"		"	26.25	41.50	67.75
1"		"	25.00	41.50	66.50
1-1/4"		"	25.00	41.50	66.50
2" x					
1/2"		EA.	33.25	44.25	77.50
3/4"		"	33.25	44.25	77.50
1"		"	33.25	44.25	77.50
1-1/4"		"	30.75	44.25	75.00
1-1/2"		"	30.75	44.25	75.00
2-1/2" x					
1"		EA.	67.00	51.00	118
2"		"	59.00	51.00	110
3" x					
1-1/2"		EA.	94.00	55.00	149
2"		"	94.00	55.00	149
2-1/2"		"	98.00	55.00	153
4" x					
2"		EA.	230	66.00	296
3"		"	230	66.00	296
Caps					
1/8"		EA.	6.46	26.50	32.96
1/4"		"	6.96	26.50	33.46
3/8"		"	6.96	30.25	37.21
1/2"		"	6.96	33.25	40.21
3/4"		"	7.34	35.00	42.34
1"		"	13.25	37.00	50.25
1-1/4"		"	19.75	39.00	58.75
1-1/2"		"	25.75	41.50	67.25
2"		"	42.25	44.25	86.50
Couplings					
1/8"		EA.	7.51	26.50	34.01
1/4"		"	7.51	26.50	34.01
3/8"		"	7.51	30.25	37.76
1/2"		"	7.51	33.25	40.76
3/4"		"	10.25	35.00	45.25

15 MECHANICAL

Plumbing	UNIT	MAT.	INST.	TOTAL
15410.14 **Brass I.p.s. Fittings** *(Cont.)*				
1"	EA.	16.25	37.00	53.25
1-1/4"	"	25.75	39.00	64.75
1-1/2"	"	32.75	41.50	74.25
2"	"	51.00	44.25	95.25
2-1/2"	"	77.00	51.00	128
Couplings, reducing, 1/4" x 1/8"	"	8.78	26.50	35.28
3/8" x				
1/8"	EA.	8.78	30.25	39.03
1/4"	"	8.78	30.25	39.03
1/2" x				
1/8"	EA.	10.75	33.25	44.00
1/4"	"	9.59	33.25	42.84
3/8"	"	9.59	33.25	42.84
3/4" x				
1/4"	EA.	15.00	35.00	50.00
3/8"	"	12.25	35.00	47.25
1/2"	"	12.25	35.00	47.25
1" x				
1/2"	EA.	17.50	35.00	52.50
3/4"	"	17.50	35.00	52.50
1-1/4" x				
1/2"	EA.	47.00	39.00	86.00
3/4"	"	32.75	39.00	71.75
1"	"	32.75	39.00	71.75
1-1/2" x				
3/4"	EA.	46.75	41.50	88.25
1"	"	42.00	41.50	83.50
1-1/4"	"	42.00	41.50	83.50
2" x				
3/4"	EA.	63.00	44.25	107
1"	"	63.00	44.25	107
1-1/4"	"	63.00	44.25	107
1-1/2"	"	63.00	44.25	107
2-1/2" x 1-1/2"	"	94.00	51.00	145
Square head plug, solid				
1/8"	EA.	7.34	26.50	33.84
1/4"	"	7.34	26.50	33.84
3/8"	"	7.34	30.25	37.59
1/2"	"	7.34	33.25	40.59
3/4"	"	8.78	35.00	43.78
Cored				
1/2"	EA.	5.84	33.25	39.09
3/4"	"	7.34	35.00	42.34
1"	"	11.75	37.00	48.75
1-1/4"	"	16.25	39.00	55.25
1-1/2"	"	16.25	41.50	57.75
2"	"	28.00	44.25	72.25
3"	"	69.00	55.00	124
4"	"	120	66.00	186
Countersunk				
1/2"	EA.	8.14	33.25	41.39
3/4"	"	8.69	35.00	43.69
1-1/2"	"	23.25	41.50	64.75

15 MECHANICAL

Plumbing

15410.14 Brass I.p.s. Fittings (Cont.)

	UNIT	MAT.	INST.	TOTAL
2"	EA.	31.00	44.25	75.25
Locknut				
3/4"	EA.	7.34	35.00	42.34
1"	"	9.16	37.00	46.16
1-1/4"	"	16.75	39.00	55.75
2"	"	28.00	44.25	72.25
Close standard red nipple, 1/8"	"	2.38	26.50	28.88
1/8" x				
1-1/2"	EA.	4.42	26.50	30.92
2"	"	4.87	26.50	31.37
2-1/2"	"	5.52	26.50	32.02
3"	"	6.24	26.50	32.74
3-1/2"	"	7.47	26.50	33.97
4"	"	7.99	26.50	34.49
4-1/2"	"	8.38	26.50	34.88
5"	"	8.84	26.50	35.34
5-1/2"	"	10.25	26.50	36.75
6"	"	11.00	26.50	37.50
1/4" x close	"	4.87	26.50	31.37
1/4" x				
1-1/2"	EA.	7.21	26.50	33.71
2"	"	7.67	26.50	34.17
2-1/2"	"	7.99	26.50	34.49
3"	"	8.38	26.50	34.88
3-1/2"	"	9.36	26.50	35.86
4"	"	9.75	26.50	36.25
4-1/2"	"	10.50	26.50	37.00
5"	"	10.75	26.50	37.25
5-1/2"	"	11.75	26.50	38.25
6"	"	12.00	26.50	38.50
3/8" x close	"	4.87	30.25	35.12
3/8" x				
1-1/2"	EA.	5.72	30.25	35.97
2"	"	6.29	30.25	36.54
2-1/2"	"	7.57	30.25	37.82
3"	"	9.22	30.25	39.47
3-1/2"	"	10.00	30.25	40.25
4"	"	13.00	30.25	43.25
4-1/2"	"	13.25	30.25	43.50
5"	"	14.25	30.25	44.50
5-1/2"	"	15.00	30.25	45.25
6"	"	17.00	30.25	47.25
1/2" x close	"	6.43	33.25	39.68
1/2" x				
1-1/2"	EA.	7.22	33.25	40.47
2"	"	8.79	33.25	42.04
2-1/2"	"	10.00	33.25	43.25
3"	"	11.50	33.25	44.75
3-1/2"	"	12.75	33.25	46.00
4"	"	13.25	33.25	46.50
4-1/2"	"	14.25	33.25	47.50
5"	"	14.50	33.25	47.75
5-1/2"	"	15.25	33.25	48.50

15 MECHANICAL

Plumbing	UNIT	MAT.	INST.	TOTAL
15410.14 **Brass I.p.s. Fittings** *(Cont.)*				
6"	EA.	16.50	33.25	49.75
7-1/2"	"	50.00	33.25	83.25
8"	"	50.00	33.25	83.25
3/4" x close	"	17.75	35.00	52.75
3/4" x				
1-1/2"	EA.	9.93	35.00	44.93
2"	"	11.50	35.00	46.50
2-1/2"	"	12.75	35.00	47.75
3"	"	13.75	35.00	48.75
3-1/2"	"	15.00	35.00	50.00
4"	"	16.00	35.00	51.00
4-1/2"	"	17.25	35.00	52.25
5"	"	17.75	35.00	52.75
5-1/2"	"	20.25	35.00	55.25
6"	"	21.00	35.00	56.00
1" x close	"	14.25	37.00	51.25
1" x				
2"	EA.	20.75	37.00	57.75
2-1/2"	"	21.00	37.00	58.00
3"	"	21.75	37.00	58.75
3-1/2"	"	23.00	37.00	60.00
4"	"	24.75	37.00	61.75
4-1/2"	"	25.25	37.00	62.25
5"	"	29.25	37.00	66.25
5-1/2"	"	30.00	37.00	67.00
6"	"	33.25	37.00	70.25
1-1/4" x close	"	21.00	39.00	60.00
1-1/4" x				
2"	EA.	23.25	39.00	62.25
2-1/2"	"	25.25	39.00	64.25
3"	"	32.50	39.00	71.50
3-1/2"	"	35.50	39.00	74.50
4"	"	35.50	39.00	74.50
4-1/2"	"	36.50	39.00	75.50
5"	"	41.50	39.00	80.50
5-1/2"	"	46.75	39.00	85.75
6"	"	50.00	39.00	89.00
1-1/2" x close	"	25.00	41.50	66.50
1-1/2" x				
2"	EA.	25.25	41.50	66.75
2-1/2"	"	26.25	41.50	67.75
3"	"	33.00	41.50	74.50
3-1/2"	"	31.50	41.50	73.00
4-1/2"	"	41.50	41.50	83.00
5"	"	45.75	41.50	87.25
5-1/2"	"	51.00	41.50	92.50
6"	"	52.00	41.50	93.50
2" x close	"	33.25	44.25	77.50
2" x				
2-1/2"	EA.	39.50	44.25	83.75
3"	"	40.50	44.25	84.75
3-1/2"	"	45.75	44.25	90.00
4"	"	46.75	44.25	91.00

15 MECHANICAL

Plumbing	UNIT	MAT.	INST.	TOTAL
15410.14 Brass I.p.s. Fittings *(Cont.)*				
4-1/2"	EA.	52.00	44.25	96.25
5"	"	54.00	44.25	98.25
5-1/2"	"	66.00	44.25	110
6"	"	71.00	44.25	115
2-1/2" x close	"	75.00	51.00	126
2-1/2" x				
3"	EA.	77.00	51.00	128
3-1/2"	"	81.00	51.00	132
4"	"	96.00	51.00	147
2-1/2" x 5"	"	110	51.00	161
3"				
Close	EA.	110	55.00	165
15410.15 Brass Fittings				
Compression fittings, union				
3/8"	EA.	4.18	11.00	15.18
1/2"	"	6.87	11.00	17.87
5/8"	"	7.89	11.00	18.89
Union elbow				
3/8"	EA.	2.90	11.00	13.90
1/2"	"	4.00	11.00	15.00
5/8"	"	5.29	11.00	16.29
Union tee				
3/8"	EA.	3.55	11.00	14.55
1/2"	"	4.91	11.00	15.91
5/8"	"	6.32	11.00	17.32
Male connector				
3/8"	EA.	2.94	11.00	13.94
1/2"	"	2.36	11.00	13.36
5/8"	"	2.07	11.00	13.07
Female connector				
3/8"	EA.	2.67	11.00	13.67
1/2"	"	3.36	11.00	14.36
5/8"	"	3.93	11.00	14.93
Brass flare fittings, union				
3/8"	EA.	2.35	10.75	13.10
1/2"	"	3.22	10.75	13.97
5/8"	"	3.47	10.75	14.22
90 deg elbow union				
3/8"	EA.	4.57	10.75	15.32
1/2"	"	7.45	10.75	18.20
5/8"	"	11.00	10.75	21.75
Three way tee				
3/8"	EA.	5.08	18.00	23.08
1/2"	"	6.73	18.00	24.73
5/8"	"	11.25	18.00	29.25
Cross				
3/8"	EA.	10.75	23.75	34.50
1/2"	"	23.00	23.75	46.75
5/8"	"	48.25	23.75	72.00
Male connector, half union				
3/8"	EA.	1.59	10.75	12.34
1/2"	"	2.79	10.75	13.54

15 MECHANICAL

Plumbing

15410.15 Brass Fittings (Cont.)

	UNIT	MAT.	INST.	TOTAL
5/8"	EA.	3.90	10.75	14.65
Female connector, half union				
3/8"	EA.	2.18	10.75	12.93
1/2"	"	2.02	10.75	12.77
5/8"	"	3.90	10.75	14.65
Long forged nut				
3/8"	EA.	1.72	10.75	12.47
1/2"	"	2.50	10.75	13.25
5/8"	"	8.65	10.75	19.40
Short forged nut				
3/8"	EA.	1.39	10.75	12.14
1/2"	"	1.89	10.75	12.64
5/8"	"	2.41	10.75	13.16
Compression elbow				
1/4"	EA.	3.43	10.75	14.18
5/16"	"	4.44	10.75	15.19
3/4"	"	19.50	10.75	30.25
Nut				
1/8"	EA.			0.31
1/4"	"			0.31
5/16"	"			0.35
3/8"	"			0.46
1/2"	"			0.68
5/8"	"			1.43
3/4"	"			2.41
7/8"	"			4.75
Sleeve				
1/8"	EA.	0.22	13.25	13.47
1/4"	"	0.07	13.25	13.32
5/16"	"	0.22	13.25	13.47
3/8"	"	0.32	13.25	13.57
1/2"	"	0.41	13.25	13.66
5/8"	"	0.59	13.25	13.84
3/4"	"	0.94	13.25	14.19
7/8"	"	1.19	13.25	14.44
Tee				
1/4"	EA.	3.58	19.00	22.58
5/16"	"	5.38	19.00	24.38
Male tee				
5/16" x 1/8"	EA.	7.40	19.00	26.40
Female union				
1/8" x 1/8"	EA.	1.79	16.50	18.29
1/4" x 3/8"	"	3.43	16.50	19.93
3/8" x 1/4"	"	2.84	16.50	19.34
3/8" x 1/2"	"	3.58	16.50	20.08
5/8" x 1/2"	"	5.60	19.00	24.60
Male union, 1/4"				
1/4" x 1/4"	EA.	1.70	16.50	18.20
3/8"	"	2.24	16.50	18.74
1/2"	"	3.36	16.50	19.86
5/16" x				
1/8"	EA.	1.64	16.50	18.14
1/4"	"	2.09	16.50	18.59

15 MECHANICAL

Plumbing

15410.15 Brass Fittings (Cont.)

	UNIT	MAT.	INST.	TOTAL
3/8"	EA.	3.19	16.50	19.69
3/8" x				
1/8"	EA.	1.94	16.50	18.44
1/4"	"	2.24	16.50	18.74
1/2"	"	3.07	16.50	19.57
3/4"	"	4.18	16.50	20.68
1/2" x				
1/4"	EA.	2.90	19.00	21.90
3/8"	"	2.99	19.00	21.99
5/8" x				
3/8"	EA.	4.47	19.00	23.47
1/2"	"	3.85	19.00	22.85
3/4"	"	5.95	19.00	24.95
1/2"	"	3.85	19.00	22.85
7/8" x				
1/2"	EA.	8.52	19.00	27.52
3/4"	"	9.40	19.00	28.40
Female elbow, 1/4" x 1/4"	"	4.23	19.00	23.23
5/16" x				
1/8"	EA.	4.63	19.00	23.63
1/4"	"	6.05	19.00	25.05
3/8" x				
3/8"	EA.	3.73	19.00	22.73
1/2"	"	2.99	19.00	21.99
Male elbow, 1/8" x 1/8"	"	4.00	19.00	23.00
3/16" x 1/4"	"	3.82	19.00	22.82
1/4" x				
1/8"	EA.	2.36	19.00	21.36
1/4"	"	2.78	19.00	21.78
3/8"	"	2.40	19.00	21.40
5/16" x				
1/8"	EA.	2.48	19.00	21.48
1/4"	"	2.90	19.00	21.90
3/8"	"	4.24	19.00	23.24
3/8" x				
1/8"	EA.	2.37	19.00	21.37
1/4"	"	3.27	19.00	22.27
3/8"	"	2.48	19.00	21.48
1/2"	"	3.30	19.00	22.30
1/2" x				
1/4"	EA.	5.20	22.25	27.45
3/8"	"	4.61	22.25	26.86
1/2"	"	4.02	22.25	26.27
5/8" x				
3/8"	EA.	5.23	22.25	27.48
1/2"	"	5.53	22.25	27.78
3/4"	"	11.25	22.25	33.50
3/4" x				
1/2"	EA.	10.00	22.25	32.25
3/4"	"	12.75	22.25	35.00
7/8" x				
3/4"	EA.	22.00	22.25	44.25
Union				

15 MECHANICAL

Plumbing	UNIT	MAT.	INST.	TOTAL

15410.15 Brass Fittings (Cont.)

	UNIT	MAT.	INST.	TOTAL
1/8"	EA.	2.39	19.00	21.39
3/16"	"	2.39	19.00	21.39
1/4"	"	2.03	19.00	21.03
5/16"	"	2.31	19.00	21.31
3/8"	"	2.61	19.00	21.61
3/4"	"	3.64	19.00	22.64
7/8"	"	9.14	19.00	28.14
Reducing union				
3/8" x 1/4"	EA.	2.85	22.25	25.10
5/8" x				
3/8"	EA.	4.52	22.25	26.77
1/2"	"	4.96	22.25	27.21

15410.17 Chrome Plated Fittings

	UNIT	MAT.	INST.	TOTAL
Fittings				
90 ell				
3/8"	EA.	32.50	16.50	49.00
1/2"	"	42.00	16.50	58.50
45 ell				
3/8"	EA.	42.00	16.50	58.50
1/2"	"	55.00	16.50	71.50
Tee				
3/8"	EA.	40.25	22.25	62.50
1/2"	"	47.75	22.25	70.00
Coupling				
3/8"	EA.	25.50	16.50	42.00
1/2"	"	25.50	16.50	42.00
Union				
3/8"	EA.	42.00	16.50	58.50
1/2"	"	43.50	16.50	60.00
Tee				
1/2" x 3/8" x 3/8"	EA.	47.75	22.25	70.00
1/2" x 3/8" x 1/2"	"	48.75	22.25	71.00

15410.18 Glass Pipe

	UNIT	MAT.	INST.	TOTAL
Glass pipe				
1-1/2" dia.	L.F.	12.25	13.25	25.50
2" dia.	"	16.50	14.75	31.25
3" dia.	"	22.00	16.50	38.50
4" dia.	"	40.25	19.00	59.25
6" dia.	"	74.00	22.25	96.25

15410.30 Pvc/cpvc Pipe

	UNIT	MAT.	INST.	TOTAL
PVC schedule 40				
1/2" pipe	L.F.	0.59	2.76	3.35
3/4" pipe	"	0.82	3.01	3.83
1" pipe	"	1.04	3.32	4.36
1-1/4" pipe	"	1.34	3.68	5.02
1-1/2" pipe	"	2.01	4.15	6.16
2" pipe	"	2.54	4.74	7.28
2-1/2" pipe	"	4.11	5.53	9.64
3" pipe	"	5.23	6.64	11.87
4" pipe	"	7.47	8.30	15.77

15 MECHANICAL

Plumbing

15410.30 Pvc/cpvc Pipe (Cont.)

	UNIT	MAT.	INST.	TOTAL
6" pipe	L.F.	13.50	16.50	30.00
8" pipe	"	19.75	22.25	42.00
Fittings, 1/2"				
90 deg ell	EA.	0.57	8.30	8.87
45 deg ell	"	0.78	8.30	9.08
Tee	"	0.58	9.48	10.06
Reducing insert	"	0.59	11.00	11.59
Threaded	"	1.40	8.30	9.70
Male adapter	"	0.55	11.00	11.55
Female adapter	"	0.59	8.30	8.89
Coupling	"	0.44	8.30	8.74
Union	"	4.68	13.25	17.93
Cap	"	0.54	11.00	11.54
Flange	"	9.88	13.25	23.13
3/4"				
90 deg elbow	EA.	0.55	11.00	11.55
45 deg elbow	"	1.32	11.00	12.32
Tee	"	0.76	13.25	14.01
Reducing insert	"	0.55	9.48	10.03
Threaded	"	0.84	11.00	11.84
1"				
90 deg elbow	EA.	0.97	13.25	14.22
45 deg elbow	"	1.43	13.25	14.68
Tee	"	1.30	14.75	16.05
Reducing insert	"	0.97	13.25	14.22
Threaded	"	1.30	14.75	16.05
Male adapter	"	0.91	16.50	17.41
Female adapter	"	0.78	16.50	17.28
Coupling	"	0.71	16.50	17.21
Union	"	7.02	22.25	29.27
Cap	"	0.78	13.25	14.03
Flange	"	9.94	22.25	32.19
1-1/4"				
90 deg elbow	EA.	1.69	19.00	20.69
45 deg elbow	"	2.01	19.00	21.01
Tee	"	1.95	22.25	24.20
Reducing insert	"	1.17	22.25	23.42
Threaded	"	1.95	22.25	24.20
Male adapter	"	1.10	22.25	23.35
Female adapter	"	1.23	22.25	23.48
Coupling	"	1.04	22.25	23.29
Union	"	16.00	26.50	42.50
Cap	"	1.10	22.25	23.35
Flange	"	10.00	26.50	36.50
1-1/2"				
90 deg elbow	EA.	1.88	19.00	20.88
45 deg elbow	"	2.79	19.00	21.79
Tee	"	2.60	22.25	24.85
Reducing insert	"	1.30	22.25	23.55
Threaded	"	2.34	22.25	24.59
Male adapter	"	1.56	22.25	23.81
Female adapter	"	1.56	22.25	23.81
Coupling	"	1.17	22.25	23.42

15 MECHANICAL

Plumbing

15410.30 Pvc/cpvc Pipe *(Cont.)*

	UNIT	MAT.	INST.	TOTAL
Union	EA.	22.00	33.25	55.25
Cap	"	1.23	22.25	23.48
Flange	"	17.00	33.25	50.25
2"				
90 deg elbow	EA.	2.99	22.25	25.24
45 deg elbow	"	3.77	22.25	26.02
Tee	"	3.96	26.50	30.46
Reducing insert	"	2.53	26.50	29.03
Threaded	"	3.38	26.50	29.88
Male adapter	"	2.08	26.50	28.58
Female adapter	"	2.14	26.50	28.64
Coupling	"	1.75	26.50	28.25
Union	"	30.00	41.50	71.50
Cap	"	1.62	26.50	28.12
Flange	"	18.00	41.50	59.50
2-1/2"				
90 deg elbow	EA.	9.03	41.50	50.53
45 deg elbow	"	13.00	41.50	54.50
Tee	"	11.75	44.25	56.00
Reducing insert	"	3.64	44.25	47.89
Threaded	"	5.20	44.25	49.45
Male adapter	"	6.04	44.25	50.29
Female adapter	"	5.00	44.25	49.25
Coupling	"	3.61	44.25	47.86
Union	"	40.25	55.00	95.25
Cap	"	4.81	41.50	46.31
Flange	"	24.00	55.00	79.00
3"				
90 deg elbow	EA.	9.75	55.00	64.75
45 deg elbow	"	12.50	55.00	67.50
Tee	"	15.50	60.00	75.50
Reducing insert	"	4.68	55.00	59.68
Threaded	"	6.04	55.00	61.04
Male adapter	"	7.41	55.00	62.41
Female adapter	"	5.98	55.00	60.98
Coupling	"	5.52	55.00	60.52
Union	"	42.00	66.00	108
Cap	"	4.81	55.00	59.81
Flange	"	22.25	66.00	88.25
4"				
90 deg elbow	EA.	17.50	66.00	83.50
45 deg elbow	"	22.75	66.00	88.75
Tee	"	25.75	74.00	99.75
Reducing insert	"	10.50	66.00	76.50
Threaded	"	13.50	66.00	79.50
Male adapter	"	9.42	66.00	75.42
Female adapter	"	10.25	66.00	76.25
Coupling	"	8.06	66.00	74.06
Union	"	51.00	83.00	134
Cap	"	11.00	66.00	77.00
Flange	"	29.75	83.00	113
PVC schedule 80 pipe				
1-1/2" pipe	L.F.	2.54	4.15	6.69

15 MECHANICAL

Plumbing

15410.30 — Pvc/cpvc Pipe (Cont.)

	UNIT	MAT.	INST.	TOTAL
2" pipe	L.F.	3.43	4.74	8.17
3" pipe	"	7.10	6.64	13.74
4" pipe	"	9.26	8.30	17.56
Fittings, 1-1/2"				
90 deg elbow	EA.	8.79	22.25	31.04
45 deg elbow	"	19.25	22.25	41.50
Tee	"	30.25	33.25	63.50
Reducing insert	"	5.57	22.25	27.82
Threaded	"	6.64	22.25	28.89
Male adapter	"	10.50	22.25	32.75
Female adapter	"	11.50	22.25	33.75
Coupling	"	12.25	22.25	34.50
Union	"	22.00	33.25	55.25
Cap	"	6.22	22.25	28.47
Flange	"	14.25	33.25	47.50
2"				
90 deg elbow	EA.	10.75	26.50	37.25
45 deg elbow	"	25.00	26.50	51.50
Tee	"	37.75	41.50	79.25
Reducing insert	"	7.93	26.50	34.43
Threaded	"	8.00	26.50	34.50
Male adapter	"	14.50	26.50	41.00
Female adapter	"	20.00	26.50	46.50
2-1/2"				
90 deg elbow	EA.	24.75	41.50	66.25
45 deg elbow	"	53.00	41.50	94.50
Tee	"	41.00	55.00	96.00
Reducing insert	"	13.75	41.50	55.25
Threaded	"	17.25	41.50	58.75
Male adapter	"	17.25	41.50	58.75
Female adapter	"	31.50	41.50	73.00
Coupling	"	17.00	41.50	58.50
Union	"	47.25	55.00	102
Cap	"	19.75	41.50	61.25
Flange	"	25.50	55.00	80.50
3"				
90 deg elbow	EA.	22.25	55.00	77.25
45 deg elbow	"	64.00	55.00	119
Tee	"	51.00	66.00	117
Reducing insert	"	21.75	55.00	76.75
Threaded	"	31.75	55.00	86.75
Male adapter	"	19.25	55.00	74.25
Female adapter	"	35.50	55.00	90.50
Coupling	"	19.50	55.00	74.50
Union	"	60.00	66.00	126
Cap	"	25.50	55.00	80.50
Flange	"	28.75	66.00	94.75
4"				
90 deg elbow	EA.	57.00	66.00	123
45 deg elbow	"	120	66.00	186
Tee	"	59.00	83.00	142
Reducing insert	"	30.25	66.00	96.25
Threaded	"	49.00	66.00	115

15 MECHANICAL

Plumbing

15410.30 Pvc/cpvc Pipe (Cont.)

	UNIT	MAT.	INST.	TOTAL
Male adapter	EA.	34.25	66.00	100
Coupling	"	24.50	66.00	90.50
Union	"	57.00	83.00	140
Cap	"	31.00	66.00	97.00
Flange	"	35.25	83.00	118
CPVC schedule 40				
1/2" pipe	L.F.	0.66	2.76	3.42
3/4" pipe	"	0.88	3.01	3.89
1" pipe	"	1.29	3.32	4.61
1-1/4" pipe	"	1.69	3.68	5.37
1-1/2" pipe	"	2.04	4.15	6.19
2" pipe	"	2.71	4.74	7.45
Fittings, CPVC, schedule 80				
1/2", 90 deg ell	EA.	4.06	6.64	10.70
Tee	"	12.25	11.00	23.25
3/4", 90 deg ell	"	5.29	6.64	11.93
Tee	"	18.25	11.00	29.25
1", 90 deg ell	"	8.38	7.37	15.75
Tee	"	19.50	12.00	31.50
1-1/4", 90 deg ell	"	15.25	7.37	22.62
Tee	"	18.25	12.00	30.25
1-1/2", 90 deg ell	"	17.00	13.25	30.25
Tee	"	20.75	16.50	37.25
2", 90 deg ell	"	18.50	13.25	31.75
Tee	"	23.50	16.50	40.00
Polypropylene, acid resistant, DWV pipe				
Schedule 40				
1-1/2" pipe	L.F.	8.97	4.74	13.71
2" pipe	"	12.00	5.53	17.53
3" pipe	"	24.75	6.64	31.39
4" pipe	"	31.50	8.30	39.80
6" pipe	"	63.00	16.50	79.50
Fittings				
1-1/2"				
1/4 bend	EA.	25.50	16.50	42.00
1/8 bend	"	25.25	16.50	41.75
Sanitary tee	"	32.00	33.25	65.25
Cleanout with plug	"	49.25	33.25	82.50
Wye	"	39.00	33.25	72.25
Combination wye & 1/8 bend	"	58.00	33.25	91.25
P-trap	"	56.00	41.50	97.50
Hub adapter	"	30.50	16.50	47.00
Coupling	"	20.00	16.50	36.50
2"				
1/4 bend	EA.	32.00	19.00	51.00
1/8 bend	"	32.00	19.00	51.00
Sanitary tee	"	38.25	39.00	77.25
Cleanout with plug	"	55.00	39.00	94.00
Wye	"	81.00	39.00	120
Combination wye & 1/8 bend	"	82.00	39.00	121
P-trap	"	76.00	55.00	131
Hub adapter	"	32.75	22.25	55.00
Mechanical joint adapter	"	32.25	22.25	54.50

15 MECHANICAL

Plumbing	UNIT	MAT.	INST.	TOTAL
15410.30 **Pvc/cpvc Pipe** *(Cont.)*				
Coupling	EA.	25.00	22.25	47.25
3"				
1/4 bend	EA.	62.00	22.25	84.25
1/8 bend	"	65.00	22.25	87.25
Sanitary tee	"	88.00	44.25	132
Cleanout with plug	"	97.00	44.25	141
Wye	"	94.00	44.25	138
Combination wye & 1/8 bend	"	120	44.25	164
P-trap	"	200	66.00	266
Hub adapter	"	52.00	22.25	74.25
Mechanical joint adapter	"	50.00	22.25	72.25
Coupling	"	37.00	22.25	59.25
4"				
1/4 bend	EA.	99.00	33.25	132
1/8 bend	"	73.00	33.25	106
Sanitary tee	"	130	66.00	196
Cleanout with plug	"	140	66.00	206
Wye	"	140	66.00	206
Combination wye & 1/8 bend	"	160	66.00	226
P-trap	"	260	110	370
Hub adapter	"	70.00	33.25	103
Mechanical joint adapter	"	67.00	33.25	100
Coupling	"	52.00	33.25	85.25
6"				
1/4 bend	EA.	240	55.00	295
1/8 bend	"	200	55.00	255
Sanitary tee	"	250	110	360
Cleanout with plug	"	260	110	370
Wye	"	350	110	460
Hub adapter	"	93.00	55.00	148
Coupling	"	83.00	55.00	138
Polyethylene pipe and fittings				
SDR-21				
3" pipe	L.F.	4.63	8.30	12.93
4" pipe	"	7.32	11.00	18.32
6" pipe	"	12.75	16.50	29.25
8" pipe	"	18.50	19.00	37.50
10" pipe	"	20.75	22.25	43.00
12" pipe	"	32.25	26.50	58.75
14" pipe	"	41.75	33.25	75.00
16" pipe	"	51.00	41.50	92.50
18" pipe	"	56.00	51.00	107
20" pipe	"	70.00	66.00	136
22" pipe	"	81.00	74.00	155
24" pipe	"	99.00	83.00	182
Fittings, 3"				
90 deg elbow	EA.	110	33.25	143
45 deg elbow	"	68.00	33.25	101
Tee	"	62.00	55.00	117
45 deg wye	"	160	55.00	215
Reducer	"	30.75	41.50	72.25
Flange assembly	"	24.50	33.25	57.75
4"				

15 MECHANICAL

Plumbing

15410.30 — Pvc/cpvc Pipe *(Cont.)*

Description	UNIT	MAT.	INST.	TOTAL
90 deg elbow	EA.	160	41.50	202
45 deg elbow	"	94.00	41.50	136
Tee	"	130	66.00	196
45 deg wye	"	250	66.00	316
Reducer	"	86.00	55.00	141
Flange assembly	"	86.00	41.50	128
8"				
90 deg elbow	EA.	430	83.00	513
45 deg elbow	"	230	83.00	313
Tee	"	410	130	540
45 deg wye	"	620	130	750
Reducer	"	210	110	320
Flange assembly	"	230	83.00	313
10"				
90 deg elbow	EA.	590	110	700
45 deg elbow	"	320	110	430
Tee	"	530	170	700
45 deg wye	"	840	170	1,010
Reducer	"	280	130	410
Flange assembly	"	270	110	380
12"				
90 deg elbow	EA.	960	130	1,090
45 deg elbow	"	580	130	710
Tee	"	750	220	970
45 deg wye	"	1,170	220	1,390
Reducer	"	440	170	610
Flange assembly	"	360	130	490
14"				
90 deg elbow	EA.	1,300	170	1,470
45 deg elbow	"	740	170	910
Tee	"	950	270	1,220
45 deg wye	"	1,760	270	2,030
Reducer	"	370	220	590
Flange assembly	"	440	170	610
16"				
90 deg elbow	EA.	1,630	170	1,800
45 deg elbow	"	970	170	1,140
Tee	"	1,180	270	1,450
45 deg wye	"	1,970	270	2,240
Reducer	"	760	220	980
Flange assembly	"	550	170	720
18"				
90 deg elbow	EA.	2,490	220	2,710
45 deg elbow	"	1,610	220	1,830
Tee	"	1,920	330	2,250
45 deg wye	"	3,320	330	3,650
Reducer	"	710	220	930
Flange assembly	"	970	220	1,190
20"				
90 deg elbow	EA.	2,000	220	2,220
45 deg elbow	"	1,190	220	1,410

15 MECHANICAL

Plumbing

15410.33 Abs Dwv Pipe

	UNIT	MAT.	INST.	TOTAL
Schedule 40 ABS				
1-1/2" pipe	L.F.	1.34	3.32	4.66
2" pipe	"	1.79	3.68	5.47
3" pipe	"	3.67	4.74	8.41
4" pipe	"	5.20	6.64	11.84
6" pipe	"	10.75	8.30	19.05
Fittings				
1/8 bend				
1-1/2"	EA.	2.14	13.25	15.39
2"	"	3.17	16.50	19.67
3"	"	7.60	22.25	29.85
4"	"	13.50	26.50	40.00
6"	"	56.00	33.25	89.25
Tee, sanitary				
1-1/2"	EA.	3.12	22.25	25.37
2"	"	4.81	26.50	31.31
3"	"	13.25	33.25	46.50
4"	"	24.00	41.50	65.50
6"	"	100	55.00	155
Tee, sanitary reducing				
2 x 1-1/2 x 1-1/2	EA.	4.42	26.50	30.92
2 x 1-1/2 x 2	"	4.55	27.75	32.30
2 x 2 x 1-1/2	"	4.22	30.25	34.47
3 x 3 x 1-1/2	"	7.67	33.25	40.92
3 x 3 x 2	"	9.55	37.00	46.55
4 x 4 x 1-1/2	"	24.00	41.50	65.50
4 x 4 x 2	"	22.25	47.50	69.75
4 x 4 x 3	"	19.25	51.00	70.25
6 x 6 x 4	"	100	55.00	155
Wye				
1-1/2"	EA.	4.55	19.00	23.55
2"	"	6.37	26.50	32.87
3"	"	14.50	33.25	47.75
4"	"	31.50	41.50	73.00
6"	"	96.00	55.00	151
Reducer				
2 x 1-1/2	EA.	3.05	16.50	19.55
3 x 1-1/2	"	7.86	22.25	30.11
3 x 2	"	6.69	22.25	28.94
4 x 2	"	13.50	26.50	40.00
4 x 3	"	13.75	26.50	40.25
6 x 4	"	28.50	33.25	61.75
P-trap				
1-1/2"	EA.	7.08	22.25	29.33
2"	"	9.55	24.50	34.05
3"	"	36.50	28.75	65.25
4"	"	75.00	33.25	108
6"	"	120	41.50	162
Double sanitary, tee				
1-1/2"	EA.	6.89	26.50	33.39
2"	"	10.00	33.25	43.25
3"	"	27.50	41.50	69.00
4"	"	43.75	55.00	98.75

15 MECHANICAL

Plumbing	UNIT	MAT.	INST.	TOTAL
15410.33 **Abs Dwv Pipe** *(Cont.)*				
Long sweep, 1/4 bend				
1-1/2"	EA.	3.57	13.25	16.82
2"	"	4.55	16.50	21.05
3"	"	11.00	22.25	33.25
4"	"	20.25	33.25	53.50
Wye, standard				
1-1/2"	EA.	4.60	22.25	26.85
2"	"	6.37	26.50	32.87
3"	"	14.50	33.25	47.75
4"	"	31.50	41.50	73.00
Wye, reducing				
2 x 1-1/2 x 1-1/2	EA.	8.51	22.25	30.76
2 x 2 x 1-1/2	"	8.12	26.50	34.62
4 x 4 x 2	"	17.25	41.50	58.75
4 x 4 x 3	"	23.75	44.25	68.00
Double wye				
1-1/2"	EA.	10.50	26.50	37.00
2"	"	12.50	33.25	45.75
3"	"	31.75	41.50	73.25
4"	"	65.00	55.00	120
2 x 2 x 1-1/2 x 1-1/2	"	12.25	33.25	45.50
3 x 3 x 2 x 2	"	26.25	41.50	67.75
4 x 4 x 3 x 3	"	61.00	55.00	116
Combination wye and 1/8 bend				
1-1/2"	EA.	7.28	22.25	29.53
2"	"	8.77	26.50	35.27
3"	"	19.00	33.25	52.25
4"	"	38.50	41.50	80.00
2 x 2 x 1-1/2	"	7.54	26.50	34.04
3 x 3 x 1-1/2	"	18.00	33.25	51.25
3 x 3 x 2	"	12.50	33.25	45.75
4 x 4 x 2	"	25.00	41.50	66.50
4 x 4 x 3	"	30.50	41.50	72.00
15410.35 **Plastic Pipe**				
Fiberglass reinforced pipe				
2" pipe	L.F.	3.31	5.10	8.41
3" pipe	"	4.71	5.53	10.24
4" pipe	"	6.16	6.03	12.19
6" pipe	"	11.75	6.64	18.39
8" pipe	"	17.25	11.00	28.25
10" pipe	"	25.75	13.25	39.00
12" pipe	"	33.50	16.50	50.00
Fittings				
90 deg elbow, flanged				
2"	EA.	170	66.00	236
3"	"	210	74.00	284
4"	"	260	83.00	343
6"	"	480	110	590
8"	"	870	130	1,000
10"	"	1,160	170	1,330
12"	"	1,550	220	1,770
45 deg elbow, flanged				

15 MECHANICAL

Plumbing

15410.35 Plastic Pipe (Cont.)

Description	UNIT	MAT.	INST.	TOTAL
2"	EA.	170	55.00	225
3"	"	210	66.00	276
4"	"	260	83.00	343
6"	"	480	110	590
8"	"	730	130	860
10"	"	980	170	1,150
12"	"	1,230	220	1,450
Tee, flanged				
2"	EA.	210	83.00	293
3"	"	310	95.00	405
4"	"	340	110	450
6"	"	580	130	710
8"	"	1,030	170	1,200
10"	"	1,680	220	1,900
12"	"	2,320	330	2,650
Wye, flanged				
2"	EA.	430	83.00	513
3"	"	590	95.00	685
4"	"	770	110	880
6"	"	980	130	1,110
8"	"	1,650	170	1,820
10"	"	2,820	220	3,040
12"	"	3,610	330	3,940
Concentric reducer, flanged				
2"	EA.	330	55.00	385
4"	"	390	66.00	456
6"	"	420	95.00	515
8"	"	720	130	850
10"	"	830	170	1,000
12"	"	1,190	220	1,410
Adapter, bell x male or female				
2"	EA.	21.25	55.00	76.25
3"	"	42.50	60.00	103
4"	"	45.00	66.00	111
6"	"	96.00	95.00	191
8"	"	140	130	270
10"	"	200	170	370
12"	"	360	220	580
Nipples				
2" x 6"	EA.	8.97	6.64	15.61
2" x 12"	"	13.50	8.30	21.80
3" x 8"	"	13.50	10.25	23.75
3" x 12"	"	13.50	11.00	24.50
4" x 8"	"	13.50	11.00	24.50
4" x 12"	"	15.25	13.25	28.50
6" x 12"	"	34.75	16.50	51.25
8" x 18"	"	99.00	16.50	116
8" x 24"	"	120	19.00	139
10" x 18"	"	120	22.25	142
10" x 24"	"	160	26.50	187
12" x 18"	"	160	30.25	190
12" x 24"	"	190	33.25	223
Sleeve coupling				

15 MECHANICAL

Plumbing

	UNIT	MAT.	INST.	TOTAL
15410.35 **Plastic Pipe** *(Cont.)*				
2"	EA.	19.75	55.00	74.75
3"	"	21.50	66.00	87.50
4"	"	29.75	95.00	125
6"	"	72.00	130	202
8"	"	110	170	280
10"	"	180	220	400
Flanges				
2"	EA.	29.25	55.00	84.25
3"	"	39.50	66.00	106
4"	"	53.00	95.00	148
6"	"	90.00	130	220
8"	"	160	170	330
10"	"	220	220	440
12"	"	270	220	490
15410.70 **Stainless Steel Pipe**				
Stainless steel, schedule 40, threaded				
1/2" pipe	L.F.	13.25	9.48	22.73
3/4" pipe	"	18.50	9.76	28.26
1" pipe	"	21.50	10.25	31.75
1-1/2" pipe	"	29.25	11.00	40.25
2" pipe	"	44.00	12.00	56.00
2-1/2" pipe	"	62.00	13.25	75.25
3" pipe	"	87.00	14.75	102
4" pipe	"	110	16.50	127
Fittings, 1/2"				
90 deg ell	EA.	21.50	83.00	105
45 deg ell	"	31.00	83.00	114
Tee	"	32.50	110	143
Cap	"	11.25	41.50	52.75
Reducer	"	17.50	55.00	72.50
Union	"	60.00	83.00	143
Flange	"	32.00	83.00	115
3/4"				
90 deg ell	EA.	24.75	83.00	108
45 deg ell	"	29.00	83.00	112
Tee	"	34.75	110	145
Cap	"	16.25	41.50	57.75
Reducer	"	22.75	55.00	77.75
Union	"	83.00	83.00	166
Flange	"	35.75	83.00	119
1"				
90 deg ell	EA.	27.75	83.00	111
45 deg ell	"	31.50	83.00	115
Tee	"	34.75	110	145
Cap	"	22.25	41.50	63.75
Reducer	"	32.75	83.00	116
Union	"	100	83.00	183
Flange	"	33.50	83.00	117
1-1/4"				
90 deg ell	EA.	43.00	83.00	126
45 deg ell	"	40.50	83.00	124
Tee	"	58.00	110	168

15 MECHANICAL

Plumbing

15410.70 Stainless Steel Pipe (Cont.)

Description	UNIT	MAT.	INST.	TOTAL
Cap	EA.	45.75	41.50	87.25
Union	"	240	83.00	323
Flange	"	42.75	83.00	126
1-1/2"				
90 deg ell	EA.	47.50	110	158
45 deg ell	"	44.25	110	154
Tee	"	72.00	130	202
Cap	"	53.00	55.00	108
Reducer	"	76.00	66.00	142
Union	"	260	95.00	355
Flange	"	42.75	95.00	138
2"				
90 deg ell	EA.	68.00	130	198
45 deg ell	"	74.00	130	204
Tee	"	89.00	220	309
Cap	"	67.00	66.00	133
Reducer	"	89.00	83.00	172
Union	"	330	130	460
Flange	"	56.00	130	186
Type 304, sch 10 pipe				
1" pipe	L.F.	10.75	8.30	19.05
1-1/4" pipe	"	15.00	11.00	26.00
1-1/2" pipe	"	18.75	12.00	30.75
2" pipe	"	20.00	14.75	34.75
2-1/2" pipe	"	23.25	16.50	39.75
3" pipe	"	27.50	19.00	46.50
4" pipe	"	41.25	22.25	63.50
6" pipe	"	61.00	26.50	87.50
Fittings, 1"				
90 deg elbow	EA.	27.25	130	157
45 deg elbow	"	29.50	130	160
Tee	"	37.00	220	257
Reducer	"	23.25	130	153
Flange	"	56.00	83.00	139
Cap	"	16.75	83.00	99.75
1-1/4"				
90 deg elbow	EA.	48.00	130	178
45 deg elbow	"	47.50	130	178
Tee	"	74.00	220	294
Reducer	"	48.75	130	179
Flange	"	72.00	83.00	155
Cap	"	30.50	83.00	114
1-1/2"				
90 deg elbow	EA.	64.00	110	174
45 deg elbow	"	47.50	110	158
Tee	"	91.00	220	311
Reducer	"	48.75	170	219
Flange	"	72.00	110	182
Cap	"	40.00	110	150
2"				
90 deg elbow	EA.	81.00	130	211
45 deg elbow	"	79.00	130	209
Tee	"	120	330	450

15 MECHANICAL

Plumbing

15410.70 Stainless Steel Pipe (Cont.)

Description	UNIT	MAT.	INST.	TOTAL
Reducer	EA.	170	220	390
Flange	"	92.00	170	262
Cap	"	51.00	170	221
2-1/2"				
90 deg elbow	EA.	180	130	310
45 deg elbow	"	220	130	350
Tee	"	250	330	580
Reducer	"	170	220	390
Flange	"	130	170	300
Cap	"	92.00	170	262
3"				
90 deg elbow	EA.	240	170	410
45 deg elbow	"	300	170	470
Tee	"	360	440	800
Reducer	"	250	330	580
Flange	"	140	220	360
Cap	"	130	220	350
4"				
90 deg elbow	EA.	350	170	520
45 deg elbow	"	250	170	420
Reducer	"	250	330	580
Flange	"	190	220	410
Cap	"	150	220	370
6"				
90 deg elbow	EA.	460	220	680
45 deg elbow	"	300	220	520
Tee	"	420	950	1,370
Reducer	"	270	330	600
Flange	"	270	330	600
Cap	"	210	330	540
Type 304 tubing				
.035 wall				
1/4"	L.F.	4.09	3.68	7.77
3/8"	"	5.12	4.15	9.27
1/2"	"	6.30	4.74	11.04
5/8"	"	8.38	5.53	13.91
3/4"	"	9.88	6.64	16.52
7/8"	"	11.00	7.37	18.37
1"	"	11.75	8.30	20.05
.049 wall				
1/4"	L.F.	5.39	3.90	9.29
3/8"	"	6.30	4.42	10.72
1/2"	"	6.82	5.10	11.92
5/8"	"	9.16	6.03	15.19
3/4"	"	10.75	7.37	18.12
7/8"	"	12.75	8.30	21.05
1"	"	13.75	9.48	23.23
.065 wall				
1/4"	L.F.	5.85	4.42	10.27
3/8"	"	7.54	5.53	13.07
1/2"	"	10.50	6.03	16.53
5/8"	"	11.25	7.37	18.62
3/4"	"	13.75	9.48	23.23

15 MECHANICAL

Plumbing

15410.70 Stainless Steel Pipe (Cont.)

	UNIT	MAT.	INST.	TOTAL
7/8"	L.F.	15.25	11.00	26.25
1"	"	17.00	13.25	30.25
Type 316 tubing				
.035 wall				
1/4"	L.F.	5.07	3.68	8.75
3/8"	"	6.50	4.15	10.65
1/2"	"	10.50	4.74	15.24
5/8"	"	13.25	5.53	18.78
3/4"	"	16.75	6.64	23.39
7/8"	"	22.75	7.37	30.12
1"	"	24.75	8.30	33.05
.049 wall				
1/4"	L.F.	6.82	4.42	11.24
3/8"	"	7.47	5.53	13.00
1/2"	"	9.42	6.03	15.45
5/8"	"	12.00	7.37	19.37
3/4"	"	12.50	9.48	21.98
7/8"	"	15.50	11.00	26.50
1"	"	17.50	13.25	30.75
.065 wall				
1/4"	L.F.	8.45	4.42	12.87
3/8"	"	11.75	5.53	17.28
1/2"	"	13.25	6.03	19.28
5/8"	"	13.75	7.37	21.12
3/4"	"	15.50	9.48	24.98
7/8"	"	20.25	11.00	31.25
1"	"	23.75	13.25	37.00
Fittings, 1/4"				
90 deg elbow	EA.	16.75	13.25	30.00
Union tee	"	29.25	22.25	51.50
Union	"	11.25	22.25	33.50
Male connector	"	7.93	16.50	24.43
3/8"				
90 deg elbow	EA.	19.75	16.50	36.25
Union tee	"	39.75	25.50	65.25
Union	"	13.75	25.50	39.25
Male connector	"	11.25	16.50	27.75
1/2"				
90 deg elbow	EA.	28.00	17.50	45.50
Union tee	"	53.00	27.75	80.75
Union	"	24.00	27.75	51.75
Male connector	"	15.00	16.50	31.50
5/8"				
90 deg elbow	EA.	34.50	22.25	56.75
Union tee	"	56.00	33.25	89.25
Union	"	32.50	33.25	65.75
Male connector	"	17.50	22.25	39.75
3/4"				
90 deg elbow	EA.	62.00	22.25	84.25
Union tee	"	73.00	33.25	106
Union	"	44.25	33.25	77.50
Male connector	"	25.25	22.25	47.50
7/8"				

15 MECHANICAL

Plumbing

15410.70 Stainless Steel Pipe (Cont.)

	UNIT	MAT.	INST.	TOTAL
90 deg elbow	EA.	94.00	23.75	118
Union tee	"	120	37.00	157
Union	"	69.00	37.00	106
Male connector	"	36.75	23.75	60.50
1"				
90 deg elbow	EA.	120	30.25	150
Union tee	"	160	41.50	202
Union	"	71.00	41.50	113
Male connector	"	47.50	33.25	80.75
Type 316 valves				
Gate valves				
1/4"	EA.	370	22.25	392
3/8"	"	370	26.50	397
1/2"	"	420	28.75	449
3/4"	"	510	33.25	543
1"	"	590	44.25	634
Globe valves				
1/4"	EA.	240	22.25	262
3/8"	"	390	26.50	417
1/2"	"	490	28.75	519
3/4"	"	540	33.25	573
1"	"	600	44.25	644
Check valves				
1/4"	EA.	160	22.25	182
3/8"	"	160	26.50	187
1/2"	"	160	28.75	189
3/4"	"	190	33.25	223
1"	"	210	44.25	254
Test and balance	"	64.00	55.00	119
Pipe identification	"	0.35	13.25	13.60
Disinfect	"	64.00	55.00	119

15410.80 Steel Pipe

	UNIT	MAT.	INST.	TOTAL
Black steel, extra heavy pipe, threaded				
1/2" pipe	L.F.	3.06	2.65	5.71
3/4" pipe	"	3.96	2.65	6.61
1" pipe	"	5.08	3.32	8.40
1-1/2" pipe	"	7.62	3.68	11.30
2-1/2" pipe	"	15.25	8.30	23.55
3" pipe	"	20.25	11.00	31.25
4" pipe	"	30.75	13.25	44.00
5" pipe	"	41.50	16.50	58.00
6" pipe	"	52.00	16.50	68.50
8" pipe	"	76.00	22.25	98.25
10" pipe	"	120	26.50	147
12" pipe	"	160	33.25	193
Fittings, malleable iron, threaded, 1/2" pipe				
90 deg ell	EA.	4.26	22.25	26.51
45 deg ell	"	5.75	22.25	28.00
Tee	"	4.64	33.25	37.89
Reducing tee	"	10.50	33.25	43.75
Cap	"	3.61	13.25	16.86
Coupling	"	4.81	26.50	31.31

15 MECHANICAL

Plumbing

15410.80 Steel Pipe (Cont.)

	UNIT	MAT.	INST.	TOTAL
Union	EA.	20.25	22.25	42.50
Nipple, 4" long	"	3.78	22.25	26.03
3/4" pipe				
90 deg ell	EA.	4.98	22.25	27.23
45 deg ell	"	7.90	33.25	41.15
Tee	"	6.70	33.25	39.95
Reducing tee	"	11.50	22.25	33.75
Cap	"	4.81	13.25	18.06
Coupling	"	5.67	22.25	27.92
Union	"	22.75	22.25	45.00
Nipple, 4" long	"	4.38	22.25	26.63
1" pipe				
90 deg ell	EA.	7.73	26.50	34.23
45 deg ell	"	10.25	26.50	36.75
Tee	"	11.50	37.00	48.50
Reducing tee	"	15.75	37.00	52.75
Cap	"	6.53	13.25	19.78
Coupling	"	8.42	26.50	34.92
Union	"	27.25	26.50	53.75
Nipple, 4" long	"	6.18	26.50	32.68
1-1/2" pipe				
90 deg ell	EA.	15.75	33.25	49.00
45 deg ell	"	19.50	33.25	52.75
Tee	"	23.00	47.50	70.50
Reducing tee	"	33.00	47.50	80.50
Cap	"	10.50	16.50	27.00
Coupling	"	14.75	33.25	48.00
Union	"	35.75	33.25	69.00
Nipple, 4" long	"	9.71	33.25	42.96
2-1/2" pipe				
90 deg ell	EA.	61.00	83.00	144
45 deg ell	"	86.00	83.00	169
Tee	"	85.00	110	195
Reducing tee	"	130	110	240
Cap	"	34.75	41.50	76.25
Coupling	"	61.00	110	171
Union	"	130	110	240
Nipple, 4" long	"	35.25	110	145
3" pipe				
90 deg ell	EA.	89.00	110	199
45 deg ell	"	110	110	220
Tee	"	130	170	300
Reducing tee	"	170	170	340
Cap	"	52.00	55.00	107
Coupling	"	82.00	110	192
Union	"	220	110	330
Nipple, 4" long	"	46.00	110	156
4" pipe				
90 deg ell	EA.	190	130	320
45 deg ell	"	220	130	350
Tee	"	300	220	520
Reducing tee	"	420	220	640
Cap	"	88.00	220	308

15 MECHANICAL

Plumbing

15410.80 Steel Pipe *(Cont.)*

	UNIT	MAT.	INST.	TOTAL
Coupling	EA.	160	66.00	226
Union	"	220	220	440
Nipple, 4" long	"	61.00	220	281
6" pipe				
90 deg ell	EA.	540	130	670
45 deg ell	"	620	130	750
Tee	"	950	220	1,170
Reducing tee	"	670	220	890
Cap	"	69.00	66.00	135
8" pipe				
90 deg ell	EA.	1,070	270	1,340
45 deg ell	"	1,200	270	1,470
Tee	"	740	420	1,160
Reducing tee	"	690	350	1,040
Cap	"	83.00	130	213
10" pipe				
90 deg ell	EA.	1,170	330	1,500
45 deg ell	"	1,320	330	1,650
Tee	"	840	420	1,260
Reducing tee	"	810	170	980
Cap	"	86.00	170	256
12" pipe				
90 deg ell	EA.	1,290	420	1,710
45 deg ell	"	1,460	420	1,880
Tee	"	980	550	1,530
Reducing tee	"	910	550	1,460
Cap	"	95.00	220	315
Butt welded, 1/2" pipe				
90 deg ell	EA.	19.50	22.25	41.75
45 deg ell	"	28.00	22.25	50.25
Tee	"	64.00	33.25	97.25
3/4" pipe				
90 deg ell	EA.	19.50	22.25	41.75
45 deg. ell	"	28.50	22.25	50.75
Tee	"	64.00	33.25	97.25
1" pipe				
90 deg ell	EA.	25.75	26.50	52.25
45 deg ell	"	28.50	26.50	55.00
Tee	"	64.00	37.00	101
1-1/2" pipe				
90 deg ell	EA.	27.75	33.25	61.00
45 deg. ell	"	29.50	33.25	62.75
Tee	"	72.00	47.50	120
Reducing tee	"	48.75	47.50	96.25
Cap	"	34.00	26.50	60.50
2-1/2" pipe				
90 deg. ell	EA.	42.75	66.00	109
45 deg. ell	"	57.00	66.00	123
Tee	"	80.00	95.00	175
Reducing tee	"	48.75	95.00	144
Cap	"	39.00	33.25	72.25
3" pipe				
90 deg ell	EA.	50.00	83.00	133

15 MECHANICAL

Plumbing

15410.80 Steel Pipe (Cont.)

	UNIT	MAT.	INST.	TOTAL
45 deg. ell	EA.	57.00	83.00	140
Tee	"	80.00	110	190
Reducing tee	"	57.00	110	167
Cap	"	38.75	55.00	93.75
4" pipe				
90 deg ell	EA.	79.00	110	189
45 deg. ell	"	64.00	110	174
Tee	"	110	170	280
Reducing tee	"	72.00	170	242
Cap	"	41.75	55.00	96.75
6" pipe				
90 deg. ell	EA.	200	130	330
45 deg. ell	"	160	130	290
Tee	"	290	220	510
Reducing tee	"	270	220	490
Cap	"	74.00	66.00	140
8" pipe				
90 deg. ell	EA.	360	220	580
45 deg. ell	"	290	220	510
Tee	"	500	330	830
Reducing tee	"	320	330	650
Cap	"	110	130	240
10" pipe				
90 deg ell	EA.	640	220	860
45 deg. ell	"	570	220	790
Tee	"	640	330	970
Reducing tee	"	420	330	750
Cap	"	200	170	370
12" pipe				
90 deg. ell	EA.	930	270	1,200
45 deg. ell	"	640	270	910
Tee	"	930	470	1,400
Reducing tee	"	480	470	950
Cap	"	300	170	470
Cast iron fittings				
1/2" pipe				
90 deg. ell	EA.	5.61	22.25	27.86
45 deg. ell	"	11.50	22.25	33.75
Tee	"	7.41	33.25	40.66
Reducing tee	"	14.00	33.25	47.25
3/4" pipe				
90 deg. ell	EA.	6.00	22.25	28.25
45 deg. ell	"	7.41	22.25	29.66
Tee	"	9.28	33.25	42.53
Reducing tee	"	12.00	33.25	45.25
1" pipe				
90 deg. ell	EA.	7.17	26.50	33.67
45 deg. ell	"	9.90	26.50	36.40
Tee	"	13.75	37.00	50.75
Reducing tee	"	11.75	37.00	48.75
1-1/2" pipe				
90 deg. ell	EA.	14.00	33.25	47.25
45 deg. ell	"	20.00	33.25	53.25

15 MECHANICAL

Plumbing

15410.80 Steel Pipe (Cont.)

Description	UNIT	MAT.	INST.	TOTAL
Tee	EA.	20.00	47.50	67.50
Reducing tee	"	27.75	47.50	75.25
2-1/2" pipe				
90 deg. ell	EA.	44.75	66.00	111
45 deg. ell	"	53.00	66.00	119
Tee	"	64.00	95.00	159
Reducing tee	"	74.00	95.00	169
3" pipe				
90 deg. ell	EA.	73.00	83.00	156
45 deg. ell	"	84.00	83.00	167
Tee	"	97.00	130	227
Reducing tee	"	110	130	240
4" pipe				
90 deg. ell	EA.	130	110	240
45 deg. ell	"	160	110	270
Tee	"	180	170	350
Reducing tee	"	230	170	400
6" pipe				
90 deg. ell	EA.	310	110	420
45 deg. ell	"	370	110	480
Tee	"	430	170	600
Reducing tee	"	520	170	690
8" pipe				
90 deg. ell	EA.	620	220	840
45 deg. ell	"	690	220	910
Tee	"	880	330	1,210
Reducing tee	"	970	330	1,300

15410.82 Galvanized Steel Pipe

Description	UNIT	MAT.	INST.	TOTAL
Galvanized pipe				
1/2" pipe	L.F.	4.16	6.64	10.80
3/4" pipe	"	5.42	8.30	13.72
1" pipe	"	8.32	9.48	17.80
1-1/4" pipe	"	10.25	11.00	21.25
1-1/2" pipe	"	11.25	13.25	24.50
2" pipe	"	16.25	16.50	32.75
2-1/2" pipe	"	23.50	22.25	45.75
3" pipe	"	32.50	23.75	56.25
4" pipe	"	44.25	27.75	72.00
6" pipe	"	80.00	55.00	135
90 degree ell, 150 lb malleable iron, galvanized				
1/2"	EA.	2.80	13.25	16.05
3/4"	"	3.71	16.50	20.21
1"	"	6.06	17.50	23.56
1-1/4"	"	8.92	19.50	28.42
1-1/2"	"	11.50	22.25	33.75
2"	"	18.25	26.50	44.75
2-1/2"	"	46.75	41.50	88.25
3"	"	74.00	51.00	125
4"	"	130	55.00	185
5"	"	390	66.00	456
6"	"	430	66.00	496
45 degree ell, 150 lb m.i., galv.				

15 MECHANICAL

Plumbing

15410.82 Galvanized Steel Pipe *(Cont.)*

	UNIT	MAT.	INST.	TOTAL
1/2"	EA.	4.46	13.25	17.71
3/4"	"	6.06	16.50	22.56
1"	"	6.80	17.50	24.30
1-1/4"	"	10.75	19.50	30.25
1-1/2"	"	13.50	22.25	35.75
2"	"	21.00	26.50	47.50
2-1/2"	"	65.00	41.50	107
3"	"	86.00	51.00	137
4"	"	150	66.00	216
5"	"	390	66.00	456
6"	"	540	83.00	623
Tees, straight, 150 lb m.i., galv.				
1/2"	EA.	3.71	16.50	20.21
3/4"	"	6.17	19.00	25.17
1"	"	9.09	22.25	31.34
1-1/4"	"	12.25	26.50	38.75
1-1/2"	"	16.75	33.25	50.00
2"	"	26.00	41.50	67.50
2-1/2"	"	70.00	55.00	125
3"	"	86.00	66.00	152
4"	"	200	83.00	283
5"	"	580	95.00	675
6"	"	670	110	780
Tees, reducing, out, 150 lb m.i., galv.				
1/2"	EA.	6.40	16.50	22.90
3/4"	"	7.43	19.00	26.43
1"	"	11.00	22.25	33.25
1-1/4"	"	20.25	26.50	46.75
1-1/2"	"	25.75	33.25	59.00
2"	"	30.00	41.50	71.50
2-1/2"	"	82.00	55.00	137
3"	"	130	66.00	196
4"	"	240	83.00	323
5"	"	530	95.00	625
6"	"	530	110	640
Couplings, straight, 150 lb m.i., galv.				
1/2"	EA.	3.43	13.25	16.68
3/4"	"	4.11	14.75	18.86
1"	"	7.03	16.50	23.53
1-1/4"	"	8.00	19.00	27.00
1-1/2"	"	10.25	22.25	32.50
2"	"	14.25	26.50	40.75
2-1/2"	"	38.50	41.50	80.00
3"	"	54.00	55.00	109
4"	"	110	60.00	170
5"	"	310	66.00	376
6"	"	310	66.00	376
Couplings, reducing, 150 lb m.i., galv				
1/2"	EA.	4.00	13.25	17.25
3/4"	"	4.46	14.75	19.21
1"	"	8.17	16.50	24.67
1-1/4"	"	10.00	19.00	29.00
1-1/2"	"	12.25	22.25	34.50

15 MECHANICAL

Plumbing

	UNIT	MAT.	INST.	TOTAL
15410.82 Galvanized Steel Pipe (Cont.)				
2"	EA.	17.25	26.50	43.75
2-1/2"	"	46.50	41.50	88.00
3"	"	58.00	55.00	113
4"	"	120	60.00	180
5"	"	290	66.00	356
6"	"	350	66.00	416
Caps, 150 lb m.i., galv.				
1/2"	EA.	2.86	6.64	9.50
3/4"	"	3.77	6.98	10.75
1"	"	5.14	7.37	12.51
1-1/4"	"	6.06	7.81	13.87
1-1/2"	"	7.77	8.30	16.07
2"	"	10.75	9.48	20.23
2-1/2"	"	26.75	12.00	38.75
3"	"	38.25	16.50	54.75
4"	"	62.00	20.75	82.75
5"	"	70.00	25.50	95.50
6"	"	74.00	33.25	107
Unions, 150 lb m.i., galv.				
1/2"	EA.	16.00	16.50	32.50
3/4"	"	18.00	19.00	37.00
1"	"	21.50	22.25	43.75
1-1/4"	"	24.50	26.50	51.00
1-1/2"	"	28.00	33.25	61.25
2"	"	34.75	37.00	71.75
2-1/2"	"	110	44.25	154
3"	"	180	55.00	235
Nipples, galvanized steel, 4" long				
1/2"	EA.	4.11	8.30	12.41
3/4"	"	5.49	8.85	14.34
1"	"	7.55	9.48	17.03
1-1/4"	"	8.58	10.25	18.83
1-1/2"	"	10.25	11.00	21.25
2"	"	12.00	12.00	24.00
2-1/2"	"	14.50	13.25	27.75
3"	"	18.75	16.50	35.25
4"	"	23.75	22.25	46.00
90 degree reducing ell, 150 lb m.i., galv.				
3/4" x 1/2"	EA.	4.46	13.25	17.71
1" x 3/4"	"	6.06	14.75	20.81
1-1/4" x 1"	"	9.55	16.50	26.05
1-1/4" x 3/4"	"	11.50	19.00	30.50
1-1/4" x 1/2"	"	12.00	22.25	34.25
1-1/2" x 1-1/4"	"	15.25	22.25	37.50
1-1/2" x 1"	"	15.25	22.25	37.50
1-1/2" x 3/4"	"	15.25	22.25	37.50
2" x 1-1/2"	"	18.00	28.75	46.75
2" x 1-1/4"	"	20.50	28.75	49.25
2" x 1"	"	21.25	28.75	50.00
2" x 3/4"	"	21.50	28.75	50.25
2-1/2" x 2"	"	60.00	41.50	102
2-1/2" x 1-1/2"	"	67.00	41.50	109
3" x 2-1/2"	"	110	55.00	165

15 MECHANICAL

Plumbing

	UNIT	MAT.	INST.	TOTAL
15410.82 Galvanized Steel Pipe *(Cont.)*				
3" x 2"	EA.	95.00	55.00	150
4" x 3"	"	240	60.00	300
Square head plug (C.I.)				
1/2"	EA.	2.83	7.37	10.20
3/4"	"	6.29	8.30	14.59
1"	"	6.60	8.85	15.45
1-1/4"	"	6.92	9.48	16.40
1-1/2"	"	8.65	10.25	18.90
2"	"	11.00	11.00	22.00
2-1/2"	"	14.50	14.75	29.25
3"	"	21.00	16.50	37.50
4"	"	30.75	22.25	53.00
5"	"	50.00	26.50	76.50
6"	"	69.00	33.25	102
Screwed flanges, galv.				
1"	EA.	49.75	33.25	83.00
1-1/4"	"	57.00	37.00	94.00
1-1/2"	"	59.00	41.50	101
2"	"	52.00	41.50	93.50
2-1/2"	"	60.00	44.25	104
3"	"	73.00	60.00	133
4"	"	100	83.00	183
5"	"	160	110	270
6"	"	170	110	280
15430.23 Cleanouts				
Cleanout, wall				
2"	EA.	250	44.25	294
3"	"	350	44.25	394
4"	"	350	55.00	405
6"	"	580	66.00	646
8"	"	810	83.00	893
Floor				
2"	EA.	220	55.00	275
3"	"	300	55.00	355
4"	"	300	66.00	366
6"	"	420	83.00	503
8"	"	790	95.00	885
15430.24 Grease Traps				
Grease traps, cast iron, 3" pipe				
35 gpm, 70 lb capacity	EA.	4,160	660	4,820
50 gpm, 100 lb capacity	"	5,300	830	6,130
15430.25 Hose Bibbs				
Hose bibb				
1/2"	EA.	10.75	22.25	33.00
3/4"	"	11.25	22.25	33.50

15 MECHANICAL

Plumbing

15430.60 Valves

	UNIT	MAT.	INST.	TOTAL
Gate valve, 125 lb, bronze, soldered				
1/2"	EA.	30.50	16.50	47.00
3/4"	"	36.50	16.50	53.00
1"	"	45.00	22.25	67.25
1-1/2"	"	79.00	26.50	106
2"	"	110	33.25	143
2-1/2"	"	250	41.50	292
Threaded				
1/4", 125 lb	EA.	35.00	26.50	61.50
1/2"				
125 lb	EA.	33.75	26.50	60.25
150 lb	"	45.00	26.50	71.50
300 lb	"	85.00	26.50	112
3/4"				
125 lb	EA.	39.50	26.50	66.00
150 lb	"	53.00	26.50	79.50
300 lb	"	100	26.50	127
1"				
125 lb	EA.	51.00	26.50	77.50
150 lb	"	70.00	26.50	96.50
300 lb	"	140	33.25	173
1-1/2"				
125 lb	EA.	88.00	33.25	121
150 lb	"	120	33.25	153
300 lb	"	260	37.00	297
2"				
125 lb	EA.	120	47.50	168
150 lb	"	180	47.50	228
300 lb	"	330	55.00	385
Cast iron, flanged				
2", 150 lb	EA.	420	55.00	475
2-1/2"				
125 lb	EA.	400	55.00	455
150 lb	"	620	55.00	675
250 lb	"	1,110	55.00	1,165
3"				
125 lb	EA.	480	66.00	546
150 lb	"	650	66.00	716
250 lb	"	1,010	66.00	1,076
4"				
125 lb	EA.	640	95.00	735
150 lb	"	1,050	95.00	1,145
250 lb	"	1,370	95.00	1,465
6"				
125 lb	EA.	1,170	130	1,300
250 lb	"	2,730	130	2,860
8"				
125 lb	EA.	1,860	170	2,030
250 lb	"	5,290	170	5,460
OS&Y, flanged				
2"				
125 lb	EA.	360	55.00	415
250 lb	"	960	55.00	1,015

15 MECHANICAL

Plumbing	UNIT	MAT.	INST.	TOTAL
15430.60 Valves *(Cont.)*				
2-1/2"				
125 lb	EA.	380	55.00	435
250 lb	"	1,190	66.00	1,256
3"				
125 lb	EA.	420	66.00	486
250 lb	"	1,230	66.00	1,296
4"				
125 lb	EA.	560	110	670
250 lb	"	1,890	110	2,000
6"				
125 lb	EA.	940	130	1,070
250 lb	"	2,990	130	3,120
Check valve, bronze, soldered, 125 lb				
1/2"	EA.	58.00	16.50	74.50
3/4"	"	72.00	16.50	88.50
1"	"	91.00	22.25	113
1-1/4"	"	130	26.50	157
1-1/2"	"	140	26.50	167
2"	"	210	33.25	243
Threaded				
1/2"				
125 lb	EA.	67.00	22.25	89.25
150 lb	"	62.00	22.25	84.25
200 lb	"	65.00	22.25	87.25
3/4"				
125 lb	EA.	50.00	26.50	76.50
150 lb	"	78.00	26.50	105
200 lb	"	86.00	26.50	113
1"				
125 lb	EA.	69.00	33.25	102
150 lb	"	110	33.25	143
200 lb	"	110	33.25	143
Flow check valve, cast iron, threaded				
1"	EA.	71.00	26.50	97.50
1-1/4"	"	89.00	33.25	122
1-1/2"				
125 lb	EA.	89.00	33.25	122
150 lb	"	110	33.25	143
200 lb	"	110	37.00	147
2"				
125 lb	EA.	100	37.00	137
150 lb	"	150	37.00	187
200 lb	"	150	41.50	192
2-1/2"				
125 lb	EA.	230	55.00	285
250 lb	"	740	66.00	806
3"				
125 lb	EA.	260	66.00	326
250 lb	"	920	83.00	1,003
4"				
125 lb	EA.	400	95.00	495
250 lb	"	1,180	110	1,290
6"				

15 MECHANICAL

Plumbing

15430.60 Valves (Cont.)

	UNIT	MAT.	INST.	TOTAL
125 lb	EA.	550	130	680
250 lb	"	2,000	130	2,130
Vertical check valve, bronze, 125 lb, threaded				
1/2"	EA.	77.00	26.50	104
3/4"	"	110	30.25	140
1"	"	130	33.25	163
1-1/4"	"	180	37.00	217
1-1/2"	"	250	41.50	292
2"	"	350	47.50	398
Cast iron, flanged				
2-1/2"	EA.	290	66.00	356
3"	"	320	83.00	403
4"	"	500	110	610
6	"	850	130	980
8"	"	1,600	170	1,770
10"	"	2,740	220	2,960
12"	"	4,250	270	4,520
Globe valve, bronze, soldered, 125 lb				
1/2"	EA.	71.00	19.00	90.00
3/4"	"	88.00	20.75	109
1"	"	120	22.25	142
1-1/4"	"	160	23.75	184
1-1/2"	"	200	27.75	228
2"	"	310	33.25	343
Threaded				
1/2"				
125 lb	EA.	65.00	22.25	87.25
150 lb	"	86.00	22.25	108
300 lb	"	160	22.25	182
3/4"				
125 lb	EA.	93.00	26.50	120
150 lb	"	100	26.50	127
300 lb	"	200	26.50	227
1"				
125 lb	EA.	110	33.25	143
150 lb	"	210	33.25	243
300 lb	"	250	33.25	283
1-1/4"				
125 lb	EA.	160	33.25	193
150 lb	"	260	33.25	293
300 lb	"	350	33.25	383
1-1/2"				
125 lb	EA.	200	37.00	237
150 lb	"	330	37.00	367
300 lb	"	430	37.00	467
2"				
125 lb	EA.	300	44.25	344
150 lb	"	510	44.25	554
300 lb	"	680	44.25	724
Cast iron flanged				
2-1/2"				
125 lb	EA.	800	66.00	866
250 lb	"	1,400	66.00	1,466

15 MECHANICAL

Plumbing

15430.60 Valves (Cont.)

	UNIT	MAT.	INST.	TOTAL
3"				
125 lb	EA.	970	83.00	1,053
250 lb	"	1,430	83.00	1,513
4"				
125 lb	EA.	1,390	110	1,500
250 lb	"	1,860	110	1,970
6"				
125 lb	EA.	2,550	130	2,680
250 lb	"	3,760	130	3,890
8"				
125 lb	EA.	5,020	170	5,190
250 lb	"	6,180	170	6,350
Butterfly valve, cast iron, wafer type				
2"				
150 lb	EA.	470	47.50	518
200 lb	"	660	55.00	715
2-1/2"				
150 lb	EA.	640	55.00	695
200 lb	"	680	60.00	740
3"				
150 lb	EA.	640	66.00	706
200 lb	"	710	74.00	784
4"				
150 lb	EA.	720	95.00	815
200 lb	"	810	110	920
6"				
150 lb	EA.	850	130	980
200 lb	"	1,060	130	1,190
8"				
150 lb	EA.	1,060	150	1,210
200 lb	"	1,390	170	1,560
10"				
150 lb	EA.	1,290	170	1,460
200 lb	"	1,870	220	2,090
Ball valve, bronze, 250 lb, threaded				
1/2"	EA.	20.00	26.50	46.50
3/4"	"	29.75	26.50	56.25
1"	"	37.75	33.25	71.00
1-1/4"	"	55.00	37.00	92.00
1-1/2"	"	88.00	41.50	130
2"	"	100	47.50	148
Angle valve, bronze, 150 lb, threaded				
1/2"	EA.	100	23.75	124
3/4"	"	140	26.50	167
1"	"	200	26.50	227
1-1/4"	"	260	33.25	293
1-1/2"	"	340	37.00	377
Balancing valve, meter connections, circuit setter				
1/2"	EA.	97.00	26.50	124
3/4"	"	100	30.25	130
1"	"	130	33.25	163
1-1/4"	"	180	37.00	217
1-1/2"	"	220	44.25	264

15 MECHANICAL

Plumbing

15430.60 Valves *(Cont.)*

	UNIT	MAT.	INST.	TOTAL
2"	EA.	310	55.00	365
2-1/2"	"	610	66.00	676
3"	"	900	83.00	983
4"	"	1,260	110	1,370
Balancing valve, straight type				
1/2"	EA.	26.25	26.50	52.75
3/4"	"	31.75	26.50	58.25
Angle type				
1/2"	EA.	35.25	26.50	61.75
3/4"	"	49.25	26.50	75.75
Square head cock, 125 lb, bronze body				
1/2"	EA.	20.75	22.25	43.00
3/4"	"	24.75	26.50	51.25
1"	"	34.50	30.25	64.75
1-1/4"	"	47.00	33.25	80.25
Radiator temp control valve, with control and sensor				
1/2" valve	EA.	140	41.50	182
1" valve	"	160	41.50	202
Pressure relief valve, 1/2", bronze				
Low pressure	EA.	33.25	26.50	59.75
High pressure	"	38.50	26.50	65.00
Pressure and temperature relief valve				
Bronze, 3/4"	EA.	120	26.50	147
Cast iron, 3/4"				
High pressure	EA.	57.00	26.50	83.50
Temperature relief	"	77.00	26.50	104
Pressure & temp relief valve	"	92.00	26.50	119
Pressure reducing valve, bronze, threaded, 250 lb				
1/2"	EA.	200	41.50	242
3/4"	"	200	41.50	242
1"	"	310	41.50	352
1-1/4"	"	440	47.50	488
1-1/2"	"	510	55.00	565
Pressure regulating valve, bronze, class 300				
1"	EA.	730	41.50	772
1-1/2"	"	980	51.00	1,031
2"	"	1,110	66.00	1,176
3"	"	1,250	95.00	1,345
4"	"	1,560	130	1,690
5"	"	2,370	170	2,540
6"	"	2,410	220	2,630
Solar water temperature regulating valve				
3/4"	EA.	750	55.00	805
1"	"	770	66.00	836
1-1/4"	"	830	74.00	904
1-1/2"	"	920	83.00	1,003
2"	"	1,140	95.00	1,235
2-1/2"	"	2,160	170	2,330
Tempering valve, threaded				
3/4"	EA.	410	22.25	432
1"	"	520	26.50	547
1-1/4"	"	770	33.25	803
1-1/2"	"	880	33.25	913

15 MECHANICAL

Plumbing	UNIT	MAT.	INST.	TOTAL
15430.60 Valves *(Cont.)*				
2"	EA.	1,200	41.50	1,242
2-1/2"	"	2,030	55.00	2,085
3"	"	2,680	66.00	2,746
4"	"	5,370	95.00	5,465
Thermostatic mixing valve, threaded				
1/2"	EA.	140	23.75	164
3/4"	"	140	26.50	167
1"	"	520	28.75	549
1-1/2"	"	580	33.25	613
2"	"	730	41.50	772
Sweat connection				
1/2"	EA.	160	23.75	184
3/4"	"	200	26.50	227
Mixing valve, sweat connection				
1/2"	EA.	85.00	23.75	109
3/4"	"	85.00	26.50	112
Liquid level gauge, aluminum body				
3/4"	EA.	430	26.50	457
4125 psi, pvc body				
3/4"	EA.	510	26.50	537
150 psi, crs body				
3/4"	EA.	400	26.50	427
1"	"	440	26.50	467
175 psi, bronze body, 1/2"	"	820	23.75	844
15430.65 Vacuum Breakers				
Vacuum breaker, atmospheric, threaded connection				
3/4"	EA.	54.00	26.50	80.50
1"	"	79.00	26.50	106
Anti-siphon, brass				
3/4"	EA.	59.00	26.50	85.50
1"	"	91.00	26.50	118
1-1/4"	"	160	33.25	193
1-1/2"	"	190	37.00	227
2"	"	290	41.50	332
Air eliminators, purger, cast iron, threaded				
1"	EA.	40.50	26.50	67.00
1-1/4"	"	42.75	33.25	76.00
1-1/2"	"	79.00	33.25	112
2"	"	90.00	41.50	132
2-1/2"	"	220	83.00	303
3"	"	250	110	360
Airtrol fitting, 3/4"	"	56.00	26.50	82.50
Air eliminator, air vents, 1/4"	"	32.75	26.50	59.25
Air vent for hot water	"	28.00	23.75	51.75
15430.68 Strainers				
Strainer, Y pattern, 125 psi, cast iron body, threaded				
3/4"	EA.	13.50	23.75	37.25
1"	"	17.25	26.50	43.75
1-1/4"	"	21.75	33.25	55.00
1-1/2"	"	27.75	33.25	61.00
2"	"	41.00	41.50	82.50

15 MECHANICAL

Plumbing	UNIT	MAT.	INST.	TOTAL
15430.68 — **Strainers** *(Cont.)*				
250 psi, brass body, threaded				
3/4"	EA.	35.25	26.50	61.75
1"	"	49.25	26.50	75.75
1-1/4"	"	61.00	33.25	94.25
1-1/2"	"	86.00	33.25	119
2"	"	150	41.50	192
Cast iron body, threaded				
3/4"	EA.	20.50	26.50	47.00
1"	"	26.25	26.50	52.75
1-1/4"	"	34.75	33.25	68.00
1-1/2"	"	46.00	33.25	79.25
2"	"	59.00	41.50	101
15430.70 — **Drains, Roof & Floor**				
Floor drain, cast iron, with cast iron top				
2"	EA.	170	55.00	225
3"	"	180	55.00	235
4"	"	380	55.00	435
6"	"	490	66.00	556
Roof drain, cast iron				
2"	EA.	270	55.00	325
3"	"	290	55.00	345
4"	"	360	55.00	415
5"	"	530	66.00	596
6"	"	540	66.00	606
15430.80 — **Traps**				
Bucket trap, threaded				
3/4"	EA.	240	41.50	282
1"	"	700	44.25	744
1-1/4"	"	830	51.00	881
1-1/2"	"	1,240	60.00	1,300
Inverted bucket steam trap, threaded				
3/4"	EA.	300	41.50	342
1"	"	590	41.50	632
1-1/4"	"	890	37.00	927
1-1/2"	"	940	55.00	995
With stainless interior				
1/2"	EA.	190	41.50	232
3/4"	"	210	41.50	252
1"	"	440	41.50	482
1-1/4"	"	540	47.50	588
Brass interior				
3/4"	EA.	330	41.50	372
1"	"	660	44.25	704
1-1/4"	"	940	47.50	988
Cast steel body, threaded, high temperature				
3/4"	EA.	860	41.50	902
1"	"	1,160	47.50	1,208
1-1/4	"	1,760	51.00	1,811
1-1/2"	"	2,220	55.00	2,275
2"	"	3,300	66.00	3,366
Float trap, 15 psi				

15 MECHANICAL

Plumbing

15430.80 Traps (Cont.)

	UNIT	MAT.	INST.	TOTAL
3/4"	EA.	210	41.50	252
1"	"	330	44.25	374
1-1/4"	"	430	47.50	478
1-1/2"	"	540	55.00	595
2"	"	940	66.00	1,006
30 psi				
3/4"	EA.	240	41.50	282
1"	"	330	44.25	374
1-1/4"	"	490	55.00	545
1-1/2"	"	730	66.00	796
75 psi				
3/4"	EA.	300	41.50	342
1"	"	340	44.25	384
1-1/4"	"	460	47.50	508
1-1/2"	"	700	55.00	755
125 psi				
3/4"	EA.	240	41.50	282
1"	"	270	44.25	314
1-1/4	"	550	47.50	598
1-1/2	"	560	55.00	615
Float and thermostatic trap, 15 psi				
3/4"	EA.	230	41.50	272
1"	"	260	44.25	304
1-1/4"	"	400	47.50	448
1-1/2"	"	510	55.00	565
2"	"	940	66.00	1,006
30 psi				
3/4"	EA.	270	41.50	312
1"	"	330	44.25	374
1-1/4"	"	370	47.50	418
1-1/2"	"	530	55.00	585
75 psi				
3/4"	EA.	300	41.50	342
1"	"	330	44.25	374
1-1/4"	"	470	47.50	518
1-1/2"	"	820	55.00	875
Steam trap, cast iron body, threaded, 125 psi				
3/4"	EA.	270	41.50	312
1"	"	310	44.25	354
1-1/4"	"	460	47.50	508
1-1/2"	"	740	55.00	795
Thermostatic trap, low pressure, angle type, 25 psi				
1/2"	EA.	83.00	41.50	125
3/4"	"	140	41.50	182
1"	"	190	44.25	234
50 psi				
1/2"	EA.	130	41.50	172
3/4"	"	170	41.50	212
1"	"	200	44.25	244
Cast iron body, threaded, 125 psi				
3/4"	EA.	190	41.50	232
1"	"	230	47.50	278
1-1/4"	"	300	51.00	351

15 MECHANICAL

Plumbing

Plumbing	UNIT	MAT.	INST.	TOTAL
15430.80		**Traps** *(Cont.)*		
1-1/2"	EA.	440	55.00	495
Low pressure, 25 psi, swivel type, 1/2"	"	88.00	41.50	130
Straightway type, 3/4"	"	120	41.50	162
Vertical type, 1/2"	"	80.00	41.50	122
Medium pressure, 50 psi, angle type, 1/2"	"	110	41.50	152
High pressure, 125 psi, angle type				
1/2"	EA.	140	41.50	182
3/4"	"	190	41.50	232
1"	"	270	47.50	318
Straightway type				
1/2"	EA.	140	41.50	182
Thermo disc trap				
3/4"	EA.	210	41.50	252
1"	"	240	44.25	284
Drip pan ell, cast iron				
2-1/2"	EA.	230	83.00	313
3"	"	240	95.00	335
4"	"	340	110	450
5"	"	440	130	570
6"	"	690	170	860
8"	"	770	220	990
Steel				
2-1/2"	EA.	1,190	83.00	1,273
3"	"	1,260	95.00	1,355
4"	"	1,630	110	1,740
5"	"	2,620	130	2,750
6"	"	2,620	170	2,790
8"	"	3,990	220	4,210

Plumbing Fixtures	UNIT	MAT.	INST.	TOTAL
15440.10		**Baths**		
Bath tub, 5' long				
Minimum	EA.	620	220	840
Average	"	1,370	330	1,700
Maximum	"	3,120	660	3,780
6' long				
Minimum	EA.	700	220	920
Average	"	1,430	330	1,760
Maximum	"	4,040	660	4,700
Square tub, whirlpool, 4'x4'				
Minimum	EA.	2,140	330	2,470
Average	"	3,040	660	3,700
Maximum	"	9,280	830	10,110
5'x5'				
Minimum	EA.	2,140	330	2,470

15 MECHANICAL

Plumbing Fixtures

	UNIT	MAT.	INST.	TOTAL
15440.10 — Baths *(Cont.)*				
Average	EA.	3,040	660	3,700
Maximum	"	9,450	830	10,280
6'x6'				
Minimum	EA.	2,610	330	2,940
Average	"	3,820	660	4,480
Maximum	"	10,960	830	11,790
For trim and rough-in				
Minimum	EA.	220	220	440
Average	"	330	330	660
Maximum	"	920	660	1,580
15440.12 — Disposals & Accessories				
Continuous feed				
Minimum	EA.	85.00	130	215
Average	"	230	170	400
Maximum	"	450	220	670
Batch feed, 1/2 hp				
Minimum	EA.	330	130	460
Average	"	650	170	820
Maximum	"	1,120	220	1,340
Hot water dispenser				
Minimum	EA.	230	130	360
Average	"	380	170	550
Maximum	"	600	220	820
Epoxy finish faucet	"	340	130	470
Lock stop assembly	"	72.00	83.00	155
Mounting gasket	"	8.32	55.00	63.32
Tailpipe gasket	"	1.22	55.00	56.22
Stopper assembly	"	28.50	66.00	94.50
Switch assembly, on/off	"	32.50	110	143
Tailpipe gasket washer	"	1.30	33.25	34.55
Stop gasket	"	2.86	37.00	39.86
Tailpipe flange	"	0.32	33.25	33.57
Tailpipe	"	3.70	41.50	45.20
15440.15 — Faucets				
Kitchen				
Minimum	EA.	98.00	110	208
Average	"	270	130	400
Maximum	"	340	170	510
Bath				
Minimum	EA.	98.00	110	208
Average	"	290	130	420
Maximum	"	440	170	610
Lavatory, domestic				
Minimum	EA.	100	110	210
Average	"	330	130	460
Maximum	"	550	170	720
Hospital, patient rooms				
Minimum	EA.	140	170	310
Average	"	450	220	670
Maximum	"	790	330	1,120
Operating room				

15 MECHANICAL

Plumbing Fixtures

15440.15 Faucets (Cont.)

	UNIT	MAT.	INST.	TOTAL
Minimum	EA.	300	170	470
Average	"	650	220	870
Maximum	"	950	330	1,280
Washroom				
Minimum	EA.	130	110	240
Average	"	330	130	460
Maximum	"	600	170	770
Handicapped				
Minimum	EA.	140	130	270
Average	"	430	170	600
Maximum	"	660	220	880
Shower				
Minimum	EA.	130	110	240
Average	"	380	130	510
Maximum	"	600	170	770
For trim and rough-in				
Minimum	EA.	91.00	130	221
Average	"	140	170	310
Maximum	"	230	330	560

15440.18 Hydrants

	UNIT	MAT.	INST.	TOTAL
Wall hydrant				
8" thick	EA.	430	110	540
12" thick	"	510	130	640
18" thick	"	550	150	700
24" thick	"	600	170	770
Ground hydrant				
2' deep	EA.	800	83.00	883
4' deep	"	920	95.00	1,015
6' deep	"	1,050	110	1,160
8' deep	"	1,170	170	1,340

15440.20 Lavatories

	UNIT	MAT.	INST.	TOTAL
Lavatory, counter top, porcelain enamel on cast iron				
Minimum	EA.	220	130	350
Average	"	340	170	510
Maximum	"	620	220	840
Wall hung, china				
Minimum	EA.	310	130	440
Average	"	360	170	530
Maximum	"	910	220	1,130
Handicapped				
Minimum	EA.	510	170	680
Average	"	580	220	800
Maximum	"	980	330	1,310
For trim and rough-in				
Minimum	EA.	260	170	430
Average	"	440	220	660
Maximum	"	550	330	880

15 MECHANICAL

Plumbing Fixtures

	UNIT	MAT.	INST.	TOTAL
15440.30 Showers				
Shower, fiberglass, 36"x34"x84"				
Minimum	EA.	680	470	1,150
Average	"	950	660	1,610
Maximum	"	1,370	660	2,030
Steel, 1 piece, 36"x36"				
Minimum	EA.	620	470	1,090
Average	"	950	660	1,610
Maximum	"	1,120	660	1,780
Receptor, molded stone, 36"x36"				
Minimum	EA.	260	220	480
Average	"	440	330	770
Maximum	"	680	550	1,230
For trim and rough-in				
Minimum	EA.	260	300	560
Average	"	440	370	810
Maximum	"	550	660	1,210
15440.40 Sinks				
Service sink, 24"x29"				
Minimum	EA.	750	170	920
Average	"	940	220	1,160
Maximum	"	1,380	330	1,710
Kitchen sink, single, stainless steel, single bowl				
Minimum	EA.	330	130	460
Average	"	380	170	550
Maximum	"	690	220	910
Double bowl				
Minimum	EA.	380	170	550
Average	"	420	220	640
Maximum	"	730	330	1,060
Porcelain enamel, cast iron, single bowl				
Minimum	EA.	230	130	360
Average	"	310	170	480
Maximum	"	480	220	700
Double bowl				
Minimum	EA.	330	170	500
Average	"	450	220	670
Maximum	"	650	330	980
Mop sink, 24"x36"x10"				
Minimum	EA.	570	130	700
Average	"	690	170	860
Maximum	"	920	220	1,140
Washing machine box				
Minimum	EA.	210	170	380
Average	"	300	220	520
Maximum	"	360	330	690
For trim and rough-in				
Minimum	EA.	340	220	560
Average	"	520	330	850
Maximum	"	660	440	1,100

15 MECHANICAL

Plumbing Fixtures	UNIT	MAT.	INST.	TOTAL
15440.50 — Urinals				
Urinal, flush valve, floor mounted				
Minimum	EA.	580	170	750
Average	"	690	220	910
Maximum	"	810	330	1,140
Wall mounted				
Minimum	EA.	480	170	650
Average	"	660	220	880
Maximum	"	860	330	1,190
For trim and rough-in				
Minimum	EA.	210	170	380
Average	"	310	330	640
Maximum	"	420	440	860
15440.60 — Water Closets				
Water closet flush tank, floor mounted				
Minimum	EA.	390	170	560
Average	"	770	220	990
Maximum	"	1,210	330	1,540
Handicapped				
Minimum	EA.	440	220	660
Average	"	790	330	1,120
Maximum	"	1,510	660	2,170
Bowl, with flush valve, floor mounted				
Minimum	EA.	550	170	720
Average	"	600	220	820
Maximum	"	1,170	330	1,500
Wall mounted				
Minimum	EA.	550	170	720
Average	"	640	220	860
Maximum	"	1,220	330	1,550
For trim and rough-in				
Minimum	EA.	250	170	420
Average	"	290	220	510
Maximum	"	390	330	720
15440.70 — Water Heaters				
Water heater, electric				
6 gal	EA.	480	110	590
10 gal	"	490	110	600
15 gal	"	490	110	600
20 gal	"	680	130	810
30 gal	"	700	130	830
40 gal	"	760	130	890
52 gal	"	860	170	1,030
66 gal	"	1,040	170	1,210
80 gal	"	1,130	170	1,300
100 gal	"	1,400	220	1,620
120 gal	"	1,790	220	2,010
Oil fired				
20 gal	EA.	1,530	330	1,860
50 gal	"	2,390	470	2,860

15 MECHANICAL

Plumbing Fixtures	UNIT	MAT.	INST.	TOTAL
15440.90 — Miscellaneous Fixtures				
Electric water cooler				
Floor mounted	EA.	1,200	220	1,420
Wall mounted	"	1,120	220	1,340
Wash fountain				
Wall mounted	EA.	2,860	330	3,190
Circular, floor supported	"	5,000	660	5,660
Deluge shower and eye wash	"	1,200	330	1,530
15440.95 — Fixture Carriers				
Water fountain, wall carrier				
Minimum	EA.	75.00	66.00	141
Average	"	100	83.00	183
Maximum	"	120	110	230
Lavatory, wall carrier				
Minimum	EA.	170	66.00	236
Average	"	250	83.00	333
Maximum	"	310	110	420
Sink, industrial, wall carrier				
Minimum	EA.	220	66.00	286
Average	"	260	83.00	343
Maximum	"	330	110	440
Toilets, water closets, wall carrier				
Minimum	EA.	330	66.00	396
Average	"	390	83.00	473
Maximum	"	510	110	620
Floor support				
Minimum	EA.	160	55.00	215
Average	"	200	66.00	266
Maximum	"	210	83.00	293
Urinals, wall carrier				
Minimum	EA.	180	66.00	246
Average	"	220	83.00	303
Maximum	"	270	110	380
Floor support				
Minimum	EA.	140	55.00	195
Average	"	210	66.00	276
Maximum	"	230	83.00	313
15450.30 — Pumps				
In-line pump, bronze, centrifugal				
5 gpm, 20' head	EA.	660	41.50	702
20 gpm, 40' head	"	1,190	41.50	1,232
50 gpm				
50' head	EA.	1,360	83.00	1,443
100' head	"	1,560	83.00	1,643
70 gpm, 100' head	"	1,940	110	2,050
100 gpm, 80' head	"	2,070	110	2,180
250 gpm, 150' head	"	6,490	170	6,660
Cast iron, centrifugal				
50 gpm, 200' head	EA.	1,290	83.00	1,373
100 gpm				
100' head	EA.	2,200	110	2,310
200' head	"	2,560	110	2,670

15 MECHANICAL

Plumbing Fixtures

15450.30 Pumps (Cont.)

	UNIT	MAT.	INST.	TOTAL
200 gpm				
100' head	EA.	3,830	170	4,000
200' head	"	5,170	170	5,340
Centrifugal, close coupled, c.i., single stage				
50 gpm, 100' head	EA.	1,540	83.00	1,623
100 gpm, 100' head	"	1,870	110	1,980
Base mounted				
50 gpm, 100' head	EA.	3,150	83.00	3,233
100 gpm, 50' head	"	3,580	110	3,690
200 gpm, 100' head	"	4,580	170	4,750
300 gpm, 175' head	"	5,940	170	6,110
Suction diffuser, flanged, strainer				
3" inlet, 2-1/2" outlet	EA.	510	83.00	593
3" outlet	"	520	83.00	603
4" inlet				
3" outlet	EA.	610	110	720
4" outlet	"	720	110	830
6" inlet				
4" outlet	EA.	830	130	960
5" outlet	"	1,000	130	1,130
6" Outlet	"	1,040	130	1,170
8" inlet				
6" outlet	EA.	1,130	170	1,300
8" outlet	"	1,960	170	2,130
10" inlet				
8" outlet	EA.	2,640	220	2,860
Vertical turbine				
Single stage, C.I., 3550 rpm, 200 gpm, 50' head	EA.	4,970	220	5,190
Multi stage, 3550 rpm				
50 gpm, 100' head	EA.	6,720	170	6,890
100 gpm				
100' head	EA.	7,030	170	7,200
200 gpm				
50' head	EA.	7,510	220	7,730
100' head	"	7,640	220	7,860
Bronze				
Single stage, 3550 rpm, 100 gpm, 50' head	EA.	6,950	170	7,120
Multi stage, 3550 rpm, 50 gpm, 100' head	"	6,550	170	6,720
100 gpm				
100' head	EA.	7,220	170	7,390
200 gpm				
50' head	EA.	7,220	220	7,440
100' head	"	7,710	220	7,930
Sump pump, bronze, 1750 rpm, 25 gpm				
20' head	EA.	640	830	1,470
150' head	"	880	1,110	1,990
50 gpm				
100' head	EA.	750	830	1,580
100 gpm				
50' head	EA.	660	830	1,490
Condensate pump, simplex				
1000 sf EDR, 2 gpm	EA.	1,610	550	2,160
2000 sf EDR, 3 gpm	"	1,640	550	2,190

15 MECHANICAL

Plumbing Fixtures

15450.30 Pumps (Cont.)

	UNIT	MAT.	INST.	TOTAL
4000 sf EDR, 6 gpm	EA.	1,650	600	2,250
6000 sf EDR, 9 gpm	"	1,680	600	2,280
Duplex, bronze				
8000 sf EDR, 12 gpm	EA.	2,300	600	2,900
10,000 sf EDR, 15 gpm	"	2,390	830	3,220
15,000 sf EDR, 23 gpm	"	2,870	950	3,820
20,000 sf EDR, 30 gpm	"	3,340	1,330	4,670
25,000 sf EDR, 38 gpm	"	3,460	1,330	4,790
30,000 sf EDR, 45 gpm	"	4,840	1,480	6,320
40,000 sf EDR, 60 gpm	"	4,840	830	5,670
50,000 sf EDR, 75 gpm	"	4,840	950	5,790
75,000 sf EDR, 112 gpm	"	7,150	1,110	8,260
100,000 sf EDR, 150 gpm	"	7,370	1,660	9,030

15450.40 Storage Tanks

	UNIT	MAT.	INST.	TOTAL
Hot water storage tank, cement lined				
10 gallon	EA.	580	220	800
70 gallon	"	1,820	330	2,150
200 gallon	"	3,450	470	3,920
900 gallon	"	7,960	830	8,790
1100 gallon	"	9,830	830	10,660
2000 gallon	"	17,790	830	18,620

15480.10 Special Systems

	UNIT	MAT.	INST.	TOTAL
Air compressor, air cooled, two stage				
5.0 cfm, 175 psi	EA.	3,200	1,330	4,530
10 cfm, 175 psi	"	3,900	1,480	5,380
20 cfm, 175 psi	"	5,380	1,580	6,960
50 cfm, 125 psi	"	7,850	1,750	9,600
80 cfm, 125 psi	"	11,230	1,900	13,130
Single stage, 125 psi				
1.0 cfm	EA.	3,090	950	4,040
1.5 cfm	"	3,140	950	4,090
2.0 cfm	"	3,210	950	4,160
Automotive, hose reel, air and water, 50' hose	"	1,440	550	1,990
Lube equipment, 3 reel, with pumps	"	7,520	2,660	10,180
Tire changer				
Truck	EA.	17,060	950	18,010
Passenger car	"	4,030	510	4,540
Air hose reel, includes, 50' hose	"	1,070	510	1,580
Hose reel, 5 reel, motor oil, gear oil, lube, air & water	"	9,950	2,660	12,610
Water hose reel, 50' hose	"	1,070	510	1,580
Pump, for motor or gear oil, fits 55 gal drum	"	1,390	66.00	1,456
For chassis lube	"	2,280	66.00	2,346
Fuel dispensing pump, lighted dial, one product				
One hose	EA.	4,970	550	5,520
Two hose	"	8,760	550	9,310
Two products, two hose	"	9,230	550	9,780

15 MECHANICAL

Heating & Ventilating

15555.10 Boilers

	UNIT	MAT.	INST.	TOTAL
Cast iron, gas fired, hot water				
115 mbh	EA.	3,070	2,050	5,120
175 mbh	"	3,670	2,240	5,910
235 mbh	"	4,690	2,460	7,150
940 mbh	"	16,150	4,920	21,070
1600 mbh	"	22,330	6,150	28,480
3000 mbh	"	36,110	8,200	44,310
6000 mbh	"	72,280	12,300	84,580
Steam				
115 mbh	EA.	3,390	2,050	5,440
175 mbh	"	4,080	2,240	6,320
235 mbh	"	4,840	2,460	7,300
940 mbh	"	16,880	4,920	21,800
1600 mbh	"	21,840	6,150	27,990
3000 mbh	"	32,530	8,200	40,730
6000 mbh	"	68,640	12,300	80,940
Electric, hot water				
115 mbh	EA.	5,500	1,230	6,730
175 mbh	"	6,080	1,230	7,310
235 mbh	"	6,940	1,230	8,170
940 mbh	"	16,890	2,460	19,350
1600 mbh	"	23,910	4,920	28,830
3000 mbh	"	35,720	6,150	41,870
6000 mbh	"	41,440	8,200	49,640
Steam				
115 mbh	EA.	6,940	1,230	8,170
175 mbh	"	8,500	1,230	9,730
235 mbh	"	9,280	1,230	10,510
940 mbh	"	18,470	2,460	20,930
1600 mbh	"	30,970	4,920	35,890
3000 mbh	"	43,990	6,150	50,140
6000 mbh	"	45,710	8,200	53,910
Oil fired, hot water				
115 mbh	EA.	4,060	1,640	5,700
175 mbh	"	5,150	1,890	7,040
235 mbh	"	7,110	2,240	9,350
940 mbh	"	13,490	4,100	17,590
1600 mbh	"	21,390	4,920	26,310
3000 mbh	"	31,010	6,150	37,160
6000 mbh	"	71,730	12,300	84,030
Steam				
115 mbh	EA.	4,060	1,640	5,700
175 mbh	"	5,150	1,890	7,040
235 mbh	"	6,570	2,240	8,810
940 mbh	"	13,100	4,100	17,200
1600 mbh	"	21,390	4,920	26,310
3000 mbh	"	28,860	6,150	35,010
6000 mbh	"	71,730	12,300	84,030

15 MECHANICAL

Heating & Ventilating

15610.10 Furnaces

	UNIT	MAT.	INST.	TOTAL
Electric, hot air				
40 mbh	EA.	960	330	1,290
60 mbh	"	1,040	350	1,390
80 mbh	"	1,130	370	1,500
100 mbh	"	1,270	390	1,660
125 mbh	"	1,560	400	1,960
160 mbh	"	2,140	420	2,560
200 mbh	"	3,120	430	3,550
400 mbh	"	5,530	440	5,970
Gas fired hot air				
40 mbh	EA.	960	330	1,290
60 mbh	"	1,030	350	1,380
80 mbh	"	1,180	370	1,550
100 mbh	"	1,230	390	1,620
125 mbh	"	1,350	400	1,750
160 mbh	"	1,610	420	2,030
200 mbh	"	2,870	430	3,300
400 mbh	"	5,140	440	5,580
Oil fired hot air				
40 mbh	EA.	1,290	330	1,620
60 mbh	"	2,130	350	2,480
80 mbh	"	2,150	370	2,520
100 mbh	"	2,180	390	2,570
125 mbh	"	2,260	400	2,660
160 mbh	"	2,600	420	3,020
200 mbh	"	3,060	430	3,490
400 mbh	"	5,070	440	5,510

Refrigeration

15670.10 Condensing Units

	UNIT	MAT.	INST.	TOTAL
Air cooled condenser, single circuit				
3 ton	EA.	1,940	110	2,050
5 ton	"	3,060	110	3,170
7.5 ton	"	5,010	320	5,330
20 ton	"	14,880	330	15,210
25 ton	"	22,430	330	22,760
30 ton	"	25,560	330	25,890
40 ton	"	33,050	470	33,520
50 ton	"	40,200	470	40,670
60 ton	"	46,290	420	46,710
With low ambient dampers				
3 ton	EA.	2,230	170	2,400
5 ton	"	3,510	170	3,680
7.5 ton	"	5,400	330	5,730
20 ton	"	14,930	440	15,370

15 MECHANICAL

Refrigeration

15670.10 Condensing Units (Cont.)

	UNIT	MAT.	INST.	TOTAL
25 ton	EA.	22,720	440	23,160
30 ton	"	26,100	440	26,540
40 ton	"	34,640	550	35,190
50 ton	"	41,790	600	42,390
60 ton	"	47,920	600	48,520
Dual circuit				
10 ton	EA.	4,770	330	5,100
15 ton	"	6,980	470	7,450
20 ton	"	14,280	470	14,750
25 ton	"	22,740	470	23,210
30 ton	"	26,450	470	26,920
40 ton	"	37,820	550	38,370
50 ton	"	41,790	550	42,340
60 ton	"	43,370	550	43,920
80 ton	"	54,220	740	54,960
100 ton	"	63,480	740	64,220
120 ton	"	75,380	740	76,120
With low ambient dampers				
15 ton	EA.	7,770	470	8,240
20 ton	"	15,170	470	15,640
25 ton	"	24,070	470	24,540
30 ton	"	27,130	470	27,600
40 ton	"	39,140	550	39,690
50 ton	"	43,120	550	43,670
60 ton	"	44,700	550	45,250
80 ton	"	56,850	740	57,590
100 ton	"	66,120	740	66,860
120 ton	"	78,820	740	79,560

15680.10 Chillers

	UNIT	MAT.	INST.	TOTAL
Chiller, reciprocal				
Air cooled, remote condenser, starter				
20 ton	EA.	35,410	820	36,230
25 ton	"	39,960	820	40,780
30 ton	"	42,250	820	43,070
40 ton	"	62,480	1,230	63,710
50 ton	"	69,350	1,370	70,720
60 ton	"	77,920	1,450	79,370
80 ton	"	96,140	2,240	98,380
100 ton	"	113,840	2,460	116,300
120 ton	"	129,030	2,730	131,760
150 ton	"	164,440	3,080	167,520
180 ton	"	188,470	3,510	191,980
200 ton	"	207,440	4,100	211,540
Water cooled, with starter				
20 ton	EA.	30,360	820	31,180
25 ton	"	33,640	820	34,460
30 ton	"	40,470	1,230	41,700
40 ton	"	55,650	1,230	56,880
50 ton	"	60,710	1,370	62,080
60 ton	"	65,770	1,450	67,220
80 ton	"	75,890	2,240	78,130
100 ton	"	88,540	2,460	91,000

15 MECHANICAL

Refrigeration

15680.10 Chillers (Cont.)

Description	UNIT	MAT.	INST.	TOTAL
120 ton	EA.	103,730	2,730	106,460
150 ton	"	129,030	3,080	132,110
180 ton	"	134,080	3,510	137,590
200 ton	"	141,660	4,100	145,760
Packaged, air cooled, with starter				
20 ton	EA.	33,900	620	34,520
25 ton	"	36,690	620	37,310
30 ton	"	42,760	620	43,380
40 ton	"	49,080	620	49,700
50 ton	"	55,410	820	56,230
60 ton	"	65,270	820	66,090
80 ton	"	76,140	1,230	77,370
100 ton	"	89,280	1,230	90,510
120 ton	"	101,440	1,230	102,670
Heat recovery, air cooled, with starter				
40 ton	EA.	56,910	1,230	58,140
50 ton	"	65,770	1,230	67,000
60 ton	"	72,100	1,640	73,740
75 ton	"	82,470	2,460	84,930
100 ton	"	94,110	2,460	96,570
Water cooled, with starter				
40 ton	EA.	56,160	1,230	57,390
50 ton	"	67,040	1,230	68,270
60 ton	"	73,870	1,640	75,510
75 ton	"	89,800	2,460	92,260
100 ton	"	101,440	2,730	104,170
Centrifugal, single bundle condenser, with starter				
80 ton	EA.	132,810	3,510	136,320
130 ton	"	134,510	4,100	138,610
160 ton	"	136,710	4,470	141,180
180 ton	"	146,430	4,920	151,350
230 ton	"	153,000	5,470	158,470
280 ton	"	168,570	6,150	174,720
360 ton	"	188,030	6,150	194,180
460 ton	"	231,820	8,200	240,020
560 ton	"	253,630	8,790	262,420
670 ton	"	309,140	9,840	318,980

15710.10 Cooling Towers

Description	UNIT	MAT.	INST.	TOTAL
Cooling tower, propeller type				
100 ton	EA.	13,780	820	14,600
200 ton	"	22,880	1,230	24,110
300 ton	"	34,580	2,050	36,630
400 ton	"	46,020	2,460	48,480
600 ton	"	69,030	3,510	72,540
800 ton	"	92,040	4,920	96,960
1000 ton	"	110,500	6,150	116,650
Centrifugal				
100 ton	EA.	19,110	820	19,930
200 ton	"	29,250	1,230	30,480
300 ton	"	41,340	2,050	43,390
400 ton	"	55,120	2,460	57,580
600 ton	"	78,900	3,510	82,410

15 MECHANICAL

Refrigeration

15710.10 Cooling Towers (Cont.)

	UNIT	MAT.	INST.	TOTAL
800 ton	EA.	105,170	4,920	110,090
1000 ton	"	128,960	6,150	135,110

Heat Transfer

15780.10 Computer Room A/c

	UNIT	MAT.	INST.	TOTAL
Air cooled, alarm, high efficiency filter, elec. heat				
3 ton	EA.	19,070	510	19,580
5 ton	"	20,360	550	20,910
7.5 ton	"	36,890	660	37,550
10 ton	"	38,560	830	39,390
15 ton	"	42,370	950	43,320
Steam heat				
3 ton	EA.	20,410	510	20,920
5 ton	"	21,710	550	22,260
7.5 ton	"	34,580	660	35,240
10 ton	"	35,620	830	36,450
15 ton	"	39,390	950	40,340
Hot water heat				
3 ton	EA.	20,410	510	20,920
5 ton	"	21,710	550	22,260
7.5 ton	"	34,580	660	35,240
10 ton	"	35,620	830	36,450
15 ton	"	39,520	950	40,470
Air cooled condenser, low ambient damper				
3 ton	EA.	1,940	130	2,070
5 ton	"	3,060	170	3,230
7.5 ton	"	4,690	330	5,020
10 ton	"	6,860	470	7,330
15 ton	"	7,580	390	7,970
Water cooled, high efficiency filter, alarm, elec. heat				
3 ton	EA.	20,150	470	20,620
5 ton	"	21,710	550	22,260
7.5 ton	"	34,580	830	35,410
10 ton	"	35,880	950	36,830
15 ton	"	41,990	1,110	43,100
Steam heat				
3 ton	EA.	23,010	470	23,480
5 ton	"	26,260	550	26,810
7.5 ton	"	37,050	830	37,880
10 ton	"	38,350	950	39,300
15 ton	"	44,590	1,110	45,700
Hot water heat				
3 ton	EA.	23,010	470	23,480
5 ton	"	24,570	550	25,120
7.5 ton	"	37,050	830	37,880

15 MECHANICAL

Heat Transfer	UNIT	MAT.	INST.	TOTAL
15780.10 — **Computer Room A/c** *(Cont.)*				
10 ton	EA.	38,350	950	39,300
15 ton	"	44,590	1,110	45,700
Chilled water, alarm, high eff. filter, elec. heat				
7.5 ton	EA.	16,120	600	16,720
10 ton	"	16,900	740	17,640
15 ton	"	19,240	830	20,070
Steam heat				
7.5 ton	EA.	18,460	600	19,060
10 ton	"	19,240	740	19,980
15 ton	"	21,580	830	22,410
Hot water heat				
7.5 ton	EA.	18,460	600	19,060
10 ton	"	19,240	740	19,980
15 ton	"	21,580	830	22,410
15780.20 — **Rooftop Units**				
Packaged, single zone rooftop unit, with roof curb				
2 ton	EA.	4,430	660	5,090
3 ton	"	4,650	660	5,310
4 ton	"	5,080	830	5,910
5 ton	"	5,510	1,110	6,620
7.5 ton	"	8,010	1,330	9,340
15820.10 — **Dehumidifiers**				
Dessicant dehumidifier, 1125 cfm	EA.			36,210
15830.10 — **Radiation Units**				
Baseboard radiation unit				
1.7 mbh/lf	L.F.	100	26.50	127
2.1 mbh/lf	"	130	33.25	163
Enclosure only				
Two tier	L.F.	52.00	11.00	63.00
Three tier	"	67.00	11.00	78.00
Copper element only, 3/4" dia.				
Two tier	L.F.	76.00	16.50	92.50
Three tier	"	120	22.25	142
Fin-tube, 16 ga, sloping cover, 1-1/4" steel				
One tier	L.F.	72.00	22.25	94.25
Two tier	"	110	26.50	137
2" steel				
Two tier	L.F.	120	26.50	147
Three tier	"	170	33.25	203
1-1/4" copper				
Two tier	L.F.	150	22.25	172
18 ga flat cover, 1-1/4" steel				
One tier	L.F.	48.25	22.25	70.50
Two tier	"	77.00	26.50	104
Three tier	"	120	33.25	153
2" steel				
One tier	L.F.	57.00	22.25	79.25
Two tier	"	91.00	26.50	118
Three tier	"	120	33.25	153
1-1/4" copper				

15 MECHANICAL

Heat Transfer	UNIT	MAT.	INST.	TOTAL
15830.10 Radiation Units *(Cont.)*				
One tier	L.F.	56.00	22.25	78.25
Two tier	"	100	26.50	127
Three tier	"	130	33.25	163
15830.20 Fan Coil Units				
Fan coil unit, 2 pipe, complete				
200 cfm ceiling hung	EA.	1,310	220	1,530
Floor mounted	"	1,230	170	1,400
300 cfm, ceiling hung	"	1,390	270	1,660
Floor mounted	"	1,330	220	1,550
400 cfm, ceiling hung	"	1,470	320	1,790
Floor mounted	"	1,400	220	1,620
500 cfm, ceiling hung	"	1,700	330	2,030
Floor mounted	"	1,640	260	1,900
600 cfm, ceiling hung	"	2,150	370	2,520
Floor mounted	"	2,000	300	2,300
800 cfm, ceiling hung	"	2,510	420	2,930
Floor mounted	"	2,000	320	2,320
1000 cfm, ceiling hung	"	2,870	470	3,340
Floor mounted	"	3,150	350	3,500
1200 cfm ceiling hung	"	3,260	550	3,810
Floor mounted	"	3,420	420	3,840
15830.70 Unit Heaters				
Steam unit heater, horizontal				
12,500 btuh, 200 cfm	EA.	640	110	750
17,000 btuh, 300 cfm	"	840	110	950
40,000 btuh, 500 cfm	"	1,020	110	1,130
60,000 btuh, 700 cfm	"	1,070	110	1,180
70,000 btuh, 1000 cfm	"	1,110	170	1,280
Vertical				
12,500 btuh, 200 cfm	EA.	640	110	750
17,000 btuh, 300 cfm	"	1,050	110	1,160
40,000 btuh, 500 cfm	"	1,020	110	1,130
60,000 btuh, 700 cfm	"	1,070	110	1,180
70,000 btuh, 1000 cfm	"	1,110	110	1,220
Gas unit heater, horizontal				
27,400 btuh	EA.	970	270	1,240
38,000 btuh	"	1,020	270	1,290
56,000 btuh	"	1,070	270	1,340
82,200 btuh	"	1,110	270	1,380
103,900 btuh	"	1,230	420	1,650
125,700 btuh	"	1,450	420	1,870
133,200 btuh	"	1,570	420	1,990
149,000 btuh	"	1,840	420	2,260
172,000 btuh	"	1,990	420	2,410
190,000 btuh	"	2,090	420	2,510
225,000 btuh	"	2,300	420	2,720
Hot water unit heater, horizontal				
12,500 btuh, 200 cfm	EA.	510	110	620
17,000 btuh, 300 cfm	"	570	110	680
25,000 btuh, 500 cfm	"	650	110	760
30,000 btuh, 700 cfm	"	770	110	880

15 MECHANICAL

Heat Transfer

	UNIT	MAT.	INST.	TOTAL
15830.70 Unit Heaters (Cont.)				
50,000 btuh, 1000 cfm	EA.	840	170	1,010
60,000 btuh, 1300 cfm	"	880	170	1,050
Vertical				
12,500 btuh, 200 cfm	EA.	750	110	860
17,000 btuh, 300 cfm	"	750	110	860
25,000 btuh, 500 cfm	"	750	110	860
30,000 btuh, 700 cfm	"	750	110	860
50,000 btuh, 1000 cfm	"	780	110	890
60,000 btuh, 1300 cfm	"	960	110	1,070
Cabinet unit heaters, ceiling, exposed, hot water				
200 cfm	EA.	1,460	220	1,680
300 cfm	"	1,570	270	1,840
400 cfm	"	1,630	320	1,950
600 cfm	"	1,680	350	2,030
800 cfm	"	2,090	420	2,510
1000 cfm	"	2,730	470	3,200
1200 cfm	"	2,930	550	3,480
2000 cfm	"	5,020	740	5,760

Air Handling

	UNIT	MAT.	INST.	TOTAL
15855.10 Air Handling Units				
Air handling unit, medium pressure, single zone				
1500 cfm	EA.	4,730	420	5,150
3000 cfm	"	6,210	740	6,950
4000 cfm	"	7,960	830	8,790
5000 cfm	"	10,030	890	10,920
6000 cfm	"	12,890	950	13,840
7000 cfm	"	14,800	1,020	15,820
8500 cfm	"	18,190	1,110	19,300
10,500 cfm	"	19,980	1,330	21,310
12,500 cfm	"	23,000	1,480	24,480
15,500 cfm	"	29,740	1,900	31,640
17,500 cfm	"	32,990	2,210	35,200
20,500 cfm	"	37,300	2,660	39,960
25,000 cfm	"	42,280	3,320	45,600
31,500 cfm	"	52,300	4,430	56,730
Rooftop air handling units				
4950 cfm	EA.	13,580	740	14,320
7370 cfm	"	17,230	950	18,180
9790 cfm	"	18,330	1,110	19,440
14,300 cfm	"	25,900	950	26,850
21,725 cfm	"	36,680	950	37,630
33,000 cfm	"	51,790	1,110	52,900

15 MECHANICAL

Air Handling

15870.20 Exhaust Fans

	UNIT	MAT.	INST.	TOTAL
Belt drive roof exhaust fans				
640 cfm, 2618 fpm	EA.	1,220	83.00	1,303
940 cfm, 2604 fpm	"	1,590	83.00	1,673
1050 cfm, 3325 fpm	"	1,420	83.00	1,503
1170 cfm, 2373 fpm	"	2,050	83.00	2,133
2440 cfm, 4501 fpm	"	1,610	83.00	1,693
2760 cfm, 4950 fpm	"	1,780	83.00	1,863
3890 cfm, 6769 fpm	"	2,030	83.00	2,113
2380 cfm, 3382 fpm	"	2,250	83.00	2,333
2880 cfm, 3859 fpm	"	2,350	83.00	2,433
3200 cfm, 4173 fpm	"	2,380	110	2,490
3660 cfm, 3437 fpm	"	2,420	110	2,530
4070 cfm, 3694 fpm	"	3,070	110	3,180
5030 cfm, 3251 fpm	"	2,140	110	2,250
5830 cfm, 6932 fpm	"	2,980	130	3,110
6380 cfm, 3817 fpm	"	2,980	130	3,110
8460 cfm, 6721 fpm	"	2,870	130	3,000
10,970 cfm, 5906 fpm	"	3,520	170	3,690
12,470 cfm, 6620 fpm	"	4,020	220	4,240
7000 cfm, 3449 fpm	"	2,870	170	3,040
13,000 cfm, 5456 fpm	"	4,250	170	4,420
11,250 cfm, 4854 fpm	"	3,890	170	4,060
18,490 cfm, 7405 fpm	"	5,680	300	5,980
11,300 cfm, 3232 fpm	"	3,690	290	3,980
18,330 cfm, 4488 fpm	"	6,150	290	6,440
21,720 cfm, 5131 fpm	"	6,460	290	6,750
31,110 cfm, 6965 fpm	"	7,110	330	7,440
Direct drive fans				
60 to 390 cfm	EA.	1,000	83.00	1,083
145 to 590 cfm	"	1,210	83.00	1,293
295 to 860 cfm	"	1,470	83.00	1,553
235 to 1300 cfm	"	1,570	83.00	1,653
415 to 1630 cfm	"	1,780	83.00	1,863
590 to 2045 cfm	"	2,050	83.00	2,133
805 cfm, 3235 fpm	"	1,350	83.00	1,433
1455 cfm, 4360 fpm	"	1,460	83.00	1,543
1385 cfm, 3655 fpm	"	1,490	83.00	1,573
2260 cfm, 4930 fpm	"	1,570	83.00	1,653
1720 cfm, 3870 fpm	"	1,720	83.00	1,803
2700 cfm, 5220 fpm	"	1,600	83.00	1,683
Terminal blenders and cooling				
400 cfm	EA.	550	130	680
800 cfm	"	600	130	730
1200 cfm	"	720	170	890
2000 cfm	"	820	170	990

15 MECHANICAL

Air Distribution

15890.10 Metal Ductwork

	UNIT	MAT.	INST.	TOTAL
Rectangular duct				
Galvanized steel				
Minimum	Lb.	1.04	6.03	7.07
Average	"	1.30	7.37	8.67
Maximum	"	1.98	11.00	12.98
Aluminum				
Minimum	Lb.	2.71	13.25	15.96
Average	"	3.61	16.50	20.11
Maximum	"	4.48	22.25	26.73
Fittings				
Minimum	EA.	8.58	22.25	30.83
Average	"	12.75	33.25	46.00
Maximum	"	19.00	66.00	85.00
For work				
10-20' high, add per pound, $.30				
30-50', add per pound, $.50				

15890.30 Flexible Ductwork

	UNIT	MAT.	INST.	TOTAL
Flexible duct, 1.25" fiberglass				
5" dia.	L.F.	3.91	3.32	7.23
6" dia.	"	4.35	3.68	8.03
7" dia.	"	5.36	3.90	9.26
8" dia.	"	5.62	4.15	9.77
10" dia.	"	7.50	4.74	12.24
12" dia.	"	8.19	5.10	13.29
14" dia.	"	10.25	5.53	15.78
16" dia.	"	15.25	6.03	21.28
Flexible duct connector, 3" wide fabric	"	2.73	11.00	13.73

15895.10 Roof Curbs

	UNIT	MAT.	INST.	TOTAL
8" high, insulated, with liner and raised can				
15" x 15"	EA.	130	33.25	163
17" x 17"	"	140	33.25	173
19" x 19"	"	140	33.25	173
21" x 21"	"	160	33.25	193
25" x 25"	"	170	41.50	212
28" x 28"	"	190	44.25	234
32" x 32"	"	200	47.50	248
36" x 36"	"	230	47.50	278
40" x 40"	"	260	47.50	308
44" x 44"	"	290	51.00	341
48" x 48"	"	640	51.00	691
52" x 52"	"	790	55.00	845
56" x 56"	"	990	55.00	1,045
60" x 60"	"	1,190	66.00	1,256
64" x 64"	"	1,430	66.00	1,496
68" x 68"	"	1,640	74.00	1,714
72" x 72"	"	1,890	83.00	1,973

15 MECHANICAL

Air Distribution

	UNIT	MAT.	INST.	TOTAL
15910.10 **Dampers**				
Horizontal parallel aluminum backdraft damper				
12" x 12"	EA.	65.00	16.50	81.50
16" x 16"	"	67.00	19.00	86.00
20" x 20"	"	86.00	23.75	110
24" x 24"	"	100	33.25	133
28" x 28"	"	150	37.00	187
32" x 32"	"	200	41.50	242
36" x 36"	"	240	47.50	288
40" x 40"	"	300	55.00	355
44" x 44"	"	350	60.00	410
48" x 48"	"	430	66.00	496
"Up", parallel dampers				
12" x 12"	EA.	110	16.50	127
16" x 16"	"	140	19.00	159
20" x 20"	"	170	23.75	194
24" x 24"	"	180	33.25	213
28" x 28"	"	260	37.00	297
32" x 32"	"	300	41.50	342
36" x 36"	"	310	47.50	358
40" x 40"	"	420	55.00	475
44" x 44"	"	490	60.00	550
48" x 48"	"	610	66.00	676
"Down", parallel dampers				
12" x 12"	EA.	110	16.50	127
16" x 16"	"	140	19.00	159
20" x 20"	"	170	23.75	194
24" x 24"	"	180	33.25	213
28" x 28"	"	260	37.00	297
32" x 32"	"	300	41.50	342
36" x 36"	"	310	47.50	358
40" x 40"	"	420	55.00	475
44" x 44"	"	490	60.00	550
48" x 48"	"	610	66.00	676
Fire damper, 1.5 hr rating				
12" x 12"	EA.	43.00	33.25	76.25
16" x 16"	"	69.00	33.25	102
20" x 20"	"	75.00	33.25	108
24" x 24"	"	87.00	33.25	120
28" x 28"	"	100	47.50	148
32" x 32"	"	120	55.00	175
36" x 36"	"	140	66.00	206
40" x 40"	"	170	74.00	244
44" x 44"	"	200	83.00	283
48" x 48"	"	280	95.00	375
15940.10 **Diffusers**				
Ceiling diffusers, round, baked enamel finish				
6" dia.	EA.	43.25	22.25	65.50
8" dia.	"	52.00	27.75	79.75
10" dia.	"	58.00	27.75	85.75
12" dia.	"	74.00	27.75	102
14" dia.	"	91.00	30.25	121
16" dia.	"	110	30.25	140

15 MECHANICAL

Air Distribution

15940.10 Diffusers (Cont.)

	UNIT	MAT.	INST.	TOTAL
18" dia.	EA.	130	33.25	163
20" dia.	"	150	33.25	183
Rectangular				
6x6"	EA.	46.25	22.25	68.50
9x9"	"	56.00	33.25	89.25
12x12"	"	81.00	33.25	114
15x15"	"	100	33.25	133
18x18"	"	130	33.25	163
21x21"	"	160	41.50	202
24x24"	"	190	41.50	232
Lay in, flush mounted, perforated face, with grid				
6x6/24x24	EA.	66.00	26.50	92.50
8x8/24x24	"	66.00	26.50	92.50
9x9/24x24	"	67.00	26.50	93.50
10x10/24x24	"	72.00	26.50	98.50
12x12/24x24	"	77.00	26.50	104
15x15/24x24	"	100	26.50	127
18x6/24x24	"	77.00	26.50	104
18x18/24x24	"	120	26.50	147
Two-way slot diffuser with balancing damper, 4'	"	74.00	66.00	140

15940.20 Relief Ventilators

	UNIT	MAT.	INST.	TOTAL
Intake ventilator, aluminum, with screen, no curbs				
12" x 12"	EA.	230	55.00	285
16" x 16"	"	320	66.00	386
20" x 20"	"	520	66.00	586
30" x 30"	"	820	95.00	915
36" x 36"	"	1,240	110	1,350
42" x 42"	"	1,690	110	1,800
48" x 48"	"	2,030	130	2,160

15940.40 Registers And Grilles

	UNIT	MAT.	INST.	TOTAL
Lay in flush mounted, perforated face, return				
6x6/24x24	EA.	57.00	26.50	83.50
8x8/24x24	"	57.00	26.50	83.50
9x9/24x24	"	62.00	26.50	88.50
10x10/24x24	"	67.00	26.50	93.50
12x12/24x24	"	67.00	26.50	93.50
Rectangular, ceiling return, single deflection				
10x10	EA.	34.50	33.25	67.75
12x12	"	40.25	33.25	73.50
14x14	"	49.00	33.25	82.25
16x8	"	40.25	33.25	73.50
16x16	"	40.25	33.25	73.50
18x8	"	46.00	33.25	79.25
20x20	"	75.00	33.25	108
24x12	"	110	33.25	143
24x18	"	140	33.25	173
36x24	"	270	37.00	307
36x30	"	390	37.00	427
Wall, return air register				
12x12	EA.	57.00	16.50	73.50
16x16	"	84.00	16.50	101

15 MECHANICAL

Air Distribution	UNIT	MAT.	INST.	TOTAL
15940.40 **Registers And Grilles** (Cont.)				
18x18	EA.	100	16.50	117
20x20	"	120	16.50	137
24x24	"	160	16.50	177
Ceiling, return air grille				
6x6	EA.	33.25	22.25	55.50
8x8	"	41.50	26.50	68.00
10x10	"	51.00	26.50	77.50
Ceiling, exhaust grille, aluminum egg crate				
6x6	EA.	22.75	22.25	45.00
8x8	"	22.75	26.50	49.25
10x10	"	25.25	26.50	51.75
12x12	"	31.25	33.25	64.50
14x14	"	40.75	33.25	74.00
16x16	"	48.00	33.25	81.25
18x18	"	57.00	33.25	90.25
15940.80 **Penthouse Louvers**				
Penthouse louvers				
12" high, extruded aluminum, 4" louver				
6' perimeter	EA.	530	170	700
8' perimeter	"	720	170	890
10' perimeter	"	900	170	1,070
12' perimeter	"	1,300	170	1,470
14' perimeter	"	1,590	220	1,810
16' perimeter	"	1,830	270	2,100
18' perimeter	"	2,140	370	2,510
20' perimeter	"	2,680	440	3,120
16" high x 4' perimeter	"	440	170	610
6' perimeter	"	620	170	790
8' perimeter	"	810	170	980
10' perimeter	"	990	170	1,160
12' perimeter	"	1,460	170	1,630
14' perimeter	"	1,790	220	2,010
16' perimeter	"	2,010	270	2,280
18' perimeter	"	2,460	370	2,830
20' perimeter	"	3,130	440	3,570
22' perimeter	"	3,350	550	3,900
24' perimeter	"	3,580	740	4,320
20" high x 4' perimeter	"	620	170	790
6' perimeter	"	660	170	830
8' perimeter	"	900	170	1,070
10' perimeter	"	1,120	170	1,290
12' perimeter	"	1,640	170	1,810
14' perimeter	"	2,010	220	2,230
16' perimeter	"	2,460	270	2,730
18' perimeter	"	2,790	370	3,160
20' perimeter	"	3,350	440	3,790
22' perimeter	"	3,600	550	4,150
24' perimeter	"	4,000	740	4,740
24" high x 4' perimeter	"	620	170	790
6' perimeter	"	750	170	920
8' perimeter	"	990	170	1,160
10' perimeter	"	1,230	170	1,400

15 MECHANICAL

Air Distribution

	UNIT	MAT.	INST.	TOTAL
15940.80 **Penthouse Louvers** *(Cont.)*				
12' perimeter	EA.	1,820	170	1,990
16' perimeter	"	2,680	270	2,950
18' perimeter	"	3,130	370	3,500
20' perimeter	"	3,780	440	4,220
22' perimeter	"	4,190	550	4,740
24' perimeter	"	4,470	740	5,210

Controls

	UNIT	MAT.	INST.	TOTAL
15950.10 **Hvac Controls**				
Pressure gauge, direct reading gage cock and siphon	EA.	140	41.50	182
Control valve, 1", modulating				
2-way	EA.	1,080	55.00	1,135
3-way	"	1,220	83.00	1,303
Self contained control valve w/ sensing elmnt, 3/4"	"	210	41.50	252
Inst air syst 2-1/2 hp comp, rcvr refrg dryer	"			9,100
Thermostat primary control device	"			200
Humidistat primary control device	"			160
Timers primary control device, indoor/outdoor, 24 hour	"			320
Thermometer, dir. reading, 3 dial	"			160
Control dampers, round				
6" dia.	EA.	150	26.50	177
8" dia	"	210	26.50	237
10" dia	"	270	26.50	297
12" dia	"	360	26.50	387
12" dia	"	530	33.25	563
18" dia	"	570	33.25	603
20" dia	"	760	33.25	793
Rectangular, parallel blade standard leakage				
12" x 12"	EA.	110	33.25	143
16" x 16"	"	160	33.25	193
20" x 20"	"	190	33.25	223
28" x 28"	"	240	41.50	282
32" x 32"	"	270	41.50	312
36" x 36"	"	410	55.00	465
40" x 40"	"	410	66.00	476
44" x 44"	"	500	83.00	583
48" x 48"	"	570	95.00	665
48" x 52"	"	600	110	710
48" x 56"	"	670	110	780
48" x 60"	"	720	110	830
48" x 64"	"	740	110	850
48" x 68"	"	810	110	920
48" x 72"	"	880	110	990
Low leakage				
12" x 12"	EA.	210	33.25	243

15 MECHANICAL

Controls

15950.10 Hvac Controls (Cont.)

	UNIT	MAT.	INST.	TOTAL
16" x 16"	EA.	260	33.25	293
20" x 20"	"	340	33.25	373
24" x 24"	"	460	33.25	493
28" x 28"	"	530	41.50	572
32" x 32"	"	600	47.50	648
36" x 36"	"	670	55.00	725
40" x 40"	"	1,080	66.00	1,146
44" x 44"	"	1,130	83.00	1,213
48" x 48"	"	1,200	95.00	1,295
48" x 56"	"	1,390	110	1,500
48" x 60"	"	1,460	110	1,570
48" x 64"	"	1,510	110	1,620
48" x 68"	"	1,580	110	1,690
48" x 72"	"	1,870	110	1,980
Rectangular, opposed horizontal blade				
12" x 12"	EA.	140	33.25	173
16" x 16"	"	190	33.25	223
20" x 20"	"	220	33.25	253
24" x 24"	"	260	33.25	293
28" x 28"	"	360	41.50	402
32" x 32"	"	390	44.25	434
36" x 36"	"	430	55.00	485
40" x 40"	"	550	66.00	616
44" x 44"	"	690	83.00	773
48" x 48"	"	860	95.00	955
48" x 52"	"	880	95.00	975
48" x 56"	"	910	110	1,020
48" x 60"	"	960	110	1,070
48" x 64"	"	1,010	110	1,120
48" x 68"	"	1,120	110	1,230
48" x 72"	"	1,200	110	1,310

16 ELECTRICAL

Basic Materials

16050.30 Bus Duct

	UNIT	MAT.	INST.	TOTAL
Bus duct, 100a, plug-in				
10', 600v	EA.	330	210	540
With ground	"	440	320	760
10', 277/480v	"	420	210	630
With ground	"	530	320	850
Cable tap box	"	190	190	380
End closure	"	230	30.75	261
Edgewise hanger	"	28.25	56.00	84.25
Flatwise hanger	"	28.25	56.00	84.25
Outside elbow	"	260	62.00	322
Inside elbow	"	260	62.00	322
Outside tee	"	340	85.00	425
Inside tee	"	340	85.00	425
Outlet cover	"	19.75	30.75	50.50
Wall flange	"	55.00	30.75	85.75
Circuit breakers, with enclosure				
1 pole				
15a-60a	EA.	330	77.00	407
70a-100a	"	380	96.00	476
2 pole				
15a-60a	EA.	490	85.00	575
70a-100a	"	580	100	680
3 pole				
15a-60a	EA.	550	89.00	639
70a-100a	"	640	120	760
Bus duct, copper feeder duct, 277/480v, 4 wire				
800a	L.F.	350	30.75	381
1000a	"	400	38.50	439
1200a	"	460	41.00	501
1350a	"	550	47.50	598
1600a	"	670	56.00	726
2000a	"	830	62.00	892
2500a	"	1,020	66.00	1,086
3000a	"	1,300	73.00	1,373
Weatherproof				
800a	L.F.	430	34.25	464
1000a	"	500	41.00	541
1350a	"	670	51.00	721
1600a	"	800	56.00	856
2000a	"	990	64.00	1,054
2500a	"	1,220	69.00	1,289
3000a	"	1,560	75.00	1,635
4000a	"	1,990	120	2,110
5000a	"	2,390	140	2,530
Plug-in feeder duct, 277/480v, 4 wire				
400a	L.F.	220	30.75	251
600a	"	230	34.25	264
800a	"	360	38.50	399
1000a	"	400	38.50	439
1200a	"	480	41.00	521
1350a	"	540	47.50	588
1600a	"	640	56.00	696
2000a	"	780	62.00	842

16 ELECTRICAL

Basic Materials

16050.30 Bus Duct *(Cont.)*

	UNIT	MAT.	INST.	TOTAL
2500a	L.F.	950	66.00	1,016
3000a	"	1,200	73.00	1,273
Ground bus				
225a	L.F.			51.00
400a	"			53.00
600a	"			54.00
800a	"			59.00
1000a	"			61.00
1200a	"			76.00
1350a	"			80.00
1600a	"			110
2000a	"			130
2500a	"			170
3000a	"			230
4000a	"			290
5000a	"			360
Copper flanged ends, 277/480v, 4 wire				
225a	EA.	700	190	890
400a	"	830	210	1,040
600a	"	1,090	230	1,320
800a	"	1,270	240	1,510
1000a	"	1,450	250	1,700
1200a	"	1,560	250	1,810
1350a	"	1,610	260	1,870
1600a	"	1,880	270	2,150
2000a	"	2,160	270	2,430
2500a	"	2,180	280	2,460
3000a	"	3,130	290	3,420
4000a	"	4,200	340	4,540
5000a	"	4,860	360	5,220
Bus duct, copper elbows, 277/480v-4w				
225a-1000a	EA.	1,650	160	1,810
1200a-3000a	"	2,840	190	3,030
4000a-5000a	"	7,040	230	7,270
Tees, 277/480v-4w				
225a-1000a	EA.	2,620	170	2,790
1200a-3000a	"	3,680	200	3,880
4000a-5000a	"	10,080	230	10,310
Crosses, 277/480v-4w				
225a-1000a	EA.	2,770	170	2,940
1200a-3000a	"	4,290	200	4,490
4000a-5000a	"	6,770	230	7,000
Copper end closures, 277/480v-4w				
225a-1000a	EA.	250	69.00	319
1200a-3000a	"	330	92.00	422
4000a-5000a	"	410	130	540
Tap boxes, 277/480v-4w				
225a	EA.	1,840	270	2,110
400a	"	1,850	340	2,190
600a	"	2,020	560	2,580
800a	"	2,200	620	2,820
1000a	"	2,460	770	3,230
1200a	"	2,510	850	3,360

16 ELECTRICAL

Basic Materials

16050.30 Bus Duct *(Cont.)*

	UNIT	MAT.	INST.	TOTAL
1350a	EA.	3,060	1,000	4,060
1600a	"	3,400	1,080	4,480
2000a	"	4,010	1,310	5,320
2500a	"	4,860	1,770	6,630
3000a	"	6,030	2,160	8,190
4000a	"	5,290	2,920	8,210
5000a	"	5,290	3,460	8,750
Circuit breaker, adapter cubicle				
225a	EA.	5,570	120	5,690
400a	"	6,570	120	6,690
600a	"	9,740	130	9,870
800a	"	11,130	140	11,270
1000a	"	12,900	150	13,050
1200a	"	15,460	150	15,610
1600a	"	18,810	160	18,970
2000a	"	22,020	170	22,190
Transformer taps, 1 phase 277/480v				
600a	EA.	960	560	1,520
800a	"	1,010	620	1,630
1000a	"	1,220	770	1,990
1200a	"	1,340	850	2,190
1350a	"	1,400	1,000	2,400
1600a	"	1,600	1,080	2,680
2000a	"	1,830	1,310	3,140
2500a	"	2,230	1,770	4,000
3000a	"	2,630	2,160	4,790
4000a	"	3,090	2,920	6,010
5000a	"	3,880	3,540	7,420
3 phase, 480v, 3 wire				
600a	EA.	2,160	560	2,720
800a	"	2,350	620	2,970
1000a	"	2,680	770	3,450
1200a	"	2,890	850	3,740
1350a	"	3,060	1,000	4,060
1600a	"	3,680	1,080	4,760
2000a	"	4,130	1,310	5,440
2500a	"	4,790	1,770	6,560
3000a	"	5,580	2,160	7,740
4000a	"	6,420	2,920	9,340
5000a	"	7,420	3,540	10,960
3 phase, 4 wire, 277/480v				
600a	EA.	2,360	560	2,920
800a	"	2,530	620	3,150
1000a	"	2,910	770	3,680
1200a	"	3,190	850	4,040
1350a	"	3,350	1,000	4,350
1600a	"	3,930	1,080	5,010
2000a	"	4,480	1,310	5,790
2500a	"	5,240	1,770	7,010
3000a	"	6,130	2,160	8,290
4000a	"	7,100	3,140	10,240
5000a	"	8,040	3,540	11,580
Transformer connection, 4 wire, 277/480v				

16 ELECTRICAL

Basic Materials

16050.30 Bus Duct (Cont.)

	UNIT	MAT.	INST.	TOTAL
600a	EA.	4,910	210	5,120
800a	"	5,100	220	5,320
1000a	"	5,350	230	5,580
1200a	"	5,480	240	5,720
1350a	"	5,580	250	5,830
1600a	"	6,000	260	6,260
2000a	"	6,240	270	6,510
2500a	"	7,050	280	7,330
3000a	"	7,730	290	8,020
4000a	"	10,240	340	10,580
5000a	"	13,690	360	14,050
Unfused reducers, 3 wire, 480v, 3 phase				
400a	EA.	840	190	1,030
600a	"	900	290	1,190
800a	"	1,130	360	1,490
1000a	"	1,330	390	1,720
1200a	"	2,250	410	2,660
1350a	"	2,870	440	3,310
1600a	"	3,130	470	3,600
2000a	"	4,200	490	4,690
2500a	"	5,200	510	5,710
3000a	"	6,270	560	6,830
4000a	"	8,150	670	8,820
5000a	"	9,830	830	10,660
4 wire, 277/480v, 3 phase, 400a	"	1,280	210	1,490
600a	"	1,660	310	1,970
800a	"	2,040	390	2,430
1000a	"	2,390	410	2,800
1200a	"	4,110	440	4,550
1350a	"	5,240	470	5,710
1600a	"	5,690	510	6,200
2000a	"	7,660	540	8,200
2500a	"	9,450	560	10,010
3000a	"	11,420	640	12,060
4000a	"	14,850	750	15,600
5000a	"	17,880	920	18,800
Circuit breaker reducers, 4 wire, 277/480v				
400a	EA.	2,820	170	2,990
600a	"	3,060	270	3,330
800a	"	3,770	320	4,090
1000a	"	4,440	360	4,800
1350a	"	9,720	410	10,130
1600a	"	10,560	440	11,000
2000a	"	14,210	470	14,680
2500a	"	17,580	510	18,090
3000a	"	21,190	540	21,730
4000a	"	27,540	560	28,100
5000a	"	33,170	770	33,940
Expansion fittings, 4 wire, 277/480v				
225a	EA.	2,040	190	2,230
400a	"	2,150	290	2,440
600a	"	2,390	360	2,750
800a	"	2,800	390	3,190

16 ELECTRICAL

Basic Materials

16050.30 Bus Duct (Cont.)

	UNIT	MAT.	INST.	TOTAL
1000a	EA.	3,170	410	3,580
1200a	"	4,320	440	4,760
1350a	"	4,320	460	4,780
1600a	"	5,560	470	6,030
2000a	"	6,300	510	6,810
2500a	"	7,620	560	8,180
3000a	"	8,890	670	9,560
4000a	"	11,580	830	12,410
5000a	"	12,170	920	13,090
Wall flanges				
225a-2500a	EA.	370	310	680
3000a-5000a	"	540	470	1,010
Weather seals	"	540	77.00	617
Roof flanges	"	1,170	310	1,480
Fire barriers	"	560	120	680
Spring hangers	"	160	130	290
Sway brace collars	"	33.25	96.00	129
Hook sticks				
8'	EA.			270
14'	"			450
Fusible switches, 240v, 3 phase				
30a	EA.	700	77.00	777
60a	"	860	96.00	956
100a	"	1,140	120	1,260
200a	"	2,000	160	2,160
400a	"	3,240	310	3,550
600a	"	6,210	470	6,680
208v, 4 wire				
30a	EA.	860	92.00	952
60a	"	880	100	980
100a	"	1,270	140	1,410
200a	"	2,240	210	2,450
400a	"	3,480	390	3,870
600a	"	4,380	620	5,000
600v				
30a	EA.	820	77.00	897
60a	"	870	96.00	966
100a	"	1,260	120	1,380
200a	"	2,210	160	2,370
400a	"	3,410	310	3,720
600a	"	4,220	470	4,690
800a	"	7,290	510	7,800
1000a	"	8,610	620	9,230
1200a	"	13,690	850	14,540
1600a	"	13,850	920	14,770
480v, 4 wire				
30a	EA.	930	92.00	1,022
60a	"	980	100	1,080
100a	"	1,450	140	1,590
200a	"	2,480	210	2,690
400a	"	3,660	390	4,050
600a	"	4,600	620	5,220
800a	"	7,550	640	8,190

16 ELECTRICAL

Basic Materials

16050.30 Bus Duct *(Cont.)*

	UNIT	MAT.	INST.	TOTAL
1000a	EA.	8,910	850	9,760
1200a	"	13,900	920	14,820
1600a	"	14,520	1,080	15,600
Fusible combination starters, 600v, 3 phase				
Size 0	EA.	2,070	99.00	2,169
Size 1	"	2,330	120	2,450
Size 2	"	2,960	140	3,100
Size 3	"	4,830	200	5,030
Circuit breaker combination starters, 600v, 3 phase				
Size 0	EA.	2,250	99.00	2,349
Size 1	"	2,350	120	2,470
Size 2	"	3,410	140	3,550
Size 3	"	4,440	200	4,640
Fusible combination contactors, 600v, 3 phase				
30a	EA.	1,570	99.00	1,669
60a	"	1,930	120	2,050
100a	"	2,900	140	3,040
200a	"	4,720	200	4,920
Circuit breaker, combination contactors, 600v, 3 phase				
30a	EA.	1,610	99.00	1,709
60a	"	1,670	120	1,790
100a	"	2,350	140	2,490
200a	"	3,090	200	3,290
Fusible contactor electrically held, 480v, 4 wire				
30a	EA.	1,570	99.00	1,669
60a	"	2,000	120	2,120
100a	"	2,870	140	3,010
200a	"	6,010	210	6,220
Mechanically held				
30a	EA.	1,700	99.00	1,799
60a	"	2,440	120	2,560
100a	"	3,380	140	3,520
200a	"	6,940	210	7,150
Circuit breakers, 240v, 3 phase				
15a-60a	EA.	810	89.00	899
70a-100a	"	900	120	1,020
600v, 3 phase				
15a-60a	EA.	880	89.00	969
125a-225a	"	2,240	180	2,420
250a-400a	"	4,630	320	4,950
500a-600a	"	6,610	410	7,020
700a-800a	"	7,950	620	8,570
900a-1000a	"	9,460	770	10,230
1200a-1600a	"	14,990	850	15,840
120/208v, 4 wire				
15a-60a	EA.	750	99.00	849
70a-100a	"	870	140	1,010
277/480v, 4 wire				
15a-60a	EA.	910	99.00	1,009
70a-100a	"	990	150	1,140
125a-225a	"	2,410	210	2,620
250a-400a	"	4,860	390	5,250
500a-600a	"	6,920	620	7,540

16 ELECTRICAL

Basic Materials

16050.30 — Bus Duct (Cont.)

	UNIT	MAT.	INST.	TOTAL
700a-800a	EA.	8,270	620	8,890
900a-1000a	"	9,760	850	10,610
1200a-1600a	"	14,890	1,080	15,970
600v, 3 phase, 65,000 aic.				
60a	EA.	1,140	89.00	1,229
70a-100a	"	1,230	120	1,350
125a-225a	"	4,020	180	4,200
250a-400a	"	6,430	320	6,750
500a-600a	"	7,610	470	8,080
700a-800a	"	9,010	620	9,630
900a-1000a	"	10,230	770	11,000
277/480v, 4 wire, 65,000 aic.				
15a-60a	EA.	1,230	99.00	1,329
70a-100a	"	1,330	150	1,480
125a-225a	"	4,140	210	4,350
250a-400a	"	6,670	390	7,060
500a-600a	"	7,950	470	8,420
700a-800a	"	9,310	620	9,930
900a-1000a	"	15,220	800	16,020
600v, 3 phase, current limiting				
15a-60a	EA.	2,960	89.00	3,049
70a-100a	"	3,650	120	3,770
125a-225a	"	6,820	180	7,000
250a-400a	"	8,100	320	8,420
500a-600a	"	12,170	470	12,640
700a-800a	"	13,720	620	14,340
900a-1000a	"	15,280	770	16,050
277/480v, 4 wire, current limiting				
15a-60a	EA.	3,030	99.00	3,129
70a-100a	"	3,900	3,080	6,980
125a-225a	"	7,000	210	7,210
250a-400a	"	8,340	390	8,730
500a-600a	"	9,300	470	9,770
700a-800a	"	14,040	620	14,660
900a-1000a	"	15,590	850	16,440
Capacitors, 3 phase, 240v				
5 kvar	EA.	1,430	390	1,820
7.5 kvar	"	1,790	470	2,260
10 kvar	"	2,090	620	2,710
15 kvar	"	2,740	710	3,450
480v				
2.5 kvar	EA.	620	210	830
5 kvar	"	920	360	1,280
7.5 kvar	"	1,120	440	1,560
10 kvar	"	1,250	620	1,870
15 kvar	"	1,500	690	2,190
20 kvar	"	1,870	890	2,760
25 kvar	"	2,340	1,000	3,340
30 kvar	"	2,760	1,090	3,850
Transformers, 3 phase, 480v				
1.0 kva	EA.	820	130	950

16 ELECTRICAL

Basic Materials

	UNIT	MAT.	INST.	TOTAL
16050.30 — **Bus Duct** *(Cont.)*				
1.5 kva	EA.	900	150	1,050
2 kva	"	970	190	1,160
3 kva	"	1,160	210	1,370
5 kva	"	1,590	310	1,900
7.5 kva	"	1,910	390	2,300
10 kva	"	2,240	410	2,650
16110.12 — **Cable Tray**				
Cable tray, 6"	L.F.	24.75	4.56	29.31
Ventilated cover	"	10.00	2.32	12.32
Solid cover	"	7.79	2.32	10.11
Flat 90	EA.	110	38.50	149
Outside 90	"	94.00	38.50	133
Inside 90	"	100	38.50	139
Flat 45	"	85.00	38.50	124
Outside 45	"	70.00	38.50	109
Inside 45	"	73.00	38.50	112
Adjustable elbow	"	110	38.50	149
Support riser	"	140	38.50	179
Adjustable riser	"	15.50	38.50	54.00
Tee	"	140	130	270
Cross	"	190	140	330
Blind end	"	11.50	22.75	34.25
Expansion joint	"	53.00	38.50	91.50
Box connector	"	35.00	190	225
Standard dropout	"	6.14	38.50	44.64
2"	"	30.00	47.50	77.50
3"	"	53.00	62.00	115
4"	"	85.00	77.00	162
Cable tray, 9"	L.F.	26.25	5.36	31.61
Ventilated cover	"	11.25	3.08	14.33
Solid cover	"	9.08	3.08	12.16
Flat 90	EA.	120	41.00	161
Outside 90	"	100	41.00	141
Inside 90	"	110	41.00	151
Flat 45	"	88.00	41.00	129
Outside 45	"	72.00	41.00	113
Inside 45	"	72.00	41.00	113
Adjustable elbow	"	110	41.00	151
Support riser	"	160	41.00	201
Adjustable riser	"	15.50	41.00	56.50
Tee	"	160	140	300
Cross	"	200	140	340
Blind end	"	12.25	24.75	37.00
Expansion joint	"	74.00	41.00	115
Box connector	"	35.50	200	236
Standard dropout	"	6.72	41.00	47.72
2"	"	30.25	51.00	81.25
3"	"	50.00	51.00	101
4"	"	85.00	84.00	169
Cable tray, 12"	L.F.	27.00	6.16	33.16
Ventilated cover	"	10.50	3.85	14.35
Solid cover	"	9.50	3.85	13.35

16 ELECTRICAL

Basic Materials

16110.12 Cable Tray (Cont.)

Description	UNIT	MAT.	INST.	TOTAL
Flat 90	EA.	130	41.00	171
Outside 90	"	100	41.00	141
Inside 90	"	110	41.00	151
Flat 45	"	93.00	41.00	134
Outside 45	"	140	41.00	181
Inside 45	"	120	41.00	161
Adjustable elbow	"	120	41.00	161
Support riser	"	160	41.00	201
Adjustable riser	"	17.50	41.00	58.50
Tee	"	170	150	320
Cross	"	210	150	360
Blind end	"	13.75	26.75	40.50
Expansion joint	"	80.00	47.50	128
Box connector	"	37.25	210	247
Standard dropout	"	7.29	47.50	54.79
2"	"	31.75	56.00	87.75
3"	"	53.00	69.00	122
4"	"	88.00	89.00	177
Cable tray, 18"	L.F.	28.25	7.70	35.95
Ventilated cover	"	14.50	4.56	19.06
Solid cover	"	12.25	4.56	16.81
Flat 90	EA.	140	56.00	196
Outside 90	"	160	56.00	216
Inside 90	"	120	56.00	176
Flat 45	"	110	56.00	166
Outside 45	"	140	56.00	196
Inside 45	"	120	56.00	176
Adjustable elbow	"	160	56.00	216
Support riser	"	190	56.00	246
Adjustable riser	"	15.50	56.00	71.50
Tee	"	200	160	360
Cross	"	300	160	460
Blind end	"	20.50	30.75	51.25
Expansion joint	"	100	56.00	156
Box connector	"	43.75	230	274
Standard dropout	"	11.00	51.00	62.00
2"	"	35.00	56.00	91.00
3"	"	55.00	73.00	128
4"	"	90.00	92.00	182
Cable tray, 24"	L.F.	27.50	9.48	36.98
Ventilated cover	"	20.00	5.36	25.36
Solid cover	"	13.75	5.36	19.11
Flat 90	EA.	190	56.00	246
Outside 90	"	160	56.00	216
Inside 90	"	130	56.00	186
Flat 45	"	120	56.00	176
Outside 45	"	160	56.00	216
Inside 45	"	130	56.00	186
Adjustable elbow	"	160	56.00	216
Support riser	"	200	56.00	256
Adjustable riser	"	17.25	56.00	73.25
Tee	"	230	170	400
Cross	"	340	170	510

16 ELECTRICAL

Basic Materials

16110.12	Cable Tray *(Cont.)*	UNIT	MAT.	INST.	TOTAL
	Blind end	EA.	22.75	34.25	57.00
	Expansion joint	"	110	56.00	166
	Box connector	"	46.75	270	317
	Standard dropout	"	11.75	51.00	62.75
	2"	"	36.50	62.00	98.50
	3"	"	58.00	75.00	133
	4"	"	93.00	96.00	189
Cable tray, 36"		L.F.	37.25	11.25	48.50
	Ventilated cover	"	23.00	6.16	29.16
	Solid cover	"	16.00	6.16	22.16
	Flat 90	EA.	210	66.00	276
	Outside 90	"	190	66.00	256
	Inside 90	"	160	66.00	226
	Flat 45	"	140	66.00	206
	Outside 45	"	170	66.00	236
	Inside 45	"	140	66.00	206
	Adjustable elbow	"	170	66.00	236
	Support riser	"	200	66.00	266
	Adjustable riser	"	18.25	66.00	84.25
	Tee	"	260	180	440
	Cross	"	390	230	620
	Blind end	"	23.50	36.25	59.75
	Expansion joint	"	110	56.00	166
	Box connector	"	54.00	290	344
	Standard dropout	"	13.25	56.00	69.25
	2"	"	38.75	62.00	101
	3"	"	60.00	77.00	137
	4"	"	93.00	99.00	192
Reducers					
	9" - 6"	EA.	84.00	38.50	123
	12" - 9"	"	88.00	38.50	127
	18" - 12"	"	93.00	47.50	141
	24" - 18"	"	100	56.00	156
	36" - 18"	"	130	62.00	192
	36" - 24"	"	130	69.00	199
Conduit dropouts					
	3/4"	EA.	11.25	26.75	38.00
	1"	"	11.50	26.75	38.25
	1-1/4"	"	13.50	30.75	44.25
	1-1/2"	"	15.50	38.50	54.00
	2"	"	17.75	41.00	58.75
	2-1/2"	"	20.25	56.00	76.25
	3"	"	23.50	62.00	85.50
Wall brackets					
	6"	EA.	43.75	11.25	55.00
	9"	"	50.00	11.25	61.25
	12"	"	53.00	15.50	68.50
	18"	"	64.00	19.25	83.25
	24"	"	74.00	22.75	96.75
	36"	"	92.00	30.75	123

16 ELECTRICAL

Basic Materials

16110.15 Fiberglass Cable Tray

	UNIT	MAT.	INST.	TOTAL
Fiberglass cable tray, 6"	L.F.	34.75	3.08	37.83
Tray cover	"	9.68	2.32	12.00
Horizontal				
90	EA.	380	21.25	401
45	"	160	21.25	181
30	"	170	21.25	191
Inside				
90	EA.	140	21.25	161
45	"	120	21.25	141
30	"	160	21.25	181
Horizontal tee	"	250	56.00	306
Horizontal cross	"	300	84.00	384
Splice plate	"	14.25	11.25	25.50
Floor flange	"	15.50	26.75	42.25
Panel flange	"	16.00	96.00	112
End plate	"	10.75	11.25	22.00
Nylon rivet	"	0.39	3.85	4.24
Barrier strip	"	5.33	3.85	9.18
Hold down clamp	"	2.34	3.85	6.19
Drop out	"	16.75	41.00	57.75
Cover stand off	"	4.48	3.85	8.33
Wall bracket	"	63.00	15.50	78.50
Sealer	"			23.00
Outside				
90	EA.	140	21.25	161
45	"	120	21.25	141
30	"	160	21.25	181
Fiberglass cable tray, 9"	L.F.	36.50	3.85	40.35
Tray cover	"	10.00	3.08	13.08
Horizontal				
90	EA.	390	21.25	411
45	"	170	21.25	191
30	"	170	21.25	191
Inside				
90	EA.	160	21.25	181
45	"	120	21.25	141
30	"	170	21.25	191
Horizontal tee	"	260	56.00	316
Horizontal cross	"	310	84.00	394
Splice plate	"	14.75	11.25	26.00
Floor flange	"	16.50	26.75	43.25
Panel flange	"	17.00	96.00	113
End plate	"	11.25	11.25	22.50
Nylon rivet	"	0.40	3.85	4.25
Barrier strap	"	5.59	3.85	9.44
Hold down clamp	"	2.47	3.85	6.32
Drop out	"	17.50	41.00	58.50
Cover stand off	"	4.68	3.85	8.53
Wall bracket	"	66.00	15.50	81.50
Sealer	"			23.00
Outside				
90	EA.	160	21.25	181
45	"	130	21.25	151

16 ELECTRICAL

Basic Materials

16110.15 Fiberglass Cable Tray (Cont.)

	UNIT	MAT.	INST.	TOTAL
30	EA.	170	21.25	191
Fiberglass cable tray, 12"	L.F.	38.25	4.56	42.81
Tray cover	"	10.75	3.85	14.60
Horizontal				
90	EA.	420	26.75	447
45	"	180	26.75	207
30	"	180	26.75	207
Inside				
90	EA.	160	26.75	187
45	"	130	26.75	157
Inside 30	"	170	26.75	197
Horizontal tee	"	270	73.00	343
Horizontal cross	"	330	99.00	429
Splice plate	"	15.50	12.25	27.75
Floor flange	"	17.25	28.00	45.25
Panel flange	"	17.75	120	138
End plate	"	11.75	12.25	24.00
Nylon rivet	"	0.41	3.85	4.26
Barrier strip	"	5.85	3.85	9.70
Hold down clamp	"	2.60	3.85	6.45
Drop out	"	26.25	41.00	67.25
Cover stand off	"	4.87	3.85	8.72
Wall bracket	"	70.00	15.50	85.50
Sealer	"			23.00
Outside				
90	EA.	160	26.75	187
45	"	130	26.75	157
30	"	180	26.75	207
Fiberglass cable tray, 18"	L.F.	40.25	5.40	45.65
Tray cover	"	19.75	4.56	24.31
Horizontal				
90	EA.	440	30.75	471
45	"	180	30.75	211
30	"	200	30.75	231
Inside				
90	EA.	170	30.75	201
45	"	130	30.75	161
30	"	180	30.75	211
Horizontal tee	"	290	120	410
Horizontal cross	"	340	100	440
Splice plate	"	16.50	12.25	28.75
Floor flange	"	18.00	29.25	47.25
Panel flange	"	18.75	130	149
End plate	"	11.75	13.00	24.75
Nylon rivet	"	0.44	3.85	4.29
Barrier strip	"	6.17	3.85	10.02
Hold down clamp	"	2.73	3.85	6.58
Drop out	"	38.25	41.00	79.25
Cover stand off	"	5.20	3.85	9.05
Wall bracket	"	73.00	15.50	88.50
Sealer	"			23.00
Outside				
90	EA.	170	30.75	201

16 ELECTRICAL

Basic Materials

16110.15 Fiberglass Cable Tray (Cont.)

	UNIT	MAT.	INST.	TOTAL
45	EA.	130	30.75	161
30	"	180	30.75	211
Fiberglass cable tray, 24"	L.F.	42.00	6.16	48.16
Tray cover	"	23.50	5.36	28.86
Horizontal				
90	EA.	440	38.50	479
45	"	200	38.50	239
30	"	200	38.50	239
Inside				
90	EA.	170	38.50	209
45	"	160	38.50	199
30	"	180	38.50	219
Horizontal tee	"	300	77.00	377
Horizontal cross	"	350	120	470
Splice plate	"	17.00	12.25	29.25
Floor flange	"	19.00	30.75	49.75
Panel flange	"	19.75	140	160
End plate	"	13.00	13.00	26.00
Nylon Rivet	"	0.45	3.85	4.30
Barrier strip	"	6.43	3.85	10.28
Hold down clamp	"	2.86	3.85	6.71
Drop out	"	50.00	41.00	91.00
Cover stand off	"	5.39	3.85	9.24
Wall bracket	"	80.00	15.50	95.50
Sealer	"			23.00
Outside				
90	EA.	170	38.50	209
45	"	140	38.50	179
30	"	180	38.50	219
Fiberglass cable tray, 30"	L.F.	48.25	6.84	55.09
Tray cover	"	33.50	6.16	39.66
Horizontal				
90	EA.	450	47.50	498
45	"	200	47.50	248
30	"	200	47.50	248
Inside				
90	EA.	180	47.50	228
45	"	140	47.50	188
30	"	200	47.50	248
Horizontal tee	"	340	84.00	424
Horizontal cross	"	400	120	520
Splice plate	"	18.00	12.25	30.25
Floor flange	"	20.00	30.75	50.75
Panel flange	"	20.75	150	171
End plate	"	13.50	15.50	29.00
Nylon rivet	"	0.49	3.85	4.34
Barrier strip	"	6.76	3.85	10.61
Hold down clamp	"	2.99	3.85	6.84
Dropout	"	58.00	47.50	106
Cover stand off	"	5.72	3.85	9.57
Wall bracket	"	90.00	15.50	106
Sealer	"			62.00
Outside				

16 ELECTRICAL

Basic Materials

16110.15 Fiberglass Cable Tray (Cont.)

Item	UNIT	MAT.	INST.	TOTAL
90	EA.	180	47.50	228
45	"	140	47.50	188
30	"	200	47.50	248
Fiberglass cable tray, 36"	L.F.	56.00	7.70	63.70
Tray cover	"	44.25	6.84	51.09
Horizontal				
90	EA.	450	51.00	501
45	"	200	51.00	251
30	"	200	51.00	251
Inside				
90	EA.	180	51.00	231
45	"	140	51.00	191
30	"	200	51.00	251
Horizontal tee	"	380	92.00	472
horizontal cross	"	440	130	570
Splice plate	"	19.00	12.25	31.25
Floor flange	"	21.00	34.25	55.25
Panel flange	"	21.75	160	182
End plate	"	14.25	15.50	29.75
Nylon rivet	"	0.54	3.85	4.39
Barrier strip	"	7.15	3.85	11.00
Hold down clamp	"	3.12	3.85	6.97
Drop out	"	67.00	51.00	118
5Cover stand off	"	5.98	3.85	9.83
Wall bracket	"	99.00	15.50	115
Sealer	"			23.00
Outside				
90	EA.	180	51.00	231
45	"	160	51.00	211
30	"	200	51.00	251
Reducers				
12" - 6"	EA.	230	22.75	253
12" - 9"	"	230	22.75	253
18" - 6"	"	210	26.75	237
18" - 9"	"	210	26.75	237
18" - 12"	"	220	30.75	251
24" - 6"	"	250	30.75	281
24" - 9"	"	260	30.75	291
24" - 12"	"	260	30.75	291
24" - 18"	"	260	34.25	294
30" - 9"	"	260	30.75	291
30" - 12"	"	260	30.75	291
30" - 18"	"	260	34.25	294
30" - 24"	"	290	38.50	329
36" - 9"	"	270	38.50	309
36" - 12"	"	290	38.50	329
36" - 18"	"	290	41.00	331
36" - 24"	"	300	47.50	348
36" - 30"	"	300	56.00	356

16 ELECTRICAL

Basic Materials

16110.20 Conduit Specialties

	UNIT	MAT.	INST.	TOTAL
Rod beam clamp, 1/2"	EA.	6.76	3.85	10.61
Hanger rod				
3/8"	L.F.	1.43	3.08	4.51
1/2"	"	3.56	3.85	7.41
All thread rod				
1/4"	L.F.	0.45	2.32	2.77
3/8"	"	0.52	3.08	3.60
1/2"	"	0.97	3.85	4.82
5/8"	"	1.71	6.16	7.87
Hanger channel, 1-1/2"				
No holes	EA.	4.49	2.32	6.81
Holes	"	5.56	2.32	7.88
Channel strap				
1/2"	EA.	1.40	3.85	5.25
3/4"	"	1.88	3.85	5.73
1"	"	2.40	3.85	6.25
1-1/4"	"	1.94	6.16	8.10
1-1/2"	"	2.33	6.16	8.49
2"	"	2.52	6.16	8.68
2-1/2"	"	4.84	9.48	14.32
3"	"	5.27	9.48	14.75
3-1/2"	"	6.56	9.48	16.04
4"	"	7.42	11.25	18.67
5"	"	12.00	11.25	23.25
6"	"	13.50	11.25	24.75
Conduit penetrations, roof and wall, 8" thick				
1/2"	EA.		47.50	47.50
3/4"	"		47.50	47.50
1"	"		62.00	62.00
1-1/4"	"		62.00	62.00
1-1/2"	"		62.00	62.00
2"	"		120	120
2-1/2"	"		120	120
3"	"		120	120
3-1/2"	"		150	150
4"	"		150	150
Plastic duct bank conduit spacer, 3" separation				
2"	EA.	1.91	3.85	5.76
3"	"	2.11	3.85	5.96
4"	"	2.37	3.85	6.22
5"	"	2.57	3.85	6.42
6"	"	4.17	3.85	8.02
Intermediate, 3" separation				
2"	EA.	1.97	3.85	5.82
3"	"	2.18	3.85	6.03
4"	"	2.44	3.85	6.29
5"	"	2.68	3.85	6.53
6"	"	4.24	3.85	8.09
Base with 1-1/2" separation				
2"	EA.	1.88	3.85	5.73
3"	"	2.06	12.25	14.31
4"	"	2.26	12.25	14.51
5"	"	2.44	12.25	14.69

16 ELECTRICAL

Basic Materials

16110.20 Conduit Specialties (Cont.)

	UNIT	MAT.	INST.	TOTAL
6"	EA.	3.94	12.25	16.19
Intermediate, 1-1/2" separation				
2"	EA.	2.02	12.25	14.27
3"	"	2.17	12.25	14.42
3-1/2"	"	2.34	12.25	14.59
4"	"	2.38	12.25	14.63
5"	"	2.52	13.25	15.77
6"	"	3.99	12.25	16.24
OD beam clamp, 1/4"	"	0.85	15.50	16.35
Threaded rod couplings				
1/4"	EA.	1.49	3.85	5.34
3/8"	"	1.57	3.85	5.42
1/2"	"	1.78	3.85	5.63
5/8"	"	2.73	3.85	6.58
3/4"	"	2.99	3.85	6.84
Hex nuts				
1/4"	EA.	0.16	3.85	4.01
3/8"	"	0.26	3.85	4.11
1/2"	"	0.55	3.85	4.40
5/8"	"	1.19	3.85	5.04
3/4"	"	1.58	3.85	5.43
Square nuts				
1/4"	EA.	0.15	3.85	4.00
3/8"	"	0.29	3.85	4.14
3/8"	"	0.49	3.85	4.34
5/8"	"	0.65	3.85	4.50
3/4"	"	1.14	3.85	4.99
Flat washers				
1/4"	EA.			0.16
3/8"	"			0.23
1/2"	"			0.32
5/8"	"			0.65
3/4"	"			0.91
Lockwashers				
1/4"	EA.			0.10
3/8"	"			0.18
1/2"	"			0.22
5/8"	"			0.39
3/4"	"			0.65
Channel closure strip	L.F.	2.47	10.25	12.72
Channel end cap	EA.	1.13	10.25	11.38
Li-channel trapeze hangers				
12" long	EA.	16.25	11.25	27.50
18" long	"	18.25	11.25	29.50
24" long	"	22.50	11.25	33.75
30" long	"	25.75	19.25	45.00
36" long	"	29.75	19.25	49.00
42" long	"	36.00	22.75	58.75
Channel spring nuts				
1/4"	EA.	1.90	4.56	6.46
3/8"	"	2.60	6.16	8.76
1/2"	"	2.77	7.70	10.47
Fireproofing, for conduit penetrations				

16 ELECTRICAL

Basic Materials

	UNIT	MAT.	INST.	TOTAL
16110.20 Conduit Specialties (Cont.)				
1/2"	EA.	3.48	38.50	41.98
3/4"	"	3.61	38.50	42.11
1"	"	3.68	38.50	42.18
1-1/4"	"	8.66	60.00	68.66
1-1/2"	"	5.11	56.00	61.11
2"	"	5.25	56.00	61.25
2-1/2"	"	10.00	69.00	79.00
3"	"	10.25	75.00	85.25
3-1/2"	"	12.00	96.00	108
4"	"	14.75	120	135
16110.21 Aluminum Conduit				
Aluminum conduit				
1/2"	L.F.	2.15	2.32	4.47
3/4"	"	2.78	3.08	5.86
1"	"	3.88	3.85	7.73
1-1/4"	"	5.20	4.56	9.76
1-1/2"	"	6.45	6.16	12.61
2"	"	8.61	6.84	15.45
2-1/2"	"	13.75	7.70	21.45
3"	"	17.75	8.21	25.96
3-1/2"	"	21.50	9.48	30.98
4"	"	25.50	11.25	36.75
5"	"	36.25	14.00	50.25
6"	"	48.00	15.50	63.50
90 deg. elbow				
1/2"	EA.	16.75	14.75	31.50
3/4"	"	22.75	19.25	42.00
1"	"	31.50	23.75	55.25
1-1/4"	"	50.00	29.25	79.25
1-1/2"	"	67.00	30.75	97.75
2"	"	98.00	34.25	132
2-1/2"	"	170	44.00	214
3"	"	260	51.00	311
3-1/2"	"	390	62.00	452
4"	"	510	68.00	578
5"	"	1,300	88.00	1,388
6"	"	1,770	170	1,940
Coupling				
1/2"	EA.	4.84	3.85	8.69
3/4"	"	7.33	4.56	11.89
1"	"	9.68	6.16	15.84
1-1/4"	"	11.75	6.84	18.59
1-1/2"	"	13.75	7.70	21.45
2"	"	19.25	8.21	27.46
2-1/2"	"	43.75	9.48	53.23
3"	"	57.00	9.48	66.48
3-1/2"	"	78.00	11.25	89.25
4"	"	95.00	12.25	107
5"	"	290	12.25	302
6"	"	450	14.75	465

16 ELECTRICAL

Basic Materials

16110.22 Emt Conduit

	UNIT	MAT.	INST.	TOTAL
EMT conduit				
1/2"	L.F.	0.72	2.32	3.04
3/4"	"	1.32	3.08	4.40
1"	"	2.21	3.85	6.06
1-1/4"	"	3.54	4.56	8.10
1-1/2"	"	4.49	6.16	10.65
2"	"	5.60	6.84	12.44
2-1/2"	"	11.25	7.70	18.95
3"	"	12.50	9.48	21.98
3-1/2"	"	17.75	11.25	29.00
4"	"	17.00	14.00	31.00
90 deg. elbow				
1/2"	EA.	6.84	6.84	13.68
3/4"	"	7.52	7.70	15.22
1"	"	11.50	8.21	19.71
1-1/4"	"	14.25	9.48	23.73
1-1/2"	"	16.75	11.25	28.00
2"	"	24.50	14.75	39.25
2-1/2"	"	60.00	16.25	76.25
3"	"	89.00	18.75	108
3-1/2"	"	120	21.50	142
4"	"	140	22.00	162
Connector, steel compression				
1/2"	EA.	2.20	6.84	9.04
3/4"	"	4.20	6.84	11.04
1"	"	6.34	6.84	13.18
1-1/4"	"	14.50	8.21	22.71
1-1/2"	"	20.75	11.25	32.00
2"	"	30.00	14.75	44.75
2-1/2"	"	84.00	19.25	103
3"	"	110	22.00	132
3-1/2"	"	160	23.75	184
4"	"	180	25.75	206
Coupling, steel, compression				
1/2"	EA.	3.73	4.56	8.29
3/4"	"	5.09	4.56	9.65
1"	"	7.70	4.56	12.26
1-1/4"	"	14.00	6.84	20.84
1-1/2"	"	21.75	8.21	29.96
2"	"	27.25	11.25	38.50
2-1/2"	"	98.00	17.00	115
3"	"	120	19.25	139
3-1/2"	"	190	22.00	212
4"	"	190	23.75	214
1 hole strap, steel				
1/2"	EA.	0.24	3.08	3.32
3/4"	"	0.31	3.08	3.39
1"	"	0.48	3.08	3.56
1-1/4"	"	0.74	3.85	4.59
1-1/2"	"	1.13	3.85	4.98
2"	"	1.79	3.85	5.64
2-1/2"	"	3.73	4.56	8.29
3"	"	4.16	4.56	8.72

16 ELECTRICAL

Basic Materials

16110.22 Emt Conduit *(Cont.)*

	UNIT	MAT.	INST.	TOTAL
3-1/2"	EA.	6.29	4.56	10.85
4"	"	7.85	4.56	12.41
Connector, steel set screw				
1/2"	EA.	1.67	5.36	7.03
3/4"	"	2.69	5.36	8.05
1"	"	4.63	5.36	9.99
1-1/4"	"	9.88	8.21	18.09
1-1/2"	"	14.25	11.25	25.50
2"	"	20.50	14.00	34.50
2-1/2"	"	66.00	18.75	84.75
3"	"	79.00	20.50	99.50
3-1/2"	"	100	22.75	123
4"	"	120	26.75	147
Insulated throat				
1/2"	EA.	2.23	5.36	7.59
3/4"	"	3.61	5.36	8.97
1"	"	5.96	5.36	11.32
1-1/4"	"	12.00	8.21	20.21
1-1/2"	"	17.50	11.25	28.75
2"	"	25.25	14.00	39.25
2-1/2"	"	110	18.75	129
3"	"	140	20.50	161
3-1/2"	"	190	22.75	213
4"	"	210	26.75	237
Connector, die cast set screw				
1/2"	EA.	1.03	4.56	5.59
3/4"	"	1.76	4.56	6.32
1"	"	3.32	4.56	7.88
1-1/4"	"	5.81	6.84	12.65
1-1/2"	"	7.89	8.21	16.10
2"	"	9.25	11.25	20.50
2-1/2"	"	21.00	15.50	36.50
3"	"	25.25	17.00	42.25
3-1/2"	"	30.00	19.25	49.25
4"	"	38.00	22.00	60.00
Insulated throat				
1/2"	EA.	2.21	4.56	6.77
3/4"	"	3.56	4.56	8.12
1"	"	6.19	4.56	10.75
1-1/4"	"	11.75	6.84	18.59
1-1/2"	"	17.00	8.21	25.21
2"	"	24.75	11.25	36.00
2-1/2"	"	110	15.50	126
3"	"	140	17.00	157
3-1/2"	"	190	20.75	211
4"	"	210	22.00	232
Coupling, steel set screw				
1/2"	EA.	2.72	3.08	5.80
3/4"	"	4.10	3.08	7.18
1"	"	6.65	3.08	9.73
1-1/4"	"	14.25	3.85	18.10
1-1/2"	"	20.50	6.16	26.66
2"	"	27.25	8.21	35.46

16 ELECTRICAL

Basic Materials

16110.22 Emt Conduit *(Cont.)*

	UNIT	MAT.	INST.	TOTAL
2-1/2"	EA.	76.00	12.25	88.25
3"	"	84.00	14.75	98.75
3-1/2"	"	95.00	17.00	112
4"	"	110	19.25	129
Diecast set screw				
1/2"	EA.	0.95	3.08	4.03
3/4"	"	1.54	3.08	4.62
1"	"	2.54	3.08	5.62
1-1/4"	"	4.43	3.85	8.28
1-1/2"	"	6.26	6.16	12.42
2"	"	8.39	8.21	16.60
2-1/2"	"	19.75	12.25	32.00
3"	"	22.75	14.25	37.00
3-1/2"	"	28.75	17.00	45.75
4"	"	34.25	19.25	53.50
1 hole malleable straps				
1/2"	EA.	0.44	3.08	3.52
3/4"	"	0.61	3.08	3.69
1"	"	1.00	3.08	4.08
1-1/4"	"	1.74	3.85	5.59
1-1/2"	"	2.11	3.85	5.96
2"	"	3.84	3.85	7.69
2-1/2"	"	8.64	4.56	13.20
3"	"	12.50	4.56	17.06
3-1/2"	"	18.50	4.56	23.06
4"	"	40.50	4.56	45.06
EMT to rigid compression coupling				
1/2"	EA.	5.13	7.70	12.83
3/4"	"	7.33	7.70	15.03
1"	"	11.25	11.50	22.75
Set screw couplings				
1/2"	EA.	1.34	7.70	9.04
3/4"	"	2.03	7.70	9.73
1"	"	3.38	11.25	14.63
Set screw offset connectors				
1/2"	EA.	3.01	7.70	10.71
3/4"	"	4.04	7.70	11.74
1"	"	7.35	11.25	18.60
Compression offset connectors				
1/2"	EA.	4.97	7.70	12.67
3/4"	"	6.29	7.70	13.99
1"	"	9.10	11.25	20.35
Type "LB" set screw condulets				
1/2"	EA.	14.25	17.50	31.75
3/4"	"	17.25	22.75	40.00
1"	"	26.25	29.25	55.50
1-1/4"	"	39.75	34.25	74.00
1-1/2"	"	52.00	41.00	93.00
2"	"	86.00	47.50	134
2-1/2"	"	140	56.00	196
3"	"	190	77.00	267
3-1/2"	"	280	100	380
4"	"	350	120	470

16 ELECTRICAL

Basic Materials

16110.22 Emt Conduit *(Cont.)*

	UNIT	MAT.	INST.	TOTAL
Type "T" set screw condulets				
1/2"	EA.	17.75	22.75	40.50
3/4"	"	22.25	30.75	53.00
1"	"	32.00	34.25	66.25
1-1/4"	"	46.00	41.00	87.00
1-1/2"	"	57.00	47.50	105
2"	"	75.00	51.00	126
Type "C" set screw condulets				
1/2"	EA.	14.75	19.25	34.00
3/4"	"	18.50	22.75	41.25
1"	"	27.75	29.25	57.00
1-1/4"	"	48.00	34.25	82.25
1-1/2"	"	58.00	41.00	99.00
2"	"	99.00	29.25	128
Type "LL" set screw condulets				
1/2"	EA.	14.75	19.25	34.00
3/4"	"	18.00	22.75	40.75
1"	"	27.50	29.25	56.75
1-1/4"	"	35.00	34.25	69.25
1-1/2"	"	44.25	41.00	85.25
2"	"	75.00	47.50	123
Type "LR" set screw condulets				
1/2"	EA.	14.75	19.25	34.00
3/4"	"	18.00	22.75	40.75
1"	"	27.50	29.25	56.75
1-1/4"	"	34.00	34.25	68.25
1-1/2"	"	55.00	41.00	96.00
2"	"	73.00	47.50	121
Type "LB" compression condulets				
1/2"	EA.	33.50	22.75	56.25
3/4"	"	49.50	38.50	88.00
1"	"	63.00	38.50	102
Type "T" compression condulets				
1/2"	EA.	44.75	30.75	75.50
3/4"	"	59.00	34.25	93.25
1"	"	91.00	47.50	139
Condulet covers				
1/2"	EA.	2.06	9.48	11.54
3/4"	"	2.50	9.48	11.98
1"	"	3.41	9.48	12.89
1-1/4"	"	3.93	9.48	13.41
1-1/2"	"	4.12	11.25	15.37
2"	"	7.25	11.25	18.50
2-1/2"	"	8.88	11.25	20.13
3"	"	11.50	14.00	25.50
3-1/2"	"	13.00	14.00	27.00
4"	"	13.50	14.00	27.50
Clamp type entrance caps				
1/2"	EA.	11.00	19.25	30.25
3/4"	"	13.00	22.75	35.75
1"	"	15.25	30.75	46.00
1-1/4"	"	17.25	41.00	58.25
1-1/2"	"	29.25	47.50	76.75

16 ELECTRICAL

Basic Materials

	UNIT	MAT.	INST.	TOTAL
16110.22 — Emt Conduit *(Cont.)*				
2"	EA.	40.00	69.00	109
2-1/2"	"	68.00	77.00	145
3"	"	100	120	220
3-1/2"	"	150	130	280
4"	"	230	170	400
Slip fitter type entrance caps				
1/2"	EA.	8.05	19.25	27.30
3/4"	"	9.62	22.75	32.37
1"	"	11.50	30.75	42.25
1-1/4"	"	13.25	41.00	54.25
1-1/2"	"	22.25	47.50	69.75
2"	"	33.50	69.00	103
2-1/2"	"	56.00	77.00	133
3"	"	85.00	120	205
3-1/2"	"	130	140	270
4"	"	200	170	370
16110.23 — Flexible Conduit				
Flexible conduit, steel				
3/8"	L.F.	0.77	2.32	3.09
1/2	"	0.88	2.32	3.20
3/4"	"	1.21	3.08	4.29
1"	"	2.29	3.08	5.37
1-1/4"	"	2.87	3.85	6.72
1-1/2"	"	4.75	4.56	9.31
2"	"	5.86	6.16	12.02
2-1/2"	"	7.12	6.84	13.96
3"	"	12.50	8.21	20.71
Flexible conduit, liquid tight				
3/8"	L.F.	2.15	2.32	4.47
1/2"	"	2.43	2.32	4.75
3/4"	"	3.31	3.08	6.39
1"	"	5.00	3.08	8.08
1-1/4"	"	6.88	3.85	10.73
1-1/2"	"	8.62	4.56	13.18
2"	EA.	11.00	6.16	17.16
2-1/2"	"	19.75	6.84	26.59
3"	"	27.25	8.21	35.46
4"	"	41.25	11.25	52.50
Connector, straight				
3/8"	EA.	4.18	6.16	10.34
1/2"	"	4.49	6.16	10.65
3/4"	"	5.70	6.84	12.54
1"	"	10.25	7.70	17.95
1-1/4"	"	14.50	8.21	22.71
1-1/2"	"	21.00	9.48	30.48
2"	"	35.25	11.25	46.50
2-1/2"	"	52.00	14.00	66.00
3"	"	76.00	14.75	90.75
Straight insulated throat connectors				
3/8"	EA.	5.07	9.48	14.55
1/2"	"	5.07	9.48	14.55
3/4"	"	7.43	11.25	18.68

16 ELECTRICAL

Basic Materials

16110.23 Flexible Conduit (Cont.)

	UNIT	MAT.	INST.	TOTAL
1"	EA.	11.50	11.25	22.75
1-1/4"	"	17.50	14.00	31.50
1-1/2"	"	25.00	16.25	41.25
2"	"	47.00	17.50	64.50
2-1/2"	"	260	20.50	281
3"	"	290	25.75	316
4"	"	340	32.50	373
90 deg connectors				
3/8"	EA.	7.09	11.50	18.59
1/2"	"	7.09	11.50	18.59
3/4"	"	11.50	13.00	24.50
1"	"	22.00	14.00	36.00
1-1/4"	"	34.00	17.50	51.50
1-1/2"	"	40.75	19.25	60.00
2"	"	61.00	20.50	81.50
2-1/2"	"	270	25.75	296
3"	"	330	29.25	359
4"	"	420	34.25	454
90 degree insulated throat connectors				
3/8"	EA.	8.77	11.25	20.02
1/2"	"	8.77	11.25	20.02
3/4"	"	13.25	13.00	26.25
1"	"	25.00	13.75	38.75
1-1/4"	"	38.25	17.50	55.75
1-1/2"	"	47.00	19.25	66.25
2"	"	70.00	20.50	90.50
2-1/2"	"	350	25.75	376
3"	"	420	29.25	449
4"	"	550	34.25	584
Flexible aluminum conduit				
3/8"	L.F.	0.52	2.32	2.84
1/2"	"	0.62	2.32	2.94
3/4"	"	0.85	3.08	3.93
1"	"	1.61	3.08	4.69
1-1/4"	"	2.18	3.85	6.03
1-1/2"	"	3.60	4.56	8.16
2"	"	4.37	6.16	10.53
2-1/2"	"	5.77	6.84	12.61
3"	"	9.83	8.21	18.04
3-1/2"	"	11.25	9.48	20.73
4"	"	12.50	11.25	23.75
Connector, straight				
3/8"	EA.	1.60	7.70	9.30
1/2"	"	2.21	7.70	9.91
3/4"	"	2.43	8.21	10.64
1"	"	8.89	9.48	18.37
1-1/4"	"	12.25	11.25	23.50
1-1/2"	"	24.00	14.00	38.00
2"	"	33.50	14.75	48.25
2-1/2"	"	47.50	17.00	64.50
3"	"	79.00	21.25	100
4"	"	260	26.75	287
Straight insulated throat connectors				

16 ELECTRICAL

Basic Materials

	UNIT	MAT.	INST.	TOTAL
16110.23 Flexible Conduit (Cont.)				
3/8"	EA.	1.55	6.84	8.39
1/2"	"	3.13	6.84	9.97
3/4"	"	3.33	6.84	10.17
1"	"	8.07	7.70	15.77
1-1/4"	"	13.00	8.21	21.21
1-1/2"	"	18.25	9.48	27.73
2"	"	29.00	11.25	40.25
2-1/2"	"	54.00	14.00	68.00
3"	"	70.00	14.75	84.75
3-1/2"	"	250	17.00	267
4"	"	300	21.25	321
90 deg connectors				
3/8"	EA.	2.55	11.25	13.80
1/2"	"	4.31	11.25	15.56
3/4"	"	6.77	11.25	18.02
1"	"	11.75	13.00	24.75
1-1/4"	"	24.00	14.00	38.00
1-1/2"	"	38.75	15.50	54.25
2"	"	48.00	16.25	64.25
2-1/2"	"	150	17.50	168
3"	"	220	20.50	241
90 deg insulated throat connectors				
3/8"	EA.	3.17	11.25	14.42
1/2"	"	4.79	11.25	16.04
3/4"	"	7.99	11.25	19.24
1"	"	13.00	13.00	26.00
1-1/4"	"	25.25	14.00	39.25
1-1/2"	"	45.00	15.50	60.50
2"	"	58.00	16.25	74.25
2-1/2"	"	150	17.50	168
3"	"	220	20.50	241
3-1/2"	"	620	25.75	646
4"	"	950	32.50	983
16110.24 Galvanized Conduit				
Galvanized rigid steel conduit				
1/2"	L.F.	3.37	3.08	6.45
3/4"	"	3.73	3.85	7.58
1"	"	5.39	4.56	9.95
1-1/4"	"	7.44	6.16	13.60
1-1/2"	"	8.76	6.84	15.60
2"	"	11.25	7.70	18.95
2-1/2"	"	20.50	11.25	31.75
3"	"	21.25	14.00	35.25
3-1/2"	"	30.75	14.75	45.50
4"	"	35.00	16.25	51.25
5"	"	65.00	22.00	87.00
6"	"	94.00	29.25	123
90 degree ell				
1/2"	EA.	12.25	19.25	31.50
3/4"	"	12.75	23.75	36.50
1"	"	19.50	29.25	48.75
1-1/4"	"	27.00	34.25	61.25

16 ELECTRICAL

Basic Materials

16110.24 Galvanized Conduit *(Cont.)*

	UNIT	MAT.	INST.	TOTAL
1-1/2"	EA.	33.25	38.50	71.75
2"	"	48.00	41.00	89.00
2-1/2"	"	100	51.00	151
3"	"	140	68.00	208
3-1/2"	"	230	77.00	307
4"	"	260	100	360
5"	"	710	170	880
6"	"	1,070	260	1,330
Couplings, with set screws				
1/2"	EA.	6.11	3.85	9.96
3/4"	"	8.06	4.56	12.62
1"	"	12.75	6.16	18.91
1-1/4"	"	21.75	7.70	29.45
1-1/2"	"	28.25	9.48	37.73
2"	"	64.00	11.25	75.25
2-1/2"	"	160	14.75	175
3"	"	190	19.25	209
3-1/2"	"	260	22.00	282
4"	"	340	23.75	364
5"	"	490	34.25	524
6"	"	650	38.50	689
Split couplings				
1/2"	EA.	5.20	14.75	19.95
3/4"	"	6.76	19.25	26.01
1"	"	9.49	21.25	30.74
1-1/4"	"	18.75	23.75	42.50
1-1/2"	"	24.25	29.25	53.50
2"	"	56.00	44.00	100
2-1/2"	"	110	44.00	154
3"	"	170	56.00	226
3-1/2"	"	260	77.00	337
4"	"	310	100	410
5"	"	550	130	680
6"	"	730	160	890
Erickson couplings				
1/2"	EA.	6.17	34.25	40.42
3/4"	"	7.54	38.50	46.04
1"	"	15.25	47.50	62.75
1-1/4"	"	27.25	68.00	95.25
1-1/2"	"	35.50	77.00	113
2"	"	69.00	100	169
2-1/2"	"	140	140	280
3"	"	200	160	360
3-1/2"	"	360	190	550
4"	"	440	210	650
5"	"	880	230	1,110
6"	"	1,270	250	1,520
Seal fittings				
1/2"	EA.	19.50	51.00	70.50
3/4"	"	21.50	62.00	83.50
1"	"	27.25	77.00	104
1-1/4"	"	32.50	88.00	121
1-1/2"	"	49.50	100	150

16 ELECTRICAL

Basic Materials

16110.24 Galvanized Conduit *(Cont.)*

	UNIT	MAT.	INST.	TOTAL
2"	EA.	62.00	120	182
2-1/2"	"	98.00	150	248
3"	"	120	160	280
3-1/2"	"	310	190	500
4"	"	470	230	700
5"	"	730	340	1,070
6"	"	810	390	1,200
Entrance fitting, (weather head), threaded				
1/2"	EA.	10.50	34.25	44.75
3/4"	"	12.75	38.50	51.25
1"	"	16.50	44.00	60.50
1-1/4"	"	21.25	56.00	77.25
1-1/2"	"	37.50	62.00	99.50
2"	"	57.00	68.00	125
2-1/2"	"	200	77.00	277
3"	"	290	100	390
3-1/2"	"	360	130	490
4"	"	470	190	660
5"	"	490	270	760
6"	"	620	340	960
Locknuts				
1/2"	EA.	0.23	3.85	4.08
3/4"	"	0.28	3.85	4.13
1"	"	0.45	3.85	4.30
1-1/4"	"	0.62	3.85	4.47
1-1/2"	"	1.02	4.56	5.58
2"	"	1.49	4.56	6.05
2-1/2"	"	4.16	6.16	10.32
3"	"	5.33	6.16	11.49
3-1/2"	"	9.00	6.16	15.16
4"	"	11.25	6.84	18.09
5"	"	24.00	6.84	30.84
6"	"	41.00	6.84	47.84
Plastic conduit bushings				
1/2"	EA.	0.51	9.48	9.99
3/4"	"	0.79	11.25	12.04
1"	"	1.11	14.75	15.86
1-1/4"	"	1.46	17.00	18.46
1-1/2"	"	1.97	19.25	21.22
2"	"	4.48	23.75	28.23
2-1/2"	"	8.59	38.50	47.09
3"	"	9.97	51.00	60.97
3-1/2"	"	11.25	62.00	73.25
4"	"	15.25	68.00	83.25
5"	"	29.50	88.00	118
6"	"	56.00	120	176
Conduit bushings, steel				
1/2"	EA.	0.68	9.48	10.16
3/4"	"	0.86	11.25	12.11
1"	"	1.31	14.75	16.06
1-1/4"	"	1.88	17.00	18.88
1-1/2"	"	2.69	19.25	21.94
2"	"	5.46	23.75	29.21

16 ELECTRICAL

Basic Materials	UNIT	MAT.	INST.	TOTAL
16110.24 Galvanized Conduit *(Cont.)*				
2-1/2"	EA.	9.56	38.50	48.06
3"	"	11.75	51.00	62.75
3-1/2"	"	24.50	62.00	86.50
4"	"	30.00	68.00	98.00
5"	"	61.00	88.00	149
6"	"	110	120	230
Pipe cap				
1/2"	EA.	0.67	3.85	4.52
3/4"	"	0.72	3.85	4.57
1"	"	1.17	3.85	5.02
1-1/4"	"	2.00	6.16	8.16
1-1/2"	"	3.12	6.16	9.28
2"	"	3.51	6.16	9.67
2-1/2"	"	5.98	6.84	12.82
3"	"	7.41	6.84	14.25
3-1/2"	"	10.25	6.84	17.09
4"	"	13.00	8.21	21.21
5"	"	17.25	11.25	28.50
6"	"	21.75	15.50	37.25
GRS elbows, 36" radius				
2"	EA.	170	51.00	221
2-1/2"	"	230	62.00	292
3"	"	310	81.00	391
3-1/2"	"	420	96.00	516
4"	"	470	120	590
5"	"	760	190	950
6"	"	810	290	1,100
42" radius				
2"	EA.	190	62.00	252
2-1/2"	"	270	77.00	347
3"	"	340	96.00	436
3-1/2"	"	480	120	600
4"	"	580	130	710
5"	"	860	220	1,080
6"	"	900	310	1,210
48" radius				
2"	EA.	220	72.00	292
2-1/2"	"	300	87.00	387
3"	"	390	110	500
3-1/2"	"	550	130	680
4"	"	660	170	830
5"	"	970	240	1,210
6"	"	1,010	340	1,350
Threaded couplings				
1/2"	EA.	2.26	3.85	6.11
3/4"	"	2.76	4.56	7.32
1"	"	4.10	6.16	10.26
1-1/4"	"	5.13	6.84	11.97
1-1/2"	"	6.29	7.70	13.99
2"	"	8.56	8.21	16.77
2-1/2"	"	21.25	9.48	30.73
3"	"	27.50	11.25	38.75
3-1/2"	"	36.75	11.25	48.00

16 ELECTRICAL

Basic Materials

16110.24 Galvanized Conduit (Cont.)

	UNIT	MAT.	INST.	TOTAL
4"	EA.	37.00	12.25	49.25
5"	"	85.00	14.00	99.00
6"	"	120	14.75	135
Threadless couplings				
1/2"	EA.	6.89	7.70	14.59
3/4"	"	7.17	9.48	16.65
1"	"	9.30	11.25	20.55
1-1/4"	"	10.75	14.75	25.50
1-1/2"	"	13.00	19.25	32.25
2"	"	18.75	23.75	42.50
2-1/2"	"	46.75	38.50	85.25
3"	"	57.00	47.50	105
3-1/2"	"	73.00	62.00	135
4"	"	85.00	77.00	162
5"	"	280	96.00	376
6"	"	310	410	720
Threadless connectors				
1/2"	EA.	3.27	7.70	10.97
3/4"	"	5.23	9.48	14.71
1"	"	8.28	11.25	19.53
1-1/4"	"	14.00	14.75	28.75
1-1/2"	"	21.50	19.25	40.75
2"	"	41.00	23.75	64.75
2-1/2"	"	95.00	38.50	134
3"	"	130	47.50	178
3-1/2"	"	160	62.00	222
4"	"	200	77.00	277
5"	"	540	96.00	636
6"	"	710	120	830
Setscrew connectors				
1/2"	EA.	3.02	6.16	9.18
3/4"	"	4.19	6.84	11.03
1"	"	6.53	7.70	14.23
1-1/4"	"	11.50	9.48	20.98
1-1/2"	"	16.75	11.25	28.00
2"	"	33.25	14.75	48.00
2-1/2"	"	99.00	19.25	118
3"	"	140	23.75	164
3-1/2"	"	180	29.25	209
4"	"	230	38.50	269
5"	"	390	47.50	438
6"	"	510	62.00	572
Clamp type entrance caps				
1/2"	EA.	10.25	23.75	34.00
3/4"	"	12.00	29.25	41.25
1"	"	16.75	34.25	51.00
1-1/4"	"	19.50	38.50	58.00
1-1/2"	"	35.00	47.50	82.50
2"	"	42.25	56.00	98.25
3-1/2"	"	160	73.00	233
3"	"	250	87.00	337
3-1/2"	"	310	110	420
4"	"	460	190	650

16 ELECTRICAL

Basic Materials

16110.24 Galvanized Conduit (Cont.)

	UNIT	MAT.	INST.	TOTAL
"LB" condulets				
1/2"	EA.	11.50	23.75	35.25
3/4"	"	13.75	29.25	43.00
1"	"	20.75	34.25	55.00
1-1/4"	"	36.00	38.50	74.50
1-1/2"	"	46.75	47.50	94.25
2"	"	77.00	56.00	133
2-1/2"	"	160	77.00	237
3"	"	210	110	320
3-1/2"	"	380	130	510
4"	"	430	160	590
"T" condulets				
1/2"	EA.	14.25	29.25	43.50
3/4"	"	17.25	34.25	51.50
1"	"	26.00	38.50	64.50
1-1/4"	"	38.00	44.00	82.00
1-1/2"	"	51.00	47.50	98.50
2"	"	78.00	56.00	134
2-1/2"	"	170	87.00	257
3"	"	220	120	340
3-1/2"	"	400	140	540
4"	"	450	170	620
"X" condulets				
1/2"	EA.	21.25	34.25	55.50
3/4"	"	22.75	38.50	61.25
1"	"	37.75	44.00	81.75
1-1/4"	"	49.25	47.50	96.75
1-1/2"	"	63.00	51.00	114
2"	"	130	68.00	198
Blank steel condulet covers				
1/2"	EA.	3.25	7.70	10.95
3/4"	"	4.03	7.70	11.73
1"	"	5.50	7.70	13.20
1-1/4"	"	6.72	9.48	16.20
1-1/2"	"	7.07	9.48	16.55
2"	"	12.00	9.48	21.48
2-1/2"	"	18.50	11.25	29.75
3"	"	20.00	11.25	31.25
3-1/2"	"	22.00	11.25	33.25
4"	"	24.00	15.50	39.50
Solid condulet gaskets				
1/2"	EA.	2.69	3.85	6.54
3/4"	"	2.91	3.85	6.76
1"	"	3.36	3.85	7.21
1-1/4"	"	4.18	6.16	10.34
1-1/2"	"	4.41	6.16	10.57
2"	"	4.93	6.16	11.09
2-1/2"	"	8.14	7.70	15.84
3"	"	8.37	7.70	16.07
3-1/2"	"	10.25	7.70	17.95
4"	"	11.25	11.25	22.50
One-hole malleable straps				
1/2"	EA.	0.44	3.08	3.52

16 ELECTRICAL

Basic Materials

16110.24 Galvanized Conduit (Cont.)

Item	UNIT	MAT.	INST.	TOTAL
3/4"	EA.	0.62	3.08	3.70
1"	"	0.89	3.08	3.97
1-1/4"	"	1.79	3.85	5.64
1-1/2"	"	2.06	3.85	5.91
2"	"	4.03	3.85	7.88
2-1/2"	"	8.35	4.56	12.91
3"	"	12.00	4.56	16.56
3-1/2"	"	17.25	4.56	21.81
4"	"	38.25	6.16	44.41
5"	"	140	6.16	146
6"	"	140	6.16	146
One-hole steel straps				
1/2"	EA.	0.11	3.08	3.19
3/4"	"	0.16	3.08	3.24
1"	"	0.28	3.08	3.36
1-1/4"	"	0.40	3.85	4.25
1-1/2"	"	0.54	3.85	4.39
2"	"	0.70	3.85	4.55
2-1/2"	"	2.46	4.56	7.02
3"	"	3.32	4.56	7.88
3-1/2"	"	4.93	4.56	9.49
4"	"	5.10	6.16	11.26
Bushed chase nipples				
1/2"	EA.	0.96	4.56	5.52
3/4"	"	1.34	5.36	6.70
1"	"	2.71	6.84	9.55
1-1/4"	"	4.07	7.70	11.77
1-1/2"	"	5.19	9.48	14.67
2"	"	6.97	11.25	18.22
2-1/2"	"	17.75	11.25	29.00
3"	"	23.00	14.00	37.00
3-1/2"	"	54.00	19.25	73.25
4"	"	77.00	22.75	99.75
Offset nipples				
1/2"	EA.	4.94	4.56	9.50
3/4"	"	5.13	5.36	10.49
1"	"	6.26	6.84	13.10
1-1/4"	"	15.25	8.21	23.46
1-1/2"	"	18.75	9.48	28.23
2"	"	29.50	11.25	40.75
3"	"	92.00	14.00	106
Short elbows				
1/2"	EA.	3.60	11.25	14.85
3/4"	"	4.92	15.50	20.42
1"	"	8.04	19.25	27.29
1-1/4"	"	24.25	22.75	47.00
1-1/2"	"	33.25	26.75	60.00
2"	"	60.00	30.75	90.75
Pulling elbows, female to female				
1/2"	EA.	8.68	19.25	27.93
3/4"	"	10.25	22.75	33.00
1"	"	17.00	30.75	47.75
1-1/4"	"	25.00	41.00	66.00

16 ELECTRICAL

Basic Materials

16110.24 Galvanized Conduit (Cont.)

	UNIT	MAT.	INST.	TOTAL
1-1/2"	EA.	46.75	56.00	103
2"	"	38.25	66.00	104
Grounding locknuts				
1/2"	EA.	2.47	6.16	8.63
3/4"	"	3.11	6.16	9.27
1"	"	4.50	6.16	10.66
1-1/4"	"	4.83	6.84	11.67
1-1/2"	"	5.04	6.84	11.88
2"	"	7.48	6.84	14.32
2-1/2"	"	13.75	7.70	21.45
3"	"	17.25	7.70	24.95
3-1/2"	"	28.25	7.70	35.95
4"	"	38.00	11.25	49.25
Insulated grounding metal bushings				
1/2"	EA.	1.75	14.75	16.50
3/4"	"	2.58	17.00	19.58
1"	"	3.68	19.25	22.93
1-1/4"	"	5.89	23.75	29.64
1-1/2"	"	7.35	29.25	36.60
2"	"	10.75	34.25	45.00
2-1/2"	"	27.75	51.00	78.75
3"	"	32.00	62.00	94.00
3-1/2"	"	46.50	73.00	120
4"	"	58.00	81.00	139
5"	"	120	120	240
6"	"	170	130	300
Nipples				
1/2" x				
4"	EA.	4.55	11.25	15.80
6"	"	6.09	11.25	17.34
8"	"	10.50	11.25	21.75
10"	"	12.25	11.25	23.50
12"	"	14.00	11.25	25.25
3/4" x				
4"	EA.	5.30	11.25	16.55
6"	"	7.12	11.25	18.37
8"	"	11.75	11.25	23.00
10"	"	14.00	11.25	25.25
12"	"	15.75	11.25	27.00
1" x				
4"	EA.	7.59	11.25	18.84
6"	"	9.43	11.25	20.68
8"	"	14.75	11.25	26.00
10"	"	19.25	11.25	30.50
12"	"	21.75	11.25	33.00
1-1/4" x				
4"	EA.	11.00	19.25	30.25
6"	"	14.25	19.25	33.50
8"	"	23.50	19.25	42.75
10"	"	29.50	19.25	48.75
12"	"	34.50	19.25	53.75
1-1/2" x				
4"	EA.	14.00	19.25	33.25

16 ELECTRICAL

Basic Materials	UNIT	MAT.	INST.	TOTAL
16110.24 **Galvanized Conduit** *(Cont.)*				
6"	EA.	19.25	19.25	38.50
8"	"	29.50	19.25	48.75
10"	"	35.50	19.25	54.75
12"	"	38.50	19.25	57.75
2" x				
4"	EA.	18.25	19.25	37.50
6"	"	24.25	19.25	43.50
8"	"	35.00	19.25	54.25
10"	"	42.00	19.25	61.25
12"	"	47.75	19.25	67.00
2-1/2" x				
6"	EA.	49.75	23.00	72.75
8"	"	65.00	23.00	88.00
10"	"	76.00	23.00	99.00
12"	"	88.00	23.00	111
3" x				
6"	EA.	60.00	23.00	83.00
8"	"	78.00	23.00	101
10"	"	92.00	23.00	115
12"	"	110	23.00	133
3-1/2" x				
6"	EA.	71.00	23.00	94.00
8"	"	89.00	23.00	112
10"	"	110	23.00	133
12"	"	130	23.00	153
4" x				
8"	EA.	100	30.75	131
10"	"	120	30.75	151
12"	"	150	30.75	181
5" x				
8"	EA.	180	30.75	211
10"	"	210	30.75	241
12"	"	260	30.75	291
6" x				
8"	EA.	220	30.75	251
10"	"	280	30.75	311
12"	"	310	30.75	341
16110.25 **Plastic Conduit**				
PVC conduit, schedule 40				
1/2"	L.F.	0.82	2.32	3.14
3/4"	"	1.02	2.32	3.34
1"	"	1.48	3.08	4.56
1-1/4"	"	2.05	3.08	5.13
1-1/2"	"	2.44	3.85	6.29
2"	"	3.13	3.85	6.98
2-1/2"	"	5.04	4.56	9.60
3"	"	6.18	4.56	10.74
3-1/2"	"	8.10	6.16	14.26
4"	"	8.88	6.16	15.04
5"	"	13.75	6.84	20.59
6"	"	18.50	7.70	26.20
Couplings				

16 ELECTRICAL

Basic Materials

16110.25	Plastic Conduit *(Cont.)*	UNIT	MAT.	INST.	TOTAL
1/2"		EA.	0.50	3.85	4.35
3/4"		"	0.61	3.85	4.46
1"		"	0.94	3.85	4.79
1-1/4"		"	1.24	4.56	5.80
1-1/2"		"	1.72	4.56	6.28
2"		"	2.27	4.56	6.83
2-1/2"		"	3.96	4.56	8.52
3"		"	6.56	6.16	12.72
3-1/2"		"	7.33	6.16	13.49
4"		"	10.75	7.70	18.45
5"		"	25.75	7.70	33.45
6"		"	33.00	7.70	40.70
90 degree elbows					
1/2"		EA.	1.96	7.70	9.66
3/4"		"	2.14	9.48	11.62
1"		"	3.39	9.48	12.87
1-1/4"		"	4.73	11.25	15.98
1-1/2"		"	6.40	14.75	21.15
2"		"	8.94	17.00	25.94
2-1/2"		"	15.75	19.25	35.00
3"		"	28.50	23.75	52.25
3-1/2"		"	37.25	29.25	66.50
4"		"	47.00	38.50	85.50
5"		"	85.00	47.50	133
6"		"	140	56.00	196
Terminal adapters					
1/2"		EA.	0.74	7.70	8.44
3/4"		"	1.19	7.70	8.89
1"		"	1.49	7.70	9.19
1-1/4"		"	1.88	12.25	14.13
1-1/2"		"	2.40	12.25	14.65
2"		"	3.31	12.25	15.56
2-1/2"		"	5.65	17.00	22.65
3"		"	7.99	17.00	24.99
3-1/2"		"	10.25	17.00	27.25
4"		"	13.25	29.25	42.50
5"		"	26.75	29.25	56.00
6"		"	32.00	29.25	61.25
End bells					
1"		EA.	4.97	7.70	12.67
1-1/4"		"	5.89	12.25	18.14
1-1/2"		"	6.12	12.25	18.37
2"		"	9.11	12.25	21.36
2-1/2"		"	10.00	17.00	27.00
3"		"	10.75	17.00	27.75
3-1/2"		"	11.75	17.00	28.75
4"		"	12.75	29.25	42.00
5"		"	20.00	29.25	49.25
6"		"	22.00	29.25	51.25
LB conduit body					
1/2"		EA.	6.42	14.75	21.17
3/4"		"	8.28	14.75	23.03
1		"	9.11	14.75	23.86

16 ELECTRICAL

Basic Materials

16110.25 Plastic Conduit (Cont.)

	UNIT	MAT.	INST.	TOTAL
1-1/4"	EA.	14.00	23.75	37.75
1-1/2"	"	16.75	23.75	40.50
2"	"	29.50	23.75	53.25
2-1/2"	"	110	34.25	144
3"	"	110	41.00	151
3-1/2"	"	120	47.50	168
4"	"	120	56.00	176
Direct burial, conduit				
2"	L.F.	1.96	3.85	5.81
3"	"	3.69	4.56	8.25
4"	"	5.85	6.16	12.01
5"	"	8.42	6.84	15.26
6"	"	11.50	7.70	19.20
Encased burial conduit				
2"	L.F.	1.24	3.85	5.09
3"	"	2.08	4.56	6.64
4"	"	3.18	6.16	9.34
5"	"	4.55	6.84	11.39
6"	"	5.91	7.70	13.61
"EB" and "DB" duct, 90 degree elbows				
1-1/2"	EA.	12.75	11.25	24.00
2"	"	14.00	17.50	31.50
3"	"	19.75	29.25	49.00
3-1/2"	"	26.00	34.25	60.25
4"	"	28.25	41.00	69.25
5"	"	70.00	51.00	121
6"	"	130	69.00	199
45 degree elbows				
1-1/2"	EA.	15.00	17.50	32.50
2"	"	15.25	17.50	32.75
3"	"	20.25	29.25	49.50
3-1/2"	"	27.25	34.25	61.50
4"	"	26.50	41.00	67.50
5"	"	50.00	51.00	101
6"	"	120	69.00	189
Couplings				
1-1/2"	EA.	1.23	4.56	5.79
2"	"	1.39	4.56	5.95
3"	"	5.03	6.16	11.19
3-1/2"	"	5.98	6.16	12.14
4"	"	7.87	7.70	15.57
5"	"	14.25	7.70	21.95
6"	"	44.00	12.25	56.25
Bell ends				
1-1/2"	EA.	7.67	12.25	19.92
2"	"	9.75	12.25	22.00
3"	"	12.25	17.00	29.25
3-1/2"	"	12.50	17.00	29.50
4"	"	14.50	29.25	43.75
5"	"	21.75	29.25	51.00
6"	"	42.50	29.25	71.75
Female adapters, 1-1/2"	"	2.34	15.50	17.84
5 degree couplings				

16 ELECTRICAL

Basic Materials

16110.25   Plastic Conduit (Cont.)

	UNIT	MAT.	INST.	TOTAL
1-1/2"	EA.	9.83	5.36	15.19
2"	"	11.00	5.36	16.36
3"	"	14.50	7.70	22.20
4"	"	15.00	11.25	26.25
5"	"	18.50	11.25	29.75
6"	"	19.50	11.25	30.75
45 degree elbows				
1/2"	EA.	1.49	9.48	10.97
3/4"	"	1.88	11.25	13.13
1"	"	2.66	11.25	13.91
1-1/4"	"	3.83	14.00	17.83
1-1/2"	"	5.26	17.50	22.76
2"	"	7.80	20.50	28.30
2-1/2"	"	14.50	22.75	37.25
3"	"	25.00	29.25	54.25
3-1/2"	"	28.25	34.25	62.50
4"	"	39.75	47.50	87.25
5"	"	63.00	56.00	119
6"	"	92.00	69.00	161
Female adapters				
1/2"	EA.	0.78	9.48	10.26
3/4"	"	1.24	9.48	10.72
1"	"	1.56	9.48	11.04
1-1/4"	"	2.01	15.50	17.51
1-1/2"	"	2.21	15.50	17.71
2"	"	3.18	15.50	18.68
2-1/2"	"	5.46	20.50	25.96
3"	"	9.10	20.50	29.60
3-1/2"	"	11.25	20.50	31.75
4"	"	14.25	34.25	48.50
5"	"	32.25	34.25	66.50
6"	"	36.50	34.25	70.75
Expansion couplings				
1/2"	EA.	33.75	9.48	43.23
3/4"	"	33.00	9.48	42.48
1"	"	34.50	11.25	45.75
1-1/4"	"	34.75	15.50	50.25
1-1/2"	"	35.00	15.50	50.50
2"	"	38.50	15.50	54.00
2-1/2"	"	53.00	22.75	75.75
3"	"	68.00	22.75	90.75
3-1/2"	"	84.00	22.75	107
4"	"	98.00	34.25	132
5"	"	150	34.25	184
6"	"	210	34.25	244
Plugs				
2"	EA.	2.49	15.50	17.99
3"	"	3.67	22.75	26.42
3-1/2"	"	3.96	22.75	26.71
4"	"	4.14	34.25	38.39
5"	"	5.61	34.25	39.86
6"	"	7.00	38.50	45.50
PVC cement				

16 ELECTRICAL

Basic Materials	UNIT	MAT.	INST.	TOTAL
16110.25 Plastic Conduit *(Cont.)*				
1 pint	EA.			17.75
1 quart	"			25.75
1 gallon	"			85.00
Type "T" condulets				
1/2"	EA.	8.77	22.75	31.52
3/4"	"	9.94	22.75	32.69
1"	"	11.25	22.75	34.00
1-1/4"	"	16.75	38.50	55.25
1-1/2"	"	22.00	38.50	60.50
2"	"	31.25	38.50	69.75
EB & DB female adapters				
2"	EA.	1.36	19.25	20.61
3"	"	3.77	29.25	33.02
3-1/2"	"	4.94	47.50	52.44
4"	"	5.07	56.00	61.07
5"	"	12.50	77.00	89.50
6"	"	16.50	120	137
16110.27 Plastic Coated Conduit				
Rigid steel conduit, plastic coated				
1/2"	L.F.	6.26	3.85	10.11
3/4"	"	7.26	4.56	11.82
1"	"	9.40	6.16	15.56
1-1/4"	"	12.00	7.70	19.70
1-1/2"	"	14.50	9.48	23.98
2"	"	18.75	11.25	30.00
2-1/2"	"	28.50	14.75	43.25
3"	"	36.00	17.00	53.00
3-1/2"	"	44.25	19.25	63.50
4"	"	53.00	23.75	76.75
5"	"	92.00	29.25	121
90 degree elbows				
1/2"	EA.	24.75	23.75	48.50
3/4"	"	25.50	29.25	54.75
1"	"	29.25	34.25	63.50
1-1/4"	"	36.00	38.50	74.50
1-1/2"	"	44.25	47.50	91.75
2"	"	62.00	62.00	124
2-1/2"	"	120	88.00	208
3"	"	190	100	290
3-1/2"	"	240	130	370
4"	"	270	150	420
5"	"	640	190	830
Couplings				
1/2"	EA.	7.03	4.56	11.59
3/4"	"	7.36	6.16	13.52
1"	"	9.75	6.84	16.59
1-1/4"	"	11.50	8.21	19.71
1-1/2"	"	16.00	9.48	25.48
2"	"	20.00	11.25	31.25
2-1/2"	"	49.75	14.00	63.75
3"	"	58.00	14.75	72.75
3-1/2"	"	80.00	15.50	95.50

16 ELECTRICAL

Basic Materials

	UNIT	MAT.	INST.	TOTAL
16110.27 Plastic Coated Conduit (Cont.)				
4"	EA.	98.00	17.00	115
5"	"	280	19.25	299
1 hole conduit straps				
3/4"	EA.	12.00	3.85	15.85
1"	"	12.25	3.85	16.10
1-1/4"	"	18.00	4.56	22.56
1-1/2"	"	19.00	4.56	23.56
2"	"	27.75	4.56	32.31
3"	"	50.00	6.16	56.16
3-1/2"	"	90.00	6.16	96.16
4"	"	96.00	7.70	104
"L.B." condulets with covers				
1/2"	EA.	59.00	38.50	97.50
3/4"	"	65.00	38.50	104
1"	"	87.00	47.50	135
1-1/4"	"	130	56.00	186
1-1/2"	"	150	68.00	218
2"	"	230	77.00	307
2-1/2"	"	420	110	530
3"	"	520	130	650
3-1/2"	"	770	170	940
4"	"	870	190	1,060
"T" condulets with covers				
1/2"	EA.	68.00	44.00	112
3/4"	"	85.00	47.50	133
1"	"	100	51.00	151
1-1/4"	"	140	62.00	202
1-1/2"	"	180	73.00	253
2"	"	260	81.00	341
2-1/2"	"	440	120	560
3-1/2"	"	800	170	970
4"	"	880	210	1,090
5"	"	940	260	1,200
16110.28 Steel Conduit				
Intermediate metal conduit (IMC)				
1/2"	L.F.	2.18	2.32	4.50
3/4"	"	2.68	3.08	5.76
1"	"	4.05	3.85	7.90
1-1/4"	"	5.19	4.56	9.75
1-1/2"	"	6.48	6.16	12.64
2"	"	8.47	6.84	15.31
2-1/2"	"	16.75	9.20	25.95
3"	"	21.50	11.25	32.75
3-1/2"	"	25.25	14.00	39.25
4"	"	28.00	14.75	42.75
90 degree ell				
1/2"	EA.	16.75	19.25	36.00
3/4"	"	17.50	23.75	41.25
1"	"	26.75	29.25	56.00
1-1/4"	"	37.00	34.25	71.25
1-1/2"	"	45.75	38.50	84.25
2"	"	66.00	44.00	110

16 ELECTRICAL

Basic Materials	UNIT	MAT.	INST.	TOTAL
16110.28 Steel Conduit *(Cont.)*				
2-1/2"	EA.	120	51.00	171
3"	"	170	68.00	238
3-1/2"	"	270	88.00	358
4"	"	300	100	400
Couplings				
1/2"	EA.	4.08	3.85	7.93
3/4"	"	5.01	4.56	9.57
1"	"	7.43	6.16	13.59
1-1/4"	"	9.30	6.84	16.14
1-1/2"	"	11.75	7.70	19.45
2"	"	15.50	8.21	23.71
2-1/2"	"	38.25	9.48	47.73
3"	"	49.50	11.25	60.75
3-1/2"	"	67.00	11.25	78.25
4"	"	67.00	12.25	79.25
16110.32 Flexible Wiring Systems				
Single circuit cables				
5'	EA.	27.25	4.56	31.81
10'	"	38.75	7.70	46.45
15'	"	48.00	11.25	59.25
20'	"	68.00	15.50	83.50
25'	"	87.00	20.50	108
30'	"	110	22.75	133
40'	"	140	30.75	171
Two circuit cables				
5'	EA.	37.00	4.56	41.56
10'	"	43.50	7.70	51.20
15'	"	54.00	11.25	65.25
20'	"	77.00	15.50	92.50
25'	"	99.00	20.50	120
30'	"	120	22.75	143
40'	"	170	30.75	201
Two wire switch and receptacle cables				
5'	EA.	15.25	4.56	19.81
10'	"	31.00	7.70	38.70
15'	"	39.75	11.25	51.00
20'	"	68.00	15.50	83.50
25'	"	91.00	20.50	112
30'	"	120	22.75	143
40'	"	160	26.75	187
Three wire switch				
5'	EA.	17.50	4.56	22.06
10'	"	35.25	7.70	42.95
15'	"	46.25	11.25	57.50
20'	"	68.00	15.50	83.50
25'	"	91.00	20.50	112
30'	"	120	22.75	143
40'	"	160	26.75	187
Distribution boxes				
2 circuit	EA.	26.25	41.00	67.25
3 circuit	"	30.25	51.00	81.25
4 circuit	"	45.50	62.00	108

16 ELECTRICAL

Basic Materials	UNIT	MAT.	INST.	TOTAL
16110.32 Flexible Wiring Systems *(Cont.)*				
6 circuit	EA.	61.00	84.00	145
12 circuit	"	94.00	170	264
18 circuit	"	140	230	370
Tap boxes				
1 single pole switch	EA.	53.00	30.75	83.75
2 single pole switches	"	58.00	41.00	99.00
1 3 way switch	"	62.00	34.25	96.25
1 4 way switch	"	77.00	41.00	118
1 receptacle	"	47.50	30.75	78.25
2 receptacles	"	59.00	41.00	100
4 receptacles	"	75.00	62.00	137
1 clock	"	59.00	30.75	89.75
2 clocks	"	69.00	41.00	110
4 clocks	"	77.00	62.00	139
Dust cap	"	7.81	7.70	15.51
Cable coupler	"	18.50	15.50	34.00
Reversing connector	"	26.25	22.75	49.00
16110.35 Surface Mounted Raceway				
Single Raceway				
3/4" x 17/32" Conduit	L.F.	1.97	3.08	5.05
Mounting Strap	EA.	0.53	4.10	4.63
Connector	"	0.71	4.10	4.81
Elbow				
45 degree	EA.	9.00	3.85	12.85
90 degree	"	2.87	3.85	6.72
internal	"	3.61	3.85	7.46
external	"	3.34	3.85	7.19
Switch	"	23.50	30.75	54.25
Utility Box	"	15.75	30.75	46.50
Receptacle	"	27.75	30.75	58.50
3/4" x 21/32" Conduit	L.F.	2.24	3.08	5.32
Mounting Strap	EA.	0.83	4.10	4.93
Connector	"	0.85	4.10	4.95
Elbow				
45 degree	EA.	11.00	3.85	14.85
90 degree	"	3.06	3.85	6.91
internal	"	4.16	3.85	8.01
external	"	4.16	3.85	8.01
Switch	"	23.50	30.75	54.25
Utility Box	"	15.75	30.75	46.50
Receptacle	"	27.75	30.75	58.50
1-1/4" x 7/8" Conduit	L.F.	5.33	3.08	8.41
Mounting Strap	EA.	0.89	4.10	4.99
Connector	"	1.43	4.10	5.53
Elbow				
90 degree	EA.	11.75	3.85	15.60
internal	"	9.28	3.85	13.13
external	"	18.25	3.85	22.10
Switch Box	"	18.75	30.75	49.50
Receptacle Box	"	18.75	30.75	49.50
1-29/32" x 7/8" Conduit	L.F.	4.99	3.08	8.07
Mounting Strap	EA.	0.83	4.10	4.93

16 ELECTRICAL

Basic Materials	UNIT	MAT.	INST.	TOTAL
16110.35 **Surface Mounted Raceway** *(Cont.)*				
Connector	EA.	2.93	4.10	7.03
Elbow				
90 degree	EA.	10.75	3.85	14.60
internal	"	26.25	3.85	30.10
external	"	16.00	3.85	19.85
Switch Box	"	16.25	30.75	47.00
Receptacle Box	"	16.25	30.75	47.00
2-3/4" x 1-15/32" Conduit	L.F.	7.98	3.08	11.06
Mounting Strap	EA.	2.52	4.10	6.62
Connector	"	4.97	4.10	9.07
Elbow				
90 degree	EA.	36.50	3.85	40.35
internal	"	24.50	3.85	28.35
external	"	31.00	3.85	34.85
Switch Cover	"	15.25	30.75	46.00
Receptacle Cover	"	14.50	30.75	45.25
Double Raceway				
5-1/2" x 2" Conduit	L.F.	14.75	3.42	18.17
Mounting Strap	EA.	2.87	5.13	8.00
Connector	"	6.14	5.13	11.27
Elbow				
90 degree	EA.	60.00	4.40	64.40
internal	"	81.00	4.40	85.40
external	"	87.00	4.40	91.40
Receptacle Cover	"	14.25	30.75	45.00
16110.40 **Underfloor Duct**				
Underfloor blank duct, insert duct				
7/8"	L.F.	21.25	3.85	25.10
1-3/8"	"	23.25	3.85	27.10
1-7/8"	"	23.50	3.85	27.35
Box opening plugs	EA.	9.79	11.25	21.04
Duct end plugs	"	11.50	11.25	22.75
Sleeve couplings	"	26.25	26.75	53.00
Expansion couplings	"	150	26.75	177
Vertical elbow	"	130	26.75	157
Offset elbow	"	100	26.75	127
Horizontal elbow	"	240	26.75	267
Adjustable elbow	"	44.50	26.75	71.25
Cabinet connector	"	36.00	92.00	128
Y-take off	"	110	47.50	158
Underfloor duct leveling legs	"	11.25	11.25	22.50
Conduit adapters				
1/2"	EA.	44.50	19.25	63.75
3/4"	"	45.25	19.25	64.50
1"	"	39.00	19.25	58.25
1-1/4"	"	51.00	22.75	73.75
2"	"	71.00	26.75	97.75
Reducer bushings				
1-1/4" x 3/4"	EA.	22.50	15.50	38.00
1-1/4" x 1"	"	26.25	15.50	41.75
2" x 1-1/2"	"	30.75	19.25	50.00
Support couplers				

16 ELECTRICAL

Basic Materials

16110.40 Underfloor Duct (Cont.)

Item	UNIT	MAT.	INST.	TOTAL
1 standard	EA.	63.00	19.25	82.25
2 standard	"	63.00	22.75	85.75
3 standard	"	63.00	26.75	89.75
Supports				
1 duct	EA.	63.00	11.25	74.25
2 duct	"	63.00	13.00	76.00
3 duct	"	63.00	14.75	77.75
4 duct	"	86.00	19.25	105
5 duct	"	93.00	22.75	116
Single level junction box				
1 standard	EA.	620	62.00	682
2 standard	"	910	120	1,030
3 standard	"	1,770	230	2,000
4 standard	"	2,270	310	2,580
Two level junction boxes				
1 standard	EA.	670	77.00	747
2 standard	"	1,050	150	1,200
Sealing compound	"			13.75
Insert adapters	"	33.50	11.25	44.75
Ellipsoids	"	37.00	26.75	63.75
Insert closing cap	"	3.91	11.25	15.16
Marker Screws	"	10.25	11.25	21.50
Access boxes	"	740	77.00	817
Closing caps	"	3.91	11.25	15.16
Afterset markers	"	10.25	11.25	21.50
Cell markers	"	22.50	11.25	33.75
Tie down straps	"	20.75	11.25	32.00
Plastic grommets	"	10.25	30.75	41.00
Metal grommets	"	72.00	30.75	103
Receptacle				
Duplex, 20a	EA.	110	38.50	149
Single				
30a	EA.	140	38.50	179
50a	"	160	47.50	208
Double duplex	"	110	47.50	158
Single, 20a	"	100	34.25	134
Double single	"	120	38.50	159
Twist lock	"	110	34.25	144
1 conduit opening	"	110	34.25	144
2 conduit openings	"	120	38.50	159
1 bushed opening	"	110	34.25	144
2 bushed openings	"	120	38.50	159
Amphenol connector				
1"	EA.	170	38.50	209
2"	"	200	41.00	241
5"	"	230	47.50	278
Standpipes				
Aluminum	EA.	130	19.25	149
Brass	"	140	19.25	159
Abandonment plates				
Aluminum	EA.	37.25	19.25	56.50
Brass	"	62.00	19.25	81.25
Split bell caps				

16 ELECTRICAL

Basic Materials

	UNIT	MAT.	INST.	TOTAL
16110.40 Underfloor Duct *(Cont.)*				
Aluminum	EA.	210	19.25	229
Brass	"	270	19.25	289
Flush floor receptacles				
Aluminum	EA.	180	56.00	236
Brass	"	190	56.00	246
Flush floor telephone				
Aluminum	EA.	130	38.50	169
Brass	"	160	38.50	199
Super underfloor duct blank duct				
1/2"	L.F.	35.50	4.56	40.06
7/8"	"	36.75	4.56	41.31
1-3/8"	"	37.50	4.56	42.06
1-7/8"	"	42.00	4.56	46.56
Box opening plugs	EA.	8.14	11.25	19.39
End plugs	"	8.66	11.25	19.91
Conduit adapters	"	34.25	56.00	90.25
Sleeve coupling	"	27.75	34.25	62.00
Expansion coupling	"	92.00	34.25	126
Reducing coupling	"	120	34.25	154
Vertical elbow	"	73.00	34.25	107
Offset elbow	"	73.00	34.25	107
Horizontal elbow	"	180	34.25	214
Adjustable elbow	"	73.00	34.25	107
Cabinet connector	"	55.00	120	175
Super underfloor duct Y-take off	"	110	56.00	166
Leveling legs	"	10.00	11.25	21.25
Support couplers				
1 super	EA.	48.75	19.25	68.00
2 super	"	66.00	22.75	88.75
1 super, 1 standard	"	60.00	22.75	82.75
2 super, 2 standard	"	71.00	26.75	97.75
Single level junction boxes				
1 super	EA.	640	120	760
2 super	"	1,380	150	1,530
4 super	"	1,770	270	2,040
Double level junction boxes				
1 super	EA.	790	92.00	882
2 super	"	1,040	120	1,160
16110.50 Wall Duct				
Lay-in wall duct, 10"	L.F.	86.00	4.56	90.56
Horizontal elbow	EA.	420	56.00	476
Edgewise elbow	"	95.00	56.00	151
Tee	"	330	69.00	399
Cross	"	200	84.00	284
Cabinet connector	"	200	120	320
Reverse elbow	"	170	41.00	211
Sweep elbow	"	300	56.00	356
Partition	"	62.00	4.56	66.56
Straight tunnel	"	86.00	21.25	107
Elbow tunnel	"	130	26.75	157
Tee kit	"	130	30.75	161
Ceiling dropout	"	500	77.00	577

16 ELECTRICAL

Basic Materials

16110.50 Wall Duct *(Cont.)*

	UNIT	MAT.	INST.	TOTAL
Coupling device	EA.	54.00	11.25	65.25
End cap	"	72.00	21.25	93.25
Lay-in wall duct, 18"	L.F.	110	6.16	116
Horizontal elbow	EA.	150	62.00	212
Edgeware elbow	"	330	77.00	407
Tee	"	340	84.00	424
Cross	"	420	96.00	516
Reverse elbow	"	210	56.00	266
Sweep elbow	"	450	62.00	512
Partition	"	86.00	7.70	93.70
Straight tunnel	"	110	30.75	141
Elbow tunnel	"	190	38.50	229
Tee kit	"	190	41.00	231
Ceiling dropout	"	500	84.00	584
Coupling device	"	55.00	15.50	70.50
Reducer coupling	"	130	30.75	161
Cabinet connector	"	140	150	290
End cap	"	89.00	30.75	120

16110.60 Trench Duct

	UNIT	MAT.	INST.	TOTAL
Trench duct, with cover				
9"	L.F.	150	13.00	163
12"	"	170	15.50	186
18"	"	220	20.50	241
24"	"	290	26.75	317
30"	"	330	30.75	361
36"	"	380	44.00	424
Tees				
9"	EA.	540	130	670
12"	"	630	150	780
18"	"	810	170	980
24"	"	1,160	190	1,350
30"	"	1,520	230	1,750
36"	"	1,980	260	2,240
Vertical elbows				
9"	EA.	190	62.00	252
12"	"	210	84.00	294
18"	"	240	100	340
24"	"	300	130	430
30"	"	330	150	480
36"	"	360	190	550
Cabinet connectors				
9"	EA.	300	150	450
12"	"	340	160	500
18"	"	390	190	580
24"	"	480	190	670
30"	"	570	210	780
36"	"	630	230	860
End closers				
9"	EA.	56.00	47.50	104
12"	"	64.00	51.00	115
18"	"	98.00	62.00	160
24"	"	130	84.00	214

16 ELECTRICAL

Basic Materials

	UNIT	MAT.	INST.	TOTAL
16110.60 Trench Duct *(Cont.)*				
30"	EA.	160	99.00	259
36"	"	190	110	300
Horizontal elbows				
9"	EA.	510	120	630
12"	"	590	130	720
18"	"	720	160	880
24"	"	1,140	190	1,330
30"	"	1,520	220	1,740
36"	"	1,980	250	2,230
Crosses				
9"	EA.	890	150	1,040
12"	"	950	170	1,120
18"	"	1,140	190	1,330
24"	"	1,450	210	1,660
30"	"	1,860	250	2,110
36"	"	2,340	270	2,610
16110.80 Wireways				
Wireway, hinge cover type				
2-1/2" x 2-1/2"				
1' section	EA.	19.75	11.75	31.50
2'	"	28.25	14.75	43.00
3'	"	38.00	19.25	57.25
5'	"	64.00	29.25	93.25
10'	"	130	51.00	181
4" x 4"				
1'	EA.	21.50	19.25	40.75
2'	"	31.50	19.25	50.75
3'	"	46.75	23.75	70.50
4'	"	64.00	23.75	87.75
10'	"	190	62.00	252
6" x 6"				
1'	EA.	41.50	29.25	70.75
2'	"	51.00	29.25	80.25
3'	"	72.00	34.25	106
4'	"	95.00	34.25	129
5'	"	100	44.00	144
10'	"	200	68.00	268
8" x 8"				
1'	EA.	67.00	34.25	101
2'	"	100	34.25	134
3'	"	140	38.50	179
4'	"	170	38.50	209
5'	"	210	47.50	258
12" x 12"				
1'	EA.	92.00	47.50	140
2'	"	140	47.50	188
3'	"	200	56.00	256
4'	"	230	56.00	286
5'	"	280	68.00	348
Fittings				
2-1/2" x 2-1/2"				
Drop hanger	EA.	22.50	9.48	31.98

16 ELECTRICAL

Basic Materials

16110.80 Wireways *(Cont.)*

Description	UNIT	MAT.	INST.	TOTAL
Bracket hanger	EA.	16.75	9.48	26.23
Panel adapter	"	20.00	38.50	58.50
End plate	"	7.48	9.48	16.96
U-connector	"	14.00	9.48	23.48
Tee	"	93.00	15.50	109
Cross	"	59.00	19.25	78.25
90 degree elbow	"	47.50	15.50	63.00
Sweep elbow	"	80.00	15.50	95.50
45 degree elbow	"	46.50	15.50	62.00
Lay-in adapter	"	20.00	11.25	31.25
4" x 4"				
Drop hanger	EA.	15.00	11.25	26.25
Bracket hanger	"	20.00	11.25	31.25
Panel adapter	"	23.50	47.50	71.00
End plate	"	7.21	11.25	18.46
U-connector	"	13.25	11.25	24.50
Tee	"	110	19.25	129
Cross	"	72.00	26.75	98.75
90 degree elbow	"	57.00	19.25	76.25
Sweep elbow	"	92.00	19.25	111
45 degree elbow	"	57.00	19.25	76.25
Lay-in adapter	"	39.75	15.50	55.25
6" x 6"				
Drop hanger	EA.	33.25	11.25	44.50
Bracket hanger	"	28.50	11.25	39.75
Reducing bushing	"	60.00	11.25	71.25
Panel adapter	"	31.25	56.00	87.25
End plate	"	8.74	11.25	19.99
U-connector	"	19.00	11.25	30.25
Tee	"	120	19.25	139
Cross	"	90.00	26.75	117
90 degree elbow	"	66.00	19.25	85.25
Sweep elbow	"	130	19.25	149
45 degree elbow	"	63.00	19.25	82.25
Lay-in adapter	"	53.00	19.25	72.25
8" x 8"				
Drop hanger	EA.	35.00	11.25	46.25
Bracket hanger	"	30.25	11.25	41.50
Reducing bushing	"	69.00	11.25	80.25
Panel adapter	"	41.75	69.00	111
End plate	"	5.82	11.25	17.07
U-connector	"	16.75	11.25	28.00
Tee	"	130	30.75	161
Cross	"	140	34.25	174
90 degree elbow	"	100	30.75	131
Sweep elbow	"	86.00	30.75	117
45 degree elbow	"	97.00	30.75	128
Lay-in adapter	"	52.00	19.25	71.25
10" x 10"				
Drop hanger	EA.	35.50	11.25	46.75
Bracket hanger	"	36.00	11.25	47.25
Reducing bushing	"	100	11.25	111
Panel adapter	"	66.00	77.00	143

16 ELECTRICAL

Basic Materials

16110.80 Wireways (Cont.)

	UNIT	MAT.	INST.	TOTAL
End plate	EA.	9.38	11.25	20.63
U-connector	"	42.25	11.25	53.50
Tee	"	130	34.25	164
Cross	"	190	38.50	229
90 degree elbow	"	110	34.25	144
Sweep elbow	"	110	34.25	144
45 degree elbow	"	130	34.25	164
Lay-in adapter	"	59.00	22.75	81.75
12" x 12"				
Drop hanger	EA.	99.00	11.25	110
Bracket hanger	"	68.00	11.25	79.25
Reducing bushing	"	100	11.25	111
Panel adapter	"	70.00	96.00	166
End plate	"	13.00	11.25	24.25
U-connector	"	63.00	47.50	111
Tee	"	260	41.00	301
Cross	"	260	47.50	308
90 degree elbow	"	170	47.50	218
Sweep elbow	"	250	47.50	298
45 degree elbow	"	260	47.50	308
Lay-in adapter	"	65.00	41.00	106
Raintight wireway, 4" x 4"				
1' section	EA.	77.00	30.75	108
5'	"	130	30.75	161
10'	"	190	77.00	267
Fittings				
90 degree elbow	EA.	74.00	30.75	105
Tee	"	140	34.25	174
Cross	"	130	38.50	169
Panel adapter	"	34.50	56.00	90.50
End plate	"	18.00	11.25	29.25
Gusset bracket	"	13.00	11.25	24.25
6" x 6"				
1' section	EA.	62.00	38.50	101
5'	"	150	56.00	206
10'	"	260	120	380
Fittings				
90 degree elbow	EA.	93.00	38.50	132
Tee	"	160	38.50	199
Cross	"	160	56.00	216
Panel adapter	"	45.75	77.00	123
End plate	"	20.25	11.25	31.50
Gusset bracket	"	16.75	11.25	28.00

16120.41 Aluminum Conductors

	UNIT	MAT.	INST.	TOTAL
Type XHHW, stranded aluminum, 600v				
#8	L.F.	0.37	0.38	0.75
#6	"	0.40	0.46	0.86
#4	"	0.48	0.61	1.09
#2	"	0.67	0.69	1.36
1/0	"	1.07	0.85	1.92
2/0	"	1.38	0.92	2.30
3/0	"	1.71	1.08	2.79

16 ELECTRICAL

Basic Materials

16120.41 Aluminum Conductors (Cont.)

	UNIT	MAT.	INST.	TOTAL
4/0	L.F.	1.91	1.15	3.06
300 MCM	"	3.21	1.54	4.75
350 MCM	"	3.27	1.76	5.03
400 MCM	"	3.83	2.16	5.99
500 MCM	"	4.21	2.56	6.77
600 MCM	"	5.33	3.08	8.41
700 MCM	"	6.19	3.62	9.81
750 MCM	"	6.24	3.97	10.21
THW, stranded				
#8	L.F.	0.37	0.38	0.75
#6	"	0.40	0.46	0.86
#4	"	0.48	0.61	1.09
#3	"	0.62	0.68	1.30
#1	"	1.07	0.77	1.84
1/0	"	1.17	0.85	2.02
2/0	"	1.38	0.91	2.29
3/0	"	1.71	0.91	2.62
4/0	"	1.91	1.15	3.06
250 MCM	"	2.33	1.38	3.71
300 MCM	"	3.21	1.54	4.75
350 MCM	"	3.27	1.76	5.03
400 MCM	"	3.83	2.16	5.99
500 MCM	"	4.21	2.56	6.77
600 MCM	"	5.33	3.08	8.41
700 MCM	"	6.19	3.62	9.81
750 MCM	"	6.24	3.97	10.21
XLP, stranded				
#6	L.F.	0.47	0.38	0.85
#4	"	0.54	0.61	1.15
#2	"	0.74	0.69	1.43
#1	"	1.02	0.77	1.79
1/0	"	1.25	0.85	2.10
2/0	"	1.48	0.92	2.40
3/0	"	1.77	1.08	2.85
4/0	"	1.95	1.15	3.10
250 MCM	"	2.63	1.24	3.87
300 MCM	"	3.41	1.54	4.95
350 MCM	"	3.48	1.76	5.24
400 MCM	"	4.27	2.16	6.43
500 MCM	"	4.69	2.56	7.25
600 MCM	"	6.03	3.08	9.11
700 MCM	"	6.92	3.62	10.54
750 MCM	"	7.02	3.97	10.99
1000 MCM	"	9.32	4.40	13.72
Bare stranded aluminum wire				
#4	L.F.	0.34	0.61	0.95
#2	"	0.48	0.69	1.17
1/0	"	0.65	0.85	1.50
2/0	"	0.81	0.92	1.73
3/0	"	1.01	1.08	2.09
4/0	"	1.27	1.15	2.42
Triplex XLP cable				
#4	L.F.	1.11	1.15	2.26

16 ELECTRICAL

Basic Materials

16120.41 Aluminum Conductors (Cont.)

	UNIT	MAT.	INST.	TOTAL
#2	L.F.	1.37	1.54	2.91
1/0	"	2.18	2.32	4.50
4/0	"	4.01	3.73	7.74
Aluminum quadruplex XLP cable				
#4	L.F.	1.48	1.38	2.86
#2	"	1.91	1.76	3.67
1/0	"	3.03	2.46	5.49
2/0	"	3.63	3.24	6.87
4/0	"	5.23	4.93	10.16
Triplexed URD-XLP cable				
#6	L.F.	0.90	0.85	1.75
#4	"	1.28	1.08	2.36
#2	"	1.65	1.38	3.03
1/0	"	2.64	2.16	4.80
2/0	"	3.03	2.56	5.59
3/0	"	3.63	3.08	6.71
4/0	"	4.24	3.62	7.86
250 MCM	"	4.86	4.25	9.11
350 MCM	"	6.29	4.40	10.69
Type S.E.U. cable				
#8/3	L.F.	1.74	1.92	3.66
#6/3	"	1.74	2.16	3.90
#4/3	"	2.24	2.68	4.92
#2/3	"	2.98	2.93	5.91
#1/3	"	4.06	3.08	7.14
1/0-3	"	4.56	3.24	7.80
2/0-3	"	5.23	3.42	8.65
3/0-3	"	7.29	3.97	11.26
4/0-3	"	7.33	4.40	11.73
Type S.E.R. cable with ground				
#8/3	L.F.	2.11	2.16	4.27
#6/3	"	2.38	2.68	5.06
#4/3	"	2.67	2.93	5.60
#2/3	"	3.93	3.08	7.01
#1/3	"	5.11	3.42	8.53
1/0-3	"	5.96	3.85	9.81
2/0-3	"	7.02	4.25	11.27
3/0-3	"	8.65	4.56	13.21
4/0-3	"	10.00	5.13	15.13
#6/4	"	4.06	2.93	6.99
#4/4	"	4.57	3.42	7.99
#2/4	"	6.67	3.42	10.09
#1/4	"	8.65	3.85	12.50
1/0-4	"	10.00	3.97	13.97
2/0-4	"	11.75	4.40	16.15
3/0-4	"	14.75	4.93	19.68
4/0-4	"	17.00	5.87	22.87

16 ELECTRICAL

Basic Materials

16120.43 Copper Conductors

	UNIT	MAT.	INST.	TOTAL
Copper conductors, type THW, solid				
#14	L.F.	0.15	0.30	0.45
#12	"	0.24	0.38	0.62
#10	"	0.37	0.46	0.83
Stranded				
#14	L.F.	0.17	0.30	0.47
#12	"	0.21	0.38	0.59
#10	"	0.32	0.46	0.78
#8	"	0.54	0.61	1.15
#6	"	0.87	0.69	1.56
#4	"	1.37	0.77	2.14
#3	"	1.73	0.77	2.50
#2	"	2.17	0.92	3.09
#1	"	2.74	1.08	3.82
1/0	"	3.28	1.23	4.51
2/0	"	4.11	1.54	5.65
3/0	"	5.19	1.92	7.11
4/0	"	6.47	2.16	8.63
250 MCM	"	7.99	2.32	10.31
300 MCM	"	9.40	2.56	11.96
350 MCM	"	11.00	3.08	14.08
400 MCM	"	12.50	3.42	15.92
500 MCM	"	15.50	3.97	19.47
600 MCM	"	20.50	4.56	25.06
750 MCM	"	25.75	5.13	30.88
1000 MCM	"	32.50	5.87	38.37
THHN-THWN, solid				
#14	L.F.	0.15	0.30	0.45
#12	"	0.24	0.38	0.62
#10	"	0.37	0.46	0.83
Stranded				
#14	L.F.	0.15	0.30	0.45
#12	"	0.24	0.38	0.62
#10	"	0.37	0.46	0.83
#8	"	0.64	0.61	1.25
#6	"	1.01	0.69	1.70
#4	"	1.60	0.77	2.37
#2	"	2.23	0.92	3.15
#1	"	2.81	1.08	3.89
1/0	"	3.47	1.23	4.70
2/0	"	4.29	1.54	5.83
3/0	"	5.39	1.92	7.31
4/0	"	6.73	2.16	8.89
250 MCM	"	8.23	2.32	10.55
350 MCM	"	9.82	3.08	12.90
XHHW				
#14	L.F.	0.27	0.30	0.57
#10	"	0.60	0.46	1.06
#8	"	0.87	0.61	1.48
#6	"	1.37	0.69	2.06
#4	"	2.13	0.69	2.82
#2	"	3.31	0.85	4.16
#1	"	4.23	1.08	5.31

16 ELECTRICAL

Basic Materials

16120.43 Copper Conductors (Cont.)

	UNIT	MAT.	INST.	TOTAL
1/0	L.F.	4.99	1.23	6.22
2/0	"	6.24	1.46	7.70
3/0	"	7.79	1.92	9.71
XLP, 600v				
#12	L.F.	0.41	0.38	0.79
#10	"	0.60	0.46	1.06
#8	"	0.78	0.61	1.39
#6	"	1.21	0.69	1.90
#4	"	1.87	0.77	2.64
#3	"	2.33	0.85	3.18
#2	"	2.88	0.92	3.80
#1	"	3.71	1.08	4.79
1/0	"	4.18	1.23	5.41
2/0	"	5.23	1.54	6.77
3/0	"	6.56	1.98	8.54
4/0	"	8.20	2.16	10.36
250 MCM	"	9.58	2.32	11.90
300 MCM	"	11.50	2.56	14.06
350 MCM	"	13.25	3.00	16.25
400 MCM	"	15.00	3.42	18.42
500 MCM	"	18.75	3.97	22.72
600 MCM	"	22.75	4.56	27.31
750 MCM	"	35.25	5.13	40.38
1000 MCM	"	46.50	5.87	52.37
Bare solid wire				
#14	L.F.	0.15	0.30	0.45
#12	"	0.27	0.38	0.65
#10	"	0.40	0.46	0.86
#8	"	0.54	0.61	1.15
#6	"	0.97	0.69	1.66
#4	"	1.60	0.77	2.37
#2	"	2.54	0.92	3.46
Bare stranded wire				
#8	L.F.	0.55	0.61	1.16
#6	"	0.92	0.77	1.69
#4	"	1.45	0.77	2.22
#2	"	2.31	0.85	3.16
#1	"	2.90	1.08	3.98
1/0	"	3.43	1.38	4.81
2/0	"	4.31	1.54	5.85
3/0	"	5.44	1.92	7.36
4/0	"	6.86	2.16	9.02
250 MCM	"	8.25	2.32	10.57
300 MCM	"	10.25	2.56	12.81
350 MCM	"	11.25	3.08	14.33
400 MCM	"	13.25	3.42	16.67
500 MCM	"	16.25	3.97	20.22
Type "BX" solid armored cable				
#14/2	L.F.	1.07	1.92	2.99
#14/3	"	1.68	2.16	3.84
#14/4	"	2.35	2.37	4.72
#12/2	"	1.10	2.16	3.26
#12/3	"	1.75	2.37	4.12

16 ELECTRICAL

Basic Materials

16120.43 Copper Conductors (Cont.)

Description	UNIT	MAT.	INST.	TOTAL
#12/4	L.F.	2.43	2.68	5.11
#10/2	"	2.03	2.37	4.40
#10/3	"	2.90	2.68	5.58
#10/4	"	4.51	3.08	7.59
#8/2	"	4.04	2.68	6.72
#8/3	"	5.69	3.08	8.77
Steel type, metal clad cable, solid, with ground				
#14/2	L.F.	0.88	1.38	2.26
#14/3	"	1.35	1.54	2.89
#14/4	"	1.82	1.76	3.58
#12/2	"	0.91	1.54	2.45
#12/3	"	1.49	1.92	3.41
#12/4	"	2.01	2.32	4.33
#10/2	"	1.87	1.76	3.63
#10/3	"	2.60	2.16	4.76
#10/4	"	4.04	2.56	6.60
Metal clad cable, stranded, with ground				
#8/2	L.F.	3.28	2.16	5.44
#8/3	"	4.70	2.68	7.38
#8/4	"	6.14	3.24	9.38
#6/2	"	4.48	2.32	6.80
#6/3	"	5.39	2.93	8.32
#6/4	"	6.43	3.42	9.85
#4/2	"	5.85	3.08	8.93
#4/3	"	6.63	3.42	10.05
#4/4	"	7.51	4.25	11.76
#3/3	"	7.68	3.85	11.53
#3/4	"	8.52	4.56	13.08
#2/3	"	6.17	4.40	10.57
#2/4	"	10.50	5.13	15.63
#1/3	"	10.75	5.87	16.62
#1/4	"	12.75	6.48	19.23

16120.45 Flat Conductor Cable

Description	UNIT	MAT.	INST.	TOTAL
Flat conductor cable, with shield, 3 conductor				
#12 awg	L.F.	9.49	4.56	14.05
#10 awg	"	11.00	4.56	15.56
4 conductor				
#12 awg	L.F.	12.75	6.16	18.91
#10 awg	"	14.50	6.16	20.66
Transition boxes				
#12 awg	L.F.	16.00	6.84	22.84
#10 awg	"	18.00	6.84	24.84
Flat conductor cable communication, with shield				
10 conductor	L.F.	6.24	4.56	10.80
16 conductor	"	7.21	5.36	12.57
24 conductor	"	8.06	7.70	15.76
Power and communication heads, duplex receptacle	EA.	85.00	62.00	147
Double duplex receptacle	"	95.00	73.00	168
Telephone	"	54.00	62.00	116
Receptacle and telephone	"	130	73.00	203
Blank cover	"	13.50	11.25	24.75
Transition boxes				

16 ELECTRICAL

Basic Materials

	UNIT	MAT.	INST.	TOTAL
16120.45 Flat Conductor Cable *(Cont.)*				
Surface	EA.	250	56.00	306
Flush	"	140	77.00	217
Flat conductor cable fittings				
End caps	EA.	2.60	11.25	13.85
Insulators	"	28.25	22.75	51.00
Splice connectors	"	1.95	34.25	36.20
Tap connectors	"	2.14	34.25	36.39
Cable connectors	"	2.40	34.25	36.65
Terminal blocks	"	16.50	47.50	64.00
Tape	"			25.25
16120.47 Sheathed Cable				
Non-metallic sheathed cable				
Type NM cable with ground				
#14/2	L.F.	0.41	1.15	1.56
#12/2	"	0.63	1.23	1.86
#10/2	"	1.01	1.36	2.37
#8/2	"	1.65	1.54	3.19
#6/2	"	2.60	1.92	4.52
#14/3	"	0.58	1.98	2.56
#12/3	"	0.91	2.05	2.96
#10/3	"	1.44	2.08	3.52
#8/3	"	2.43	2.12	4.55
#6/3	"	3.92	2.16	6.08
#4/3	"	8.12	2.46	10.58
#2/3	"	12.25	2.68	14.93
Type U.F. cable with ground				
#14/2	L.F.	0.48	1.23	1.71
#12/2	"	0.72	1.46	2.18
#10/2	"	1.15	1.54	2.69
#8/2	"	1.98	1.76	3.74
#6/2	"	3.10	2.08	5.18
#14/3	"	0.67	1.54	2.21
#12/3	"	1.02	1.68	2.70
#10/3	"	1.59	1.92	3.51
#8/3	"	3.01	2.16	5.17
#6/3	"	4.87	2.46	7.33
Type S.F.U. cable, 3 conductor				
#8	L.F.	2.08	2.16	4.24
#6	"	3.62	2.37	5.99
#3	"	7.05	3.08	10.13
#2	"	8.76	3.42	12.18
#1	"	11.25	3.85	15.10
#1/0	"	14.00	4.25	18.25
#2/0	"	17.50	4.93	22.43
#3/0	"	22.00	5.36	27.36
#4/0	"	24.25	5.87	30.12
Type SER cable, 4 conductor				
#6	L.F.	5.18	2.80	7.98
#4	"	7.26	3.00	10.26
#3	"	9.80	3.42	13.22
#2	"	11.25	3.73	14.98
#1	"	14.25	4.25	18.50

16 ELECTRICAL

Basic Materials	UNIT	MAT.	INST.	TOTAL
16120.47 Sheathed Cable *(Cont.)*				
#1/0	L.F.	18.00	4.93	22.93
#2/0	"	22.50	5.13	27.63
#3/0	"	28.00	5.87	33.87
#4/0	"	35.25	6.48	41.73
Flexible cord, type STO cord				
#18/2	L.F.	0.88	0.30	1.18
#18/3	"	1.02	0.38	1.40
#18/4	"	1.43	0.46	1.89
#16/2	"	1.01	0.30	1.31
#16/3	"	0.85	0.34	1.19
#16/4	"	1.19	0.38	1.57
#14/2	"	1.58	0.38	1.96
#14/3	"	1.44	0.47	1.91
#14/4	"	1.79	0.53	2.32
#12/2	"	2.01	0.46	2.47
#12/3	"	1.52	0.51	2.03
#12/4	"	2.19	0.61	2.80
#10/2	"	2.49	0.53	3.02
#10/3	"	2.39	0.61	3.00
#10/4	"	3.70	0.69	4.39
#8/2	"	4.16	0.61	4.77
#8/3	"	4.61	0.68	5.29
#8/4	"	6.47	0.77	7.24
16130.10 Floor Boxes				
Adjustable floor boxes, steel	EA.	28.50	41.00	69.50
Cast bronze round	"	36.50	56.00	92.50
1 gang	"	42.25	62.00	104
2 gang	"	78.00	73.00	151
3 gang	"	110	77.00	187
Aluminum round	"	63.00	56.00	119
1 gang	"	48.00	62.00	110
2 gang	"	59.00	73.00	132
3 gang	"	70.00	77.00	147
Steel plate single recept	"	16.00	11.25	27.25
Duplex receptacle	"	15.25	14.00	29.25
Twist lock receptacle	"	16.25	14.00	30.25
Plug, 3/4"	"	20.25	11.25	31.50
1" plug	"	19.00	11.25	30.25
Carpet flange	"	24.75	11.25	36.00
Adjustable bronze plates for round cast boxes				
1/2" plug	EA.	8.32	11.25	19.57
3/4" plug	"	8.32	11.25	19.57
1" plug	"	10.50	11.25	21.75
1-1/4" plug	"	11.75	14.00	25.75
2" plug	"	15.50	15.50	31.00
Combination plug	"	18.25	15.50	33.75
Duplex receptacle plug	"	31.00	15.50	46.50
Adjustable aluminum plates for round cast boxes				
1/2" plug	EA.	24.75	11.25	36.00
3/4" plug	"	25.00	11.25	36.25
1" plug	"	25.75	11.25	37.00
1-1/4" plug	"	27.50	14.00	41.50

16 ELECTRICAL

Basic Materials	UNIT	MAT.	INST.	TOTAL
16130.10 **Floor Boxes** *(Cont.)*				
2" plug	EA.	28.00	15.50	43.50
Combination plug	"	25.00	15.50	40.50
Duplex receptacle plug	"	42.25	15.50	57.75
Adjustable bronze plates for gang type boxes				
1/2" plug	EA.	26.00	11.25	37.25
3/4" plug	"	26.25	11.25	37.50
1" plug	"	26.75	11.25	38.00
1-1/4" plug	"	27.75	14.00	41.75
2" plug	"	29.00	15.50	44.50
Carpet plate				
1 gang	EA.	23.00	11.25	34.25
2 gang	"	34.75	11.25	46.00
3 gang	"	46.50	15.50	62.00
Adjustable aluminum plates for gang type boxes				
1/2" plug	EA.	23.25	11.25	34.50
3/4" plug	"	23.75	11.25	35.00
1" plug	"	24.00	11.25	35.25
1-1/4" plug	"	27.50	14.00	41.50
2" plug	"	27.50	15.50	43.00
Duplex recept	"	23.50	15.50	39.00
Carpet plate				
1 gang	EA.	51.00	11.25	62.25
2 gang	"	71.00	11.25	82.25
3 gang	"	110	15.50	126
4 gang carpet plate	"	37.75	44.00	81.75
Telephone	"	34.50	38.50	73.00
Floor box nozzles, horizontal				
Duplex recept	EA.	59.00	41.00	100
Single recept	"	78.00	41.00	119
Double duplex recept	"	64.00	56.00	120
Vertical with duplex recept	"	52.00	47.50	99.50
Double duplex recept	"	55.00	56.00	111
Floor box bell nozzles split bell	"	18.75	19.25	38.00
One piece bell	"	59.00	19.25	78.25
Floor box standpipe				
1/2" x 3"	EA.	16.50	11.25	27.75
1/2" x 1"	"	16.25	11.25	27.50
Poke thru floor outlets				
2" floor	EA.	51.00	77.00	128
3" floor	"	55.00	92.00	147
4" floor	"	56.00	99.00	155
7" floor	"	59.00	120	179
9" floor	"	85.00	120	205
11" floor	"	85.00	140	225
13" floor	"	91.00	150	241
16130.40 **Boxes**				
Round cast box, type SEH				
1/2"	EA.	23.75	26.75	50.50
3/4"	"	23.75	32.50	56.25
SEHC				
1/2"	EA.	28.25	26.75	55.00
3/4"	"	28.25	32.50	60.75

16 ELECTRICAL

Basic Materials

16130.40 Boxes (Cont.)

	UNIT	MAT.	INST.	TOTAL
SEHL				
1/2"	EA.	29.00	26.75	55.75
3/4"	"	28.25	34.25	62.50
SEHT				
1/2"	EA.	31.25	32.50	63.75
3/4"	"	31.25	38.50	69.75
SEHX				
1/2"	EA.	33.50	38.50	72.00
3/4"	"	33.50	47.50	81.00
Blank cover	"	5.72	11.25	16.97
1/2", hub cover	"	5.46	11.25	16.71
Cover with gasket	"	5.98	13.75	19.73
Rectangle, type FS boxes				
1/2"	EA.	12.25	26.75	39.00
3/4"	"	13.00	30.75	43.75
1"	"	14.00	38.50	52.50
FSA				
1/2"	EA.	21.75	26.75	48.50
3/4"	"	20.25	30.75	51.00
FSC				
1/2"	EA.	13.50	26.75	40.25
3/4"	"	14.75	32.50	47.25
1"	"	18.50	38.50	57.00
FSL				
1/2"	EA.	21.75	26.75	48.50
3/4"	"	21.75	30.75	52.50
FSR				
1/2"	EA.	22.50	26.75	49.25
3/4"	"	23.00	30.75	53.75
FSS				
1/2"	EA.	13.50	26.75	40.25
3/4"	"	14.75	30.75	45.50
FSLA				
1/2"	EA.	9.25	26.75	36.00
3/4"	"	10.50	30.75	41.25
FSCA				
1/2"	EA.	27.25	26.75	54.00
3/4"	"	26.50	30.75	57.25
FSCC				
1/2"	EA.	16.50	30.75	47.25
3/4"	"	24.75	38.50	63.25
FSCT				
1/2"	EA.	16.50	30.75	47.25
3/4"	"	20.75	38.50	59.25
1"	"	16.75	44.00	60.75
FST				
1/2"	EA.	24.25	38.50	62.75
3/4"	"	24.25	44.00	68.25
FSX				
1/2"	EA.	27.75	47.50	75.25
3/4"	"	25.75	56.00	81.75
FSCD boxes				
1/2"	EA.	23.00	47.50	70.50

16 ELECTRICAL

Basic Materials

16130.40 Boxes (Cont.)

	UNIT	MAT.	INST.	TOTAL
3/4"	EA.	24.25	56.00	80.25
Rectangle, type FS, 2 gang boxes				
1/2"	EA.	26.00	26.75	52.75
3/4"	"	26.75	30.75	57.50
1"	"	28.25	38.50	66.75
FSC, 2 gang boxes				
1/2"	EA.	27.50	26.75	54.25
3/4"	"	30.50	30.75	61.25
1"	"	37.00	38.50	75.50
FSS, 2 gang boxes				
3/4"	EA.	28.75	30.75	59.50
FS, tandem boxes				
1/2"	EA.	28.75	30.75	59.50
3/4"	"	29.50	34.25	63.75
FSC, tandem boxes				
1/2"	EA.	38.75	30.75	69.50
3/4"	"	41.25	34.25	75.50
FS, three gang boxes				
3/4"	EA.	42.25	34.25	76.50
1"	"	46.50	38.50	85.00
FSS, three gang boxes, 3/4"	"	54.00	38.50	92.50
Weatherproof cast aluminum boxes, 1 gang, 3 outlets				
1/2"	EA.	7.73	30.75	38.48
3/4"	"	8.38	38.50	46.88
2 gang, 3 outlets				
1/2"	EA.	14.75	38.50	53.25
3/4"	"	15.75	41.00	56.75
1 gang, 4 outlets				
1/2"	EA.	13.50	47.50	61.00
3/4"	"	14.75	56.00	70.75
2 gang, 4 outlets				
1/2"	EA.	14.00	47.50	61.50
3/4"	"	15.75	56.00	71.75
1 gang, 5 outlets				
1/2"	EA.	11.25	56.00	67.25
3/4"	"	13.25	62.00	75.25
2 gang, 5 outlets				
1/2"	EA.	20.00	56.00	76.00
3/4"	"	24.50	62.00	86.50
2 gang, 6 outlets				
1/2"	EA.	22.75	66.00	88.75
3/4"	"	24.50	69.00	93.50
2 gang, 7 outlets				
1/2"	EA.	24.25	77.00	101
3/4"	"	30.25	84.00	114
Weatherproof and type FS box covers, blank, 1 gang	"	3.52	11.25	14.77
Tumbler switch, 1 gang	"	7.22	11.25	18.47
1 gang, single recept	"	4.55	11.25	15.80
Duplex recept	"	5.81	11.25	17.06
Despard	"	5.83	11.25	17.08
Red pilot light	"	27.50	11.25	38.75
SW and				
Single recept	EA.	12.25	15.50	27.75

16 ELECTRICAL

Basic Materials

16130.40 Boxes (Cont.)

	UNIT	MAT.	INST.	TOTAL
Duplex recept	EA.	10.00	15.50	25.50
2 gang				
Blank	EA.	3.66	14.00	17.66
Tumbler switch	"	4.81	14.00	18.81
Single recept	"	4.81	14.00	18.81
Duplex recept	"	4.81	14.00	18.81
3 gang				
Blank	EA.	8.38	15.50	23.88
Tumbler switch	"	10.50	15.50	26.00
4 gang				
Tumbler switch	EA.	13.25	19.25	32.50
Explosion proof boxes type E				
1/2"	EA.	47.50	26.75	74.25
3/4"	"	45.75	30.75	76.50
1"	"	47.25	38.50	85.75
1-1/4"	"	82.00	44.00	126
1-1/2"	"	190	47.50	238
Type L.B.				
1/2"	EA.	45.25	30.75	76.00
3/4"	"	48.00	38.50	86.50
1"	"	50.00	44.00	94.00
1-1/4"	"	88.00	51.00	139
1-1/2"	"	170	56.00	226
2"	"	200	62.00	262
Type C				
1/2"	EA.	40.50	30.75	71.25
3/4"	"	42.50	38.50	81.00
1"	"	43.75	44.00	87.75
1-1/4"	"	73.00	51.00	124
1-1/2"	"	160	56.00	216
2"	"	170	62.00	232
Type CA				
1/2"	EA.	39.50	44.00	83.50
3/4"	"	44.75	56.00	101
Type L				
1/2"	EA.	41.25	30.75	72.00
3/4"	"	43.75	38.50	82.25
1"	"	45.75	44.00	89.75
1-1/4"	"	77.00	51.00	128
1-1/2"	"	160	56.00	216
2"	"	160	62.00	222
Type N				
1/2"	EA.	41.75	30.75	72.50
3/4"	"	43.50	38.50	82.00
1"	"	44.75	47.50	92.25
1-1/4"	"	78.00	51.00	129
Type T				
1/2"	EA.	40.00	41.00	81.00
3/4"	"	42.75	56.00	98.75
1"	"	44.00	66.00	110
1-1/4"	"	75.00	77.00	152
1-1/2"	"	160	89.00	249
2"	"	160	99.00	259

16 ELECTRICAL

Basic Materials	UNIT	MAT.	INST.	TOTAL
16130.40 — **Boxes** *(Cont.)*				
Type TA				
1/2"	EA.	46.25	56.00	102
3/4"	"	48.75	62.00	111
Type X				
1/2"	EA.	41.75	56.00	97.75
3/4"	"	44.75	66.00	111
1"	"	49.00	77.00	126
1-1/4"	"	88.00	89.00	177
1-1/2"	"	170	99.00	269
2"	"	180	110	290
With union hubs				
1/2"	EA.	86.00	56.00	142
3/4"	"	90.00	62.00	152
Box covers				
Surface	EA.	18.50	15.50	34.00
Sealing	"	20.00	15.50	35.50
Dome	"	27.75	15.50	43.25
1/2" nipple	"	35.50	15.50	51.00
3/4" nipple	"	36.50	15.50	52.00
16130.45 — **Explosion Proof Fittings**				
Flexible couplings with female unions				
1/2" x 18"	EA.	280	15.50	296
3/4" x 18"	"	350	21.25	371
1" x 18"	"	640	26.75	667
1-1/4" x 18"	"	1,070	32.50	1,103
1-1/2" x 18"	"	1,300	38.50	1,339
2" x 18"	"	1,800	44.00	1,844
1/2" x 24"	"	370	19.25	389
3/4" x 24"	"	470	22.75	493
1" x 24"	"	850	30.75	881
1-1/4" x 24"	"	1,360	34.25	1,394
1-1/2" x 24"	"	1,730	44.00	1,774
2" x 24"	"	2,410	47.50	2,458
Female seal-offs				
1/2"	EA.	22.25	44.00	66.25
3/4"	"	26.25	51.00	77.25
1"	"	33.75	56.00	89.75
1-1/4"	"	41.00	66.00	107
1-1/2"	"	62.00	77.00	139
2"	"	81.00	89.00	170
2-1/2"	"	120	130	250
3"	"	160	170	330
4"	"	620	210	830
Conduit plugs				
1/2"	EA.	3.30	11.25	14.55
3/4"	"	3.76	11.25	15.01
1"	"	4.51	11.25	15.76
1-1/4"	"	4.93	19.25	24.18
1-1/2"	"	7.00	19.25	26.25
2"	"	12.00	22.75	34.75
2-1/2"	"	19.00	22.75	41.75
3"	"	27.00	26.75	53.75

16 ELECTRICAL

Basic Materials

16130.45 Explosion Proof Fittings (Cont.)

	UNIT	MAT.	INST.	TOTAL
4"	EA.	45.00	26.75	71.75
Sealing cement				
1 pound	EA.			17.50
5 pound	"			48.50
Fibre				
1 ounce	EA.			8.12
8 ounce	"			58.00
Male unions				
1/2"	EA.	15.00	15.50	30.50
3/4"	"	21.00	18.75	39.75
1"	"	37.25	21.25	58.50
1-1/4"	"	56.00	22.75	78.75
1-1/2"	"	73.00	26.75	99.75
2"	"	93.00	32.50	126
2-1/2"	"	150	38.50	189
3"	"	200	56.00	256
4"	"	240	69.00	309
Female unions				
1/2"	EA.	10.25	15.50	25.75
3/4"	"	14.25	18.75	33.00
1"	"	25.75	21.25	47.00
1-1/4"	"	38.75	22.75	61.50
1-1/2"	"	49.75	26.75	76.50
2"	"	64.00	32.50	96.50
2-1/2"	"	92.00	38.50	131
3"	"	130	56.00	186
4"	"	190	69.00	259
Male elbows				
1/2"	EA.	17.50	19.25	36.75
3/4"	"	19.75	22.75	42.50
1"	"	29.50	26.75	56.25
1-1/4"	"	33.75	34.25	68.00
Female elbows				
1/2"	EA.	14.75	19.25	34.00
3/4"	"	16.75	22.75	39.50
1"	"	22.75	26.75	49.50
1-1/4"	"	31.50	34.25	65.75
Pulling elbows				
1/2"	EA.	83.00	26.75	110
3/4"	"	89.00	34.25	123
1"	"	210	38.50	249
1-1/4"	"	250	47.50	298
1-1/2"	"	330	56.00	386
2"	"	350	150	500
2-1/2"	"	780	190	970
3"	"	740	230	970
3-1/2"	"	1,400	270	1,670
4"	"	1,420	320	1,740
Male expansion couplings				
1/2"	EA.	22.00	19.25	41.25
3/4"	"	28.75	22.75	51.50
1"	"	54.00	34.25	88.25
Female expansion couplings				

16 ELECTRICAL

Basic Materials	UNIT	MAT.	INST.	TOTAL
16130.45 Explosion Proof Fittings *(Cont.)*				
1/2"	EA.	20.00	19.25	39.25
3/4"	"	31.25	22.75	54.00
1"	"	54.00	34.25	88.25
16130.60 Pull And Junction Boxes				
4"				
Octagon box	EA.	4.33	8.80	13.13
Box extension	"	7.29	4.56	11.85
Plaster ring	"	4.00	4.56	8.56
Cover blank	"	1.77	4.56	6.33
Square box	"	6.23	8.80	15.03
Box extension	"	6.10	4.56	10.66
Plaster ring	"	3.34	4.56	7.90
Cover blank	"	1.71	4.56	6.27
4-11/16"				
Square box	EA.	12.50	8.80	21.30
Box extension	"	13.75	4.56	18.31
Plaster ring	"	8.32	4.56	12.88
Cover blank	"	3.08	4.56	7.64
Switch and device boxes				
2 gang	EA.	18.75	8.80	27.55
3 gang	"	33.25	8.80	42.05
4 gang	"	44.50	12.25	56.75
Device covers				
2 gang	EA.	15.00	4.56	19.56
3 gang	"	15.50	4.56	20.06
4 gang	"	21.00	4.56	25.56
Handy box	"	4.64	8.80	13.44
Extension	"	4.37	4.56	8.93
Switch cover	"	2.31	4.56	6.87
Switch box with knockout	"	6.97	11.25	18.22
Weatherproof cover, spring type	"	13.00	6.16	19.16
Cover plate, dryer receptacle 1 gang plastic	"	1.98	7.70	9.68
For 4" receptacle, 2 gang	"	3.53	7.70	11.23
Duplex receptacle cover plate, plastic	"	0.87	4.56	5.43
4", vertical bracket box, 1-1/2" with				
RMX clamps	EA.	8.98	11.25	20.23
BX clamps	"	9.63	11.25	20.88
4", octagon device cover				
1 switch	EA.	5.27	4.56	9.83
1 duplex recept	"	5.27	4.56	9.83
4", octagon swivel hanger box, 1/2" hub	"	14.00	4.56	18.56
3/4" hub	"	16.00	4.56	20.56
4" octagon adjustable bar hangers				
18-1/2"	EA.	6.54	3.85	10.39
26-1/2"	"	7.15	3.85	11.00
With clip				
18-1/2"	EA.	4.84	3.85	8.69
26-1/2"	"	5.43	3.85	9.28
4", square face bracket boxes, 1-1/2"				
RMX	EA.	10.75	11.25	22.00
BX	"	11.75	11.25	23.00
4" square to round plaster rings	"	3.57	4.56	8.13

16 ELECTRICAL

Basic Materials

16130.60 Pull And Junction Boxes *(Cont.)*

	UNIT	MAT.	INST.	TOTAL
2 gang device plaster rings	EA.	3.68	4.56	8.24
Surface covers				
1 gang switch	EA.	3.21	4.56	7.77
2 gang switch	"	3.28	4.56	7.84
1 single recept	"	4.84	4.56	9.40
1 20a twist lock recept	"	6.06	4.56	10.62
1 30a twist lock recept	"	7.76	4.56	12.32
1 duplex recept	"	3.00	4.56	7.56
2 duplex recept	"	3.00	4.56	7.56
Switch and duplex recept	"	5.00	4.56	9.56
4-11/16" square to round plaster rings	"	8.32	4.56	12.88
2 gang device plaster rings	"	6.86	4.56	11.42
Surface covers				
1 gang switch	EA.	9.23	4.56	13.79
2 gang switch	"	14.25	4.56	18.81
1 single recept	"	12.75	4.56	17.31
1 20a twist lock recept	"	12.50	4.56	17.06
1 30a twist lock recept	"	15.75	4.56	20.31
1 duplex recept	"	13.75	4.56	18.31
2 duplex recept	"	12.00	4.56	16.56
Switch and duplex recept	"	20.75	4.56	25.31
4" plastic round boxes, ground straps				
Box only	EA.	2.28	11.25	13.53
Box w/clamps	"	2.65	15.50	18.15
Box w/16" bar	"	5.64	17.50	23.14
Box w/24" bar	"	5.63	19.25	24.88
4" plastic round box covers				
Blank cover	EA.	1.49	4.56	6.05
Plaster ring	"	2.43	4.56	6.99
4" plastic square boxes				
Box only	EA.	1.76	11.25	13.01
Box w/clamps	"	2.18	15.50	17.68
Box w/hanger	"	2.68	19.25	21.93
Box w/nails and clamp	"	3.85	19.25	23.10
4" plastic square box covers				
Blank cover	EA.	1.44	4.56	6.00
1 gang ring	"	1.76	4.56	6.32
2 gang ring	"	2.46	4.56	7.02
Round ring	"	1.96	4.56	6.52

16130.65 Pull Boxes And Cabinets

	UNIT	MAT.	INST.	TOTAL
Galvanized pull boxes, screw cover				
4x4x4	EA.	9.88	14.75	24.63
4x6x4	"	11.75	14.75	26.50
6x6x4	"	15.00	14.75	29.75
6x8x4	"	17.75	14.75	32.50
8x8x4	"	22.00	19.25	41.25
8x10x4	"	25.50	18.75	44.25
8x12x4	"	28.00	19.25	47.25
Screw cover				
10x10x4	EA.	28.00	23.75	51.75
12x12x6	"	41.25	34.25	75.50
12x15x6	"	49.25	34.25	83.50

16 ELECTRICAL

Basic Materials

16130.65 Pull Boxes And Cabinets *(Cont.)*

Description	UNIT	MAT.	INST.	TOTAL
12x18x6	EA.	55.00	38.50	93.50
15x18x6	"	61.00	44.00	105
18x24x6	"	110	47.50	158
18x30x6	"	130	56.00	186
24x36x6	"	200	56.00	256
Cast iron junction box, unflanged				
6x6x4				
3/4" tap	EA.	93.00	38.50	132
1" tap	"	100	38.50	139
Two 1/2" taps	"	100	38.50	139
3/4" taps	"	100	38.50	139
6" adapter plate	"	40.75	26.75	67.50
6" exterior collar	"	66.00	26.75	92.75
Screw cover cabinet				
12x12x4	EA.	78.00	47.50	126
12x16x4	"	100	47.50	148
12x16x6	"	120	47.50	168
12x18x4	"	110	51.00	161
12x18x6	"	140	51.00	191
18x18x4	"	140	77.00	217
18x18x6	"	150	77.00	227
18x24x6	"	210	88.00	298
24x24x6	"	250	100	350
24x36x6	"	350	130	480
36x48x6	"	720	190	910
NEMA 3R, rain tight screw cover enclosures				
6x6x4	EA.	26.75	16.25	43.00
8x6x4	"	30.50	22.75	53.25
8x8x4	"	35.25	22.75	58.00
10x8x4	"	39.75	30.75	70.50
10x10x4	"	45.25	30.75	76.00
12x8x4	"	51.00	34.25	85.25
12x12x4	"	57.00	34.25	91.25
15x12x4	"	66.00	41.00	107
8x8x6	"	41.25	30.75	72.00
10x8x6	"	46.75	34.25	81.00
10x10x6	"	53.00	34.25	87.25
12x8x6	"	52.00	41.00	93.00
12x10x6	"	59.00	42.25	101
12x12x6	"	66.00	42.25	108
18x12x6	"	94.00	54.00	148

16130.80 Receptacles

Description	UNIT	MAT.	INST.	TOTAL
Contractor grade duplex receptacles, 15a 120v				
Duplex	EA.	1.90	15.50	17.40
125 volt, 20a, duplex, standard grade	"	14.00	15.50	29.50
Ground fault interrupter type	"	45.75	22.75	68.50
250 volt, 20a, 2 pole, single, ground type	"	23.50	15.50	39.00
120/208v, 4 pole, single receptacle, twist lock				
20a	EA.	28.00	26.75	54.75
50a	"	53.00	26.75	79.75
125/250v, 3 pole, flush receptacle				
30a	EA.	28.50	22.75	51.25

16 ELECTRICAL

Basic Materials

	UNIT	MAT.	INST.	TOTAL
16130.80 Receptacles *(Cont.)*				
50a	EA.	35.00	22.75	57.75
60a	"	91.00	26.75	118
277v, 20a, 2 pole, grounding type, twist lock	"	15.25	15.50	30.75
Dryer receptacle, 250v, 30a/50a, 3 wire	"	21.25	22.75	44.00
Clock receptacle, 2 pole, grounding type	"	14.00	15.50	29.50
125v, 20a single recept. grounding type				
Standard grade	EA.	15.25	15.50	30.75
Specification	"	18.50	15.50	34.00
Hospital	"	19.25	15.50	34.75
Isolated ground orange	"	65.00	19.25	84.25
Duplex				
Specification grade	EA.	15.25	15.50	30.75
Hospital	"	31.50	15.50	47.00
Isolated ground orange	"	65.00	19.25	84.25
250v, 20a, duplex, 2 pole, grounding, spec. grade	"	25.00	15.50	40.50
Combination recepts, 20a, 125v and 250v, duplex	"	33.25	15.50	48.75
GFI hospital grade recepts, 20a, 125v, duplex	"	69.00	22.75	91.75
125/250v, 3 pole, 3 wire surface recepts				
30a	EA.	24.00	22.75	46.75
50a	"	26.75	22.75	49.50
60a	"	59.00	26.75	85.75
Cord set, 3 wire, 6' cord				
30a	EA.	21.50	22.75	44.25
50a	"	30.25	22.75	53.00
125/250v, 3 pole, 3 wire cap				
30a	EA.	21.25	30.75	52.00
50a	"	38.75	30.75	69.50
60a	"	49.75	34.25	84.00
16198.10 Electric Manholes				
Precast, handhole, 4' deep				
2'x2'	EA.	520	270	790
3'x3'	"	690	430	1,120
4'x4'	"	1,480	790	2,270
Power manhole, complete, precast, 8' deep				
4'x4'	EA.	2,120	1,080	3,200
6'x6'	"	2,830	1,540	4,370
8'x8'	"	3,360	1,620	4,980
6' deep, 9' x 12'	"	3,710	1,930	5,640
Cast in place, power manhole, 8' deep				
4'x4'	EA.	2,510	1,080	3,590
6'x6'	"	3,240	1,540	4,780
8'x8'	"	3,600	1,620	5,220
16199.10 Utility Poles & Fittings				
Wood pole, creosoted				
25'	EA.	530	180	710
30'	"	640	230	870
35'	"	840	270	1,110
40'	"	1,010	290	1,300
45'	"	1,170	540	1,710
50'	"	1,380	560	1,940
55'	"	1,590	580	2,170

16 ELECTRICAL

Basic Materials

16199.10 Utility Poles & Fittings (Cont.)

	UNIT	MAT.	INST.	TOTAL
Treated, wood preservative, 6"x6"				
8'	EA.	110	38.50	149
10'	"	170	62.00	232
12'	"	170	68.00	238
14'	"	220	100	320
16'	"	260	120	380
18'	"	300	150	450
20'	"	380	150	530
Aluminum, brushed, no base				
8'	EA.	700	150	850
10'	"	810	210	1,020
15'	"	900	210	1,110
20'	"	1,090	250	1,340
25'	"	1,460	290	1,750
30'	"	2,180	340	2,520
35'	"	2,560	390	2,950
40'	"	3,290	480	3,770
Steel, no base				
10'	EA.	820	190	1,010
15'	"	900	230	1,130
20'	"	1,200	290	1,490
25'	"	1,350	350	1,700
30'	"	1,730	390	2,120
35'	"	2,010	480	2,490
Concrete, no base				
13'	EA.	1,050	430	1,480
16'	"	1,460	560	2,020
18'	"	1,760	680	2,440
25'	"	2,160	770	2,930
30'	"	2,870	930	3,800
35'	"	3,700	1,080	4,780
40'	"	4,310	1,230	5,540
45'	"	5,130	1,310	6,440
50'	"	6,360	1,400	7,760
55'	"	7,080	1,470	8,550
60'	"	8,110	1,540	9,650
Pole line hardware				
Wood crossarm				
4'	EA.	94.00	100	194
8'	"	190	130	320
10'	"	360	160	520
Angle steel brace				
1 piece	EA.	12.25	19.25	31.50
2 piece	"	26.50	26.75	53.25
Eye nut, 5/8"	"	3.04	3.85	6.89
Bolt (14-16"), 5/8"	"	22.75	15.50	38.25
Transformer, ground connection	"	7.21	19.25	26.46
Stirrup	"	16.00	23.75	39.75
Secondary lead support	"	24.50	30.75	55.25
Spool insulator	"	5.25	15.50	20.75
Guy grip, preformed				
7/16"	EA.	3.12	11.25	14.37
1/2"	"	5.39	11.25	16.64

16 ELECTRICAL

Basic Materials

16199.10 Utility Poles & Fittings (Cont.)

	UNIT	MAT.	INST.	TOTAL
Hook	EA.	3.90	19.25	23.15
Strain insulator	"	31.50	28.00	59.50
Wire				
5/16"	L.F.	1.74	0.38	2.12
7/16"	"	1.72	0.46	2.18
1/2"	"	4.16	0.61	4.77
Soft drawn ground, copper, #8	"	0.53	0.61	1.14
Ground clamp	EA.	4.34	23.75	28.09
Perforated strapping for conduit, 1-1/2"	L.F.	3.25	11.25	14.50
Hot line clamp	EA.	17.75	62.00	79.75
Lightning arrester				
3kv	EA.	570	77.00	647
10kv	"	880	120	1,000
30kv	"	1,610	150	1,760
36kv	"	3,280	190	3,470
Fittings				
Plastic molding	L.F.	3.82	11.25	15.07
Molding staples	EA.	0.91	3.85	4.76
Ground wires staples	"	0.39	2.32	2.71
Copper butt plate	"	1.09	22.75	23.84
Anchor bond clamp	"	4.78	11.25	16.03
Guy wire				
1/4"	L.F.	0.54	2.32	2.86
3/8"	"	0.84	3.85	4.69
Guy grip				
1/4"	EA.	9.29	3.85	13.14
3/8"	"	10.75	3.85	14.60

Power Generation

16210.10 Generators

	UNIT	MAT.	INST.	TOTAL
Diesel generator, with auto transfer switch				
30kw	EA.	32,120	2,370	34,490
50kw	"	40,680	2,370	43,050
75kw	"	51,650	3,240	54,890
100kw	"	57,320	3,630	60,950
125kw	"	61,260	3,850	65,110
150kw	"	70,620	4,400	75,020
175kw	"	73,590	5,140	78,730
200kw	"	76,340	6,160	82,500
250kw	"	83,420	6,850	90,270
300kw	"	100,330	7,700	108,030
350kw	"	107,420	8,810	116,230
400kw	"	131,870	10,270	142,140
450kw	"	141,250	11,210	152,460
500kw	"	153,010	12,330	165,340

16 ELECTRICAL

Power Generation

16210.10 Generators (Cont.)

	UNIT	MAT.	INST.	TOTAL
600kw	EA.	209,570	15,410	224,980
750kw	"	292,540	15,410	307,950

16230.10 Capacitors

	UNIT	MAT.	INST.	TOTAL
Three phase capacitors				
240v				
1.5 kvar	EA.	610	190	800
2.5 kvar	"	680	250	930
3.0 kvar	"	750	310	1,060
4 kvar	"	820	390	1,210
5 kvar	"	870	410	1,280
6 kvar	"	980	440	1,420
7.5 kvar	"	1,080	470	1,550
10 kvar	"	1,200	620	1,820
15 kvar	"	1,630	730	2,360
20 kvar	"	1,980	920	2,900
25 kvar	"	2,310	1,000	3,310
40 kvar	"	4,120	1,390	5,510
50 kvar	"	4,990	1,620	6,610
60 kvar	"	6,290	1,660	7,950
75 kvar	"	7,160	1,930	9,090
100 kvar	"	8,480	2,310	10,790
480v				
1.5 kvar	EA.	520	190	710
2.5 kvar	"	550	250	800
3 kvar	"	580	310	890
4 kvar	"	640	390	1,030
5 kvar	"	690	410	1,100
6 kvar	"	730	440	1,170
7.5 kvar	"	740	470	1,210
10 kvar	"	840	620	1,460
12.5 kvar	"	950	730	1,680
15 kvar	"	1,040	920	1,960
18 kvar	"	1,080	960	2,040
20 kvar	"	1,130	1,000	2,130
22.5 kvar	"	1,170	1,040	2,210
25 kvar	"	1,260	1,140	2,400
30 kvar	"	1,480	1,140	2,620
35 kvar	"	1,660	1,230	2,890
40 kvar	"	1,870	1,390	3,260
45 kvar	"	1,950	1,540	3,490
50 kvar	"	2,500	1,620	4,120
60 kvar	"	2,930	1,690	4,620
70 kvar	"	2,820	1,850	4,670
75 kvar	"	3,040	1,930	4,970
80 kvar	"	3,470	2,080	5,550
90 kvar	"	3,690	2,230	5,920
100 kvar	"	3,910	2,310	6,220
125 kvar	"	4,990	2,550	7,540
150 kvar	"	5,860	2,850	8,710

16 ELECTRICAL

Power Generation

16320.10 Transformers

Description	UNIT	MAT.	INST.	TOTAL
Floor mtd, one phase, int. dry, 480v-120/240v				
3 kva	EA.	740	140	880
5 kva	"	990	240	1,230
7.5 kva	"	1,340	270	1,610
10 kva	"	1,660	290	1,950
15 kva	"	2,230	330	2,560
25 kva	"	4,080	580	4,660
37.5 kva	"	4,360	730	5,090
50 kva	"	5,180	790	5,970
75 kva	"	6,820	820	7,640
100 kva	"	8,960	890	9,850
Three phase, 480v-120/208v				
15 kva	EA.	2,440	460	2,900
30 kva	"	2,930	730	3,660
45 kva	"	3,890	830	4,720
75 kva	"	5,850	840	6,690
112.5 kva	"	7,670	980	8,650
150 kva	"	9,130	1,040	10,170
225 kva	"	13,980	1,190	15,170
Single phase, dry type, 2400v				
167 kva	EA.	24,910	1,730	26,640
250 kva	"	32,480	2,310	34,790
333 kva	"	39,900	2,890	42,790
5000v				
167 kva	EA.	27,100	1,730	28,830
250 kva	"	33,550	2,310	35,860
333 kva	"	41,960	2,890	44,850
8660v				
167 kva	EA.	28,040	2,120	30,160
250 kva	"	37,910	2,690	40,600
333 kva	"	45,020	5,220	50,240
1500v				
167 kva	EA.	31,620	2,120	33,740
250 kva	"	40,940	2,690	43,630
333 kva	"	48,910	3,280	52,190
Three phase, dry type transformer, 2400v				
225 kva	EA.	32,810	1,930	34,740
300 kva	"	39,590	2,120	41,710
500 kva	"	46,240	3,280	49,520
750 kva	"	57,900	4,060	61,960
5000v				
225.0 kva	EA.	35,500	1,930	37,430
300 kva	"	44,010	2,120	46,130
500 kva	"	53,920	3,280	57,200
750 kva	"	74,260	4,060	78,320
8660v				
225.0 kva	EA.	42,080	2,310	44,390
300 kva	"	50,770	2,510	53,280
500 kva	"	59,290	3,670	62,960
750 kva	"	74,240	4,430	78,670
1500v				
225 kva	EA.	47,240	2,310	49,550
300 kva	"	55,430	2,510	57,940

16 ELECTRICAL

Power Generation

16320.10 Transformers (Cont.)

	UNIT	MAT.	INST.	TOTAL
500 kva	EA.	68,900	3,670	72,570
750 kva	"	84,340	4,430	88,770
Buck boost transformers				
.25 kva	EA.	210	77.00	287
.50 kva	"	280	96.00	376
.75 kva	"	360	120	480
1.00 kva	"	430	130	560
1.50 kva	"	550	150	700
2.00 kva	"	690	190	880
3.00 kva	"	920	230	1,150

16350.10 Circuit Breakers

	UNIT	MAT.	INST.	TOTAL
Molded case, 240v, 15-60a, bolt-on				
1 pole	EA.	19.50	19.25	38.75
2 pole	"	41.50	26.75	68.25
70-100a, 2 pole	"	120	41.00	161
15-60a, 3 pole	"	140	30.75	171
70-100a, 3 pole	"	240	47.50	288
480v, 2 pole				
15-60a	EA.	300	22.75	323
70-100a	"	390	30.75	421
3 pole				
15-60a	EA.	390	30.75	421
70-100a	"	460	34.25	494
70-225a	"	940	47.50	988
Draw out air circuit breakers				
600a	EA.	15,030	1,230	16,260
800a	"	19,420	1,400	20,820
1600a	"	31,190	1,870	33,060
2000a	"	41,790	2,130	43,920
3000a	"	72,530	2,470	75,000
4000a	"	111,400	2,940	114,340
Load center circuit breakers, 240v				
1 pole, 10-60a	EA.	19.50	19.25	38.75
2 pole				
10-60a	EA.	45.50	30.75	76.25
70-100a	"	140	51.00	191
110-150a	"	290	56.00	346
3 pole				
10-60a	EA.	130	38.50	169
70-100a	"	200	56.00	256
Load center, G.F.I. breakers, 240v				
1 pole, 15-30a	EA.	170	22.75	193
2 pole, 15-30a	"	300	30.75	331
Key operated breakers, 240v, 1 pole, 10-30a	"	110	22.75	133
Tandem breakers, 240v				
1 pole, 15-30a	EA.	37.00	30.75	67.75
2 pole, 15-30a	"	68.00	41.00	109
Bolt-on, G.F.I. breakers, 240v, 1 pole, 15-30a	"	160	26.75	187
Enclosed breaker, 120v, 1 pole, 15-50a, NEMA 1	"	180	62.00	242
240v, 2 pole				
15-60a, NEMA 1	EA.	260	96.00	356
70-100a, NEMA 1	"	330	130	460

16 ELECTRICAL

Power Generation

16350.10 Circuit Breakers (Cont.)

	UNIT	MAT.	INST.	TOTAL
3 pole				
15-60a, NEMA 1	EA.	330	120	450
70-100a, NEMA 1	"	430	170	600
Enclosed circuit breakers				
120v, 1 pole, NEMA 3R, 15-50a	EA.	350	69.00	419
240v, 2 pole, NEMA 3R				
15-60a	EA.	430	96.00	526
70-100a	"	490	130	620
3 pole, NEMA 3R				
15-60a	EA.	480	120	600
70-100a	"	580	170	750
480v, NEMA 1				
1 pole, 15-50a	EA.	220	62.00	282
2 pole, 15-60a	"	390	96.00	486
70-100a	"	480	120	600
3 pole, NEMA 1				
15-60a	EA.	470	120	590
70-100a	"	550	170	720
480v, 1 pole, 15-50a, NEMA 3R	"	360	77.00	437
2 pole				
2 pole, 15-60a, NEMA 3R	EA.	550	96.00	646
70-100a, NEMA 3R	"	620	130	750
3 pole				
15-60a, NEMA 3R	EA.	610	120	730
70-100a, NEMA 3R	"	690	170	860
70-100a, NEMA 1	"	520	130	650
3 pole				
15-60a, NEMA 1	EA.	520	130	650
70-100a, NEMA 1	"	620	170	790
Enclosed breakers, 600v, 2 phase, NEMA 3R				
15-60a	EA.	560	96.00	656
70-100a	"	690	130	820
3 phase, NEMA 3R				
15-60a	EA.	650	120	770
70-100a	"	740	170	910
600v, 3 phase, NEMA 1				
125a	EA.	910	170	1,080
150a	"	1,400	230	1,630
175a	"	1,400	230	1,630
200a	"	2,110	230	2,340
225a	"	2,110	230	2,340
250a	"	2,580	470	3,050
300a	"	2,580	470	3,050
350a	"	3,580	470	4,050
400a	"	3,580	470	4,050
500a	"	4,910	750	5,660
600a	"	4,910	750	5,660
700a	"	5,390	830	6,220
800a	"	6,800	830	7,630
900a	"	6,800	1,160	7,960
1000a	"	8,560	1,160	9,720
1200a	"	10,810	1,430	12,240
1400a	"	15,010	1,430	16,440

16 ELECTRICAL

Power Generation

	UNIT	MAT.	INST.	TOTAL
16350.10 Circuit Breakers *(Cont.)*				
1600a	EA.	15,010	1,850	16,860
1800a	"	21,060	2,310	23,370
2000a	"	22,460	2,310	24,770
600v, 3 phase, NEMA 3R				
125-225a	EA.	1,760	170	1,930
250-400a	"	3,290	440	3,730
500-600a	"	4,930	750	5,680
700-800a	"	6,570	850	7,420
900-1000a	"	7,460	1,160	8,620
1000-1200a	"	17,610	1,460	19,070
1400-1600a	"	17,830	1,850	19,680
1800-2000a	"	18,050	2,310	20,360
16360.10 Safety Switches				
Fused, 3 phase, 30 amp, 600v, heavy duty				
NEMA 1	EA.	310	88.00	398
NEMA 3r	"	710	88.00	798
NEMA 4	"	1,990	120	2,110
NEMA 12	"	630	130	760
60a				
NEMA 1	EA.	440	88.00	528
NEMA 3r	"	840	88.00	928
NEMA 4	"	2,190	120	2,310
NEMA 12	"	750	130	880
100a				
NEMA 1	EA.	740	130	870
NEMA 3r	"	1,300	130	1,430
NEMA 4	"	4,660	150	4,810
NEMA 12	"	1,130	190	1,320
200a				
NEMA 1	EA.	1,110	190	1,300
NEMA 3r	"	1,800	190	1,990
NEMA 4	"	6,120	210	6,330
NEMA 12	"	1,670	270	1,940
400a				
NEMA 1	EA.	2,680	430	3,110
NEMA 3r	"	5,160	430	5,590
NEMA 4	"	12,120	440	12,560
NEMA 12	"	3,950	540	4,490
600a				
NEMA 1	EA.	4,650	620	5,270
NEMA 3r	"	6,900	620	7,520
NEMA 4	"	11,190	690	11,880
NEMA 12	"	7,270	950	8,220
Non-fused, 240-600v, heavy duty, 3 phase, 30 amp				
NEMA 1	EA.	220	88.00	308
NEMA 3r	"	350	88.00	438
NEMA 4	"	1,400	130	1,530
NEMA 12	"	420	130	550
60a				
NEMA1	EA.	290	88.00	378
NEMA 3r	"	530	88.00	618
NEMA 4	"	1,520	130	1,650

16 ELECTRICAL

Power Generation

16360.10 Safety Switches *(Cont.)*

	UNIT	MAT.	INST.	TOTAL
NEMA 12	EA.	510	130	640
100a				
NEMA 1	EA.	470	130	600
NEMA 3r	"	740	130	870
NEMA 4	"	3,080	190	3,270
NEMA 12	"	720	190	910
200a, NEMA 1	"	730	190	920
600a, NEMA 12	"	4,050	950	5,000
Bolt-on hubs				
3/4" - 1-1/2"	EA.	22.75	19.25	42.00
2"	"	41.50	22.75	64.25
2-1/2"	"	65.00	22.75	87.75
3"	"	120	26.75	147
3-1/2"	"	180	30.75	211
4"	"	230	30.75	261
Watertight hubs				
1/2"	EA.	18.25	19.25	37.50
3/4"	"	27.25	22.75	50.00
1"	"	28.00	30.75	58.75
1-1/4"	"	32.25	34.25	66.50
1-1/2"	"	47.25	36.25	83.50
2"	"	70.00	38.50	109
2-1/2"	"	86.00	41.00	127
3"	"	110	47.50	158
3-1/2"	"	160	62.00	222
4"	"	230	66.00	296
Non-fused, 600v, 3 pole, NEMA 7				
600a	EA.	2,030	170	2,200
100a	"	2,540	250	2,790
225a	"	5,530	310	5,840
NEMA 9				
60a	EA.	1,710	190	1,900
100a	"	2,140	260	2,400
225a	"	4,790	320	5,110
Fusible bolted pressure switches, 600v/3 pole, NEMA 1				
800a	EA.	12,150	1,230	13,380
1200a	"	14,710	1,690	16,400
1600a	"	15,880	1,930	17,810
2000a	"	16,340	2,310	18,650
2500a	"	18,680	2,690	21,370
3000a	"	25,240	3,460	28,700
4000a	"	33,660	4,000	37,660
Non-fusible				
800a	EA.	11,670	1,120	12,790
1200a	"	13,080	1,540	14,620
1600a	"	14,020	1,770	15,790
2000a	"	14,850	2,160	17,010
2500a	"	17,050	2,690	19,740
3000a	"	24,300	3,460	27,760
4000a	"	32,720	4,000	36,720
Fusible load interrupter switches, 4.16 kv, NEMA 1				
200a	EA.	11,340	2,310	13,650
600a	"	14,300	5,410	19,710

16 ELECTRICAL

Power Generation

16360.10 Safety Switches (Cont.)

	UNIT	MAT.	INST.	TOTAL
Fusible load interrupter switch, 13.8 kv				
NEMA 1, 600a	EA.	15,860	7,700	23,560
NEMA 3R, 600a	"	19,030	7,700	26,730
4.16 kv, NEMA 3R				
200a	EA.	9,910	2,310	12,220
600a	"	17,850	5,410	23,260
Non-fused load interrupter switch, 4.16 kv, NEMA 1				
200a	EA.	7,920	2,310	10,230
600a	"	13,670	5,410	19,080
13.8 kv, NEMA 1, 600a	"	14,670	7,700	22,370
4.16 kv, NEMA 3R				
200a	EA.	8,330	2,310	10,640
600a	"	16,680	5,410	22,090
13.8 kv, NEMA 3R, 600a	"	17,850	7,700	25,550
Interrupter switch accessories, strip heater	"			1,080
Cable lugs	"			200
Key interlock	"			1,380
Auxiliary switch	"			770
Lightning arrester				
5 kva	EA.			7,360
15 kv	"			8,040

16365.10 Fuses

	UNIT	MAT.	INST.	TOTAL
Fuse, one-time, 250v				
30a	EA.	2.97	3.85	6.82
60a	"	5.03	3.85	8.88
100a	"	21.00	3.85	24.85
200a	"	51.00	3.85	54.85
400a	"	110	3.85	114
600a	"	200	3.85	204
600v				
30a	EA.	15.00	3.85	18.85
60a	"	23.75	3.85	27.60
100a	"	45.00	3.85	48.85
200a	"	120	3.85	124
400a	"	250	3.85	254
Fusetron, 600v				
200a	EA.	100	3.85	104
400a	"	210	3.85	214
Fuse, amp-trap, K1, 250v				
30a	EA.	9.68	3.85	13.53
60a	"	17.75	3.85	21.60
100a	"	39.75	3.85	43.60
200a	"	87.00	3.85	90.85
400a	"	200	3.85	204
600a	"	260	3.85	264
600v				
30a	EA.	29.00	3.85	32.85
60a	"	52.00	3.85	55.85
100a	"	100	3.85	104
200a	"	160	3.85	164
400a	"	320	3.85	324
K5, 250v				

16 ELECTRICAL

Power Generation

16365.10 Fuses (Cont.)

	UNIT	MAT.	INST.	TOTAL
30a	EA.	7.26	3.85	11.11
60a	"	13.25	3.85	17.10
100a	"	29.75	3.85	33.60
200a	"	66.00	3.85	69.85
400a	"	120	3.85	124
600a	"	190	3.85	194
600v				
30a	EA.	16.00	3.85	19.85
60a	"	27.50	3.85	31.35
100a	"	57.00	3.85	60.85
200a	"	110	3.85	114
400a	"	230	3.85	234
600a	"	330	3.85	334
J, 600v				
30a	EA.	26.25	3.85	30.10
60a	"	43.75	3.85	47.60
100a	"	77.00	3.85	80.85
200a	"	150	3.85	154
400a	"	320	3.85	324
L, 600v				
1200a	EA.	750	30.75	781
1600a	"	970	30.75	1,001
2000a	"	1,300	30.75	1,331
2500a	"	1,730	30.75	1,761
3000a	"	1,980	30.75	2,011
4000a	"	2,710	30.75	2,741
5000a	"	4,270	30.75	4,301
Fuse cl-ay 250v				
600a	EA.	630	22.75	653
1200a	"	630	22.75	653
1600a	"	780	22.75	803
2000a	"	1,010	22.75	1,033
600v				
1200a	EA.	750	22.75	773
1600a	"	870	22.75	893
2000a	"	1,050	22.75	1,073
Reducers, 600v				
60a-30a	EA.	16.00	11.25	27.25
100a-30a	"	56.00	11.25	67.25
100a-60a	"	36.00	11.25	47.25
200a-60a	"	140	19.25	159
200a-100a	"	53.00	19.25	72.25
400a-100a	"	250	26.75	277
400a-200a	"	220	26.75	247
600a-100a	"	340	30.75	371
600a-200a	"	380	30.75	411
600a-400a	"	340	30.75	371

16 ELECTRICAL

Power Generation

16395.10 Grounding

	UNIT	MAT.	INST.	TOTAL
Ground rods, copper clad, 1/2" x				
6'	EA.	15.75	51.00	66.75
8'	"	21.75	56.00	77.75
10'	"	27.25	77.00	104
5/8" x				
5'	EA.	19.75	47.50	67.25
6'	"	21.00	56.00	77.00
8'	"	27.25	77.00	104
10'	"	33.75	96.00	130
3/4" x				
8'	EA.	48.25	56.00	104
10'	"	53.00	62.00	115
Ground rod clamp				
5/8"	EA.	6.46	9.48	15.94
3/4"	"	9.13	9.48	18.61
Coupling, on threaded rods, 3/4"	"	19.00	3.85	22.85
Ground receptacles	"	24.00	19.25	43.25
Bus bar, copper, 2" x 1/4"	L.F.	6.96	11.25	18.21
Copper braid, 1" x 1/8", for door ground	EA.	5.40	7.70	13.10
Brazed connection for				
#6 wire	EA.	22.50	38.50	61.00
#2 wire	"	28.50	62.00	90.50
#2/0 wire	"	37.75	77.00	115
#4/0 wire	"	52.00	88.00	140
Ground rod couplings				
1/2"	EA.	11.75	7.70	19.45
5/8"	"	16.50	7.70	24.20
Ground rod, driving stud				
1/2"	EA.	9.49	7.70	17.19
5/8"	"	11.25	7.70	18.95
3/4"	"	12.75	7.70	20.45
Ground rod clamps, #8-2 to				
1" pipe	EA.	10.25	15.50	25.75
2" pipe	"	12.75	19.25	32.00
3" pipe	"	51.00	22.75	73.75
5" pipe	"	83.00	26.75	110
6" pipe	"	110	34.25	144
#4-4/0 to				
1" pipe	EA.	24.50	15.50	40.00
2" pipe	"	38.50	19.25	57.75
3" pipe	"	59.00	22.75	81.75
3" pipe	"	87.00	26.75	114
8 pipe	"	130	34.25	164
8 pipe	"	140	51.00	191
10 pipe	"	170	73.00	243
12 pipe	"	200	99.00	299

16 ELECTRICAL

Service And Distribution

16425.10 Switchboards

	UNIT	MAT.	INST.	TOTAL
Switchboard, 90" high, no main disconnect, 208/120v				
400a	EA.	3,110	610	3,720
600a	"	4,820	620	5,440
1000a	"	6,070	620	6,690
1200a	"	6,420	770	7,190
1600a	"	7,060	920	7,980
2000a	"	7,580	1,080	8,660
2500a	"	7,670	1,230	8,900
277/480v				
600a	EA.	5,540	630	6,170
800a	"	6,070	630	6,700
1600a	"	7,640	920	8,560
2000a	"	8,180	1,080	9,260
2500a	"	8,710	1,230	9,940
3000a	"	10,020	2,130	12,150
4000a	"	12,140	2,280	14,420
Main breaker sections, 600v				
1200a, GFI	EA.	23,000	1,310	24,310
1600a, GFI	"	27,160	1,500	28,660
2000a, GFI	"	28,830	1,540	30,370
2500a, GFI	"	31,870	1,930	33,800
3000a, GFI	"	42,120	2,310	44,430
4000a, GFI	"	38,300	2,690	40,990
Switchboard meter sections, 600v				
400a	EA.	4,150	620	4,770
600a	"	6,370	770	7,140
800a	"	7,750	850	8,600
1000a	"	9,420	1,040	10,460
2000a	"	11,640	1,230	12,870
2500a	"	13,020	1,540	14,560
3000a	"	13,580	1,930	15,510
4000a	"	15,250	2,310	17,560
Insulated case, draw out compartment, 208/120v				
800a	EA.	2,640	190	2,830
1600a	"	3,160	230	3,390
2000a	"	3,950	270	4,220
2500a	"	4,750	270	5,020
3000a	"	7,920	310	8,230
4000a	"	11,090	370	11,460
Accessories for power trip breakers				
Shunt trip	EA.	1,260	38.50	1,299
Key interlock	"	610	170	780
Lifting and transport truck	"	3,950	350	4,300
Lifting device	"	400	100	500
Bus duct connection, 3 phase, 4 wire				
225a	EA.	570	230	800
400a	"	710	230	940
600a	"	930	260	1,190
800a	"	1,010	310	1,320
2500a	"	1,880	460	2,340
3000a	"	2,840	580	3,420
4000a	"	5,090	680	5,770
Provision for mounting current transformers				

16 ELECTRICAL

Service And Distribution

16425.10 Switchboards *(Cont.)*

	UNIT	MAT.	INST.	TOTAL
800a & below primary	EA.	2,130	230	2,360
1000 to 1500a primary	"	2,650	230	2,880
2000 to 6000a primary	"	3,190	230	3,420
Provision for mounting potential transformers				
2000a max	EA.	7,620	290	7,910
Switchboard instruments				
Voltmeter	EA.	2,600	77.00	2,677
Ammeter, incoming line	"	2,470	77.00	2,547
Wattmeter	"	4,130	77.00	4,207
Varmeter	"	4,280	77.00	4,357
Power factor meter	"	4,940	77.00	5,017
Frequency meter	"	5,540	77.00	5,617
Recording voltmeter	"	11,050	150	11,200
Wattmeter	"	11,770	150	11,920
Power factor meter	"	13,730	150	13,880
Frequency meter	"	13,730	150	13,880
Instrument phase select switch	"	520	38.50	559
Enclosure, 90" high, 3 phase, 4 wire				
1000a	EA.	4,490	530	5,020
1200a	"	4,750	540	5,290
1600a	"	5,280	660	5,940
2000a	"	5,540	1,030	6,570
5500a	"	6,860	1,210	8,070
3000a	"	7,640	1,400	9,040
4000a	"	10,020	1,810	11,830
Circuit breakers, 600v, 100a, frame				
15-30a, 1 pole	EA.	170	22.75	193
15-60a, 2 pole	"	450	26.75	477
70-100a, 2 pole	"	560	30.75	591
15-60a, 3 pole	"	580	34.25	614
70-100a, 3 pole	"	670	38.50	709
Bolt on breakers, 600v, 225a frame, 110-225a				
2 pole	EA.	1,310	47.50	1,358
3 pole	"	1,640	84.00	1,724
400a frame, 250-400a, 2 pole	"	2,400	96.00	2,496
800a frame				
450-600a, 2 pole	EA.	4,230	150	4,380
700-800a, 2 pole	"	4,930	190	5,120
450-600a, 3 pole	"	4,840	320	5,160
700-800a, 3 pole	"	6,260	340	6,600
Bolt on branch breakers, 600v				
1000-2000a, 2 pole	EA.	8,780	410	9,190
2500a, 2 pole	"	16,200	830	17,030
1000-2000a, 3 pole	"	11,190	620	11,810
2500a, 3 pole	"	19,760	850	20,610
3000a, 3 pole	"	37,610	1,540	39,150
Metal clad substation switch board, selector switch				
600a, 5kv	EA.	29,030	3,240	32,270
15kv	"	32,200	3,690	35,890
Fused switch, 600a				
5kv	EA.	21,130	2,690	23,820
15kv	"	29,290	2,690	31,980
1200a				

16 ELECTRICAL

Service And Distribution

	UNIT	MAT.	INST.	TOTAL
16425.10 **Switchboards** *(Cont.)*				
5kv	EA.	23,490	3,080	26,570
15kv	"	31,470	3,080	34,550
Oil cutout switch				
5 kv	EA.	8,180	1,160	9,340
15 kv	"	11,350	1,390	12,740
Liquid air terminal section	"	1,050	620	1,670
Dry air terminal section	"	2,110	660	2,770
Auxiliary compartment	"	15,040	2,310	17,350
16430.20 **Metering**				
Outdoor wp meter sockets, 1 gang, 240v, 1 phase				
Includes sealing ring, 100a	EA.	56.00	120	176
150a	"	74.00	140	214
200a	"	95.00	150	245
Die cast hubs, 1-1/4"	"	8.65	24.75	33.40
1-1/2"	"	9.93	24.75	34.68
2"	"	12.00	24.75	36.75
Indoor meter center, main switch single phase, 240v				
400a	EA.	2,310	620	2,930
600a	"	4,050	850	4,900
800a	"	6,340	900	7,240
Main breaker				
400a	EA.	4,150	620	4,770
600a	"	5,590	850	6,440
800a	"	6,520	900	7,420
1000a	"	9,000	1,230	10,230
1200a	"	9,840	1,270	11,110
1600a	"	14,770	1,390	16,160
Terminal box				
800a	EA.	640	770	1,410
1600a	"	2,230	1,390	3,620
Main switch, three phase, 208v				
400a	EA.	2,370	660	3,030
600a	"	4,260	920	5,180
800a	"	8,560	1,040	9,600
Main breaker				
400a	EA.	4,140	660	4,800
600a	"	6,530	920	7,450
800a	"	8,670	1,040	9,710
1000a	"	11,100	1,310	12,410
1200a	"	15,460	1,390	16,850
1600a	"	22,770	1,620	24,390
Terminal box				
800a	EA.	740	1,000	1,740
1600a	"	2,350	1,620	3,970
Indoor meter center				
2 meters	EA.	680	390	1,070
3 meters	"	860	470	1,330
4 meters	"	1,110	560	1,670
5 meters	"	1,400	620	2,020
6 meters	"	1,790	690	2,480
Plug on breakers, single phase, 208v				
60a	EA.	38.00	19.25	57.25

16 ELECTRICAL

Service And Distribution

16430.20 Metering (Cont.)

	UNIT	MAT.	INST.	TOTAL
70a	EA.	76.00	19.25	95.25
80a	"	100	19.25	119
90a	"	110	19.25	129
100a	"	120	26.75	147
Indoor meter center, single phase, 125a breakers				
3 meters	EA.	980	470	1,450
4 meters	"	1,180	560	1,740
5 meters	"	1,470	620	2,090
6 meters	"	1,700	660	2,360
7 meters	"	2,160	770	2,930
8 meters	"	2,370	850	3,220
10 meters	"	2,950	920	3,870
150a breakers				
3 meters	EA.	3,350	470	3,820
4 meters	"	4,470	560	5,030
6 meters	"	6,700	620	7,320
7 meters	"	7,830	770	8,600
8 meters	"	8,950	850	9,800
200a breakers				
3 meters	EA.	2,760	470	3,230
4 meters	"	3,730	560	4,290
6 meters	"	5,540	620	6,160
7 meters	"	6,510	770	7,280
8 meters	"	7,470	850	8,320
Indoor meter center, three phase, 125a breakers				
3 meters	EA.	970	470	1,440
4 meters	"	1,160	560	1,720
5 meters	"	1,450	620	2,070
6 meters	"	1,660	690	2,350
7 meters	"	2,120	770	2,890
8 meters	"	2,340	850	3,190
10 meters	"	2,900	920	3,820
150a breakers				
3 meters	EA.	4,100	470	4,570
4 meters	"	5,470	560	6,030
6 meters	"	8,310	660	8,970
7 meters	"	9,590	850	10,440
8 meters	"	10,950	920	11,870
200a breakers				
3 meters	EA.	4,880	510	5,390
4 meters	"	6,520	560	7,080
6 meters	"	8,590	690	9,280
7 meters	"	10,040	850	10,890
8 meters	"	11,460	920	12,380
NEMA 3R, meter center, main switch, 1 phase, 240v				
400a	EA.	2,270	620	2,890
600a	"	4,400	770	5,170
800a	"	6,760	850	7,610
Main breaker				
400a	EA.	5,630	620	6,250
600a	"	7,190	770	7,960
800a	"	8,830	950	9,780
1000a	"	10,420	1,160	11,580

16 ELECTRICAL

Service And Distribution

16430.20 Metering *(Cont.)*

	UNIT	MAT.	INST.	TOTAL
1200a	EA.	14,700	1,230	15,930
Terminal box				
225a	EA.	550	560	1,110
800a	"	810	890	1,700
1600a	"	2,680	1,390	4,070
NEMA 3R, three phase, 280v				
400a	EA.	2,620	660	3,280
600a	"	4,950	920	5,870
800a	"	7,550	1,000	8,550
Main breaker				
400a	EA.	6,510	660	7,170
600a	"	8,360	920	9,280
800a	"	10,590	1,000	11,590
1000a	"	11,890	1,310	13,200
1200a	"	16,290	1,390	17,680
Terminal box				
225a	EA.	620	620	1,240
800a	"	910	1,000	1,910
1600a	"	2,790	1,620	4,410
NEMA 3R meter center, single phase, 208v, 100a				
2 meters	EA.	680	390	1,070
3 meters	"	810	470	1,280
4 meters	"	1,190	560	1,750
5 meters	"	1,430	620	2,050
6 meters	"	2,270	690	2,960
125a, 3 meters	"	1,800	470	2,270
4 meters	"	2,300	560	2,860
6 meters	"	2,870	630	3,500
7 meters	"	3,360	770	4,130
8 meters	"	4,110	850	4,960
150a, 3 meters	"	3,550	470	4,020
4 meters	"	4,750	560	5,310
6 meters	"	8,550	660	9,210
7 meters	"	9,990	770	10,760
8 meters	"	11,400	850	12,250
NEMA 3R center, 3 phase, 208v, 125a breakers				
3 meters	EA.	940	470	1,410
4 meters	"	1,280	560	1,840
6 meters	"	1,930	660	2,590
7 meters	"	2,250	770	3,020
8 meters	"	2,570	850	3,420
150a				
3 meters	EA.	4,250	510	4,760
4 meters	"	5,570	560	6,130
6 meters	"	8,340	690	9,030
7 meters	"	9,740	810	10,550
8 meters	"	11,140	890	12,030
200a				
3 meters	EA.	3,290	510	3,800
4 meters	"	4,390	560	4,950
6 meters	"	6,570	690	7,260
7 meters	"	7,690	850	8,540
8 meters	"	8,790	890	9,680

16 ELECTRICAL

Service And Distribution

16430.20 Metering (Cont.)

	UNIT	MAT.	INST.	TOTAL
NEMA 3R, center plug-on breakers, 208v, 1 phase				
60a	EA.	50.00	19.25	69.25
70a	"	100	19.25	119
90a	"	140	19.25	159
100a	"	150	26.75	177
125a	"	300	30.75	331

16460.10 Transformers

	UNIT	MAT.	INST.	TOTAL
Pad mounted, single phase, dry type, 480v-120/240v				
15 kva	EA.	2,250	620	2,870
25 kva	"	3,030	690	3,720
37.5 kva	"	4,420	770	5,190
50 kva	"	5,230	840	6,070
3 phase				
225 kva	EA.	13,100	1,930	15,030
300 kva	"	16,850	2,370	19,220
500 kva	"	26,520	2,940	29,460
750 kva	"	42,750	3,630	46,380
1000 kva	"	51,790	3,850	55,640
1500 kva	"	61,080	4,400	65,480
Substation transformers, outdoor, 5 kv - 208v				
112.5 kva	EA.	20,180	1,690	21,870
150 kva	"	21,940	1,850	23,790
225 kva	"	24,870	2,160	27,030
300 kva	"	28,660	2,310	30,970
500 kva	"	38,020	3,460	41,480
750 kva	"	52,660	4,250	56,910
1000 kva	"	64,940	5,010	69,950
15 kv, 208v				
112 kva	EA.	25,270	2,160	27,430
150 kva	"	25,840	2,310	28,150
225 kva	"	26,700	2,690	29,390
300 kva	"	30,100	3,080	33,180
500 kva	"	42,310	3,850	46,160
750 kva	"	42,310	4,630	46,940
1000 kva	"	63,610	5,410	69,020
5kv, 480v				
112kva	EA.	19,310	1,690	21,000
150 kva	"	20,450	1,850	22,300
225 kva	"	23,010	2,160	25,170
300 kva	"	25,840	2,310	28,150
500 kva	"	34,640	3,460	38,100
750 kva	"	47,150	4,250	51,400
1000 kva	"	55,380	5,010	60,390
1500 kva	"	72,420	5,760	78,180
2000 kva	"	89,170	6,930	96,100
2500 kva	"	105,930	8,440	114,370
15 kv, 480v				
112.5 kva	EA.	24,700	2,160	26,860
150 kva	"	25,270	2,310	27,580
225 kva	"	25,550	2,690	28,240
300 kva	"	27,830	3,080	30,910
500 kva	"	39,470	3,850	43,320

16 ELECTRICAL

Service And Distribution	UNIT	MAT.	INST.	TOTAL
16460.10 — **Transformers** *(Cont.)*				
750 kva	EA.	47,990	4,630	52,620
1000 kva	"	55,380	5,410	60,790
1500 kva	"	72,700	6,160	78,860
2000 kva	"	89,460	6,930	96,390
2500 kva	"	106,220	9,200	115,420
Pad mounted 3 phase, 15 kv outdoor				
50 kva	EA.	6,260	790	7,050
75 kva	"	8,790	910	9,700
112 kva	"	11,330	990	12,320
150 kva	"	16,090	1,120	17,210
225 kva	"	19,380	1,190	20,570
300 kva	"	23,850	1,310	25,160
500 kva	"	29,230	2,130	31,360
750 kva	"	37,270	2,800	40,070
1000 kva	"	44,130	3,420	47,550
1500 kva	"	56,360	4,110	60,470
Dry type, for power gear, 5 kv indoor				
75 kva	EA.	17,300	1,230	18,530
112.5 kva	"	19,380	1,430	20,810
150 kva	"	22,360	1,620	23,980
225 kva	"	27,730	1,810	29,540
300 kva	"	32,790	1,930	34,720
500 kva	"	43,330	2,130	45,460
750 kva	"	56,940	2,800	59,740
16470.10 — **Panelboards**				
Indoor load center, 1 phase 240v main lug only				
30a - 2 spaces	EA.	34.50	150	185
100a - 8 spaces	"	110	190	300
150a - 16 spaces	"	290	230	520
200a - 24 spaces	"	590	270	860
200a - 42 spaces	"	620	310	930
Main circuit breaker				
100a - 8 spaces	EA.	350	190	540
100a - 16 spaces	"	380	210	590
150a - 16 spaces	"	620	230	850
150a - 24 spaces	"	740	250	990
200a - 24 spaces	"	690	270	960
200a - 42 spaces	"	980	280	1,260
3 phase, 480/277v, main lugs only, 120a, 30 circuits	"	1,790	270	2,060
277/480v, 4 wire, flush surface				
225a, 30 circuits	EA.	2,930	310	3,240
400a, 30 circuits	"	3,960	390	4,350
600a, 42 circuits	"	7,640	460	8,100
208/120v, main circuit breaker, 3 phase, 4 wire				
100a				
12 circuits	EA.	1,550	390	1,940
20 circuits	"	1,930	490	2,420
30 circuits	"	2,830	540	3,370
225a				
30 circuits	EA.	2,420	600	3,020
42 circuits	"	4,380	730	5,110
400a				

16 ELECTRICAL

Service And Distribution

16470.10 Panelboards (Cont.)

	UNIT	MAT.	INST.	TOTAL
30 circuits	EA.	5,960	1,140	7,100
42 circuits	"	7,150	1,230	8,380
600a, 42 circuits	"	13,910	1,400	15,310
120/208v, flush, 3 ph., 4 wire, main only				
100a				
12 circuits	EA.	1,090	390	1,480
20 circuits	"	1,510	490	2,000
30 circuits	"	2,250	540	2,790
225a				
30 circuits	EA.	2,280	600	2,880
42 circuits	"	2,890	730	3,620
400a				
30 circuits	EA.	4,380	1,140	5,520
42 circuits	"	6,380	1,230	7,610
600a, 42 circuits	"	9,950	1,400	11,350
Panelboard accessories				
Grounding bus	EA.	51.00	26.75	77.75
Handle lock device	"	24.75	11.25	36.00
Factory assembled panel				
1 pole space	EA.	31.75	26.75	58.50
2 pole space	"	68.00	11.25	79.25
3 pole space	"	100	10.25	110
Panelboards 1 phase, 240/120v main circuit breaker				
Single phase, 3 wire, 120/240v flush				
100a, 20 circuits	EA.	1,600	270	1,870
225a, 30 circuits	"	3,060	310	3,370
240/120v, main lugs only				
100a				
8 circuits	EA.	770	230	1,000
12 circuits	"	830	230	1,060
20 circuits	"	870	230	1,100
225a				
24 circuits	EA.	920	270	1,190
30 circuits	"	1,060	290	1,350
42 circuits	"	1,170	290	1,460
Distribution panelboards, 3 ph, main breaker				
225a	EA.	2,930	1,230	4,160
400a	"	5,360	1,390	6,750
600a	"	8,360	1,690	10,050
800a	"	10,740	1,850	12,590
1000a	"	13,930	2,160	16,090
1200a	"	17,120	2,310	19,430
Single phase				
225a	EA.	2,590	1,080	3,670
400a	"	4,780	1,230	6,010
600a	"	6,760	1,540	8,300
800a	"	9,150	1,850	11,000
1000a	"	11,930	2,160	14,090
1200a	"	15,120	2,310	17,430
Fusible distribution panelboards, 3 phase, 600v				
100a	EA.	2,770	1,080	3,850
200a	"	3,170	1,230	4,400
400a	"	5,570	1,540	7,110

16 ELECTRICAL

Service And Distribution

16470.10 Panelboards (Cont.)

	UNIT	MAT.	INST.	TOTAL
600a	EA.	7,140	1,850	8,990
800a	"	10,740	2,160	12,900
Single phase				
100a	EA.	2,380	920	3,300
200a	"	2,770	1,080	3,850
400a	"	5,960	1,390	7,350
600a	"	6,360	1,690	8,050
800a	"	8,760	2,000	10,760
Hospital panels, operating room				
3kv - 208v	EA.	8,420	470	8,890
3kv - 277v	"	8,800	470	9,270
5kv - 208v	"	8,800	470	9,270
5kv - 277v	"	9,570	470	10,040
Coronary care				
3kv - 208v	EA.	9,950	560	10,510
3kv - 277v	"	10,720	560	11,280
5kv - 208v	"	10,720	560	11,280
5kv - 277v	"	11,480	560	12,040
Intensive care				
3kv - 208v	EA.	11,100	620	11,720
3kv - 277v	"	11,480	620	12,100
5kv - 208v	"	11,480	620	12,100
5kv - 277v	"	12,250	620	12,870
15kv - 208v	"	21,820	920	22,740
15kv - 277v	"	22,580	920	23,500
25kv - 208v	"	23,350	1,230	24,580
25kv - 277v	"	24,100	1,230	25,330
Explosion proof, 240v, m.l.b. 20a, single phase				
6 breakers	EA.	6,120	850	6,970
8 breakers	"	6,890	910	7,800
10 breakers	"	8,040	960	9,000
12 breakers	"	9,170	1,020	10,190
14 breakers	"	11,100	1,080	12,180
16 breakers	"	13,400	1,080	14,480
18 breakers	"	14,530	1,190	15,720
20 breakers	"	15,330	1,250	16,580
22 breakers	"	15,690	1,310	17,000
24 breakers	"	16,460	1,370	17,830

16480.10 Motor Controls

	UNIT	MAT.	INST.	TOTAL
Motor generator set, 3 phase, 480/277v, w/controls				
10kw	EA.	13,810	2,130	15,940
15kw	"	18,010	2,370	20,380
20kw	"	19,980	2,470	22,450
25kw	"	23,050	2,680	25,730
30kw	"	25,790	2,800	28,590
40kw	"	28,130	2,940	31,070
50kw	"	31,390	3,080	34,470
60kw	"	35,570	3,420	38,990
75kw	"	39,950	3,850	43,800
100kw	"	46,100	4,740	50,840
125kw	"	83,420	5,140	88,560
150kw	"	92,210	5,140	97,350

16 ELECTRICAL

Service And Distribution

16480.10 Motor Controls *(Cont.)*

	UNIT	MAT.	INST.	TOTAL
200kw	EA.	105,380	5,600	110,980
250kw	"	114,150	5,600	119,750
300kw	"	131,720	6,160	137,880
2 pole, 230 volt starter, w/NEMA-1				
1 hp, 9a, size 00	EA.	170	77.00	247
2 hp, 18a, size 0	"	190	77.00	267
3 hp, 27a, size 1	"	270	77.00	347
5 hp, 45a, size 1p	"	270	77.00	347
7-1/2 hp, 45a, size 2	"	660	77.00	737
15 hp, 90a, size 3	"	990	77.00	1,067
2 pole, w/NEMA-4 enclosure				
2 hp, 18a, size 1	EA.	560	120	680
5 hp, 45a, size 1p	"	720	120	840
7-1/2 hp, 45a, size 2	"	1,120	120	1,240
3 pole, 2 hp, 9a, 200-575v starter				
W/NEMA-1, size 00	EA.	350	100	450
W/NEMA-4 enclosure, size 00	"	560	130	690
5hp, 18a				
W/NEMA-1 enclosure, size 0	EA.	430	100	530
W/NEMA-4 enclosure, size 0	"	850	130	980
7.5-10hp, 27a				
7.5-10hp 27a, w/NEMA-1 enclosure, size 1	EA.	490	100	590
W/NEMA-4 enclosure size 1	"	920	130	1,050
10-25hp, 45a				
W/NEMA-1 enclosure, size 2	EA.	970	100	1,070
W/NEMA-4 enclosure, size 2	"	1,820	130	1,950
25-50hp, 90a				
W/NEMA-1 enclosure, size 3	EA.	1,520	130	1,650
W/NEMA-4 enclosure, size 3	"	1,430	190	1,620
40-100hp, 135a				
W/NEMA-1 enclosure, size 4	EA.	2,440	190	2,630
W/NEMA-4 enclosure, size 4	"	3,770	270	4,040
75-200hp, 270a				
W/NEMA-1 enclosure, size 5	EA.	5,760	430	6,190
W/NEMA-4 enclosure, size 5	"	7,310	540	7,850
Magnetic starter accessories				
On-off-auto selector switch kit	EA.	39.75	24.75	64.50
With pilot light	"	74.00	26.75	101
Control center main lug only, 208v, 3 phase				
600a	EA.	2,310	920	3,230
1200a	"	4,940	1,230	6,170
Main circuit breakers, 208v, 3 phase				
400a	EA.	4,610	770	5,380
600a	"	5,140	1,080	6,220
800a	"	5,900	1,230	7,130
1000a	"	6,420	1,390	7,810
1200a	"	11,800	1,540	13,340
Non-reversing starters				
Size 1	EA.	540	56.00	596
Size 2	"	1,060	96.00	1,156
Size 3	"	1,670	120	1,790
Size 4	"	3,430	130	3,560
Reversing starters				

16 ELECTRICAL

Service And Distribution

16480.10   Motor Controls *(Cont.)*

	UNIT	MAT.	INST.	TOTAL
Size 1	EA.	1,220	56.00	1,276
Size 2	"	2,830	84.00	2,914
Fusible switch, non-revolving starters				
Size 1	EA.	830	56.00	886
Size 2	"	1,040	96.00	1,136
Size 3	"	1,340	120	1,460
Size 4	"	2,140	130	2,270
Reversing starters				
Size 1	EA.	1,220	56.00	1,276
Size 2	"	1,290	84.00	1,374
Two speed, non-reversing starter				
Size 1	EA.	1,230	56.00	1,286
Size 2	"	1,800	84.00	1,884
Magnetic starter, 600v, 2 pole, NEMA 3R				
Size 0, 2 hp	EA.	310	77.00	387
Size 1, 5hp	"	410	84.00	494
NEMA 3R				
Size 2, 7.5 hp	EA.	860	88.00	948
Size 3, 15 hp	"	1,900	92.00	1,992
NEMA 7				
Size 0, 2 hp	EA.	1,570	130	1,700
Size 1, 5 hp	"	1,770	150	1,920
Size 2, 7.5 hp	"	3,070	170	3,240
NEMA 12				
Size 0, 2 hp	EA.	400	120	520
Size 1, 5 hp	"	530	130	660
Size 2, 7.5 hp	"	990	150	1,140
Size 3, 15 hp	"	1,570	170	1,740
3 pole, NEMA 1				
Size 6	EA.	17,160	770	17,930
Size 7	"	23,120	920	24,040
Size 8	"	32,090	1,230	33,320
NEMA 4				
Size 6	EA.	21,640	1,080	22,720
Size 7	"	27,600	1,230	28,830
Size 8	"	36,570	1,540	38,110
NEMA 3R				
Size 0	EA.	430	96.00	526
Size 1	"	490	100	590
Size 2	"	920	140	1,060
Size 3	"	1,430	150	1,580
Size 4	"	3,580	210	3,790
NEMA 7				
Size 0	EA.	1,750	150	1,900
Size 1	"	1,850	170	2,020
Size 2	"	2,960	170	3,130
Size 3	"	4,450	210	4,660
Size 4	"	7,440	340	7,780
Size 5	"	17,230	750	17,980
Size 6	"	40,570	1,230	41,800
NEMA 12				
Size 00	EA.	300	130	430
Size 0	"	330	140	470

16 ELECTRICAL

Service And Distribution

16480.10 Motor Controls *(Cont.)*

	UNIT	MAT.	INST.	TOTAL
Size 1	EA.	460	150	610
Size 2	"	830	150	980
Size 3	"	1,340	190	1,530
Size 4	"	3,070	270	3,340
Size 5	"	7,440	560	8,000
Size 6	"	17,910	960	18,870
Size 7	"	25,690	1,080	26,770
Size 8	"	37,870	1,540	39,410
Reversing magnetic starters, 600v, 3 pole, NEMA 1				
Size 00	EA.	680	96.00	776
Size 0	"	790	99.00	889
Size 1	"	890	100	990
Size 2	"	1,780	120	1,900
Size 3	"	3,080	130	3,210
Size 4	"	7,050	150	7,200
Size 5	"	15,220	410	15,630
Size 6	"	34,090	730	34,820
Size 7	"	46,500	850	47,350
Size 8	"	66,220	1,390	67,610
NEMA 4				
Size 0	EA.	1,140	130	1,270
Size 4	"	1,400	140	1,540
Size 2	"	2,500	150	2,650
Size 3	"	3,830	150	3,980
Size 4	"	8,710	190	8,900
Size 5	"	15,870	560	16,430
Size 6	"	35,610	960	36,570
Size 7	"	46,620	1,160	47,780
Size 8	"	64,560	1,540	66,100
NEMA 7				
Size 0	EA.	2,360	150	2,510
Size 1	"	2,460	170	2,630
Size 2	"	4,230	190	4,420
Size 3	"	6,650	230	6,880
NEMA 12				
Size 0	EA.	1,020	130	1,150
Size 1	"	1,150	150	1,300
Size 2	"	2,140	150	2,290
Size 3	"	3,830	170	4,000
Size 4	"	8,270	190	8,460
Size 5	"	18,250	560	18,810
Size 6	"	38,600	1,080	39,680
Size 7	"	51,550	1,230	52,780
Size 8	"	72,170	1,540	73,710
Electrically held lighting contactors, NEMA 1, 20a				
2 pole	EA.	410	77.00	487
3 pole	"	450	96.00	546
4 pole	"	570	120	690
6 pole	"	570	150	720
8 pole	"	740	190	930
10 pole	"	850	230	1,080
12 pole	"	990	270	1,260
30a				

16 ELECTRICAL

Service And Distribution

16480.10 Motor Controls (Cont.)

	UNIT	MAT.	INST.	TOTAL
2 pole	EA.	3,450	210	3,660
3 pole	"	3,690	230	3,920
4 pole	"	4,910	250	5,160
300a				
2 pole	EA.	6,510	320	6,830
3 pole	"	6,820	410	7,230
400a				
2 pole	EA.	13,280	320	13,600
3 pole	"	15,020	410	15,430
600a				
2 pole	EA.	16,340	510	16,850
3 pole	"	18,190	710	18,900
800a				
2 pole	EA.	19,510	620	20,130
3 pole	"	21,610	850	22,460
Mechanically held lighting contactors, NEMA 1, 20a				
2 pole	EA.	440	77.00	517
3 pole	"	470	96.00	566
4 pole	"	500	120	620
6 pole	"	820	150	970
8 pole	"	890	190	1,080
10 pole	"	1,000	230	1,230
30a				
2 pole	EA.	470	77.00	547
3 pole	"	500	96.00	596
4 pole	"	510	120	630
5 pole	"	650	130	780
60a				
2 pole	EA.	940	77.00	1,017
3 pole	"	990	96.00	1,086
4 pole	"	1,190	120	1,310
5 pole	"	1,530	130	1,660
100a				
2 pole	EA.	1,320	96.00	1,416
3 pole	"	1,390	130	1,520
4 pole	"	1,670	150	1,820
5 pole	"	2,250	190	2,440
200a				
2 pole	EA.	3,400	130	3,530
3 pole	"	3,850	190	4,040
4 pole	"	4,680	250	4,930
300a				
2 pole	EA.	6,060	320	6,380
3 pole	"	6,590	410	7,000
400a				
2 pole	EA.	14,490	320	14,810
3 pole	"	16,340	410	16,750
600a				
2 pole	EA.	17,400	510	17,910
3 pole	"	19,510	680	20,190
800a				
2 pole	EA.	20,570	620	21,190
3 pole	"	22,940	880	23,820

16 ELECTRICAL

Service And Distribution

	UNIT	MAT.	INST.	TOTAL
16480.10 Motor Controls *(Cont.)*				
AC relays, control type open, 15a, 600v				
2 pole	EA.	160	77.00	237
3 pole	"	170	96.00	266
4 pole	"	210	120	330
6 pole	"	260	150	410
8 pole	"	310	190	500
10 pole	"	500	230	730
12 pole	"	530	270	800
16490.10 Switches				
Oil switches, medium voltage, bus components				
Switches, 277/120v, toggle device only	EA.	600	120	720
With oil 35kv, g&w gram 44, 4 way switch	"	19,790	620	20,410
Weatherproof enclosure				
3 way switch	EA.	24,440	770	25,210
4 way switch	"	23,270	840	24,110
Fused interrupter load, 35kv				25,600
20A				
1 pole	EA.	27,940	1,230	29,170
2 pole	"	30,260	1,310	31,570
3 way	"	32,590	1,310	33,900
4 way	"	34,920	1,400	36,320
30a, 1 pole	"	27,940	1,230	29,170
3 way	"	32,590	1,310	33,900
4 way	"	34,920	1,400	36,320
Weatherproof switch, including box & cover, 20a				
1 pole	EA.	30,250	1,230	31,480
2 pole	"	32,590	1,310	33,900
3 way	"	34,920	1,400	36,320
4 way	"	37,260	1,400	38,660
3 way, oil switch, 15kv enclosure	"	25,600	920	26,520
Pedestal for 35kv double breaker switch	"	1,010	390	1,400
Bus terminal connector, 2	"	880	190	1,070
2 to 3	"	1,080	190	1,270
Support connector, 3	"	570	120	690
Tee connector, 2 to 3	"	820	150	970
Flexible bus stud connector	"	700	130	830
End cap 3	"	750	100	850
Weldment connection, 3	"	310	77.00	387
Plate switch, 1 gang	"	0.76	3.85	4.61
Start stop stations, manual motor starters	"	76.00	56.00	132
Lockout switch	"	12.75	19.25	32.00
Forward-reverse switch	"	94.00	56.00	150
On-off switch	"	95.00	56.00	151
Open-close switch	"	94.00	56.00	150
Forward-reverse-stop switch	"	150	77.00	227
Standard 3 button switch any standard legend	"	100	77.00	177
Standard 3 button with lockout	"	100	77.00	177
Manual motor starters, tog, 115/230v				
Size 1 gp	EA.	160	77.00	237
Size 2	"	210	77.00	287
Button				
Size 0	EA.	120	77.00	197

16 ELECTRICAL

Service And Distribution

16490.10 Switches (Cont.)

	UNIT	MAT.	INST.	TOTAL
Size 1	EA.	180	77.00	257
Size 2	"	220	77.00	297
3-phase				
Size 0	EA.	260	100	360
Size 1	"	320	100	420
Time & float switches	"	400	120	520
Astronomical time switch, 40a, 240v	"	620	77.00	697
Timer switch 0-5 minute, with box	"	30.50	38.50	69.00
Single pole/single throw time, 277v, NEMA-1	"	110	56.00	166
Single toggle switch, 20a, 120v, with pilot	"	27.25	19.25	46.50
3-way toggle	"	75.00	22.75	97.75
Photo electric switches				
1000 watt				
105-135v	EA.	39.75	56.00	95.75
208-277v	"	53.00	56.00	109
3000 watt, 105-130v				
Double throw	EA.	160	77.00	237
Single throw	"	140	77.00	217
Double pole/single throw, 210-250v	"	180	100	280
Dimmer switch and switch plate				
600w	EA.	36.50	23.75	60.25
1000w	"	60.00	26.75	86.75
Dimmer switch incandescent				
1500w	EA.	120	54.00	174
2000w	"	150	58.00	208
Fluorescent				
12 lamps	EA.	78.00	38.50	117
20 lamps	"	180	42.50	223
30 lamps	"	330	46.25	376
40 lamps	"	310	54.00	364
Time clocks with skip, 40a, 120v				
SPST	EA.	110	58.00	168
SPDT	"	240	58.00	298
DPST	"	160	58.00	218
DPDT	"	180	77.00	257
SPST	"	200	77.00	277
Astronomic time clocks with skip, 40a, 120v				
DPST	EA.	160	58.00	218
SPST	"	230	77.00	307
SPDT	"	170	58.00	228
Raintight time clocks, 40a, 120v				
SPDT	EA.	170	77.00	247
DPST	"	160	77.00	237
Contractor grade wall switch 15a, 120v				
Single pole	EA.	1.92	12.25	14.17
Three way	"	3.51	15.50	19.01
Four way	"	11.75	20.50	32.25
Specification grade toggle switches, 20a, 120-277v				
Single pole	EA.	4.22	15.50	19.72
Double pole	"	10.25	22.75	33.00
3 way	"	11.00	19.25	30.25
4 way	"	33.25	22.75	56.00
30a, 120-277v				

16 ELECTRICAL

Service And Distribution

16490.10 Switches (Cont.)

	UNIT	MAT.	INST.	TOTAL
Single pole	EA.	27.50	15.50	43.00
Double pole	"	38.25	22.75	61.00
3 way	"	38.25	19.25	57.50
Specification grade key switches, 20a, 120-277v				
Single pole	EA.	29.00	15.50	44.50
Double pole	"	36.75	22.75	59.50
3 way	"	31.50	19.25	50.75
4 way	"	61.00	22.75	83.75
Red pilot light handle switches, 20a, 120-277v				
Single pole	EA.	31.00	15.50	46.50
Double pole	"	36.00	22.75	58.75
3 way	"	55.00	19.25	74.25
30a, 120-277v				
Single pole	EA.	38.50	15.50	54.00
Double pole	"	47.25	22.75	70.00
3 way	"	72.00	19.25	91.25
Momentary contact switches, 20a				
SPDT, ivory	EA.	36.25	19.25	55.50
SPDT, locking	"	48.00	22.75	70.75
Maintained contact switches				
SPDT ivory	EA.	73.00	19.25	92.25
DPDT ivory	"	74.00	19.25	93.25
SPDT locking	"	86.00	22.75	109
DPDT locking	"	86.00	26.75	113
Mercury switch, 3 way	"	18.25	19.25	37.50
Door switches, open on or off	"	45.00	38.50	83.50
Combination switch and pilot light, single pole	"	14.50	22.75	37.25
3 way	"	18.00	26.75	44.75
Combination switch and receptacle, single pole	"	21.00	22.75	43.75
3 way	"	25.75	22.75	48.50
Combination two switches, single pole/single pole	"	17.25	19.25	36.50
3 way	"	21.25	30.75	52.00
Switch plates, plastic ivory				
1 gang	EA.	0.44	6.16	6.60
2 gang	"	1.04	7.70	8.74
3 gang	"	1.62	9.20	10.82
4 gang	"	4.16	11.25	15.41
5 gang	"	4.35	12.25	16.60
6 gang	"	5.13	14.00	19.13
Stainless steel				
1 gang	EA.	3.74	6.16	9.90
2 gang	"	5.20	7.70	12.90
3 gang	"	7.98	9.48	17.46
4 gang	"	13.75	11.25	25.00
5 gang	"	16.00	12.25	28.25
6 gang	"	20.00	14.00	34.00
Brass				
1 gang	EA.	6.98	6.16	13.14
2 gang	"	15.00	7.70	22.70
3 gang	"	23.25	9.48	32.73
4 gang	"	26.75	11.25	38.00
5 gang	"	33.00	12.25	45.25
6 gang	"	39.75	14.00	53.75

16 ELECTRICAL

Service And Distribution

	UNIT	MAT.	INST.	TOTAL
16490.20 — Transfer Switches				
Automatic transfer switch 600v, 3 pole				
30a	EA.	2,620	270	2,890
60a	"	3,170	270	3,440
100a	"	3,470	370	3,840
150a	"	4,630	460	5,090
225a	"	5,790	620	6,410
260a	"	6,320	620	6,940
400a	"	7,840	770	8,610
600a	"	11,400	1,160	12,560
800a	"	14,430	1,400	15,830
1000a	"	20,570	1,620	22,190
1200a	"	23,780	1,760	25,540
1600a	"	30,010	1,930	31,940
2000a	"	30,360	2,280	32,640
2600a	"	62,310	3,240	65,550
3000a	"	97,920	3,850	101,770
16490.80 — Safety Switches				
Safety switch, 600v, 3 pole, heavy duty, NEMA-1				
30a	EA.	230	77.00	307
60a	"	310	88.00	398
100a	"	600	120	720
200a	"	940	190	1,130
400a	"	2,410	430	2,840
600a	"	4,280	620	4,900
800a	"	9,680	810	10,490
1200a	"	12,030	1,100	13,130

Lighting

	UNIT	MAT.	INST.	TOTAL
16510.05 — Interior Lighting				
Recessed fluorescent fixtures, 2'x2'				
2 lamp	EA.	82.00	56.00	138
4 lamp	"	110	56.00	166
2 lamp w/flange	"	100	77.00	177
4 lamp w/flange	"	130	77.00	207
1'x4'				
2 lamp	EA.	84.00	51.00	135
3 lamp	"	110	51.00	161
2 lamp w/flange	"	100	56.00	156
3 lamp w/flange	"	140	56.00	196
2'x4'				
2 lamp	EA.	100	56.00	156
3 lamp	"	130	56.00	186
4 lamp	"	110	56.00	166
2 lamp w/flange	"	130	77.00	207

16 ELECTRICAL

Lighting

16510.05 Interior Lighting (Cont.)

	UNIT	MAT.	INST.	TOTAL
3 lamp w/flange	EA.	140	77.00	217
4 lamp w/flange	"	140	77.00	217
4'x4'				
4 lamp	EA.	410	77.00	487
6 lamp	"	490	77.00	567
8 lamp	"	530	77.00	607
4 lamp w/flange	"	510	120	630
6 lamp w/flange	"	630	120	750
8 lamp, w/flange	"	710	120	830
Surface mounted incandescent fixtures				
40w	EA.	120	51.00	171
75w	"	130	51.00	181
100w	"	140	51.00	191
150w	"	190	51.00	241
Pendant				
40w	EA.	100	62.00	162
75w	"	110	62.00	172
100w	"	130	62.00	192
150w	"	140	62.00	202
Contractor grade recessed down lights				
100 watt housing only	EA.	84.00	77.00	161
150 watt housing only	"	120	77.00	197
100 watt trim	"	69.00	38.50	108
150 watt trim	"	110	38.50	149
Recessed incandescent fixtures				
40w	EA.	170	120	290
75w	"	190	120	310
100w	"	200	120	320
150w	"	210	120	330
Exit lights, 120v				
Recessed	EA.	57.00	96.00	153
Back mount	"	93.00	56.00	149
Universal mount	"	97.00	56.00	153
Emergency battery units, 6v-120v, 50 unit	"	200	120	320
With 1 head	"	230	120	350
With 2 heads	"	260	120	380
Mounting bucket	"	40.00	56.00	96.00
Light track single circuit				
2'	EA.	50.00	38.50	88.50
4'	"	59.00	38.50	97.50
8'	"	81.00	77.00	158
12'	"	110	120	230
Fittings and accessories				
Dead end	EA.	19.75	11.25	31.00
Starter kit	"	26.50	19.25	45.75
Conduit feed	"	25.75	11.25	37.00
Straight connector	"	22.75	11.25	34.00
Center feed	"	36.50	11.25	47.75
L-connector	"	25.75	11.25	37.00
T-connector	"	34.50	11.25	45.75
X-connector	"	41.50	15.50	57.00
Cord and plug	"	41.75	7.70	49.45
Rigid corner	"	55.00	11.25	66.25

16 ELECTRICAL

Lighting

16510.05 Interior Lighting (Cont.)

	UNIT	MAT.	INST.	TOTAL
Flex connector	EA.	43.00	11.25	54.25
2 way connector	"	120	15.50	136
Spacer clip	"	1.85	3.85	5.70
Grid box	"	10.25	11.25	21.50
T-bar clip	"	2.77	3.85	6.62
Utility hook	"	8.03	11.25	19.28
Fixtures, square				
R-20	EA.	51.00	11.25	62.25
R-30	"	79.00	11.25	90.25
40w flood	"	130	11.25	141
40w spot	"	130	11.25	141
100w flood	"	150	11.25	161
100w spot	"	110	11.25	121
Mini spot	"	48.25	11.25	59.50
Mini flood	"	110	11.25	121
Quartz, 500w	"	280	11.25	291
R-20 sphere	"	85.00	11.25	96.25
R-30 sphere	"	45.00	11.25	56.25
R-20 cylinder	"	60.00	11.25	71.25
R-30 cylinder	"	70.00	11.25	81.25
R-40 cylinder	"	71.00	11.25	82.25
R-30 wall wash	"	110	11.25	121
R-40 wall wash	"	140	11.25	151
Explosion proof, incan., surface mounted				
100w - 200w	EA.	660	130	790
300w	"	930	130	1,060
500w	"	1,370	130	1,500
With guard				
100w-200w	EA.	630	170	800
300w	"	880	170	1,050
500w	"	1,290	170	1,460
Reflectors for incan. light fixtures, dome	"	110	19.25	129
Angle	"	130	19.25	149
Highbay	"	240	22.75	263
Explosion proof fluor. fixtures, 800 ms.				
1 lamp	EA.	2,570	170	2,740
2 lamp	"	3,180	210	3,390
3 lamp	"	4,760	230	4,990
4 lamp	"	6,160	250	6,410
Explosion proof hp sodium fixtures				
50w-70w	EA.	1,740	170	1,910
100w	"	1,780	170	1,950
150w	"	1,850	190	2,040
200w	"	1,870	190	2,060
250w	"	1,970	190	2,160
310w	"	2,030	190	2,220
400w	"	3,060	190	3,250
With guard				
50w-70w	EA.	1,760	190	1,950
100w	"	1,800	190	1,990
150w	"	1,930	210	2,140
200w	"	1,970	210	2,180
250w	"	2,030	210	2,240

16 ELECTRICAL

Lighting

16510.05 Interior Lighting (Cont.)

	UNIT	MAT.	INST.	TOTAL
310w	EA.	2,110	210	2,320
400w	"	3,180	210	3,390
Explosion proof metal halide fixtures				
175w	EA.	1,500	190	1,690
250w	"	1,640	190	1,830
400w	"	2,430	190	2,620
With guard, 175w	"	1,540	210	1,750
250w	"	1,680	210	1,890
400w	"	2,530	210	2,740
Energy saving rapid start fluor. lamps				
F30 cw	EA.	9.52	7.70	17.22
F40 cw	"	9.52	7.70	17.22
F40 cwx	"	10.75	7.70	18.45
F30 ww	"	14.50	7.70	22.20
F40 ww	"	7.85	7.70	15.55
F40 wwx	"	10.75	7.70	18.45
Slimline				
F48 cw	EA.	16.00	11.25	27.25
F96 cwx	"	28.50	11.25	39.75
F48 ww	"	21.25	11.25	32.50
F96 ww	"	19.50	11.25	30.75
F96 wwx	"	18.25	11.25	29.50
High output				
F96 cwx	"	12.00	11.25	23.25
F96 cw	"	17.75	11.25	29.00
Power groove, F48 cw	"	13.50	11.25	24.75
Circle	"	30.75	11.25	42.00
Fc6 cw	EA.	13.00	7.70	20.70
Fc8 cw	"	11.00	7.70	18.70
Fc12 cw	"	11.75	7.70	19.45
Fc16 cw	"	17.25	7.70	24.95
Fc6 ww	"	14.00	7.70	21.70
Fc8 ww	"	12.00	7.70	19.70
Fc12 ww	"	14.25	7.70	21.95
Fc16 ww	"	21.00	7.70	28.70
Incandescent lamps				
200w	EA.	6.24	7.70	13.94
300w	"	6.82	7.70	14.52
500w	"	15.25	7.70	22.95
750w	"	37.25	11.25	48.50
1000w	"	40.50	15.50	56.00
1500w	"	59.00	15.50	74.50
Energy saving reflector floodlight lamps				
25w	EA.	11.00	7.70	18.70
30w	"	11.75	7.70	19.45
50w	"	11.75	7.70	19.45
75w	"	13.50	7.70	21.20
120w	"	13.50	7.70	21.20
150w	"	18.75	7.70	26.45
200w	"	19.50	7.70	27.20
300w	"	26.25	7.70	33.95
500w	"	43.50	11.25	54.75
750w	"	57.00	11.25	68.25

16 ELECTRICAL

Lighting

16510.05 Interior Lighting (Cont.)

	UNIT	MAT.	INST.	TOTAL
Reflector spotlight				
75w	EA.	10.25	7.70	17.95
100w	"	12.25	7.70	19.95
125w	"	21.00	7.70	28.70
150w	"	22.25	7.70	29.95
250w	"	36.75	7.70	44.45
300w	"	58.00	7.70	65.70
400w	"	64.00	11.25	75.25
500w	"	80.00	11.25	91.25
1000w	"	170	15.50	186
Medium par flood lamps				
75w	EA.	10.00	7.70	17.70
100w	"	11.25	7.70	18.95
150w	"	12.00	7.70	19.70
200w	"	46.50	7.70	54.20
300w	"	52.00	7.70	59.70
500w	"	100	11.25	111
Medium par spot lamps				
75w	EA.	12.50	7.70	20.20
120w	"	12.75	7.70	20.45
150w	"	13.50	7.70	21.20
Tubular quartz lamps				
100w	EA.	83.00	11.25	94.25
150w	"	83.00	11.25	94.25
200w	"	98.00	11.25	109
400w	"	98.00	11.25	109
500w	"	100	15.50	116
750w	"	100	15.50	116
1000w	"	130	19.25	149
1250w	"	140	19.25	159
1500w	"	170	19.25	189
Ballast replacements rapid start fluor				
1f-40-120v	EA.	27.25	56.00	83.25
1f-40-277v	"	45.75	56.00	102
1f-96-120v	"	140	56.00	196
1f-96-277v	"	170	56.00	226
2f-40-120v	"	38.75	56.00	94.75
2f-40-277v	"	45.75	56.00	102
2f-96-120v	"	120	56.00	176
2f-96-277v	"	130	56.00	186
Circline, 1fc6-1fc16	"	35.75	56.00	91.75
Very high output, 1500ma				
1f48-120v	EA.	210	56.00	266
1f48-277v	"	210	56.00	266
1f96-120v	"	180	56.00	236
1f96-277v	"	210	56.00	266
2f48-120v	"	180	56.00	236
2f48-277v	"	210	56.00	266
2f96-120v	"	210	56.00	266
2f96-277v	"	230	56.00	286
Mercury, multi tap				
475w	EA.	170	77.00	247
100w	"	170	77.00	247

16 ELECTRICAL

Lighting

	UNIT	MAT.	INST.	TOTAL
16510.05 Interior Lighting *(Cont.)*				
175w	EA.	190	77.00	267
250w	"	230	77.00	307
400w	"	260	77.00	337
1000w	"	640	77.00	717
Metal halide, multi tap				
175w	EA.	150	77.00	227
250w	"	200	77.00	277
400w	"	240	77.00	317
1000w	"	440	77.00	517
1500w	"	570	77.00	647
High pressure sodium				
70w	EA.	260	77.00	337
100w	"	270	77.00	347
150w	"	290	77.00	367
250w	"	430	77.00	507
400w	"	490	77.00	567
1000w	"	670	77.00	747
16510.08 Energy Efficient Interior Lighting				
Ballast				
Fluorescent, 12 VDC				
Min.	EA.	120	51.00	171
Ave.	"	140	51.00	191
Max.	"	170	51.00	221
24 VDC				
Min.	EA.	120	51.00	171
Ave.	"	150	51.00	201
Max.	"	180	51.00	231
Pressure sodium, 12 VDC				
Min.	EA.	140	51.00	191
Ave.	"	160	51.00	211
Max.	"	180	51.00	231
24 VDC				
Min.	EA.	150	51.00	201
Ave.	"	170	51.00	221
Max.	"	180	51.00	231
Lamps				
Photovoltaic source, Fluorescent, 12 VDC, 7 Watt				
Min.	EA.	26.00	20.50	46.50
Ave.	"	31.25	20.50	51.75
Max.	"	36.50	20.50	57.00
11 Watt				
Min.	EA.	28.50	20.50	49.00
Ave.	"	33.75	20.50	54.25
Max.	"	39.00	20.50	59.50
15 Watt				
Min.	EA.	31.25	20.50	51.75
Ave.	"	36.50	20.50	57.00
Max.	"	41.50	20.50	62.00
25 Watt				
Min.	EA.	35.00	20.50	55.50
Ave.	"	39.00	20.50	59.50
Max.	"	45.50	20.50	66.00

16 ELECTRICAL

Lighting

	UNIT	MAT.	INST.	TOTAL
16510.08 Energy Efficient Interior Lighting *(Cont.)*				
30 Watt				
Min.	EA.	39.00	20.50	59.50
Ave.	"	44.25	20.50	64.75
Max.	"	49.50	20.50	70.00
LED, 85-265 V, 300 lumens				
Min.	EA.	56.00	20.50	76.50
Ave.	"	61.00	20.50	81.50
Max.	"	65.00	20.50	85.50
600 lumens				
Min.	EA.	140	20.50	161
Ave.	"	300	20.50	321
Max.	"	160	20.50	181
12-24 V, 2500 lumens				
Min.	EA.	570	51.00	621
Ave.	"	580	51.00	631
Max.	"	580	51.00	631
Compact fluorescent, 13 Watt, 900 lumens				
Min.	EA.	41.50	20.50	62.00
Ave.	"	46.75	20.50	67.25
Max.	"	52.00	20.50	72.50
36 Watt, 2800 lumens				
Min.	EA.	32.50	20.50	53.00
Ave.	"	36.50	20.50	57.00
Max.	"	39.00	20.50	59.50
40 Watt, 3150 lumens				
Min.	EA.	35.00	20.50	55.50
Ave.	"	39.00	20.50	59.50
Max.	"	41.50	20.50	62.00
Low Pressure Sodium, 18 Watt, 1800 lumens				
Min.	EA.	78.00	20.50	98.50
Ave.	"	85.00	20.50	106
Max.	"	91.00	20.50	112
35 Watt, 5000 lumens				
Min.	EA.	100	20.50	121
Ave.	"	110	20.50	131
Max.	"	110	20.50	131
40 Watt, 3150 lumens				
Min.	EA.	35.00	20.50	55.50
Ave.	"	39.00	20.50	59.50
Max.	"	41.50	20.50	62.00
16510.10 Lighting Industrial				
Surface mounted fluorescent, wrap around lens				
1 lamp	EA.	100	62.00	162
2 lamps	"	170	68.00	238
4 lamps	"	160	77.00	237
Wall mounted fluorescent				
2-20w lamps	EA.	100	38.50	139
2-30w lamps	"	120	38.50	159
2-40w lamps	"	120	51.00	171
Indirect, with wood shielding, 2049w lamps				
4'	EA.	120	77.00	197
8'	"	160	120	280

16 ELECTRICAL

Lighting

16510.10 Lighting Industrial (Cont.)

Description	UNIT	MAT.	INST.	TOTAL
Industrial fluorescent, 2 lamp				
4'	EA.	81.00	56.00	137
8'	"	130	100	230
Strip fluorescent				
4'				
1 lamp	EA.	51.00	51.00	102
2 lamps	"	62.00	51.00	113
8'				
1 lamp	EA.	75.00	56.00	131
2 lamps	"	110	68.00	178
Wire guard for strip fixture, 4' long	"	11.50	26.75	38.25
Strip fluorescent, 8' long, two 4' lamps	"	160	100	260
With four 4' lamps	"	200	120	320
Wet location fluorescent, plastic housing				
4' long				
1 lamp	EA.	170	77.00	247
2 lamps	"	180	100	280
8' long				
2 lamps	EA.	310	120	430
4 lamps	"	420	130	550
Parabolic troffer, 2'x2'				
With 2 "U" lamps	EA.	140	77.00	217
With 3 "U" lamps	"	170	88.00	258
2'x4'				
With 2 40w lamps	EA.	170	88.00	258
With 3 40w lamps	"	170	100	270
With 4 40w lamps	"	180	100	280
1'x4'				
With 1 T-12 lamp, 9 cell	EA.	88.00	56.00	144
With 2 T-12 lamps	"	99.00	68.00	167
With 1 T-12 lamp, 20 cell	"	99.00	56.00	155
With 2 T-12 lamps	"	110	68.00	178
Steel sided surface fluorescent, 2'x4'				
3 lamps	EA.	170	100	270
4 lamps	"	200	100	300
Outdoor sign fluor., 1 lamp, remote ballast				
4' long	EA.	3,560	460	4,020
6' long	"	4,280	620	4,900
Recess mounted, commercial, 2'x2', 13" high				
100w	EA.	1,180	310	1,490
250w	"	1,300	350	1,650
High pressure sodium, hi-bay open				
400w	EA.	510	130	640
1000w	"	880	190	1,070
Enclosed				
400w	EA.	830	190	1,020
1000w	"	1,160	230	1,390
Metal halide hi-bay, open				
400w	EA.	320	130	450
1000w	"	650	190	840
Enclosed				
400w	EA.	720	190	910
1000w	"	690	230	920

16 ELECTRICAL

Lighting

16510.10 Lighting Industrial (Cont.)

	UNIT	MAT.	INST.	TOTAL
High pressure sodium, low bay, surface mounted				
100w	EA.	260	77.00	337
150w	"	290	88.00	378
250w	"	320	100	420
400w	"	410	120	530
Metal halide, low bay, pendant mounted				
175w	EA.	420	100	520
250w	"	570	120	690
400w	"	620	170	790
Indirect luminare, square, metal halide, freestanding				
175w	EA.	510	77.00	587
250w	"	560	77.00	637
400w	"	580	77.00	657
High pressure sodium				
150w	EA.	940	77.00	1,017
250w	"	1,000	77.00	1,077
400w	"	1,110	77.00	1,187
Round, metal halide				
175w	EA.	1,080	77.00	1,157
250w	"	1,120	77.00	1,197
400w	"	1,170	77.00	1,247
High pressure sodium				
150w	EA.	1,030	77.00	1,107
250w	"	1,200	77.00	1,277
400w	"	1,250	77.00	1,327
Wall mounted, metal halide				
175w	EA.	470	190	660
250w	"	460	190	650
400w	"	510	250	760
High pressure sodium				
150w	EA.	440	190	630
250w	"	460	190	650
400w	"	480	250	730
Wall pack lithonia, high pressure sodium				
35w	EA.	67.00	68.00	135
55w	"	85.00	77.00	162
150w	"	220	120	340
250w	"	230	130	360
Low pressure sodium				
35w	EA.	360	130	490
55w	"	490	150	640
Wall pack hubbell, high pressure sodium				
35w	EA.	300	68.00	368
150w	"	380	120	500
250w	"	480	130	610
Compact fluorescent				
2-7w	EA.	180	77.00	257
2-13w	"	210	100	310
1-18w	"	250	100	350
Handball & racquet ball court, 2'x2', metal halide				
250w	EA.	650	190	840
400w	"	780	210	990
High pressure sodium				

16 ELECTRICAL

Lighting

16510.10 Lighting Industrial (Cont.)

	UNIT	MAT.	INST.	TOTAL
250w	EA.	720	190	910
400w	"	780	210	990
Bollard light, 42" w/found., high pressure sodium				
70w	EA.	1,090	200	1,290
100w	"	1,120	200	1,320
150w	"	1,130	200	1,330
Light fixture lamps				
Lamp				
20w med. bipin base, cool white, 24"	EA.	8.76	11.25	20.01
30w cool white, rapid start, 36"	"	11.00	11.25	22.25
40w cool white "U", 3"	"	24.25	11.25	35.50
40w cool white, rapid start, 48"	"	10.25	11.25	21.50
70w high pressure sodium, mogul base	"	80.00	15.50	95.50
75w slimline, 96"	"	23.75	15.50	39.25
100w				
Incandescent, 100a, inside frost	EA.	4.16	7.70	11.86
Mercury vapor, clear, mogul base	"	72.00	15.50	87.50
High pressure sodium, mogul base	"	110	15.50	126
150w				
Par 38 flood or spot, incandescent	EA.	24.50	7.70	32.20
High pressure sodium, 1/2 mogul base	"	98.00	15.50	114
175w				
Mercury vapor, clear, mogul base	EA.	44.00	15.50	59.50
Metal halide, clear, mogul base	"	88.00	15.50	104
High pressure sodium, mogul base	"	98.00	15.50	114
250w				
Mercury vapor, clear, mogul base	EA.	61.00	15.50	76.50
Metal halide, clear, mogul base	"	88.00	15.50	104
High pressure sodium, mogul base	"	100	15.50	116
400w				
Mercury vapor, clear, mogul base	EA.	67.00	15.50	82.50
Metal halide, clear, mogul base	"	88.00	15.50	104
High pressure sodium, mogul base	"	110	15.50	126
1000w				
Mercury vapor, clear, mogul base	EA.	160	19.25	179
High pressure sodium, mogul base	"	280	19.25	299

16510.30 Exterior Lighting

	UNIT	MAT.	INST.	TOTAL
Exterior light fixtures				
Rectangle, high pressure sodium				
70w	EA.	350	190	540
100w	"	360	200	560
150w	"	390	200	590
250w	"	530	210	740
400w	"	590	270	860
Flood, rectangular, high pressure sodium				
70w	EA.	260	190	450
100w	"	300	200	500
150w	"	280	200	480
400w	"	350	270	620
1000w	"	580	350	930
Round				
400w	EA.	650	270	920

16 ELECTRICAL

Lighting	UNIT	MAT.	INST.	TOTAL
16510.30 Exterior Lighting (Cont.)				
1000w	EA.	1,010	350	1,360
Round, metal halide				
400w	EA.	720	270	990
1000w	"	1,070	350	1,420
Light fixture arms, cobra head, 6', high press. sodium				
100w	EA.	390	150	540
150w	"	620	190	810
250w	"	640	190	830
400w	"	660	230	890
Flood, metal halide				
400w	EA.	640	270	910
1000w	"	880	350	1,230
1500w	"	1,110	460	1,570
Mercury vapor				
250w	EA.	420	210	630
400w	"	480	270	750
Incandescent				
300w	EA.	98.00	130	228
500w	"	180	150	330
1000w	"	200	250	450
16510.40 Energy Efficient Exterior Lighting				
Solar Powered, led area light, 100 Watt, Zone 4				
Min.	EA.	1,280	100	1,380
Ave.	"	1,430	120	1,550
Max.	"	1,560	150	1,710
Zone 2				
Min.	EA.	1,680	100	1,780
Ave.	"	1,750	120	1,870
Max.	"	1,820	150	1,970
Zone 4DD				
Min.	EA.	1,930	100	2,030
Ave.	"	2,070	120	2,190
Max.	"	2,210	150	2,360
Zone 2DD				
Min.	EA.	2,140	100	2,240
Ave.	"	2,300	120	2,420
Max.	"	2,470	150	2,620
16510.90 Power Line Filters				
Heavy duty power line filter, 240v				
100a	EA.	5,330	770	6,100
300a	"	17,690	1,230	18,920
600a	"	24,520	1,870	26,390
16600.20 Central Inverter Systems				
Central inverter systems				
500va	EA.	11,620	230	11,850
1000va	"	12,700	310	13,010
1500va	"	14,860	410	15,270
2400va	"	19,160	510	19,670
3000va	"	20,240	660	20,900

16 ELECTRICAL

Lighting

	UNIT	MAT.	INST.	TOTAL
16600.20 Central Inverter Systems *(Cont.)*				
4500va	EA.	30,800	770	31,570
6000va	"	38,770	850	39,620
7500va	"	44,800	1,080	45,880
10,000va	"	51,690	1,230	52,920
16,600va	"	89,600	1,770	91,370
25,000va	"	106,820	2,690	109,510
16610.30 Uninterruptible Power				
Uninterruptible power systems, (U.P.S.), 3kva	EA.	8,310	620	8,930
5 kva	"	9,320	850	10,170
7.5 kva	"	11,180	1,230	12,410
10 kva	"	13,980	1,690	15,670
15 kva	"	16,770	1,760	18,530
20 kva	"	23,290	1,850	25,140
25 kva	"	29,820	1,930	31,750
30 kva	"	30,750	2,000	32,750
35 kva	"	32,610	2,080	34,690
40 kva	"	35,410	2,160	37,570
45 kva	"	37,270	2,230	39,500
50 kva	"	40,070	2,310	42,380
62.5 kva	"	47,520	2,470	49,990
75 kva	"	54,970	2,690	57,660
100 kva	"	73,610	2,780	76,390
150 kva	"	111,820	3,850	115,670
200 kva	"	149,090	4,250	153,340
300 kva	"	223,630	5,760	229,390
400 kva	"	334,550	6,930	341,480
500 kva	"	418,190	8,440	426,630
16670.10 Lightning Protection				
Lightning protection				
Copper point, nickel plated, 12'				
1/2" dia.	EA.	52.00	77.00	129
5/8" dia.	"	59.00	77.00	136

Communications

	UNIT	MAT.	INST.	TOTAL
16720.10 Fire Alarm Systems				
Master fire alarm box, pedestal mounted	EA.	8,400	1,230	9,630
Master fire alarm box	"	4,320	460	4,780
Box light	"	150	38.50	189
Ground assembly for box	"	120	51.00	171
Bracket for pole type box	"	160	56.00	216
Pull station				
Waterproof	EA.	77.00	38.50	116
Manual	"	57.00	30.75	87.75

16 ELECTRICAL

Communications	UNIT	MAT.	INST.	TOTAL
16720.10 **Fire Alarm Systems** (Cont.)				
Horn, waterproof	EA.	110	77.00	187
Interior alarm	"	72.00	56.00	128
Coded transmitter, automatic	"	1,110	150	1,260
Control panel, 8 zone	"	2,540	620	3,160
Battery charger and cabinet	"	840	150	990
Batteries, nickel cadmium or lead calcium	"	640	390	1,030
CO2 pressure switch connection	"	120	56.00	176
Annunciator panels				
Fire detection annunciator, remote type, 8 zone	EA.	430	140	570
12 zone	"	550	150	700
16 zone	"	690	190	880
Fire alarm systems				
Bell	EA.	130	47.50	178
Weatherproof bell	"	84.00	51.00	135
Horn	"	77.00	56.00	133
Siren	"	780	150	930
Chime	"	96.00	47.50	144
Audio/visual	"	140	56.00	196
Strobe light	"	130	56.00	186
Smoke detector	"	210	51.00	261
Heat detection	"	36.00	38.50	74.50
Thermal detector	"	33.50	38.50	72.00
Ionization detector	"	170	41.00	211
Duct detector	"	560	210	770
Test switch	"	96.00	38.50	135
Remote indicator	"	60.00	44.00	104
Door holder	"	210	56.00	266
Telephone jack	"	3.90	22.75	26.65
Fireman phone	"	510	77.00	587
Speaker	"	100	62.00	162
Remote fire alarm annunciator panel				
24 zone	EA.	2,640	510	3,150
48 zone	"	5,280	1,000	6,280
Control panel				
12 zone	EA.	1,790	230	2,020
16 zone	"	2,350	340	2,690
24 zone	"	3,600	510	4,110
48 zone	"	6,720	1,230	7,950
Power supply	"	420	120	540
Status command	"	11,390	390	11,780
Printer	"	3,630	120	3,750
Transponder	"	180	69.00	249
Transformer	"	260	51.00	311
Transceiver	"	360	56.00	416
Relays	"	140	38.50	179
Flow switch	"	470	150	620
Tamper switch	"	290	230	520
End of line resistor	"	20.25	26.75	47.00
Printed ckt. card	"	180	38.50	219
Central processing unit	"	12,450	470	12,920
UPS backup to c.p.u.	"	22,800	690	23,490
Smoke detector, fixed temp. & rate of rise comb.	"	360	120	480

16 ELECTRICAL

Communications

16720.50 Security Systems

	UNIT	MAT.	INST.	TOTAL
Sensors				
Balanced magnetic door switch, surface mounted	EA.	180	38.50	219
With remote test	"	230	77.00	307
Flush mounted	"	170	140	310
Mounted bracket	"	13.00	26.75	39.75
Mounted bracket spacer	"	11.75	26.75	38.50
Photoelectric sensor, for fence				
6 beam	EA.	18,980	210	19,190
9 beam	"	23,210	330	23,540
Photoelectric sensor, 12 volt dc				
500' range	EA.	560	120	680
800' range	"	620	150	770
Capacitance wire grid kit				
Surface	EA.	160	77.00	237
Duct	"	120	120	240
Tube grid kit	"	200	38.50	239
Vibration sensor, 30 max per zone	"	230	38.50	269
Audio sensor, 30 max per zone	"	250	38.50	289
Inertia sensor				
Outdoor	EA.	180	56.00	236
Indoor	"	120	38.50	159
Ultrasonic transmitter, 20 max per zone				
Omni-directional	EA.	130	120	250
Directional	"	140	100	240
Transceiver				
Omni-directional	EA.	140	77.00	217
Directional	"	160	77.00	237
Passive infra-red sensor, 20 max per zone	"	1,000	120	1,120
Access/secure unit, balanced magnetic switch	"	620	120	740
Photoelectric sensor	"	1,030	120	1,150
Photoelectric fence sensor	"	1,050	120	1,170
Capacitance sensor	"	1,220	130	1,350
Audio and vibration sensor	"	1,080	120	1,200
Inertia sensor	"	1,440	120	1,560
Ultrasonic sensor	"	1,650	130	1,780
Infra-red sensor	"	1,030	150	1,180
Monitor panel, with access/secure tone, standard	"	700	130	830
High security	"	1,030	150	1,180
Emergency power indicator	"	420	38.50	459
Monitor rack with 115v power supply				
1 zone	EA.	580	77.00	657
10 zone	"	2,980	190	3,170
Monitor cabinet, wall mounted				
1 zone	EA.	900	77.00	977
5 zone	"	3,240	120	3,360
10 zone	"	1,480	130	1,610
20 zone	"	4,510	150	4,660
Floor mounted, 50 zone	"	4,690	310	5,000
Security system accessories				
Tamper assembly for monitor cabinet	EA.	120	34.25	154
Monitor panel blank	"	15.50	26.75	42.25
Audible alarm	"	130	38.50	169
Audible alarm control	"	550	26.75	577

16 ELECTRICAL

Communications

	UNIT	MAT.	INST.	TOTAL
16720.50 Security Systems (Cont.)				
Termination screw, terminal cabinet				
25 pair	EA.	390	120	510
50 pair	"	620	190	810
150 pair	"	1,010	390	1,400
Universal termination, cabinets & panel				
Remote test	EA.	91.00	130	221
No remote test	"	65.00	56.00	121
High security line supervision termination	"	450	77.00	527
Door cord for capacitance sensor, 12"	"	15.50	38.50	54.00
Insulation block kit for capacitance sensor	"	73.00	26.75	99.75
Termination block for capacitance sensor	"	15.75	26.75	42.50
Guard alert display	"	1,650	47.50	1,698
Uninterrupted power supply	"	1,680	620	2,300
Plug-in 40kva transformer				
12 volt	EA.	66.00	26.75	92.75
18 volt	"	44.00	26.75	70.75
24 volt	"	31.25	26.75	58.00
Test relay	"	100	26.75	127
Coaxial cable, 50 ohm	L.F.	0.45	0.46	0.91
Door openers	EA.	120	38.50	159
Push buttons				
Standard	EA.	26.00	26.75	52.75
Weatherproof	"	38.75	34.25	73.00
Bells	"	99.00	56.00	155
Horns				
Standard	EA.	100	77.00	177
Weatherproof	"	200	96.00	296
Chimes	"	160	51.00	211
Flasher	"	110	47.50	158
Motion detectors	"	450	120	570
Intercom units	"	100	56.00	156
Remote annunciator	"	4,670	390	5,060
16730.20 Clock Systems				
Clock systems				
Single face	EA.	150	62.00	212
Double face	"	420	62.00	482
Skeleton	"	330	210	540
Master	"	3,380	390	3,770
Signal generator	"	2,930	310	3,240
Elapsed time indicator	"	580	62.00	642
Controller	"	130	41.00	171
Clock and speaker	"	250	84.00	334
Bell				
Standard	EA.	120	41.00	161
Weatherproof	"	140	62.00	202
Horn				
Standard	EA.	74.00	56.00	130
Weatherproof	"	90.00	73.00	163
Chime	"	82.00	41.00	123
Buzzer	"	29.50	41.00	70.50

16 ELECTRICAL

Communications

	UNIT	MAT.	INST.	TOTAL
16730.20 Clock Systems (Cont.)				
Flasher	EA.	120	47.50	168
Control Board	"	420	270	690
Program unit	"	450	390	840
Block back box	"	25.75	38.50	64.25
Double clock back box	"	57.00	51.00	108
Wire guard	"	17.25	15.50	32.75
16740.10 Telephone Systems				
Communication cable				
25 pair	L.F.	1.10	1.98	3.08
100 pair	"	5.26	2.20	7.46
150 pair	"	7.99	2.56	10.55
200 pair	"	10.75	3.08	13.83
300 pair	"	13.75	3.24	16.99
400 pair	"	18.75	3.42	22.17
Cable tap in manhole or junction box				
25 pair cable	EA.	7.67	290	298
50 pair cable	"	15.50	580	596
75 pair cable	"	23.25	870	893
100 pair cable	"	31.00	1,160	1,191
150 pair cable	"	46.50	1,710	1,757
200 pair cable	"	62.00	2,280	2,342
300 pair cable	"	93.00	3,420	3,513
400 pair cable	"	120	4,740	4,860
Cable terminations, manhole or junction box				
25 pair cable	EA.	7.67	290	298
50 pair cable	"	15.50	580	596
100 pair cable	"	31.00	1,160	1,191
150 pair cable	"	46.50	1,710	1,757
200 pair cable	"	62.00	2,280	2,342
300 pair cable	"	93.00	3,420	3,513
400 pair cable	"	96.00	4,740	4,836
Telephones, standard				
1 button	EA.	140	230	370
2 button	"	210	270	480
6 button	"	310	410	720
12 button	"	790	590	1,380
18 button	"	830	680	1,510
Hazardous area				
Desk	EA.	2,380	560	2,940
Wall	"	1,110	390	1,500
Accessories				
Standard ground	EA.	39.25	120	159
Push button	"	40.00	120	160
Buzzer	"	41.75	120	162
Interface device	"	22.25	62.00	84.25
Long cord	"	23.50	62.00	85.50
Interior jack	"	14.25	30.75	45.00
Exterior jack	"	28.75	47.50	76.25
Hazardous area				
Selector switch	EA.	230	250	480
Bell	"	350	250	600
Horn	"	530	320	850

16 ELECTRICAL

Communications	UNIT	MAT.	INST.	TOTAL
16740.10 — **Telephone Systems** *(Cont.)*				
Horn relay	EA.	440	240	680
16740.30 — **Call Systems**				
Call systems, single bed station	EA.	230	41.00	271
Double bed station	"	430	56.00	486
Call-in cord	"	82.00	15.50	97.50
Pull cord	"	130	15.50	146
Pillow speaker	"	260	21.25	281
Dome light	"	68.00	41.00	109
Zone light	"	60.00	41.00	101
Stake station	"	180	47.50	228
Duty station	"	230	38.50	269
Utility station	"	180	47.50	228
Nurses station	"	200	41.00	241
Surgical station	"	440	56.00	496
Master station	"	4,630	190	4,820
Control station	"	1,630	620	2,250
Annunciator	"	610	150	760
Power supply	"	450	120	570
Speakers	"	68.00	62.00	130
Foot switch	"	82.00	22.75	105
Code blue systems				
Bed station	EA.	440	56.00	496
Dome light	"	68.00	51.00	119
Zone light	"	120	56.00	176
Pull cord	"	130	19.25	149
Nurses station	"	200	41.00	241
Annunciator	"	610	150	760
Power supply	"	450	110	560
Nurse station indicator, alarm annunciators, flush				
4 circuit	EA.	3,950	310	4,260
6 circuit	"	4,630	620	5,250
12 circuit	"	7,880	930	8,810
Desktop				
4 circuit	EA.	4,350	270	4,620
6 circuit	"	5,430	270	5,700
12 circuit	"	8,420	390	8,810
16750.20 — **Signaling Systems**				
Signaling systems				
4" bell	EA.	140	46.25	186
6" bell	"	170	50.00	220
10" bell	"	200	58.00	258
Buzzer				
Size 0	EA.	37.75	34.25	72.00
Size 1	"	40.25	34.25	74.50
Size 2	"	42.75	38.50	81.25
Size 3	"	45.25	41.00	86.25
Horn	"	130	47.50	178
Chime	"	200	41.00	241
Push button				
Standard	EA.	42.75	30.75	73.50
Weatherproof	"	65.00	38.50	104

16 ELECTRICAL

Communications	UNIT	MAT.	INST.	TOTAL
16750.20 Signaling Systems (Cont.)				
Door opener				
Mortise	EA.	45.25	38.50	83.75
Rim	"	64.00	30.75	94.75
Transformer	"	23.00	34.25	57.25
Contractor grade doorbell chime kit				
Chime	EA.	43.50	77.00	121
Doorbutton	"	6.04	24.75	30.79
Transformer	"	19.25	38.50	57.75
16770.30 Sound Systems				
Power amplifiers	EA.	1,120	270	1,390
Pre-amplifiers	"	890	210	1,100
Tuner	"	570	110	680
Horn				
Equalizer	EA.	1,450	120	1,570
Mixer	"	600	170	770
Tape recorder	"	1,900	140	2,040
Microphone	"	160	77.00	237
Cassette Player	"	960	170	1,130
Record player	"	84.00	150	234
Equipment rack	"	110	99.00	209
Speaker				
Wall	EA.	600	310	910
Paging	"	210	62.00	272
Column	"	320	41.00	361
Single	"	72.00	47.50	120
Double	"	220	340	560
Volume control	"	72.00	41.00	113
Plug-in	"	220	62.00	282
Desk	"	180	30.75	211
Outlet	"	36.00	30.75	66.75
Stand	"	72.00	22.75	94.75
Console	"	16,930	620	17,550
Power supply	"	320	99.00	419
16780.10 Antennas And Towers				
Guy cable, alumaweld				
1x3, 7/32"	L.F.	0.53	3.85	4.38
1x3, 1/4"	"	0.62	3.85	4.47
1x3, 25/64"	"	0.88	4.56	5.44
1x19, 1/2"	"	2.21	5.36	7.57
1x7, 35/64"	"	2.53	6.16	8.69
1x19, 13/16"	"	2.73	7.70	10.43
Preformed alumaweld end grip				
1/4" cable	EA.	4.03	7.70	11.73
3/8" cable	"	5.33	7.70	13.03
1/2" cable	"	6.69	11.25	17.94
9/16" cable	"	8.32	15.50	23.82
5/8" cable	"	10.25	19.25	29.50
Fiberglass guy rod, white epoxy coated				
1/4" dia.	L.F.	2.73	11.25	13.98
3/8" dia	"	4.09	11.25	15.34
1/2" dia	"	5.46	15.50	20.96

16 ELECTRICAL

Communications	UNIT	MAT.	INST.	TOTAL
16780.10 — Antennas And Towers *(Cont.)*				
5/8" dia	L.F.	6.82	19.25	26.07
Preformed glass grip end grip, guy rod				
1/4" dia.	EA.	15.00	11.25	26.25
3/8" dia.	"	17.50	15.50	33.00
1/2" dia.	"	21.50	19.25	40.75
5/8" dia.	"	24.00	19.25	43.25
Spelter socket end grip, 1/4" dia. guy rod				
Standard strength	EA.	44.50	38.50	83.00
High performance	"	55.00	38.50	93.50
3/8" dia. guy rod				
Standard strength	EA.	43.50	26.75	70.25
High performance	"	55.00	38.50	93.50
Timber pole, Douglas Fir				
80-85 ft	EA.	4,170	1,500	5,670
90-95 ft	"	5,230	1,710	6,940
Southern yellow pine				
35-45 ft	EA.	2,500	840	3,340
50-55 ft	"	3,480	1,080	4,560
16780.50 — Television Systems				
TV outlet, self terminating, w/cover plate	EA.	7.05	23.75	30.80
Thru splitter	"	15.25	120	135
End of line	"	12.75	100	113
In line splitter multitap				
4 way	EA.	25.75	140	166
2 way	"	19.25	130	149
Equipment cabinet	"	64.00	120	184
Antenna				
Broad band uhf	EA.	130	270	400
Lightning arrester	"	39.00	56.00	95.00
TV cable	L.F.	0.58	0.38	0.96
Coaxial cable rg	"	0.39	0.38	0.77
Cable drill, with replacement tip	EA.	6.43	38.50	44.93
Cable blocks for in-line taps	"	12.75	56.00	68.75
In-line taps ptu-series 36 tv system	"	15.50	88.00	104
Control receptacles	"	10.00	34.50	44.50
Coupler	"	19.25	190	209
Head end equipment	"	2,420	510	2,930
TV camera	"	1,310	130	1,440
TV power bracket	"	120	62.00	182
TV monitor	"	1,010	110	1,120
Video recorder	"	1,950	160	2,110
Console	"	4,040	660	4,700
Selector switch	"	630	110	740
TV controller	"	300	110	410

16 ELECTRICAL

Resistance Heating

16850.10 Electric Heating

	UNIT	MAT.	INST.	TOTAL
Baseboard heater				
2', 375w	EA.	49.50	77.00	127
3', 500w	"	59.00	77.00	136
4', 750w	"	65.00	88.00	153
5', 935w	"	92.00	100	192
6', 1125w	"	110	120	230
7', 1310w	"	120	140	260
8', 1500w	"	140	150	290
9', 1680w	"	150	170	320
10', 1875w	"	210	180	390
Unit heater, wall mounted				
750w	EA.	200	120	320
1500w	"	260	130	390
2000w	"	270	130	400
2500w	"	280	140	420
3000w	"	340	150	490
4000w	"	380	180	560
Thermostat				
Integral	EA.	44.25	38.50	82.75
Line voltage	"	45.50	38.50	84.00
Electric heater connection	"	1.95	19.25	21.20
Fittings				
Inside corner	EA.	28.50	30.75	59.25
Outside corner	"	31.25	30.75	62.00
Receptacle section	"	32.50	30.75	63.25
Blank section	"	40.25	30.75	71.00
Infrared heaters				
600w	EA.	170	77.00	247
2000w	"	190	92.00	282
3000w	"	280	150	430
4000w	"	400	190	590
Controller	"	77.00	51.00	128
Wall bracket	"	150	56.00	206
Radiant ceiling heater panels				
500w	EA.	340	77.00	417
750w	"	380	77.00	457
Unit heaters, suspended, single phase				
3.0 kw	EA.	560	210	770
5.0 kw	"	570	210	780
7.5 kw	"	940	250	1,190
10.0 kw	"	1,010	290	1,300
Three phase				
5 kw	EA.	570	210	780
7.5 kw	"	750	250	1,000
10 kw	"	800	290	1,090
15 kw	"	1,360	320	1,680
20 kw	"	1,820	410	2,230
25 kw	"	2,180	490	2,670
30 kw	"	2,550	620	3,170
35 kw	"	3,100	620	3,720
Unit heater thermostat	"	59.00	41.00	100
Mounting bracket	"	60.00	56.00	116

16 ELECTRICAL

Resistance Heating

16850.10 Electric Heating (Cont.)

	UNIT	MAT.	INST.	TOTAL
Relay	EA.	77.00	47.50	125
Duct heaters, three phase				
10 kw	EA.	1,100	290	1,390
15 kw	"	1,310	290	1,600
17.5 kw	"	1,380	310	1,690
20 kw	"	1,480	470	1,950

Controls

16910.40 Control Cable

	UNIT	MAT.	INST.	TOTAL
Control cable, 600v, #14 THWN, PVC jacket				
2 wire	L.F.	0.41	0.61	1.02
4 wire	"	0.70	0.77	1.47
6 wire	"	1.21	10.00	11.21
8 wire	"	1.50	11.25	12.75
10 wire	"	1.71	12.25	13.96
12 wire	"	2.07	14.00	16.07
14 wire	"	2.43	16.25	18.68
16 wire	"	2.64	17.00	19.64
18 wire	"	2.86	18.75	21.61
20 wire	"	3.36	19.25	22.61
22 wire	"	3.36	22.00	25.36
Audio cables, shielded, #24 gauge				
3 conductor	L.F.	0.34	0.30	0.64
4 conductor	"	0.41	0.46	0.87
5 conductor	"	0.48	0.53	1.01
6 conductor	"	0.54	0.69	1.23
7 conductor	"	0.61	0.84	1.45
8 conductor	"	0.68	0.92	1.60
9 conductor	"	0.71	1.08	1.79
10 conductor	"	0.80	1.15	1.95
15 conductor	"	1.37	1.38	2.75
20 conductor	"	1.83	1.76	3.59
25 conductor	"	2.23	2.08	4.31
30 conductor	"	2.74	2.32	5.06
40 conductor	"	3.54	2.80	6.34
50 conductor	"	4.46	3.24	7.70
#22 gauge				
3 conductor	L.F.	0.50	0.30	0.80
4 conductor	"	0.64	0.46	1.10
#20 gauge				
3 conductor	L.F.	0.37	0.30	0.67
10 conductor	"	1.20	1.15	2.35
15 conductor	"	1.54	1.38	2.92
#18 gauge				
3 conductor	L.F.	0.48	0.30	0.78

16 ELECTRICAL

Controls

16910.40 Control Cable (Cont.)

	UNIT	MAT.	INST.	TOTAL
4 conductor	L.F.	0.71	0.46	1.17
Microphone cables, #24 gauge				
2 conductor	L.F.	0.48	0.30	0.78
3 conductor	"	0.54	0.38	0.92
#20 gauge				
1 conductor	L.F.	0.52	0.30	0.82
2 conductor	"	0.82	0.30	1.12
2 conductor	"	0.94	0.30	1.24
3 conductor	"	1.14	0.46	1.60
4 conductor	"	1.60	0.53	2.13
5 conductor	"	1.94	0.69	2.63
7 conductor	"	2.17	0.84	3.01
8 conductor	"	2.40	0.92	3.32
Computer cables shielded, #24 gauge				
1 pair	L.F.	0.31	0.30	0.61
2 pair	"	0.42	0.30	0.72
3 pair	"	0.52	0.46	0.98
4 pair	"	0.61	0.53	1.14
5 pair	"	0.80	0.69	1.49
6 pair	"	0.94	0.85	1.79
7 pair	"	0.98	0.92	1.90
8 pair	"	1.14	1.08	2.22
50 pair	"	6.57	3.00	9.57
Coaxial cables				
RG 6/u	L.F.	0.48	0.46	0.94
RG 6a/u	"	0.75	0.46	1.21
RG 8/u	"	0.87	0.46	1.33
RG 8a/u	"	1.05	0.46	1.51
RG 9/u	"	1.88	0.46	2.34
RG 11/u	"	2.34	0.46	2.80
RG 58/u	"	2.57	0.46	3.03
RG 59/u	"	2.80	0.46	3.26
RG 62/u	"	3.08	0.46	3.54
RG 174/u	"	3.31	0.46	3.77
RG 213/u	"	3.54	0.46	4.00
MATV and CCTV camera cables				
1 conductor	L.F.	0.48	0.30	0.78
2 conductor	"	0.61	0.38	0.99
4 conductor	"	1.31	0.46	1.77
7 conductor	"	1.83	0.69	2.52
12 conductor	"	2.80	1.15	3.95
13 conductor	"	3.03	1.23	4.26
14 conductor	"	3.08	1.38	4.46
28 conductor	"	7.37	2.08	9.45
Fire alarm cables, #22 gauge				
6 conductor	L.F.	2.40	0.77	3.17
9 conductor	"	3.08	1.15	4.23
12 conductor	"	3.54	1.23	4.77
#18 gauge				
2 conductor	L.F.	2.40	0.38	2.78
4 conductor	"	3.08	0.53	3.61
#16 gauge				
2 conductor	L.F.	2.40	0.53	2.93

16 ELECTRICAL

Controls	UNIT	MAT.	INST.	TOTAL
16910.40 **Control Cable** *(Cont.)*				
4 conductor	L.F.	3.31	0.61	3.92
#14 gauge				
2 conductor	L.F.	3.54	0.61	4.15
#12 gauge				
2 conductor	L.F.	4.34	0.77	5.11
Plastic jacketed thermostat cable				
2 conductor	L.F.	0.15	0.30	0.45
3 conductor	"	0.21	0.38	0.59
4 conductor	"	0.28	0.46	0.74
5 conductor	"	0.35	0.61	0.96
6 conductor	"	0.42	0.69	1.11
7 conductor	"	0.47	0.92	1.39
8 conductor	"	0.68	1.00	1.68

Man-Hour Tables

The man-hour productivities used to develop the labor costs are listed in the following section of this book. These productivities represent typical installation labor for thousands of construction items. The data takes into account all activities involved in normal construction under commonly experienced working conditions. As with the Costbook pages, these items are listed according to the 16 divisions. In order to best use the information in this book, please review this sample page and read the "Features in this Book" section.

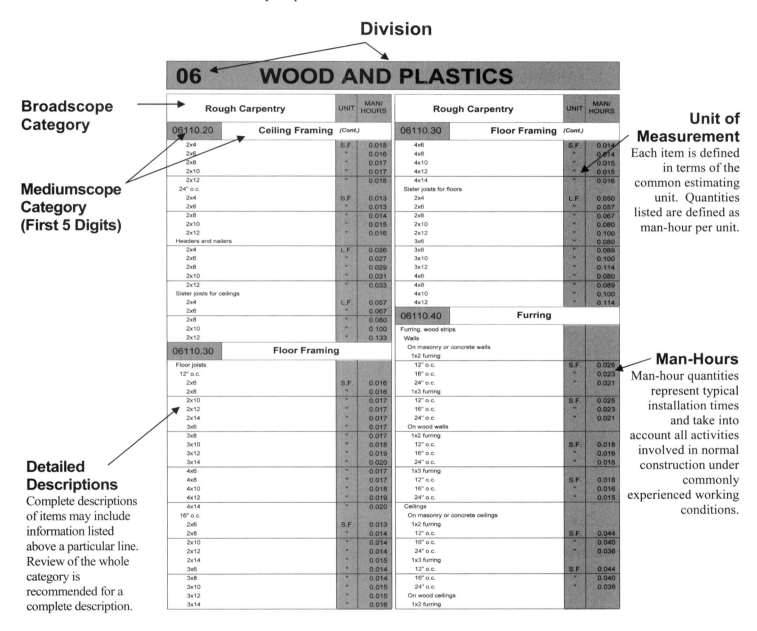

Division

Broadscope Category

Mediumscope Category (First 5 Digits)

Detailed Descriptions
Complete descriptions of items may include information listed above a particular line. Review of the whole category is recommended for a complete description.

Unit of Measurement
Each item is defined in terms of the common estimating unit. Quantities listed are defined as man-hour per unit.

Man-Hours
Man-hour quantities represent typical installation times and take into account all activities involved in normal construction under commonly experienced working conditions.

453

BNi Building News

02 SITE CONSTRUCTION

Site Remediation	UNIT	MAN/HOURS
02115.60 Underground Storage Tank		
Remove underground storage tank, and backfill		
50 to 250 gals	EA.	8.000
600 gals	"	8.000
1000 gals	"	12.000
4000 gals	"	19.200
5000 gals	"	19.200
10,000 gals	"	32.000
12,000 gals	"	40.000
15,000 gals	"	48.000
20,000 gals	"	60.000
02115.66 Septic Tank Removal		
Remove septic tank		
1000 gals	EA.	2.000
2000 gals	"	2.400
5000 gals	"	3.000
15,000 gals	"	24.000
25,000 gals	"	32.000
40,000 gals	"	48.000

Site Preparation	UNIT	MAN/HOURS
02210.10 Soil Boring		
Borings, uncased, stable earth		
2-1/2" dia.		
Minimum	L.F.	0.200
Average	"	0.300
Maximum	"	0.480
4" dia.		
Minimum	L.F.	0.218
Average	"	0.343
Maximum	"	0.600
Cased, including samples		
2-1/2" dia.		
Minimum	L.F.	0.240
Average	"	0.400
Maximum	"	0.800
4" dia.		
Minumum	L.F.	0.480
Average	"	0.686
Maximum	"	0.960
Drilling in rock		
No sampling		
Minimum	L.F.	0.436
Average	"	0.632
Maximum	"	0.857

Site Preparation	UNIT	MAN/HOURS
02210.10 Soil Boring *(Cont.)*		
With casing and sampling		
Minimum	L.F.	0.600
Average	"	0.800
Maximum	"	1.200
Test pits		
Light soil		
Minimum	EA.	3.000
Average	"	4.000
Maximum	"	8.000
Heavy soil		
Minimum	EA.	4.800
Average	"	6.000
Maximum	"	12.000

Demolition	UNIT	MAN/HOURS
02220.10 Complete Building Demolition		
Wood frame	C.F.	0.003
Concrete	"	0.004
Steel frame	"	0.005
02220.15 Selective Building Demolition		
Partition removal		
Concrete block partitions		
4" thick	S.F.	0.040
8" thick	"	0.053
12" thick	"	0.073
Brick masonry partitions		
4" thick	S.F.	0.040
8" thick	"	0.050
12" thick	"	0.067
16" thick	"	0.100
Cast in place concrete partitions		
Unreinforced		
6" thick	S.F.	0.160
8" thick	"	0.171
10" thick	"	0.200
12" thick	"	0.240
Reinforced		
6" thick	S.F.	0.185
8" thick	"	0.240
10" thick	"	0.267
12" thick	"	0.320
Terra cotta		
To 6" thick	S.F.	0.040
Stud partitions		

02 SITE CONSTRUCTION

Demolition

02220.15 Selective Building Demolition (Cont.)

	UNIT	MAN/HOURS
Metal or wood, with drywall both sides	S.F.	0.040
Metal studs, both sides, lath and plaster	"	0.053
Door and frame removal		
Hollow metal in masonry wall		
Single		
2'6"x6'8"	EA.	1.000
3'x7'	"	1.333
Double		
3'x7'	EA.	1.600
4'x8'	"	1.600
Wood in framed wall		
Single		
2'6"x6'8"	EA.	0.571
3'x6'8"	"	0.667
Double		
2'6"x6'8"	EA.	0.800
3'x6'8"	"	0.889
Remove for re-use		
Hollow metal	EA.	2.000
Wood	"	1.333
Floor removal		
Brick flooring	S.F.	0.032
Ceramic or quarry tile	"	0.018
Terrazzo	"	0.036
Heavy wood	"	0.021
Residential wood	"	0.023
Resilient tile or linoleum	"	0.008
Ceiling removal		
Acoustical tile ceiling		
Adhesive fastened	S.F.	0.008
Furred and glued	"	0.007
Suspended grid	"	0.005
Drywall ceiling		
Furred and nailed	S.F.	0.009
Nailed to framing	"	0.008
Plastered ceiling		
Furred on framing	S.F.	0.020
Suspended system	"	0.027
Roofing removal		
Steel frame		
Corrugated metal roofing	S.F.	0.016
Built-up roof on metal deck	"	0.027
Wood frame		
Built up roof on wood deck	S.F.	0.025
Roof shingles	"	0.013
Roof tiles	"	0.027
Concrete frame	C.F.	0.053
Concrete plank	S.F.	0.040
Built-up roof on concrete	"	0.023
Cut-outs		
Concrete, elevated slabs, mesh reinforcing		
Under 5 cf	C.F.	0.800
Over 5 cf	"	0.667

02220.15 Selective Building Demolition (Cont.)

	UNIT	MAN/HOURS
Bar reinforcing		
Under 5 cf	C.F.	1.333
Over 5 cf	"	1.000
Window removal		
Metal windows, trim included		
2'x3'	EA.	0.800
2'x4'	"	0.889
2'x6'	"	1.000
3'x4'	"	1.000
3'x6'	"	1.143
3'x8'	"	1.333
4'x4'	"	1.333
4'x6'	"	1.600
4'x8'	"	2.000
Wood windows, trim included		
2'x3'	EA.	0.444
2'x4'	"	0.471
2'x6'	"	0.500
3'x4'	"	0.533
3'x6'	"	0.571
3'x8'	"	0.615
6'x4'	"	0.667
6'x6'	"	0.727
6'x8'	"	0.800
Walls, concrete, bar reinforcing		
Small jobs	C.F.	0.533
Large jobs	"	0.444
Brick walls, not including toothing		
4" thick	S.F.	0.040
8" thick	"	0.050
12" thick	"	0.067
16" thick	"	0.100
Concrete block walls, not including toothing		
4" thick	S.F.	0.044
6" thick	"	0.047
8" thick	"	0.050
10" thick	"	0.057
12" thick	"	0.067
Rubbish handling		
Load in dumpster or truck		
Minimum	C.F.	0.018
Maximum	"	0.027
For use of elevators, add		
Minimum	C.F.	0.004
Maximum	"	0.008
Rubbish hauling		
Hand loaded on trucks, 2 mile trip	C.Y.	0.320
Machine loaded on trucks, 2 mile trip	"	0.240

02 SITE CONSTRUCTION

Selective Site Demolition	UNIT	MAN/HOURS
02225.10 Catch Basin / Manhole Demolition		
Abandon catch basin or manhole (fill with sand)		
Minimum	EA.	3.000
Average	"	4.800
Maximum	"	8.000
Remove and reset frame and cover		
Minimum	EA.	1.600
Average	"	2.400
Maximum	"	4.000
Remove catch basin, to 10' deep		
Masonry		
Minumum	EA.	4.800
Average	"	6.000
Maximum	"	8.000
Concrete		
Minimum	EA.	6.000
Average	"	8.000
Maximum	"	9.600
02225.13 Core Drilling		
Concrete		
6" thick		
3" dia.	EA.	0.571
4" dia.	"	0.667
6" dia.	"	0.800
8" dia.	"	1.333
8" thick		
3" dia.	EA.	0.800
4" dia.	"	1.000
6" dia.	"	1.143
8" dia.	"	1.600
10" thick		
3" dia.	EA.	1.000
4" dia.	"	1.143
6" dia.	"	1.333
8" dia.	"	2.000
12" thick		
3" dia.	EA.	1.333
4" dia.	"	1.600
6" dia.	"	2.000
8" dia.	"	2.667
02225.15 Curb & Gutter Demolition		
Curb removal		
Concrete, unreinforced		
Minimum	L.F.	0.048
Average	"	0.060
Maximum	"	0.075
Reinforced		
Minimum	L.F.	0.077
Average	"	0.086
Maximum	"	0.096
Combination curb and 2' gutter		
Unreinforced		

Selective Site Demolition	UNIT	MAN/HOURS
02225.15 Curb & Gutter Demolition (Cont.)		
Minimum	L.F.	0.063
Average	"	0.083
Maximum	"	0.120
Reinforced		
Minimum	L.F.	0.100
Average	"	0.133
Maximum	"	0.240
Granite curb		
Minimum	L.F.	0.069
Average	"	0.080
Maximum	"	0.092
Asphalt curb		
Minimum	L.F.	0.040
Average	"	0.048
Maximum	"	0.057
02225.20 Fence Demolition		
Remove fencing		
Chain link, 8' high		
For disposal	L.F.	0.040
For reuse	"	0.100
Wood		
4' high	S.F.	0.027
6' high	"	0.032
8' high	"	0.040
Masonry		
8" thick		
4' high	S.F.	0.080
6' high	"	0.100
8' high	"	0.114
12" thick		
4' high	S.F.	0.133
6' high	"	0.160
8' high	"	0.200
12' high	"	0.267
02225.25 Guardrail Demolition		
Remove standard guardrail		
Steel		
Minimum	L.F.	0.060
Average	"	0.080
Maximum	"	0.120
Wood		
Minimum	L.F.	0.052
Average	"	0.062
Maximum	"	0.100
02225.30 Hydrant Demolition		
Remove fire hydrant		
Minimum	EA.	3.000
Average	"	4.000
Maximum	"	6.000
Remove and reset fire hydrant		

02 SITE CONSTRUCTION

Selective Site Demolition	UNIT	MAN/HOURS
02225.30 Hydrant Demolition *(Cont.)*		
Minimum	EA.	8.000
Average	"	12.000
Maximum	"	24.000
02225.40 Pavement And Sidewalk		
Bituminous pavement, up to 3" thick		
On streets		
Minimum	S.Y.	0.069
Average	"	0.096
Maximum	"	0.160
On pipe trench		
Minimum	S.Y.	0.096
Average	"	0.120
Maximum	"	0.240
Concrete pavement, 6" thick		
No reinforcement		
Minimum	S.Y.	0.120
Average	"	0.160
Maximum	"	0.240
With wire mesh		
Minimum	S.Y.	0.185
Average	"	0.240
Maximum	"	0.300
With rebars		
Minimum	S.Y.	0.240
Average	"	0.300
Maximum	"	0.400
9" thick		
No reinforcement		
Minimum	S.Y.	0.160
Average	"	0.200
Maximum	"	0.240
With wire mesh		
Minimum	S.Y.	0.253
Average	"	0.300
Maximum	"	0.369
With rebars		
Minimum	S.Y.	0.320
Average	"	0.400
Maximum	"	0.533
12" thick		
No reinforcement		
Minimum	S.Y.	0.200
Average	"	0.240
Maximum	"	0.300
With wire mesh		
Minimum	S.Y.	0.282
Average	"	0.343
Maximum	"	0.436
With rebars		
Minimum	S.Y.	0.400
Average	"	0.480
Maximum	"	0.600

Selective Site Demolition	UNIT	MAN/HOURS
02225.40 Pavement And Sidewalk *(Cont.)*		
Sidewalk, 4" thick, with disposal		
Minimum	S.Y.	0.057
Average	"	0.080
Maximum	"	0.114
Removal of pavement markings by waterblasting		
Minimum	S.F.	0.003
Average	"	0.004
Maximum	"	0.008
02225.42 Drainage Piping Demolition		
Remove drainage pipe, not including excavation		
12" dia.		
Minimum	L.F.	0.080
Average	"	0.100
Maximum	"	0.126
18" dia.		
Minimum	L.F.	0.109
Average	"	0.126
Maximum	"	0.160
24" dia.		
Minimum	L.F.	0.133
Average	"	0.160
Maximum	"	0.200
36" dia.		
Minimum	L.F.	0.160
Average	"	0.200
Maximum	"	0.253
02225.43 Gas Piping Demolition		
Remove welded steel pipe, not including excavation		
4" dia.		
Minimum	L.F.	0.120
Average	"	0.150
Maximum	"	0.200
5" dia.		
Minimum	L.F.	0.200
Average	"	0.240
Maximum	"	0.300
6" dia.		
Minimum	L.F.	0.253
Average	"	0.300
Maximum	"	0.400
8" dia.		
Minimum	L.F.	0.369
Average	"	0.480
Maximum	"	0.632
10" dia.		
Minimum	L.F.	0.480
Average	"	0.600
Maximum	"	0.800

02 SITE CONSTRUCTION

Selective Site Demolition	UNIT	MAN/HOURS
02225.45 — **Sanitary Piping Demolition**		
Remove sewer pipe, not including excavation		
4" dia.		
Minimum	L.F.	0.067
Average	"	0.096
Maximum	"	0.160
6" dia.		
Minimum	L.F.	0.075
Average	"	0.109
Maximum	"	0.200
8" dia.		
Minimum	L.F.	0.080
Average	"	0.120
Maximum	"	0.240
10" dia.		
Minimum	L.F.	0.086
Average	"	0.126
Maximum	"	0.267
12" dia.		
Minimum	L.F.	0.092
Average	"	0.133
Maximum	"	0.300
15" dia.		
Minimum	L.F.	0.100
Average	"	0.141
Maximum	"	0.343
18" dia.		
Minimum	L.F.	0.109
Average	"	0.160
Maximum	"	0.400
24" dia.		
Minimum	L.F.	0.120
Average	"	0.200
Maximum	"	0.480
30" dia.		
Minimum	L.F.	0.133
Average	"	0.240
Maximum	"	0.600
36" dia.		
Minimum	L.F.	0.160
Average	"	0.300
Maximum	"	0.800
02225.48 — **Water Piping Demolition**		
Remove water pipe, not including excavation		
4" dia.		
Minimum	L.F.	0.096
Average	"	0.109
Maximum	"	0.126
6" dia.		
Minimum	L.F.	0.100
Average	"	0.114
Maximum	"	0.133
8" dia.		

Selective Site Demolition	UNIT	MAN/HOURS
02225.48 — **Water Piping Demolition** *(Cont.)*		
Minimum	L.F.	0.109
Average	"	0.126
Maximum	"	0.150
10" dia.		
Minimum	L.F.	0.114
Average	"	0.133
Maximum	"	0.160
12" dia.		
Minimum	L.F.	0.120
Average	"	0.141
Maximum	"	0.171
14" dia.		
Minimum	L.F.	0.126
Average	"	0.150
Maximum	"	0.185
16" dia.		
Minimum	L.F.	0.133
Average	"	0.160
Maximum	"	0.200
18" dia.		
Minimum	L.F.	0.141
Average	"	0.171
Maximum	"	0.218
20" dia.		
Minimum	L.F.	0.150
Average	"	0.185
Maximum	"	0.240
Remove valves		
6"	EA.	1.200
10"	"	1.333
14"	"	1.500
18"	"	2.000
02225.50 — **Saw Cutting Pavement**		
Pavement, bituminous		
2" thick	L.F.	0.016
3" thick	"	0.020
4" thick	"	0.025
5" thick	"	0.027
6" thick	"	0.029
Concrete pavement, with wire mesh		
4" thick	L.F.	0.031
5" thick	"	0.033
6" thick	"	0.036
8" thick	"	0.040
10" thick	"	0.044
Plain concrete, unreinforced		
4" thick	L.F.	0.027
5" thick	"	0.031
6" thick	"	0.033
8" thick	"	0.036
10" thick	"	0.040

02 SITE CONSTRUCTION

Selective Site Demolition	UNIT	MAN/HOURS
02225.80 Wall, Exterior, Demolition		
Concrete wall		
Light reinforcing		
6" thick	S.F.	0.120
8" thick	"	0.126
10" thick	"	0.133
12" thick	"	0.150
Medium reinforcing		
6" thick	S.F.	0.126
8" thick	"	0.133
10" thick	"	0.150
12" thick	"	0.171
Heavy reinforcing		
6" thick	S.F.	0.141
8" thick	"	0.150
10" thick	"	0.171
12" thick	"	0.200
Masonry		
No reinforcing		
8" thick	S.F.	0.053
12" thick	"	0.060
16" thick	"	0.069
Horizontal reinforcing		
8" thick	S.F.	0.060
12" thick	"	0.065
16" thick	"	0.077
Vertical reinforcing		
8" thick	S.F.	0.077
12" thick	"	0.089
16" thick	"	0.109
Remove concrete headwall		
15" pipe	EA.	1.714
18" pipe	"	2.000
24" pipe	"	2.182
30" pipe	"	2.400
36" pipe	"	2.667
48" pipe	"	3.429
60" pipe	"	4.800

Site Clearing	UNIT	MAN/HOURS
02230.10 Clear Wooded Areas		
Clear wooded area		
Light density	ACRE	60.000
Medium density	"	80.000
Heavy density	"	96.000

Site Clearing	UNIT	MAN/HOURS
02230.50 Tree Cutting & Clearing		
Cut trees and clear out stumps		
9" to 12" dia.	EA.	4.800
To 24" dia.	"	6.000
24" dia. and up	"	8.000
Loading and trucking		
For machine load, per load, round trip		
1 mile	EA.	0.960
3 mile	"	1.091
5 mile	"	1.200
10 mile	"	1.600
20 mile	"	2.400
Hand loaded, round trip		
1 mile	EA.	2.000
3 mile	"	2.286
5 mile	"	2.667
10 mile	"	3.200
20 mile	"	4.000
Tree trimming for pole line construction		
Light cutting	L.F.	0.012
Medium cutting	"	0.016
Heavy cutting	"	0.024

Dewatering	UNIT	MAN/HOURS
02240.10 Wellpoint Systems		
Pumping, gas driven, 50' hose		
3" header pipe	DAY	8.000
6" header pipe	"	10.000
Wellpoint system per job; 150' length of PVC header		
6" header pipe, 2" wellpoints, 5' centers	L.F.	0.032
8" header pipe	"	0.040
10" header pipe	"	0.053
Jetting wellpoint system		
14' long	EA.	0.533
18' long	"	0.667
Sand filter for wellpoints	L.F.	0.013
Replacement of wellpoint components	EA.	0.160

02 SITE CONSTRUCTION

Shoring And Underpinning	UNIT	MAN/HOURS
02250.10 Trench Sheeting		
Closed timber, including pull and salvage, excavation		
8' deep	S.F.	0.064
10' deep	"	0.067
12' deep	"	0.071
14' deep	"	0.075
16' deep	"	0.080
18' deep	"	0.091
20' deep	"	0.098
02260.10 Cofferdams		
Cofferdam, steel, driven from shore		
15' deep	S.F.	0.137
20' deep	"	0.128
25' deep	"	0.120
30' deep	"	0.113
40' deep	"	0.107
Driven from barge		
20' deep	S.F.	0.148
30' deep	"	0.137
40' deep	"	0.128
50' deep	"	0.120
02260.70 Steel Sheet Piling		
Steel sheet piling, 12" wide		
20' long	S.F.	0.096
35' long	"	0.069
50' long	"	0.048
Over 50' long	"	0.044

Earthwork, Excavation & Fill	UNIT	MAN/HOURS
02315.10 Base Course		
Base course, crushed stone		
3" thick	S.Y.	0.004
4" thick	"	0.004
6" thick	"	0.005
8" thick	"	0.005
10" thick	"	0.006
12" thick	"	0.007
Base course, bank run gravel		
4" deep	S.Y.	0.004
6" deep	"	0.005
8" deep	"	0.005
10" deep	"	0.005
12" deep	"	0.006
Prepare and roll sub base		

Earthwork, Excavation & Fill	UNIT	MAN/HOURS
02315.10 Base Course *(Cont.)*		
Minimum	S.Y.	0.004
Average	"	0.005
Maximum	"	0.007
02315.20 Borrow		
Borrow fill, F.O.B. at pit		
Sand, haul to site, round trip		
10 mile	C.Y.	0.080
20 mile	"	0.133
30 mile	"	0.200
Place borrow fill and compact		
Less than 1 in 4 slope	C.Y.	0.040
Greater than 1 in 4 slope	"	0.053
02315.30 Bulk Excavation		
Excavation, by small dozer		
Large areas	C.Y.	0.016
Small areas	"	0.027
Trim banks	"	0.040
Drag line		
1-1/2 cy bucket		
Sand or gravel	C.Y.	0.040
Light clay	"	0.053
Heavy clay	"	0.060
Unclassified	"	0.064
2 cy bucket		
Sand or gravel	C.Y.	0.037
Light clay	"	0.048
Heavy clay	"	0.053
Unclassified	"	0.056
2-1/2 cy bucket		
Sand or gravel	C.Y.	0.034
Light clay	"	0.044
Heavy clay	"	0.048
Unclassified	"	0.051
3 cy bucket		
Sand or gravel	C.Y.	0.030
Light clay	"	0.040
Heavy clay	"	0.044
Unclassified	"	0.046
Hydraulic excavator		
1 cy capacity		
Light material	C.Y.	0.040
Medium material	"	0.048
Wet material	"	0.060
Blasted rock	"	0.069
1-1/2 cy capacity		
Light material	C.Y.	0.010
Medium material	"	0.013
Wet material	"	0.016
Blasted rock	"	0.020
2 cy capacity		
Light material	C.Y.	0.009

02 SITE CONSTRUCTION

Earthwork, Excavation & Fill	UNIT	MAN/HOURS
02315.30 Bulk Excavation *(Cont.)*		
Medium material	C.Y.	0.011
Wet material	"	0.013
Blasted rock	"	0.016
Wheel mounted front-end loader		
7/8 cy capacity		
Light material	C.Y.	0.020
Medium material	"	0.023
Wet material	"	0.027
Blasted rock	"	0.032
1-1/2 cy capacity		
Light material	C.Y.	0.011
Medium material	"	0.012
Wet material	"	0.013
Blasted rock	"	0.015
2-1/2 cy capacity		
Light material	C.Y.	0.009
Medium material	"	0.010
Wet material	"	0.011
Blasted rock	"	0.011
3-1/2 cy capacity		
Light material	C.Y.	0.009
Medium material	"	0.009
Wet material	"	0.010
Blasted rock	"	0.011
6 cy capacity		
Light material	C.Y.	0.005
Medium material	"	0.006
Wet material	"	0.006
Blasted rock	"	0.007
Track mounted front-end loader		
1-1/2 cy capacity		
Light material	C.Y.	0.013
Medium material	"	0.015
Wet material	"	0.016
Blasted rock	"	0.018
2-3/4 cy capacity		
Light material	C.Y.	0.008
Medium material	"	0.009
Wet material	"	0.010
Blasted rock	"	0.011
02315.40 Building Excavation		
Structural excavation, unclassified earth		
3/8 cy backhoe	C.Y.	0.107
3/4 cy backhoe	"	0.080
1 cy backhoe	"	0.067
Foundation backfill and compaction by machine	"	0.160

Earthwork, Excavation & Fill	UNIT	MAN/HOURS
02315.45 Hand Excavation		
Excavation		
To 2' deep		
Normal soil	C.Y.	0.889
Sand and gravel	"	0.800
Medium clay	"	1.000
Heavy clay	"	1.143
Loose rock	"	1.333
To 6' deep		
Normal soil	C.Y.	1.143
Sand and gravel	"	1.000
Medium clay	"	1.333
Heavy clay	"	1.600
Loose rock	"	2.000
Backfilling foundation without compaction, 6" lifts	"	0.500
Compaction of backfill around structures or in trench		
By hand with air tamper	C.Y.	0.571
By hand with vibrating plate tamper	"	0.533
1 ton roller	"	0.400
Miscellaneous hand labor		
Trim slopes, sides of excavation	S.F.	0.001
Trim bottom of excavation	"	0.002
Excavation around obstructions and services	C.Y.	2.667
02315.50 Roadway Excavation		
Roadway excavation		
1/4 mile haul	C.Y.	0.016
2 mile haul	"	0.027
5 mile haul	"	0.040
Excavation of open ditches	"	0.011
Trim banks, swales or ditches	S.Y.	0.013
Bulk swale excavation by dragline		
Small jobs	C.Y.	0.060
Large jobs	"	0.034
Spread base course	"	0.020
Roll and compact	"	0.027
02315.60 Trenching		
Trenching and continuous footing excavation		
By gradall		
1 cy capacity		
Light soil	C.Y.	0.023
Medium soil	"	0.025
Heavy/wet soil	"	0.027
Loose rock	"	0.029
Blasted rock	"	0.031
By hydraulic excavator		
1/2 cy capacity		
Light soil	C.Y.	0.027
Medium soil	"	0.029
Heavy/wet soil	"	0.032
Loose rock	"	0.036
Blasted rock	"	0.040
1 cy capacity		

02 SITE CONSTRUCTION

Earthwork, Excavation & Fill	UNIT	MAN/HOURS
02315.60 Trenching *(Cont.)*		
Light soil	C.Y.	0.019
Medium soil	"	0.020
Heavy/wet soil	"	0.021
Loose rock	"	0.023
Blasted rock	"	0.025
1-1/2 cy capacity		
Light soil	C.Y.	0.017
Medium soil	"	0.018
Heavy/wet soil	"	0.019
Loose rock	"	0.020
Blasted rock	"	0.021
2 cy capacity		
Light soil	C.Y.	0.016
Medium soil	"	0.017
Heavy/wet soil	"	0.018
Loose rock	"	0.019
Blasted rock	"	0.020
2-1/2 cy capacity		
Light soil	C.Y.	0.015
Medium soil	"	0.015
Heavy/wet soil	"	0.016
Loose rock	"	0.017
Blasted rock	"	0.018
Trencher, chain, 1' wide to 4' deep		
Light soil	C.Y.	0.020
Medium soil	"	0.023
Heavy soil	"	0.027
Hand excavation		
Bulk, wheeled 100'		
Normal soil	C.Y.	0.889
Sand or gravel	"	0.800
Medium clay	"	1.143
Heavy clay	"	1.600
Loose rock	"	2.000
Trenches, up to 2' deep		
Normal soil	C.Y.	1.000
Sand or gravel	"	0.889
Medium clay	"	1.333
Heavy clay	"	2.000
Loose rock	"	2.667
Trenches, to 6' deep		
Normal soil	C.Y.	1.143
Sand or gravel	"	1.000
Medium clay	"	1.600
Heavy clay	"	2.667
Loose rock	"	4.000
Backfill trenches		
With compaction		
By hand	C.Y.	0.667
By 60 hp tracked dozer	"	0.020
By 200 hp tracked dozer	"	0.009
By small front-end loader	"	0.023
Spread dumped fill or gravel, no compaction		

Earthwork, Excavation & Fill	UNIT	MAN/HOURS
02315.60 Trenching *(Cont.)*		
6" layers	S.Y.	0.013
12" layers	"	0.016
Compaction in 6" layers		
By hand with air tamper	S.Y.	0.016
Backfill trenches, sand bedding, no compaction		
By hand	C.Y.	0.667
By small front-end loader	"	0.023
02315.70 Utility Excavation		
Trencher, sandy clay, 8" wide trench		
18" deep	L.F.	0.018
24" deep	"	0.020
36" deep	"	0.023
Trench backfill, 95% compaction		
Tamp by hand	C.Y.	0.500
Vibratory compaction	"	0.400
Trench backfilling, with borrow sand, place & compact	"	0.400
02315.80 Hauling Material		
Haul material by 10 cy dump truck, round trip distance		
1 mile	C.Y.	0.044
2 mile	"	0.053
5 mile	"	0.073
10 mile	"	0.080
20 mile	"	0.089
30 mile	"	0.107
Site grading, cut & fill, sandy clay, 200' haul, 75 hp dozer	"	0.032
Spread topsoil by equipment on site	"	0.036
Site grading (cut and fill to 6") less than 1 acre		
75 hp dozer	C.Y.	0.053
1.5 cy backhoe/loader	"	0.080

Soil Stabilization & Treatment	UNIT	MAN/HOURS
02340.05 Soil Stabilization		
Straw bale secured with rebar	L.F.	0.027
Filter barrier, 18" high filter fabric	"	0.080
Sediment fence, 36" fabric with 6" mesh	"	0.100
Soil stabilization with tar paper, burlap, straw and stakes	S.F.	0.001
02340.30 Geotextile		
Filter cloth, light reinforcement		
Woven		
12'-6" wide x 50' long	S.F.	0.001
Various lengths	"	0.001
Non-woven		

02 SITE CONSTRUCTION

Soil Stabilization & Treatment	UNIT	MAN/HOURS
02340.30 Geotextile *(Cont.)*		
14'-8" wide x 430' long	S.F.	0.001
Various lengths	"	0.001
02360.20 Soil Treatment		
Soil treatment, termite control pretreatment		
Under slabs	S.F.	0.004
By walls	"	0.005
02370.10 Slope Protection		
Gabions, stone filled		
6" deep	S.Y.	0.200
9" deep	"	0.229
12" deep	"	0.267
18" deep	"	0.320
36" deep	"	0.533
02370.40 Riprap		
Riprap		
Crushed stone blanket, max size 2-1/2"	TON	0.533
Stone, quarry run, 300 lb. stones	"	0.492
400 lb. stones	"	0.457
500 lb. stones	"	0.427
750 lb. stones	"	0.400
Dry concrete riprap in bags 3" thick, 80 lb. per bag	BAG	0.027

Tunneling, Boring & Jacking	UNIT	MAN/HOURS
02445.10 Pipe Jacking		
Pipe casing, horizontal jacking		
18" dia.	L.F.	0.711
21" dia.	"	0.762
24" dia.	"	0.800
27" dia.	"	0.800
30" dia.	"	0.842
36" dia.	"	0.914
42" dia.	"	1.000
48" dia.	"	1.067

Piles And Caissons	UNIT	MAN/HOURS
02455.60 Steel Piles		
H-section piles		
8x8		
36 lb/ft		
30' long	L.F.	0.080
40' long	"	0.064
50' long	"	0.053
10x10		
42 lb/ft		
30' long	L.F.	0.080
40' long	"	0.064
50' long	"	0.053
57 lb/ft		
30' long	L.F.	0.080
40' long	"	0.064
50' long	"	0.053
12x12		
53 lb/ft		
30' long	L.F.	0.087
40' long	"	0.069
50' long	"	0.053
74 lb/ft		
30' long	L.F.	0.087
40' long	"	0.069
50' long	"	0.053
14x14		
73 lb/ft		
40' long	L.F.	0.087
50' long	"	0.069
60' long	"	0.053
89 lb/ft		
40' long	L.F.	0.087
50' long	"	0.069
60' long	"	0.053
102 lb/ft		
40' long	L.F.	0.087
50' long	"	0.069
60' long	"	0.053
117 lb/ft		
40' long	L.F.	0.091
50' long	"	0.071
60' long	"	0.055
Splice		
8"	EA.	1.333
10"	"	1.600
12"	"	1.600
14"	"	2.000
Driving cap		
8"	EA.	0.800
10"	"	1.000
12"	"	1.000
14"	"	1.143
Standard point		
8"	EA.	0.800

02 SITE CONSTRUCTION

Piles And Caissons	UNIT	MAN/HOURS
02455.60 **Steel Piles** *(Cont.)*		
10"	EA.	1.000
12"	"	1.143
14"	"	1.333
Heavy duty point		
8"	EA.	0.889
10"	"	1.143
12"	"	1.333
14"	"	1.600
Tapered friction piles, fluted casing, up to 50'		
With 4000 psi concrete no reinforcing		
12" dia.	L.F.	0.048
14" dia.	"	0.049
16" dia.	"	0.051
18" dia.	"	0.056
02455.65 **Steel Pipe Piles**		
Concrete filled, 3000# concrete, up to 40'		
8" dia.	L.F.	0.069
10" dia.	"	0.071
12" dia.	"	0.074
14" dia.	"	0.077
16" dia.	"	0.080
18" dia.	"	0.083
Pipe piles, non-filled		
8" dia.	L.F.	0.053
10" dia.	"	0.055
12" dia.	"	0.056
14" dia.	"	0.060
16" dia.	"	0.062
18" dia.	"	0.064
Splice		
8" dia.	EA.	1.600
10" dia.	"	1.600
12" dia.	"	2.000
14" dia.	"	2.000
16" dia.	"	2.667
18" dia.	"	2.667
Standard point		
8" dia.	EA.	1.600
10" dia.	"	1.600
12" dia.	"	2.000
14" dia.	"	2.000
16" dia.	"	2.667
18" dia.	"	2.667
Heavy duty point		
8" dia.	EA.	2.000
10" dia.	"	2.000
12" dia.	"	2.667
14" dia.	"	2.667
16" dia.	"	3.200
18" dia.	"	3.200

Piles And Caissons	UNIT	MAN/HOURS
02455.80 **Wood And Timber Piles**		
Treated wood piles, 12" butt, 8" tip		
25' long	L.F.	0.096
30' long	"	0.080
35' long	"	0.069
40' long	"	0.060
12" butt, 7" tip		
40' long	L.F.	0.060
45' long	"	0.053
50' long	"	0.048
55' long	"	0.044
60' long	"	0.040
02465.50 **Prestressed Piling**		
Prestressed concrete piling, less than 60' long		
10" sq.	L.F.	0.040
12" sq.	"	0.042
14" sq.	"	0.043
16" sq.	"	0.044
18" sq.	"	0.047
20" sq.	"	0.048
24" sq.	"	0.049
More than 60' long		
12" sq.	L.F.	0.034
14" sq.	"	0.035
16" sq.	"	0.036
18" sq.	"	0.036
20" sq.	"	0.037
24" sq.	"	0.038
Straight cylinder, less than 60' long		
12" dia.	L.F.	0.044
14" dia.	"	0.045
16" dia.	"	0.046
18" dia.	"	0.047
20" dia.	"	0.048
24" dia.	"	0.049
More than 60' long		
12" dia.	L.F.	0.035
14" dia.	"	0.036
16" dia.	"	0.036
18" dia.	"	0.037
20" dia.	"	0.038
24" dia.	"	0.038
Concrete sheet piling		
12" thick x 20' long	S.F.	0.096
25' long	"	0.087
30' long	"	0.080
35' long	"	0.074
40' long	"	0.069
16" thick x 40' long	"	0.053
45' long	"	0.051
50' long	"	0.048
55' long	"	0.046
60' long	"	0.044

02 SITE CONSTRUCTION

Piles And Caissons	UNIT	MAN/HOURS
02475.10 Caissons (includes Casing)		
Caisson, 3000# conc., 60 # reinf./CY, stable ground		
18" dia., 0.065 CY/ LF	L.F.	0.192
24" dia., 0.116 CY/ LF	"	0.200
30" dia., 0.182 CY/ LF	"	0.240
36" dia., 0.262 CY/ LF	"	0.274
48" dia., 0.465 CY/ LF	"	0.320
60" dia., 0.727 CY/ LF	"	0.436
72" dia., 1.05 CY/ LF	"	0.533
84" dia., 1.43 CY/ LF	"	0.686
Wet ground, casing required but pulled		
18" dia.	L.F.	0.240
24" dia.	"	0.267
30" dia.	"	0.300
36" dia.	"	0.320
48" dia.	"	0.400
60" dia.	"	0.533
72" dia.	"	0.800
84" dia.	"	1.200
Soft rock		
18" dia.	L.F.	0.686
24" dia.	"	1.200
30" dia.	"	1.600
36" dia.	"	2.400
48" dia.	"	3.200
60" dia.	"	4.800
72" dia.	"	5.333
84" dia.	"	6.000

Utility Services	UNIT	MAN/HOURS
02510.10 Wells		
Domestic water, drilled and cased		
4" dia.	L.F.	0.480
6" dia.	"	0.533
8" dia.	"	0.600
02510.13 Gate Valves		
Gate valve, (AWWA) mechanical joint, with adjustable box		
4" valve	EA.	0.800
6" valve	"	0.960
8" valve	"	1.200
10" valve	"	1.412
12" valve	"	1.714
14" valve	"	2.000
16" valve	"	2.182
18" valve	"	2.400

Utility Services	UNIT	MAN/HOURS
02510.13 Gate Valves (Cont.)		
Flanged, with box, post indicator (AWWA)		
4" valve	EA.	0.960
6" valve	"	1.091
8" valve	"	1.333
10" valve	"	1.600
12" valve	"	2.000
14" valve	"	2.400
16" valve	"	3.000
02510.15 Water Meters		
Water meter, displacement type		
1"	EA.	0.800
1-1/2"	"	0.889
2"	"	1.000
02510.17 Corporation Stops		
Stop for flared copper service pipe		
3/4"	EA.	0.400
1"	"	0.444
1-1/4"	"	0.533
1-1/2"	"	0.667
2"	"	0.800
02510.19 Thrust Blocks		
Thrust block, 3000# concrete		
1/4 c.y.	EA.	1.333
1/2 c.y.	"	1.600
3/4 c.y.	"	2.667
1 c.y.	"	5.333
02510.20 Tapping Saddles & Sleeves		
Tapping saddle, tap size to 2"		
4" saddle	EA.	0.400
6" saddle	"	0.500
8" saddle	"	0.667
10" saddle	"	0.800
12" saddle	"	1.143
14" saddle	"	1.600
Tapping sleeve		
4x4	EA.	0.533
6x4	"	0.615
6x6	"	0.615
8x4	"	0.800
8x6	"	0.800
10x4	"	0.960
10x6	"	0.960
10x8	"	0.960
10x10	"	1.000
12x4	"	1.000
12x6	"	1.091
12x8	"	1.200
12x10	"	1.333
12x12	"	1.500

02 SITE CONSTRUCTION

Utility Services	UNIT	MAN/HOURS
02510.20 Tapping Saddles & Sleeves (Cont.)		
Tapping valve, mechanical joint		
4" valve	EA.	3.000
6" valve	"	4.000
8" valve	"	6.000
10" valve	"	8.000
12" valve	"	12.000
Tap hole in pipe		
4" hole	EA.	1.000
6" hole	"	1.600
8" hole	"	2.667
10" hole	"	3.200
12" hole	"	4.000
02510.25 Valve Boxes		
Valve box, adjustable, for valves up to 20"		
3' deep	EA.	0.267
4' deep	"	0.320
5' deep	"	0.400
02510.30 Fire Hydrants		
Standard, 3 way post, 6" mechanical joint		
2' deep	EA.	8.000
4' deep	"	9.600
6' deep	"	12.000
8' deep	"	13.714
02510.35 Chilled Water Systems		
Chilled water pipe, 2" thick insulation, w/casing		
Align and tack weld on sleepers		
1-1/2" dia.	L.F.	0.022
3" dia.	"	0.034
4" dia.	"	0.048
6" dia.	"	0.060
8" dia.	"	0.069
10" dia.	"	0.080
12" dia.	"	0.096
14" dia.	"	0.104
16" dia.	"	0.120
Align and tack weld on trench bottom		
18" dia.	L.F.	0.133
20" dia.	"	0.150
Preinsulated fittings		
Align and tack weld on sleepers		
Elbows		
1-1/2"	EA.	0.500
3"	"	0.800
4"	"	1.000
6"	"	1.333
8"	"	1.600
Tees		
1-1/2"	EA.	0.533
3"	"	0.889
4"	"	1.143

Utility Services	UNIT	MAN/HOURS
02510.35 Chilled Water Systems (Cont.)		
6"	EA.	1.600
8"	"	2.000
Reducers		
3"	EA.	0.667
4"	"	0.800
6"	"	1.000
8"	"	1.333
Anchors, not including concrete		
4"	EA.	1.000
6"	"	1.000
Align and tack weld on trench bottom		
Elbows		
10"	EA.	1.500
12"	"	1.714
14"	"	1.846
16"	"	2.000
18"	"	2.182
20"	"	2.400
Tees		
10"	EA.	1.500
12"	"	1.714
14"	"	1.846
16"	"	2.000
18"	"	2.182
20"	"	2.400
Reducers		
10"	EA.	1.000
12"	"	1.091
14"	"	1.200
16"	"	1.333
18"	"	1.500
20"	"	1.714
Anchors, not including concrete		
10"	EA.	1.000
12"	"	1.091
14"	"	1.200
16"	"	1.333
18"	"	1.500
20"	"	1.714
02510.40 Ductile Iron Pipe		
Ductile iron pipe, cement lined, slip-on joints		
4"	L.F.	0.067
6"	"	0.071
8"	"	0.075
10"	"	0.080
12"	"	0.096
14"	"	0.120
16"	"	0.133
18"	"	0.150
20"	"	0.171
Mechanical joint pipe		
4"	L.F.	0.092

02 SITE CONSTRUCTION

Utility Services	UNIT	MAN/HOURS
02510.40 **Ductile Iron Pipe** (Cont.)		
6"	L.F.	0.100
8"	"	0.109
10"	"	0.120
12"	"	0.160
14"	"	0.185
16"	"	0.218
18"	"	0.240
20"	"	0.267
Fittings, mechanical joint		
90 degree elbow		
4"	EA.	0.533
6"	"	0.615
8"	"	0.800
10"	"	1.143
12"	"	1.600
14"	"	2.000
16"	"	2.667
18"	"	3.200
20"	"	4.000
45 degree elbow		
4"	EA.	0.533
6"	"	0.615
8"	"	0.800
10"	"	1.143
12"	"	1.600
14"	"	2.000
16"	"	2.667
18"	"	4.000
20"	"	4.000
Tee		
4"x3"	EA.	1.000
4"x4"	"	1.000
6"x3"	"	1.143
6"x4"	"	1.143
6"x6"	"	1.143
8"x4"	"	1.333
8"x6"	"	1.333
8"x8"	"	1.333
10"x4"	"	1.600
10"x6"	"	1.600
10"x8"	"	1.600
10"x10"	"	1.600
12"x4"	"	2.000
12"x6"	"	2.000
12"x8"	"	2.000
12"x10"	"	2.000
12"x12"	"	2.133
14"x4"	"	2.286
14"x6"	"	2.286
14"x8"	"	2.286
14"x10"	"	2.286
14"x12"	"	2.462
14"x14"	"	2.462

Utility Services	UNIT	MAN/HOURS
02510.40 **Ductile Iron Pipe** (Cont.)		
16"x4"	EA.	2.667
16"x6"	"	2.667
16"x8"	"	2.667
16"x10"	"	2.667
16"x12"	"	2.667
16"x14"	"	2.667
16"x16"	"	2.667
18"x6"	"	2.909
18"x8"	"	2.909
18"x10"	"	2.909
18"x12"	"	2.909
18"x14"	"	2.909
18"x16"	"	2.909
18"x18"	"	2.909
20"x6"	"	3.200
20"x8"	"	3.200
20"x10"	"	3.200
20"x12"	"	3.200
20"x14"	"	3.200
20"x16"	"	3.200
20"x18"	"	3.200
20"x20"	"	3.200
Cross		
4"x3"	EA.	1.333
4"x4"	"	1.333
6"x3"	"	1.600
6"x4"	"	1.600
6"x6"	"	1.600
8"x4"	"	1.778
8"x6"	"	1.778
8"x8"	"	1.778
10"x4"	"	2.000
10"x6"	"	2.000
10"x8"	"	2.000
10"x10"	"	2.000
12"x4"	"	2.286
12"x6"	"	2.286
12"x8"	"	2.286
12"x10"	"	2.462
12"x12"	"	2.462
14"x4"	"	2.667
14"x6"	"	2.667
14"x8"	"	2.667
14"x10"	"	2.667
14"x12"	"	2.909
14"x14"	"	2.909
16"x4"	"	3.200
16"x6"	"	3.200
16"x8"	"	3.200
16"x10"	"	3.200
16"x12"	"	3.200
16"x14"	"	3.200
16"x16"	"	3.200

02 SITE CONSTRUCTION

Utility Services	UNIT	MAN/HOURS
02510.40 Ductile Iron Pipe *(Cont.)*		
18"x6"	EA.	3.556
18"x8"	"	3.556
18"x10"	"	3.556
18"x12"	"	3.556
18"x14"	"	3.556
18"x16"	"	3.556
18"x18"	"	3.556
20"x6"	"	3.810
20"x8"	"	3.810
20"x10"	"	3.810
20"x12"	"	3.810
20"x14"	"	3.810
20"x16"	"	3.810
20"x18"	"	4.000
20"x20"	"	4.000
02510.60 Plastic Pipe		
PVC, class 150 pipe		
4" dia.	L.F.	0.060
6" dia.	"	0.065
8" dia.	"	0.069
10" dia.	"	0.075
12" dia.	"	0.080
Schedule 40 pipe		
1-1/2" dia.	L.F.	0.047
2" dia.	"	0.050
2-1/2" dia.	"	0.053
3" dia.	"	0.057
4" dia.	"	0.067
6" dia.	"	0.080
90 degree elbows		
1"	EA.	0.133
1-1/2"	"	0.133
2"	"	0.145
2-1/2"	"	0.160
3"	"	0.178
4"	"	0.200
6"	"	0.267
45 degree elbows		
1"	EA.	0.133
1-1/2"	"	0.133
2"	"	0.145
2-1/2"	"	0.160
3"	"	0.178
4"	"	0.200
6"	"	0.267
Tees		
1"	EA.	0.160
1-1/2"	"	0.160
2"	"	0.178
2-1/2"	"	0.200
3"	"	0.229
4"	"	0.267

Utility Services	UNIT	MAN/HOURS
02510.60 Plastic Pipe *(Cont.)*		
6"	EA.	0.320
Couplings		
1"	EA.	0.133
1-1/2"	"	0.133
2"	"	0.145
2-1/2"	"	0.160
3"	"	0.178
4"	"	0.200
6"	"	0.267
Drainage pipe		
PVC schedule 80		
1" dia.	L.F.	0.047
1-1/2" dia.	"	0.047
ABS, 2" dia.	"	0.050
2-1/2" dia.	"	0.053
3" dia.	"	0.057
4" dia.	"	0.067
6" dia.	"	0.080
8" dia.	"	0.063
10" dia.	"	0.075
12" dia.	"	0.080
90 degree elbows		
1"	EA.	0.133
1-1/2"	"	0.133
2"	"	0.145
2-1/2"	"	0.160
3"	"	0.178
4"	"	0.200
6"	"	0.267
45 degree elbows		
1"	EA.	0.133
1-1/2"	"	0.133
2"	"	0.145
2-1/2"	"	0.160
3"	"	0.178
4"	"	0.200
6"	"	0.267
Tees		
1"	EA.	0.160
1-1/2"	"	0.160
2"	"	0.178
2-1/2"	"	0.200
3"	"	0.229
4"	"	0.267
6"	"	0.320
Couplings		
1"	EA.	0.133
1-1/2"	"	0.133
2"	"	0.145
2-1/2"	"	0.160
3"	"	0.178
4"	"	0.200
6"	"	0.267

02 SITE CONSTRUCTION

Utility Services		UNIT	MAN/HOURS
02510.60	**Plastic Pipe** (Cont.)		
Pressure pipe			
PVC, class 200 pipe			
3/4"		L.F.	0.040
1"		"	0.042
1-1/4"		"	0.044
1-1/2"		"	0.047
2"		"	0.050
2-1/2"		"	0.053
3"		"	0.057
4"		"	0.067
6"		"	0.080
8"		"	0.069
90 degree elbows			
3/4"		EA.	0.133
1"		"	0.133
1-1/4"		"	0.133
1-1/2"		"	0.133
2"		"	0.145
2-1/2"		"	0.160
3"		"	0.178
4"		"	0.200
6"		"	0.267
8"		"	0.400
45 degree elbows			
3/4"		EA.	0.133
1"		"	0.133
1-1/4"		"	0.133
1-1/2"		"	0.133
2"		"	0.145
2-1/2"		"	0.160
3"		"	0.178
4"		"	0.200
6"		"	0.267
8"		"	0.400
Tees			
3/4"		EA.	0.160
1"		"	0.160
1-1/4"		"	0.160
1-1/2"		"	0.160
2"		"	0.178
2-1/2"		"	0.200
3"		"	0.229
4"		"	0.267
6"		"	0.320
8"		"	0.444
Couplings			
3/4"		EA.	0.133
1"		"	0.133
1-1/4"		"	0.133
1-1/2"		"	0.133
2"		"	0.145
2-1/2"		"	0.160
3"		"	0.178

Utility Services		UNIT	MAN/HOURS
02510.60	**Plastic Pipe** (Cont.)		
4"		EA.	0.178
6"		"	0.200
8"		"	0.267

Sanitary Sewer		UNIT	MAN/HOURS
02530.10	**Cast Iron Flanged Pipe**		
Cast iron flanged sections			
4" pipe, with one bolt set			
3' section		EA.	0.218
4' section		"	0.240
5' section		"	0.267
6' section		"	0.300
8' section		"	0.343
10' section		"	0.480
12' section		"	0.800
15' section		"	1.200
18' section		"	1.600
6" pipe, with one bolt set			
3' section		EA.	0.240
4' section		"	0.282
5' section		"	0.320
6' section		"	0.369
8' section		"	0.533
10' section		"	0.600
12' section		"	0.800
15' section		"	1.200
18' section		"	1.714
8" pipe, with one bolt set			
3' section		EA.	0.300
4' section		"	0.343
5' section		"	0.400
6' section		"	0.480
8' section		"	0.686
10' section		"	0.800
12' section		"	1.200
15' section		"	1.600
18' section		"	2.000
10" pipe, with one bolt set			
3' section		EA.	0.308
4' section		"	0.353
5' section		"	0.414
6' section		"	0.500
8' section		"	0.727
10' section		"	0.857
12' section		"	1.333

02 SITE CONSTRUCTION

Sanitary Sewer	UNIT	MAN/HOURS
02530.10 **Cast Iron Flanged Pipe** *(Cont.)*		
15' section	EA.	1.714
18' section	"	2.400
12" pipe, with one bolt set		
3' section	EA.	0.333
4' section	"	0.387
5' section	"	0.462
6' section	"	0.545
8' section	"	0.800
10' section	"	0.923
12' section	"	1.500
15' section	"	2.000
18' section	"	2.667
02530.11 **Cast Iron Fittings**		
Mechanical joint, with 2 bolt kits		
90 deg bend		
4"	EA.	0.533
6"	"	0.615
8"	"	0.800
10"	"	1.143
12"	"	1.600
14"	"	2.000
16"	"	2.667
45 deg bend		
4"	EA.	0.533
6"	"	0.615
8"	"	0.800
10"	"	1.143
12"	"	1.600
14"	"	2.000
16"	"	2.667
Tee, with 3 bolt kits		
4" x 4"	EA.	0.800
6" x 6"	"	1.000
8" x 8"	"	1.333
10" x 10"	"	2.000
12" x 12"	"	2.667
Wye, with 3 bolt kits		
6" x 6"	EA.	1.000
8" x 8"	"	1.333
10" x 10"	"	2.000
12" x 12"	"	2.667
Reducer, with 2 bolt kits		
6" x 4"	EA.	1.000
8" x 6"	"	1.333
10" x 8"	"	2.000
12" x 10"	"	2.667
Flanged, 90 deg bend, 125 lb.		
4"	EA.	0.667
6"	"	0.800
8"	"	1.000
10"	"	1.333
12"	"	2.000

Sanitary Sewer	UNIT	MAN/HOURS
02530.11 **Cast Iron Fittings** *(Cont.)*		
14"	EA.	2.667
16"	"	2.667
Tee		
4"	EA.	1.000
6"	"	1.143
8"	"	1.333
10"	"	1.600
12"	"	2.000
14"	"	2.667
16"	"	4.000
02530.20 **Vitrified Clay Pipe**		
Vitrified clay pipe, extra strength		
6" dia.	L.F.	0.109
8" dia.	"	0.114
10" dia.	"	0.120
12" dia.	"	0.160
15" dia.	"	0.240
18" dia.	"	0.267
24" dia.	"	0.343
30" dia.	"	0.480
36" dia.	"	0.686
02530.30 **Manholes**		
Precast sections, 48" dia.		
Base section	EA.	2.000
1'0" riser	"	1.600
1'4" riser	"	1.714
2'8" riser	"	1.846
4'0" riser	"	2.000
2'8" cone top	"	2.400
Precast manholes, 48" dia.		
4' deep	EA.	4.800
6' deep	"	6.000
7' deep	"	6.857
8' deep	"	8.000
10' deep	"	9.600
Cast-in-place, 48" dia., with frame and cover		
5' deep	EA.	12.000
6' deep	"	13.714
8' deep	"	16.000
10' deep	"	19.200
Brick manholes, 48" dia. with cover, 8" thick		
4' deep	EA.	8.000
6' deep	"	8.889
8' deep	"	10.000
10' deep	"	11.429
12' deep	"	13.333
14' deep	"	16.000
Inverts for manholes		
Single channel	EA.	3.200
Triple channel	"	4.000
Frames and covers, 24" diameter		

02 SITE CONSTRUCTION

Sanitary Sewer	UNIT	MAN/HOURS
02530.30 Manholes *(Cont.)*		
300 lb	EA.	0.800
400 lb	"	0.889
500 lb	"	1.143
Watertight, 350 lb	"	2.667
For heavy equipment, 1200 lb	"	4.000
Steps for manholes		
7" x 9"	EA.	0.160
8" x 9"	"	0.178
Curb inlet, 4' throat, cast in place		
12"-30" pipe	EA.	12.000
36"-48" pipe	"	13.714
Raise exist frame and cover, when repaving	"	4.800
02530.40 Sanitary Sewers		
Clay		
6" pipe	L.F.	0.080
8" pipe	"	0.086
10" pipe	"	0.092
12" pipe	"	0.100
PVC		
4" pipe	L.F.	0.060
6" pipe	"	0.063
8" pipe	"	0.067
10" pipe	"	0.071
12" pipe	"	0.075
Cleanout		
4" pipe	EA.	1.000
6" pipe	"	1.000
8" pipe	"	1.000
Connect new sewer line		
To existing manhole	EA.	2.667
To new manhole	"	1.600
02540.10 Drainage Fields		
Perforated PVC pipe, for drain field		
4" pipe	L.F.	0.053
6" pipe	"	0.057
02540.50 Septic Tanks		
Septic tank, precast concrete		
1000 gals	EA.	4.000
2000 gals	"	6.000
5000 gals	"	12.000
25,000 gals	"	48.000
40,000 gals	"	80.000
Leaching pit, precast concrete, 72" diameter		
3' deep	EA.	3.000
6' deep	"	3.429
8' deep	"	4.000

Piped Energy Distribution	UNIT	MAN/HOURS
02550.10 Gas Distribution		
Gas distribution lines		
Polyethylene, 60 psi coils		
1-1/4" dia.	L.F.	0.053
1-1/2" dia.	"	0.057
2" dia.	"	0.067
3" dia.	"	0.080
30' pipe lengths		
3" dia.	L.F.	0.089
4" dia.	"	0.100
6" dia.	"	0.133
8" dia.	"	0.160
Steel, schedule 40, plain end		
1" dia.	L.F.	0.067
2" dia.	"	0.073
3" dia.	"	0.080
4" dia.	"	0.160
5" dia.	"	0.171
6" dia.	"	0.200
8" dia.	"	0.218
Natural gas meters, direct digital reading, threaded		
250 cfh @ 5 lbs	EA.	1.600
425 cfh @ 10 lbs	"	1.600
800 cfh @ 20 lbs	"	2.000
1,000 cfh @ 25 lbs	"	2.000
1,400 cfh @ 100 lbs	"	2.667
2,300 cfh @ 100 lbs	"	4.000
5,000 cfh @ 100 lbs	"	8.000
Gas pressure regulators		
Threaded		
3/4"	EA.	1.000
1"	"	1.333
1-1/4"	"	1.333
1-1/2"	"	1.333
2"	"	1.600
Flanged		
3"	EA.	2.000
4"	"	2.667
02550.40 Steel Pipe		
Steel pipe, extra heavy, A 53, grade B, seamless		
1/2" dia.	L.F.	0.080
3/4" dia.	"	0.084
1" dia.	"	0.089
1-1/4" dia.	"	0.100
1-1/2" dia.	"	0.114
2" dia.	"	0.133
3" dia.	"	0.120
4" dia.	"	0.133
6" dia.	"	0.150
8" dia.	"	0.171
10" dia.	"	0.200
12" dia.	"	0.240

02 SITE CONSTRUCTION

Piped Energy Distribution	UNIT	MAN/HOURS
02550.80 Steam Meters		
In-line turbine, direct reading, 300 lb, flanged		
2"	EA.	1.000
3"	"	1.333
4"	"	1.600
Threaded, 2"		
5" line	EA.	8.000
6" line	"	8.000
8" line	"	8.000
10" line	"	8.000
12" line	"	8.000
14" line	"	8.000
16" line	"	8.000

Power & Communications	UNIT	MAN/HOURS
02580.20 High Voltage Cable		
High voltage XLP copper cable, shielded, 5000v		
#6 awg	L.F.	0.013
#4 awg	"	0.016
#2 awg	"	0.019
#1 awg	"	0.021
#1/0 awg	"	0.024
#2/0 awg	"	0.029
#3/0 awg	"	0.034
#4/0 awg	"	0.036
#250 awg	"	0.043
#300 awg	"	0.048
#350 awg	"	0.053
#500 awg	"	0.073
#750 awg	"	0.080
Ungrounded, 15,000v		
#1 awg	L.F.	0.031
#1/0 awg	"	0.034
#2/0 awg	"	0.036
#3/0 awg	"	0.040
#4/0 awg	"	0.046
#250 awg	"	0.048
#300 awg	"	0.053
#350 awg	"	0.062
#500 awg	"	0.080
#750 awg	"	0.098
#1000 awg	"	0.123
Aluminum cable, shielded, 5000v		
#6 awg	L.F.	0.011
#4 awg	"	0.013
#2 awg	"	0.015

Power & Communications	UNIT	MAN/HOURS
02580.20 High Voltage Cable *(Cont.)*		
#1 awg	L.F.	0.017
#1/0 awg	"	0.019
#2/0 awg	"	0.020
#3/0 awg	"	0.021
#4/0 awg	"	0.024
#250 awg	"	0.026
#300 awg	"	0.031
#350 awg	"	0.034
#500 awg	"	0.036
#750 awg	"	0.044
#1000 awg	"	0.050
Ungrounded, 15,000v		
#1 awg	L.F.	0.021
#1/0 awg	"	0.025
#2/0 awg	"	0.027
#3/0 awg	"	0.028
#4/0 awg	"	0.029
#250 awg	"	0.031
#300 awg	"	0.032
#350 awg	"	0.036
#500 awg	"	0.043
#750 awg	"	0.052
#1000 awg	"	0.064
Indoor terminations, 5000v		
#6 - #4	EA.	0.157
#2 - #2/0	"	0.157
#3/0 - #250	"	0.157
#300 - #750	"	2.759
#1000	"	3.810
In-line splice, 5000v		
#6 - #4/0	EA.	3.810
#250 - #500	"	10.000
#750 - #1000	"	13.008
T-splice, 5000v		
#2 - #4/0	EA.	11.994
#250 - #500	"	20.000
#750 - #1000	"	25.000
Indoor terminations, 15,000v		
#2 - #2/0	EA.	3.478
#3/0 - #500	"	5.333
#750 - #1000	"	6.154
In-line splice, 15,000v		
#2 - #4/0	EA.	8.999
#250 - #500	"	11.994
#750 - #1000	"	18.018
T-splice, 15,000v		
#4	EA.	18.018
#250 - #500	"	29.963
#750 - #1000	"	44.944
Compression lugs, 15,000v		
#4	EA.	0.400
#2	"	0.533
#1	"	0.533

02 SITE CONSTRUCTION

Power & Communications	UNIT	MAN/HOURS
02580.20 High Voltage Cable *(Cont.)*		
#1/0	EA.	0.667
#2/0	"	0.667
#3/0	"	0.851
#4/0	"	0.851
#250	"	0.952
#300	"	0.952
#350	"	1.159
#500	"	1.250
#750	"	1.509
#1000	"	1.905
Compression splices, 15,000v		
#4	EA.	0.667
#2	"	0.727
#1	"	0.899
#1/0	"	1.000
#2/0	"	1.159
#3/0	"	1.250
#4/0	"	1.404
#250	"	1.509
#350	"	1.739
#500	"	2.000
#750	"	2.500
02580.40 Supports & Connectors		
Cable supports for conduit		
1-1/2"	EA.	0.348
2"	"	0.348
2-1/2"	"	0.400
3"	"	0.400
3-1/2"	"	0.500
4"	"	0.500
5"	"	0.667
6"	"	0.727
Split bolt connectors		
#10	EA.	0.200
#8	"	0.200
#6	"	0.200
#4	"	0.400
#3	"	0.400
#2	"	0.400
#1/0	"	0.667
#2/0	"	0.667
#3/0	"	0.667
#4/0	"	0.667
#250	"	1.000
#350	"	1.000
#500	"	1.000
#750	"	1.509
#1000	"	1.509
Single barrel lugs		
#6	EA.	0.250
#1/0	"	0.500
#250	"	0.667

Power & Communications	UNIT	MAN/HOURS
02580.40 Supports & Connectors *(Cont.)*		
#350	EA.	0.667
#500	"	0.667
#600	"	0.899
#800	"	0.899
#1000	"	0.899
Double barrel lugs		
#1/0	EA.	0.899
#250	"	1.290
#350	"	1.290
#600	"	1.905
#800	"	1.905
#1000	"	1.905
Three barrel lugs		
#2/0	EA.	1.290
#250	"	1.905
#350	"	1.905
#600	"	2.667
#800	"	2.667
#1000	"	2.667
Four barrel lugs		
#250	EA.	2.759
#350	"	2.759
#600	"	3.478
#800	"	3.478
Compression conductor adapters		
#6	EA.	0.296
#4	"	0.348
#2	"	0.444
#1	"	0.444
#1/0	"	0.533
#250	"	0.800
#350	"	0.851
#500	"	1.096
#750	"	1.143
Terminal blocks, 2 screw		
3 circuit	EA.	0.200
6 circuit	"	0.200
8 circuit	"	0.200
10 circuit	"	0.296
12 circuit	"	0.296
18 circuit	"	0.296
24 circuit	"	0.348
36 circuit	"	0.348
Compression splice		
#8 awg	EA.	0.381
#6 awg	"	0.276
#4 awg	"	0.276
#2 awg	"	0.533
#1 awg	"	0.533
#1/0 awg	"	0.533
#2/0 awg	"	0.851
#3/0 awg	"	0.851
#4/0 awg	"	0.851

02 SITE CONSTRUCTION

Power & Communications	UNIT	MAN/HOURS
02580.40 Supports & Connectors (Cont.)		
#250 awg	EA.	1.356
#300 awg	"	1.356
#350 awg	"	1.404
#400 awg	"	1.404
#500 awg	"	1.509
#600 awg	"	1.509
#750 awg	"	1.739
#1000 awg	"	1.739

Drainage And Containment	UNIT	MAN/HOURS
02630.10 Catch Basins		
Standard concrete catch basin		
Cast in place, 3'8" x 3'8", 6" thick wall		
2' deep	EA.	6.000
3' deep	"	6.000
4' deep	"	8.000
5' deep	"	8.000
6' deep	"	9.600
4'x4', 8" thick wall, cast in place		
2' deep	EA.	6.000
3' deep	"	6.000
4' deep	"	8.000
5' deep	"	8.000
6' deep	"	9.600
Frames and covers, cast iron		
Round		
24" dia.	EA.	2.000
26" dia.	"	2.000
28" dia.	"	2.000
Rectangular		
23"x23"	EA.	2.000
27"x20"	"	2.000
24"x24"	"	2.000
26"x26"	"	2.000
Curb inlet frames and covers		
27"x27"	EA.	2.000
24"x36"	"	2.000
24"x25"	"	2.000
24"x22"	"	2.000
20"x22"	"	2.000
Airfield catch basin frame and grating, galvanized		
2'x4'	EA.	2.000
2'x2'	"	2.000

Drainage And Containment	UNIT	MAN/HOURS
02630.40 Storm Drainage		
Concrete pipe		
Plain, bell and spigot joint, Class II		
6" pipe	L.F.	0.109
8" pipe	"	0.120
10" pipe	"	0.126
12" pipe	"	0.133
15" pipe	"	0.141
18" pipe	"	0.150
21" pipe	"	0.160
24" pipe	"	0.171
Reinforced, class III, tongue and groove joint		
12" pipe	L.F.	0.133
15" pipe	"	0.141
18" pipe	"	0.150
21" pipe	"	0.160
24" pipe	"	0.171
27" pipe	"	0.185
30" pipe	"	0.200
36" pipe	"	0.218
42" pipe	"	0.240
48" pipe	"	0.267
54" pipe	"	0.300
60" pipe	"	0.343
66" pipe	"	0.400
72" pipe	"	0.480
Flared end-section, concrete		
12" pipe	L.F.	0.133
15" pipe	"	0.141
18" pipe	"	0.150
24" pipe	"	0.171
30" pipe	"	0.200
36" pipe	"	0.218
42" pipe	"	0.240
48" pipe	"	0.267
54" pipe	"	0.300
Porous concrete pipe standard strength		
4" pipe	L.F.	0.092
6" pipe	"	0.096
8" pipe	"	0.100
10" pipe	"	0.104
12" pipe	"	0.109
Corrugated metal pipe, coated, paved invert		
16 ga.		
8" pipe	L.F.	0.080
10" pipe	"	0.083
12" pipe	"	0.086
15" pipe	"	0.092
18" pipe	"	0.100
21" pipe	"	0.109
24" pipe	"	0.120
30" pipe	"	0.133
36" pipe	"	0.150
12 ga., 48" pipe	"	0.171

02 SITE CONSTRUCTION

Drainage And Containment	UNIT	MAN/HOURS
02630.40 Storm Drainage (Cont.)		
10 ga.		
60" pipe	L.F.	0.200
72" pipe	"	0.240
Galvanized or aluminum, plain		
16 ga.		
8" pipe	L.F.	0.080
10" pipe	"	0.083
12" pipe	"	0.086
15" pipe	"	0.092
18" pipe	"	0.100
24" pipe	"	0.120
30" pipe	"	0.133
36" pipe	"	0.150
12 ga., 48" pipe	"	0.171
10 ga., 60" pipe	"	0.200
Galvanized or aluminum, coated oval arch		
16 ga.		
17" x 13"	L.F.	0.109
21" x 15"	"	0.120
14 ga.		
28" x 20"	L.F.	0.133
35" x 24"	"	0.171
12 ga.		
42" x 29"	L.F.	0.200
57" x 38"	"	0.240
64" x 43"	"	0.253
Oval arch culverts, plain		
16 ga.		
17" x 13"	L.F.	0.109
21" x 15"	"	0.120
14 ga.		
28" x 20"	L.F.	0.133
35" x 24"	"	0.171
12 ga.		
57" x 38"	L.F.	0.200
64" x 43"	"	0.240
71" x 47"	"	0.253
Nestable corrugated metal pipe		
16 ga.		
10" pipe	L.F.	0.083
12" pipe	"	0.086
15" pipe	"	0.092
18" pipe	"	0.100
24" pipe	"	0.120
30" pipe	"	0.133
14 ga., 36" pipe	"	0.150
Headwalls, cast in place, 30 deg wingwall		
12" pipe	EA.	2.000
15" pipe	"	2.000
18" pipe	"	2.286
24" pipe	"	2.286
30" pipe	"	2.667
36" pipe	"	4.000

Drainage And Containment	UNIT	MAN/HOURS
02630.40 Storm Drainage (Cont.)		
42" pipe	EA.	4.000
48" pipe	"	5.333
54" pipe	"	6.667
60" pipe	"	8.000
4" cleanout for storm drain		
4" pipe	EA.	1.000
6" pipe	"	1.000
8" pipe	"	1.000
Connect new drain line		
To existing manhole	EA.	2.667
To new manhole	"	1.600
02630.70 Underdrain		
Drain tile, clay		
6" pipe	L.F.	0.053
8" pipe	"	0.056
12" pipe	"	0.060
Porous concrete, standard strength		
6" pipe	L.F.	0.053
8" pipe	"	0.056
12" pipe	"	0.060
15" pipe	"	0.067
18" pipe	"	0.080
Corrugated metal pipe, perforated type		
6" pipe	L.F.	0.060
8" pipe	"	0.063
10" pipe	"	0.067
12" pipe	"	0.071
18" pipe	"	0.075
Perforated clay pipe		
6" pipe	L.F.	0.069
8" pipe	"	0.071
12" pipe	"	0.073
Drain tile, concrete		
6" pipe	L.F.	0.053
8" pipe	"	0.056
12" pipe	"	0.060
Perforated rigid PVC underdrain pipe		
4" pipe	L.F.	0.040
6" pipe	"	0.048
8" pipe	"	0.053
10" pipe	"	0.060
12" pipe	"	0.069
Underslab drainage, crushed stone		
3" thick	S.F.	0.008
4" thick	"	0.009
6" thick	"	0.010
8" thick	"	0.010
Plastic filter fabric for drain lines	"	0.008
Gravel fill in trench, crushed or bank run, 1/2" to 3/4"	C.Y.	0.600

02 SITE CONSTRUCTION

Base Courses And Ballasts

02720.10 Railroad Ballast, Rail,

Item	UNIT	MAN/HOURS
Rail		
90 lb	L.F.	0.010
100 lb	"	0.010
115 lb	"	0.010
132 lb	"	0.010
Rail relay		
90 lb	L.F.	0.010
100 lb	"	0.010
115 lb	"	0.010
132 lb	"	0.010
New angle bars, per pair		
90 lb	EA.	0.012
100 lb	"	0.012
115 lb	"	0.012
132 lb	"	0.012
Angle bar relay		
90 lb	EA.	0.012
100 lb	"	0.012
115 lb	"	0.012
132 lb	"	0.012
New tie plates		
90 lb	EA.	0.009
100 lb	"	0.009
115 lb	"	0.009
132 lb	"	0.009
Tie plate relay		
90 lb	EA.	0.009
100 lb	"	0.009
115 lb	"	0.009
132 lb	"	0.009
Track accessories		
Wooden cross ties, 8'	EA.	0.060
Concrete cross ties, 8'	"	0.120
Tie plugs, 5"	"	0.006
Track bolts and nuts, 1"	"	0.006
Lockwashers, 1"	"	0.004
Track spikes, 6"	"	0.024
Wooden switch ties	B.F.	0.006
Rail anchors	EA.	0.022
Ballast	TON	0.120
Gauge rods	EA.	0.096
Compromise splice bars	"	0.160
Turnout		
90 lb	EA.	24.000
100 lb	"	24.000
110 lb	"	24.000
115 lb	"	24.000
132 lb	"	24.000
Turnout relay		
90 lb	EA.	24.000
100 lb	"	24.000
110 lb	"	24.000
115 lb	"	24.000

02720.10 Railroad Ballast, Rail (Cont.)

Item	UNIT	MAN/HOURS
132 lb	EA.	24.000
Railroad track in place, complete		
New rail		
90 lb	L.F.	0.240
100 lb	"	0.240
110 lb	"	0.240
115 lb	"	0.240
132 lb	"	0.240
Rail relay		
90 lb	L.F.	0.240
100 lb	"	0.240
110 lb	"	0.240
115 lb	"	0.240
132 lb	"	0.240
No. 8 turnout		
90 lb	EA.	32.000
100 lb	"	32.000
110 lb	"	32.000
115 lb	"	32.000
132 lb	"	32.000
No. 8 turnout relay		
90 lb	EA.	32.000
100 lb	"	32.000
110 lb	"	32.000
115 lb	"	32.000
132 lb	"	32.000
Railroad crossings, asphalt, based on 8" thick x 20'		
Including track and approach		
12' roadway	EA.	6.000
15' roadway	"	6.857
18' roadway	"	8.000
21' roadway	"	9.600
24' roadway	"	12.000
Precast concrete inserts		
12' roadway	EA.	2.400
15' roadway	"	3.000
18' roadway	"	4.000
21' roadway	"	4.800
24' roadway	"	5.333
Molded rubber, with headers		
12' roadway	EA.	2.400
15' roadway	"	3.000
18' roadway	"	4.000
21' roadway	"	4.800
24' roadway	"	5.333

02 SITE CONSTRUCTION

Flexible Surfaces	UNIT	MAN/HOURS
02740.10 Asphalt Repair		
Coal tar seal coat, rubber add., fuel resist.	S.Y.	0.011
Bituminous surface treatment, single	"	0.008
Double	"	0.001
Bituminous prime coat	"	0.001
Tack coat	"	0.001
Crack sealing, concrete paving	L.F.	0.005
Bituminous paving for pipe trench, 4" thick	S.Y.	0.160
Polypropylene, nonwoven paving fabric	"	0.004
Rubberized asphalt	"	0.073
Asphalt slurry seal	"	0.047
02740.20 Asphalt Surfaces		
Asphalt wearing surface, flexible pavement		
1" thick	S.Y.	0.016
1-1/2" thick	"	0.019
2" thick	"	0.024
3" thick	"	0.032
Binder course		
1-1/2" thick	S.Y.	0.018
2" thick	"	0.022
3" thick	"	0.029
4" thick	"	0.032
5" thick	"	0.036
6" thick	"	0.040
Bituminous sidewalk, no base		
2" thick	S.Y.	0.028
3" thick	"	0.030

Rigid Pavement	UNIT	MAN/HOURS
02750.10 Concrete Paving		
Concrete paving, reinforced, 5000 psi concrete		
6" thick	S.Y.	0.150
7" thick	"	0.160
8" thick	"	0.171
9" thick	"	0.185
10" thick	"	0.200
11" thick	"	0.218
12" thick	"	0.240
15" thick	"	0.300
Concrete paving, for pipe trench, reinforced		
7" thick	S.Y.	0.240
8" thick	"	0.267
9" thick	"	0.300
10" thick	"	0.343
Fibrous concrete		

Rigid Pavement	UNIT	MAN/HOURS
02750.10 Concrete Paving *(Cont.)*		
5" thick	S.Y.	0.185
8" thick	"	0.200
Roller comp.conc., (RCC), place and compact		
8" thick	S.Y.	0.240
12" thick	"	0.300
Steel edge forms up to		
12" deep	L.F.	0.027
15" deep	"	0.032
Paving finishes		
Belt dragged	S.Y.	0.040
Curing	"	0.008
02760.10 Pavement Markings		
Pavement line marking, paint		
4" wide	L.F.	0.002
6" wide	"	0.004
8" wide	"	0.007
Reflective paint, 4" wide	"	0.007
Airfield markings, retro-reflective		
White	L.F.	0.007
Yellow	"	0.007
Preformed tape, 4" wide		
Inlaid reflective	L.F.	0.001
Reflective paint	"	0.002
Thermoplastic		
White	L.F.	0.004
Yellow	"	0.004
12" wide, thermoplastic, white	"	0.011
Directional arrows, reflective preformed tape	EA.	0.800
Messages, reflective preformed tape (per letter)	"	0.400
Handicap symbol, preformed tape	"	0.800
Parking stall painting	"	0.160

Site Improvements	UNIT	MAN/HOURS
02810.40 Lawn Irrigation		
Minimum	ACRE	
02820.10 Chain Link Fence		
Chain link fence, 9 ga., galvanized, with posts 10' o.c.		
4' high	L.F.	0.057
5' high	"	0.073
6' high	"	0.100
7' high	"	0.123
8' high	"	0.160
For barbed wire with hangers, add		

02 SITE CONSTRUCTION

Site Improvements

02820.10 Chain Link Fence (Cont.)

Item	UNIT	MAN/HOURS
3 strand	L.F.	0.040
6 strand	"	0.067
Corner or gate post, 3" post		
4' high	EA.	0.267
5' high	"	0.296
6' high	"	0.348
7' high	"	0.400
8' high	"	0.444
4" post		
4' high	EA.	0.296
5' high	"	0.348
6' high	"	0.400
7' high	"	0.444
8' high	"	0.500
Gate with gate posts, galvanized, 3' wide		
4' high	EA.	2.000
5' high	"	2.667
6' high	"	2.667
7' high	"	4.000
8' high	"	4.000
Fabric, galvanized chain link, 2" mesh, 9 ga.		
4' high	L.F.	0.027
5' high	"	0.032
6' high	"	0.040
8' high	"	0.053
Line post, no rail fitting, galvanized, 2-1/2" dia.		
4' high	EA.	0.229
5' high	"	0.250
6' high	"	0.267
7' high	"	0.320
8' high	"	0.400
1-7/8" H beam		
4' high	EA.	0.229
5' high	"	0.250
6' high	"	0.267
7' high	"	0.320
8' high	"	0.400
2-1/4" H beam		
4' high	EA.	0.229
5' high	"	0.250
6' high	"	0.267
7' high	"	0.320
8' high	"	0.400
Vinyl coated, 9 ga., with posts 10' o.c.		
4' high	L.F.	0.057
5' high	"	0.073
6' high	"	0.100
7' high	"	0.123
8' high	"	0.160
For barbed wire w/hangers, add		
3 strand	L.F.	0.040
6 Strand	"	0.067
Corner, or gate post, 4' high		

02820.10 Chain Link Fence (Cont.)

Item	UNIT	MAN/HOURS
3" dia.	EA.	0.267
4" dia.	"	0.267
6" dia.	"	0.320
Gate, with posts, 3' wide		
4' high	EA.	2.000
5' high	"	2.667
6' high	"	2.667
7' high	"	4.000
8' high	"	4.000
Line post, no rail fitting, 2-1/2" dia.		
4' high	EA.	0.229
5' high	"	0.250
6' high	"	0.267
7' high	"	0.320
8' high	"	0.400
Corner post, no top rail fitting, 4" dia.		
4' high	EA.	0.267
5' high	"	0.296
6' high	"	0.348
7' high	"	0.400
8' high	"	0.444
Fabric, vinyl, chain link, 2" mesh, 9 ga.		
4' high	L.F.	0.027
5' high	"	0.032
6' high	"	0.040
8' high	"	0.053
Swing gates, galvanized, 4' high		
Single gate		
3' wide	EA.	2.000
4' wide	"	2.000
Double gate		
10' wide	EA.	3.200
12' wide	"	3.200
14' wide	"	3.200
16' wide	"	3.200
18' wide	"	4.571
20' wide	"	4.571
22' wide	"	4.571
24' wide	"	5.333
26' wide	"	5.333
28' wide	"	6.400
30' wide	"	6.400
5' high		
Single gate		
3' wide	EA.	2.667
4' wide	"	2.667
Double gate		
10' wide	EA.	4.000
12' wide	"	4.000
14' wide	"	4.000
16' wide	"	4.000
18' wide	"	4.571
20' wide	"	4.571

02 SITE CONSTRUCTION

Site Improvements	UNIT	MAN/HOURS
02820.10 **Chain Link Fence** *(Cont.)*		
22' wide	EA.	4.571
24' wide	"	5.333
26' wide	"	5.333
28' wide	"	6.400
30' wide	"	6.400
6' high		
Single gate		
3' wide	EA.	2.667
4' wide	"	2.667
Double gate		
10' wide	EA.	4.000
12' wide	"	4.000
14' wide	"	4.000
16' wide	"	4.000
18' wide	"	4.571
20' wide	"	4.571
22' wide	"	4.571
24' wide	"	5.333
26' wide	"	5.333
28' wide	"	6.400
30' wide	"	6.400
7' high		
Single gate		
3' wide	EA.	4.000
4' wide	"	4.000
Double gate		
10' wide	EA.	5.333
12' wide	"	5.333
14' wide	"	5.333
16' wide	"	5.333
18' wide	"	6.400
20' wide	"	6.400
22' wide	"	6.400
24' wide	"	8.000
26' wide	"	8.000
28' wide	"	10.000
30' wide	"	10.000
8' high		
Single gate		
3' wide	EA.	4.000
4' wide	"	4.000
Double gate		
10' wide	EA.	5.333
12' wide	"	5.333
14' wide	"	5.333
16' wide	"	5.333
18' wide	"	6.400
20' wide	"	6.400
22' wide	"	6.400
24' wide	"	8.000
26' wide	"	8.000
28' wide	"	10.000
30' wide	"	10.000

Site Improvements	UNIT	MAN/HOURS
02820.10 **Chain Link Fence** *(Cont.)*		
Vinyl coated swing gates, 4' high		
Single gate		
3' wide	EA.	2.000
4' wide	"	2.000
Double gate		
10' wide	EA.	3.200
12' wide	"	3.200
14' wide	"	3.200
16' wide	"	3.200
18' wide	"	4.571
20' wide	"	4.571
22' wide	"	4.571
24' wide	"	5.333
26' wide	"	5.333
28' wide	"	6.400
30' wide	"	6.400
5' high		
Single gate		
3' wide	EA.	2.667
4' wide	"	2.667
Double gate		
10' wide	EA.	4.000
12' wide	"	4.000
14' wide	"	4.000
16' wide	"	4.000
18' wide	"	4.571
20' wide	"	4.571
22' wide	"	4.571
24' wide	"	5.333
26' wide	"	5.333
28' wide	"	6.400
30' wide	"	6.400
6' high		
Single gate		
3' wide	EA.	2.667
4' wide	"	2.667
Double gate		
10' wide	EA.	4.000
12' wide	"	4.000
14' wide	"	4.000
16' wide	"	4.000
18' wide	"	4.571
20' wide	"	4.571
22' wide	"	4.571
24' wide	"	5.333
26' wide	"	5.333
28' wide	"	6.400
30' wide	"	6.400
7' high		
Single gate		
3' wide	EA.	4.000
4' wide	"	4.000
Double gate		

02 SITE CONSTRUCTION

Site Improvements	UNIT	MAN/HOURS
02820.10 Chain Link Fence *(Cont.)*		
10' wide	EA.	5.333
12' wide	"	5.333
14' wide	"	5.333
16' wide	"	5.333
18' wide	"	6.400
20' wide	"	6.400
22' wide	"	6.400
24' wide	"	8.000
26' wide	"	8.000
28' wide	"	10.000
30' wide	"	10.000
8' high		
Single gate		
3' wide	EA.	4.000
4' wide	"	4.000
Double gate		
10' wide	EA.	5.333
12' wide	"	5.333
14' wide	"	5.333
16' wide	"	5.333
18' wide	"	6.400
20' wide	"	6.400
22' wide	"	6.400
24' wide	"	8.000
28' wide	"	8.000
30' wide	"	10.000
Drilling fence post holes		
In soil		
By hand	EA.	0.400
By machine auger	"	0.200
In rock		
By jackhammer	EA.	2.667
By rock drill	"	0.800
Aluminum privacy slats, installed vertically	S.F.	0.020
Post hole, dig by hand	EA.	0.533
Set fence post in concrete	"	0.400
02840.30 Guardrails		
Pipe bollard, steel pipe, concrete filled, painted		
6" dia.	EA.	0.667
8" dia.	"	1.000
12" dia.	"	2.667
Corrugated steel, guardrail, galvanized	L.F.	0.040
End section, wrap around or flared	EA.	0.800
Timber guardrail, 4" x 8"	L.F.	0.030
Guard rail, 3 cables, 3/4" dia.		
Steel posts	L.F.	0.120
Wood posts	"	0.096
Steel box beam		
6" x 6"	L.F.	0.133
6" x 8"	"	0.150
Concrete posts	EA.	0.400
Barrel type impact barrier	"	0.800

Site Improvements	UNIT	MAN/HOURS
02840.30 Guardrails *(Cont.)*		
Light shield, 6' high	L.F.	0.160
02840.40 Parking Barriers		
Timber, treated, 4' long		
4" x 4"	EA.	0.667
6" x 6"	"	0.800
Precast concrete, 6' long, with dowels		
12" x 6"	EA.	0.400
12" x 8"	"	0.444
02840.60 Signage		
Traffic signs		
Reflectorized per OSHA stds., incl. post		
Stop, 24"x24"	EA.	0.533
Yield, 30" triangle	"	0.533
Speed limit, 12"x18"	"	0.533
Directional, 12"x18"	"	0.533
Exit, 12"x18"	"	0.533
Entry, 12"x18"	"	0.533
Warning, 24"x24"	"	0.533
Informational, 12"x18"	"	0.533
Handicap parking, 12"x18"	"	0.533
02880.40 Recreational Facilities		
Bleachers, outdoor, portable, per seat		
10 tiers		
Minimum	EA.	0.150
Maximum	"	0.200
20 tiers		
Minimum	EA.	0.141
Maximum	"	0.185
Grandstands, fixed, wood seat, steel frame		
Per seat, 15 tiers		
Minimum	EA.	0.240
Maximum	"	0.400
30 tiers		
Minimum	EA.	0.218
Maximum	"	0.343
Seats		
Seat backs only		
Fiberglass	EA.	0.080
Steel and wood seat	"	0.080
Seat restoration, fiberglass on wood		
Seats	EA.	0.160
Plain bench, no backs	"	0.067
Benches		
Park, precast concrete with backs		
4' long	EA.	2.667
8' long	"	4.000
Fiberglass, with backs		
4' long	EA.	2.000
8' long	"	2.667
Wood, with backs and fiberglass supports		

02 SITE CONSTRUCTION

Site Improvements	UNIT	MAN/HOURS
02880.40 Recreational Facilities (Cont.)		
4' long	EA.	2.000
8' long	"	2.667
Steel frame, 6' long		
All steel	EA.	2.000
Hardwood boards	"	2.000
Players bench, steel frame, fir seat, 10' long	"	2.667
Soccer goal posts	PAIR	
Running track		
Gravel and cinders over stone base	S.Y.	0.060
Rubber-cork base resilient pavement	"	0.480
For colored surfaces, add	"	0.048
Colored rubberized asphalt	"	0.600
Artificial resilient mat over asphalt	"	1.200
Tennis courts		
Bituminous pavement, 2-1/2" thick	S.Y.	0.150
Colored sealer, acrylic emulsion		
3 coats	S.Y.	0.053
For 2 color seal coating, add	"	0.008
For preparing old courts, add	"	0.005
Net, nylon, 42' long	EA.	1.000
Paint markings on asphalt, 2 coats	"	8.000
Playground equipment		
Basketball backboard		
Minimum	EA.	2.000
Maximum	"	2.286
Bike rack, 10' long	"	1.600
Golf shelter, fiberglass	"	2.000
Ground socket for movable posts		
Minimum	EA.	0.500
Maximum	"	0.500
Horizontal monkey ladder, 14' long	"	1.333
Posts, tether ball	"	0.400
Multiple purpose, 10' long	"	0.800
See-saw, steel		
Minimum	EA.	3.200
Average	"	4.000
Maximum	"	5.333
Slide		
Minimum	EA.	6.400
Maximum	"	7.273
Swings, plain seats		
8' high		
Minimum	EA.	5.333
Maximum	"	6.154
12' high		
Minimum	EA.	6.154
Maximum	"	8.889

Site Improvements	UNIT	MAN/HOURS
02880.70 Recreational Courts		
Walls, galvanized steel		
8' high	L.F.	0.160
10' high	"	0.178
12' high	"	0.211
Vinyl coated		
8' high	L.F.	0.160
10' high	"	0.178
12' high	"	0.211
Gates, galvanized steel		
Single, 3' transom		
3'x7'	EA.	4.000
4'x7'	"	4.571
5'x7'	"	5.333
6'x7'	"	6.400
Double, 3' transom		
10'x7'	EA.	16.000
12'x7'	"	17.778
14'x7'	"	20.000
Double, no transom		
10'x10'	EA.	13.333
12'x10'	"	16.000
14'x10'	"	17.778
Vinyl coated		
Single, 3' transom		
3'x7'	EA.	4.000
4'x7'	"	4.571
5'x7'	"	5.333
6'x7'	"	6.400
Double, 3'		
10'x7'	EA.	16.000
12'x7'	"	17.778
14'x7'	"	20.000
Double, no transom		
10'x10'	EA.	13.333
12'x10'	"	16.000
14'x10'	"	17.778
Wire and miscellaneous metal fences		
Chicken wire, post 4' o.c.		
2" mesh		
4' high	L.F.	0.040
6' high	"	0.053
Galvanized steel		
12 gauge, 2" by 4" mesh, posts 5' o.c.		
3' high	L.F.	0.040
5' high	"	0.050
14 gauge, 1" by 2" mesh, posts 5' o.c.		
3' high	L.F.	0.040
5' high	"	0.050

02 SITE CONSTRUCTION

Planting

02910.10 Topsoil

	UNIT	MAN/HOURS
Spread topsoil, with equipment		
Minimum	C.Y.	0.080
Maximum	"	0.100
By hand		
Minimum	C.Y.	0.800
Maximum	"	1.000
Area prep. seeding (grade, rake and clean)		
Square yard	S.Y.	0.006
By acre	ACRE	32.000
Remove topsoil and stockpile on site		
4" deep	C.Y.	0.067
6" deep	"	0.062
Spreading topsoil from stock pile		
By loader	C.Y.	0.073
By hand	"	0.800
Top dress by hand	S.Y.	0.008
Place imported top soil		
By loader		
4" deep	S.Y.	0.008
6" deep	"	0.009
By hand		
4" deep	S.Y.	0.089
6" deep	"	0.100
Plant bed preparation, 18" deep		
With backhoe/loader	S.Y.	0.020
By hand	"	0.133

02920.10 Fertilizing

	UNIT	MAN/HOURS
Fertilizing (23#/1000 sf)		
By square yard	S.Y.	0.002
By acre	ACRE	10.000
Liming (70#/1000 sf)		
By square yard	S.Y.	0.003
By acre	ACRE	13.333

02920.30 Seeding

	UNIT	MAN/HOURS
Mechanical seeding, 175 lb/acre		
By square yard	S.Y.	0.002
By acre	ACRE	8.000
450 lb/acre		
By square yard	S.Y.	0.002
By acre	ACRE	10.000
Seeding by hand, 10 lb per 100 s.y.		
By square yard	S.Y.	0.003
By acre	ACRE	13.333
Reseed disturbed areas	S.F.	0.004

02930.10 Plants

	UNIT	MAN/HOURS
Euonymus coloratus, 18" (Purple Wintercreeper)	EA.	0.133
Hedera Helix, 2-1/4" pot (English ivy)	"	0.133
Liriope muscari, 2" clumps	"	0.080

02930.10 Plants (Cont.)

	UNIT	MAN/HOURS
Santolina, 12"	EA.	0.080
Vinca major or minor, 3" pot	"	0.080
Cortaderia argentia, 2 gallon (Pampas Grass)	"	0.080
Ophiopogan japonicus, 1 quart (4" pot)	"	0.080
Ajuga reptans, 2-3/4" pot (carpet bugle)	"	0.080
Pachysandra terminalis, 2-3/4" pot (Japanese Spurge)	"	0.080

02930.30 Shrubs

	UNIT	MAN/HOURS
Juniperus conferia litoralis, 18"-24" (Shore Juniper)	EA.	0.320
Horizontalis plumosa, 18"-24" (Andorra Juniper)	"	0.320
Sabina tamar-iscfolia-tamarix juniper, 18"-24"	"	0.320
Chin San Jose, 18"-24" (San Jose Juniper)	"	0.320
Sargenti, 18"-24" (Sargent's Juniper)	"	0.320
Nandina domestica, 18"-24" (Heavenly Bamboo)	"	0.320
Raphiolepis Indica Springtime, 18"-24"	"	0.320
Osmanthus Heterophyllus Gulftide, 18"-24"	"	0.320
Ilex Cornuta Burfordi Nana, 18"-24"	"	0.320
Glabra, 18"-24" (Inkberry Holly)	"	0.320
Azalea, Indica types, 18"-24"	"	0.320
Kurume types, 18"-24"	"	0.320
Berberis Julianae, 18"-24" (Wintergreen Barberry)	"	0.320
Pieris Japonica Japanese, 18"-24"	"	0.320
Ilex Cornuta Rotunda, 18"-24"	"	0.320
Juniperus Horiz. Plumosa, 24"-30"	"	0.400
Rhodopendrow Hybrids, 24"-30"	"	0.400
Aucuba Japonica Varigata, 24"-30"	"	0.400
Ilex Crenata Willow Leaf, 24"-30"	"	0.400
Cleyera Japonica, 30"-36"	"	0.500
Pittosporum Tobira, 30"-36"	"	0.500
Prumus Laurocerasus, 30"-36"	"	0.500
Ilex Cornuta Burfordi, 30"-36" (Burford Holly)	"	0.500
Abelia Grandiflora, 24"-36" (Yew Podocarpus)	"	0.400
Podocarpos Macrophylla, 24"-36"	"	0.400
Pyracantha Coccinea Lalandi, 3'-4' (Firethorn)	"	0.500
Photinia Frazieri, 3'-4' (Red Photinia)	"	0.500
Forsythia Suspensa, 3'-4' (Weeping Forsythia)	"	0.500
Camellia Japonica, 3'-4' (Common Camellia)	"	0.500
Juniperus Chin Torulosa, 3'-4' (Hollywood Juniper)	"	0.500
Cupressocyparis Leylandi, 3'-4'	"	0.500
Ilex Opaca Fosteri, 5'-6' (Foster's Holly)	"	0.667
Opaca, 5'-6' (American Holly)	"	0.667
Nyrica Cerifera, 4'-5' (Southern Wax Myrtles)	"	0.571
Ligustrum Japonicum, 4'-5' (Japanese Privet)	"	0.571

02930.60 Trees

	UNIT	MAN/HOURS
Cornus Florida, 5'-6' (White flowering Dogwood)	EA.	0.667
Prunus Serrulata Kwanzan, 6'-8' (Kwanzan Cherry)	"	0.800
Caroliniana, 6'-8' (Carolina Cherry Laurel)	"	0.800
Cercis Canadensis, 6'-8' (Eastern Redbud)	"	0.800
Koelreuteria Paniculata, 8'-10' (Goldenrain Tree)	"	1.000
Acer Platanoides, 1-3/4"-2" (11'-13')	"	1.333
Rubrum, 1-3/4"-2" (11'-13') (Red Maple)	"	1.333
Saccharum, 1-3/4"-2" (Sugar Maple)	"	1.333

02 SITE CONSTRUCTION

Planting

02930.60 Trees (Cont.)

Description	UNIT	MAN/HOURS
Fraxinus Pennsylvanica, 1-3/4"-2"	EA.	1.333
Celtis Occidentalis, 1-3/4"-2"	"	1.333
Glenditsia Triacantos Inermis, 2"	"	1.333
Prunus Cerasifera 'Thundercloud', 6'-8'	"	0.800
Yeodensis, 6'-8' (Yoshino Cherry)	"	0.800
Lagerstroemia Indica, 8'-10' (Crapemyrtle)	"	1.000
Crataegus Phaenopyrum, 8'-10'	"	1.000
Quercus Borealis, 1-3/4"-2" (Northern Red Oak)	"	1.333
Quercus Acutissima, 1-3/4"-2" (8'-10')	"	1.333
Saliz Babylonica, 1-3/4"-2" (Weeping Willow)	"	1.333
Tilia Cordata Greenspire, 1-3/4"-2" (10'-12')	"	1.333
Malus, 2"-2-1/2" (8'-10') (Flowering Crabapple)	"	1.333
Platanus Occidentalis, (12'-14')	"	1.600
Pyrus Calleryana Bradford, 2"-2-1/2"	"	1.333
Quercus Palustris, 2"-2-1/2" (12'-14') (Pin Oak)	"	1.333
Phellos, 2-1/2"-3" (Willow Oak)	"	1.600
Nigra, 2"-2-1/2" (Water Oak)	"	1.333
Magnolia Soulangeana, 4'-5' (Saucer Magnolia)	"	0.667
Grandiflora, 6'-8' (Southern Magnolia)	"	0.800
Cedrus Deodara, 10'-12' (Deodare Cedar)	"	1.333
Gingko Biloba, 10'-12' (2"-2-1/2")	"	1.333
Pinus Thunbergi, 5'-6' (Japanese Black Pine)	"	0.667
Strobus, 6'-8' (White Pine)	"	0.800
Taeda, 6'-8' (Loblolly Pine)	"	0.800
Quercus Virginiana, 2"-2-1/2" (Live Oak)	"	1.600

02935.10 Shrub & Tree Maintenance

Description	UNIT	MAN/HOURS
Moving shrubs on site		
12" ball	EA.	1.000
24" ball	"	1.333
3' high	"	0.800
4' high	"	0.889
5' high	"	1.000
18" spread	"	1.143
30" spread	"	1.333
Moving trees on site		
24" ball	EA.	1.200
48" ball	"	1.600
Trees		
3' high	EA.	0.480
6' high	"	0.533
8' high	"	0.600
10' high	"	0.800
Palm trees		
7' high	EA.	0.600
10' high	"	0.800
20' high	"	2.400
40' high	"	4.800
Guying trees		
4" dia.	EA.	0.400
8" dia.	"	0.500

Planting

02935.30 Weed Control

Description	UNIT	MAN/HOURS
Weed control, bromicil, 15 lb./acre, wettable powder	ACRE	4.000
Vegetation control, by application of plant killer	S.Y.	0.003
Weed killer, lawns and fields	"	0.002

02945.10 Prefabricated Planters

Description	UNIT	MAN/HOURS
Concrete precast, circular		
24" dia., 18" high	EA.	0.800
42" dia., 30" high	"	1.000
Fiberglass, circular		
36" dia., 27" high	EA.	0.400
60" dia., 39" high	"	0.444
Tapered, circular		
24" dia., 36" high	EA.	0.364
40" dia., 36" high	"	0.400
Square		
2' by 2', 17" high	EA.	0.364
4' by 4', 39" high	"	0.444
Rectangular		
4' by 1', 18" high	EA.	0.400

02945.20 Landscape Accessories

Description	UNIT	MAN/HOURS
Steel edging, 3/16" x 4"	L.F.	0.010
Landscaping stepping stones, 15"x15", white	EA.	0.040
Wood chip mulch	C.Y.	0.533
2" thick	S.Y.	0.016
4" thick	"	0.023
6" thick	"	0.029
Gravel mulch, 3/4" stone	C.Y.	0.800
White marble chips, 1" deep	S.F.	0.008
Peat moss		
2" thick	S.Y.	0.018
4" thick	"	0.027
6" thick	"	0.033
Landscaping timbers, treated lumber		
4" x 4"	L.F.	0.027
6" x 6"	"	0.029
8" x 8"	"	0.033

Site Restoration

02955.10 Pipeline Restoration

Description	UNIT	MAN/HOURS
Relining existing water main		
6" dia.	L.F.	0.240
8" dia.	"	0.253
10" dia.	"	0.267
12" dia.	"	0.282

02 SITE CONSTRUCTION

Site Restoration		UNIT	MAN/HOURS
02955.10	**Pipeline Restoration** *(Cont.)*		
14" dia.		L.F.	0.300
16" dia.		"	0.320
18" dia.		"	0.343
20" dia.		"	0.369
24" dia.		"	0.400
36" dia.		"	0.480
48" dia.		"	0.533
72" dia.		"	0.600
Replacing in line gate valves			
6" valve		EA.	3.200
8" valve		"	4.000
10" valve		"	4.800
12" valve		"	6.000
16" valve		"	6.857
18" valve		"	8.000
20" valve		"	9.600
24" valve		"	12.000
36" valve		"	16.000

03 CONCRETE

Formwork	UNIT	MAN/HOURS
03110.05 Beam Formwork		
Beam forms, job built		
Beam bottoms		
1 use	S.F.	0.133
2 uses	"	0.127
3 uses	"	0.123
4 uses	"	0.118
5 uses	"	0.114
Beam sides		
1 use	S.F.	0.089
2 uses	"	0.084
3 uses	"	0.080
4 uses	"	0.076
5 uses	"	0.073
03110.10 Box Culvert Formwork		
Box culverts, job built		
6' x 6'		
1 use	S.F.	0.080
2 uses	"	0.076
3 uses	"	0.073
4 uses	"	0.070
5 uses	"	0.067
8' x 12'		
1 use	S.F.	0.067
2 uses	"	0.064
3 uses	"	0.062
4 uses	"	0.059
5 uses	"	0.057
03110.15 Column Formwork		
Column, square forms, job built		
8" x 8" columns		
1 use	S.F.	0.160
2 uses	"	0.154
3 uses	"	0.148
4 uses	"	0.143
5 uses	"	0.138
12" x 12" columns		
1 use	S.F.	0.145
2 uses	"	0.140
3 uses	"	0.136
4 uses	"	0.131
5 uses	"	0.127
16" x 16" columns		
1 use	S.F.	0.133
2 uses	"	0.129
3 uses	"	0.125
4 uses	"	0.121
5 uses	"	0.118
24" x 24" columns		
1 use	S.F.	0.123
2 uses	"	0.119
3 uses	"	0.116

Formwork	UNIT	MAN/HOURS
03110.15 Column Formwork (Cont.)		
4 uses	S.F.	0.113
5 uses	"	0.110
36" x 36" columns		
1 use	S.F.	0.114
2 uses	"	0.111
3 uses	"	0.108
4 uses	"	0.105
5 uses	"	0.103
Round fiber forms, 1 use		
10" dia.	L.F.	0.160
12" dia.	"	0.163
14" dia.	"	0.170
16" dia.	"	0.178
18" dia.	"	0.190
24" dia.	"	0.205
30" dia.	"	0.222
36" dia.	"	0.242
42" dia.	"	0.267
03110.18 Curb Formwork		
Curb forms		
Straight, 6" high		
1 use	L.F.	0.080
2 uses	"	0.076
3 uses	"	0.073
4 uses	"	0.070
5 uses	"	0.067
Curved, 6" high		
1 use	L.F.	0.100
2 uses	"	0.094
3 uses	"	0.089
4 uses	"	0.085
5 uses	"	0.082
03110.20 Elevated Slab Formwork		
Elevated slab formwork		
Slab, with drop panels		
1 use	S.F.	0.064
2 uses	"	0.062
3 uses	"	0.059
4 uses	"	0.057
5 uses	"	0.055
Floor slab, hung from steel beams		
1 use	S.F.	0.062
2 uses	"	0.059
3 uses	"	0.057
4 uses	"	0.055
5 uses	"	0.053
Floor slab, with pans or domes		
1 use	S.F.	0.073
2 uses	"	0.070
3 uses	"	0.067
4 uses	"	0.064

03 CONCRETE

Formwork	UNIT	MAN/HOURS
03110.20 Elevated Slab Formwork *(Cont.)*		
5 uses	S.F.	0.062
Equipment curbs, 12" high		
1 use	L.F.	0.080
2 uses	"	0.076
3 uses	"	0.073
4 uses	"	0.070
5 uses	"	0.067
03110.25 Equipment Pad Formwork		
Equipment pad, job built		
1 use	S.F.	0.100
2 uses	"	0.094
3 uses	"	0.089
4 uses	"	0.084
5 uses	"	0.080
03110.35 Footing Formwork		
Wall footings, job built, continuous		
1 use	S.F.	0.080
2 uses	"	0.076
3 uses	"	0.073
4 uses	"	0.070
5 uses	"	0.067
Column footings, spread		
1 use	S.F.	0.100
2 uses	"	0.094
3 uses	"	0.089
4 uses	"	0.084
5 uses	"	0.080
03110.50 Grade Beam Formwork		
Grade beams, job built		
1 use	S.F.	0.080
2 uses	"	0.076
3 uses	"	0.073
4 uses	"	0.070
5 uses	"	0.067
03110.53 Pile Cap Formwork		
Pile cap forms, job built		
Square		
1 use	S.F.	0.100
2 uses	"	0.094
3 uses	"	0.089
4 uses	"	0.084
5 uses	"	0.080
Triangular		
1 use	S.F.	0.114
2 uses	"	0.107
3 uses	"	0.100
4 uses	"	0.094
5 uses	"	0.089

Formwork	UNIT	MAN/HOURS
03110.55 Slab / Mat Formwork		
Mat foundations, job built		
1 use	S.F.	0.100
2 uses	"	0.094
3 uses	"	0.089
4 uses	"	0.084
5 uses	"	0.080
Edge forms		
6" high		
1 use	L.F.	0.073
2 uses	"	0.070
3 uses	"	0.067
4 uses	"	0.064
5 uses	"	0.062
12" high		
1 use	L.F.	0.080
2 uses	"	0.076
3 uses	"	0.073
4 uses	"	0.070
5 uses	"	0.067
Formwork for openings		
1 use	S.F.	0.160
2 uses	"	0.145
3 uses	"	0.133
4 uses	"	0.123
5 uses	"	0.114
03110.60 Stair Formwork		
Stairway forms, job built		
1 use	S.F.	0.160
2 uses	"	0.145
3 uses	"	0.133
4 uses	"	0.123
5 uses	"	0.114
Stairs, elevated		
1 use	S.F.	0.160
2 uses	"	0.133
3 uses	"	0.114
4 uses	"	0.107
5 uses	"	0.100
03110.65 Wall Formwork		
Wall forms, exterior, job built		
Up to 8' high wall		
1 use	S.F.	0.080
2 uses	"	0.076
3 uses	"	0.073
4 uses	"	0.070
5 uses	"	0.067
Over 8' high wall		
1 use	S.F.	0.100
2 uses	"	0.094
3 uses	"	0.089
4 uses	"	0.084

03 CONCRETE

Formwork

03110.65 Wall Formwork (Cont.)

Description	Unit	Man/Hours
5 uses	S.F.	0.080
Over 16' high wall		
1 use	S.F.	0.114
2 uses	"	0.107
3 uses	"	0.100
4 uses	"	0.094
5 uses	"	0.089
Radial wall forms		
1 use	S.F.	0.123
2 uses	"	0.114
3 uses	"	0.107
4 uses	"	0.100
5 uses	"	0.094
Retaining wall forms		
1 use	S.F.	0.089
2 uses	"	0.084
3 uses	"	0.080
4 uses	"	0.076
5 uses	"	0.073
Radial retaining wall forms		
1 use	S.F.	0.133
2 uses	"	0.123
3 uses	"	0.114
4 uses	"	0.107
5 uses	"	0.100
Column pier and pilaster		
1 use	S.F.	0.160
2 uses	"	0.145
3 uses	"	0.133
4 uses	"	0.123
5 uses	"	0.114
Interior wall forms		
Up to 8' high		
1 use	S.F.	0.073
2 uses	"	0.070
3 uses	"	0.067
4 uses	"	0.064
5 uses	"	0.062
Over 8' high		
1 use	S.F.	0.089
2 uses	"	0.084
3 uses	"	0.080
4 uses	"	0.076
5 uses	"	0.073
Over 16' high		
1 use	S.F.	0.100
2 uses	"	0.094
3 uses	"	0.089
4 uses	"	0.084
5 uses	"	0.080
Radial wall forms		
1 use	S.F.	0.107
2 uses	"	0.100

03110.65 Wall Formwork (Cont.)

Description	Unit	Man/Hours
3 uses	S.F.	0.094
4 uses	"	0.089
5 uses	"	0.084
Curved wall forms, 24" sections		
1 use	S.F.	0.160
2 uses	"	0.145
3 uses	"	0.133
4 uses	"	0.123
5 uses	"	0.114
PVC form liner, per side, smooth finish		
1 use	S.F.	0.067
2 uses	"	0.064
3 uses	"	0.062
4 uses	"	0.057
5 uses	"	0.053

03110.90 Miscellaneous Formwork

Description	Unit	Man/Hours
Keyway forms (5 uses)		
2 x 4	L.F.	0.040
2 x 6	"	0.044
Bulkheads		
Walls, with keyways		
2 piece	L.F.	0.073
3 piece	"	0.080
Elevated slab, with keyway		
2 piece	L.F.	0.067
3 piece	"	0.073
Ground slab, with keyway		
2 piece	L.F.	0.057
3 piece	"	0.062
Chamfer strips		
Wood		
1/2" wide	L.F.	0.018
3/4" wide	"	0.018
1" wide	"	0.018
PVC		
1/2" wide	L.F.	0.018
3/4" wide	"	0.018
1" wide	"	0.018
Radius		
1"	L.F.	0.019
1-1/2"	"	0.019
Reglets		
Galvanized steel, 24 ga.	L.F.	0.032
Metal formwork		
Straight edge forms		
4" high	L.F.	0.050
6" high	"	0.053
8" high	"	0.057
12" high	"	0.062
16" high	"	0.067

03 CONCRETE

Formwork	UNIT	MAN/HOURS
03110.90 Miscellaneous Formwork (Cont.)		
Curb form, S-shape		
12" x		
1'-6"	L.F.	0.114
2'	"	0.107
2'-6"	"	0.100
3'	"	0.089

Reinforcement	UNIT	MAN/HOURS
03210.05 Beam Reinforcing		
Beam-girders		
#3 - #4	TON	20.000
#5 - #6	"	16.000
#7 - #8	"	13.333
#9 - #10	"	11.429
#11	"	10.667
#14	"	10.000
Galvanized		
#3 - #4	TON	20.000
#5 - #6	"	16.000
#7 - #8	"	13.333
#9 - #10	"	11.429
#11	"	10.667
#14	"	10.000
Bond Beams		
#3 - #4	TON	26.667
#5 - #6	"	20.000
#7 - #8	"	17.778
Galvanized		
#3 - #4	TON	26.667
#5 - #6	"	20.000
#7 - #8	"	17.778
03210.10 Box Culvert Reinforcing		
Box culverts		
#3 - #4	TON	10.000
#5 - #6	"	8.889
#7 - #8	"	8.000
#9 - #10	"	7.273
#11	"	6.667
Galvanized		
#3 - #4	TON	10.000
#5 - #6	"	8.889
#7 - #8	"	8.000
#9 - #10	"	7.273
#11	"	6.667

Reinforcement	UNIT	MAN/HOURS
03210.15 Column Reinforcing		
Columns		
#3 - #4	TON	22.857
#5 - #6	"	17.778
#7 - #8	"	16.000
#9 - #10	"	14.545
#11	"	13.333
#14	"	12.308
#18	"	11.429
Galvanized		
#3 - #4	TON	22.857
#5 - #6	"	17.778
#7 - #8	"	16.000
#9 - #10	"	14.545
#11	"	13.333
#14	"	12.308
#18	"	11.429
Spirals		
8" to 24" dia.	TON	20.000
24" to 48" dia.	"	17.778
48" to 84" dia.	"	16.000
03210.20 Elevated Slab Reinforcing		
Elevated slab		
#3 - #4	TON	10.000
#5 - #6	"	8.889
#7 - #8	"	8.000
#9 - #10	"	7.273
#11	"	6.667
Galvanized		
#3 - #4	TON	10.000
#5 - #6	"	8.889
#7 - #8	"	8.000
#9 - #10	"	7.273
#11	"	6.667
03210.25 Equip. Pad Reinforcing		
Equipment pad		
#3 - #4	TON	16.000
#5 - #6	"	14.545
#7 - #8	"	13.333
#9 - #10	"	12.308
#11	"	11.429
03210.35 Footing Reinforcing		
Footings		
Grade 50		
#3 - #4	TON	13.333
#5 - #6	"	11.429
#7 - #8	"	10.000
#9 - #10	"	8.889
Grade 60		
#3 - #4	TON	13.333
#5 - #6	"	11.429

03 CONCRETE

Reinforcement	UNIT	MAN/HOURS
03210.35 Footing Reinforcing (Cont.)		
#7 - #8	TON	10.000
#9 - #10	"	8.889
Grade 70		
#3 - #4	TON	13.333
#5 - #6	"	11.429
#7 - #8	"	10.000
#9 - #10	"	8.889
#11	"	8.000
Straight dowels, 24" long		
1" dia. (#8)	EA.	0.080
3/4" dia. (#6)	"	0.080
5/8" dia. (#5)	"	0.067
1/2" dia. (#4)	"	0.057
03210.45 Foundation Reinforcing		
Foundations		
#3 - #4	TON	13.333
#5 - #6	"	11.429
#7 - #8	"	10.000
#9 - #10	"	8.889
#11	"	8.000
Galvanized		
#3 - #4	TON	13.333
#5 - #6	"	11.429
#7 - #8	"	10.000
#9 - #10	"	8.889
#11	"	8.000
03210.50 Grade Beam Reinforcing		
Grade beams		
#3 - #4	TON	12.308
#5 - #6	"	10.667
#7 - #8	"	9.412
#9 - #10	"	8.421
#11	"	7.619
Galvanized		
#3 - #4	TON	12.308
#5 - #6	"	10.667
#7 - #8	"	9.412
#9 - #10	"	8.421
#11	"	7.619
03210.53 Pile Cap Reinforcing		
Pile caps		
#3 - #4	TON	20.000
#5 - #6	"	17.778
#7 - #8	"	16.000
#9 - #10	"	14.545
#11	"	13.333
Galvanized		
#3 - #4	TON	20.000
#5 - #6	"	17.778
#7 - #8	"	16.000

Reinforcement	UNIT	MAN/HOURS
03210.53 Pile Cap Reinforcing (Cont.)		
#9 - #10	TON	14.545
#11	"	13.333
03210.55 Slab / Mat Reinforcing		
Bars, slabs		
#3 - #4	TON	13.333
#5 - #6	"	11.429
#7 - #8	"	10.000
#9 - #10	"	8.889
#11	"	8.000
Galvanized		
#3 - #4	TON	13.333
#5 - #6	"	11.429
#7 - #8	"	10.000
#9 - #10	"	8.889
#11	"	8.000
Wire mesh, slabs		
Galvanized		
4x4		
W1.4xW1.4	S.F.	0.005
W2.0xW2.0	"	0.006
W2.9xW2.9	"	0.006
W4.0xW4.0	"	0.007
6x6		
W1.4xW1.4	S.F.	0.004
W2.0xW2.0	"	0.004
W2.9xW2.9	"	0.005
W4.0xW4.0	"	0.005
Standard		
2x2		
W.9xW.9	S.F.	0.005
4x4		
W1.4xW1.4	S.F.	0.005
W2.0xW2.0	"	0.006
W2.9xW2.9	"	0.006
W4.0xW4.0	"	0.007
6x6		
W1.4xW1.4	S.F.	0.004
W2.0xW2.0	"	0.004
W2.9xW2.9	"	0.005
W4.0xW4.0	"	0.005
03210.60 Stair Reinforcing		
Stairs		
#3 - #4	TON	16.000
#5 - #6	"	13.333
#7 - #8	"	11.429
#9 - #10	"	10.000
Galvanized		
#3 - #4	TON	16.000
#5 - #6	"	13.333
#7 - #8	"	11.429
#9 - #10	"	10.000

03 CONCRETE

Reinforcement

03210.65 Wall Reinforcing

	UNIT	MAN/HOURS
Walls		
#3 - #4	TON	11.429
#5 - #6	"	10.000
#7 - #8	"	8.889
#9 - #10	"	8.000
Galvanized		
#3 - #4	TON	11.429
#5 - #6	"	10.000
#7 - #8	"	8.889
#9 - #10	"	8.000
Masonry wall (horizontal)		
#3 - #4	TON	32.000
#5 - #6	"	26.667
Galvanized		
#3 - #4	TON	32.000
#5 - #6	"	26.667
Masonry wall (vertical)		
#3 - #4	TON	40.000
#5 - #6	"	32.000
Galvanized		
#3 - #4	TON	40.000
#5 - #6	"	32.000

Accessories

03250.40 Concrete Accessories

	UNIT	MAN/HOURS
Expansion joint, poured		
Asphalt		
1/2" x 1"	L.F.	0.016
1" x 2"	"	0.017
Liquid neoprene, cold applied		
1/2" x 1"	L.F.	0.016
1" x 2"	"	0.018
Polyurethane, 2 parts		
1/2" x 1"	L.F.	0.027
1" x 2"	"	0.029
Rubberized asphalt, cold		
1/2" x 1"	L.F.	0.016
1" x 2"	"	0.017
Hot, fuel resistant		
1/2" x 1"	L.F.	0.016
1" x 2"	"	0.017
Expansion joint, premolded, in slabs		
Asphalt		
1/2" x 6"	L.F.	0.020
1" x 12"	"	0.027

03250.40 Concrete Accessories (Cont.)

	UNIT	MAN/HOURS
Cork		
1/2" x 6"	L.F.	0.020
1" x 12"	"	0.027
Neoprene sponge		
1/2" x 6"	L.F.	0.020
1" x 12"	"	0.027
Polyethylene foam		
1/2" x 6"	L.F.	0.020
1" x 12"	"	0.027
Polyurethane foam		
1/2" x 6"	L.F.	0.020
1" x 12"	"	0.027
Polyvinyl chloride foam		
1/2" x 6"	L.F.	0.020
1" x 12"	"	0.027
Rubber, gray sponge		
1/2" x 6"	L.F.	0.020
1" x 12"	"	0.027
Asphalt felt control joints or bond breaker, screed joints		
4" slab	L.F.	0.016
6" slab	"	0.018
8" slab	"	0.020
10" slab	"	0.023
Keyed cold expansion and control joints, 24 ga.		
4" slab	L.F.	0.050
5" slab	"	0.050
6" slab	"	0.053
8" slab	"	0.057
10" slab	"	0.062
Waterstops		
Polyvinyl chloride		
Ribbed		
3/16" thick x		
4" wide	L.F.	0.040
6" wide	"	0.044
1/2" thick x		
9" wide	L.F.	0.050
Ribbed with center bulb		
3/16" thick x 9" wide	L.F.	0.050
3/8" thick x 9" wide	"	0.050
Dumbbell type, 3/8" thick x 6" wide	"	0.044
Plain, 3/8" thick x 9" wide	"	0.050
Center bulb, 3/8" thick x 9" wide	"	0.050
Rubber		
Flat dumbbell		
3/8" thick x		
6" wide	L.F.	0.044
9" wide	"	0.050
Center bulb		
3/8" thick x		
6" wide	L.F.	0.044
9" wide	"	0.050
Vapor barrier		

03 CONCRETE

Accessories	UNIT	MAN/HOURS
03250.40 Concrete Accessories (Cont.)		
4 mil polyethylene	S.F.	0.003
6 mil polyethylene	"	0.003
Gravel porous fill, under floor slabs, 3/4" stone	C.Y.	1.333
Reinforcing accessories		
Beam bolsters		
1-1/2" high, plain	L.F.	0.008
Galvanized	"	0.008
3" high		
Plain	L.F.	0.010
Galvanized	"	0.010
Slab bolsters		
1" high		
Plain	L.F.	0.004
Galvanized	"	0.004
2" high		
Plain	L.F.	0.004
Galvanized	"	0.004
Chairs, high chairs		
3" high		
Plain	EA.	0.020
Galvanized	"	0.020
5" high		
Plain	EA.	0.021
Galvanized	"	0.021
8" high		
Plain	EA.	0.023
Galvanized	"	0.023
12" high		
Plain	EA.	0.027
Galvanized	"	0.027
Continuous, high chair		
3" high		
Plain	L.F.	0.005
Galvanized	"	0.005
5" high		
Plain	L.F.	0.006
Galvanized	"	0.006
8" high		
Plain	L.F.	0.006
Galvanized	"	0.006
12" high		
Plain	L.F.	0.007
Galvanized	"	0.007

Cast-in-place Concrete	UNIT	MAN/HOURS
03300.10 Concrete Admixtures		
Floor finishes		
Broom	S.F.	0.011
Screed	"	0.010
Darby	"	0.010
Steel float	"	0.013
Granolithic topping		
1/2" thick	S.F.	0.036
1" thick	"	0.040
2" thick	"	0.044
Wall finishes		
Burlap rub, with cement paste	S.F.	0.013
Float finish	"	0.020
Etch with acid	"	0.013
Sandblast		
Minimum	S.F.	0.016
Maximum	"	0.016
Bush hammer		
Green concrete	S.F.	0.040
Cured concrete	"	0.062
Break ties and patch holes	"	0.016
Carborundum		
Dry rub	S.F.	0.027
Wet rub	"	0.040
Floor hardeners		
Metallic		
Light service	S.F.	0.010
Heavy service	"	0.013
Non-metallic		
Light service	S.F.	0.010
Heavy service	"	0.013
Rusticated concrete finish		
Beveled edge	L.F.	0.044
Square edge	"	0.057
Solid board concrete finish		
Standard	S.F.	0.067
Rustic	"	0.080
03360.10 Pneumatic Concrete		
Pneumatic applied concrete (gunite)		
2" thick	S.F.	0.030
3" thick	"	0.040
4" thick	"	0.048
Finish surface		
Minimum	S.F.	0.040
Maximum	"	0.080
03370.10 Curing Concrete		
Sprayed membrane		
Slabs	S.F.	0.002
Walls	"	0.002
Curing paper		
Slabs	S.F.	0.002
Walls	"	0.002

03 CONCRETE

Cast-in-place Concrete

03370.10 Curing Concrete (Cont.)

	UNIT	MAN/HOURS
Burlap		
7.5 oz.	S.F.	0.003
12 oz.	"	0.003

Placing Concrete

03380.05 Beam Concrete

	UNIT	MAN/HOURS
Beams and girders		
2500# or 3000# concrete		
By crane	C.Y.	0.960
By pump	"	0.873
By hand buggy	"	0.800
3500# or 4000# concrete		
By crane	C.Y.	0.960
By pump	"	0.873
By hand buggy	"	0.800
5000# concrete		
By crane	C.Y.	0.960
By pump	"	0.873
By hand buggy	"	0.800
Bond beam, 3000# concrete		
By pump		
8" high		
4" wide	L.F.	0.019
6" wide	"	0.022
8" wide	"	0.024
10" wide	"	0.027
12" wide	"	0.030
16" high		
8" wide	L.F.	0.030
10" wide	"	0.034
12" wide	"	0.040
By crane		
8" high		
4" wide	L.F.	0.021
6" wide	"	0.023
8" wide	"	0.024
10" wide	"	0.027
12" wide	"	0.030
16" high		
8" wide	L.F.	0.030
10" wide	"	0.032
12" wide	"	0.037

Placing Concrete

03380.15 Column Concrete

	UNIT	MAN/HOURS
Columns		
2500# or 3000# concrete		
By crane	C.Y.	0.873
By pump	"	0.800
3500# or 4000# concrete		
By crane	C.Y.	0.873
By pump	"	0.800
5000# concrete		
By crane	C.Y.	0.873
By pump	"	0.800

03380.20 Elevated Slab Concrete

	UNIT	MAN/HOURS
Elevated slab		
2500# or 3000# concrete		
By crane	C.Y.	0.480
By pump	"	0.369
By hand buggy	"	0.800
3500# or 4000# concrete		
By crane	C.Y.	0.480
By pump	"	0.369
By hand buggy	"	0.800
5000# concrete		
By crane	C.Y.	0.480
By pump	"	0.369
By hand buggy	"	0.800
Topping		
2500# or 3000# concrete		
By crane	C.Y.	0.480
By pump	"	0.369
By hand buggy	"	0.800
3500# or 4000# concrete		
By crane	C.Y.	0.480
By pump	"	0.369
By hand buggy	"	0.800
5000# concrete		
By crane	C.Y.	0.480
By pump	"	0.369
By hand buggy	"	0.800

03380.25 Equipment Pad Concrete

	UNIT	MAN/HOURS
Equipment pad		
2500# or 3000# concrete		
By chute	C.Y.	0.267
By pump	"	0.686
By crane	"	0.800
3500# or 4000# concrete		
By chute	C.Y.	0.267
By pump	"	0.686
By crane	"	0.800
5000# concrete		
By chute	C.Y.	0.267
By pump	"	0.686
By crane	"	0.800

03 CONCRETE

Placing Concrete

03380.35 Footing Concrete

	Unit	Man/Hours
Continuous footing		
2500# or 3000# concrete		
By chute	C.Y.	0.267
By pump	"	0.600
By crane	"	0.686
3500# or 4000# concrete		
By chute	C.Y.	0.267
By pump	"	0.600
By crane	"	0.686
5000# concrete		
By chute	C.Y.	0.267
By pump	"	0.600
By crane	"	0.686
Spread footing		
2500# or 3000# concrete		
Under 5 cy		
By chute	C.Y.	0.267
By pump	"	0.640
By crane	"	0.738
Over 5 cy		
By chute	C.Y.	0.200
By pump	"	0.565
By crane	"	0.640
3500# or 4000# concrete		
Under 5 c.y.		
By chute	C.Y.	0.267
By pump	"	0.640
By crane	"	0.738
Over 5 c.y.		
By pump	C.Y.	0.565
By crane	"	0.640
5000# concrete		
Under 5 c.y.		
By chute	C.Y.	0.267
By pump	"	0.640
By crane	"	0.738
Over 5 c.y.		
By chute	C.Y.	0.200
By pump	"	0.565
By crane	"	0.640

03380.50 Grade Beam Concrete

	Unit	Man/Hours
Grade beam		
2500# or 3000# concrete		
By chute	C.Y.	0.267
By crane	"	0.686
By pump	"	0.600
By hand buggy	"	0.800
3500# or 4000# concrete		
By chute	C.Y.	0.267
By crane	"	0.686
By pump	"	0.600

03380.50 Grade Beam Concrete (Cont.)

	Unit	Man/Hours
By hand buggy	C.Y.	0.800
5000# concrete		
By chute	C.Y.	0.267
By crane	"	0.686
By pump	"	0.600
By hand buggy	"	0.800

03380.53 Pile Cap Concrete

	Unit	Man/Hours
Pile cap		
2500# or 3000 concrete		
By chute	C.Y.	0.267
By crane	"	0.800
By pump	"	0.686
By hand buggy	"	0.800
3500# or 4000# concrete		
By chute	C.Y.	0.267
By crane	"	0.800
By pump	"	0.686
By hand buggy	"	0.800
5000# concrete		
By chute	C.Y.	0.267
By crane	"	0.800
By pump	"	0.686
By hand buggy	"	0.800

03380.55 Slab / Mat Concrete

	Unit	Man/Hours
Slab on grade		
2500# or 3000# concrete		
By chute	C.Y.	0.200
By crane	"	0.400
By pump	"	0.343
By hand buggy	"	0.533
3500# or 4000# concrete		
By chute	C.Y.	0.200
By crane	"	0.400
By pump	"	0.343
By hand buggy	"	0.533
5000# concrete		
By chute	C.Y.	0.200
By crane	"	0.400
By pump	"	0.343
By hand buggy	"	0.533
Foundation mat		
2500# or 3000# concrete, over 20 cy		
By chute	C.Y.	0.160
By crane	"	0.343
By pump	"	0.300
By hand buggy	"	0.400

03 CONCRETE

Placing Concrete	UNIT	MAN/HOURS
03380.58 Sidewalks		
Walks, cast in place with wire mesh, base not incl.		
4" thick	S.F.	0.027
5" thick	"	0.032
6" thick	"	0.040
03380.60 Stair Concrete		
Stairs		
2500# or 3000# concrete		
By chute	C.Y.	0.267
By crane	"	0.800
By pump	"	0.686
By hand buggy	"	0.800
3500# or 4000# concrete		
By chute	C.Y.	0.267
By crane	"	0.800
By pump	"	0.686
By hand buggy	"	0.800
5000# concrete		
By chute	C.Y.	0.267
By crane	"	0.800
By pump	"	0.686
By hand buggy	"	0.800
03380.65 Wall Concrete		
Walls		
2500# or 3000# concrete		
To 4'		
By chute	C.Y.	0.229
By crane	"	0.800
By pump	"	0.738
To 8'		
By crane	C.Y.	0.873
By pump	"	0.800
To 16'		
By crane	C.Y.	0.960
By pump	"	0.873
Over 16'		
By crane	C.Y.	1.067
By pump	"	0.960
3500# or 4000# concrete		
To 4'		
By chute	C.Y.	0.229
By crane	"	0.800
By pump	"	0.738
To 8'		
By crane	C.Y.	0.873
By pump	"	0.800
To 16'		
By crane	C.Y.	0.960
By pump	"	0.873
Over 16'		
By crane	C.Y.	1.067
By pump	"	0.960

Placing Concrete	UNIT	MAN/HOURS
03380.65 Wall Concrete *(Cont.)*		
5000# concrete		
To 4'		
By chute	C.Y.	0.229
By crane	"	0.800
By pump	"	0.738
To 8'		
By crane	C.Y.	0.873
By pump	"	0.800
To 16'		
By crane	C.Y.	0.960
By pump	"	0.873
Filled block (CMU)		
3000# concrete, by pump		
4" wide	S.F.	0.034
6" wide	"	0.040
8" wide	"	0.048
10" wide	"	0.056
12" wide	"	0.069
Pilasters, 3000# concrete	C.F.	0.960
Wall cavity, 2" thick, 3000# concrete	S.F.	0.032

Precast Concrete	UNIT	MAN/HOURS
03400.10 Precast Beams		
Prestressed, double tee, 24" deep, 8' wide		
35' span		
115 psf	S.F.	0.008
140 psf	"	0.008
40' span		
80 psf	S.F.	0.009
143 psf	"	0.009
45' span		
50 psf	S.F.	0.007
70 psf	"	0.007
100 psf	"	0.007
130 psf	"	0.007
50' span		
75 psf	S.F.	0.007
100 psf	"	0.007
Precast beams, girders and joists		
1000 lb/lf live load		
10' span	L.F.	0.160
20' span	"	0.096
30' span	"	0.080
3000 lb/lf live load		
10' span	L.F.	0.160

03 CONCRETE

Precast Concrete	UNIT	MAN/HOURS
03400.10 Precast Beams *(Cont.)*		
20' span	L.F.	0.096
30' span	"	0.080
5000 lb/lf live load		
10' span	L.F.	0.160
20' span	"	0.096
30' span	"	0.080
03400.20 Precast Columns		
Prestressed concrete columns		
10" x 10"		
10' long	EA.	0.960
15' long	"	1.000
20' long	"	1.067
25' long	"	1.143
30' long	"	1.200
12" x 12"		
20' long	EA.	1.200
25' long	"	1.297
30' long	"	1.371
16" x 16"		
20' long	EA.	1.200
25' long	"	1.297
30' long	"	1.371
20" x 20"		
20' long	EA.	1.263
25' long	"	1.333
30' long	"	1.412
24" x 24"		
20' long	EA.	1.333
25' long	"	1.412
30' long	"	1.500
28" x 28"		
20' long	EA.	1.500
25' long	"	1.600
30' long	"	1.714
32" x 32"		
20' long	EA.	1.600
25' long	"	1.714
30' long	"	1.846
36" x 36"		
20' long	EA.	1.714
25' long	"	1.846
30' long	"	2.000
03400.30 Precast Slabs		
Prestressed flat slab		
6" thick, 4' wide		
20' span		
80 psf	S.F.	0.020
110 psf	"	0.020
25' span		
80 psf	S.F.	0.019
Cored slab		

Precast Concrete	UNIT	MAN/HOURS
03400.30 Precast Slabs *(Cont.)*		
6" thick, 4' wide		
20' span		
80 psf	S.F.	0.020
100 psf	"	0.020
130 psf	"	0.020
8" thick, 4' wide		
25' span		
70 psf	S.F.	0.019
125 psf	"	0.019
170 psf	"	0.019
30' span		
70 psf	S.F.	0.016
90 psf	"	0.016
35' span		
70 psf	S.F.	0.015
10" thick, 4' wide		
30' span		
75 psf	S.F.	0.016
100 psf	"	0.016
130 psf	"	0.016
35' span		
60 psf	S.F.	0.015
80 psf	"	0.015
120 psf	"	0.015
40' span		
65 psf	S.F.	0.012
Slabs, roof and floor members, 4' wide		
6" thick, 25' span	S.F.	0.019
8" thick, 30' span	"	0.015
10" thick, 40' span	"	0.013
Tee members		
Multiple tee, roof and floor		
Minimum	S.F.	0.012
Maximum	"	0.024
Double tee wall member		
Minimum	S.F.	0.014
Maximum	"	0.027
Single tee		
Short span, roof members		
Minimum	S.F.	0.015
Maximum	"	0.030
Long span, roof members		
Minimum	S.F.	0.012
Maximum	"	0.024
03400.40 Precast Walls		
Wall panel, 8' x 20'		
Gray cement		
Liner finish		
4" wall	S.F.	0.014
5" wall	"	0.014
6" wall	"	0.015
8" wall	"	0.015

03 CONCRETE

Precast Concrete	UNIT	MAN/HOURS
03400.40 Precast Walls *(Cont.)*		
Sandblast finish		
4" wall	S.F.	0.014
5" wall	"	0.014
6" wall	"	0.015
8" wall	"	0.015
White cement		
Liner finish		
4" wall	S.F.	0.014
5" wall	"	0.014
6" wall	"	0.015
8" wall	"	0.015
Sandblast finish		
4" wall	S.F.	0.014
5" wall	"	0.014
6" wall	"	0.015
8" wall	"	0.015
Double tee wall panel, 24" deep		
Gray cement		
Liner finish	S.F.	0.016
Sandblast finish	"	0.016
White cement		
Form liner finish	S.F.	0.016
Sandblast finish	"	0.016
Partition panels		
4" wall	S.F.	0.016
5" wall	"	0.016
6" wall	"	0.016
8" wall	"	0.016
Cladding panels		
4" wall	S.F.	0.017
5" wall	"	0.017
6" wall	"	0.017
8" wall	"	0.017
Sandwich panel, 2.5" cladding panel, 2" insulation		
5" wall	S.F.	0.017
6" wall	"	0.017
8" wall	"	0.017
Adjustable tilt-up brace	EA.	0.200
03400.90 Precast Specialties		
Precast concrete, coping, 4' to 8' long		
12" wide	L.F.	0.060
10" wide	"	0.069
Splash block, 30"x12"x4"	EA.	0.400
Stair unit, per riser	"	0.400
Sun screen and trellis, 8' long, 12" high		
4" thick blades	EA.	0.300
5" thick blades	"	0.300
6" thick blades	"	0.320
8" thick blades	"	0.320
Bearing pads for precast members, 2" wide strips		

Precast Concrete	UNIT	MAN/HOURS
03400.90 Precast Specialties *(Cont.)*		
1/8" thick	L.F.	0.003
1/4" thick	"	0.003
1/2" thick	"	0.003
3/4" thick	"	0.004
1" thick	"	0.004
1-1/2" thick	"	0.004

Cementitous Toppings	UNIT	MAN/HOURS
03550.10 Concrete Toppings		
Gypsum fill		
2" thick	S.F.	0.005
2-1/2" thick	"	0.005
3" thick	"	0.005
3-1/2" thick	"	0.005
4" thick	"	0.006
Formboard		
Mineral fiber board		
1" thick	S.F.	0.020
1-1/2" thick	"	0.023
Cement fiber board		
1" thick	S.F.	0.027
1-1/2" thick	"	0.031
Glass fiber board		
1" thick	S.F.	0.020
1-1/2" thick	"	0.023
Poured deck		
Vermiculite or perlite		
1 to 4 mix	C.Y.	0.800
1 to 6 mix	"	0.738
Vermiculite or perlite		
2" thick		
1 to 4 mix	S.F.	0.005
1 to 6 mix	"	0.005
3" thick		
1 to 4 mix	S.F.	0.007
1 to 6 mix	"	0.007
Concrete plank, lightweight		
2" thick	S.F.	0.024
2-1/2" thick	"	0.024
3-1/2" thick	"	0.027
4" thick	"	0.027
Channel slab, lightweight, straight		
2-3/4" thick	S.F.	0.024
3-1/2" thick	"	0.024
3-3/4" thick	"	0.024

03 CONCRETE

Cementitous Toppings	UNIT	MAN/HOURS
03550.10 Concrete Toppings (Cont.)		
4-3/4" thick	S.F.	0.027
Gypsum plank		
2" thick	S.F.	0.024
3" thick	"	0.024
Cement fiber, T and G planks		
1" thick	S.F.	0.022
1-1/2" thick	"	0.022
2" thick	"	0.024
2-1/2" thick	"	0.024
3" thick	"	0.024
3-1/2" thick	"	0.027
4" thick	"	0.027

Grout	UNIT	MAN/HOURS
03600.10 Grouting		
Grouting for bases		
Nonshrink		
Metallic grout		
1" deep	S.F.	0.160
2" deep	"	0.178
Non-metallic grout		
1" deep	S.F.	0.160
2" deep	"	0.178
Fluid type		
Non-metallic		
1" deep	S.F.	0.160
2" deep	"	0.178
Grouting for joints		
Portland cement grout (1 cement to 3 sand)		
1/2" joint thickness		
6" wide joints	L.F.	0.027
8" wide joints	"	0.032
1" joint thickness		
4" wide joints	L.F.	0.025
6" wide joints	"	0.028
8" wide joints	"	0.033
Nonshrink, nonmetallic grout		
1/2" joint thickness		
4" wide joint	L.F.	0.023
6" wide joint	"	0.027
8" wide joint	"	0.032
1" joint thickness		
4" wide joint	L.F.	0.025
6" wide joint	"	0.028
8" wide joint	"	0.033

Concrete Restoration	UNIT	MAN/HOURS
03730.10 Concrete Repair		
Epoxy grout floor patch, 1/4" thick	S.F.	0.080
Epoxy gel grout	"	0.800
Injection valve, 1 way, threaded plastic	EA.	0.160
Grout crack seal, 2 component	C.F.	0.800
Grout, non shrink	"	0.800
Concrete, epoxy modified		
Sand mix	C.F.	0.320
Gravel mix	"	0.296
Concrete repair		
Soffit repair		
16" wide	L.F.	0.160
18" wide	"	0.167
24" wide	"	0.178
30" wide	"	0.190
32" wide	"	0.200
Edge repair		
2" spall	L.F.	0.200
3" spall	"	0.211
4" spall	"	0.216
6" spall	"	0.222
8" spall	"	0.235
9" spall	"	0.267
Crack repair, 1/8" crack	"	0.080
Reinforcing steel repair		
1 bar, 4 ft		
#4 bar	L.F.	0.100
#5 bar	"	0.100
#6 bar	"	0.107
#8 bar	"	0.107
#9 bar	"	0.114
#11 bar	"	0.114
Pile repairs		
Polyethylene wrap		
30 mil thick		
60" wide	S.F.	0.267
72" wide	"	0.320
60 mil thick		
60" wide	S.F.	0.267
80" wide	"	0.364
Pile spall, average repair 3'		
18" x 18"	EA.	0.667
20" x 20"	"	0.800

04 MASONRY

Mortar And Grout	UNIT	MAN/HOURS
04100.10 Masonry Grout		
Grout, non shrink, non-metallic, trowelable	C.F.	0.016
Grout door frame, hollow metal		
Single	EA.	0.600
Double	"	0.632
Grout-filled concrete block (CMU)		
4" wide	S.F.	0.020
6" wide	"	0.022
8" wide	"	0.024
12" wide	"	0.025
Grout-filled individual CMU cells		
4" wide	L.F.	0.012
6" wide	"	0.012
8" wide	"	0.012
10" wide	"	0.014
12" wide	"	0.014
Bond beams or lintels, 8" deep		
6" thick	L.F.	0.022
8" thick	"	0.024
10" thick	"	0.027
12" thick	"	0.030
Cavity walls		
2" thick	S.F.	0.032
3" thick	"	0.032
4" thick	"	0.034
6" thick	"	0.040
04150.10 Masonry Accessories		
Foundation vents	EA.	0.320
Bar reinforcing		
Horizontal		
#3 - #4	Lb.	0.032
#5 - #6	"	0.027
Vertical		
#3 - #4	Lb.	0.040
#5 - #6	"	0.032
Horizontal joint reinforcing		
Truss type		
4" wide, 6" wall	L.F.	0.003
6" wide, 8" wall	"	0.003
8" wide, 10" wall	"	0.003
10" wide, 12" wall	"	0.004
12" wide, 14" wall	"	0.004
Ladder type		
4" wide, 6" wall	L.F.	0.003
6" wide, 8" wall	"	0.003
8" wide, 10" wall	"	0.003
10" wide, 12" wall	"	0.003
Rectangular wall ties		
3/16" dia., galvanized		
2" x 6"	EA.	0.013
2" x 8"	"	0.013
2" x 10"	"	0.013
2" x 12"	"	0.013

Mortar And Grout	UNIT	MAN/HOURS
04150.10 Masonry Accessories *(Cont.)*		
4" x 6"	EA.	0.016
4" x 8"	"	0.016
4" x 10"	"	0.016
4" x 12"	"	0.016
1/4" dia., galvanized		
2" x 6"	EA.	0.013
2" x 8"	"	0.013
2" x 10"	"	0.013
2" x 12"	"	0.013
4" x 6"	"	0.016
4" x 8"	"	0.016
4" x 10"	"	0.016
4" x 12"	"	0.016
"Z" type wall ties, galvanized		
6" long		
1/8" dia.	EA.	0.013
3/16" dia.	"	0.013
1/4" dia.	"	0.013
8" long		
1/8" dia.	EA.	0.013
3/16" dia.	"	0.013
1/4" dia.	"	0.013
10" long		
1/8" dia.	EA.	0.013
3/16" dia.	"	0.013
1/4" dia.	"	0.013
Dovetail anchor slots		
Galvanized steel, filled		
24 ga.	L.F.	0.020
20 ga.	"	0.020
16 oz. copper, foam filled	"	0.020
Dovetail anchors		
16 ga.		
3-1/2" long	EA.	0.013
5-1/2" long	"	0.013
12 ga.		
3-1/2" long	EA.	0.013
5-1/2" long	"	0.013
Dovetail, triangular galvanized ties, 12 ga.		
3" x 3"	EA.	0.013
5" x 5"	"	0.013
7" x 7"	"	0.013
7" x 9"	"	0.013
Brick anchors		
Corrugated, 3-1/2" long		
16 ga.	EA.	0.013
12 ga.	"	0.013
Non-corrugated, 3-1/2" long		
16 ga.	EA.	0.013
12 ga.	"	0.013
Cavity wall anchors, corrugated, galvanized		
5" long		
16 ga.	EA.	0.013

04 MASONRY

Mortar And Grout	UNIT	MAN/HOURS
04150.10 **Masonry Accessories** (Cont.)		
12 ga.	EA.	0.013
7" long		
28 ga.	EA.	0.013
24 ga.	"	0.013
22 ga.	"	0.013
16 ga.	"	0.013
Mesh ties, 16 ga., 3" wide		
8" long	EA.	0.013
12" long	"	0.013
20" long	"	0.013
24" long	"	0.013
04150.20 **Masonry Control Joints**		
Control joint, cross shaped PVC	L.F.	0.020
Closed cell joint filler		
1/2"	L.F.	0.020
3/4"	"	0.020
Rubber, for		
4" wall	L.F.	0.020
6" wall	"	0.021
8" wall	"	0.022
PVC, for		
4" wall	L.F.	0.020
6" wall	"	0.021
8" wall	"	0.022
04150.50 **Masonry Flashing**		
Through-wall flashing		
5 oz. coated copper	S.F.	0.067
0.030" elastomeric	"	0.053

Unit Masonry	UNIT	MAN/HOURS
04210.10 **Brick Masonry**		
Standard size brick, running bond		
Face brick, red (6.4/sf)		
Veneer	S.F.	0.133
Cavity wall	"	0.114
9" solid wall	"	0.229
Common brick (6.4/sf)		
Select common for veneers	S.F.	0.133
Back-up		
4" thick	S.F.	0.100
8" thick	"	0.160
Firewall		
12" thick	S.F.	0.267

Unit Masonry	UNIT	MAN/HOURS
04210.10 **Brick Masonry** (Cont.)		
16" thick	S.F.	0.364
Glazed brick (7.4/sf)		
Veneer	S.F.	0.145
Buff or gray face brick (6.4/sf)		
Veneer	S.F.	0.133
Cavity wall	"	0.114
Jumbo or oversize brick (3/sf)		
4" veneer	S.F.	0.080
4" back-up	"	0.067
8" back-up	"	0.114
12" firewall	"	0.200
16" firewall	"	0.267
Norman brick, red face, (4.5/sf)		
4" veneer	S.F.	0.100
Cavity wall	"	0.089
Chimney, standard brick, including flue		
16" x 16"	L.F.	0.800
16" x 20"	"	0.800
16" x 24"	"	0.800
20" x 20"	"	1.000
20" x 24"	"	1.000
20" x 32"	"	1.143
Window sill, face brick on edge	"	0.200
04210.20 **Structural Tile**		
Structural glazed tile		
6T series, 5-1/2" x 12"		
Glazed on one side		
2" thick	S.F.	0.080
4" thick	"	0.080
6" thick	"	0.089
8" thick	"	0.100
Glazed on two sides		
4" thick	S.F.	0.100
6" thick	"	0.114
Special shapes		
Group 1	S.F.	0.160
Group 2	"	0.160
Group 3	"	0.160
Group 4	"	0.160
Group 5	"	0.160
Fire rated		
4" thick, 1 hr rating	S.F.	0.080
6" thick, 2 hr rating	"	0.089
8W series, 8" x 16"		
Glazed on one side		
2" thick	S.F.	0.053
4" thick	"	0.053
6" thick	"	0.062
8" thick	"	0.062
Glazed on two sides		
4" thick	S.F.	0.067
6" thick	"	0.080

04 MASONRY

Unit Masonry	UNIT	MAN/HOURS
04210.20 Structural Tile *(Cont.)*		
8" thick	S.F.	0.080
Special shapes		
Group 1	S.F.	0.114
Group 2	"	0.114
Group 3	"	0.114
Group 4	"	0.114
Group 5	"	0.114
Fire rated		
4" thick, 1 hr rating	S.F.	0.114
6" thick, 2 hr rating	"	0.114
04210.60 Pavers, Masonry		
Brick walk laid on sand, sand joints		
Laid flat, (4.5 per sf)	S.F.	0.089
Laid on edge, (7.2 per sf)	"	0.133
Precast concrete patio blocks		
2" thick		
Natural	S.F.	0.027
Colors	"	0.027
Exposed aggregates, local aggregate		
Natural	S.F.	0.027
Colors	"	0.027
Granite or limestone aggregate	"	0.027
White tumblestone aggregate	"	0.027
Stone pavers, set in mortar		
Bluestone		
1" thick		
Irregular	S.F.	0.200
Snapped rectangular	"	0.160
1-1/2" thick, random rectangular	"	0.200
2" thick, random rectangular	"	0.229
Slate		
Natural cleft		
Irregular, 3/4" thick	S.F.	0.229
Random rectangular		
1-1/4" thick	S.F.	0.200
1-1/2" thick	"	0.222
Granite blocks		
3" thick, 3" to 6" wide		
4" to 12" long	S.F.	0.267
6" to 15" long	"	0.229
Crushed stone, white marble, 3" thick	"	0.016
04220.10 Concrete Masonry Units		
Hollow, load bearing		
4"	S.F.	0.059
6"	"	0.062
8"	"	0.067
10"	"	0.073
12"	"	0.080
Solid, load bearing		
4"	S.F.	0.059
6"	"	0.062

Unit Masonry	UNIT	MAN/HOURS
04220.10 Concrete Masonry Units *(Cont.)*		
8"	S.F.	0.067
10"	"	0.073
12"	"	0.080
Back-up block, 8" x 16"		
2"	S.F.	0.046
4"	"	0.047
6"	"	0.050
8"	"	0.053
10"	"	0.057
12"	"	0.062
Foundation wall, 8" x 16"		
6"	S.F.	0.057
8"	"	0.062
10"	"	0.067
12"	"	0.073
Solid		
6"	S.F.	0.062
8"	"	0.067
10"	"	0.073
12"	"	0.080
Exterior, styrofoam inserts, std weight, 8" x 16"		
6"	S.F.	0.062
8"	"	0.067
10"	"	0.073
12"	"	0.080
Lightweight		
6"	S.F.	0.062
8"	"	0.067
10"	"	0.073
12"	"	0.080
Acoustical slotted block		
4"	S.F.	0.073
6"	"	0.073
8"	"	0.080
Filled cavities		
4"	S.F.	0.089
6"	"	0.094
8"	"	0.100
Hollow, split face		
4"	S.F.	0.059
6"	"	0.062
8"	"	0.067
10"	"	0.073
12"	"	0.080
Split rib profile		
4"	S.F.	0.073
6"	"	0.073
8"	"	0.080
10"	"	0.080
12"	"	0.080
High strength block, 3500 psi		
2"	S.F.	0.059
4"	"	0.062

04 MASONRY

Unit Masonry

04220.10 Concrete Masonry Units *(Cont.)*

	UNIT	MAN/HOURS
6"	S.F.	0.062
8"	"	0.067
10"	"	0.073
12"	"	0.080
Solar screen concrete block		
4" thick		
6" x 6"	S.F.	0.178
8" x 8"	"	0.160
12" x 12"	"	0.123
8" thick		
8" x 16"	S.F.	0.114
Glazed block		
Cove base, glazed 1 side, 2"	L.F.	0.089
4"	"	0.089
6"	"	0.100
8"	"	0.100
Single face		
2"	S.F.	0.067
4"	"	0.067
6"	"	0.073
8"	"	0.080
10"	"	0.089
12"	"	0.094
Double face		
4"	S.F.	0.084
6"	"	0.089
8"	"	0.100
Corner or bullnose		
2"	EA.	0.100
4"	"	0.114
6"	"	0.114
8"	"	0.133
10"	"	0.145
12"	"	0.160
Gypsum unit masonry		
Partition blocks (12"x30")		
Solid		
2"	S.F.	0.032
Hollow		
3"	S.F.	0.032
4"	"	0.033
6"	"	0.036
Vertical reinforcing		
4' o.c., add 5% to labor		
2'8" o.c., add 15% to labor		
Interior partitions, add 10% to labor		

04220.90 Bond Beams & Lintels

	UNIT	MAN/HOURS
Bond beam, no grout or reinforcement		
8" x 16" x		
4" thick	L.F.	0.062
6" thick	"	0.064
8" thick	"	0.067

04220.90 Bond Beams & Lintels *(Cont.)*

	UNIT	MAN/HOURS
10" thick	L.F.	0.070
12" thick	"	0.073
Beam lintel, no grout or reinforcement		
8" x 16" x		
10" thick	L.F.	0.080
12" thick	"	0.089
Precast masonry lintel		
6 lf, 8" high x		
4" thick	L.F.	0.133
6" thick	"	0.133
8" thick	"	0.145
10" thick	"	0.145
10 lf, 8" high x		
4" thick	L.F.	0.080
6" thick	"	0.080
8" thick	"	0.089
10" thick	"	0.089
Steel angles and plates		
Minimum	Lb.	0.011
Maximum	"	0.020
Various size angle lintels		
1/4" stock		
3" x 3"	L.F.	0.050
3" x 3-1/2"	"	0.050
3/8" stock		
3" x 4"	L.F.	0.050
3-1/2" x 4"	"	0.050
4" x 4"	"	0.050
5" x 3-1/2"	"	0.050
6" x 3-1/2"	"	0.050
1/2" stock		
6" x 4"	L.F.	0.050

04240.10 Clay Tile

	UNIT	MAN/HOURS
Hollow clay tile, for back-up, 12" x 12"		
Scored face		
Load bearing		
4" thick	S.F.	0.057
6" thick	"	0.059
8" thick	"	0.062
10" thick	"	0.064
12" thick	"	0.067
Non-load bearing		
3" thick	S.F.	0.055
4" thick	"	0.057
6" thick	"	0.059
8" thick	"	0.062
12" thick	"	0.067
Partition, 12" x 12"		
In walls		
3" thick	S.F.	0.067
4" thick	"	0.067
6" thick	"	0.070

04 MASONRY

Unit Masonry	UNIT	MAN/HOURS
04240.10 Clay Tile *(Cont.)*		
8" thick	S.F.	0.073
10" thick	"	0.076
12" thick	"	0.080
Clay tile floors		
4" thick	S.F.	0.044
6" thick	"	0.047
8" thick	"	0.050
10" thick	"	0.053
12" thick	"	0.057
Terra cotta		
Coping, 10" or 12" wide, 3" thick	L.F.	0.160
04270.10 Glass Block		
Glass block, 4" thick		
6" x 6"	S.F.	0.267
8" x 8"	"	0.200
12" x 12"	"	0.160
Replacement glass blocks, 4" x 8" x 8"		
Minimum	S.F.	0.800
Maximum	"	1.600
04295.10 Parging / Masonry Plaster		
Parging		
1/2" thick	S.F.	0.053
3/4" thick	"	0.067
1" thick	"	0.080

Stone	UNIT	MAN/HOURS
04400.10 Stone		
Rubble stone		
Walls set in mortar		
8" thick	S.F.	0.200
12" thick	"	0.320
18" thick	"	0.400
24" thick	"	0.533
Dry set wall		
8" thick	S.F.	0.133
12" thick	"	0.200
18" thick	"	0.267
24" thick	"	0.320
Cut stone		
Imported marble		
Facing panels		
3/4" thick	S.F.	0.320
1-1/2" thick	"	0.364

Stone	UNIT	MAN/HOURS
04400.10 Stone *(Cont.)*		
2-1/4" thick	S.F.	0.444
Base		
1" thick		
4" high	L.F.	0.400
6" high	"	0.400
Columns, solid		
Plain faced	C.F.	5.333
Fluted	"	5.333
Flooring, travertine, minimum	S.F.	0.123
Average	"	0.160
Maximum	"	0.178
Domestic marble		
Facing panels		
7/8" thick	S.F.	0.320
1-1/2" thick	"	0.364
2-1/4" thick	"	0.444
Stairs		
12" treads	L.F.	0.400
6" risers	"	0.267
Thresholds, 7/8" thick, 3' long, 4" to 6" wide		
Plain	EA.	0.667
Beveled	"	0.667
Window sill		
6" wide, 2" thick	L.F.	0.320
Stools		
5" wide, 7/8" thick	L.F.	0.320
Limestone panels up to 12' x 5', smooth finish		
2" thick	S.F.	0.096
3" thick	"	0.096
4" thick	"	0.096
Miscellaneous limestone items		
Steps, 14" wide, 6" deep	L.F.	0.533
Coping, smooth finish	C.F.	0.267
Sills, lintels, jambs, smooth finish	"	0.320
Granite veneer facing panels, polished		
7/8" thick		
Black	S.F.	0.320
Gray	"	0.320
Base		
4" high	L.F.	0.160
6" high	"	0.178
Curbing, straight, 6" x 16"	"	0.400
Radius curbs, radius over 5'	"	0.533
Ashlar veneer		
4" thick, random	S.F.	0.320
Pavers, 4" x 4" split		
Gray	S.F.	0.160
Pink	"	0.160
Black	"	0.160
Slate, panels		
1" thick	S.F.	0.320
2" thick	"	0.364
Sills or stools		

04 MASONRY

Stone	UNIT	MAN/HOURS
04400.10 Stone *(Cont.)*		
1" thick		
6" wide	L.F.	0.320
10" wide	"	0.348
2" thick		
6" wide	L.F.	0.364
10" wide	"	0.400

Masonry Restoration	UNIT	MAN/HOURS
04520.10 Restoration And Cleaning		
Masonry cleaning		
Washing brick		
Smooth surface	S.F.	0.013
Rough surface	"	0.018
Steam clean masonry		
Smooth face		
Minimum	S.F.	0.010
Maximum	"	0.015
Rough face		
Minimum	S.F.	0.013
Maximum	"	0.020
Sandblast masonry		
Minimum	S.F.	0.016
Maximum	"	0.027
Pointing masonry		
Brick	S.F.	0.032
Concrete block	"	0.023
Cut and repoint		
Brick		
Minimum	S.F.	0.040
Maximum	"	0.080
Stone work	L.F.	0.062
Cut and recaulk		
Oil base caulks	L.F.	0.053
Butyl caulks	"	0.053
Polysulfides and acrylics	"	0.053
Silicones	"	0.053
Cement and sand grout on walls, to 1/8" thick		
Minimum	S.F.	0.032
Maximum	"	0.040
Brick removal and replacement		
Minimum	EA.	0.100
Average	"	0.133
Maximum	"	0.400

Masonry Restoration	UNIT	MAN/HOURS
04550.10 Refractories		
Flue liners		
Rectangular		
8" x 12"	L.F.	0.133
12" x 12"	"	0.145
12" x 18"	"	0.160
16" x 16"	"	0.178
18" x 18"	"	0.190
20" x 20"	"	0.200
24" x 24"	"	0.229
Round		
18" dia.	L.F.	0.190
24" dia.	"	0.229

05 METALS

Metal Fastening	UNIT	MAN/HOURS
05050.10 Structural Welding		
Welding		
Single pass		
1/8"	L.F.	0.040
3/16"	"	0.053
1/4"	"	0.067
Miscellaneous steel shapes		
Plain	Lb.	0.002
Galvanized	"	0.003
Plates		
Plain	Lb.	0.002
Galvanized	"	0.003
05050.95 Metal Lintels		
Lintels, steel		
Plain	Lb.	0.020
Galvanized	"	0.020
05120.10 Structural Steel		
Beams and girders, A-36		
Welded	TON	4.800
Bolted	"	4.364
Columns		
Pipe		
6" dia.	Lb.	0.005
12" dia.	"	0.004
Purlins and girts		
Welded	TON	8.000
Bolted	"	6.857
Column base plates		
Up to 150 lb each	Lb.	0.005
Over 150 lb each	"	0.007
Structural pipe		
3" to 5" o.d.	TON	9.600
6" to 12" o.d.	"	6.857
Structural tube		
6" square		
Light sections	TON	9.600
Heavy sections	"	6.857
6" wide rectangular		
Light sections	TON	8.000
Heavy sections	"	6.000
Greater than 6" wide rectangular		
Light sections	TON	8.000
Heavy sections	"	6.000
Miscellaneous structural shapes		
Steel angle	TON	12.000
Steel plate	"	8.000
Trusses, field welded		
60 lb/lf	TON	6.000
100 lb/lf	"	4.800
150 lb/lf	"	4.000
Bolted		
60 lb/lf	TON	5.333

Metal Fastening	UNIT	MAN/HOURS
05120.10 Structural Steel *(Cont.)*		
100 lb/lf	TON	4.364
150 lb/lf	"	3.692
05200.10 Metal Joists		
Joist		
DLH series	TON	3.200
K series	"	3.200
LH series	"	3.200
05300.10 Metal Decking		
Roof, 1-1/2" deep, non-composite		
16 ga.		
Primed	S.F.	0.008
Galvanized	"	0.008
18 ga.		
Primed	S.F.	0.008
Galvanized	"	0.008
20 ga.		
Primed	S.F.	0.008
Galvanized	"	0.008
22 ga.		
Primed	S.F.	0.008
Galvanized	"	0.008
Open type decking, galvanized		
1-1/2" deep		
18 ga.	S.F.	0.008
20 ga.	"	0.008
22 ga.	"	0.008
3" deep		
16 ga.	S.F.	0.009
18 ga.	"	0.009
20 ga.	"	0.009
22 ga.	"	0.009
4-1/2" deep		
16 ga.	S.F.	0.010
18 ga.	"	0.010
6" deep		
16 ga.	S.F.	0.011
18 ga.	"	0.011
7-1/2" deep		
16 ga.	S.F.	0.011
18 ga.	"	0.011
Cellular type		
1-1/2" deep, galvanized		
18-18 ga.	S.F.	0.010
22-18 ga.	"	0.010
3" deep, galvanized		
16-16 ga.	S.F.	0.011
18-16 ga.	"	0.011
18-18 ga.	"	0.011
20-18 ga.	"	0.011
4-1/2" deep, galvanized		
16-16 ga.	S.F.	0.011

05 METALS

Metal Fastening	UNIT	MAN/HOURS
05300.10 Metal Decking *(Cont.)*		
18-16 ga.	S.F.	0.011
18-18 ga.	"	0.011
20-18 ga.	"	0.011
Composite deck, non-cellular, galvanized		
1-1/2" deep		
18 ga.	S.F.	0.009
20 ga.	"	0.009
22 ga.	"	0.009
3" deep		
18 ga.	S.F.	0.009
20 ga.	"	0.009
22 ga.	"	0.009
Slab form floor deck		
9/16" deep		
28 ga.	S.F.	0.009
1-5/16" deep		
24 ga.	S.F.	0.009
22 ga.	"	0.009

Cold Formed Framing	UNIT	MAN/HOURS
05410.10 Metal Framing		
Furring channel, galvanized		
Beams and columns, 3/4"		
12" o.c.	S.F.	0.080
16" o.c.	"	0.073
Walls, 3/4"		
12" o.c.	S.F.	0.040
16" o.c.	"	0.033
24" o.c.	"	0.027
1-1/2"		
12" o.c.	S.F.	0.040
16" o.c.	"	0.033
24" o.c.	"	0.027
Stud, load bearing		
16" o.c.		
16 ga.		
2-1/2"	S.F.	0.036
3-5/8"	"	0.036
4"	"	0.036
6"	"	0.040
18 ga.		
2-1/2"	S.F.	0.036
3-5/8"	"	0.036
4"	"	0.036
6"	"	0.040

Cold Formed Framing	UNIT	MAN/HOURS
05410.10 Metal Framing *(Cont.)*		
8"	S.F.	0.040
20 ga.		
2-1/2"	S.F.	0.036
3-5/8"	"	0.036
4"	"	0.036
6"	"	0.040
8"	"	0.040
24" o.c.		
16 ga.		
2-1/2"	S.F.	0.031
3-5/8"	"	0.031
4"	"	0.031
6"	"	0.033
8"	"	0.033
18 ga.		
2-1/2"	S.F.	0.031
3-5/8"	"	0.031
4"	"	0.031
6"	"	0.033
8"	"	0.033
20 ga.		
2-1/2"	S.F.	0.031
3-5/8"	"	0.031
4"	"	0.031
6"	"	0.033
8"	"	0.033

Metal Fabrications	UNIT	MAN/HOURS
05510.10 Stairs		
Stock unit, steel, complete, per riser		
Tread		
3'-6" wide	EA.	1.000
4' wide	"	1.143
5' wide	"	1.333
Metal pan stair, cement filled, per riser		
3'-6" wide	EA.	0.800
4' wide	"	0.889
5' wide	"	1.000
Landing, steel pan	S.F.	0.200
Cast iron tread, steel stringers, stock units, per riser		
Tread		
3'-6" wide	EA.	1.000
4' wide	"	1.143
5' wide	"	1.333
Stair treads, abrasive, 12" x 3'-6"		

05 METALS

Metal Fabrications	UNIT	MAN/HOURS
05510.10 Stairs (Cont.)		
Cast iron		
3/8"	EA.	0.400
1/2"	"	0.400
Cast aluminum		
5/16"	EA.	0.400
3/8"	"	0.400
1/2"	"	0.400
05515.10 Ladders		
Ladder, 18" wide		
With cage	L.F.	0.533
Without cage	"	0.400
05520.10 Railings		
Railing, pipe		
1-1/4" diameter, welded steel		
2-rail		
Primed	L.F.	0.160
Galvanized	"	0.160
3-rail		
Primed	L.F.	0.200
Galvanized	"	0.200
Wall mounted, single rail, welded steel		
Primed	L.F.	0.123
Galvanized	"	0.123
1-1/2" diameter, welded steel		
2-rail		
Primed	L.F.	0.160
Galvanized	"	0.160
3-rail		
Primed	L.F.	0.200
Galvanized	"	0.200
Wall mounted, single rail, welded steel		
Primed	L.F.	0.123
Galvanized	"	0.123
2" diameter, welded steel		
2-rail		
Primed	L.F.	0.178
Galvanized	"	0.178
3-rail		
Primed	L.F.	0.229
Galvanized	"	0.229
Wall mounted, single rail, welded steel		
Primed	L.F.	0.133
Galvanized	"	0.133
05530.10 Metal Grating		
Floor plate, checkered, steel		
1/4"		
Primed	S.F.	0.011
Galvanized	"	0.011
3/8"		
Primed	S.F.	0.012

Metal Fabrications	UNIT	MAN/HOURS
05530.10 Metal Grating (Cont.)		
Galvanized	S.F.	0.012
Aluminum grating, pressure-locked bearing bars		
3/4" x 1/8"	S.F.	0.020
1" x 1/8"	"	0.020
1-1/4" x 1/8"	"	0.020
1-1/4" x 3/16"	"	0.020
1-1/2" x 1/8"	"	0.020
1-3/4" x 3/16"	"	0.020
Miscellaneous expenses		
Cutting		
Minimum	L.F.	0.053
Maximum	"	0.080
Banding		
Minimum	L.F.	0.133
Maximum	"	0.160
Toe plates		
Minimum	L.F.	0.160
Maximum	"	0.200
Steel grating, primed		
3/4" x 1/8"	S.F.	0.027
1" x 1/8"	"	0.027
1-1/4" x 1/8"	"	0.027
1-1/4" x 3/16"	"	0.027
1-1/2" x 1/8"	"	0.027
1-3/4" x 3/16"	"	0.027
Galvanized		
3/4" x 1/8"	S.F.	0.027
1" x 1/8"	"	0.027
1-1/4" x 1/8"	"	0.027
1-1/4" x 3/16"	"	0.027
1-1/2" x 1/8"	"	0.027
1-3/4" x 3/16"	"	0.027
Miscellaneous expenses		
Cutting		
Minimum	L.F.	0.057
Maximum	"	0.089
Banding		
Minimum	L.F.	0.145
Maximum	"	0.178
Toe plates		
Minimum	L.F.	0.178
Maximum	"	0.229
05540.10 Castings		
Miscellaneous castings		
Light sections	Lb.	0.016
Heavy sections	"	0.011
Manhole covers and frames		
Regular, city type		
18" dia.		
100 lb	EA.	1.600
24" dia.		
200 lb	EA.	1.600

05 METALS

Metal Fabrications	UNIT	MAN/HOURS
05540.10 Castings *(Cont.)*		
300 lb	EA.	1.778
400 lb	"	1.778
26" dia., 475 lb	"	2.000
30" dia., 600 lb	"	2.286
8" square, 75 lb	"	0.320
24" square		
126 lb	EA.	1.600
500 lb	"	2.000
Watertight type		
20" dia., 200 lb	EA.	2.000
24" dia., 350 lb	"	2.667
Steps, cast iron		
7" x 9"	EA.	0.160
8" x 9"	"	0.178
Manhole covers and frames, aluminum		
12" x 12"	EA.	0.320
18" x 18"	"	0.320
24" x 24"	"	0.400
Corner protection		
Steel angle guard with anchors		
2" x 2" x 3/16"	L.F.	0.114
2" x 3" x 1/4"	"	0.114
3" x 3" x 5/16"	"	0.114
3" x 4" x 5/16"	"	0.123
4" x 4" x 5/16"	"	0.123

Misc. Fabrications	UNIT	MAN/HOURS
05580.10 Metal Specialties		
Kick plate		
4" high x 1/4" thick		
Primed	L.F.	0.160
Galvanized	"	0.160
6" high x 1/4" thick		
Primed	L.F.	0.178
Galvanized	"	0.178
05700.10 Ornamental Metal		
Railings, square bars, 6" o.c., shaped top rails		
Steel	L.F.	0.400
Aluminum	"	0.400
Bronze	"	0.533
Stainless steel	"	0.533
Laminated metal or wood handrails		
2-1/2" round or oval shape	L.F.	0.400
Grilles and louvers		

Misc. Fabrications	UNIT	MAN/HOURS
05700.10 Ornamental Metal *(Cont.)*		
Fixed type louvers		
4 through 10 sf	S.F.	0.133
Over 10 sf	"	0.100
Movable type louvers		
4 through 10 sf	S.F.	0.133
Over 10 sf	"	0.100
Aluminum louvers		
Residential use, fixed type, with screen		
8" x 8"	EA.	0.400
12" x 12"	"	0.400
12" x 18"	"	0.400
14" x 24"	"	0.400
18" x 24"	"	0.400
30" x 24"	"	0.444
05800.10 Expansion Control		
Expansion joints with covers, floor assembly type		
With 1" space		
Aluminum	L.F.	0.133
Bronze	"	0.133
Stainless steel	"	0.133
With 2" space		
Aluminum	L.F.	0.133
Bronze	"	0.133
Stainless steel	"	0.133
Ceiling and wall assembly type		
With 1" space		
Aluminum	L.F.	0.160
Bronze	"	0.160
Stainless steel	"	0.160
With 2" space		
Aluminum	L.F.	0.160
Bronze	"	0.160
Stainless steel	"	0.160
Exterior roof and wall, aluminum		
Roof to roof		
With 1" space	L.F.	0.133
With 2" space	"	0.133
Roof to wall		
With 1" space	L.F.	0.145
With 2" space	"	0.145
Flat wall to wall		
With 1" space	L.F.	0.133
With 2" space	"	0.133
Corner to flat wall		
With 1" space	L.F.	0.160
With 2" in space	"	0.160

06 WOOD AND PLASTICS

Fasteners And Adhesives	UNIT	MAN/HOURS
06050.10 Accessories		
Column/post base, cast aluminum		
4" x 4"	EA.	0.200
6" x 6"	"	0.200
Bridging, metal, per pair		
12" o.c.	EA.	0.080
16" o.c.	"	0.073
Anchors		
Bolts, threaded two ends, with nuts and washers		
1/2" dia.		
4" long	EA.	0.050
7-1/2" long	"	0.050
3/4" dia.		
7-1/2" long	EA.	0.050
15" long	"	0.050
Framing anchors		
10 gauge	EA.	0.067
Bolts, carriage		
1/4 x 4	EA.	0.080
5/16 x 6	"	0.084
3/8 x 6	"	0.084
1/2 x 6	"	0.084
Joist and beam hangers		
18 ga.		
2 x 4	EA.	0.080
2 x 6	"	0.080
2 x 8	"	0.080
2 x 10	"	0.089
2 x 12	"	0.100
16 ga.		
3 x 6	EA.	0.089
3 x 8	"	0.089
3 x 10	"	0.094
3 x 12	"	0.107
3 x 14	"	0.114
4 x 6	"	0.089
4 x 8	"	0.089
4 x 10	"	0.094
4 x 12	"	0.107
4 x 14	"	0.114
Rafter anchors, 18 ga., 1-1/2" wide		
5-1/4" long	EA.	0.067
10-3/4" long	"	0.067
Shear plates		
2-5/8" dia.	EA.	0.062
4" dia.	"	0.067
Sill anchors		
Embedded in concrete	EA.	0.080
Split rings		
2-1/2" dia.	EA.	0.089
4" dia.	"	0.100
Strap ties, 14 ga., 1-3/8" wide		
12" long	EA.	0.067

Fasteners And Adhesives	UNIT	MAN/HOURS
06050.10 Accessories (Cont.)		
18" long	EA.	0.073
24" long	"	0.080
36" long	"	0.089
Toothed rings		
2-5/8" dia.	EA.	0.133
4" dia.	"	0.160

Rough Carpentry	UNIT	MAN/HOURS
06110.10 Blocking		
Steel construction		
Walls		
2x4	L.F.	0.053
2x6	"	0.062
2x8	"	0.067
2x10	"	0.073
2x12	"	0.080
Ceilings		
2x4	L.F.	0.062
2x6	"	0.073
2x8	"	0.080
2x10	"	0.089
2x12	"	0.100
Wood construction		
Walls		
2x4	L.F.	0.044
2x6	"	0.050
2x8	"	0.053
2x10	"	0.057
2x12	"	0.062
Ceilings		
2x4	L.F.	0.050
2x6	"	0.057
2x8	"	0.062
2x10	"	0.067
2x12	"	0.073
06110.20 Ceiling Framing		
Ceiling joists		
12" o.c.		
2x4	S.F.	0.019
2x6	"	0.020
2x8	"	0.021
2x10	"	0.022
2x12	"	0.024
16" o.c.		

06 WOOD AND PLASTICS

Rough Carpentry		UNIT	MAN/HOURS
06110.20	**Ceiling Framing** (Cont.)		
2x4		S.F.	0.015
2x6		"	0.016
2x8		"	0.017
2x10		"	0.017
2x12		"	0.018
24" o.c.			
2x4		S.F.	0.013
2x6		"	0.013
2x8		"	0.014
2x10		"	0.015
2x12		"	0.016
Headers and nailers			
2x4		L.F.	0.026
2x6		"	0.027
2x8		"	0.029
2x10		"	0.031
2x12		"	0.033
Sister joists for ceilings			
2x4		L.F.	0.057
2x6		"	0.067
2x8		"	0.080
2x10		"	0.100
2x12		"	0.133
06110.30	**Floor Framing**		
Floor joists			
12" o.c.			
2x6		S.F.	0.016
2x8		"	0.016
2x10		"	0.017
2x12		"	0.017
2x14		"	0.017
3x6		"	0.017
3x8		"	0.017
3x10		"	0.018
3x12		"	0.019
3x14		"	0.020
4x6		"	0.017
4x8		"	0.017
4x10		"	0.018
4x12		"	0.019
4x14		"	0.020
16" o.c.			
2x6		S.F.	0.013
2x8		"	0.014
2x10		"	0.014
2x12		"	0.014
2x14		"	0.015
3x6		"	0.014
3x8		"	0.014
3x10		"	0.015
3x12		"	0.015
3x14		"	0.016

Rough Carpentry		UNIT	MAN/HOURS
06110.30	**Floor Framing** (Cont.)		
4x6		S.F.	0.014
4x8		"	0.014
4x10		"	0.015
4x12		"	0.015
4x14		"	0.016
Sister joists for floors			
2x4		L.F.	0.050
2x6		"	0.057
2x8		"	0.067
2x10		"	0.080
2x12		"	0.100
3x6		"	0.080
3x8		"	0.089
3x10		"	0.100
3x12		"	0.114
4x6		"	0.080
4x8		"	0.089
4x10		"	0.100
4x12		"	0.114
06110.40	**Furring**		
Furring, wood strips			
Walls			
On masonry or concrete walls			
1x2 furring			
12" o.c.		S.F.	0.025
16" o.c.		"	0.023
24" o.c.		"	0.021
1x3 furring			
12" o.c.		S.F.	0.025
16" o.c.		"	0.023
24" o.c.		"	0.021
On wood walls			
1x2 furring			
12" o.c.		S.F.	0.018
16" o.c.		"	0.016
24" o.c.		"	0.015
1x3 furring			
12" o.c.		S.F.	0.018
16" o.c.		"	0.016
24" o.c.		"	0.015
Ceilings			
On masonry or concrete ceilings			
1x2 furring			
12" o.c.		S.F.	0.044
16" o.c.		"	0.040
24" o.c.		"	0.036
1x3 furring			
12" o.c.		S.F.	0.044
16" o.c.		"	0.040
24" o.c.		"	0.036
On wood ceilings			
1x2 furring			

06 WOOD AND PLASTICS

Rough Carpentry	UNIT	MAN/HOURS
06110.40 Furring (Cont.)		
12" o.c.	S.F.	0.030
16" o.c.	"	0.027
24" o.c.	"	0.024
1x3		
12" o.c.	S.F.	0.030
16" o.c.	"	0.027
24" o.c.	"	0.024
06110.50 Roof Framing		
Roof framing		
Rafters, gable end		
0-2 pitch (flat to 2-in-12)		
12" o.c.		
2x4	S.F.	0.017
2x6	"	0.017
2x8	"	0.018
2x10	"	0.019
2x12	"	0.020
16" o.c.		
2x6	S.F.	0.014
2x8	"	0.015
2x10	"	0.015
2x12	"	0.016
24" o.c.		
2x6	S.F.	0.012
2x8	"	0.013
2x10	"	0.013
2x12	"	0.013
4-6 pitch (4-in-12 to 6-in-12)		
12" o.c.		
2x4	S.F.	0.017
2x6	"	0.018
2x8	"	0.019
2x10	"	0.020
2x12	"	0.021
16" o.c.		
2x6	S.F.	0.015
2x8	"	0.015
2x10	"	0.016
2x12	"	0.017
24" o.c.		
2x6	S.F.	0.013
2x8	"	0.013
2x10	"	0.014
2x12	"	0.015
8-12 pitch (8-in-12 to 12-in-12)		
12" o.c.		
2x4	S.F.	0.018
2x6	"	0.019
2x8	"	0.020
2x10	"	0.021
2x12	"	0.022
16" o.c.		

Rough Carpentry	UNIT	MAN/HOURS
06110.50 Roof Framing (Cont.)		
2x6	S.F.	0.015
2x8	"	0.016
2x10	"	0.017
2x12	"	0.017
24" o.c.		
2x6	S.F.	0.013
2x8	"	0.013
2x10	"	0.014
2x12	"	0.014
Ridge boards		
2x6	L.F.	0.040
2x8	"	0.044
2x10	"	0.050
2x12	"	0.057
Hip rafters		
2x6	L.F.	0.029
2x8	"	0.030
2x10	"	0.031
2x12	"	0.032
Jack rafters		
4-6 pitch (4-in-12 to 6-in-12)		
16" o.c.		
2x6	S.F.	0.024
2x8	"	0.024
2x10	"	0.026
2x12	"	0.027
24" o.c.		
2x6	S.F.	0.018
2x8	"	0.019
2x10	"	0.020
2x12	"	0.020
8-12 pitch (8-in-12 to 12-in-12)		
16" o.c.		
2x6	S.F.	0.025
2x8	"	0.026
2x10	"	0.027
2x12	"	0.028
24" o.c.		
2x6	S.F.	0.019
2x8	"	0.020
2x10	"	0.020
2x12	"	0.021
Sister rafters		
2x4	L.F.	0.057
2x6	"	0.067
2x8	"	0.080
2x10	"	0.100
2x12	"	0.133
Fascia boards		
2x4	L.F.	0.040
2x6	"	0.040
2x8	"	0.044
2x10	"	0.044

06 WOOD AND PLASTICS

Rough Carpentry	UNIT	MAN/HOURS
06110.50 Roof Framing (Cont.)		
2x12	L.F.	0.050
Cant strips		
Fiber		
3x3	L.F.	0.023
4x4	"	0.024
Wood		
3x3	L.F.	0.024
06110.60 Sleepers		
Sleepers, over concrete		
12" o.c.		
1x2	S.F.	0.018
1x3	"	0.019
2x4	"	0.022
2x6	"	0.024
16" o.c.		
1x2	S.F.	0.016
1x3	"	0.016
2x4	"	0.019
2x6	"	0.020
06110.65 Soffits		
Soffit framing		
2x3	L.F.	0.057
2x4	"	0.062
2x6	"	0.067
2x8	"	0.073
06110.70 Wall Framing		
Framing wall, studs		
12" o.c.		
2x3	S.F.	0.015
2x4	"	0.015
2x6	"	0.016
2x8	"	0.017
16" o.c.		
2x3	S.F.	0.013
2x4	"	0.013
2x6	"	0.013
2x8	"	0.014
24" o.c.		
2x3	S.F.	0.011
2x4	"	0.011
2x6	"	0.011
2x8	"	0.012
Plates, top or bottom		
2x3	L.F.	0.024
2x4	"	0.025
2x6	"	0.027
2x8	"	0.029
Headers, door or window		
2x6		
Single		

Rough Carpentry	UNIT	MAN/HOURS
06110.70 Wall Framing (Cont.)		
3' long	EA.	0.400
6' long	"	0.500
Double		
3' long	EA.	0.444
6' long	"	0.571
2x8		
Single		
4' long	EA.	0.500
8' long	"	0.615
Double		
4' long	EA.	0.571
8' long	"	0.727
2x10		
Single		
5' long	EA.	0.615
10' long	"	0.800
Double		
5' long	EA.	0.667
10' long	"	0.800
2x12		
Single		
6' long	EA.	0.615
12' long	"	0.800
Double		
6' long	EA.	0.727
12' long	"	0.889
06115.10 Floor Sheathing		
Sub-flooring, plywood, CDX		
1/2" thick	S.F.	0.010
5/8" thick	"	0.011
3/4" thick	"	0.013
Structural plywood		
1/2" thick	S.F.	0.010
5/8" thick	"	0.011
3/4" thick	"	0.012
Board type subflooring		
1x6		
Minimum	S.F.	0.018
Maximum	"	0.020
1x8		
Minimum	S.F.	0.017
Maximum	"	0.019
1x10		
Minimum	S.F.	0.016
Maximum	"	0.018
Underlayment		
Hardboard, 1/4" tempered	S.F.	0.010
Plywood, CDX		
3/8" thick	S.F.	0.010
1/2" thick	"	0.011
5/8" thick	"	0.011
3/4" thick	"	0.012

06 WOOD AND PLASTICS

Rough Carpentry	UNIT	MAN/HOURS
06115.20 Roof Sheathing		
Sheathing		
Plywood, CDX		
3/8" thick	S.F.	0.010
1/2" thick	"	0.011
5/8" thick	"	0.011
3/4" thick	"	0.012
Structural plywood		
3/8" thick	S.F.	0.010
1/2" thick	"	0.011
5/8" thick	"	0.011
3/4" thick	"	0.012
06115.30 Wall Sheathing		
Sheathing		
Plywood, CDX		
3/8" thick	S.F.	0.012
1/2" thick	"	0.012
5/8" thick	"	0.013
3/4" thick	"	0.015
Waferboard		
3/8" thick	S.F.	0.012
1/2" thick	"	0.012
5/8" thick	"	0.013
3/4" thick	"	0.015
Structural plywood		
3/8" thick	S.F.	0.012
1/2" thick	"	0.012
5/8" thick	"	0.013
3/4" thick	"	0.015
Gypsum, 1/2" thick	"	0.012
Asphalt impregnated fiberboard, 1/2" thick	"	0.012
06125.10 Wood Decking		
Decking, T&G solid		
Cedar		
3" thick	S.F.	0.020
4" thick	"	0.021
Fir		
3" thick	S.F.	0.020
4" thick	"	0.021
Southern yellow pine		
3" thick	S.F.	0.023
4" thick	"	0.025
White pine		
3" thick	S.F.	0.020
4" thick	"	0.021
06130.10 Heavy Timber		
Mill framed structures		
Beams to 20' long		
Douglas fir		
6x8	L.F.	0.080
6x10	"	0.083

Rough Carpentry	UNIT	MAN/HOURS
06130.10 Heavy Timber *(Cont.)*		
6x12	L.F.	0.089
6x14	"	0.092
6x16	"	0.096
8x10	"	0.083
8x12	"	0.089
8x14	"	0.092
8x16	"	0.096
Southern yellow pine		
6x8	L.F.	0.080
6x10	"	0.083
6x12	"	0.089
6x14	"	0.092
6x16	"	0.096
8x10	"	0.083
8x12	"	0.089
8x14	"	0.092
8x16	"	0.096
Columns to 12' high		
Douglas fir		
6x6	L.F.	0.120
8x8	"	0.120
10x10	"	0.133
12x12	"	0.133
Southern yellow pine		
6x6	L.F.	0.120
8x8	"	0.120
10x10	"	0.133
12x12	"	0.133
Posts, treated		
4x4	L.F.	0.032
6x6	"	0.040
06190.20 Wood Trusses		
Truss, fink, 2x4 members		
3-in-12 slope		
24' span	EA.	0.686
26' span	"	0.686
28' span	"	0.727
30' span	"	0.727
34' span	"	0.774
38' span	"	0.774
5-in-12 slope		
24' span	EA.	0.706
28' span	"	0.727
30' span	"	0.750
32' span	"	0.750
40' span	"	0.800
Gable, 2x4 members		
5-in-12 slope		
24' span	EA.	0.706
26' span	"	0.706
28' span	"	0.727
30' span	"	0.750

06 WOOD AND PLASTICS

Rough Carpentry	UNIT	MAN/HOURS
06190.20 **Wood Trusses** *(Cont.)*		
32' span	EA.	0.750
36' span	"	0.774
40' span	"	0.800
King post type, 2x4 members		
4-in-12 slope		
16' span	EA.	0.649
18' span	"	0.667
24' span	"	0.706
26' span	"	0.706
30' span	"	0.750
34' span	"	0.750
38' span	"	0.774
42' span	"	0.828

Finish Carpentry	UNIT	MAN/HOURS
06200.10 **Finish Carpentry**		
Mouldings and trim		
Apron, flat		
9/16 x 2	L.F.	0.040
9/16 x 3-1/2	"	0.042
Base		
Colonial		
7/16 x 2-1/4	L.F.	0.040
7/16 x 3	"	0.040
7/16 x 3-1/4	"	0.040
9/16 x 3	"	0.042
9/16 x 3-1/4	"	0.042
11/16 x 2-1/4	"	0.044
Ranch		
7/16 x 2-1/4	L.F.	0.040
7/16 x 3-1/4	"	0.040
9/16 x 2-1/4	"	0.042
9/16 x 3	"	0.042
9/16 x 3-1/4	"	0.042
Casing		
11/16 x 2-1/2	L.F.	0.036
11/16 x 3-1/2	"	0.038
Chair rail		
9/16 x 2-1/2	L.F.	0.040
9/16 x 3-1/2	"	0.040
Closet pole		
1-1/8" dia.	L.F.	0.053
1-5/8" dia.	"	0.053
Cove		
9/16 x 1-3/4	L.F.	0.040

Finish Carpentry	UNIT	MAN/HOURS
06200.10 **Finish Carpentry** *(Cont.)*		
11/16 x 2-3/4	L.F.	0.040
Crown		
9/16 x 1-5/8	L.F.	0.053
9/16 x 2-5/8	"	0.062
11/16 x 3-5/8	"	0.067
11/16 x 4-1/4	"	0.073
11/16 x 5-1/4	"	0.080
Drip cap		
1-1/16 x 1-5/8	L.F.	0.040
Glass bead		
3/8 x 3/8	L.F.	0.050
1/2 x 9/16	"	0.050
5/8 x 5/8	"	0.050
3/4 x 3/4	"	0.050
Half round		
1/2	L.F.	0.032
5/8	"	0.032
3/4	"	0.032
Lattice		
1/4 x 7/8	L.F.	0.032
1/4 x 1-1/8	"	0.032
1/4 x 1-3/8	"	0.032
1/4 x 1-3/4	"	0.032
1/4 x 2	"	0.032
Ogee molding		
5/8 x 3/4	L.F.	0.040
11/16 x 1-1/8	"	0.040
11/16 x 1-3/8	"	0.040
Parting bead		
3/8 x 7/8	L.F.	0.050
Quarter round		
1/4 x 1/4	L.F.	0.032
3/8 x 3/8	"	0.032
1/2 x 1/2	"	0.032
11/16 x 11/16	"	0.035
3/4 x 3/4	"	0.035
1-1/16 x 1-1/16	"	0.036
Railings, balusters		
1-1/8 x 1-1/8	L.F.	0.080
1-1/2 x 1-1/2	"	0.073
Screen moldings		
1/4 x 3/4	L.F.	0.067
5/8 x 5/16	"	0.067
Shoe		
7/16 x 11/16	L.F.	0.032
Sash beads		
1/2 x 3/4	L.F.	0.067
1/2 x 7/8	"	0.067
1/2 x 1-1/8	"	0.073
5/8 x 7/8	"	0.073
Stop		
5/8 x 1-5/8		
Colonial	L.F.	0.050

06 WOOD AND PLASTICS

Finish Carpentry	UNIT	MAN/HOURS
06200.10 Finish Carpentry *(Cont.)*		
Ranch	L.F.	0.050
Stools		
11/16 x 2-1/4	L.F.	0.089
11/16 x 2-1/2	"	0.089
11/16 x 5-1/4	"	0.100
Exterior trim, casing, select pine, 1x3	"	0.040
Douglas fir		
1x3	L.F.	0.040
1x4	"	0.040
1x6	"	0.044
1x8	"	0.050
Cornices, white pine, #2 or better		
1x2	L.F.	0.040
1x4	"	0.040
1x6	"	0.044
1x8	"	0.047
1x10	"	0.050
1x12	"	0.053
Shelving, pine		
1x8	L.F.	0.062
1x10	"	0.064
1x12	"	0.067
Plywood shelf, 3/4", with edge band, 12" wide	"	0.080
Adjustable shelf, and rod, 12" wide		
3' to 4' long	EA.	0.200
5' to 8' long	"	0.267
Prefinished wood shelves with brackets and supports		
8" wide		
3' long	EA.	0.200
4' long	"	0.200
6' long	"	0.200
10" wide		
3' long	EA.	0.200
4' long	"	0.200
6' long	"	0.200
06220.10 Millwork		
Countertop, laminated plastic		
25" x 7/8" thick		
Minimum	L.F.	0.200
Average	"	0.267
Maximum	"	0.320
25" x 1-1/4" thick		
Minimum	L.F.	0.267
Average	"	0.320
Maximum	"	0.400
Add for cutouts	EA.	0.500
Backsplash, 4" high, 7/8" thick	L.F.	0.160
Plywood, sanded, A-C		
1/4" thick	S.F.	0.027
3/8" thick	"	0.029
1/2" thick	"	0.031
A-D		

Finish Carpentry	UNIT	MAN/HOURS
06220.10 Millwork *(Cont.)*		
1/4" thick	S.F.	0.027
3/8" thick	"	0.029
1/2" thick	"	0.031
Base cab., 34-1/2" high, 24" deep, hardwood		
Minimum	L.F.	0.320
Average	"	0.400
Maximum	"	0.533
Wall cabinets		
Minimum	L.F.	0.267
Average	"	0.320
Maximum	"	0.400
Oil borne		
Water borne		

Architectural Woodwork	UNIT	MAN/HOURS
06420.10 Panel Work		
Hardboard, tempered, 1/4" thick		
Natural faced	S.F.	0.020
Plastic faced	"	0.023
Pegboard, natural	"	0.020
Plastic faced	"	0.023
Untempered, 1/4" thick		
Natural faced	S.F.	0.020
Plastic faced	"	0.023
Pegboard, natural	"	0.020
Plastic faced	"	0.023
Plywood unfinished, 1/4" thick		
Birch		
Natural	S.F.	0.027
Select	"	0.027
Knotty pine	"	0.027
Cedar (closet lining)		
Standard boards T&G	S.F.	0.027
Particle board	"	0.027
Plywood, prefinished, 1/4" thick, premium grade		
Birch veneer	S.F.	0.032
Cherry veneer	"	0.032
Chestnut veneer	"	0.032
Lauan veneer	"	0.032
Mahogany veneer	"	0.032
Oak veneer (red)	"	0.032
Pecan veneer	"	0.032
Rosewood veneer	"	0.032
Teak veneer	"	0.032
Walnut veneer	"	0.032

06 WOOD AND PLASTICS

Architectural Woodwork	UNIT	MAN/HOURS
06430.10 Stairwork		
Risers, 1x8, 42" wide		
White oak	EA.	0.400
Pine	"	0.400
Treads, 1-1/16" x 9-1/2" x 42"		
White oak	EA.	0.500
06440.10 Columns		
Column, hollow, round wood		
12" diameter		
10' high	EA.	0.800
12' high	"	0.857
14' high	"	0.960
16' high	"	1.200
24" diameter		
16' high	EA.	1.200
18' high	"	1.263
20' high	"	1.263
22' high	"	1.333
24' high	"	1.333

07 THERMAL AND MOISTURE

Moisture Protection	UNIT	MAN/HOURS
07100.10 Waterproofing		
Membrane waterproofing, elastomeric		
Butyl		
1/32" thick	S.F.	0.032
1/16" thick	"	0.033
Butyl with nylon		
1/32" thick	S.F.	0.032
1/16" thick	"	0.033
Neoprene		
1/32" thick	S.F.	0.032
1/16" thick	"	0.033
Neoprene with nylon		
1/32" thick	S.F.	0.032
1/16" thick	"	0.033
Plastic vapor barrier (polyethylene)		
4 mil	S.F.	0.003
6 mil	"	0.003
10 mil	"	0.004
Bituminous membrane, asphalt felt, 15 lb.		
One ply	S.F.	0.020
Two ply	"	0.024
Three ply	"	0.029
Four ply	"	0.033
Five ply	"	0.042
Modified asphalt membrane, fibrous asphalt		
One ply	S.F.	0.033
Two ply	"	0.040
Three ply	"	0.044
Four ply	"	0.053
Five ply	"	0.064
Asphalt coated protective board		
1/8" thick	S.F.	0.020
1/4" thick	"	0.020
3/8" thick	"	0.020
1/2" thick	"	0.021
Cement protective board		
3/8" thick	S.F.	0.027
1/2" thick	"	0.027
Fluid applied, neoprene		
50 mil	S.F.	0.027
90 mil	"	0.027
Tab extended polyurethane		
.050" thick	S.F.	0.020
Fluid applied rubber based polyurethane		
6 mil	S.F.	0.025
15 mil	"	0.020
Bentonite waterproofing, panels		
3/16" thick	S.F.	0.020
1/4" thick	"	0.020
5/8" thick	"	0.021
Granular admixtures, trowel on, 3/8" thick	"	0.020
Metallic oxide, iron compound, troweled		
5/8" thick	S.F.	0.020
3/4" thick	"	0.023

Moisture Protection	UNIT	MAN/HOURS
07150.10 Dampproofing		
Silicone dampproofing, sprayed on		
Concrete surface		
1 coat	S.F.	0.004
2 coats	"	0.006
Concrete block		
1 coat	S.F.	0.005
2 coats	"	0.007
Brick		
1 coat	S.F.	0.006
2 coats	"	0.008
07160.10 Bituminous Dampproofing		
Building paper, asphalt felt		
15 lb	S.F.	0.032
30 lb	"	0.033
Asphalt, troweled, cold, primer plus		
1 coat	S.F.	0.027
2 coats	"	0.040
3 coats	"	0.050
Fibrous asphalt, hot troweled, primer plus		
1 coat	S.F.	0.032
2 coats	"	0.044
3 coats	"	0.057
Asphaltic paint dampproofing, per coat		
Brush on	S.F.	0.011
Spray on	"	0.009
07190.10 Vapor Barriers		
Vapor barrier, polyethylene		
2 mil	S.F.	0.004
6 mil	"	0.004
8 mil	"	0.004
10 mil	"	0.004

Insulation	UNIT	MAN/HOURS
07210.10 Batt Insulation		
Ceiling, fiberglass, unfaced		
3-1/2" thick, R11	S.F.	0.009
6" thick, R19	"	0.011
9" thick, R30	"	0.012
Suspended ceiling, unfaced		
3-1/2" thick, R11	S.F.	0.009
6" thick, R19	"	0.010
9" thick, R30	"	0.011
Crawl space, unfaced		

07 THERMAL AND MOISTURE

Insulation	UNIT	MAN/HOURS
07210.10 Batt Insulation *(Cont.)*		
3-1/2" thick, R11	S.F.	0.012
6" thick, R19	"	0.013
9" thick, R30	"	0.015
Wall, fiberglass		
Paper backed		
2" thick, R7	S.F.	0.008
3" thick, R8	"	0.009
4" thick, R11	"	0.009
6" thick, R19	"	0.010
Foil backed, 1 side		
2" thick, R7	S.F.	0.008
3" thick, R11	"	0.009
4" thick, R14	"	0.009
6" thick, R21	"	0.010
Foil backed, 2 sides		
2" thick, R7	S.F.	0.009
3" thick, R11	"	0.010
4" thick, R14	"	0.011
6" thick, R21	"	0.011
Unfaced		
2" thick, R7	S.F.	0.008
3" thick, R9	"	0.009
4" thick, R11	"	0.009
6" thick, R19	"	0.010
Mineral wool batts		
Paper backed		
2" thick, R6	S.F.	0.008
4" thick, R12	"	0.009
6" thick, R19	"	0.010
Fasteners, self adhering, attached to ceiling deck		
2-1/2" long	EA.	0.013
4-1/2" long	"	0.015
Capped, self-locking washers	"	0.008
07210.20 Board Insulation		
Insulation, rigid		
Fiberglass, roof		
0.75" thick, R2.78	S.F.	0.007
1.06" thick, R4.17	"	0.008
1.31" thick, R5.26	"	0.008
1.63" thick, R6.67	"	0.008
2.25" thick, R8.33	"	0.009
Composite board, roof		
1-1/2" thick, R6.67	S.F.	0.008
1-5/8" thick, R7.69	"	0.008
2" thick, R10.0	"	0.009
2-1/4" thick, R12.50	"	0.009
2-1/2" thick, R14.29	"	0.010
2-3/4" thick, R16.67	"	0.011
3-1/4" thick, R20.00	"	0.011
Perlite board, roof		
1.00" thick, R2.78	S.F.	0.007
1.50" thick, R4.17	"	0.007

Insulation	UNIT	MAN/HOURS
07210.20 Board Insulation *(Cont.)*		
2.00" thick, R5.92	S.F.	0.007
2.50" thick, R6.67	"	0.008
3.00" thick, R8.33	"	0.008
4.00" thick, R10.00	"	0.008
5.25" thick, R14.29	"	0.009
Rigid urethane		
Roof		
1" thick, R6.67	S.F.	0.007
1.20" thick, R8.33	"	0.007
1.50" thick, R11.11	"	0.007
2" thick, R14.29	"	0.007
2.25" thick, R16.67	"	0.008
Wall		
1" thick, R6.67	S.F.	0.008
1.5" thick, R11.11	"	0.009
2" thick, R14.29	"	0.009
Polystyrene		
Roof		
1.0" thick, R4.17	S.F.	0.007
1.5" thick, R6.26	"	0.007
2.0" thick, R8.33	"	0.007
Wall		
1.0" thick, R4.17	S.F.	0.008
1.5" thick, R6.26	"	0.009
2.0" thick, R8.33	"	0.009
Rigid board insulation, deck		
Mineral fiberboard		
1" thick, R3.0	S.F.	0.007
2" thick, R5.26	"	0.007
Fiberglass		
1" thick, R4.3	S.F.	0.007
2" thick, R8.5	"	0.007
Polystyrene		
1" thick, R5.4	S.F.	0.007
2" thick, R10.8	"	0.007
Urethane		
.75" thick, R5.4	S.F.	0.007
1" thick, R6.4	"	0.007
1.5" thick, R10.7	"	0.007
2" thick, R14.3	"	0.007
Foamglass		
1" thick, R1.8	S.F.	0.007
2" thick, R5.26	"	0.007
Wood fiber		
1" thick, R3.85	S.F.	0.007
2" thick, R7.7	"	0.007
Particle board		
3/4" thick, R2.08	S.F.	0.007
1" thick, R2.77	"	0.007
2" thick, R5.50	"	0.007

07 THERMAL AND MOISTURE

Insulation	UNIT	MAN/HOURS
07210.60 Loose Fill Insulation		
Blown-in type		
Fiberglass		
5" thick, R11	S.F.	0.007
6" thick, R13	"	0.008
9" thick, R19	"	0.011
Rockwool, attic application		
6" thick, R13	S.F.	0.008
8" thick, R19	"	0.010
10" thick, R22	"	0.012
12" thick, R26	"	0.013
15" thick, R30	"	0.016
Poured type		
Fiberglass		
1" thick, R4	S.F.	0.005
2" thick, R8	"	0.006
3" thick, R12	"	0.007
4" thick, R16	"	0.008
Mineral wool		
1" thick, R3	S.F.	0.005
2" thick, R6	"	0.006
3" thick, R9	"	0.007
4" thick, R12	"	0.008
Vermiculite or perlite		
2" thick, R4.8	S.F.	0.006
3" thick, R7.2	"	0.007
4" thick, R9.6	"	0.008
Masonry, poured vermiculite or perlite		
4" block	S.F.	0.004
6" block	"	0.005
8" block	"	0.006
10" block	"	0.006
12" block	"	0.007
07210.70 Sprayed Insulation		
Foam, sprayed on		
Polystyrene		
1" thick, R4	S.F.	0.008
2" thick, R8	"	0.011
Urethane		
1" thick, R4	S.F.	0.008
2" thick, R8	"	0.011
07250.10 Fireproofing		
Sprayed on		
1" thick		
On beams	S.F.	0.018
On columns	"	0.016
On decks		
Flat surface	S.F.	0.008
Fluted surface	"	0.010
1-1/2" thick		
On beams	S.F.	0.023
On columns	"	0.020

Insulation	UNIT	MAN/HOURS
07250.10 Fireproofing (Cont.)		
On decks		
Flat surface	S.F.	0.010
Fluted surface	"	0.013

Shingles And Tiles	UNIT	MAN/HOURS
07310.10 Asphalt Shingles		
Standard asphalt shingles, strip shingles		
210 lb/square	SQ.	0.800
235 lb/square	"	0.889
240 lb/square	"	1.000
260 lb/square	"	1.143
300 lb/square	"	1.333
385 lb/square	"	1.600
Roll roofing, mineral surface		
90 lb	SQ.	0.571
110 lb	"	0.667
140 lb	"	0.800
07310.50 Metal Shingles		
Aluminum, .020" thick		
Plain	SQ.	1.600
Colors	"	1.600
Steel, galvanized		
26 ga.		
Plain	SQ.	1.600
Colors	"	1.600
24 ga.		
Plain	SQ.	1.600
Colors	"	1.600
Porcelain enamel, 22 ga.		
Minimum	SQ.	2.000
Average	"	2.000
Maximum	"	2.000
07310.60 Slate Shingles		
Slate shingles		
Pennsylvania		
Ribbon	SQ.	4.000
Clear	"	4.000
Vermont		
Black	SQ.	4.000
Gray	"	4.000
Green	"	4.000
Red	"	4.000
Replacement shingles		

07 THERMAL AND MOISTURE

Shingles And Tiles	UNIT	MAN/HOURS
07310.60 Slate Shingles *(Cont.)*		
Small jobs	EA.	0.267
Large jobs	S.F.	0.133
07310.70 Wood Shingles		
Wood shingles, on roofs		
White cedar, #1 shingles		
4" exposure	SQ.	2.667
5" exposure	"	2.000
#2 shingles		
4" exposure	SQ.	2.667
5" exposure	"	2.000
Resquared and rebutted		
4" exposure	SQ.	2.667
5" exposure	"	2.000
On walls		
White cedar, #1 shingles		
4" exposure	SQ.	4.000
5" exposure	"	3.200
6" exposure	"	2.667
#2 shingles		
4" exposure	SQ.	4.000
5" exposure	"	3.200
6" exposure	"	2.667
07310.80 Wood Shakes		
Shakes, hand split, 24" red cedar, on roofs		
5" exposure	SQ.	4.000
7" exposure	"	3.200
9" exposure	"	2.667
On walls		
6" exposure	SQ.	4.000
8" exposure	"	3.200
10" exposure	"	2.667

Roofing And Siding	UNIT	MAN/HOURS
07410.10 Manufactured Roofs		
Aluminum roof panels, for steel framing		
Corrugated		
Unpainted finish		
.024"	S.F.	0.020
.030"	"	0.020
Painted finish		
.024"	S.F.	0.020
.030"	"	0.020
V-beam		

Roofing And Siding	UNIT	MAN/HOURS
07410.10 Manufactured Roofs *(Cont.)*		
Unpainted finish		
.032"	S.F.	0.020
.040"	"	0.020
.050"	"	0.020
Painted finish		
.032"	S.F.	0.020
.040"	"	0.020
.050"	"	0.020
Steel roof panels, for structural steel framing		
Corrugated, painted		
18 ga.	S.F.	0.020
20 ga.	"	0.020
22 ga.	"	0.020
Box rib, painted		
18 ga.	S.F.	0.021
20 ga.	"	0.021
22 ga.	"	0.021
4" rib, painted		
18 ga.	S.F.	0.022
20 ga.	"	0.022
22 ga.	"	0.022
Standing seam roof		
2" high seam, painted		
22 ga.	S.F.	0.032
24 ga.	"	0.032
26 ga.	"	0.032
07410.30 Manufactured Walls		
Sandwich panels with 1-1/2" fiberglass insulation		
Galvanized 18 ga. steel interior panels		
Exterior panels		
16 ga. aluminum	S.F.	0.107
18 ga. galvanized steel	"	0.107
20 ga. painted steel	"	0.107
20 ga. stainless steel	"	0.107
Metal liner panels, 1-3/8" thick, 24" wide		
Galvanized		
22 ga.	S.F.	0.027
20 ga.	"	0.027
18 ga.	"	0.027
Primed		
22 ga.	S.F.	0.027
20 ga.	"	0.027
18 ga.	"	0.027
07440.10 Aggregate Coated Panels		
Dryvit type system		
1" thick	S.F.	0.027
1-1/2" thick	"	0.029
2" thick	"	0.033

07 THERMAL AND MOISTURE

Roofing And Siding	UNIT	MAN/HOURS
07460.10 Metal Siding Panels		
Aluminum siding panels		
Corrugated		
Plain finish		
.024"	S.F.	0.032
.032"	"	0.032
Painted finish		
.024"	S.F.	0.032
.032"	"	0.032
V. beam		
Plain finish		
.032"	S.F.	0.032
.040"	"	0.032
.050"	"	0.032
Painted finish		
.032"	S.F.	0.032
.040"	"	0.032
.050"	"	0.032
4" rib		
Plain finish		
.032"	S.F.	0.036
.040"	"	0.036
.050"	"	0.036
Painted finish		
.032"	S.F.	0.036
.040"	"	0.036
.050"	"	0.036
Steel siding panels		
Corrugated		
22 ga.	S.F.	0.053
24 ga.	"	0.053
26 ga.	"	0.053
Box rib		
20 ga.	S.F.	0.053
22 ga.	"	0.053
24 ga.	"	0.053
26 ga.	"	0.053
07460.50 Plastic Siding		
Horizontal vinyl siding, solid		
8" wide		
Standard	S.F.	0.031
Insulated	"	0.031
10" wide		
Standard	S.F.	0.029
Insulated	"	0.029
Vinyl moldings for doors and windows	L.F.	0.032
07460.60 Plywood Siding		
Rough sawn cedar, 3/8" thick	S.F.	0.027
Fir, 3/8" thick	"	0.027
Texture 1-11, 5/8" thick		
Cedar	S.F.	0.029
Fir	"	0.029

Roofing And Siding	UNIT	MAN/HOURS
07460.60 Plywood Siding (Cont.)		
Redwood	S.F.	0.029
Southern Yellow Pine	"	0.029
07460.70 Steel Siding		
Ribbed, sheets, galvanized		
22 ga.	S.F.	0.032
24 ga.	"	0.032
26 ga.	"	0.032
28 ga.	"	0.032
Primed		
24 ga.	S.F.	0.032
26 ga.	"	0.032
28 ga.	"	0.032
07460.80 Wood Siding		
Beveled siding, cedar		
A grade		
1/2 x 6	S.F.	0.040
1/2 x 8	"	0.032
3/4 x 10	"	0.027
Clear		
1/2 x 6	S.F.	0.040
1/2 x 8	"	0.032
3/4 x 10	"	0.027
B grade		
1/2 x 6	S.F.	0.040
1/2 x 8	"	0.320
3/4 x 10	"	0.027
Board and batten		
Cedar		
1x6	S.F.	0.040
1x8	"	0.032
1x10	"	0.029
1x12	"	0.026
Pine		
1x6	S.F.	0.040
1x8	"	0.032
1x10	"	0.029
1x12	"	0.026
Redwood		
1x6	S.F.	0.040
1x8	"	0.032
1x10	"	0.029
1x12	"	0.026
Tongue and groove		
Cedar		
1x4	S.F.	0.044
1x6	"	0.042
1x8	"	0.040
1x10	"	0.038
Pine		
1x4	S.F.	0.044
1x6	"	0.042

07 THERMAL AND MOISTURE

Roofing And Siding	UNIT	MAN/HOURS
07460.80 Wood Siding (Cont.)		
1x8	S.F.	0.040
1x10	"	0.038
Redwood		
1x4	S.F.	0.044
1x6	"	0.042
1x8	"	0.040
1x10	"	0.038

Membrane Roofing	UNIT	MAN/HOURS
07510.10 Built-up Asphalt Roofing		
Built-up roofing, asphalt felt, including gravel		
2 ply	SQ.	2.000
3 ply	"	2.667
4 ply	"	3.200
Walkway, for built-up roofs		
3' x 3' x		
1/2" thick	S.F.	0.027
3/4" thick	"	0.027
1" thick	"	0.027
Cant strip, 4" x 4"		
Treated wood	L.F.	0.023
Foamglass	"	0.020
Mineral fiber	"	0.020
New gravel for built-up roofing, 400 lb/sq	SQ.	1.600
Roof gravel (ballast)	C.Y.	4.000
Aluminum coating, top surfacing, for built-up roofing	SQ.	1.333
Remove 4-ply built-up roof (includes gravel)	"	4.000
Remove & replace gravel, includes flood coat	"	2.667
07530.10 Single-ply Roofing		
Elastic sheet roofing		
Neoprene, 1/16" thick	S.F.	0.010
EPDM rubber		
45 mil	S.F.	0.010
60 mil	"	0.010
PVC		
45 mil	S.F.	0.010
60 mil	"	0.010
Flashing		
Pipe flashing, 90 mil thick		
1" pipe	EA.	0.200
2" pipe	"	0.200
3" pipe	"	0.211
4" pipe	"	0.211
5" pipe	"	0.222

Membrane Roofing	UNIT	MAN/HOURS
07530.10 Single-ply Roofing (Cont.)		
6" pipe	EA.	0.222
8" pipe	"	0.235
10" pipe	"	0.267
12" pipe	"	0.267
Neoprene flashing, 60 mil thick strip		
6" wide	L.F.	0.067
12" wide	"	0.100
18" wide	"	0.133
24" wide	"	0.200
Adhesives		
Mastic sealer, applied at joints only		
1/4" bead	L.F.	0.004
Fluid applied roofing		
Urethane, 2 part, elastomeric membrane		
1" thick	S.F.	0.013
Vinyl liquid roofing, 2 coats, 2 mils per coat	"	0.011
Silicone roofing, 2 coats sprayed, 16 mil per coat	"	0.013
Inverted roof system		
Insulated membrane with coarse gravel ballast		
3 ply with 2" polystyrene	S.F.	0.013
Ballast, 3/4" through 1-1/2" gravel, 100lb/sf	"	0.800
Walkway for membrane roofs, 1/2" thick	"	0.027

Flashing And Sheet Metal	UNIT	MAN/HOURS
07610.10 Metal Roofing		
Sheet metal roofing, copper, 16 oz, batten seam	SQ.	5.333
Standing seam	"	5.000
Aluminum roofing, natural finish		
Corrugated, on steel frame		
.0175" thick	SQ.	2.286
.0215" thick	"	2.286
.024" thick	"	2.286
.032" thick	"	2.286
V-beam, on steel frame		
.032" thick	SQ.	2.286
.040" thick	"	2.286
.050" thick	"	2.286
Ridge cap		
.019" thick	L.F.	0.027
Corrugated galvanized steel roofing, on steel frame		
28 ga.	SQ.	2.286
26 ga.	"	2.286
24 ga.	"	2.286
22 ga.	"	2.286
26 ga., factory insulated with 1" polystyrene	"	3.200

07 THERMAL AND MOISTURE

07610.10 Metal Roofing (Cont.)

Flashing And Sheet Metal	UNIT	MAN/HOURS
Ridge roll		
10" wide	L.F.	0.027
20" wide	"	0.032

07620.10 Flashing And Trim

Flashing And Sheet Metal	UNIT	MAN/HOURS
Counter flashing		
Aluminum, .032"	S.F.	0.080
Stainless steel, .015"	"	0.080
Copper		
16 oz.	S.F.	0.080
20 oz.	"	0.080
24 oz.	"	0.080
32 oz.	"	0.080
Valley flashing		
Aluminum, .032"	S.F.	0.050
Stainless steel, .015	"	0.050
Copper		
16 oz.	S.F.	0.050
20 oz.	"	0.067
24 oz.	"	0.050
32 oz.	"	0.050
Base flashing		
Aluminum, .040"	S.F.	0.067
Stainless steel, .018"	"	0.067
Copper		
16 oz.	S.F.	0.067
20 oz.	"	0.050
24 oz.	"	0.067
32 oz.	"	0.067
Waterstop, "T" section, 22 ga.		
1-1/2" x 3"	L.F.	0.040
2" x 2"	"	0.040
4" x 3"	"	0.040
6" x 4"	"	0.040
8" x 4"	"	0.040
Scupper outlets		
10" x 10" x 4"	EA.	0.200
22" x 4" x 4"	"	0.200
8" x 8" x 5"	"	0.200
Flashing and trim, aluminum		
.019" thick	S.F.	0.057
.032" thick	"	0.057
.040" thick	"	0.062
Neoprene sheet flashing, .060" thick	"	0.050
Copper, paper backed		
2 oz.	S.F.	0.080
5 oz.	"	0.080
Drainage boots, roof, cast iron		
2 x 3	L.F.	0.100
3 x 4	"	0.100
4 x 5	"	0.107
4 x 6	"	0.107
5 x 7	"	0.114

07620.10 Flashing And Trim (Cont.)

Flashing And Sheet Metal	UNIT	MAN/HOURS
Pitch pocket, copper, 16 oz.		
4 x 4	EA.	0.200
6 x 6	"	0.200
8 x 8	"	0.200
8 x 10	"	0.200
8 x 12	"	0.200
Reglets, copper 10 oz.	L.F.	0.053
Stainless steel, .020"	"	0.053
Gravel stop		
Aluminum, .032"		
4"	L.F.	0.027
6"	"	0.027
8"	"	0.031
10"	"	0.031
Copper, 16 oz.		
4"	L.F.	0.027
6"	"	0.027
8"	"	0.031
10"	"	0.031

07620.20 Gutters And Downspouts

Flashing And Sheet Metal	UNIT	MAN/HOURS
Copper gutter and downspout		
Downspouts, 16 oz. copper		
Round		
3" dia.	L.F.	0.053
4" dia.	"	0.053
Rectangular, corrugated		
2" x 3"	L.F.	0.050
3" x 4"	"	0.050
Rectangular, flat surface		
2" x 3"	L.F.	0.053
3" x 4"	"	0.053
Lead-coated copper downspouts		
Round		
3" dia.	L.F.	0.050
4" dia.	"	0.057
Rectangular, corrugated		
2" x 3"	L.F.	0.053
3" x 4"	"	0.053
Rectangular, plain		
2" x 3"	L.F.	0.053
3" x 4"	"	0.053
Gutters, 16 oz. copper		
Half round		
4" wide	L.F.	0.080
5" wide	"	0.089
Type K		
4" wide	L.F.	0.080
5" wide	"	0.089
Lead-coated copper gutters		
Half round		
4" wide	L.F.	0.080
6" wide	"	0.089

07 THERMAL AND MOISTURE

Flashing And Sheet Metal	UNIT	MAN/HOURS
07620.20 Gutters And Downspouts (Cont.)		
Type K		
4" wide	L.F.	0.080
5" wide	"	0.089
Aluminum gutter and downspout		
Downspouts		
2" x 3"	L.F.	0.053
3" x 4"	"	0.057
4" x 5"	"	0.062
Round		
3" dia.	L.F.	0.053
4" dia.	"	0.057
Gutters, stock units		
4" wide	L.F.	0.084
5" wide	"	0.089
Galvanized steel gutter and downspout		
Downspouts, round corrugated		
3" dia.	L.F.	0.053
4" dia.	"	0.053
5" dia.	"	0.057
6" dia.	"	0.057
Rectangular		
2" x 3"	L.F.	0.053
3" x 4"	"	0.050
4" x 4"	"	0.050
Gutters, stock units		
5" wide		
Plain	L.F.	0.089
Painted	"	0.089
6" wide		
Plain	L.F.	0.094
Painted	"	0.094

Roofing Specialties	UNIT	MAN/HOURS
07700.10 Manufactured Specialties		
Moisture relief vent		
Aluminum	EA.	0.114
Copper	"	0.114
Expansion joint		
Aluminum		
Opening to 2.5"	L.F.	0.057
Opening to 3.5"	"	0.062
Copper, 16 oz.		
Opening to 2.5"	L.F.	0.057
Opening to 3.5"	"	0.062
Butyl or neoprene		

Roofing Specialties	UNIT	MAN/HOURS
07700.10 Manufactured Specialties (Cont.)		
4" wide		
16 oz. copper bellows	L.F.	0.067
28 ga. stainless steel bellows	"	0.067
6" wide		
Copper bellows	L.F.	0.073
Stainless steel		
Opening to 2.5"	L.F.	0.057
Opening to 3.5"	"	0.062
Smoke vent, 48" x 48"		
Aluminum	EA.	2.000
Galvanized steel	"	2.000
Heat/smoke vent, 48" x 96"		
Aluminum	EA.	2.667
Galvanized steel	"	2.667
Ridge vent strips		
Mill finish	L.F.	0.053
Connectors	EA.	0.200
End cap	"	0.229
Soffit vents		
Mill finish		
2-1/2" wide	L.F.	0.032
3" wide	"	0.032
6" wide	"	0.032
Roof hatches		
Steel, plain, primed		
2'6" x 3'0"	EA.	2.000
2'6" x 4'6"	"	2.667
2'6" x 8'0"	"	4.000
Galvanized steel		
2'6" x 3'0"	EA.	2.000
2'6" x 4'6"	"	2.667
2'6" x 8'0"	"	4.000
Aluminum		
2'6" x 3'0"	EA.	2.000
2'6" x 4'6"	"	2.667
2'6" x 8'0"	"	4.000
Ceiling access doors		
Swing up model, metal frame		
Steel door		
2'6" x 2'6"	EA.	0.800
2'6" x 3'0"	"	0.800
Aluminum door		
2'6" x 2'6"	EA.	0.800
2'6" x 3'0"	"	0.800
Swing down model, metal frame		
Steel door		
2'6" x 2'6"	EA.	0.800
2'6" x 3'0"	"	0.800
Aluminum door		
2'6" x 2'6"	EA.	0.800
2'6" x 3'0"	"	0.800
Gravity ventilators, with curb, base, damper and screen		
Stationary siphon		

07 THERMAL AND MOISTURE

Roofing Specialties	UNIT	MAN/HOURS
07700.10 Manufactured Specialties (Cont.)		
6" dia.	EA.	0.533
12" dia.	"	0.533
24" dia.	"	0.800
36" dia.	"	0.800
Wind driven spinner		
6" dia.	EA.	0.533
12" dia.	"	0.533
24" dia.	"	0.800
36" dia.	"	0.800
Stationary mushroom		
16" dia.	EA.	0.800
30" dia.	"	1.000
36" dia.	"	1.333
42" dia.	"	1.600

Skylights	UNIT	MAN/HOURS
07810.10 Plastic Skylights		
Single thickness, not including mounting curb		
2' x 4'	EA.	1.000
4' x 4'	"	1.333
5' x 5'	"	2.000
6' x 8'	"	2.667
Double thickness, not including mounting curb		
2' x 4'	EA.	1.000
4' x 4'	"	1.333
5' x 5'	"	2.000
6' x 8'	"	2.667
Metal framed skylights		
Translucent panels, 2-1/2" thick	S.F.	0.080
Continuous vaults, 8' wide		
Single glazed	S.F.	0.100
Double glazed	"	0.114

Joint Sealers	UNIT	MAN/HOURS
07920.10 Caulking		
Caulk exterior, two component		
1/4 x 1/2	L.F.	0.040
3/8 x 1/2	"	0.044
1/2 x 1/2	"	0.050
Caulk interior, single component		
1/4 x 1/2	L.F.	0.038
3/8 x 1/2	"	0.042
1/2 x 1/2	"	0.047
Butyl rubber fillers		
1/4" x 1/4"	L.F.	0.016
1/2" x 1/2"	"	0.027
1/2" x 3/4"	"	0.032
3/4" x 3/4"	"	0.032
1" x 1"	"	0.036
Seals, "O" ring type cord		
1/4" dia.	L.F.	0.020
1/2" dia.	"	0.021
1" dia.	"	0.022
1-1/4" dia.	"	0.024
1-1/2" dia.	"	0.025
1-3/4" dia.	"	0.026
2" dia.	"	0.027
Polyvinyl chloride, closed cell		
1/4" x 2"	L.F.	0.029
1/4" x 6"	"	0.036
Silicon foam penetration seal		
1/4" x 1/2"	L.F.	0.010
1/2" x 1/2"	"	0.013
1/2" x 3/4"	"	0.016
3/4" x 3/4"	"	0.020
1/8" x 1"	"	0.010
1/8" x 3"	"	0.016
1/4" x 3"	"	0.020
1/4" x 6"	"	0.027
1/2" x 6"	"	0.062
1/2" x 9"	"	0.100
1/2" x 12"	"	0.145
Oil base sealants and caulking		
1/4" x 1/4"	L.F.	0.020
1/4" x 3/8"	"	0.021
1/4" x 1/2"	"	0.022
3/8" x 3/8"	"	0.023
3/8" x 1/2"	"	0.024
3/8" x 5/8"	"	0.026
3/8" x 3/4"	"	0.028
1/2" x 1/2"	"	0.031
1/2" x 5/8"	"	0.035
1/2" x 3/4"	"	0.040
1/2" x 7/8"	"	0.041
1/2" x 1"	"	0.042
3/4" x 3/4"	"	0.043
1" x 1"	"	0.044
Polyurethane compounds		

07 THERMAL AND MOISTURE

Joint Sealers	UNIT	MAN/HOURS
07920.10 **Caulking** *(Cont.)*		
1/4" x 1/4"	L.F.	0.020
1/4" x 3/8"	"	0.021
1/4" x 1/2"	"	0.022
3/8" x 3/8"	"	0.023
3/8" x 1/2"	"	0.024
3/8" x 5/8"	"	0.026
3/8" x 3/4"	"	0.028
1/2" x 1/2"	"	0.031
1/2" x 5/8"	"	0.035
1/2" x 3/4"	"	0.040
1/2" x 7/8"	"	0.041
1/2" x 1"	"	0.044
3/4" x 3/4"	"	0.043
3/4" x 1"	"	0.044
Backer rod, polyethylene		
1/4"	L.F.	0.020
1/2"	"	0.021
3/4"	"	0.022
1"	"	0.024

08 DOORS AND WINDOWS

Metal	UNIT	MAN/HOURS
08110.10 Metal Doors		
Flush hollow metal, std. duty, 20 ga., 1-3/8" thick		
2-6 x 6-8	EA.	0.889
2-8 x 6-8	"	0.889
3-0 x 6-8	"	0.889
1-3/4" thick		
2-6 x 6-8	EA.	0.889
2-8 x 6-8	"	0.889
3-0 x 6-8	"	0.889
2-6 x 7-0	"	0.889
2-8 x 7-0	"	0.889
3-0 x 7-0	"	0.889
Heavy duty, 20 ga., unrated, 1-3/4"		
2-8 x 6-8	EA.	0.889
3-0 x 6-8	"	0.889
2-8 x 7-0	"	0.889
3-0 x 7-0	"	0.889
3-4 x 7-0	"	0.889
18 ga., 1-3/4", unrated door		
2-0 x 7-0	EA.	0.889
2-4 x 7-0	"	0.889
2-6 x 7-0	"	0.889
2-8 x 7-0	"	0.889
3-0 x 7-0	"	0.889
3-4 x 7-0	"	0.889
2", unrated door		
2-0 x 7-0	EA.	1.000
2-4 x 7-0	"	1.000
2-6 x 7-0	"	1.000
2-8 x 7-0	"	1.000
3-0 x 7-0	"	1.000
3-4 x 7-0	"	1.000
Galvanized metal door		
3-0 x 7-0	EA.	1.000
08110.40 Metal Door Frames		
Hollow metal, stock, 18 ga., 4-3/4" x 1-3/4"		
2-0 x 7-0	EA.	1.000
2-4 x 7-0	"	1.000
2-6 x 7-0	"	1.000
2-8 x 7-0	"	1.000
3-0 x 7-0	"	1.000
4-0 x 7-0	"	1.333
5-0 x 7-0	"	1.333
6-0 x 7-0	"	1.333
16 ga., 6-3/4" x 1-3/4"		
2-0 x 7-0	EA.	1.000
2-4 x 7-0	"	1.000
2-6 x 7-0	"	1.000
2-8 x 7-0	"	1.000
3-0 x 7-0	"	1.000
4-0 x 7-0	"	1.333
6-0 x 7-0	"	1.333
Transom frame		

Metal	UNIT	MAN/HOURS
08110.40 Metal Door Frames *(Cont.)*		
3-4 x 1-6	EA.	1.000
3-8 x 1-6	"	1.000
6-4 x 1-6	"	1.000
Transom sash		
3-0 x 1-4	EA.	1.000
3-4 x 1-4	"	1.000
6-0 x 1-4	"	1.000
Sidelights, complete		
1-0 x 7-2	EA.	1.000
1-4 x 7-2	"	1.000
1-0 x 8-8	"	1.000
1-6 x 8-8	"	1.000
16 ga., 4-3/4" x 1-3/4"		
2-0 x 7-0	EA.	1.000
2-4 x 7-0	"	1.000
2-6 x 7-0	"	1.000
2-8 x 7-0	"	1.000
3-0 x 7-0	"	1.000
4-0 x 7-0	"	1.333
6-0 x 7-0	"	1.333
3-4 x 1-6	"	1.000
3-8 x 1-6	"	1.000
6-4 x 1-6	"	1.000
Transom sash		
3-0 x 1-4	EA.	1.000
3-4 x 1-4	"	1.000
6-0 x 1-4	"	1.000
Sidelights, complete		
1-0 x 7-2	EA.	1.000
1-4 x 7-2	"	1.000
1-0 x 8-8	"	1.000
1-4 x 8-8	"	1.000
16 ga., 5-3/4" x 1-3/4"		
2-0 x 7-0	EA.	1.000
2-4 x 7-0	"	1.000
2-6 x 7-0	"	1.000
2-8 x 7-0	"	1.000
3-0 x 7-0	"	1.000
4-0 x 7-0	"	1.333
5-0 x 7-0	"	1.333
6-0 x 7-0	"	1.333
Mullions, vertical		
5-1/4" x 1-3/4"	L.F.	0.100
5-1/4" x 2"	"	0.100
Horizontal		
5-1/4" x 1-3/4"	L.F.	0.100
5-1/4" x 2"	"	0.100

08 DOORS AND WINDOWS

Metal	UNIT	MAN/HOURS
08120.10 — Aluminum Doors		
Aluminum doors, commercial		
Narrow stile		
2-6 x 7-0	EA.	4.000
3-0 x 7-0	"	4.000
3-6 x 7-0	"	4.000
Pair		
5-0 x 7-0	EA.	8.000
6-0 x 7-0	"	8.000
7-0 x 7-0	"	8.000
Wide stile		
2-6 x 7-0	EA.	4.000
3-0 x 7-0	"	4.000
3-6 x 7-0	"	4.000
Pair		
5-0 x 7-0	EA.	8.000
6-0 x 7-0	"	8.000
7-0 x 7-0	"	8.000

Wood And Plastic	UNIT	MAN/HOURS
08210.10 — Wood Doors		
Solid core, 1-3/8" thick		
Birch faced		
2-4 x 7-0	EA.	1.000
2-8 x 7-0	"	1.000
3-0 x 7-0	"	1.000
3-4 x 7-0	"	1.000
2-4 x 6-8	"	1.000
2-6 x 6-8	"	1.000
2-8 x 6-8	"	1.000
3-0 x 6-8	"	1.000
Lauan faced		
2-4 x 6-8	EA.	1.000
2-8 x 6-8	"	1.000
3-0 x 6-8	"	1.000
3-4 x 6-8	"	1.000
Tempered hardboard faced		
2-4 x 7-0	EA.	1.000
2-8 x 7-0	"	1.000
3-0 x 7-0	"	1.000
3-4 x 7-0	"	1.000
Hollow core, 1-3/8" thick		
Birch faced		
2-4 x 7-0	EA.	1.000
2-8 x 7-0	"	1.000
3-0 x 7-0	"	1.000

Wood And Plastic	UNIT	MAN/HOURS
08210.10 — Wood Doors *(Cont.)*		
3-4 x 7-0	EA.	1.000
Lauan faced		
2-4 x 6-8	EA.	1.000
2-6 x 6-8	"	1.000
2-8 x 6-8	"	1.000
3-0 x 6-8	"	1.000
3-4 x 6-8	"	1.000
Tempered hardboard faced		
2-4 x 7-0	EA.	1.000
2-6 x 7-0	"	1.000
2-8 x 7-0	"	1.000
3-0 x 7-0	"	1.000
3-4 x 7-0	"	1.000
Solid core, 1-3/4" thick		
Birch faced		
2-4 x 7-0	EA.	1.000
2-6 x 7-0	"	1.000
2-8 x 7-0	"	1.000
3-0 x 7-0	"	1.000
3-4 x 7-0	"	1.000
Lauan faced		
2-4 x 7-0	EA.	1.000
2-6 x 7-0	"	1.000
2-8 x 7-0	"	1.000
3-4 x 7-0	"	1.000
3-0 x 7-0	"	1.000
Tempered hardboard faced		
2-4 x 7-0	EA.	1.000
2-6 x 7-0	"	1.000
2-8 x 7-0	"	1.000
3-0 x 7-0	"	1.000
3-4 x 7-0	"	1.000
Hollow core, 1-3/4" thick		
Birch faced		
2-4 x 7-0	EA.	1.000
2-6 x 7-0	"	1.000
2-8 x 7-0	"	1.000
3-0 x 7-0	"	1.000
3-4 x 7-0	"	1.000
Lauan faced		
2-4 x 6-8	EA.	1.000
2-6 x 6-8	"	1.000
2-8 x 6-8	"	1.000
3-0 x 6-8	"	1.000
3-4 x 6-8	"	1.000
Tempered hardboard		
2-4 x 7-0	EA.	1.000
2-6 x 7-0	"	1.000
2-8 x 7-0	"	1.000
3-0 x 7-0	"	1.000
3-4 x 7-0	"	1.000
Add-on, louver	"	0.800
Glass	"	0.800

08 DOORS AND WINDOWS

08210.10 Wood Doors (Cont.)

Wood And Plastic	UNIT	MAN/HOURS
Exterior doors, 3-0 x 7-0 x 2-1/2", solid core		
Carved		
One face	EA.	2.000
Two faces	"	2.000
Closet doors, 1-3/4" thick		
Bi-fold or bi-passing, includes frame and trim		
Paneled		
4-0 x 6-8	EA.	1.333
6-0 x 6-8	"	1.333
Louvered		
4-0 x 6-8	EA.	1.333
6-0 x 6-8	"	1.333
Flush		
4-0 x 6-8	EA.	1.333
6-0 x 6-8	"	1.333
Primed		
4-0 x 6-8	EA.	1.333
6-0 x 6-8	"	1.333

08210.90 Wood Frames

Wood And Plastic	UNIT	MAN/HOURS
Frame, interior, pine		
2-6 x 6-8	EA.	1.143
2-8 x 6-8	"	1.143
3-0 x 6-8	"	1.143
5-0 x 6-8	"	1.143
6-0 x 6-8	"	1.143
2-6 x 7-0	"	1.143
2-8 x 7-0	"	1.143
3-0 x 7-0	"	1.143
5-0 x 7-0	"	1.600
6-0 x 7-0	"	1.600
Exterior, custom, with threshold, including trim		
Walnut		
3-0 x 7-0	EA.	2.000
6-0 x 7-0	"	2.000
Oak		
3-0 x 7-0	EA.	2.000
6-0 x 7-0	"	2.000
Pine		
2-4 x 7-0	EA.	1.600
2-6 x 7-0	"	1.600
2-8 x 7-0	"	1.600
3-0 x 7-0	"	1.600
3-4 x 7-0	"	1.600
6-0 x 7-0	"	2.667

08300.10 Special Doors

Wood And Plastic	UNIT	MAN/HOURS
Vault door and frame, class 5, steel	EA.	8.000
Overhead door, coiling insulated		
Chain gear, no frame, 12' x 12'	EA.	10.000
Aluminum, bronze glass panels, 12-9 x 13-0	"	8.000
Garage, flush, ins. metal, primed, 9-0 x 7-0	"	2.667
Sliding fire doors, motorized, fusible link, 3 hr.		

08300.10 Special Doors (Cont.)

Wood And Plastic	UNIT	MAN/HOURS
3-0 x 6-8	EA.	16.000
3-8 x 6-8	"	16.000
4-0 x 8-0	"	16.000
5-0 x 8-0	"	16.000
Metal clad doors, including electric motor		
Light duty		
Minimum	S.F.	0.133
Maximum	"	0.320
Heavy duty		
Minimum	S.F.	0.400
Maximum	"	0.500
Hangar doors, based on 150' openings		
To 20' high	S.F.	0.096
20' to 40' high	"	0.060
40' to 60' high	"	0.040
60' to 80' high	"	0.024
Over 80' high	"	0.019
Counter doors, (roll-up shutters), std, manual		
Opening, 4' high		
4' wide	EA.	6.667
6' wide	"	6.667
8' wide	"	7.273
10' wide	"	10.000
14' wide	"	10.000
6' high		
4' wide	EA.	6.667
6' wide	"	7.273
8' wide	"	8.000
10' wide	"	10.000
14' wide	"	11.429
Service doors, (roll up shutters), std, manual		
Opening		
8' high x 8' wide	EA.	4.444
10' high x 10' wide	"	6.667
12' high x 12' wide	"	10.000
14' high x 14' wide	"	13.333
16' high x 14' wide	"	13.333
20' high x 14' wide	"	20.000
24' high x 16' wide	"	17.778
Roll-up doors		
13-0 high x 14-0 wide	EA.	11.429
12-0 high x 14-0 wide	"	11.429
Top coiling grilles, manual, steel or aluminum		
Opening, 4' high x		
4' wide	EA.	3.200
6' wide	"	3.200
8' wide	"	4.444
12' wide	"	4.444
16' wide	"	6.667
6' high x		
4' wide	EA.	6.667
6' wide	"	7.273
8' wide	"	8.000

08 DOORS AND WINDOWS

Wood And Plastic	UNIT	MAN/HOURS
08300.10 **Special Doors** *(Cont.)*		
12' wide	EA.	8.889
16' wide	"	11.429
Side coiling grilles, manual, aluminum		
Opening, 8' high x		
18' wide	EA.	60.000
24' wide	"	68.571
12' high x		
12' wide	EA.	60.000
18' wide	"	68.571
24' wide	"	80.000
Accordion folding, tracks and fittings included		
Vinyl covered, 2 layers	S.F.	0.320
Woven mahogany and vinyl	"	0.320
Economy vinyl	"	0.320
Rigid polyvinyl chloride	"	0.320
Sectional wood overhead, frames not incl.		
Commercial grade, HD, 1-3/4" thick, manual		
8' x 8'	EA.	6.667
10' x 10'	"	7.273
12' x 12'	"	8.000
Chain hoist		
12' x 16' high	EA.	13.333
14' x 14' high	"	10.000
20' x 8' high	"	16.000
16' high	"	20.000
Sectional metal overhead doors, complete		
Residential grade, manual		
9' x 7'	EA.	3.200
16' x 7'	"	4.000
Commercial grade		
8' x 8'	EA.	6.667
10' x 10'	"	7.273
12' x 12'	"	8.000
20' x 14', with chain hoist	"	16.000
Sliding glass doors		
Tempered plate glass, 1/4" thick		
6' wide		
Economy grade	EA.	2.667
Premium grade	"	2.667
12' wide		
Economy grade	EA.	4.000
Premium grade	"	4.000
Insulating glass, 5/8" thick		
6' wide		
Economy grade	EA.	2.667
Premium grade	"	2.667
12' wide		
Economy grade	EA.	4.000
Premium grade	"	4.000
1" thick		
6' wide		
Economy grade	EA.	2.667
Premium grade	"	2.667

Wood And Plastic	UNIT	MAN/HOURS
08300.10 **Special Doors** *(Cont.)*		
12' wide		
Economy grade	EA.	4.000
Premium grade	"	4.000
Vertical lift doors, channel frame construction		
20' high x		
10' wide	EA.	9.600
15' wide	"	9.600
20' wide	"	17.143
25' wide	"	17.143
25' high x		
20' wide	EA.	17.143
25' wide	"	20.000
30' high x		
25' wide	EA.	20.000
30' wide	"	20.000
35' wide	"	20.000
Residential storm door		
Minimum	EA.	1.333
Average	"	1.333
Maximum	"	2.000

Storefronts	UNIT	MAN/HOURS
08410.10 **Storefronts**		
Storefront, aluminum and glass		
Minimum	S.F.	0.100
Average	"	0.114
Maximum	"	0.133
Entrance doors, premium, closers, panic dev.,etc.		
1/2" thick glass		
3' x 7'	EA.	6.667
6' x 7'	"	10.000
3/4" thick glass		
3' x 7'	EA.	6.667
6' x 7'	"	10.000
1" thick glass		
3' x 7'	EA.	6.667
6' x 7'	"	10.000
Revolving doors		
7' diameter, 7' high		
Minimum	EA.	60.000
Average	"	96.000
Maximum	"	120.000

08 DOORS AND WINDOWS

Metal Windows	UNIT	MAN/HOURS
08510.10 Steel Windows		
Steel windows, primed		
Casements		
Operable		
Minimum	S.F.	0.047
Maximum	"	0.053
Fixed sash	"	0.040
Double hung	"	0.044
Industrial windows		
Horizontally pivoted sash	S.F.	0.053
Fixed sash	"	0.044
Security sash		
Operable	S.F.	0.053
Fixed	"	0.044
Picture window	"	0.044
Projecting sash		
Minimum	S.F.	0.050
Maximum	"	0.050
Mullions	L.F.	0.040
08520.10 Aluminum Windows		
Jalousie		
3-0 x 4-0	EA.	1.000
3-0 x 5-0	"	1.000
Fixed window		
6 sf to 8 sf	S.F.	0.114
12 sf to 16 sf	"	0.089
Projecting window		
6 sf to 8 sf	S.F.	0.200
12 sf to 16 sf	"	0.133
Horizontal sliding		
6 sf to 8 sf	S.F.	0.100
12 sf to 16 sf	"	0.080
Double hung		
6 sf to 8 sf	S.F.	0.160
10 sf to 12 sf	"	0.133
Storm window, 0.5 cfm, up to		
60 u.i. (united inches)	EA.	0.400
70 u.i.	"	0.400
80 u.i.	"	0.400
90 u.i.	"	0.444
100 u.i.	"	0.444
2.0 cfm, up to		
60 u.i.	EA.	0.400
70 u.i.	"	0.400
80 u.i.	"	0.400
90 u.i.	"	0.444
100 u.i.	"	0.444

Wood And Plastic	UNIT	MAN/HOURS
08600.10 Wood Windows		
Double hung		
24" x 36"		
Minimum	EA.	0.800
Average	"	1.000
Maximum	"	1.333
24" x 48"		
Minimum	EA.	0.800
Average	"	1.000
Maximum	"	1.333
30" x 48"		
Minimum	EA.	0.889
Average	"	1.143
Maximum	"	1.600
30" x 60"		
Minimum	EA.	0.889
Average	"	1.143
Maximum	"	1.600
Casement		
1 leaf, 22" x 38" high		
Minimum	EA.	0.800
Average	"	1.000
Maximum	"	1.333
2 leaf, 50" x 50" high		
Minimum	EA.	1.000
Average	"	1.333
Maximum	"	2.000
3 leaf, 71" x 62" high		
Minimum	EA.	1.000
Average	"	1.333
Maximum	"	2.000
4 leaf, 95" x 75" high		
Minimum	EA.	1.143
Average	"	1.600
Maximum	"	2.667
5 leaf, 119" x 75" high		
Minimum	EA.	1.143
Average	"	1.600
Maximum	"	2.667
Picture window, fixed glass, 54" x 54" high		
Minimum	EA.	1.000
Average	"	1.143
Maximum	"	1.333
68" x 55" high		
Minimum	EA.	1.000
Average	"	1.143
Maximum	"	1.333
Sliding, 40" x 31" high		
Minimum	EA.	0.800
Average	"	1.000
Maximum	"	1.333
52" x 39" high		
Minimum	EA.	1.000
Average	"	1.143

08 DOORS AND WINDOWS

Wood And Plastic	UNIT	MAN/HOURS
08600.10 **Wood Windows** (Cont.)		
Maximum	EA.	1.333
64" x 72" high		
Minimum	EA.	1.000
Average	"	1.333
Maximum	"	1.600
Awning windows		
34" x 21" high		
Minimum	EA.	0.800
Average	"	1.000
Maximum	"	1.333
40" x 21" high		
Minimum	EA.	0.889
Average	"	1.143
Maximum	"	1.600
48" x 27" high		
Minimum	EA.	0.889
Average	"	1.143
Maximum	"	1.600
60" x 36" high		
Minimum	EA.	1.000
Average	"	1.333
Maximum	"	1.600
Window frame, milled		
Minimum	L.F.	0.160
Average	"	0.200
Maximum	"	0.267

Hardware	UNIT	MAN/HOURS
08710.20 **Locksets**		
Latchset, heavy duty		
Cylindrical	EA.	0.500
Mortise	"	0.800
Lockset, heavy duty		
Cylindrical	EA.	0.500
Mortise	"	0.800
Mortise locks and latchsets, chrome		
Latchset passage or closet latch	EA.	0.667
Privacy (bath or bedroom)	"	0.667
Entry lockset	"	0.667
Classroom lockset (outside key operated)	"	0.667
Storeroom lock	"	0.667
Front door lock	"	0.667
Dormitory or exit lock	"	0.667
Preassembled locks and latches, brass		
Latchset, passage or closet latch	EA.	0.667

Hardware	UNIT	MAN/HOURS
08710.20 **Locksets** (Cont.)		
Lockset		
Privacy (bath or bathroom)	EA.	0.667
Entry lock	"	0.667
Classroom lock (outside key, operated)	"	0.667
Storeroom lock	"	0.667
Bored locks and latches, satin chrome plated		
Latchset passage or closet latch	EA.	0.667
Lockset		
Privacy (bath or bedroom)	EA.	0.667
Entry lock	"	0.667
Classroom lock	"	0.667
Corridor lock	"	0.667
Miscellaneous locks		
Exit lock with alarm, single door	EA.	3.200
Electric strike		
Rim mounted wrought steel	EA.	2.000
Mortised, wrought steel with bronze plating	"	3.200
Dead bolt		
Bored, wrought brass, keyed both sides	EA.	1.333
Mortised, cast brass	"	1.333
Lockset, cipher, mechanical	"	0.800
08710.30 **Closers**		
Door closers		
Surface mounted, traditional type, parallel arm		
Standard	EA.	1.000
Heavy duty	"	1.000
Modern type, parallel arm, standard duty	"	1.000
Overhead, concealed, pivot hung, single acting		
Interior	EA.	1.000
Exterior	"	1.000
Floor concealed, single acting, offset, pivoted		
Interior	EA.	2.667
Exterior	"	2.667
08710.40 **Door Trim**		
Door bumper, bronze, wall type	EA.	0.160
Wall type, 4" dia. with convex rubber pad, aluminum	"	0.160
Floor type		
Aluminum	EA.	0.160
Brass	"	0.160
Door holders		
Wall type, bronze	EA.	0.160
Overhead	"	0.400
Floor type	"	0.400
Plunger type	"	0.400
Wall type, aluminum	"	0.400
Surface bolt	"	0.160
Panic device		
Rim type with thumb piece	EA.	2.000
Mortise	"	2.000
Vertical rod	"	2.000
Labeled, rim type	"	2.000

08 DOORS AND WINDOWS

Hardware	UNIT	MAN/HOURS
08710.40 Door Trim *(Cont.)*		
Mortise	EA.	2.000
Vertical rod	"	2.000
Silencers, rubber type	"	0.016
Dust proof strike with plate, brass	"	0.267
Flush bolt, lever extension, brass, rated	"	0.160
Surface bolt with strike, brass, 6" long	"	0.160
Door coordinator, labeled, brass, satin chrome	"	0.571
Door plates		
Kick plate, aluminum, 3 beveled edges		
10" x 28"	EA.	0.400
10" x 30"	"	0.400
10" x 34"	"	0.400
10" x 38"	"	0.400
Push plate, 4" x 16"		
Aluminum	EA.	0.160
Bronze	"	0.160
Stainless steel	"	0.160
Armor plate, 40" x 34"	"	0.320
Pull handle, 4" x 16"		
Aluminum	EA.	0.160
Bronze	"	0.160
Stainless steel	"	0.160
Hasp assembly		
3"	EA.	0.133
4-1/2"	"	0.178
6"	"	0.229
Electro-magnetic door holder		
Wall mounted	EA.	2.667
Floor mounted	"	2.667
Smoke detector door holder		
Photo electric type	EA.	2.667
Ionization type	"	2.667
Pneumatic operators, activated by rubber mats		
Swing		
Single	EA.	6.667
Double	"	10.000
Sliding		
Single	EA.	6.667
Double	"	10.000
08710.60 Weatherstripping		
Weatherstrip, head and jamb, metal strip, neoprene bulb		
Standard duty	L.F.	0.044
Heavy duty	"	0.050
Spring type		
Metal doors	EA.	2.000
Wood doors	"	2.667
Sponge type with adhesive backing	"	0.800
Astragal		
1-3/4" x 13 ga., aluminum	L.F.	0.067
1-3/8" x 5/8", oak	"	0.053
Thresholds		
Bronze	L.F.	0.200

Hardware	UNIT	MAN/HOURS
08710.60 Weatherstripping *(Cont.)*		
Aluminum		
Plain	L.F.	0.200
Vinyl insert	"	0.200
Aluminum with grit	"	0.200
Steel		
Plain	L.F.	0.200
Interlocking	"	0.667

Glazing	UNIT	MAN/HOURS
08810.10 Glazing		
Sheet glass, 1/8" thick	S.F.	0.044
Plate glass, bronze or grey, 1/4" thick	"	0.073
Clear	"	0.073
Polished	"	0.073
Plexiglass		
1/8" thick	S.F.	0.073
1/4" thick	"	0.044
Float glass, clear		
3/16" thick	S.F.	0.067
1/4" thick	"	0.073
5/16" thick	"	0.080
3/8" thick	"	0.100
1/2" thick	"	0.133
5/8" thick	"	0.160
3/4" thick	"	0.200
1" thick	"	0.267
Tinted glass, polished plate, twin ground		
3/16" thick	S.F.	0.067
1/4" thick	"	0.073
3/8" thick	"	0.100
1/2" thick	"	0.133
Total, full vision, all glass window system		
To 10' high		
Minimum	S.F.	0.200
Average	"	0.200
Maximum	"	0.200
10' to 20' high		
Minimum	S.F.	0.200
Average	"	0.200
Maximum	"	0.200
Insulated glass, bronze or gray		
1/2" thick	S.F.	0.133
1" thick	"	0.200
Spandrel, polished, 1 side, 1/4" thick	"	0.073
Tempered glass (safety)		

08 DOORS AND WINDOWS

Glazing	UNIT	MAN/HOURS
08810.10 **Glazing** *(Cont.)*		
Clear sheet glass		
1/8" thick	S.F.	0.044
3/16" thick	"	0.062
Clear float glass		
1/4" thick	S.F.	0.067
5/16" thick	"	0.080
3/8" thick	"	0.100
1/2" thick	"	0.133
5/8" thick	"	0.160
3/4" thick	"	0.267
Tinted float glass		
3/16" thick	S.F.	0.062
1/4" thick	"	0.067
3/8" thick	"	0.100
1/2" thick	"	0.133
Laminated glass		
Float safety glass with polyvinyl plastic layer		
1/4", sheet or float		
Two lites, 1/8" thick, clear glass	S.F.	0.067
1/2" thick, float glass		
Two lites, 1/4" thick, clear glass	S.F.	0.133
Tinted glass	"	0.133
Insulating glass, two lites, clear float glass		
1/2" thick	S.F.	0.133
5/8" thick	"	0.160
3/4" thick	"	0.200
7/8" thick	"	0.229
1" thick	"	0.267
Glass seal edge		
3/8" thick	S.F.	0.133
Tinted glass		
1/2" thick	S.F.	0.133
1" thick	"	0.267
Tempered, clear		
1" thick	S.F.	0.267
Wire reinforced	"	0.267
Plate mirror glass		
1/4" thick		
15 sf	S.F.	0.080
Over 15 sf	"	0.073
Door type, 1/4" thick	"	0.080
Transparent, one way vision, 1/4" thick	"	0.080
Sheet mirror glass		
3/16" thick	S.F.	0.080
1/4" thick	"	0.067
Wall tiles, 12" x 12"		
Clear glass	S.F.	0.044
Veined glass	"	0.044
Wire glass, 1/4" thick		
Clear	S.F.	0.267
Hammered	"	0.267
Obscure	"	0.267
Bullet resistant, plate, with inter-leaved vinyl		

Glazing	UNIT	MAN/HOURS
08810.10 **Glazing** *(Cont.)*		
1-3/16" thick		
To 15 sf	S.F.	0.400
Over 15 sf	"	0.400
2" thick		
To 15 sf	S.F.	0.667
Over 15 sf	"	0.667
Glazing accessories		
Neoprene glazing gaskets		
1/4" glass	L.F.	0.032
3/8" glass	"	0.033
1/2" glass	"	0.035
3/4" glass	"	0.036
1" glass	"	0.040
Mullion section		
1/4" glass	L.F.	0.016
3/8" glass	"	0.020
1/2" glass	"	0.023
3/4" glass	"	0.027
1" glass	"	0.032
Molded corners	EA.	0.533

Glazed Curtain Walls	UNIT	MAN/HOURS
08910.10 **Glazed Curtain Walls**		
Curtain wall, aluminum system, framing sections		
2" x 3"		
Jamb	L.F.	0.067
Horizontal	"	0.067
Mullion	"	0.067
2" x 4"		
Jamb	L.F.	0.100
Horizontal	"	0.100
Mullion	"	0.100
3" x 5-1/2"		
Jamb	L.F.	0.100
Horizontal	"	0.100
Mullion	"	0.100
4" corner mullion	"	0.133
Coping sections		
1/8" x 8"	L.F.	0.133
1/8" x 9"	"	0.133
1/8" x 12-1/2"	"	0.160
Sill section		
1/8" x 6"	L.F.	0.080
1/8" x 7"	"	0.080
1/8" x 8-1/2"	"	0.080

08 DOORS AND WINDOWS

Glazed Curtain Walls	UNIT	MAN/HOURS
08910.10 **Glazed Curtain Walls** *(Cont.)*		
Column covers, aluminum		
1/8" x 26"	L.F.	0.200
1/8" x 34"	"	0.211
1/8" x 38"	"	0.211
Doors		
Aluminum framed, standard hardware		
Narrow stile		
2-6 x 7-0	EA.	4.000
3-0 x 7-0	"	4.000
3-6 x 7-0	"	4.000
Wide stile		
2-6 x 7-0	EA.	4.000
3-0 x 7-0	"	4.000
3-6 x 7-0	"	4.000
Flush panel doors, to match adjacent wall panels		
2-6 x 7-0	EA.	5.000
3-0 x 7-0	"	5.000
3-6 x 7-0	"	5.000
Wall panel, insulated		
"U"=.08	S.F.	0.067
"U"=.10	"	0.067
"U"=.15	"	0.067
Window wall system, complete		
Minimum	S.F.	0.080
Average	"	0.089
Maximum	"	0.114

09 FINISHES

Support Systems	UNIT	MAN/HOURS
09110.10 Metal Studs		
Studs, non load bearing, galvanized		
2-1/2", 20 ga.		
12" o.c.	S.F.	0.017
16" o.c.	"	0.013
25 ga.		
12" o.c.	S.F.	0.017
16" o.c.	"	0.013
24" o.c.	"	0.011
3-5/8", 20 ga.		
12" o.c.	S.F.	0.020
16" o.c.	"	0.016
24" o.c.	"	0.013
25 ga.		
12" o.c.	S.F.	0.020
16" o.c.	"	0.016
24" o.c.	"	0.013
4", 20 ga.		
12" o.c.	S.F.	0.020
16" o.c.	"	0.016
24" o.c.	"	0.013
25 ga.		
12" o.c.	S.F.	0.020
16" o.c.	"	0.016
24" o.c.	"	0.013
6", 20 ga.		
12" o.c.	S.F.	0.025
16" o.c.	"	0.020
24" o.c.	"	0.017
25 ga.		
12" o.c.	S.F.	0.025
16" o.c.	"	0.020
24" o.c.	"	0.017
Load bearing studs, galvanized		
3-5/8", 16 ga.		
12" o.c.	S.F.	0.020
16" o.c.	"	0.016
18 ga.		
12" o.c.	S.F.	0.013
16" o.c.	"	0.016
4", 16 ga.		
12" o.c.	S.F.	0.020
16" o.c.	"	0.016
6", 16 ga.		
12" o.c.	S.F.	0.025
16" o.c.	"	0.020
Furring		
On beams and columns		
7/8" channel	L.F.	0.053
1-1/2" channel	"	0.062
On ceilings		
3/4" furring channels		
12" o.c.	S.F.	0.033
16" o.c.	"	0.032

Support Systems	UNIT	MAN/HOURS
09110.10 Metal Studs (Cont.)		
24" o.c.	S.F.	0.029
1-1/2" furring channels		
12" o.c.	S.F.	0.036
16" o.c.	"	0.033
24" o.c.	"	0.031
On walls		
3/4" furring channels		
12" o.c.	S.F.	0.027
16" o.c.	"	0.025
24" o.c.	"	0.024
1-1/2" furring channels		
12" o.c.	S.F.	0.029
16" o.c.	"	0.027
24" o.c.	"	0.025

Lath And Plaster	UNIT	MAN/HOURS
09205.10 Gypsum Lath		
Gypsum lath, 1/2" thick		
Clipped	S.Y.	0.044
Nailed	"	0.050
09205.20 Metal Lath		
Diamond expanded, galvanized		
2.5 lb., on walls		
Nailed	S.Y.	0.100
Wired	"	0.114
On ceilings		
Nailed	S.Y.	0.114
Wired	"	0.133
3.4 lb., on walls		
Nailed	S.Y.	0.100
Wired	"	0.114
On ceilings		
Nailed	S.Y.	0.114
Wired	"	0.133
Flat rib		
2.75 lb., on walls		
Nailed	S.Y.	0.100
Wired	"	0.114
On ceilings		
Nailed	S.Y.	0.114
Wired	"	0.133
3.4 lb., on walls		
Nailed	S.Y.	0.100
Wired	"	0.114

09 FINISHES

Lath And Plaster

09205.20 Metal Lath (Cont.)

	UNIT	MAN/HOURS
On ceilings		
Nailed	S.Y.	0.114
Wired	"	0.133
Stucco lath		
1.8 lb.	S.Y.	0.100
3.6 lb.	"	0.100
Paper backed		
Minimum	S.Y.	0.080
Maximum	"	0.114

09205.60 Plaster Accessories

	UNIT	MAN/HOURS
Expansion joint, 3/4", 26 ga., galv.	L.F.	0.020
Plaster corner beads, 3/4", galvanized	"	0.023
Casing bead, expanded flange, galvanized	"	0.020
Expanded wing, 1-1/4" wide, galvanized	"	0.020
Joint clips for lath	EA.	0.004
Metal base, galvanized, 2-1/2" high	L.F.	0.027
Stud clips for gypsum lath	EA.	0.004
Sound deadening board, 1/4"	S.F.	0.013

09210.10 Plaster

	UNIT	MAN/HOURS
Gypsum plaster, trowel finish, 2 coats		
Ceilings	S.Y.	0.250
Walls	"	0.235
3 coats		
Ceilings	S.Y.	0.348
Walls	"	0.308
Vermiculite plaster		
2 coats		
Ceilings	S.Y.	0.381
Walls	"	0.348
3 coats		
Ceilings	S.Y.	0.471
Walls	"	0.421
Keenes cement plaster		
2 coats		
Ceilings	S.Y.	0.308
Walls	"	0.267
3 coats		
Ceilings	S.Y.	0.348
Walls	"	0.308
On columns, add to installation, 50%	"	
Chases, fascia, and soffits, add to installation, 50%	"	
Beams, add to installation, 50%	"	
Patch holes, average size holes		
1 sf to 5 sf		
Minimum	S.F.	0.133
Average	"	0.160
Maximum	"	0.200
Over 5 sf		
Minimum	S.F.	0.080

09210.10 Plaster (Cont.)

	UNIT	MAN/HOURS
Average	S.F.	0.114
Maximum	"	0.133
Patch cracks		
Minimum	S.F.	0.027
Average	"	0.040
Maximum	"	0.080

09220.10 Portland Cement Plaster

	UNIT	MAN/HOURS
Stucco, portland, gray, 3 coat, 1" thick		
Sand finish	S.Y.	0.348
Trowel finish	"	0.364
White cement		
Sand finish	S.Y.	0.364
Trowel finish	"	0.400
Scratch coat		
For ceramic tile	S.Y.	0.080
For quarry tile	"	0.080
Portland cement plaster		
2 coats, 1/2"	S.Y.	0.160
3 coats, 7/8"	"	0.200

09250.10 Gypsum Board

	UNIT	MAN/HOURS
Drywall, plasterboard, 3/8" clipped to		
Metal furred ceiling	S.F.	0.009
Columns and beams	"	0.020
Walls	"	0.008
Nailed or screwed to		
Wood framed ceiling	S.F.	0.008
Columns and beams	"	0.018
Walls	"	0.007
1/2", clipped to		
Metal furred ceiling	S.F.	0.009
Columns and beams	"	0.020
Walls	"	0.008
Nailed or screwed to		
Wood framed ceiling	S.F.	0.008
Columns and beams	"	0.018
Walls	"	0.007
5/8", clipped to		
Metal furred ceiling	S.F.	0.010
Columns and beams	"	0.022
Walls	"	0.009
Nailed or screwed to		
Wood framed ceiling	S.F.	0.010
Columns and beams	"	0.022
Walls	"	0.009
Vinyl faced, clipped to metal studs		
1/2"	S.F.	0.010
5/8"	"	0.010
Taping and finishing joints		
Minimum	S.F.	0.005
Average	"	0.007
Maximum	"	0.008

09 FINISHES

Lath And Plaster

09250.10 Gypsum Board (Cont.)

	UNIT	MAN/HOURS
Casing bead		
Minimum	L.F.	0.023
Average	"	0.027
Maximum	"	0.040
Corner bead		
Minimum	L.F.	0.023
Average	"	0.027
Maximum	"	0.040

Tile

09310.10 Ceramic Tile

	UNIT	MAN/HOURS
Glazed wall tile, 4-1/4" x 4-1/4"		
Minimum	S.F.	0.057
Average	"	0.067
Maximum	"	0.080
6" x 6"		
Minimum	S.F.	0.050
Average	"	0.057
Maximum	"	0.067
Base, 4-1/4" high		
Minimum	L.F.	0.100
Average	"	0.100
Maximum	"	0.100
Glazed modlings and trim, 12" x 12"		
Minimum	L.F.	0.080
Average	"	0.080
Maximum	"	0.080
Unglazed floor tile		
Portland cem., cushion edge, face mtd		
1" x 1"	S.F.	0.073
2" x 2"	"	0.067
4" x 4"	"	0.067
6" x 6"	"	0.057
12" x 12"	"	0.050
16" x 16"	"	0.044
18" x 18"	"	0.040
Adhesive bed, with white grout		
1" x 1"	S.F.	0.073
2" x 2"	"	0.067
4" x 4"	"	0.067
6" x 6"	"	0.057
12" x 12"	"	0.050
16" x 16"	"	0.044
18" x 18"	"	0.040
Organic adhesive bed, thin set, back mounted		

Tile

09310.10 Ceramic Tile (Cont.)

	UNIT	MAN/HOURS
1" x 1"	S.F.	0.073
2" x 2"	"	0.067
Porcelain floor tile		
1" x 1"	S.F.	0.073
2" x 2"	"	0.070
4" x 4"	"	0.067
6" x 6"	"	0.057
12" x 12"	"	0.050
16" x 16"	"	0.044
18" x 18"	"	0.040
Unglazed wall tile		
Organic adhesive, face mounted cushion edge		
1" x 1"		
Minimum	S.F.	0.067
Average	"	0.073
Maximum	"	0.080
2" x 2"		
Minimum	S.F.	0.062
Average	"	0.067
Maximum	"	0.073
Back mounted		
1" x 1"		
Minimum	S.F.	0.067
Average	"	0.073
Maximum	"	0.080
2" x 2"		
Minimum	S.F.	0.062
Average	"	0.067
Maximum	"	0.073
Conductive floor tile, unglazed square edged		
Portland cement bed		
1 x 1	S.F.	0.100
1-9/16 x 1-9/16	"	0.100
Dry set		
1 x 1	S.F.	0.100
1-9/16 x 1-9/16	"	0.100
Epoxy bed with epoxy joints		
1 x 1	S.F.	0.100
1-9/16 x 1-9/16	"	0.100
Ceramic accessories		
Towel bar, 24" long		
Minimum	EA.	0.320
Average	"	0.400
Maximum	"	0.533
Soap dish		
Minimum	EA.	0.533
Average	"	0.667
Maximum	"	0.800

09 FINISHES

Tile	UNIT	MAN/HOURS
09330.10 Quarry Tile		
Floor		
4 x 4 x 1/2"	S.F.	0.107
6 x 6 x 1/2"	"	0.100
6 x 6 x 3/4"	"	0.100
12 x 12 x 3/4"	"	0.089
16 x 16 x 3/4"	"	0.080
18 x 18 x 3/4"	"	0.067
Medallion		
36" dia.	EA.	2.000
48" dia.	"	2.000
Wall, applied to 3/4" portland cement bed		
4 x 4 x 1/2"	S.F.	0.160
6 x 6 x 3/4"	"	0.133
Cove base		
5 x 6 x 1/2" straight top	L.F.	0.133
6 x 6 x 3/4" round top	"	0.133
Moldings		
2 x 12	L.F.	0.080
4 x 12	"	0.080
Stair treads 6 x 6 x 3/4"	"	0.200
Window sill 6 x 8 x 3/4"	"	0.160
For abrasive surface, add to material, 25%		
09410.10 Terrazzo		
Floors on concrete, 1-3/4" thick, 5/8" topping		
Gray cement	S.F.	0.114
White cement	"	0.114
Sand cushion, 3" thick, 5/8" top, 1/4"		
Gray cement	S.F.	0.133
White cement	"	0.133
Monolithic terrazzo, 3-1/2" base slab, 5/8" topping	"	0.100
Terrazzo wainscot, cast-in-place, 1/2" thick	"	0.200
Base, cast in place, terrazzo cove type, 6" high	L.F.	0.114
Curb, cast in place, 6" wide x 6" high, polished top	"	0.400
Stairs, cast-in-place, topping on concrete or metal		
1-1/2" thick treads, 12" wide	L.F.	0.400
Combined tread and riser	"	1.000
Precast terrazzo, thin set		
Terrazzo tiles, non-slip surface		
9" x 9" x 1" thick	S.F.	0.114
12" x 12"		
1" thick	S.F.	0.107
1-1/2" thick	"	0.114
18" x 18" x 1-1/2" thick	"	0.114
24" x 24" x 1-1/2" thick	"	0.094
Terrazzo wainscot		
12" x 12" x 1" thick	S.F.	0.200
18" x 18" x 1-1/2" thick	"	0.229
Base		
6" high		
Straight	L.F.	0.062
Coved	"	0.062
8" high		

Tile	UNIT	MAN/HOURS
09410.10 Terrazzo *(Cont.)*		
Straight	L.F.	0.067
Coved	"	0.067
Terrazzo curbs		
8" wide x 8" high	L.F.	0.320
6" wide x 6" high	"	0.267
Precast terrazzo stair treads, 12" wide		
1-1/2" thick		
Diamond pattern	L.F.	0.145
Non-slip surface	"	0.145
2" thick		
Diamond pattern	L.F.	0.145
Non-slip surface	"	0.160
Stair risers, 1" thick to 6" high		
Straight sections	L.F.	0.080
Cove sections	"	0.080
Combined tread and riser		
Straight sections		
1-1/2" tread, 3/4" riser	L.F.	0.229
3" tread, 1" riser	"	0.229
Curved sections		
2" tread, 1" riser	L.F.	0.267
3" tread, 1" riser	"	0.267
Stair stringers, notched for treads and risers		
1" thick	L.F.	0.200
2" thick	"	0.267
Landings, structural, nonslip		
1-1/2" thick	S.F.	0.133
3" thick	"	0.160
Conductive terrazzo, spark proof industrial floor		
Epoxy terrazzo		
Floor	S.F.	0.050
Base	"	0.067
Polyacrylate		
Floor	S.F.	0.050
Base	"	0.067
Polyester		
Floor	S.F.	0.032
Base	"	0.040
Synthetic latex mastic		
Floor	S.F.	0.050
Base	"	0.067

09 FINISHES

Acoustical Treatment

09510.10 Ceilings And Walls

	UNIT	MAN/HOURS
Acoustical panels, suspension system not included		
Fiberglass panels		
5/8" thick		
2' x 2'	S.F.	0.011
2' x 4'	"	0.009
3/4" thick		
2' x 2'	S.F.	0.011
2' x 4'	"	0.009
Glass cloth faced fiberglass panels		
3/4" thick	S.F.	0.013
1" thick	"	0.013
Mineral fiber panels		
5/8" thick		
2' x 2'	S.F.	0.011
2' x 4'	"	0.009
3/4" thick		
2' x 2'	S.F.	0.011
2' x 4'	"	0.009
Wood fiber panels		
1/2" thick		
2' x 2'	S.F.	0.011
2' x 4'	"	0.009
5/8" thick		
2' x 2'	S.F.	0.011
2' x 4'	"	0.009
3/4" thick		
2' x 2'	S.F.	0.011
2' x 4'	"	0.009
2" thick		
2' x 2'	S.F.	0.013
2' x 4'	"	0.010
Air distributing panels		
3/4" thick	S.F.	0.020
5/8" thick	"	0.016
Acoustical tiles, suspension system not included		
Fiberglass tile, 12" x 12"		
5/8" thick	S.F.	0.015
3/4" thick	"	0.018
Glass cloth faced fiberglass tile		
3/4" thick	S.F.	0.018
3" thick	"	0.020
Mineral fiber tile, 12" x 12"		
5/8" thick		
Standard	S.F.	0.016
Vinyl faced	"	0.016
3/4" thick		
Standard	S.F.	0.016
Vinyl faced	"	0.016
Fire rated	"	0.016
Aluminum or mylar faced	"	0.016
Wood fiber tile, 12" x 12"		
1/2" thick	S.F.	0.016
3/4" thick	"	0.016

Acoustical Treatment

09510.10 Ceilings And Walls (Cont.)

	UNIT	MAN/HOURS
Metal pan units, 24 ga. steel		
12" x 12"	S.F.	0.032
12" x 24"	"	0.027
Aluminum, .025" thick		
12" x 12"	S.F.	0.032
12" x 24"	"	0.027
Anodized aluminum, 0.25" thick		
12" x 12"	S.F.	0.032
12" x 24"	"	0.027
Stainless steel, 24 ga.		
12" x 12"	S.F.	0.032
12" x 24"	"	0.027
Metal ceiling systems		
.020" thick panels		
10', 12', and 16' lengths	S.F.	0.023
Custom lengths, 3' to 20'	"	0.023
.025" thick panels		
32 sf, 38 sf, and 52 sf pieces	S.F.	0.027
Custom lengths, 10 sf to 65 sf	"	0.027
Sound absorption walls, with fabric cover		
2-6" x 9' x 3/4"	S.F.	0.027
2' x 9' x 1"	"	0.027
Starter spline	L.F.	0.020
Internal spline	"	0.020
Acoustical treatment		
Barriers for plenums		
Leaded vinyl		
0.48 lb per sf	S.F.	0.038
0.87 lb per sf	"	0.040
Aluminum foil, fiberglass reinforcement		
Minimum	S.F.	0.027
Maximum	"	0.040
Aluminum mesh, paper backed	"	0.027
Fibered cement sheet, 3/16" thick	"	0.029
Sheet lead, 1/64" thick	"	0.020
Sound attenuation blanket		
1" thick	S.F.	0.080
1-1/2" thick	"	0.080
2" thick	"	0.080
3" thick	"	0.089
Ceiling suspension systems		
T bar system		
2' x 4'	S.F.	0.008
2' x 2'	"	0.009
Concealed Z bar suspension system, 12" module	"	0.013

09 FINISHES

Flooring	UNIT	MAN/HOURS
09550.10 Wood Flooring		
Wood strip flooring, unfinished		
Fir floor		
C and better		
Vertical grain	S.F.	0.027
Flat grain	"	0.027
Oak floor		
Minimum	S.F.	0.038
Average	"	0.038
Maximum	"	0.038
Maple floor		
25/32" x 2-1/4"		
Minimum	S.F.	0.038
Maximum	"	0.038
33/32" x 3-1/4"		
Minimum	S.F.	0.038
Maximum	"	0.038
Wood block industrial flooring		
Creosoted		
2" thick	S.F.	0.021
2-1/2" thick	"	0.025
3" thick	"	0.027
Parquet, 5/16", white oak		
Finished	S.F.	0.040
Unfinished	"	0.040
Gym floor, 2 ply felt, 25/32" maple, finished, in mastic	"	0.044
Over wood sleepers	"	0.050
Finishing, sand, fill, finish, and wax	"	0.020
Refinish sand, seal, and 2 coats of polyurethane	"	0.027
Clean and wax floors	"	0.004
09630.10 Unit Masonry Flooring		
Clay brick		
9 x 4-1/2 x 3" thick		
Glazed	S.F.	0.067
Unglazed	"	0.067
8 x 4 x 3/4" thick		
Glazed	S.F.	0.070
Unglazed	"	0.070
09660.10 Resilient Tile Flooring		
Solid vinyl tile, 1/8" thick, 12" x 12"		
Marble patterns	S.F.	0.020
Solid colors	"	0.020
Travertine patterns	"	0.020
Conductive resilient flooring, vinyl tile		
1/8" thick, 12" x 12"	S.F.	0.023
09665.10 Resilient Sheet Flooring		
Vinyl sheet flooring		
Minimum	S.F.	0.008
Average	"	0.010
Maximum	"	0.013
Cove, to 6"	L.F.	0.016

Flooring	UNIT	MAN/HOURS
09665.10 Resilient Sheet Flooring *(Cont.)*		
Fluid applied resilient flooring		
Polyurethane, poured in place, 3/8" thick	S.F.	0.067
Vinyl sheet goods, backed		
0.070" thick	S.F.	0.010
0.093" thick	"	0.010
0.125" thick	"	0.010
0.250" thick	"	0.010
09678.10 Resilient Base And Accessories		
Wall base, vinyl		
Group 1		
4" high	L.F.	0.027
6" high	"	0.027
Group 2		
4" high	L.F.	0.027
6" high	"	0.027
Group 3		
4" high	L.F.	0.027
6" high	"	0.027
Stair accessories		
Treads, 1/4" x 12", rubber diamond surface		
Marbled	L.F.	0.067
Plain	"	0.067
Grit strip safety tread, 12" wide, colors		
3/16" thick	L.F.	0.067
5/16" thick	"	0.067
Risers, 7" high, 1/8" thick, colors		
Flat	L.F.	0.040
Coved	"	0.040
Nosing, rubber		
3/16" thick, 3" wide		
Black	L.F.	0.040
Colors	"	0.040
6" wide		
Black	L.F.	0.067
Colors	"	0.067

Carpet	UNIT	MAN/HOURS
09680.10 Floor Leveling		
Repair and level floors to receive new flooring		
Minimum	S.Y.	0.027
Average	"	0.067
Maximum	"	0.080

09 FINISHES

Carpet	UNIT	MAN/HOURS
09682.10 Carpet Padding		
Carpet padding		
Foam rubber, waffle type, 0.3" thick	S.Y.	0.040
Jute padding		
Minimum	S.Y.	0.036
Average	"	0.040
Maximum	"	0.044
Sponge rubber cushion		
Minimum	S.Y.	0.036
Average	"	0.040
Maximum	"	0.044
Urethane cushion, 3/8" thick		
Minimum	S.Y.	0.036
Average	"	0.040
Maximum	"	0.044
09685.10 Carpet		
Carpet, acrylic		
24 oz., light traffic	S.Y.	0.089
28 oz., medium traffic	"	0.089
Residential		
Nylon		
15 oz., light traffic	S.Y.	0.089
28 oz., medium traffic	"	0.089
Commercial		
Nylon		
28 oz., medium traffic	S.Y.	0.089
35 oz., heavy traffic	"	0.089
Wool		
30 oz., medium traffic	S.Y.	0.089
36 oz., medium traffic	"	0.089
42 oz., heavy traffic	"	0.089
Carpet tile		
Foam backed		
Minimum	S.F.	0.016
Average	"	0.018
Maximum	"	0.020
Tufted loop or shag		
Minimum	S.F.	0.016
Average	"	0.018
Maximum	"	0.020
Clean and vacuum carpet		
Minimum	S.Y.	0.004
Average	"	0.005
Maximum	"	0.008
09700.10 Special Flooring		
Epoxy flooring, marble chips		
Epoxy with colored quartz chips in 1/4" base	S.F.	0.044
Heavy duty epoxy topping, 3/16" thick	"	0.044
Epoxy terrazzo		
1/4" thick chemical resistant	S.F.	0.050

Painting	UNIT	MAN/HOURS
09905.10 Painting Preparation		
Dropcloths		
Minimum	S.F.	0.001
Average	"	0.001
Maximum	"	0.001
Masking		
Paper and tape		
Minimum	L.F.	0.008
Average	"	0.010
Maximum	"	0.013
Doors		
Minimum	EA.	0.100
Average	"	0.133
Maximum	"	0.178
Windows		
Minimum	EA.	0.100
Average	"	0.133
Maximum	"	0.178
Sanding		
Walls and flat surfaces		
Minimum	S.F.	0.005
Average	"	0.007
Maximum	"	0.008
Doors and windows		
Minimum	EA.	0.133
Average	"	0.200
Maximum	"	0.267
Trim		
Minimum	L.F.	0.010
Average	"	0.013
Maximum	"	0.018
Puttying		
Minimum	S.F.	0.012
Average	"	0.016
Maximum	"	0.020
Chemical Preparation		
Concrete floors		
Acid Etch		
Minimum	S.F.	0.002
Average	"	0.003
Maximum	"	0.004
Chemical Stripping		
Minimum	S.F.	0.013
Average	"	0.020
Maximum	"	0.032
Stone		
Chemical cleaning		
Minimum	S.F.	0.040
Average	"	0.053
Maximum	"	0.067
Wood		
Bleaching		
Minimum	S.F.	0.009
Average	"	0.011

09 FINISHES

Painting	UNIT	MAN/HOURS
09905.10 Painting Preparation (Cont.)		
Maximum	S.F.	0.015
Chemical stripping		
Minimum	S.F.	0.032
Average	"	0.080
Maximum	"	0.133
Water cleaning/preparation		
Washing (General)		
Minimum	S.F.	0.001
Average	"	0.001
Maximum	"	0.001
Mildew eradication		
Minimum	S.F.	0.001
Average	"	0.002
Maximum	"	0.003
Remove loose paint		
Minimum	S.F.	0.002
Average	"	0.003
Maximum	"	0.004
Steam clean		
Minimum	S.F.	0.002
Average	"	0.003
Maximum	"	0.004
09910.05 Ext. Painting, Sitework		
Benches		
Brush		
First Coat		
Minimum	S.F.	0.008
Average	"	0.010
Maximum	"	0.013
Second Coat		
Minimum	S.F.	0.005
Average	"	0.006
Maximum	"	0.007
Roller		
First Coat		
Minimum	S.F.	0.004
Average	"	0.004
Maximum	"	0.005
Second Coat		
Minimum	S.F.	0.003
Average	"	0.003
Maximum	"	0.004
Brickwork		
Brush		
First Coat		
Minimum	S.F.	0.005
Average	"	0.007
Maximum	"	0.010
Second Coat		
Minimum	S.F.	0.004
Average	"	0.005
Maximum	"	0.007

Painting	UNIT	MAN/HOURS
09910.05 Ext. Painting, Sitework (Cont.)		
Roller		
First Coat		
Minimum	S.F.	0.004
Average	"	0.005
Maximum	"	0.007
Second Coat		
Minimum	S.F.	0.003
Average	"	0.004
Maximum	"	0.005
Spray		
First Coat		
Minimum	S.F.	0.002
Average	"	0.003
Maximum	"	0.004
Second Coat		
Minimum	S.F.	0.002
Average	"	0.003
Maximum	"	0.003
Concrete Block		
Roller		
First Coat		
Minimum	S.F.	0.004
Average	"	0.005
Maximum	"	0.008
Second Coat		
Minimum	S.F.	0.003
Average	"	0.004
Maximum	"	0.007
Spray		
First Coat		
Minimum	S.F.	0.002
Average	"	0.003
Maximum	"	0.003
Second Coat		
Minimum	S.F.	0.001
Average	"	0.002
Maximum	"	0.003
Fences, Chain Link		
Brush		
First Coat		
Minimum	S.F.	0.008
Average	"	0.009
Maximum	"	0.010
Second Coat		
Minimum	S.F.	0.005
Average	"	0.006
Maximum	"	0.007
Roller		
First Coat		
Minimum	S.F.	0.006
Average	"	0.007
Maximum	"	0.008
Second Coat		

09 FINISHES

Painting	UNIT	MAN/HOURS
09910.05 Ext. Painting, Sitework *(Cont.)*		
Minimum	S.F.	0.003
Average	"	0.004
Maximum	"	0.005
Spray		
First Coat		
Minimum	S.F.	0.003
Average	"	0.003
Maximum	"	0.003
Second Coat		
Minimum	S.F.	0.002
Average	"	0.002
Maximum	"	0.003
Fences, Wood or Masonry		
Brush		
First Coat		
Minimum	S.F.	0.008
Average	"	0.010
Maximum	"	0.013
Second Coat		
Minimum	S.F.	0.005
Average	"	0.006
Maximum	"	0.008
Roller		
First Coat		
Minimum	S.F.	0.004
Average	"	0.005
Maximum	"	0.006
Second Coat		
Minimum	S.F.	0.003
Average	"	0.004
Maximum	"	0.005
Spray		
First Coat		
Minimum	S.F.	0.003
Average	"	0.004
Maximum	"	0.005
Second Coat		
Minimum	S.F.	0.002
Average	"	0.003
Maximum	"	0.003
Storage Tanks		
Roller		
First Coat		
Minimum	S.F.	0.003
Average	"	0.004
Maximum	"	0.005
Second Coat		
Minimum	S.F.	0.003
Average	"	0.003
Maximum	"	0.004
Spray		
First Coat		
Minimum	S.F.	0.002

Painting	UNIT	MAN/HOURS
09910.05 Ext. Painting, Sitework *(Cont.)*		
Average	S.F.	0.002
Maximum	"	0.003
Second Coat		
Minimum	S.F.	0.002
Average	"	0.002
Maximum	"	0.002
09910.15 Ext. Painting, Buildings		
Decks, Metal		
Spray		
First Coat		
Minimum	S.F.	0.004
Average	"	0.004
Maximum	"	0.004
Second Coat		
Minimum	S.F.	0.003
Average	"	0.003
Maximum	"	0.003
Decks, Wood, Stained		
Brush		
First Coat		
Minimum	S.F.	0.004
Average	"	0.004
Maximum	"	0.005
Second Coat		
Minimum	S.F.	0.003
Average	"	0.003
Maximum	"	0.003
Roller		
First Coat		
Minimum	S.F.	0.003
Average	"	0.003
Maximum	"	0.003
Second Coat		
Minimum	S.F.	0.003
Average	"	0.003
Maximum	"	0.003
Spray		
First Coat		
Minimum	S.F.	0.003
Average	"	0.003
Maximum	"	0.003
Second Coat		
Minimum	S.F.	0.002
Average	"	0.002
Maximum	"	0.003
Doors, Metal		
Roller		
First Coat		
Minimum	S.F.	0.006
Average	"	0.007
Maximum	"	0.008
Second Coat		

09 FINISHES

Painting	UNIT	MAN/HOURS
09910.15 Ext. Painting, Buildings *(Cont.)*		
Minimum	S.F.	0.004
Average	"	0.004
Maximum	"	0.005
Spray		
First Coat		
Minimum	S.F.	0.005
Average	"	0.006
Maximum	"	0.007
Second Coat		
Minimum	S.F.	0.004
Average	"	0.004
Maximum	"	0.004
Door Frames, Metal		
Brush		
First Coat		
Minimum	L.F.	0.010
Average	"	0.013
Maximum	"	0.015
Second Coat		
Minimum	L.F.	0.006
Average	"	0.007
Maximum	"	0.008
Spray		
First Coat		
Minimum	L.F.	0.004
Average	"	0.006
Maximum	"	0.008
Second Coat		
Minimum	L.F.	0.004
Average	"	0.004
Maximum	"	0.004
Doors, Wood		
Brush		
First Coat		
Minimum	S.F.	0.012
Average	"	0.016
Maximum	"	0.020
Second Coat		
Minimum	S.F.	0.010
Average	"	0.011
Maximum	"	0.013
Roller		
First Coat		
Minimum	S.F.	0.005
Average	"	0.007
Maximum	"	0.010
Second Coat		
Minimum	S.F.	0.004
Average	"	0.004
Maximum	"	0.007
Spray		
First Coat		
Minimum	S.F.	0.003

Painting	UNIT	MAN/HOURS
09910.15 Ext. Painting, Buildings *(Cont.)*		
Average	S.F.	0.003
Maximum	"	0.004
Second Coat		
Minimum	S.F.	0.002
Average	"	0.002
Maximum	"	0.003
Gutters and Downspouts		
Brush		
First Coat		
Minimum	L.F.	0.010
Average	"	0.011
Maximum	"	0.013
Second Coat		
Minimum	L.F.	0.007
Average	"	0.008
Maximum	"	0.010
Siding, Metal		
Roller		
First Coat		
Minimum	S.F.	0.003
Average	"	0.004
Maximum	"	0.004
Second Coat		
Minimum	S.F.	0.003
Average	"	0.003
Maximum	"	0.004
Spray		
First Coat		
Minimum	S.F.	0.003
Average	"	0.003
Maximum	"	0.003
Second Coat		
Minimum	S.F.	0.002
Average	"	0.002
Maximum	"	0.003
Siding, Wood		
Roller		
First Coat		
Minimum	S.F.	0.003
Average	"	0.003
Maximum	"	0.004
Second Coat		
Minimum	S.F.	0.003
Average	"	0.004
Maximum	"	0.004
Spray		
First Coat		
Minimum	S.F.	0.003
Average	"	0.003
Maximum	"	0.003
Second Coat		
Minimum	S.F.	0.002
Average	"	0.003

09 FINISHES

Painting

09910.15 Ext. Painting, Buildings (Cont.)

Description	Unit	Man/Hours
Maximum	S.F.	0.004
Stucco		
Roller		
First Coat		
Minimum	S.F.	0.004
Average	"	0.004
Maximum	"	0.005
Second Coat		
Minimum	S.F.	0.003
Average	"	0.003
Maximum	"	0.004
Spray		
First Coat		
Minimum	S.F.	0.003
Average	"	0.003
Maximum	"	0.003
Second Coat		
Minimum	S.F.	0.002
Average	"	0.002
Maximum	"	0.003
Trim		
Brush		
First Coat		
Minimum	L.F.	0.003
Average	"	0.004
Maximum	"	0.005
Second Coat		
Minimum	L.F.	0.003
Average	"	0.003
Maximum	"	0.005
Walls		
Roller		
First Coat		
Minimum	S.F.	0.003
Average	"	0.003
Maximum	"	0.003
Second Coat		
Minimum	S.F.	0.003
Average	"	0.003
Maximum	"	0.003
Spray		
First Coat		
Minimum	S.F.	0.001
Average	"	0.002
Maximum	"	0.002
Second Coat		
Minimum	S.F.	0.001
Average	"	0.001
Maximum	"	0.002
Windows		
Brush		
First Coat		
Minimum	S.F.	0.013

09910.15 Ext. Painting, Buildings (Cont.)

Description	Unit	Man/Hours
Average	S.F.	0.016
Maximum	"	0.020
Second Coat		
Minimum	S.F.	0.011
Average	"	0.013
Maximum	"	0.016

09910.25 Ext. Painting, Misc.

Description	Unit	Man/Hours
Gratings, Metal		
Roller		
First Coat		
Minimum	S.F.	0.023
Average	"	0.027
Maximum	"	0.032
Second Coat		
Minimum	S.F.	0.016
Average	"	0.020
Maximum	"	0.027
Spray		
First Coat		
Minimum	S.F.	0.011
Average	"	0.013
Maximum	"	0.016
Second Coat		
Minimum	S.F.	0.009
Average	"	0.010
Maximum	"	0.011
Ladders		
Brush		
First Coat		
Minimum	L.F.	0.020
Average	"	0.023
Maximum	"	0.027
Second Coat		
Minimum	L.F.	0.016
Average	"	0.018
Maximum	"	0.020
Spray		
First Coat		
Minimum	L.F.	0.013
Average	"	0.015
Maximum	"	0.016
Second Coat		
Minimum	L.F.	0.011
Average	"	0.012
Maximum	"	0.013
Shakes		
Spray		
First Coat		
Minimum	S.F.	0.003
Average	"	0.004
Maximum	"	0.004
Second Coat		

09 FINISHES

Painting

09910.25 Ext. Painting, Misc. (Cont.)

Description	UNIT	MAN/HOURS
Minimum	S.F.	0.003
Average	"	0.003
Maximum	"	0.004
Shingles, Wood		
Roller		
First Coat		
Minimum	S.F.	0.004
Average	"	0.005
Maximum	"	0.006
Second Coat		
Minimum	S.F.	0.003
Average	"	0.003
Maximum	"	0.004
Spray		
First Coat		
Minimum	L.F.	0.003
Average	"	0.003
Maximum	"	0.004
Second Coat		
Minimum	L.F.	0.002
Average	"	0.003
Maximum	"	0.003
Shutters and Louvres		
Brush		
First Coat		
Minimum	EA.	0.160
Average	"	0.200
Maximum	"	0.267
Second Coat		
Minimum	EA.	0.100
Average	"	0.123
Maximum	"	0.160
Spray		
First Coat		
Minimum	EA.	0.053
Average	"	0.064
Maximum	"	0.080
Second Coat		
Minimum	EA.	0.040
Average	"	0.053
Maximum	"	0.064
Stairs, metal		
Brush		
First Coat		
Minimum	S.F.	0.009
Average	"	0.010
Maximum	"	0.011
Second Coat		
Minimum	S.F.	0.005
Average	"	0.006
Maximum	"	0.007
Spray		
First Coat		

09910.25 Ext. Painting, Misc. (Cont.)

Description	UNIT	MAN/HOURS
Minimum	S.F.	0.004
Average	"	0.006
Maximum	"	0.006
Second Coat		
Minimum	S.F.	0.003
Average	"	0.004
Maximum	"	0.005
Steel, Structural, Light		
Brush		
First Coat		
Minimum	S.F.	0.016
Average	"	0.020
Maximum	"	0.027
Second Coat		
Minimum	S.F.	0.011
Average	"	0.013
Maximum	"	0.016
Roller		
First Coat		
Minimum	S.F.	0.011
Average	"	0.013
Maximum	"	0.016
Second Coat		
Minimum	S.F.	0.007
Average	"	0.008
Maximum	"	0.010
Spray		
First Coat		
Minimum	S.F.	0.007
Average	"	0.008
Maximum	"	0.010
Second Coat		
Minimum	S.F.	0.006
Average	"	0.007
Maximum	"	0.008
Steel, Medium to Heavy		
Brush		
First Coat		
Minimum	S.F.	0.008
Average	"	0.009
Maximum	"	0.010
Second Coat		
Minimum	S.F.	0.007
Average	"	0.008
Maximum	"	0.009
Roller		
First Coat		
Minimum	S.F.	0.007
Average	"	0.008
Maximum	"	0.009
Second Coat		
Minimum	S.F.	0.005
Average	"	0.006

09 FINISHES

Painting	UNIT	MAN/HOURS
09910.25 Ext. Painting, Misc. *(Cont.)*		
Maximum	S.F.	0.007
Spray		
First Coat		
Minimum	S.F.	0.004
Average	"	0.005
Maximum	"	0.006
Second Coat		
Minimum	S.F.	0.004
Average	"	0.004
Maximum	"	0.004
09910.35 Int. Painting, Buildings		
Acoustical Ceiling		
Roller		
First Coat		
Minimum	S.F.	0.005
Average	"	0.007
Maximum	"	0.010
Second Coat		
Minimum	S.F.	0.004
Average	"	0.005
Maximum	"	0.007
Spray		
First Coat		
Minimum	S.F.	0.002
Average	"	0.003
Maximum	"	0.003
Second Coat		
Minimum	S.F.	0.002
Average	"	0.002
Maximum	"	0.002
Cabinets and Casework		
Brush		
First Coat		
Minimum	S.F.	0.008
Average	"	0.009
Maximum	"	0.010
Second Coat		
Minimum	S.F.	0.007
Average	"	0.007
Maximum	"	0.008
Spray		
First Coat		
Minimum	S.F.	0.004
Average	"	0.005
Maximum	"	0.006
Second Coat		
Minimum	S.F.	0.003
Average	"	0.003
Maximum	"	0.004
Ceilings		
Roller		
First Coat		

Painting	UNIT	MAN/HOURS
09910.35 Int. Painting, Buildings *(Cont.)*		
Minimum	S.F.	0.003
Average	"	0.004
Maximum	"	0.004
Second Coat		
Minimum	S.F.	0.003
Average	"	0.003
Maximum	"	0.003
Spray		
First Coat		
Minimum	S.F.	0.002
Average	"	0.002
Maximum	"	0.003
Second Coat		
Minimum	S.F.	0.002
Average	"	0.002
Maximum	"	0.002
Doors, Metal		
Roller		
First Coat		
Minimum	L.F.	0.005
Average	"	0.006
Maximum	"	0.007
Second Coat		
Minimum	L.F.	0.004
Average	"	0.004
Maximum	"	0.005
Spray		
First Coat		
Minimum	L.F.	0.004
Average	"	0.005
Maximum	"	0.006
Second Coat		
Minimum	L.F.	0.003
Average	"	0.004
Maximum	"	0.004
Doors, Wood		
Brush		
First Coat		
Minimum	S.F.	0.011
Average	"	0.015
Maximum	"	0.018
Second Coat		
Minimum	S.F.	0.009
Average	"	0.010
Maximum	"	0.011
Spray		
First Coat		
Minimum	S.F.	0.002
Average	"	0.003
Maximum	"	0.004
Second Coat		
Minimum	S.F.	0.002
Average	"	0.002

09 FINISHES

Painting		UNIT	MAN/HOURS
09910.35	**Int. Painting, Buildings** (Cont.)		
Maximum		S.F.	0.003
Ductwork			
Brush			
Minimum		L.F.	0.010
Average		"	0.011
Maximum		"	0.013
Roller			
Minimum		S.F.	0.007
Average		"	0.007
Maximum		"	0.008
Spray			
Minimum		S.F.	0.003
Average		"	0.003
Maximum		"	0.003
Floors			
Roller			
First Coat			
Minimum		S.F.	0.003
Average		"	0.003
Maximum		"	0.003
Second Coat			
Minimum		S.F.	0.002
Average		"	0.002
Maximum		"	0.002
Spray			
First Coat			
Minimum		S.F.	0.002
Average		"	0.002
Maximum		"	0.002
Second Coat			
Minimum		S.F.	0.002
Average		"	0.002
Maximum		"	0.002
Pipes to 6" diameter			
Brush			
Minimum		L.F.	0.010
Average		"	0.011
Maximum		"	0.013
Spray			
Minimum		L.F.	0.003
Average		"	0.004
Maximum		"	0.005
Pipes to 12" diameter			
Brush			
Minimum		L.F.	0.020
Average		"	0.023
Maximum		"	0.027
Spray			
Minimum		L.F.	0.007
Average		"	0.008
Maximum		"	0.010
Trim			
Brush			

Painting		UNIT	MAN/HOURS
09910.35	**Int. Painting, Buildings** (Cont.)		
First Coat			
Minimum		L.F.	0.003
Average		"	0.004
Maximum		"	0.004
Second Coat			
Minimum		L.F.	0.002
Average		"	0.003
Maximum		"	0.004
Walls			
Roller			
First Coat			
Minimum		S.F.	0.003
Average		"	0.003
Maximum		"	0.003
Second Coat			
Minimum		S.F.	0.003
Average		"	0.003
Maximum		"	0.003
Spray			
First Coat			
Minimum		S.F.	0.001
Average		"	0.002
Maximum		"	0.002
Second Coat			
Minimum		S.F.	0.001
Average		"	0.001
Maximum		"	0.002
09955.10	**Wall Covering**		
Vinyl wall covering			
Medium duty		S.F.	0.011
Heavy duty		"	0.013
Over pipes and irregular shapes			
Lightweight, 13 oz.		S.F.	0.016
Medium weight, 25 oz.		"	0.018
Heavy weight, 34 oz.		"	0.020
Cork wall covering			
1' x 1' squares			
1/4" thick		S.F.	0.020
1/2" thick		"	0.020
3/4" thick		"	0.020
Wall fabrics			
Natural fabrics, grass cloths			
Minimum		S.F.	0.012
Average		"	0.013
Maximum		"	0.016
Flexible gypsum coated wall fabric, fire resistant		"	0.008
Vinyl corner guards			
3/4" x 3/4" x 8'		EA.	0.100
2-3/4" x 2-3/4" x 4'		"	0.100

10 SPECIALTIES

Specialties	UNIT	MAN/HOURS
10110.10 Chalkboards		
Chalkboard, metal frame, 1/4" thick		
48"x60"	EA.	0.800
48"x96"	"	0.889
48"x144"	"	1.000
48"x192"	"	1.143
Liquid chalkboard		
48"x60"	EA.	0.800
48"x96"	"	0.889
48"x144"	"	1.000
48"x192"	"	1.143
Map rail, deluxe	L.F.	0.040
10165.10 Toilet Partitions		
Toilet partition, plastic laminate		
Ceiling mounted	EA.	2.667
Floor mounted	"	2.000
Metal		
Ceiling mounted	EA.	2.667
Floor mounted	"	2.000
Wheel chair partition, plastic laminate		
Ceiling mounted	EA.	2.667
Floor mounted	"	2.000
Painted metal		
Ceiling mounted	EA.	2.667
Floor mounted	"	2.000
Urinal screen, plastic laminate		
Wall hung	EA.	1.000
Floor mounted	"	1.000
Porcelain enameled steel, floor mounted	"	1.000
Painted metal, floor mounted	"	1.000
Stainless steel, floor mounted	"	1.000
Metal toilet partitions		
Front door and side divider, floor mounted		
Porcelain enameled steel	EA.	2.000
Painted steel	"	2.000
Stainless steel	"	2.000
10185.10 Shower Stalls		
Shower receptors		
Precast, terrazzo		
32" x 32"	EA.	0.667
32" x 48"	"	0.800
Concrete		
32" x 32"	EA.	0.667
48" x 48"	"	0.889
Shower door, trim and hardware		
Economy, 24" wide, chrome, tempered glass	EA.	0.800
Porcelain enameled steel, flush	"	0.800
Baked enameled steel, flush	"	0.800
Aluminum, tempered glass, 48" wide, sliding	"	1.000
Folding	"	1.000
Aluminum and tempered glass, molded plastic		
Complete with receptor and door		

Specialties	UNIT	MAN/HOURS
10185.10 Shower Stalls (Cont.)		
32" x 32"	EA.	2.000
36" x 36"	"	2.000
40" x 40"	"	2.286
Shower compartment, precast concrete receptor		
Single entry type		
Porcelain enameled steel	EA.	8.000
Baked enameled steel	"	8.000
Stainless steel	"	8.000
Double entry type		
Porcelain enameled steel	EA.	10.000
Baked enameled steel	"	10.000
Stainless steel	"	10.000
10190.10 Cubicles		
Hospital track		
Ceiling hung	L.F.	0.089
Suspended	"	0.114
Hospital metal dividers, galvanized steel		
Baked enamel finish		
54" high		
10" glass light	L.F.	0.400
14" glass light	"	0.400
24" glass light	"	0.400
60" high		
10" glass light	L.F.	0.444
14" glass light	"	0.444
24" glass light	"	0.444
Stainless steel		
54" high		
10" glass light	L.F.	0.444
14" glass light	"	0.444
24" glass light	"	0.444
60" high		
14" glass light	L.F.	0.500
14" glass light	"	0.500
24" glass light	"	0.500
10210.10 Vents And Wall Louvers		
Block vent, 8"x16"x4" alum., w/screen, mill finish	EA.	0.267
Standard	"	0.250
Vents w/screen, 4" deep, 8" wide, 5" high		
Modular	EA.	0.250
Aluminum gable louvers	S.F.	0.133
Vent screen aluminum, 4" wide, continuous	L.F.	0.027
Louvers, aluminum, anodized, fixed blade		
Horizontal line	S.F.	0.200
Vertical line	"	0.200
Wall louvre, aluminum mill finish		
Under, 2 sf	S.F.	0.100
2 to 4 sf	"	0.089
5 to 10 sf	"	0.089
Galvanized steel		
Under 2 sf	S.F.	0.100

10 SPECIALTIES

Specialties	UNIT	MAN/HOURS
10210.10 Vents And Wall Louvers *(Cont.)*		
2 to 4 sf	S.F.	0.089
5 to 10 sf	"	0.089
10225.10 Door Louvers		
Fixed, 1" thick, enameled steel		
8"x8"	EA.	0.100
12"x8"	"	0.100
12"x12"	"	0.114
16"x12"	"	0.123
18"x12"	"	0.200
20"x8"	"	0.114
20"x12"	"	0.229
20"x16"	"	0.267
20"x20"	"	0.320
24"x12"	"	0.267
24"x16"	"	0.286
24"x18"	"	0.308
24"x20"	"	0.333
24"x24"	"	0.364
26"x26"	"	0.500
10270.40 Access & Pedestal Floor		
Panels, no covering, 2'x2'		
Plain	S.F.	0.010
Perforated	"	0.400
Pedestals		
For 6" to 12" clearance	EA.	0.080
Stringers		
2'	L.F.	0.038
6'	"	0.027
Accessories		
Ramp assembly	S.F.	0.032
Elevated floor assembly	"	0.030
Handrail	L.F.	0.400
Fascia plate	"	0.200
RF shielding components, floor liner		
Hot rolled steel sheet		
14 ga.	S.F.	0.020
11 ga.	"	0.062
10290.10 Pest Control		
Termite control		
Under slab spraying		
Minimum	S.F.	0.002
Average	"	0.004
Maximum	"	0.008
10350.10 Flagpoles		
Installed in concrete base		
Fiberglass		
25' high	EA.	5.333
50' high	"	13.333
Aluminum		

Specialties	UNIT	MAN/HOURS
10350.10 Flagpoles *(Cont.)*		
25' high	EA.	5.333
50' high	"	13.333
Bonderized steel		
25' high	EA.	6.154
50' high	"	16.000
Freestanding tapered, fiberglass		
30' high	EA.	5.714
40' high	"	7.273
50' high	"	8.000
60' high	"	9.412
Wall mounted, with collar, brushed aluminum finish		
15' long	EA.	4.000
18' long	"	4.000
20' long	"	4.211
24' long	"	4.706
Outrigger, wall, including base		
10' long	EA.	5.333
20' long	"	6.667
10400.10 Identifying Devices		
Directory and bulletin boards		
Open face boards		
Chrome plated steel frame	S.F.	0.400
Aluminum framed	"	0.400
Bronze framed	"	0.400
Stainless steel framed	"	0.400
Tack board, aluminum framed	"	0.400
Visual aid board, aluminum framed	"	0.400
Glass encased boards, hinged and keyed		
Aluminum framed	S.F.	1.000
Bronze framed	"	1.000
Stainless steel framed	"	1.000
Chrome plated steel framed	"	1.000
Metal plaque		
Cast bronze	S.F.	0.667
Aluminum	"	0.667
Metal engraved plaque		
Porcelain steel	S.F.	0.667
Stainless steel	"	0.667
Brass	"	0.667
Aluminum	"	0.667
Metal built-up plaque		
Bronze	S.F.	0.800
Copper and bronze	"	0.800
Copper and aluminum	"	0.800
Metal nameplate plaques		
Cast bronze	S.F.	0.500
Cast aluminum	"	0.500
Engraved, 1-1/2" x 6"		
Bronze	EA.	0.500
Aluminum	"	0.500
Letters, on masonry, aluminum, satin finish		
1/2" thick		

10 SPECIALTIES

Specialties	UNIT	MAN/HOURS
10400.10 Identifying Devices *(Cont.)*		
2" high	EA.	0.320
4" high	"	0.400
6" high	"	0.444
3/4" thick		
8" high	EA.	0.500
10" high	"	0.571
1" thick		
12" high	EA.	0.667
14" high	"	0.800
16" high	"	1.000
3/8" thick		
2" high	EA.	0.320
4" high	"	0.400
1/2" thick, 6" high	"	0.444
5/8" thick, 8" high	"	0.500
1" thick		
10" high	EA.	0.571
12" high	"	0.667
14" high	"	0.800
16" high	"	1.000
Interior door signs, adhesive, flexible		
2" x 8"	EA.	0.200
4" x 4"	"	0.200
6" x 7"	"	0.200
6" x 9"	"	0.200
10" x 9"	"	0.200
10" x 12"	"	0.200
Hard plastic type, no frame		
3" x 8"	EA.	0.200
4" x 4"	"	0.200
4" x 12"	"	0.200
Hard plastic type, with frame		
3" x 8"	EA.	0.200
4" x 4"	"	0.200
4" x 12"	"	0.200
10450.10 Control		
Access control, 7' high, indoor or outdoor impenetrability		
Remote or card control, type B	EA.	10.667
Free passage, type B	"	10.667
Remote or card control, type AA	"	10.667
Free passage, type AA	"	10.667
10500.10 Lockers		
Locker bench, floor mounted, laminated maple		
4'	EA.	0.667
6'	"	0.667
Wardrobe locker, 12" x 60" x 15", baked on enamel		
1-tier	EA.	0.400
2-tier	"	0.400
3-tier	"	0.421
4-tier	"	0.421
12" x 72" x 15", baked on enamel		

Specialties	UNIT	MAN/HOURS
10500.10 Lockers *(Cont.)*		
1-tier	EA.	0.400
2-tier	"	0.400
4-tier	"	0.421
5-tier	"	0.421
15" x 60" x 15", baked on enamel		
1-tier	EA.	0.400
4-tier	"	0.421
Wardrobe locker, single tier type		
12" x 15" x 72"	EA.	0.800
18" x 15" x 72"	"	0.842
12" x 18" x 72"	"	0.889
18" x 18" x 72"	"	0.941
Double tier type		
12" x 15" x 36"	EA.	0.400
18" x 15" x 36"	"	0.400
12" x 18" x 36"	"	0.400
18" x 18" x 36"	"	0.400
Two person unit		
18" x 15" x 72"	EA.	1.333
18" x 18" x 72"	"	1.600
Duplex unit		
15" x 15" x 72"	EA.	0.800
15" x 21" x 72"	"	0.800
Basket lockers, basket sets with baskets		
24 basket set	SET	4.000
30 basket set	"	5.000
36 basket set	"	6.667
42 basket set	"	8.000
10520.10 Fire Protection		
Portable fire extinguishers		
Water pump tank type		
2.5 gal.		
Red enameled galvanized	EA.	0.533
Red enameled copper	"	0.533
Polished copper	"	0.533
Carbon dioxide type, red enamel steel		
Squeeze grip with hose and horn		
2.5 lb	EA.	0.533
5 lb	"	0.615
10 lb	"	0.800
15 lb	"	1.000
20 lb	"	1.000
Wheeled type		
125 lb	EA.	1.600
250 lb	"	1.600
500 lb	"	1.600
Dry chemical, pressurized type		
Red enameled steel		
2.5 lb	EA.	0.533
5 lb	"	0.615
10 lb	"	0.800
20 lb	"	1.000

10 SPECIALTIES

Specialties	UNIT	MAN/HOURS
10520.10 Fire Protection (Cont.)		
30 lb	EA.	1.000
Chrome plated steel, 2.5 lb	"	0.533
Other type extinguishers		
2.5 gal, stainless steel, pressurized water tanks	EA.	0.533
Soda and acid type	"	0.533
Cartridge operated, water type	"	0.533
Loaded stream, water type	"	0.533
Foam type	"	0.533
40 gal, wheeled foam type	"	1.600
Fire extinguisher cabinets		
Enameled steel		
8" x 12" x 27"	EA.	1.600
8" x 16" x 38"	"	1.600
Aluminum		
8" x 12" x 27"	EA.	1.600
8" x 16" x 38"	"	1.600
8" x 12" x 27"	"	1.600
Stainless steel		
8" x 16" x 38"	EA.	1.600
10550.10 Postal Specialties		
Mail chutes		
Single mail chute		
Finished aluminum	L.F.	2.000
Bronze	"	2.000
Single mail chute receiving box		
Finished aluminum	EA.	4.000
Bronze	"	4.000
Twin mail chute, double parallel		
Finished aluminum	FLR	4.000
Bronze	"	4.000
Receiving box, 36" x 20" x 12"		
Finished aluminum	EA.	6.667
Bronze	"	6.667
Locked receiving mail box		
Finished aluminum	EA.	4.000
Bronze	"	4.000
Commercial postal accessories for mail chutes		
Letter slot, brass	EA.	1.333
Bulk mail slot, brass	"	1.333
Mail boxes		
Residential postal accessories		
Letter slot	EA.	0.400
Rural letter box	"	1.000
Apartment house, keyed, 3.5" x 4.5" x 16"	"	0.267
Ranch style	"	0.400
Commercial postal accessories		
Letter box, with combination lock	EA.	0.286
Key lock	"	0.286
Mail box, aluminum w/glass front, 4x5		
Horizontal rear load	EA.	0.229
Vertical front load	"	0.229

Specialties	UNIT	MAN/HOURS
10600.10 Movable Partitions		
Partition, movable, 2-1/2" thick, vinyl-gypsum	S.F.	0.040
Enameled steel frame, with 1/4" thick clear glass	"	0.040
Door frame and hardware for movable partitions	EA.	2.667
Cased opening for movable partitions	"	1.333
Add for acoustic movable partition	S.F.	
Accordion partition, 12' high		
Vinyl	S.F.	0.133
Acoustical	"	0.133
Standard office cubicles, 8' high, steel framed		
Baked enamel finish		
100% flush	L.F.	0.200
75% flush and 25% glass	"	0.222
50% flush and 50% glass	"	0.222
100% glass	"	0.267
Natural hardwood panels		
100% flush	L.F.	0.267
50% flush and 50% glass	"	0.286
Plastic laminated panels		
100% flush	L.F.	0.267
75% flush and 25% glass	"	0.286
50% flush and 50% glass	"	0.286
Vinyl covered panels		
100% flush	L.F.	0.276
75% flush and 25% glass	"	0.296
50% and 50% glass	"	0.296
Aluminum framed		
Enameled or anodized aluminum panels		
100% flush	L.F.	0.200
75% flush and 25% glass	"	0.222
50% flush and 50% glass	"	0.222
Vinyl covered panels		
100% flush	L.F.	0.276
75% flush and 25% glass	"	0.296
50% flush and 50% glass	"	0.296
60" high partitions, steel framed		
Enameled panels	L.F.	0.178
Natural hardwood panels, two sides	"	0.186
Plastic laminated panels	"	0.186
Vinyl covered panels	"	0.178
Aluminum framed		
Anodized or baked enamel panels	L.F.	0.178
Natural hardwood panels	"	0.186
Plastic laminated panels	"	0.186
Vinyl covered panels	"	0.178
Wire mesh partitions		
Wall panels		
4' x 7'	EA.	0.500
4' x 8'	"	0.533
4' x 10'	"	0.615
Wall filler panels		
1' x 7'	EA.	0.500
1' x 8'	"	0.533
1' x 10'	"	0.571

10 SPECIALTIES

Specialties	UNIT	MAN/HOURS
10600.10 Movable Partitions *(Cont.)*		
2' x 7'	EA.	0.500
2' x 8'	"	0.533
2' x 10'	"	0.571
3' x 7'	"	0.500
3' x 8'	"	0.533
3' x 10'	"	0.571
Ceiling panels		
10' x 2'	EA.	1.143
10' x 4'	"	1.600
Wall panel with service window		
5' wide		
7' high	EA.	0.500
8' high	"	0.533
10' high	"	0.571
Doors		
Sliding		
3' x 7'	EA.	2.000
3' x 8'	"	2.286
3' x 10'	"	3.200
4' x 7'	"	2.286
4' x 8'	"	3.200
4' x 10'	"	4.000
5' x 7'	"	3.200
5' x 8'	"	4.000
5' x 10'	"	4.000
Swing door		
3' x 7'	EA.	2.000
4' x 7'	"	2.286
Swing door, with 1' transom		
3' x 7'	EA.	2.286
4' x 7'	"	3.200
Swing door, with 3' transom		
3' x 7'	EA.	3.200
4' x 7'	"	4.000
10670.10 Shelving		
Shelving, enamel, closed side and back, 12" x 36"		
5 shelves	EA.	1.333
8 shelves	"	1.778
Open		
5 shelves	EA.	1.333
8 shelves	"	1.778
Metal storage shelving, baked enamel		
7 shelf unit, 72" or 84" high		
10" shelf	L.F.	0.800
12" shelf	"	0.842
15" shelf	"	0.889
18" shelf	"	0.941
24" shelf	"	1.000
30" shelf	"	1.067
36" shelf	"	1.143
4 shelf unit, 40" high		
10" shelf	L.F.	0.667

Specialties	UNIT	MAN/HOURS
10670.10 Shelving *(Cont.)*		
12" shelf	L.F.	0.727
15" shelf	"	0.800
18" shelf	"	0.842
24" shelf	"	0.889
3 shelf unit, 32" high		
10" shelf	L.F.	0.400
12" shelf	"	0.421
15" shelf	"	0.444
18" shelf	"	0.471
24" shelf	"	0.500
Single shelf unit, attached to masonry		
10" shelf	L.F.	0.133
12" shelf	"	0.145
15" shelf	"	0.154
18" shelf	"	0.163
24" shelf	"	0.174
Built-in wood shelves		
Posts and trimmed plywood	L.F.	0.114
Solid clear pine	"	0.123
Closet shelf, pine with rod	"	0.123
10750.10 Telephone Enclosures		
Enclosure, wall mounted, shelf, 28" x 30" x 15"	EA.	2.000
Directory shelf, stainless steel, 3 binders	"	1.333
10800.10 Bath Accessories		
Ash receiver, wall mounted, aluminum	EA.	0.400
Grab bar, 1-1/2" dia., stainless steel, wall mounted		
24" long	EA.	0.400
36" long	"	0.421
42" long	"	0.444
48" long	"	0.471
52" long	"	0.500
1" dia., stainless steel		
12" long	EA.	0.348
18" long	"	0.364
24" long	"	0.400
30" long	"	0.421
36" long	"	0.444
48" long	"	0.471
Hand dryer, surface mounted, 110 volt	"	1.000
Medicine cabinet, 16 x 22, baked enamel, lighted	"	0.320
With mirror, lighted	"	0.533
Mirror, 1/4" plate glass, up to 10 sf	S.F.	0.080
Mirror, stainless steel frame		
18"x24"	EA.	0.267
18"x32"	"	0.320
18"x36"	"	0.400
24"x30"	"	0.400
24"x36"	"	0.444
24"x48"	"	0.667
24"x60"	"	0.800
30"x30"	"	0.800

10 SPECIALTIES

Specialties	UNIT	MAN/HOURS
10800.10 Bath Accessories *(Cont.)*		
30"x72"	EA.	1.000
48"x72"	"	1.333
With shelf, 18"x24"	"	0.320
Sanitary napkin dispenser, stainless steel	"	0.533
Shower rod, 1" diameter		
Chrome finish over brass	EA.	0.400
Stainless steel	"	0.400
Soap dish, stainless steel, wall mounted	"	0.533
Toilet tissue dispenser, stainless, wall mounted		
Single roll	EA.	0.200
Double roll	"	0.229
Towel dispenser, stainless steel		
Flush mounted	EA.	0.444
Surface mounted	"	0.400
Combination towel and waste receptacle	"	0.533
Towel bar, stainless steel		
18" long	EA.	0.320
24" long	"	0.364
30" long	"	0.400
36" long	"	0.444
Toothbrush and tumbler holder	"	0.267
Waste receptacle, stainless steel, wall mounted	"	0.667
10900.10 Wardrobe Specialties		
Hospital wardrobe units, 24" x 24" x 76", with door		
Baked enameled steel	EA.	4.444
Hardwood	"	4.444
Stainless steel	"	4.444
Plastic laminated	"	4.444
Dormitory wardrobe units, 24" x 76", with door		
Hardwood	EA.	4.444
Plastic laminated	"	4.444
Hat and coat rack		
Single tier		
Baked enameled steel	L.F.	0.200
Stainless steel	"	0.200
Aluminum	"	0.200
Double tier		
Baked enameled steel	L.F.	0.229
Stainless steel	"	0.229
Aluminum	"	0.229

11 EQUIPMENT

Architectural Equipment	UNIT	MAN/HOURS
11010.10 Maintenance Equipment		
Vacuum cleaning system		
3 valves		
1.5 hp	EA.	8.889
2.5 hp	"	11.429
5 valves	"	16.000
7 valves	"	20.000
11020.10 Security Equipment		
Bulletproof teller window		
4' x 4'	EA.	13.333
5' x 4'	"	16.000
Bulletproof partitions		
Up to 12' high, 2.5" thick	S.F.	0.053
Counter for banks		
Minimum	L.F.	1.600
Maximum	"	2.667
Drive-up window		
Minimum	EA.	11.429
Maximum	"	26.667
Night depository		
Minimum	EA.	11.429
Maximum	"	26.667
Office safes, 30" x 20" x 20", 1 hr rating	"	2.000
30" x 16" x 15", 2 hr rating	"	1.600
30" x 28" x 20", H&G rating	"	1.000
Service windows, pass through painted steel		
24" x 36"	EA.	8.000
48" x 40"	"	10.000
72" x 40"	"	16.000
Special doors and windows		
3' x 7' bulletproof door with frame	EA.	11.429
12" x 12" vision panel	"	5.714
Surveillance system		
Minimum	EA.	16.000
Maximum	"	80.000
Vault door, 3' wide, 6'6" high		
3-1/2" thick	EA.	100.000
7" thick	"	133.333
10" thick	"	160.000
Insulated vault door		
2 hr rating		
32" wide	EA.	8.000
40" wide	"	8.421
4 hr rating		
32" wide	EA.	8.889
40" wide	"	10.000
6 hr rating		
32" wide	EA.	8.889
40" wide	"	10.000
Insulated file room door		
1 hr rating		
32" wide	EA.	8.000
40" wide	"	8.889

Architectural Equipment	UNIT	MAN/HOURS
11060.10 Theater Equipment		
Roll out stage, steel frame, wood floor		
Manual	S.F.	0.050
Electric	"	0.080
Portable stages		
8" high	S.F.	0.040
18" high	"	0.044
36" high	"	0.047
48" high	"	0.050
Band risers		
Minimum	S.F.	0.040
Maximum	"	0.040
Chairs for risers		
Minimum	EA.	0.036
Maximum	"	0.036
11080.10 Police Equipment		
Firing range equipment, rifle		
3 position	EA.	26.667
4 position	"	40.000
5 position	"	44.444
6 position	"	47.059
11090.10 Checkroom Equipment		
Motorized checkroom equipment		
No shelf system, 6'4" height		
7'6" length	EA.	8.000
14'6" length	"	8.000
28' length	"	8.000
One shelf, 6'8" height		
7'6" length	EA.	8.000
14'6" length	"	8.000
28' length	"	8.000
Two shelves, 7'5" height		
7'6" length	EA.	8.000
14'6" length	"	8.000
28' length	"	8.000
Three shelves, 8' height		
7'6" length	EA.	16.000
14'6" length	"	16.000
28' length	"	16.000
Four shelves, 8'7" height		
7'6" length	EA.	16.000
14'6" length	"	16.000
28' length	"	16.000
11110.10 Laundry Equipment		
High capacity, heavy duty		
Washer extractors		
135 lb		
Standard	EA.	6.667
Pass through	"	6.667
200 lb		
Standard	EA.	6.667

11 EQUIPMENT

Architectural Equipment	UNIT	MAN/HOURS
11110.10 Laundry Equipment *(Cont.)*		
Pass through	EA.	6.667
110 lb dryer	"	6.667
Hand operated presser	"	8.889
Mushroom press	"	8.889
Spreader feeders		
2 station	EA.	8.889
4 station	"	16.000
Delivery carts		
12 bushel	EA.	0.100
16 bushel	"	0.107
18 bushel	"	0.114
30 bushel	"	0.133
40 bushel	"	0.160
Low capacity		
Pressers		
Air operated	EA.	3.200
Hand operated	"	3.200
Extractor, low capacity	"	3.200
Ironer, 48"	"	1.600
Coin washers		
10 lb capacity	EA.	1.600
20 lb capacity	"	1.600
Coin dryer	"	1.000
Coin dry cleaner, 20 lb	"	3.200
11161.10 Loading Dock Equipment		
Dock leveler, 10 ton capacity		
6' x 8'	EA.	8.000
7' x 8'	"	8.000
Bumpers, laminated rubber		
4-1/2" thick		
6" x 14"	EA.	0.160
6" x 36"	"	0.178
10" x 14"	"	0.200
10" x 24"	"	0.229
10" x 36"	"	0.267
12" x 14"	"	0.211
12" x 24"	"	0.250
12" x 36"	"	0.296
6" thick		
10" x 14"	EA.	0.229
10" x 24"	"	0.276
10" x 36"	"	0.400
Extruded rubber bumpers		
T-section, 22" x 22" x 3"	EA.	0.160
Molded rubber bumpers		
24" x 12" x 3" thick	EA.	0.400
Door seal, 12" x 12", vinyl covered	L.F.	0.200
Dock boards, heavy duty, 5' x 5'		
5000 lb		
Minimum	EA.	6.667
Maximum	"	6.667
9000 lb		

Architectural Equipment	UNIT	MAN/HOURS
11161.10 Loading Dock Equipment *(Cont.)*		
Minimum	EA.	6.667
Maximum	"	7.273
15,000 lb	"	7.273
Truck shelters		
Minimum	EA.	6.154
Maximum	"	11.429
11170.10 Waste Handling		
Incinerator, electric		
100 lb/hr		
Minimum	EA.	8.000
Maximum	"	8.000
400 lb/hr		
Minimum	EA.	16.000
Maximum	"	16.000
1000 lb/hr		
Minimum	EA.	24.242
Maximum	"	24.242
Incinerator, medical-waste		
25 lb/hr, 2-7 x 4-0	EA.	16.000
50 lb/hr, 2-11 x 4-11	"	16.000
75 lb/hr, 3-8 x 5-0	"	32.000
100 lb/hr, 3-8 x 6-0	"	32.000
Industrial compactor		
1 cy	EA.	8.889
3 cy	"	11.429
5 cy	"	16.000
Trash chutes steel, including sprinklers		
18" dia.	L.F.	4.000
24" dia.	"	4.211
30" dia.	"	4.444
36" dia.	"	4.706
Refuse bottom hopper	EA.	4.444
11400.10 Food Service Equipment		
Unit kitchens		
30" compact kitchen		
Refrigerator, with range, sink	EA.	4.000
Sink only	"	2.667
Range only	"	2.000
Cabinet for upper wall section	"	1.143
Stainless shield, for rear wall	"	0.320
Side wall	"	0.320
42" compact kitchen		
Refrigerator with range, sink	EA.	4.444
Sink only	"	4.000
Cabinet for upper wall section	"	1.333
Stainless shield, for rear wall	"	0.333
Side wall	"	0.333
54" compact kitchen		
Refrigerator, oven, range, sink	EA.	5.714
Cabinet for upper wall section	"	1.600
Stainless shield, for		

11 EQUIPMENT

Architectural Equipment	UNIT	MAN/HOURS
11400.10 Food Service Equipment *(Cont.)*		
Rear wall	EA.	0.364
Side wall	"	0.364
60" compact kitchen		
Refrigerator, oven, range, sink	EA.	5.714
Cabinet for upper wall section	"	1.600
Stainless shield, for		
Rear wall	EA.	0.364
Side wall	"	0.364
72" compact kitchen		
Refrigerator, oven, range, sink	EA.	6.667
Cabinet for upper wall section	"	1.600
Stainless shield for		
Rear wall	EA.	0.400
Side wall	"	0.400
Bake oven		
Single deck		
Minimum	EA.	1.000
Maximum	"	2.000
Double deck		
Minimum	EA.	1.333
Maximum	"	2.000
Triple deck		
Minimum	EA.	1.333
Maximum	"	2.667
Convection type oven, electric, 40" x 45" x 57"		
Minimum	EA.	1.000
Maximum	"	2.000
Broiler, without oven, 69" x 26" x 39"		
Minimum	EA.	1.000
Maximum	"	1.333
Coffee urns, 10 gallons		
Minimum	EA.	2.667
Maximum	"	4.000
Fryer, with submerger		
Single		
Minimum	EA.	1.600
Maximum	"	2.667
Double		
Minimum	EA.	2.000
Maximum	"	2.667
Griddle, counter		
3' long		
Minimum	EA.	1.333
Maximum	"	1.600
5' long		
Minimum	EA.	2.000
Maximum	"	2.667
Kettles, steam, jacketed		
20 gallons		
Minimum	EA.	2.000
Maximum	"	4.000
40 gallons		
Minimum	EA.	2.000

Architectural Equipment	UNIT	MAN/HOURS
11400.10 Food Service Equipment *(Cont.)*		
Maximum	EA.	4.000
60 gallons		
Minimum	EA.	2.000
Maximum	"	4.000
Range		
Heavy duty, single oven, open top		
Minimum	EA.	1.000
Maximum	"	2.667
Fry top		
Minimum	EA.	1.000
Maximum	"	2.667
Steamers, electric		
27 kw		
Minimum	EA.	2.000
Maximum	"	2.667
18 kw		
Minimum	EA.	2.000
Maximum	"	2.667
Dishwasher, rack type		
Single tank, 190 racks/hr	EA.	4.000
Double tank		
234 racks/hr	EA.	4.444
265 racks/hr	"	5.333
Dishwasher, automatic 100 meals/hr	"	2.667
Disposals		
100 gal/hr	EA.	2.667
120 gal/hr	"	2.759
250 gal/hr	"	2.857
Exhaust hood for dishwasher, gutter 4 sides		
4'x4'x2'	EA.	2.963
4'x7'x2'	"	3.200
Food preparation machines		
Vertical cutter mixers		
25 quart	EA.	2.667
40 quart	"	2.667
80 quart	"	4.000
130 quart	"	6.667
Choppers		
5 lb	EA.	2.000
16 lb	"	2.667
40 lb	"	4.000
Mixers, floor models		
20 quart	EA.	1.000
60 quart	"	1.000
80 quart	"	1.143
140 quart	"	1.600
Ice cube maker		
50 lb per day		
Minimum	EA.	8.000
Maximum	"	8.000
500 lb per day		
Minimum	EA.	13.333
Maximum	"	13.333

11 EQUIPMENT

Architectural Equipment

11400.10 Food Service Equipment (Cont.)

	UNIT	MAN/HOURS
Ice flakers		
300 lb per day	EA.	8.000
600 lb per day	"	13.333
1000 lb per day	"	17.778
2000 lb per day	"	20.000
Refrigerated cases		
Dairy products		
Multi deck type	L.F.	0.533
Delicatessen case, service deli		
Single deck	L.F.	4.000
Multi deck	"	5.000
Meat case		
Single deck	L.F.	4.706
Multi deck	"	5.000
Produce case		
Single deck	L.F.	4.706
Multi deck	"	5.000
Bottle coolers		
6' long		
Minimum	EA.	16.000
Maximum	"	16.000
10' long		
Minimum	EA.	26.667
Maximum	"	26.667
Frozen food cases		
Chest type	L.F.	4.706
Reach-in, glass door	"	5.000
Island case, single	"	4.706
Multi deck	"	5.000
Ice storage bins		
500 lb capacity	EA.	11.429
1000 lb capacity	"	22.857

11450.10 Residential Equipment

	UNIT	MAN/HOURS
Compactor, 4 to 1 compaction	EA.	2.000
Dishwasher, built-in		
2 cycles	EA.	4.000
4 or more cycles	"	4.000
Disposal		
Garbage disposer	EA.	2.667
Heaters, electric, built-in		
Ceiling type	EA.	2.667
Wall type		
Minimum	EA.	2.000
Maximum	"	2.667
Hood for range, 2-speed, vented		
30" wide	EA.	2.667
42" wide	"	2.667
Ice maker, automatic		
30 lb per day	EA.	1.143
50 lb per day	"	4.000
Folding access stairs, disappearing metal stair		
8' long	EA.	1.143

Architectural Equipment

11450.10 Residential Equipment (Cont.)

	UNIT	MAN/HOURS
11' long	EA.	1.143
12' long	"	1.143
Wood frame, wood stair		
22" x 54" x 8'9" long	EA.	0.800
25" x 54" x 10' long	"	0.800
Ranges electric		
Built-in, 30", 1 oven	EA.	2.667
2 oven	"	2.667
Counter top, 4 burner, standard	"	2.000
With grill	"	2.000
Free standing, 21", 1 oven	"	2.667
30", 1 oven	"	1.600
2 oven	"	1.600
Water softener		
30 grains per gallon	EA.	2.667
70 grains per gallon	"	4.000

11470.10 Darkroom Equipment

	UNIT	MAN/HOURS
Dryers		
36" x 25" x 68"	EA.	4.000
48" x 25" x 68"	"	4.000
Processors, film		
Black and white	EA.	4.000
Color negatives	"	4.000
Prints	"	4.000
Transparencies	"	4.000
Sinks with cabinet and/or stand		
5" sink with stand		
24" x 48"	EA.	2.000
32" x 64"	"	2.667
38" x 52"	"	2.667
42" x 132"	"	4.000
48" x 52"	"	4.000
5" sink with cabinet		
24" x 48"	EA.	2.000
32" x 64"	"	2.667
38" x 52"	"	2.667
42" x 132"	"	4.000
48" x 52"	"	4.000
10" sink with stand		
24" x 48"	EA.	2.000
32" x 64"	"	2.667
38" x 52"	"	2.667
10" sink with cabinet		
24" x 48"	EA.	2.000
38" x 52"	"	2.667

11480.10 Athletic Equipment

	UNIT	MAN/HOURS
Basketball backboard		
Fixed	EA.	10.000
Swing-up	"	16.000
Portable, hydraulic	"	4.000
Suspended type, standard	"	16.000

11 EQUIPMENT

Architectural Equipment	UNIT	MAN/HOURS
11480.10 Athletic Equipment *(Cont.)*		
Bleacher, telescoping, manual		
15 tier, minimum	SEAT	0.160
Maximum	"	0.160
20 tier, minimum	"	0.178
Maximum	"	0.178
30 tier, minimum	"	0.267
Maximum	"	0.267
Boxing ring elevated, complete, 22' x 22'	EA.	114.286
Gym divider curtain		
Minimum	S.F.	0.011
Maximum	"	0.011
Scoreboards, single face		
Minimum	EA.	8.000
Maximum	"	40.000
Parallel bars		
Minimum	EA.	8.000
Maximum	"	13.333
11500.10 Industrial Equipment		
Vehicular paint spray booth, solid back, 14'4" x 9'6"		
24' deep	EA.	8.000
26'6" deep	"	8.000
28'6" deep	"	8.000
Drive through, 14'9" x 9'6"		
24' deep	EA.	8.000
26'6" deep	"	8.000
28'6" deep	"	8.000
Water wash, paint spray booth		
5' x 11'2" x 10'8"	EA.	8.000
6' x 11'2" x 10'8"	"	8.000
8' x 11'2" x 10'8"	"	8.000
10' x 11'2" x 11'2"	"	8.000
12' x 12'2" x 11'2"	"	8.000
14' x 12'2" x 11'2"	"	8.000
16' x 12'2" x 11'2"	"	8.000
20' x 12'2" x 11'2"	"	8.000
Dry type spray booth, with paint arrestors		
5'4" x 7'2" x 6'8"	EA.	8.000
6'4" x 7'2" x 6'8"	"	8.000
8'4" x 7'2" x 9'2"	"	8.000
10'4" x 7'2" x 9'2"	"	8.000
12'4" x 7'6" x 9'2"	"	8.000
14'4" x 7'6" x 9'8"	"	8.000
16'4" x 7'7" x 9'8"	"	8.000
20'4" x 7'7" x 10'8"	"	8.000
Air compressor, electric		
1 hp		
115 volt	EA.	5.333
7.5 hp		
115 volt	EA.	8.000
230 volt	"	8.000
Hydraulic lifts		
8,000 lb capacity	EA.	20.000

Architectural Equipment	UNIT	MAN/HOURS
11500.10 Industrial Equipment *(Cont.)*		
11,000 lb capacity	EA.	32.000
24,000 lb capacity	"	53.333
Power tools		
Band saws		
10"	EA.	0.667
14"	"	0.800
Motorized shaper	"	0.615
Motorized lathe	"	0.667
Bench saws		
9" saw	EA.	0.533
10" saw	"	0.571
12" saw	"	0.667
Electric grinders		
1/3 hp	EA.	0.320
1/2 hp	"	0.348
3/4 hp	"	0.348
11600.10 Laboratory Equipment		
Cabinets, base		
Minimum	L.F.	0.667
Maximum	"	0.667
Full storage, 7' high		
Minimum	L.F.	0.667
Maximum	"	0.667
Wall		
Minimum	L.F.	0.800
Maximum	"	0.800
Counter tops		
Minimum	S.F.	0.100
Average	"	0.114
Maximum	"	0.133
Tables		
Open underneath	S.F.	0.400
Doors underneath	"	0.500
Medical laboratory equipment		
Analyzer		
Chloride	EA.	0.400
Blood	"	0.667
Bath, water, utility, countertop unit	"	0.800
Hot plate, lab, countertop	"	0.727
Stirrer	"	0.727
Incubator, anaerobic, 23x23x36"	"	4.000
Dry heat bath	"	1.333
Incinerator, for sterilizing	"	0.080
Meter, serum protein	"	0.100
Ph analog, general purpose	"	0.114
Refrigerator, blood bank, undercounter type 153 litres	"	1.333
5.4 cf, undercounter type	"	1.333
Refrigerator/freezer, 4.4 cf, undercounter type	"	1.333
Sealer, impulse, free standing, 20x12x4"	"	0.267
Timer, electric, 1-60 minutes, bench or wall mounted	"	0.444
Glassware washer - dryer, undercounter	"	10.000
Balance, torsion suspension, tabletop, 4.5 lb capacity	"	0.444

11 EQUIPMENT

Architectural Equipment	UNIT	MAN/HOURS
11600.10 Laboratory Equipment *(Cont.)*		
Binocular microscope, with in base illuminator	EA.	0.308
Centrifuge, table model, 19x16x13"	"	0.320
Clinical model, with four place head	"	0.178
11700.10 Medical Equipment		
Hospital equipment, lights		
Examination, portable	EA.	0.667
Meters		
Air flow meter	EA.	0.444
Oxygen flow meters	"	0.333
Racks		
40 chart, revolving open frame; mobile caddy	EA.	0.667
Scales.		
Clinical, metric with measure rod, 350 lb	EA.	0.727
Physical therapy		
Chair, hydrotherapy	EA.	0.133
Diathermy, shortwave, portable, on casters	"	0.320
Exercise bicycle, floor standing, 35" x 15"	"	0.267
Hydrocollator, 4 pack, portable, 129 x 90 x 160"	"	0.114
Lamp, infrared, mobile with variable heat control	"	0.615
Ultra violet, base mounted	"	0.615
Mirror, posture training, 27" wide and 72" high	"	0.200
Parallel bars, adjustable	"	1.000
Platform mat 10'x6', 1" thick	"	0.200
Pulley, duplex, wall mounted	"	2.667
Rack, crutch, wall mounted, 66 x 16 x 13"	"	0.800
Stimulator, galvanic-faradic, hand held	"	0.053
Ultrasound, stimulator, portable, 13x13x8"	"	0.067
Sandbag set, velcro straps, saddle bag type	"	0.114
Whirlpool, 85 gallon	"	4.000
65 gallon capacity	"	4.000
Radiology		
Radiographic table, motor driven tilting table	EA.	80.000
Fluoroscope image/tv system	"	160.000
Processor for washing and drying radiographs		
Water filter unit, 30" x 48-1/2" x 37-1/2"	EA.	13.333
Cassette transfer cabinet	"	0.667
Base storage cabinets, sectional design		
With back splash, 24" deep and 35" high	L.F.	0.667
Wall storage cabinets	"	1.000
Steam sterilizers		
For heat and moisture stable materials	EA.	0.800
For fast drying after sterilization	"	1.000
Compact unit	"	1.000
Semi-automatic	"	4.000
Floor loading		
Single door	EA.	6.667
Double door	"	8.000
Utensil washer, sanitizer	"	6.154
Automatic washer/sterilizer	"	16.000
16 x 16 x 26", including accessories	"	26.667
Steam generator, elec., 10 kw to 180 kw	"	16.000
Surgical scrub		

Architectural Equipment	UNIT	MAN/HOURS
11700.10 Medical Equipment *(Cont.)*		
Minimum	EA.	2.667
Maximum	"	2.667
Gas sterilizers		
Automatic, free standing, 21x19x29"	EA.	8.000
Surgical tables		
Minimum	EA.	11.429
Maximum	"	16.000
Surgical lights, ceiling mounted		
Minimum	EA.	13.333
Maximum	"	16.000
Water stills		
4 liters/hr	EA.	2.667
8 liters/hr	"	2.667
19 liters/hr	"	6.667
X-ray equipment		
Mobile unit		
Minimum	EA.	4.000
Maximum	"	8.000
Film viewers		
Minimum	EA.	1.333
Maximum	"	2.667
Autopsy table		
Minimum	EA.	8.000
Maximum	"	8.000
Incubators		
15 cf	EA.	4.000
29 cf	"	6.667
Infant transport, portable	"	4.211
Beds		
Stretcher, with pad, 30" x 78"	EA.	2.000
Transfer, for patient transport	"	2.000
Headwall		
Aluminum, with back frame and console	EA.	4.000
Hospital ground detection system		
Power ground module	EA.	2.286
Ground slave module	"	1.739
Master ground module	"	1.509
Remote indicator	"	1.600
X-ray indicator	"	1.739
Micro ammeter	"	2.000
Supervisory module	"	1.739
Ground cords	"	0.296
Hospital isolation monitors, 5 ma		
120v	EA.	3.478
208v	"	3.478
240v	"	3.478
Digital clock-timers separate display	"	1.600
One display	"	1.600
Remote control	"	1.250
Battery pack	"	1.250
Surgical chronometer clock and 3 timers	"	2.500
Auxilary control	"	1.159

11 EQUIPMENT

Architectural Equipment	UNIT	MAN/HOURS
11700.20 **Dental Equipment**		
Dental care equipment		
Drill console with accessories	EA.	13.333
Amalgamator	"	0.400
Lathe	"	0.267
Finish polisher	"	0.533
Model trimmer	"	0.364
Motor, wall mounted	"	0.364
Cleaner, ultrasonic	"	0.800
Curing unit, bench mounted	"	1.333
Oral evacuation system, dual pump	"	1.000
Sterilizer, table top, self contained	"	0.444
Dental lights		
Light, floor or ceiling mounted	EA.	4.000
X-ray unit		
Portable	EA.	2.000
Wall mounted with remote control	"	6.667
Illuminator, single panel	"	11.429
X-ray film processor	"	6.667
Shield, portable x-ray, lead lined	"	0.533

12 FURNISHINGS

Interior	UNIT	MAN/HOURS
12302.10 — Casework		
Kitchen base cabinet, standard, 24" deep, 35" high		
12" wide	EA.	0.800
18" wide	"	0.800
24" wide	"	0.889
27" wide	"	0.889
36" wide	"	1.000
48" wide	"	1.000
Drawer base, 24" deep, 35" high		
15" wide	EA.	0.800
18" wide	"	0.800
24" wide	"	0.889
27" wide	"	0.889
30" wide	"	0.889
Sink-ready, base cabinet		
30" wide	EA.	0.889
36" wide	"	0.889
42" wide	"	0.889
60" wide	"	1.000
Corner cabinet, 36" wide	"	1.000
Wall cabinet, 12" deep, 12" high		
30" wide	EA.	0.800
36" wide	"	0.800
15" high		
30" wide	EA.	0.889
36" wide	"	0.889
24" high		
30" wide	EA.	0.889
36" wide	"	0.889
30" high		
12" wide	EA.	1.000
18" wide	"	1.000
24" wide	"	1.000
27" wide	"	1.000
30" wide	"	1.143
36" wide	"	1.143
Corner cabinet, 30" high		
24" wide	EA.	1.333
30" wide	"	1.333
36" wide	"	1.333
Wardrobe	"	2.000
Vanity with top, laminated plastic		
24" wide	EA.	2.000
30" wide	"	2.000
36" wide	"	2.667
48" wide	"	3.200
12390.10 — Counter Tops		
Stainless steel, counter top, with backsplash	S.F.	0.200
Acid-proof, kemrock surface	"	0.133

Interior	UNIT	MAN/HOURS
12500.10 — Window Treatment		
Drapery tracks, wall or ceiling mounted		
Basic traverse rod		
50 to 90"	EA.	0.400
84 to 156"	"	0.444
136 to 250"	"	0.444
165 to 312"	"	0.500
Traverse rod with stationary curtain rod		
30 to 50"	EA.	0.400
50 to 90"	"	0.400
84 to 156"	"	0.444
136 to 250"	"	0.500
Double traverse rod		
30 to 50"	EA.	0.400
50 to 84"	"	0.400
84 to 156"	"	0.444
136 to 250"	"	0.500
12510.10 — Blinds		
Venetian blinds		
2" slats	S.F.	0.020
1" slats	"	0.020
12690.40 — Floor Mats		
Recessed entrance mat, 3/8" thick, aluminum link	S.F.	0.400
Steel, flexible	"	0.400

13 SPECIAL

Construction	UNIT	MAN/HOURS
13056.10 Vaults		
Floor safes		
Class C		
1.0 cf	EA.	0.667
1.3 cf	"	1.000
1.9 cf	"	1.333
5.2 cf	"	1.333
13121.10 Pre-engineered Buildings		
Pre-engineered metal building, 40'x100'		
14' eave height	S.F.	0.032
16' eave height	"	0.037
20' eave height	"	0.048
60'x100'		
14' eave height	S.F.	0.032
16' eave height	"	0.037
20' eave height	"	0.048
80'x100'		
14' eave height	S.F.	0.032
16' eave height	"	0.037
20' eave height	"	0.048
100'x100'		
14' eave height	S.F.	0.032
16' eave height	"	0.037
20' eave height	"	0.048
100'x150'		
14' eave height	S.F.	0.032
16' eave height	"	0.037
20' eave height	"	0.048
120'x150'		
14' eave height	S.F.	0.032
16' eave height	"	0.037
20' eave height	"	0.048
140'x150'		
14' eave height	S.F.	0.032
16' eave height	"	0.037
20' eave height	"	0.048
160'x200'		
14' eave height	S.F.	0.032
16' eave height	"	0.037
20' eave height	"	0.048
200'x200'		
14' eave height	S.F.	0.032
16' eave height	"	0.037
20' eave height	"	0.048
Liner panel, 26 ga, painted steel	"	0.020
Wall panel insulated, 26 ga. steel, foam core	"	0.020
Roof panel, 26 ga. painted steel	"	0.011
Plastic (sky light)	"	0.011
Insulation, 3-1/2" thick blanket, R11	"	0.005

Construction	UNIT	MAN/HOURS
13152.10 Swimming Pool Equipment		
Diving boards		
14' long		
Aluminum	EA.	4.444
Fiberglass	"	4.444
Ladders, heavy duty		
2 steps		
Minimum	EA.	1.600
Maximum	"	1.600
4 steps		
Minimum	EA.	2.000
Maximum	"	2.000
Lifeguard chair		
Minimum	EA.	8.000
Maximum	"	8.000
Lights, underwater		
12 volt, with transformer, 100 watt		
Incandescent	EA.	2.000
Halogen	"	2.000
LED	"	2.000
110 volt		
Minimum	EA.	2.000
Maximum	"	2.000
Ground fault interrupter for 110 volt, each light	"	0.667
Pool cover		
Reinforced polyethylene	S.F.	0.062
Vinyl water tube		
Minimum	S.F.	0.062
Maximum	"	0.062
Slides with water tube		
Minimum	EA.	6.667
Maximum	"	6.667
13200.10 Storage Tanks		
Oil storage tank, underground, single wall, no excv.		
Steel		
500 gals	EA.	3.000
1,000 gals	"	4.000
4,000 gals	"	8.000
5,000 gals	"	12.000
10,000 gals	"	24.000
Fiberglass, double wall		
550 gals	EA.	4.000
1,000 gals	"	4.000
2,000 gals	"	6.000
4,000 gals	"	12.000
6,000 gals	"	16.000
8,000 gals	"	24.000
10,000 gals	"	30.000
12,000 gals	"	40.000
15,000 gals	"	53.333
20,000 gals	"	60.000
Above ground		
Steel, single wall		

13 SPECIAL

Construction	UNIT	MAN/HOURS
13200.10 Storage Tanks *(Cont.)*		
275 gals	EA.	2.400
500 gals	"	4.000
1,000 gals	"	4.800
1,500 gals	"	6.000
2,000 gals	"	8.000
5,000 gals	"	12.000
Fill cap	"	0.800
Vent cap	"	0.800
Level indicator	"	0.800

Hazardous Waste	UNIT	MAN/HOURS
13280.10 Asbestos Removal		
Enclosure using wood studs & poly, install & remove	S.F.	0.020
13280.12 Duct Insulation Removal		
Remove duct insulation, duct size		
6" x 12"	L.F.	0.044
x 18"	"	0.062
x 24"	"	0.089
8" x 12"	"	0.067
x 18"	"	0.073
x 24"	"	0.100
12" x 12"	"	0.067
x 18"	"	0.089
x 24"	"	0.114
13280.15 Pipe Insulation Removal		
Removal, asbestos insulation		
2" thick, pipe		
1" to 3" dia.	L.F.	0.067
4" to 6" dia.	"	0.076
3" thick		
7" to 8" dia.	L.F.	0.080
9" to 10" dia.	"	0.084
11" to 12" dia.	"	0.089
13" to 14" dia.	"	0.094
15" to 18" dia.	"	0.100

14 CONVEYING

Elevators

14210.10 Elevators

	UNIT	MAN/HOURS
Passenger elevators, electric, geared		
Based on a shaft of 6 stops and 6 openings		
50 fpm, 2000 lb	EA.	24.000
100 fpm, 2000 lb	"	26.667
150 fpm		
2000 lb	EA.	30.000
3000 lb	"	34.286
4000 lb	"	40.000
200 fpm		
2500 lb	EA.	34.286
3000 lb	"	36.923
4000 lb	"	40.000
250 fpm		
2500 lb	EA.	34.286
3000 lb	"	36.923
4000 lb	"	40.000
300 fpm		
2500 lb	EA.	34.286
3000 lb	"	36.923
4000 lb	"	24.000
Based on a shaft of 8 stops and 8 openings		
300 fpm		
3000 lb	EA.	48.000
3500 lb	"	48.000
4000 lb	"	53.333
5000 lb	"	57.143
400 fpm		
3000 lb	EA.	48.000
3500 lb	"	48.000
4000 lb	"	53.333
5000 lb	"	57.143
600 fpm		
3000 lb	EA.	53.333
3500 lb	"	57.143
4000 lb	"	58.537
5000 lb	"	60.000
800 fpm		
3000 lb	EA.	53.333
3500 lb	"	57.143
4000 lb	"	58.537
5000 lb	"	60.000
Hydraulic, based on a shaft of 3 stops, 3 openings		
50 fpm		
2000 lb	EA.	20.000
2500 lb	"	20.000
3000 lb	"	20.870
100 fpm		
2000 lb	EA.	20.000
2500 lb	"	20.870
3000 lb	"	21.818
150 fpm		
2000 lb	EA.	20.000
2500 lb	"	20.870

14210.10 Elevators (Cont.)

	UNIT	MAN/HOURS
3000 lb	EA.	22.857
Small elevators, 4 to 6 passenger capacity		
Electric, push		
2 stops	EA.	20.000
3 stops	"	21.818
4 stops	"	24.000
Freight elevators, electric		
Based on a shaft of 6 stops and 6 openings		
50 fpm		
3500 lb	EA.	26.667
4000 lb	"	26.667
5000 lb	"	30.000
100 fpm		
3500 lb	EA.	30.000
4000 lb	"	30.000
5000 lb	"	34.286
200 fpm		
3500 lb	EA.	34.286
4000 lb	"	34.286
5000 lb	"	40.000
Based on shaft of 8 stops and 8 openings		
100 fpm		
4000 lb	EA.	30.000
6000 lb	"	30.769
8000 lb	"	32.432
150 fpm		
4000 lb	EA.	34.286
6000 lb	"	35.294
8000 lb	"	37.500
200 fpm		
4000 lb	EA.	40.000
6000 lb	"	41.379
8000 lb	"	43.636
Hydraulic, based on 3 stops and 3 openings		
50 fpm		
3000 lb	EA.	17.143
4000 lb	"	17.778
6000 lb	"	18.462
100 fpm		
3000 lb	EA.	17.143
4000 lb	"	17.778
6000 lb	"	18.462
150 fpm		
3000 lb	EA.	17.143
4000 lb	"	17.778
6000 lb	"	18.462

14300.10 Escalators

	UNIT	MAN/HOURS
Escalators		
32" wide, floor to floor		
12' high	EA.	40.000
15' high	"	48.000
18' high	"	60.000

14 CONVEYING

Elevators

	UNIT	MAN/HOURS
14300.10 Escalators *(Cont.)*		
22' high	EA.	80.000
25' high	"	96.000
48" wide		
12' high	EA.	41.379
15' high	"	50.000
18' high	"	63.158
22' high	"	85.714
25' high	"	96.000

Lifts

	UNIT	MAN/HOURS
14410.10 Personnel Lifts		
Residential stair climber, per story	EA.	6.667
curved	"	8.000
14410.20 Wheelchair Lifts		
600 lb, Residential	EA.	8.000
Commercial	"	8.000
14450.10 Vehicle Lifts		
Automotive hoist, one post, semi-hydraulic, 8,000 lb	EA.	24.000
Full hydraulic, 8,000 lb	"	24.000
2 post, semi-hydraulic, 10,000 lb	"	34.286
Full hydraulic		
10,000 lb	EA.	34.286
13,000 lb	"	60.000
18,500 lb	"	60.000
24,000 lb	"	60.000
26,000 lb	"	60.000
Pneumatic hoist, fully hydraulic		
11,000 lb	EA.	80.000
24,000 lb	"	80.000

Material Handling

	UNIT	MAN/HOURS
14560.10 Chutes		
Linen chutes, stainless steel, with supports		
18" dia.	L.F.	0.057
24" dia.	"	0.062
30" dia.	"	0.067
Hopper	EA.	0.533
Skylight	"	0.800
Sprinkler unit at top	"	0.889
14580.10 Pneumatic Systems		
Pneumatic message tube system		
Average, 20 station job		
3" round system	E.A.	72.727
4" round system	"	80.000
6" round system	"	88.889
4" x 7" oval system	"	160.000
Trash and linen tube system		
10 stations	EA.	120.000
15 stations	"	160.000
20 stations	"	184.615
30 stations	"	218.182

Hoists And Cranes

	UNIT	MAN/HOURS
14600.10 Industrial Hoists		
Industrial hoists, electric, light to medium duty		
500 lb	EA.	4.000
1000 lb	"	4.211
2000 lb	"	4.444
3000 lb	"	4.706
4000 lb	"	5.000
5000 lb	"	5.333
6000 lb	"	5.517
7500 lb	"	5.714
10,000 lb	"	5.926
15,000 lb	"	6.154
20,000 lb	"	6.667
25,000 lb	"	7.273
30,000 lb	"	8.000
Heavy duty		
500 lb	EA.	4.000
1000 lb	"	4.211
2000 lb	"	4.444
3000 lb	"	4.706
4000 lb	"	5.000
5000 lb	"	5.333
6000 lb	"	5.517

14 CONVEYING

Hoists And Cranes	UNIT	MAN/HOURS
14600.10 **Industrial Hoists** *(Cont.)*		
7500 lb	EA.	5.714
10,000 lb	"	5.926
15,000 lb	"	6.154
20,000 lb	"	6.667
25,000 lb	"	7.273
30,000 lb	"	8.000
Air powered hoists		
500 lb	EA.	4.000
1000 lb	"	4.000
2000 lb	"	4.211
4000 lb	"	4.706
6000 lb	"	6.154
Overhead traveling bridge crane		
Single girder, 20' span		
3 ton	EA.	12.000
5 ton	"	12.000
7.5 ton	"	12.000
10 ton	"	15.000
15 ton	"	15.000
30' span		
3 ton	EA.	12.000
5 ton	"	12.000
10 ton	"	15.000
15 ton	"	15.000
Double girder, 40' span		
3 ton	EA.	26.667
5 ton	"	26.667
7.5 ton	"	26.667
10 ton	"	34.286
15 ton	"	34.286
25 ton	"	34.286
50' span		
3 ton	EA.	26.667
5 ton	"	26.667
7.5 ton	"	26.667
10 ton	"	34.286
15 ton	"	34.286
25 ton	"	34.286
14650.10 **Jib Cranes**		
Self supporting, swinging 8' boom, 200 deg rotation		
1000 lb	EA.	6.667
2000 lb	"	6.667
3000 lb	"	13.333
4000 lb	"	13.333
6000 lb	"	13.333
10,000 lb	"	13.333
Wall mounted, 180 deg rotation		
2000 lb	EA.	6.667
3000 lb	"	6.667
4000 lb	"	13.333
6000 lb	"	13.333
10,000 lb	"	13.333

15 MECHANICAL

Basic Materials	UNIT	MAN/HOURS
15100.10 Specialties		
Wall penetration		
Concrete wall, 6" thick		
2" dia.	EA.	0.267
4" dia.	"	0.400
8" dia.	"	0.571
12" thick		
2" dia.	EA.	0.364
4" dia.	"	0.571
8" dia.	"	0.889
Non-destructive testing, piping systems		
X-ray of welds		
3" dia. pipe	EA.	0.800
4" dia. pipe	"	0.800
6" dia. pipe	"	0.800
8" dia. pipe	"	1.000
10" dia. pipe	"	1.000
Liquid penetration of welds		
2" dia. pipe	EA.	0.500
3" dia. pipe	"	0.500
4" dia. pipe	"	0.500
6" dia. pipe	"	0.500
8" dia. pipe	"	0.500
10" dia. pipe	"	0.500
15120.10 Backflow Preventers		
Backflow preventer, flanged, cast iron, with valves		
3" pipe	EA.	4.000
4" pipe	"	4.444
6" pipe	"	6.667
8" pipe	"	8.000
Threaded		
3/4" pipe	EA.	0.500
2" pipe	"	0.800
Reduced pressure assembly, bronze, threaded		
3/4"	EA.	0.500
1"	"	0.571
1-1/4"	"	0.667
1-1/2"	"	0.800
15140.10 Pipe Hangers, Heavy		
Hangers		
1/2" pipe, clevis pipe hanger		
Black steel	EA.	0.267
Galvanized	"	0.267
U bolt	"	0.080
3/4" pipe, clevis pipe hanger		
Black steel	EA.	0.267
Galvanized	"	0.267
U bolt	"	0.080
1" pipe, clevis pipe hanger		
Black steel	EA.	0.267
Galvanized	"	0.267
U bolt	"	0.080

Basic Materials	UNIT	MAN/HOURS
15140.10 Pipe Hangers, Heavy *(Cont.)*		
1-1/4" pipe, clevis pipe hanger		
Black steel	EA.	0.267
Galvanized	"	0.267
U bolt	"	0.080
1-1/2" pipe, clevis pipe hanger		
Black steel	EA.	0.267
Galvanized	"	0.267
U bolt	"	0.080
2" pipe, clevis pipe hanger		
Black steel	EA.	0.267
Galvanized	"	0.267
Adjustable pipe roll stand	"	1.000
U bolt	"	0.080
Adjustable roller hanger	"	0.267
2-1/2" pipe, clevis pipe hanger		
Black steel	EA.	0.267
Galvanized	"	0.267
Adjustable pipe roll stand	"	0.267
Adjustable roller hanger	"	0.267
3" pipe, clevis pipe hanger		
Black steel	EA.	0.267
Galvanized	"	0.267
Adjustable pipe roll stand	"	0.267
U bolt	"	0.080
Adjustable roller hanger	"	0.267
3-1/2" pipe, clevis pipe hanger		
Black steel	EA.	0.267
Galvanized	"	0.267
Adjustable pipe roll stand	"	0.267
U bolt	"	0.080
Adjustable roller hanger	"	0.267
4" pipe, clevis pipe hanger		
Black steel	EA.	0.267
Galvanized	"	0.267
Adjustable pipe roll stand	"	0.267
U bolt	"	0.080
Adjustable roller hanger	"	0.267
5" pipe, clevis pipe hanger		
Black steel	EA.	0.320
Galvanized	"	0.320
Adjustable pipe roll stand	"	0.320
U bolt	"	0.080
Adjustable roller hanger	"	0.320
6" pipe, clevis pipe hanger		
Black steel	EA.	0.320
Galvanized	"	0.320
Adjustable pipe roll stand	"	0.320
U bolt	"	0.100
Adjustable roller hanger	"	0.320
8" pipe, clevis pipe hanger		
Black steel	EA.	0.320
Galvanized	"	0.320
Adjustable pipe roll stand	"	0.320

15 MECHANICAL

Basic Materials	UNIT	MAN/HOURS
15140.10 Pipe Hangers, Heavy *(Cont.)*		
U bolt	EA.	0.100
Adjustable roller hanger	"	0.320
10" clevis pipe hanger		
Black steel	EA.	0.320
Galvanized	"	0.320
Adjustable pipe roll stand	"	0.320
Adjustable roller hanger	"	0.320
12" pipe, clevis pipe hanger		
Black steel	EA.	0.320
Galvanized	"	0.320
Adjustable pipe roll stand	"	0.320
Adjustable roller hanger	"	0.320
14" pipe, clevis pipe hanger		
Black steel	EA.	0.364
Galvanized	"	0.364
Threaded rod, galvanized		
C-clamp, steel, with lock nut		
3/8"	EA.	0.100
1/2"	"	0.100
5/8"	"	0.100
3/4"	"	0.100
7/8"	"	0.100
Angle support, medium, welded steel		
12"x18"	EA.	0.800
18"x24"	"	0.800
24"x30"	"	0.800
Heavy, welded steel		
12"x18"	EA.	0.800
18"x24"	"	0.800
24"x30"	"	0.800
15140.11 Pipe Hangers, Light		
A band, black iron		
1/2"	EA.	0.057
1"	"	0.059
1-1/4"	"	0.062
1-1/2"	"	0.067
2"	"	0.073
2-1/2"	"	0.080
3"	"	0.089
4"	"	0.100
5"	"	0.107
6"	"	0.114
8"	"	0.133
Copper		
1/2"	EA.	0.057
3/4"	"	0.059
1"	"	0.059
1-1/4"	"	0.062
1-1/2"	"	0.067
2"	"	0.073
2-1/2"	"	0.080
3"	"	0.089

Basic Materials	UNIT	MAN/HOURS
15140.11 Pipe Hangers, Light *(Cont.)*		
4"	EA.	0.100
Black riser friction hangers		
3/4"	EA.	0.067
1"	"	0.070
1-1/4"	"	0.073
1-1/2"	"	0.076
2"	"	0.080
2-1/2"	"	0.089
3"	"	0.100
4"	"	0.114
5"	"	0.123
6"	"	0.133
8"	"	0.145
10"	"	0.160
Short pattern black riser clamps		
1-1/2"	EA.	0.073
2"	"	0.076
3"	"	0.080
4"	"	0.089
Copper riser friction hanger		
1/2"	EA.	0.062
3/4"	"	0.064
1"	"	0.067
1-1/4"	"	0.070
1-1/2"	"	0.073
2"	"	0.076
2-1/2"	"	0.080
3"	"	0.080
4"	"	0.089
Auto grip hangers, galvanized		
1/2"	EA.	0.057
3/4"	"	0.062
1"	"	0.064
1-1/4"	"	0.067
1-1/2"	"	0.070
2"	"	0.073
2-1/2"	"	0.076
3"	"	0.080
4"	"	0.089
Copper		
1/2"	EA.	0.057
3/4"	"	0.062
1"	"	0.064
1-1/4"	"	0.067
1-1/2"	"	0.070
2"	"	0.073
2-1/2"	"	0.076
3"	"	0.080
4"	"	0.089
Split rings (F&M), galvanized		
3/8"	EA.	0.057
1/2"	"	0.062
3/4"	"	0.064

15 MECHANICAL

15140.11 Pipe Hangers, Light (Cont.)

Basic Materials	UNIT	MAN/HOURS
1"	EA.	0.067
1-1/4"	"	0.070
1-1/2"	"	0.073
2"	"	0.076
2-1/2"	"	0.080
3"	"	0.084
4"	"	0.089
Copper		
1/4"	EA.	0.057
3/8"	"	0.057
1/2"	"	0.062
1"	"	0.067
1-1/4"	"	0.070
1-1/2"	"	0.073
2"	"	0.076
2-1/2"	"	0.080
3"	"	0.084
4"	"	0.089
F&M plates, galvanized		
3/8"	EA.	0.057
1/2"	"	0.059
Copper		
3/8"	EA.	0.057
1/2"	"	0.059
2 hole clips, galvanized		
3/4"	EA.	0.053
1"	"	0.055
1-1/4"	"	0.057
1-1/2"	"	0.059
2"	"	0.062
2-1/2"	"	0.064
3"	"	0.067
4"	"	0.073
Perforated strap		
3/4"		
Galvanized, 20 ga.	L.F.	0.040
Copper, 22 ga.	"	0.040
Threaded rod-couplings		
1/4"	EA.	0.050
3/4"	"	0.053
1/2"	"	0.057
5/8"	"	0.062
Reducing rod coupling, 1/2" x 3/8"	"	0.057
C-clamps		
3/4"	EA.	0.080
Top beam clamp		
3/8"	EA.	0.067
1/2"	"	0.073
Side beam connector		
3/8"	EA.	0.067
1/2"	"	0.073
J-Hooks		
1/2"	EA.	0.036

15140.11 Pipe Hangers, Light (Cont.)

Basic Materials	UNIT	MAN/HOURS
3/4"	EA.	0.036
1"	"	0.038
1-1/4"	"	0.039
1-1/2"	"	0.040
2"	"	0.040
3"	"	0.042
4"	"	0.042
PVC coated hangers, galvanized, 28 ga.		
1-1/2" x 12"	EA.	0.053
2" x 12"	"	0.057
3" x 12"	"	0.062
4" x 12"	"	0.067
Copper, 30 ga.		
1-1/2" x 12"	EA.	0.053
2" x 12"	"	0.057
3" x 12"	"	0.062
4" x 12"	"	0.067
2" x 24"	"	0.062
3" x 24"	"	0.067
4" x 24"	"	0.073
Milford hangers		
1/2" x 6"	EA.	0.057
1/2" x 12"	"	0.062
3/4" x 6"	"	0.059
3/4" x 12"	"	0.064
1" x 6"	"	0.062
1" x 12"	"	0.064
1-1/4" x 6"	"	0.064
1-1/4" x 12"	"	0.067
1-1/2" x 6"	"	0.067
1-1/2" x 12"	"	0.070
2" x 6"	"	0.070
2" x 12"	"	0.073
Wire hook hangers		
Black wire, 1/2" x		
4"	EA.	0.040
6"	"	0.042
8"	"	0.044
10"	"	0.044
12"	"	0.047
3/4" x		
4"	EA.	0.042
6"	"	0.044
8"	"	0.047
10"	"	0.050
12"	"	0.053
1" x		
4"	EA.	0.044
6"	"	0.047
8"	"	0.050
10"	"	0.053
12"	"	0.057
1-1/4" x		

15 MECHANICAL

Basic Materials	UNIT	MAN/HOURS
15140.11 Pipe Hangers, Light *(Cont.)*		
4"	EA.	0.047
6"	"	0.050
8"	"	0.053
10"	"	0.057
12"	"	0.062
1-1/2" x		
6"	EA.	0.053
8"	"	0.057
10"	"	0.062
12"	"	0.067
2" x		
6"	EA.	0.057
8"	"	0.062
10"	"	0.067
12"	"	0.073
Copper wire hooks		
1/2" x		
4"	EA.	0.040
6"	"	0.042
8"	"	0.044
10"	"	0.047
12"	"	0.050
3/4" x		
4"	EA.	0.042
6"	"	0.044
8"	"	0.047
10"	"	0.050
12"	"	0.053
1" x		
4"	EA.	0.044
6"	"	0.047
8"	"	0.050
10"	"	0.053
12"	"	0.057
1-1/4" x		
6"	EA.	0.047
8"	"	0.050
10"	"	0.053
12"	"	0.057
1-1/2" x		
6"	EA.	0.053
8"	"	0.057
10"	"	0.062
12"	"	0.067
2" x		
6"	EA.	0.057
8"	"	0.062
10"	"	0.067
12"	"	0.073

Basic Materials	UNIT	MAN/HOURS
15175.60 Expansion Tanks		
Expansion tank, 125 psi, steel		
20 gallon	EA.	1.000
30 gallon	"	1.333
65 gallon	"	2.000
80 gallon	"	2.286
15240.10 Vibration Control		
Vibration isolator, in-line, stainless connector		
1/2"	EA.	0.444
3/4"	"	0.471
1"	"	0.500
1-1/4"	"	0.533
1-1/2"	"	0.571
2"	"	0.615
2-1/2"	"	0.667
3"	"	0.727
4"	"	0.800
6"	"	0.889
Flanged		
8"	EA.	1.000
10"	"	1.143
12"	"	1.333

Insulation	UNIT	MAN/HOURS
15260.10 Fiberglass Pipe Insulation		
Fiberglass insulation on 1/2" pipe		
1" thick	L.F.	0.027
1-1/2" thick	"	0.033
3/4" pipe		
1" thick	L.F.	0.027
1-1/2" thick	"	0.033
1" pipe		
1" thick	L.F.	0.027
1-1/2" thick	"	0.033
2" thick	"	0.040
1-1/4" pipe		
1" thick	L.F.	0.033
1-1/2" thick	"	0.036
2" thick	"	0.040
1-1/2" pipe		
1" thick	L.F.	0.033
1-1/2" thick	"	0.036
2" thick	"	0.038
2-1/2" thick	"	0.040
3-1/2" thick	"	0.042

15 MECHANICAL

Insulation		UNIT	MAN/HOURS
15260.10	**Fiberglass Pipe Insulation** (Cont.)		
2" pipe			
	1" thick	L.F.	0.033
	1-1/2" thick	"	0.036
	2" thick	"	0.040
	2-1/2" thick	"	0.044
	3-1/2" thick	"	0.050
2-1/2" pipe			
	1" thick	L.F.	0.033
	1-1/2" thick	"	0.036
	2" thick	"	0.040
	2-1/2" thick	"	0.044
	3" thick	"	0.050
	3-1/2" thick	"	0.057
3" pipe			
	1" thick	L.F.	0.038
	1-1/2" thick	"	0.040
	2" thick	"	0.044
	2-1/2" thick	"	0.050
	3" thick	"	0.057
	3-1/2" thick	"	0.067
4" pipe			
	1" thick	L.F.	0.038
	1-1/2" thick	"	0.040
	2" thick	"	0.044
	2-1/2" thick	"	0.050
	3" thick	"	0.057
	3-1/2" thick	"	0.067
5" pipe			
	1" thick	L.F.	0.038
	2" thick	"	0.040
	3" thick	"	0.047
	4" thick	"	0.062
6" pipe			
	1" thick	L.F.	0.042
	2" thick	"	0.044
	4" thick	"	0.053
	6" thick	"	0.073
8" pipe			
	2" thick	L.F.	0.042
	3" thick	"	0.044
	4" thick	"	0.053
	6" thick	"	0.073
10" pipe			
	2" thick	L.F.	0.042
	3" thick	"	0.044
	4" thick	"	0.053
	6" thick	"	0.073
12" pipe			
	2" thick	L.F.	0.042
	3" thick	"	0.044
	4" thick	"	0.053
	6" thick	"	0.073

Insulation		UNIT	MAN/HOURS
15260.20	**Calcium Silicate**		
Calcium silicate insulation, 6" pipe			
	2" thick	L.F.	0.057
	2-1/2" thick	"	0.062
	3" thick	"	0.067
	4" thick	"	0.073
	6" thick	"	0.080
8" pipe			
	2" thick	L.F.	0.062
	2-1/2" thick	"	0.067
	3" thick	"	0.073
	4" thick	"	0.080
	6" thick	"	0.089
10" pipe			
	2" thick	L.F.	0.062
	2-1/2" thick	"	0.067
	3" thick	"	0.073
	4" thick	"	0.080
	6" thick	"	0.089
12" pipe			
	2" thick	L.F.	0.062
	2-1/2" thick	"	0.067
	3" thick	"	0.073
	4" thick	"	0.080
	6" thick	"	0.089
15260.60	**Exterior Pipe Insulation**		
Fiberglass insulation, aluminum jacket			
1/2" pipe			
	1" thick	L.F.	0.062
	1-1/2" thick	"	0.067
3/4" pipe			
	1" thick	L.F.	0.062
	1-1/2" thick	"	0.067
1" pipe			
	1" thick	L.F.	0.062
	1-1/2" thick	"	0.067
	3-1/2" thick	"	0.089
1-1/4" pipe			
	1" thick	L.F.	0.073
	1-1/2" thick	"	0.076
	3-1/2" thick	"	0.089
1-1/2" pipe			
	1" thick	L.F.	0.073
	1-1/2" thick	"	0.076
	2" thick	"	0.080
	2-1/2" thick	"	0.084
	3-1/2" thick	"	0.089
2" pipe			
	1" thick	L.F.	0.073
	1-1/2" thick	"	0.076
	2" thick	"	0.080
	2-1/2" thick	"	0.084
	3-1/2" thick	"	0.089

15 MECHANICAL

Insulation	UNIT	MAN/HOURS
15260.60 **Exterior Pipe Insulation** *(Cont.)*		
2-1/2" pipe		
1" thick	L.F.	0.073
1-1/2" thick	"	0.076
2" thick	"	0.080
2-1/2" thick	"	0.084
3" thick	"	0.089
3-1/2" thick	"	0.094
3" pipe		
1" thick	L.F.	0.080
1-1/2" thick	"	0.084
2" thick	"	0.089
2-1/2" thick	"	0.094
3" thick	"	0.100
3-1/2" thick	"	0.107
4" pipe		
1" thick	L.F.	0.080
1-1/2" thick	"	0.084
2" thick	"	0.089
2-1/2" thick	"	0.094
3" thick	"	0.100
3-1/2" thick	"	0.107
5" pipe		
1" thick	L.F.	0.080
2" thick	"	0.084
2-1/2" thick	"	0.089
3" thick	"	0.094
3-1/2" thick	"	0.100
6" pipe		
1" thick	L.F.	0.089
2" thick	"	0.094
2-1/2" thick	"	0.100
3" thick	"	0.107
3-1/2" thick	"	0.114
4" thick	"	0.123
6" thick	"	0.133
8" pipe		
2" thick	L.F.	0.089
3" thick	"	0.094
3-1/2" thick	"	0.100
4" thick	"	0.107
6" thick	"	0.114
10" pipe		
2" thick	L.F.	0.089
3" thick	"	0.094
3-1/2" thick	"	0.100
4" thick	"	0.107
6" thick	"	0.114
12" pipe		
2" thick	L.F.	0.089
3" thick	"	0.094
3-1/2" thick	"	0.100
4" thick	"	0.107
6" thick	"	0.114

Insulation	UNIT	MAN/HOURS
15260.60 **Exterior Pipe Insulation** *(Cont.)*		
Calcium silicate with aluminum jacket, 6" pipe		
2" thick	L.F.	0.100
2-1/2" thick	"	0.107
3" thick	"	0.114
4" thick	"	0.133
8" pipe		
2" thick	L.F.	0.100
2-1/2" thick	"	0.107
3" thick	"	0.114
4" thick	"	0.123
6" thick	"	0.133
10" pipe		
2" thick	L.F.	0.114
2-1/2" thick	"	0.123
3" thick	"	0.133
4" thick	"	0.145
6" thick	"	0.160
12" pipe		
2" thick	L.F.	0.114
2-1/2" thick	"	0.123
3" thick	"	0.133
4" thick	"	0.145
6" thick	"	0.160
15260.90 **Pipe Insulation Fittings**		
Insulation protection saddle		
1" thick covering		
1/2" pipe	EA.	0.320
3/4" pipe	"	0.320
1" pipe	"	0.320
1-1/4" pipe	"	0.320
1-1/2" pipe	"	0.320
2" pipe	"	0.320
2-1/2" pipe	"	0.320
3" pipe	"	0.364
4" pipe	"	0.400
6" pipe	"	0.500
1-1/2" thick covering		
3/4" pipe	EA.	0.320
1" pipe	"	0.320
1-1/4" pipe	"	0.320
1-1/2" pipe	"	0.320
2" pipe	"	0.320
2-1/2" pipe	"	0.320
3" pipe	"	0.320
4" pipe	"	0.400
6" pipe	"	0.500
8" pipe	"	0.667
10" pipe	"	0.667
12" pipe	"	0.667
2" thick covering		
3/4" pipe	EA.	0.320
1" pipe	"	0.320

15 MECHANICAL

Insulation	UNIT	MAN/HOURS
15260.90 Pipe Insulation Fittings (Cont.)		
1-1/4" pipe	EA.	0.320
1-1/2" pipe	"	0.320
2" pipe	"	0.320
2-1/2" pipe	"	0.320
3" pipe	"	0.320
4" pipe	"	0.400
6" pipe	"	0.500
8" pipe	"	0.667
10" pipe	"	0.667
12" pipe	"	0.667
3" thick covering		
2" pipe	EA.	0.320
2-1/2" pipe	"	0.320
3" pipe	"	0.320
4" pipe	"	0.400
6" pipe	"	0.500
8" pipe	"	0.667
10" pipe	"	0.667
12" pipe	"	0.667
15280.10 Equipment Insulation		
Equipment insulation, 2" thick, cellular glass	S.F.	0.050
Urethane, rigid, jacket, plastered finish	"	0.100
Fiberglass, rigid, with vapor barrier	"	0.044
15290.10 Ductwork Insulation		
Fiberglass duct insulation, plain blanket		
1-1/2" thick	S.F.	0.010
2" thick	"	0.013
With vapor barrier		
1-1/2" thick	S.F.	0.010
2" thick	"	0.013
Rigid with vapor barrier		
2" thick	S.F.	0.027
3" thick	"	0.032
4" thick	"	0.040
6" thick	"	0.053
Weatherproof, poly, 3" thick, w/vapor barrier	"	0.080
Urethane board with vapor barrier	"	0.100

Fire Protection	UNIT	MAN/HOURS
15330.10 Wet Sprinkler System		
Sprinkler head, 212 deg, brass, exposed piping	EA.	0.320
Chrome, concealed piping	"	0.444
Water motor alarm	"	1.333
Fire department inlet connection	"	1.600
Wall plate for fire dept connection	"	0.667
Swing check valve flanged iron body, 4"	"	2.667
Check valve, 6"	"	4.000
Wet pipe valve, flange to groove, 4"	"	0.889
Flange to flange		
6"	EA.	1.333
8"	"	2.667
Alarm valve, flange to flange, (wet valve)		
4"	EA.	0.889
8"	"	6.667
Inspector's test connection	"	0.667
Wall hydrant, polished brass, 2-1/2" x 2-1/2", single	"	0.571
2-way	"	0.571
3-way	"	0.571
Wet valve trim, includes retard chamber & gauges, 4"-6"	"	0.667
Retard pressure switch for wet systems	"	1.600
Air maintenance device	"	0.667
Wall hydrant non-freeze, 8" thick wall, vacuum breaker	"	0.400
12" thick wall	"	0.400
15330.50 Dry Sprinkler System		
Dry pipe valve, flange to flange		
4"	EA.	1.600
6"	"	2.000
Trim, 4" and 6", includes gauges	"	0.667
Field testing and flushing	"	6.667
Disinfection	"	6.667
Pressure switch double circuit, open/close contacts	"	2.000
Low air		
Supervisory unit	EA.	1.333
Pressure switch	"	0.667
15330.70 CO_2 System		
CO_2 system, high pressure, 75# cylinder with		
Valve assemblies	EA.	1.600
Storage rack	"	1.143
Manifold	"	5.714
Flexible loops	"	0.100
Beam scale for cylinders	"	1.333
Mechanically control head	"	0.533
Electrically control head	"	0.533
Stop valves	"	0.800
Check valves	"	1.000
Activation station	"	0.800
Nozzles	"	0.667
Hose reel with 75' of 3/4" hose	"	4.000
Main/reserve transfer switch	"	1.333
Pressure switch	"	0.800
Heat responsive device	"	1.333

15 MECHANICAL

Fire Protection	UNIT	MAN/HOURS
15330.70	**CO₂ System** *(Cont.)*	
Battery and charger	EA.	4.000
Low pressure		
Battery and charger	EA.	4.000
Pressure switch	"	0.889
Nozzles	"	0.727
Master selector valve	"	1.333
Selector valve	"	1.333
Low pressure hose reel with 75' of 3/4" hose	"	4.000
Tank fill lines	"	1.000
Activation stations	"	0.667
Electro manual pilot panels	"	1.000

Plumbing	UNIT	MAN/HOURS
15410.05	**C.i. Pipe, Above Ground**	
No hub pipe		
1-1/2" pipe	L.F.	0.057
2" pipe	"	0.067
3" pipe	"	0.080
4" pipe	"	0.133
6" pipe	"	0.160
8" pipe	"	0.267
10" pipe	"	0.320
No hub fittings, 1-1/2" pipe		
1/4 bend	EA.	0.267
1/8 bend	"	0.267
Sanitary tee	"	0.400
Sanitary cross	"	0.400
Wye	"	0.400
Tapped tee	"	0.267
P-trap	"	0.267
Tapped cross	"	0.267
2" pipe		
1/4 bend	EA.	0.320
1/8 bend	"	0.320
Sanitary tee	"	0.533
Sanitary cross	"	0.533
Wye	"	0.667
Double wye	"	0.667
2x1-1/2" wye & 1/8 bend	"	0.500
Double wye & 1/8 bend	"	0.667
Test tee less 2" plug	"	0.320
Tapped tee		
2"x2"	EA.	0.320
2"x1-1/2"	"	0.320
P-trap		

Plumbing	UNIT	MAN/HOURS
15410.05	**C.i. Pipe, Above Ground** *(Cont.)*	
2"x2"	EA.	0.320
Tapped cross		
2"x1-1/2"	EA.	0.320
3" pipe		
1/4 bend	EA.	0.400
1/8 bend	"	0.400
Sanitary tee	"	0.500
3"x2" sanitary tee	"	0.500
3"x1-1/2" sanitary tee	"	0.500
Sanitary cross	"	0.667
3x2" sanitary cross	"	0.667
Wye	"	0.667
3x2" wye	"	0.667
Double wye	"	0.667
3x2" double wye	"	0.667
3x2" wye & 1/8 bend	"	0.571
3x1-1/2" wye & 1/8 bend	"	0.571
Double wye & 1/8 bend	"	0.667
3x2" double wye & 1/8 bend	"	0.667
3x2" reducer	"	0.364
Test tee, less 3" plug	"	0.400
3x3" tapped tee	"	0.400
3x2" tapped tee	"	0.400
3x1-1/2" tapped tee	"	0.400
P-trap	"	0.400
3x2" tapped cross	"	0.400
3x1-1/2" tapped cross	"	0.400
Closet flange, 3-1/2" deep	"	0.200
4" pipe		
1/4 bend	EA.	0.400
1/8 bend	"	0.400
Sanitary tee	"	0.667
4x3" sanitary tee	"	0.667
4x2" sanitary tee	"	0.667
Sanitary cross	"	0.800
4x3" sanitary cross	"	0.800
4x2" sanitary cross	"	0.800
Wye	"	0.667
4x3" wye	"	0.667
4x2" wye	"	0.667
Double wye	"	0.800
4x3" double wye	"	0.800
4x2" double wye	"	0.800
Wye & 1/8 bend	"	0.667
4x3" wye & 1/8 bend	"	0.667
4x2" wye & 1/8 bend	"	0.667
Double wye & 1/8 bend	"	0.800
4x3" double wye & 1/8 bend	"	0.800
4x2" double wye & 1/8 bend	"	0.800
4x3" reducer	"	0.400
4x2" reducer	"	0.400
Test tee, less 4" plug	"	0.400
4x2" tapped tee	"	0.400

15 MECHANICAL

Plumbing	UNIT	MAN/HOURS
15410.05 **C.i. Pipe, Above Ground** *(Cont.)*		
4x1-1/2" tapped tee	EA.	0.400
P-trap	"	0.400
4x2" tapped cross	"	0.400
4x1-1/2" tapped cross	"	0.400
Closet flange		
3" deep	EA.	0.400
8" deep	"	0.400
6" pipe		
1/4 bend	EA.	0.667
1/8 bend	"	0.667
Sanitary tee	"	0.800
6x4" sanitary tee	"	0.800
Wye	"	0.800
6x4" wye	"	0.800
6x3" wye	"	0.800
6x2" wye	"	0.800
Double wye	"	1.000
6x4" double wye	"	1.000
Wye & 1/8 bend	"	0.800
6x4" wye & 1/8 bend	"	0.800
6x3" wye & 1/8 bend	"	0.800
6x2" wye & 1/8 bend	"	0.800
6x4" reducer	"	0.444
6x3" reducer	"	0.444
6x2" reducer	"	0.400
Test tee		
Less 6" plug	EA.	0.500
Plug	"	
P-trap	"	0.500
8" pipe		
1/4 bend	EA.	0.667
1/8 bend	"	0.667
Sanitary tee	"	1.000
8x6" sanitary tee	"	1.000
Wye	"	0.800
8x6" wye	"	0.667
8x4" wye	"	0.667
Double wye	"	0.667
Wye & 1/8 bend	"	0.667
8x6" wye & 1/8 bend	"	0.667
8x4" wye & 1/8 bend	"	0.667
8x6" reducer	"	0.667
8x4" reducer	"	0.667
8x3" reducer	"	0.667
8x2" reducer	"	0.444
Test tee		
Less 8" plug	EA.	0.667
Plug	"	
10" pipe		
1/4 bend	EA.	0.667
1/8 bend	"	0.667
Wye	"	1.333

Plumbing	UNIT	MAN/HOURS
15410.05 **C.i. Pipe, Above Ground** *(Cont.)*		
10x8" wye	EA.	1.333
10x6" wye	"	1.333
10x4" wye	"	1.333
10x8" reducer	"	0.667
10x6" reducer	"	0.667
10x4" reducer	"	0.667
15410.06 **C.i. Pipe, Below Ground**		
No hub pipe		
1-1/2" pipe	L.F.	0.040
2" pipe	"	0.044
3" pipe	"	0.050
4" pipe	"	0.067
6" pipe	"	0.073
8" pipe	"	0.089
10" pipe	"	0.100
Fittings, 1-1/2"		
1/4 bend	EA.	0.229
1/8 bend	"	0.229
Wye	"	0.320
Wye & 1/8 bend	"	0.229
P-trap	"	0.229
2"		
1/4 bend	EA.	0.267
1/8 bend	"	0.267
Double wye	"	0.500
Wye & 1/8 bend	"	0.400
Double wye & 1/8 bend	"	0.500
P-trap	"	0.267
3"		
1/4 bend	EA.	0.320
1/8 bend	"	0.320
Wye	"	0.500
3x2" wye	"	0.500
Wye & 1/8 bend	"	0.500
Double wye & 1/8 bend	"	0.500
3x2" double wye & 1/8 bend	"	0.500
3x2" reducer	"	0.320
P-trap	"	0.320
4"		
1/4 bend	EA.	0.320
1/8 bend	"	0.320
Wye	"	0.500
4x3" wye	"	0.500
4x2" wye	"	0.500
Double wye	"	0.667
4x3" double wye	"	0.667
4x2" double wye	"	0.667
Wye & 1/8 bend	"	0.500
4x3" wye & 1/8 bend	"	0.500
4x2" wye & 1/8 bend	"	0.500
Double wye & 1/8 bend	"	0.667
4x3" double wye & 1/8 bend	"	0.667

15 MECHANICAL

Plumbing	UNIT	MAN/HOURS
15410.06 C.i. Pipe, Below Ground (Cont.)		
4x2" double wye & 1/8 bend	EA.	0.667
4x3" reducer	"	0.320
4x2" reducer	"	0.320
6"		
1/4 bend	EA.	0.500
1/8 bend	"	0.500
Wye & 1/8 bend	"	0.667
6x4" wye & 1/8 bend	"	0.667
6x3" wye & 1/8 bend	"	0.667
6x2" wye & 1/8 bend	"	0.667
6x3" reducer	"	0.364
P-trap	"	0.400
8"		
1/4 bend	EA.	0.500
1/8 bend	"	0.500
Wye	"	0.667
8x6" wye	"	0.500
8x4" wye	"	0.500
8x6" wye & 1/8 bend	"	0.500
8x4" reducer	"	0.400
8x3" reducer	"	0.400
8x2" reducer	"	0.400
10"		
1/4 bend	EA.	0.500
1/8 bend	"	0.500
Wye	"	1.000
10x8" wye	"	1.000
10x6" wye	"	1.000
10x4" wye	"	1.000
10x8" reducer	"	0.500
10x6" reducer	"	0.500
15410.08 Extra Heavy Soil Pipe		
Extra heavy soil pipe, single hub		
2" x 5'	EA.	0.160
3" x 5'	"	0.170
4" x 5'	"	0.186
6" x 5'	"	0.200
Double hub		
2" x 5'	EA.	0.200
4" x 5'	"	0.211
5" x 5'	"	0.222
6" x 5'	"	0.229
Single hub		
3" x 10'	EA.	0.133
4" x 10'	"	0.138
6" x 10'	"	0.145
8" x 10'	"	0.154
"Mini", single hub		
2" x 42"	EA.	0.160
3" x 42"	"	0.170
4" x 42"	"	0.186
6" x 42"	"	0.200

Plumbing	UNIT	MAN/HOURS
15410.08 Extra Heavy Soil Pipe (Cont.)		
8" x 42"	EA.	0.211
Fittings, 1/4" bend		
2"	EA.	0.267
3"	"	0.320
4"	"	0.320
1/8 bend		
2"	EA.	0.267
3"	"	0.320
4"	"	0.320
5"	"	0.400
6"	"	0.500
8"	"	0.500
Long sweep		
2"	EA.	0.267
3"	"	0.320
4"	"	0.320
Straight T		
4" x 2"	EA.	0.500
4" x 3"	"	0.500
4"	"	0.571
Sanitary T		
3" x 2"	EA.	0.500
3"	"	0.500
4" x 2"	"	0.571
4" x 3"	"	0.571
4"	"	0.615
Wye		
4" x 2"	EA.	0.571
4" x 3"	"	0.571
4"	"	0.615
Combination Y and 1/8 bend, 4"	"	0.615
Double wye, 4"	"	0.615
Tapped sanitary T, 4" x 2"	"	0.667
P trap		
2"	EA.	0.320
4"	"	0.533
Dandy		
4", with 3" brass plug	EA.	0.444
6", 4" brass plug	"	0.500
15410.09 Service Weight Pipe		
Service weight pipe, single hub		
2" x 5'	EA.	0.160
3" x 5'	"	0.170
4" x 5'	"	0.178
5" x 5'	"	0.190
6" x 5'	"	0.200
8" x 5'	"	0.211
10" x 5'	"	0.222
12" x 5'	"	0.250
Double hub		
2" x 5'	EA.	0.200
3" x 5'	"	0.216

15 MECHANICAL

Plumbing	UNIT	MAN/HOURS
15410.09 Service Weight Pipe *(Cont.)*		
4" x 5'	EA.	0.229
5" x 5'	"	0.250
6" x 5'	"	0.267
10" x 5'	"	0.286
12' x 5'	"	0.308
Single hub		
2" x 10'	EA.	0.200
3" x 10'	"	0.216
4" x 10'	"	0.229
5" x 10'	"	0.250
6" x 10'	"	0.267
8" x 10'	"	0.286
10" x 10'	"	0.308
12" x 10'	"	0.333
Shorty		
2" x 42"	EA.	0.160
3" x 42"	"	0.170
4" x 42"	"	0.178
5" x 42"	"	0.190
6" x 42"	"	0.200
8" x 42"	"	0.211
10" x 42"	"	0.222
Soil plug		
2"	EA.	0.267
3"	"	0.286
4"	"	0.308
5"	"	0.320
6"	"	0.333
1/5 bend		
3"	EA.	0.320
4"	"	0.364
1/6 bend		
2"	EA.	0.267
3"	"	0.320
4"	"	0.364
1/8 bend		
2"	EA.	0.267
3"	"	0.320
4"	"	0.364
5"	"	0.381
6"	"	0.400
8"	"	0.500
10"	"	0.571
12"	"	0.667
1/16 bend		
2"	EA.	0.267
3"	"	0.320
4"	"	0.364
5"	"	0.381
6"	"	0.400
1/4 bend		
2"	EA.	0.267
3"	"	0.320

Plumbing	UNIT	MAN/HOURS
15410.09 Service Weight Pipe *(Cont.)*		
4"	EA.	0.364
5"	"	0.381
6"	"	0.400
8"	"	0.500
10"	"	0.571
12"	"	0.667
1/4 bend, long		
2" x 12"	EA.	0.286
4" x 18"	"	0.400
4" x 12"	"	0.400
Sweep		
2"	EA.	0.267
3"	"	0.320
4"	"	0.364
5"	"	0.381
6"	"	0.400
8"	"	0.500
Reducing long sweep		
3" x 2"	EA.	0.320
4" x 3"	"	0.400
Straight T		
2"	EA.	0.500
3"	"	0.533
4" x 2"	"	0.571
4" x 3"	"	0.615
4"	"	0.667
Sanitary T		
2"	EA.	0.500
3" x 2"	"	0.533
3"	"	0.571
4" x 2"	"	0.615
4" x 3"	"	0.667
4"	"	0.667
5"	"	0.727
6"	"	0.727
Wye		
2"	EA.	0.400
3" x 2"	"	0.444
3"	"	0.444
4" x 2"	"	0.471
4" x 3"	"	0.471
4"	"	0.471
5" x 2"	"	0.500
5" x 3"	"	0.500
5" x 4"	"	0.500
5"	"	0.500
6" x 2"	"	0.500
6" x 3"	"	0.500
6" x 4"	"	0.533
6" x 5"	"	0.533
6"	"	0.571
8" x 4"	"	0.571
8" x 6"	"	0.615

15 MECHANICAL

Plumbing	UNIT	MAN/HOURS
15410.09 Service Weight Pipe *(Cont.)*		
8"	EA.	0.667
10" x 4"	"	0.667
10" x 6"	"	0.727
10" x 8"	"	0.800
Service wye		
10"	EA.	1.000
12" x 4"	"	1.000
12" x 6"	"	1.000
12" x 8"	"	1.143
12"	"	1.333
Combination wye and 1/8 bend		
2"	EA.	0.533
3" x 2"	"	0.571
3"	"	0.571
4" x 2"	"	0.615
4" x 3"	"	0.615
4"	"	0.667
5" x 4"	"	0.667
6" x 4"	"	0.800
8"	"	1.000
Straight cross, 4"	"	0.667
Sanitary cross		
2"	EA.	0.571
3"	"	0.615
3" x 2"	"	0.615
4"	"	0.667
4" x 3"	"	0.667
4" x 2"	"	0.727
Double wye		
2"	EA.	0.571
3" x 2"	"	0.615
3"	"	0.615
4" x 2"	"	0.667
4" x 3"	"	0.667
4"	"	0.667
5"	"	0.800
6" x 4"	"	1.000
Combination double wye and 1/8 bend		
2"	EA.	0.571
3" x 2"	"	0.615
3"	"	0.615
4" x 3"	"	0.667
4"	"	0.667
Tapped sanitary T		
2" x 1-1/2"	EA.	0.571
2" x 2"	"	0.571
3" x 1-1/2"	"	0.615
3" x 2"	"	0.615
4" x 1-1/2"	"	0.667
4" x 2"	"	0.667
Tapped straight T		
3" x 1-1/2"	EA.	0.615
3" x 2"	"	0.615

Plumbing	UNIT	MAN/HOURS
15410.09 Service Weight Pipe *(Cont.)*		
4" x 2"	EA.	0.667
4" x 1-1/2"	"	0.667
Tapped Y		
4" x 2"	EA.	0.667
Tapped sanitary cross		
2" x 1-1/2"	EA.	0.571
2" x 2"	"	0.615
3" x 1-1/2"	"	0.615
3" x 2"	"	0.615
3" x 3"	"	0.615
4" x 1-1/2"	"	0.667
4" x 2"	"	0.667
Cleanout, dandy, with brass plug		
2", 1-1/2" plug	EA.	0.571
3", 2" plug	"	0.615
4", 3" plug	"	0.667
5", 4" plug	"	0.727
6", 4" plug	"	0.800
8", 6" plug	"	1.000
Reducer, 3" x 2"	"	0.571
4" x		
2"	EA.	0.615
3"	"	0.615
5" x		
2"	EA.	0.667
3"	"	0.667
4"	"	0.667
6" x		
2"	EA.	0.800
3"	"	0.800
4"	"	0.800
5"	"	0.800
8" x		
4"	EA.	1.000
6"	"	1.000
10" x		
4"	EA.	1.143
6"	"	1.143
8"	"	1.143
12" x		
4"	EA.	1.333
6"	"	1.333
8"	"	1.333
10"	"	1.333
P trap		
2"	EA.	0.571
3"	"	0.615
4"	"	0.667
5"	"	0.800
6"	"	1.000

15 MECHANICAL

Plumbing		UNIT	MAN/HOURS
15410.10	**Copper Pipe**		
Type "K" copper			
1/2"		L.F.	0.025
3/4"		"	0.027
1"		"	0.029
1-1/4"		"	0.031
1-1/2"		"	0.033
2"		"	0.036
2-1/2"		"	0.040
3"		"	0.042
4"		"	0.044
DWV, copper			
1-1/4"		L.F.	0.033
1-1/2"		"	0.036
2"		"	0.040
3"		"	0.044
4"		"	0.050
6"		"	0.057
Refrigeration tubing, copper, sealed			
1/8"		L.F.	0.032
3/16"		"	0.033
1/4"		"	0.035
5/16"		"	0.036
3/8"		"	0.038
1/2"		"	0.040
7/8"		"	0.046
1-1/8"		"	0.053
1-3/8"		"	0.062
Type "L" copper			
1/4"		L.F.	0.024
3/8"		"	0.024
1/2"		"	0.025
3/4"		"	0.027
1"		"	0.029
1-1/4"		"	0.031
1-1/2"		"	0.033
2"		"	0.036
2-1/2"		"	0.040
3"		"	0.042
3-1/2"		"	0.043
4"		"	0.044
Type "M" copper			
1/2"		L.F.	0.025
3/4"		"	0.027
1"		"	0.029
1-1/4"		"	0.031
2"		"	0.036
2-1/2"		"	0.040
3"		"	0.042
4"		"	0.044

Plumbing		UNIT	MAN/HOURS
15410.11	**Copper Fittings**		
Coupling, with stop			
1/4"		EA.	0.267
3/8"		"	0.320
1/2"		"	0.348
5/8"		"	0.400
3/4"		"	0.444
1"		"	0.471
3"		"	0.800
4"		"	1.000
Reducing coupling			
1/4" x 1/8"		EA.	0.320
3/8" x 1/4"		"	0.348
1/2" x			
3/8"		EA.	0.400
1/4"		"	0.400
1/8"		"	0.400
3/4" x			
3/8"		EA.	0.444
1/2"		"	0.444
1" x			
3/8"		EA.	0.500
1" x 1/2"		"	0.500
1" x 3/4"		"	0.500
1-1/4" x			
1/2"		EA.	0.533
3/4"		"	0.533
1"		"	0.533
1-1/2" x			
1/2"		EA.	0.571
3/4"		"	0.571
1"		"	0.571
1-1/4"		"	0.571
2" x			
1/2"		EA.	0.667
3/4"		"	0.667
1"		"	0.667
1-1/4"		"	0.667
1-1/2"		"	0.667
2-1/2" x			
1"		EA.	0.800
1-1/4"		"	0.800
1-1/2"		"	0.800
2"		"	0.800
3" x			
1-1/2"		EA.	1.000
2"		"	1.000
2-1/2"		"	1.000
4" x			
2"		EA.	1.143
2-1/2"		"	1.143
3"		"	1.143
Slip coupling			
1/4"		EA.	0.267

15 MECHANICAL

Plumbing	UNIT	MAN/HOURS
15410.11 Copper Fittings (Cont.)		
1/2"	EA.	0.320
3/4"	"	0.400
1"	"	0.444
1-1/4"	"	0.500
1-1/2"	"	0.533
2"	"	0.667
2-1/2"	"	0.667
3"	"	0.800
4"	"	1.000
Coupling with drain		
1/2"	EA.	0.400
3/4"	"	0.444
1"	"	0.500
Reducer		
3/8" x 1/4"	EA.	0.320
1/2" x 3/8"	"	0.320
3/4" x		
1/4"	EA.	0.364
3/8"	"	0.364
1/2"	"	0.364
1" x		
1/2"	EA.	0.400
3/4"	"	0.400
1-1/4" x		
1/2"	EA.	0.444
3/4"	"	0.444
1"	"	0.444
1-1/2" x		
1/2"	EA.	0.500
3/4"	"	0.500
1"	"	0.500
1-1/4"	"	0.500
2" x		
1/2"	EA.	0.571
3/4"	"	0.571
1"	"	0.571
1-1/4"	"	0.571
1-1/2"	"	0.571
2-1/2" x		
1"	EA.	0.667
1-1/4"	"	0.667
1-1/2"	"	0.667
2"	"	0.667
3" x		
1-1/4"	EA.	0.800
1-1/2"	"	0.800
2"	"	0.800
2-1/2"	"	0.800
4" x		
2"	EA.	1.000
3"	"	1.000
Female adapters		
1/4"	EA.	0.320

Plumbing	UNIT	MAN/HOURS
15410.11 Copper Fittings (Cont.)		
3/8"	EA.	0.364
1/2"	"	0.400
3/4"	"	0.444
1"	"	0.444
1-1/4"	"	0.500
1-1/2"	"	0.500
2"	"	0.533
2-1/2"	"	0.571
3"	"	0.667
4"	"	0.800
Increasing female adapters		
1/8" x		
3/8"	EA.	0.320
1/2"	"	0.320
1/4" x 1/2"	"	0.348
3/8" x 1/2"	"	0.364
1/2" X		
3/4"	EA.	0.400
1"	"	0.400
3/4" X		
1"	EA.	0.444
1-1/4"	"	0.444
1" x		
1-1/4"	EA.	0.444
1-1/2"	"	0.444
1-1/4" x		
1-1/2"	EA.	0.500
2"	"	0.500
1-1/2" x 2"	"	0.533
Reducing female adapters		
3/8" x 1/4"	EA.	0.364
1/2" x		
1/4"	EA.	0.400
3/8"	"	0.400
3/4" x 1/2"	"	0.444
1" x		
1/2"	EA.	0.444
3/4"	"	0.444
1-1/4" x		
1/2"	EA.	0.500
3/4"	"	0.500
1"	"	0.500
1-1/2" x		
1"	EA.	0.533
1-1/4"	"	0.533
2" x		
1"	EA.	0.571
1-1/4"	"	0.571
1-1/2"	"	0.571
Female fitting adapters		
1/2"	EA.	0.400
3/4"	"	0.400
3/4" x 1/2"	"	0.421

15 MECHANICAL

Plumbing	UNIT	MAN/HOURS
15410.11 Copper Fittings *(Cont.)*		
1"	EA.	0.444
1-1/4"	"	0.471
1-1/2"	"	0.500
2"	"	0.533
Male adapters		
1/4"	EA.	0.364
3/8"	"	0.364
3"	"	0.667
4"	"	0.800
Increasing male adapters		
3/8" x 1/2"	EA.	0.364
1/2" x		
3/4"	EA.	0.400
1"	"	0.400
3/4" x		
1"	EA.	0.421
1-1/4"	"	0.421
1" x 1-1/4"	"	0.444
1-1/2" x		
3/4"	EA.	0.471
1"	"	0.471
1-1/4"	"	0.471
2" x		
1"	EA.	0.500
1-1/4"	"	0.500
1-1/2"	"	0.500
2" x 2-1/2"	"	0.533
Reducing male adapters		
1/2" x		
1/4"	EA.	0.400
3/8"	"	0.400
3/4" x 1/2"	"	0.421
1" x		
1/2"	EA.	0.444
3/4"	"	0.444
1-1/4" x		
3/4"	EA.	0.471
1"	"	0.471
1-1/2" x		
3/4"	EA.	0.500
1"	"	0.500
1-1/4'	"	0.500
2" x		
3/4"	EA.	0.533
1"	"	0.533
1-1/4"	"	0.533
1-1/2"	"	0.533
2-1/2" x		
2"	EA.	0.615
Fitting x male adapters		
1/2"	EA.	0.400
3/4"	"	0.421
1"	"	0.444

Plumbing	UNIT	MAN/HOURS
15410.11 Copper Fittings *(Cont.)*		
1-1/4"	EA.	0.471
1-1/2"	"	0.500
2"	"	0.533
90 ells		
1/8"	EA.	0.320
1/4"	"	0.320
3/8"	"	0.364
1/2"	"	0.400
3/4"	"	0.421
1"	"	0.444
1-1/4"	"	0.471
1-1/2"	"	0.500
2"	"	0.533
2-1/2"	"	0.615
3"	"	0.667
4"	"	0.800
Reducing 90 ell		
3/8" x 1/4"	EA.	0.364
1/2" x		
1/4"	EA.	0.400
3/8"	"	0.400
3/4" x 1/2"	"	0.421
1" x		
1/2"	EA.	0.444
3/4"	"	0.444
1-1/4" x		
1/2"	EA.	0.471
3/4"	"	0.471
1"	"	0.471
1-1/2" x		
1/2"	EA.	0.500
3/4"	"	0.500
1"	"	0.500
1-1/4"	"	0.500
2" x		
3/4"	EA.	0.533
1"	"	0.533
1-1/4"	"	0.533
1-1/2"	"	0.533
2-1/2" x		
1-1/2"	EA.	0.615
2"	"	0.615
3" x		
2"	EA.	0.667
2-1/2"	"	0.667
Street ells, copper		
1/4"	EA.	0.320
3/8"	"	0.364
1/2"	"	0.400
3/4"	"	0.421
1"	"	0.444
1-1/4"	"	0.471
1-1/2"	"	0.500

15 MECHANICAL

Plumbing

15410.11 Copper Fittings (Cont.)

	UNIT	MAN/HOURS
2"	EA.	0.533
2-1/2"	"	0.615
3"	"	0.667
4"	"	0.800
Female, 90 ell		
1/2"	EA.	0.400
3/4"	"	0.421
1"	"	0.444
1-1/4"	"	0.471
1-1/2"	"	0.500
2"	"	0.533
Female increasing, 90 ell		
3/8" x 1/2"	EA.	0.364
1/2" x		
3/4"	EA.	0.400
1"	"	0.400
3/4" x 1"	"	0.421
1" x 1-1/4"	"	0.444
1-1/4" x 1-1/2"	"	0.471
Female reducing, 90 ell		
1/2" x 3/8"	EA.	0.400
3/4" x 1/2"	"	0.421
1" x		
1/2"	EA.	0.444
3/4"	"	0.444
1-1/4" x		
3/4"	EA.	0.471
1"	"	0.471
1-1/2"	"	0.471
1-1/2" x		
1"	EA.	0.500
1-1/4"	"	0.500
2" x 1-1/2"	"	0.533
Male, 90 ell		
1/4"	EA.	0.320
3/8"	"	0.364
1/2"	"	0.400
3/4"	"	0.421
1"	"	0.444
1-1/4"	"	0.471
1-1/2"	"	0.500
2"	"	0.533
Male, increasing 90 ell		
1/2" x		
3/4"	EA.	0.400
1"	"	0.400
3/4" x 1"	"	0.421
1" x 1-1/4"	"	0.444
1-1/4" x 1-1/2"	"	0.471
Male, reducing 90 ell		
1/2" x 3/8"	EA.	0.400
3/4" x 1/2"	"	0.421
1" x		

15410.11 Copper Fittings (Cont.)

	UNIT	MAN/HOURS
1/2"	EA.	0.444
3/4"	"	0.444
1-1/4" x 1"	"	0.471
Drop ear ells		
1/2"	EA.	0.400
Female drop ear ells		
1/2"	EA.	0.400
1/2" x 3/8"	"	0.400
3/4"	"	0.421
Female flanged sink ell		
1/2"	EA.	0.400
45 ells		
1/4"	EA.	0.320
3/8"	"	0.364
3"	"	0.667
4"	"	0.800
45 street ell		
1/4"	EA.	0.320
3/8"	"	0.364
1/2"	"	0.400
3/4"	"	0.421
1"	"	0.444
1-1/2"	"	0.500
2"	"	0.533
2-1/2"	"	0.615
3"	"	0.667
4"	"	0.800
Tee		
1/8"	EA.	0.320
1/4"	"	0.320
3/8"	"	0.364
3"	"	0.667
4"	"	0.800
Caps		
1/4"	EA.	0.320
3/8"	"	0.364
3"	"	0.667
4"	"	0.800
Test caps		
1/2"	EA.	0.400
3/4"	"	0.421
1"	"	0.444
1-1/4"	"	0.471
1-1/2"	"	0.500
2"	"	0.533
3"	"	0.667
Flush bushing		
1/4" x 1/8"	EA.	0.320
1/2" x		
1/4"	EA.	0.400
3/8"	"	0.400
3/4" x		
3/8"	EA.	0.421

15 MECHANICAL

Plumbing	UNIT	MAN/HOURS
15410.11 **Copper Fittings** (Cont.)		
1/2"	EA.	0.421
1" x		
1/2"	EA.	0.444
3/4"	"	0.444
1-1/4" x		
1/2"	EA.	0.471
3/4"	"	0.471
x 1"	"	0.471
1-1/2" x		
1/2"	EA.	0.500
1-1/4"	"	0.500
2" x		
1"	EA.	0.533
1-1/4"	"	0.533
1-1/2"	"	0.533
Female flush bushing		
1/2" x		
1/2" x 1/8"	EA.	0.400
1/4"	"	0.400
Union		
1/4"	EA.	0.320
3/8"	"	0.364
3"	"	0.667
Female		
1/2"	EA.	0.400
3/4"	"	0.421
Male		
1/2"	EA.	0.400
3/4"	"	0.421
1"	"	0.444
45 degree wye		
1/2"	EA.	0.400
3/4"	"	0.421
1"	"	0.444
1" x 3/4" x 3/4"	"	0.444
1-1/4"	"	0.471
1-1/4" x 1" x 1"	"	0.471
Twin ells		
1" x 3/4" x 3/4"	EA.	0.444
1" x 1" x 1"	"	0.444
1-1/4" x 1" x 1"	"	0.471
Companion flanges, 125#		
2" x 6"	EA.	0.533
2-1/2"	"	0.615
3" x 7-1/2"	"	0.667
4" x 9"	"	0.800
90 union ells, male		
1/2"	EA.	0.400
3/4"	"	0.421
1"	"	0.444
DWV fittings, coupling with stop		
1-1/4"	EA.	0.471
1-1/2"	"	0.500

Plumbing	UNIT	MAN/HOURS
15410.11 **Copper Fittings** (Cont.)		
1-1/2" x 1-1/4"	EA.	0.500
2"	"	0.533
2" x 1-1/4"	"	0.533
2" x 1-1/2"	"	0.533
3"	"	0.667
3" x 1-1/2"	"	0.667
3" x 2"	"	0.667
4"	"	0.800
Slip coupling		
1-1/2"	EA.	0.500
2"	"	0.533
3"	"	0.667
90 ells		
1-1/2"	EA.	0.500
1-1/2" x 1-1/4"	"	0.500
2"	"	0.533
2" x 1-1/2"	"	0.533
3"	"	0.667
4"	"	0.800
Street, 90 elbows		
1-1/2"	EA.	0.500
2"	"	0.533
3"	"	0.667
4"	"	0.800
Female, 90 elbows		
1-1/2"	EA.	0.500
2"	"	0.533
Male, 90 elbows		
1-1/2"	EA.	0.500
2"	"	0.533
90 with side inlet		
3" x 3" x 1"	EA.	0.667
3" x 3" x 1-1/2"	"	0.667
3" x 3" x 2"	"	0.667
45 ells		
1-1/4"	EA.	0.471
1-1/2"	"	0.500
2"	"	0.533
3"	"	0.667
4"	"	0.800
Street, 45 ell		
1-1/2"	EA.	0.500
2"	"	0.533
3"	"	0.667
60 ell		
1-1/2"	EA.	0.500
2"	"	0.533
3"	"	0.667
22-1/2 ell		
1-1/2"	EA.	0.500
2"	"	0.533
3"	"	0.667
11-1/4 ell		

15 MECHANICAL

Plumbing		UNIT	MAN/HOURS
15410.11	**Copper Fittings** (Cont.)		
1-1/2"		EA.	0.500
2"		"	0.533
3"		"	0.667
Wye			
1-1/4"		EA.	0.471
1-1/2"		"	0.500
2"		"	0.533
2" x 1-1/2" x 1-1/2"		"	0.533
2" x 1-1/2" x 2"		"	0.533
2" x 1-1/2" x 2"		"	0.533
3"		"	0.667
3" x 3" x 1-1/2"		"	0.667
3" x 3" x 2"		"	0.667
4"		"	0.800
4" x 4" x 2"		"	0.800
4" x 4" x 3"		"	0.800
Sanitary tee			
1-1/4"		EA.	0.471
1-1/2"		"	0.500
2"		"	0.533
2" x 1-1/2" x 1-1/2"		"	0.533
2" x 1-1/2" x 2"		"	0.533
2" x 2" x 1-1/2"		"	0.533
3"		"	0.667
3" x 3" x 1-1/2"		"	0.667
3" x 3" x 2"		"	0.667
4"		"	0.800
4" x 4" x 3"		"	0.800
Female sanitary tee			
1-1/2"		EA.	0.500
Long turn tee			
1-1/2"		EA.	0.500
2"		"	0.533
3" x 1-1/2"		"	0.667
Double wye			
1-1/2"		EA.	0.500
2"		"	0.533
2" x 2" x 1-1/2" x 1-1/2"		"	0.533
3"		"	0.667
3" x 3" x 1-1/2" x 1-1/2"		"	0.667
3" x 3" x 2" x 2"		"	0.667
4" x 4" x 1-1/2" x 1-1/2"		"	0.800
Double sanitary tee			
1-1/2"		EA.	0.500
2"		"	0.533
2" x 2" x 1-1/2"		"	0.533
3"		"	0.667
3" x 3" x 1-1/2"		"	0.667
3" x 3" x 2" x 2"		"	0.667
4" x 4" x 1-1/2" x 1-1/2"		"	0.800
Long			
2" x 1-1/2"		EA.	0.533
Twin elbow			

Plumbing		UNIT	MAN/HOURS
15410.11	**Copper Fittings** (Cont.)		
1-1/2"		EA.	0.500
2"		"	0.533
2" x 1-1/2" x 1-1/2"		"	0.533
Spigot adapter, manoff			
1-1/2" x 2"		EA.	0.500
1-1/2" x 3"		"	0.500
2"		"	0.533
2" x 3"		"	0.533
2" x 4"		"	0.533
3"		"	0.667
3" x 4"		"	0.667
4"		"	0.800
No-hub adapters			
1-1/2" x 2"		EA.	0.500
2"		"	0.533
2" x 3"		"	0.533
3"		"	0.667
3" x 4"		"	0.667
4"		"	0.800
Fitting reducers			
1-1/2" x 1-1/4"		EA.	0.500
2" x 1-1/2"		"	0.533
3" x 1-1/2"		"	0.667
3" x 2"		"	0.667
Slip joint (Desanco)			
1-1/4"		EA.	0.471
1-1/2"		"	0.500
1-1/2" x 1-1/4"		"	0.500
Street x slip joint (Desanco)			
1-1/2"		EA.	0.500
1-1/2" x 1-1/4"		"	0.500
Flush bushing			
1-1/2" x 1-1/4"		EA.	0.500
2" x 1-1/2"		"	0.533
3" x 1-1/2"		"	0.667
3" x 2"		"	0.667
Male hex trap bushing			
1-1/4" x 1-1/2"		EA.	0.471
1-1/2"		"	0.500
1-1/2" x 2"		"	0.500
2"		"	0.533
Round trap bushing			
1-1/2"		EA.	0.500
2"		"	0.533
Female adapter			
1-1/4"		EA.	0.471
1-1/2"		"	0.500
1-1/2" x 2"		"	0.500
2"		"	0.533
2" x 1-1/2"		"	0.533
3"		"	0.667
Fitting x female adapter			
1-1/2"		EA.	0.500

15 MECHANICAL

Plumbing	UNIT	MAN/HOURS
15410.11 Copper Fittings *(Cont.)*		
2"	EA.	0.533
Male adapters		
1-1/4"	EA.	0.471
1-1/4" x 1-1/2"	"	0.471
1-1/2"	"	0.500
1-1/2" x 2"	"	0.500
2"	"	0.533
2" x 1-1/2"	"	0.533
3"	"	0.667
Male x slip joint adapters		
1-1/2" x 1-1/4"	EA.	0.500
Dandy cleanout		
1-1/2"	EA.	0.500
2"	"	0.533
3"	"	0.667
End cleanout, flush pattern		
1-1/2" x 1"	EA.	0.500
2" x 1-1/2"	"	0.533
3" x 2-1/2"	"	0.667
Copper caps		
1-1/2"	EA.	0.500
2"	"	0.533
Closet flanges		
3"	EA.	0.667
4"	"	0.800
Drum traps, with cleanout		
1-1/2" x 3" x 6"	EA.	0.500
P-trap, swivel, with cleanout		
1-1/2"	EA.	0.500
P-trap, solder union		
1-1/2"	EA.	0.500
2"	"	0.533
With cleanout		
1-1/2"	EA.	0.500
2"	"	0.533
2" x 1-1/2"	"	0.533
Swivel joint, with cleanout		
1-1/2" x 1-1/4"	EA.	0.500
1-1/2"	"	0.500
2" x 1-1/2"	"	0.533
Estabrook TY, with inlets		
3", with 1-1/2" inlet	EA.	0.667
Fine thread adapters		
1/2"	EA.	0.400
1/2" x 1/2" IPS	"	0.400
1/2" x 3/4" IPS	"	0.400
1/2" x male	"	0.400
1/2" x female	"	0.400
Copper pipe fittings		
1/2"		
90 deg ell	EA.	0.178
45 deg ell	"	0.178
Tee	"	0.229

Plumbing	UNIT	MAN/HOURS
15410.11 Copper Fittings *(Cont.)*		
Cap	EA.	0.089
Coupling	"	0.178
Union	"	0.200
3/4"		
90 deg ell	EA.	0.200
45 deg ell	"	0.200
Tee	"	0.267
Cap	"	0.094
Coupling	"	0.200
Union	"	0.229
1"		
90 deg ell	EA.	0.267
45 deg ell	"	0.267
Tee	"	0.320
Cap	"	0.133
Coupling	"	0.267
Union	"	0.267
1-1/4"		
90 deg ell	EA.	0.229
45 deg ell	"	0.229
Tee	"	0.400
Cap	"	0.133
Union	"	0.286
1-1/2"		
90 deg ell	EA.	0.286
45 deg ell	"	0.286
Tee	"	0.444
Cap	"	0.133
Coupling	"	0.267
Union	"	0.364
2"		
90 deg ell	EA.	0.320
45 deg ell	"	0.500
Tee	"	0.500
Cap	"	0.160
Coupling	"	0.320
Union	"	0.400
2-1/2"		
90 deg ell	EA.	0.400
45 deg ell	"	0.400
Tee	"	0.571
Cap	"	0.200
Coupling	"	0.400
Union	"	0.444
15410.14 Brass I.p.s. Fittings		
Fittings, iron pipe size, 45 deg ell		
1/8"	EA.	0.320
1/4"	"	0.320
3/8"	"	0.364
1/2"	"	0.400
3/4"	"	0.421
1"	"	0.444

15 MECHANICAL

Plumbing	UNIT	MAN/HOURS
15410.14 Brass I.p.s. Fittings *(Cont.)*		
1-1/4"	EA.	0.471
1-1/2"	"	0.500
2"	"	0.533
90 deg ell		
1/8"	EA.	0.320
1/4"	"	0.320
3/8"	"	0.364
1/2"	"	0.400
3/4"	"	0.421
1"	"	0.444
1-1/4"	"	0.471
1-1/2"	"	0.500
2"	"	0.533
2-1/2"	"	0.615
90 deg ell, reducing		
1/4" x 1/8"	EA.	0.320
3/8" x 1/8"	"	0.364
3/8" x 1/4"	"	0.364
1/2" x 1/4"	"	0.400
1/2" x 3/8"	"	0.400
3/4" x 1/2"	"	0.421
1" x 3/8"	"	0.444
1" x 1/2"	"	0.444
1" x 3/4"	"	0.444
1-1/4" x 3/4"	"	0.471
1-1/4" x 1"	"	0.471
1-1/2" x 1/2"	"	0.500
1-1/2" x 3/4"	"	0.500
1-1/2" x 1"	"	0.500
2" x 1"	"	0.533
2" x 1-1/4"	"	0.533
2" x 1-1/2"	"	0.533
Street ell, 45 deg		
1/2"	EA.	0.400
3/4"	"	0.421
2"	"	0.533
90 deg		
1/8"	EA.	0.320
1/4"	"	0.320
3/8"	"	0.364
1/2"	"	0.400
3/4"	"	0.421
1"	"	0.444
1-1/4"	"	0.471
1-1/2"	"	0.500
2"	"	0.533
Tee, 1/8"	"	0.320
1/4"	"	0.320
3/8"	"	0.364
1/2"	"	0.400
3/4"	"	0.421
1"	"	0.444
1-1/4"	"	0.471

Plumbing	UNIT	MAN/HOURS
15410.14 Brass I.p.s. Fittings *(Cont.)*		
1-1/2"	EA.	0.500
2"	"	0.533
2-1/2"	"	0.615
Tee, reducing, 3/8" x		
1/4"	EA.	0.364
1/2"	"	0.364
1/2" x		
1/4"	EA.	0.400
3/8"	"	0.400
3/4"	"	0.400
3/4" x		
1/4"	EA.	0.421
1/2"	"	0.421
1"	"	0.421
1" x		
1/2"	EA.	0.444
3/4"	"	0.444
1-1/4" x		
1/2"	EA.	0.471
1"	"	0.471
1-1/2" x		
1/2"	EA.	0.500
3/4"	"	0.500
1"	"	0.500
1-1/2"	"	0.500
2" x		
1/2"	EA.	0.533
3/4"	"	0.533
1"	"	0.533
1-1/4"	"	0.533
1-1/2"	"	0.533
2-1/2" x 2"	"	0.615
Tee, reducing		
1/2" x 3/8" x 1/2"	EA.	0.400
3/4" x 1/2" x 1/2"	"	0.421
3/4" x 1/2" x 3/4"	"	0.421
1" x 1/2" x 1/2"	"	0.444
1" x 1/2" x 3/4"	"	0.444
1" x 3/4" x 1/2"	"	0.444
1" x 3/4" x 3/4"	"	0.444
1-1/4" x 1/2" x 1-1/4"	"	0.471
1-1/4" x 1" x 1"	"	0.471
1-1/4" x 1-1/4" x 3/4"	"	0.471
1-1/2" x 1-1/4" x 1-1/4"	"	0.500
Union		
1/8"	EA.	0.320
1/4"	"	0.320
3/8"	"	0.364
1/2"	"	0.400
3/4"	"	0.421
1"	"	0.444
1-1/4"	"	0.471
1-1/2"	"	0.500

15 MECHANICAL

Plumbing	UNIT	MAN/HOURS
15410.14 Brass I.p.s. Fittings *(Cont.)*		
2"	EA.	0.533
Brass face bushing		
3/8" x 1/4"	EA.	0.364
1/2" x 3/8"	"	0.400
3/4" x 1/2"	"	0.421
1" x 3/4"	"	0.444
1-1/4" x 1"	"	0.471
Hex bushing, 1/4" x 1/8"	"	0.320
1/2" x		
1/4"	EA.	0.400
3/8"	"	0.400
5/8" x		
1/8"	EA.	0.400
1/4"	"	0.400
3/4" x		
1/8"	EA.	0.421
1/4"	"	0.421
3/8"	"	0.421
1/2"	"	0.421
1" x		
1/4"	EA.	0.444
3/8"	"	0.444
1/2"	"	0.444
3/4"	"	0.444
1-1/4" x		
3/8"	EA.	0.471
1/2"	"	0.471
3/4"	"	0.471
1"	"	0.471
1-1/2" x		
1/4"	EA.	0.500
1/2"	"	0.500
3/4"	"	0.500
1"	"	0.500
1-1/4"	"	0.500
2" x		
1/2"	EA.	0.533
3/4"	"	0.533
1"	"	0.533
1-1/4"	"	0.533
1-1/2"	"	0.533
2-1/2" x		
1"	EA.	0.615
2"	"	0.615
3" x		
1-1/2"	EA.	0.667
2"	"	0.667
2-1/2"	"	0.667
4" x		
2"	EA.	0.800
3"	"	0.800
Caps		
1/8"	EA.	0.320

Plumbing	UNIT	MAN/HOURS
15410.14 Brass I.p.s. Fittings *(Cont.)*		
1/4"	EA.	0.320
3/8"	"	0.364
1/2"	"	0.400
3/4"	"	0.421
1"	"	0.444
1-1/4"	"	0.471
1-1/2"	"	0.500
2"	"	0.533
Couplings		
1/8"	EA.	0.320
1/4"	"	0.320
3/8"	"	0.364
1/2"	"	0.400
3/4"	"	0.421
1"	"	0.444
1-1/4"	"	0.471
1-1/2"	"	0.500
2"	"	0.533
2-1/2"	"	0.615
Couplings, reducing, 1/4" x 1/8"	"	0.320
3/8" x		
1/8"	EA.	0.364
1/4"	"	0.364
1/2" x		
1/8"	EA.	0.400
1/4"	"	0.400
3/8"	"	0.400
3/4" x		
1/4"	EA.	0.421
3/8"	"	0.421
1/2"	"	0.421
1" x		
1/2"	EA.	0.421
3/4"	"	0.421
1-1/4" x		
1/2"	EA.	0.471
3/4"	"	0.471
1"	"	0.471
1-1/2" x		
3/4"	EA.	0.500
1"	"	0.500
1-1/4"	"	0.500
2" x		
3/4"	EA.	0.533
1"	"	0.533
1-1/4"	"	0.533
1-1/2"	"	0.533
2-1/2" x 1-1/2"	"	0.615
Square head plug, solid		
1/8"	EA.	0.320
1/4"	"	0.320
3/8"	"	0.364
1/2"	"	0.400

15 MECHANICAL

Plumbing	UNIT	MAN/HOURS
15410.14 **Brass I.p.s. Fittings** *(Cont.)*		
3/4"	EA.	0.421
Cored		
1/2"	EA.	0.400
3/4"	"	0.421
1"	"	0.444
1-1/4"	"	0.471
1-1/2"	"	0.500
2"	"	0.533
3"	"	0.667
4"	"	0.800
Countersunk		
1/2"	EA.	0.400
3/4"	"	0.421
1-1/2"	"	0.500
2"	"	0.533
Locknut		
3/4"	EA.	0.421
1"	"	0.444
1-1/4"	"	0.471
2"	"	0.533
Close standard red nipple, 1/8"	"	0.320
1/8" x		
1-1/2"	EA.	0.320
2"	"	0.320
2-1/2"	"	0.320
3"	"	0.320
3-1/2"	"	0.320
4"	"	0.320
4-1/2"	"	0.320
5"	"	0.320
5-1/2"	"	0.320
6"	"	0.320
1/4" x close	"	0.320
1/4" x		
1-1/2"	EA.	0.320
2"	"	0.320
2-1/2"	"	0.320
3"	"	0.320
3-1/2"	"	0.320
4"	"	0.320
4-1/2"	"	0.320
5"	"	0.320
5-1/2"	"	0.320
6"	"	0.320
3/8" x close	"	0.364
3/8" x		
1-1/2"	EA.	0.364
2"	"	0.364
2-1/2"	"	0.364
3"	"	0.364
3-1/2"	"	0.364
4"	"	0.364
4-1/2"	"	0.364

Plumbing	UNIT	MAN/HOURS
15410.14 **Brass I.p.s. Fittings** *(Cont.)*		
5"	EA.	0.364
5-1/2"	"	0.364
6"	"	0.364
1/2" x close	"	0.400
1/2" x		
1-1/2"	EA.	0.400
2"	"	0.400
2-1/2"	"	0.400
3"	"	0.400
3-1/2"	"	0.400
4"	"	0.400
4-1/2"	"	0.400
5"	"	0.400
5-1/2"	"	0.400
6"	"	0.400
7-1/2"	"	0.400
8"	"	0.400
3/4" x close	"	0.421
3/4" x		
1-1/2"	EA.	0.421
2"	"	0.421
2-1/2"	"	0.421
3"	"	0.421
3-1/2"	"	0.421
4"	"	0.421
4-1/2"	"	0.421
5"	"	0.421
5-1/2"	"	0.421
6"	"	0.421
1" x close	"	0.444
1" x		
2"	EA.	0.444
2-1/2"	"	0.444
3"	"	0.444
3-1/2"	"	0.444
4"	"	0.444
4-1/2"	"	0.444
5"	"	0.444
5-1/2"	"	0.444
6"	"	0.444
1-1/4" x close	"	0.471
1-1/4" x		
2"	EA.	0.471
2-1/2"	"	0.471
3"	"	0.471
3-1/2"	"	0.471
4"	"	0.471
4-1/2"	"	0.471
5"	"	0.471
5-1/2"	"	0.471
6"	"	0.471
1-1/2" x close	"	0.500
1-1/2" x		

15 MECHANICAL

Plumbing	UNIT	MAN/HOURS
15410.14 Brass I.p.s. Fittings (Cont.)		
2"	EA.	0.500
2-1/2"	"	0.500
3"	"	0.500
3-1/2"	"	0.500
4-1/2"	"	0.500
5"	"	0.500
5-1/2"	"	0.500
6"	"	0.500
2" x close	"	0.533
2" x		
2-1/2"	EA.	0.533
3"	"	0.533
3-1/2"	"	0.533
4"	"	0.533
4-1/2"	"	0.533
5"	"	0.533
5-1/2"	"	0.533
6"	"	0.533
2-1/2" x close	"	0.615
2-1/2" x		
3"	EA.	0.615
3-1/2"	"	0.615
4"	"	0.615
2-1/2" x 5"	"	0.615
3"		
Close	EA.	0.667
15410.15 Brass Fittings		
Compression fittings, union		
3/8"	EA.	0.133
1/2"	"	0.133
5/8"	"	0.133
Union elbow		
3/8"	EA.	0.133
1/2"	"	0.133
5/8"	"	0.133
Union tee		
3/8"	EA.	0.133
1/2"	"	0.133
5/8"	"	0.133
Male connector		
3/8"	EA.	0.133
1/2"	"	0.133
5/8"	"	0.133
Female connector		
3/8"	EA.	0.133
1/2"	"	0.133
5/8"	"	0.133
Brass flare fittings, union		
3/8"	EA.	0.129
1/2"	"	0.129
5/8"	"	0.129
90 deg elbow union		

Plumbing	UNIT	MAN/HOURS
15410.15 Brass Fittings (Cont.)		
3/8"	EA.	0.129
1/2"	"	0.129
5/8"	"	0.129
Three way tee		
3/8"	EA.	0.216
1/2"	"	0.216
5/8"	"	0.216
Cross		
3/8"	EA.	0.286
1/2"	"	0.286
5/8"	"	0.286
Male connector, half union		
3/8"	EA.	0.129
1/2"	"	0.129
5/8"	"	0.129
Female connector, half union		
3/8"	EA.	0.129
1/2"	"	0.129
5/8"	"	0.129
Long forged nut		
3/8"	EA.	0.129
1/2"	"	0.129
5/8"	"	0.129
Short forged nut		
3/8"	EA.	0.129
1/2"	"	0.129
5/8"	"	0.129
Compression elbow		
1/4"	EA.	0.129
5/16"	"	0.129
3/4"	"	0.129
Sleeve		
1/8"	EA.	0.160
1/4"	"	0.160
5/16"	"	0.160
3/8"	"	0.160
1/2"	"	0.160
5/8"	"	0.160
3/4"	"	0.160
7/8"	"	0.160
Tee		
1/4"	EA.	0.229
5/16"	"	0.229
Male tee		
5/16" x 1/8"	EA.	0.229
Female union		
1/8" x 1/8"	EA.	0.200
1/4" x 3/8"	"	0.200
3/8" x 1/4"	"	0.200
3/8" x 1/2"	"	0.200
5/8" x 1/2"	"	0.229
Male union, 1/4"		
1/4" x 1/4"	EA.	0.200

15 MECHANICAL

Plumbing	UNIT	MAN/HOURS
15410.15 Brass Fittings (Cont.)		
3/8"	EA.	0.200
1/2"	"	0.200
5/16" x		
1/8"	EA.	0.200
1/4"	"	0.200
3/8"	"	0.200
3/8" x		
1/8"	EA.	0.200
1/4"	"	0.200
1/2"	"	0.200
3/4"	"	0.200
1/2" x		
1/4"	EA.	0.229
3/8"	"	0.229
5/8" x		
3/8"	EA.	0.229
1/2"	"	0.229
3/4"	"	0.229
1/2"	"	0.229
7/8" x		
1/2"	EA.	0.229
3/4"	"	0.229
Female elbow, 1/4" x 1/4"	"	0.229
5/16" x		
1/8"	EA.	0.229
1/4"	"	0.229
3/8" x		
3/8"	EA.	0.229
1/2"	"	0.229
Male elbow, 1/8" x 1/8"	"	0.229
3/16" x 1/4"	"	0.229
1/4" x		
1/8"	EA.	0.229
1/4"	"	0.229
3/8"	"	0.229
5/16" x		
1/8"	EA.	0.229
1/4"	"	0.229
3/8"	"	0.229
3/8" x		
1/8"	EA.	0.229
1/4"	"	0.229
3/8"	"	0.229
1/2"	"	0.229
1/2" x		
1/4"	EA.	0.267
3/8"	"	0.267
1/2"	"	0.267
5/8" x		
3/8"	EA.	0.267
1/2"	"	0.267
3/4"	"	0.267
3/4" x		

Plumbing	UNIT	MAN/HOURS
15410.15 Brass Fittings (Cont.)		
1/2"	EA.	0.267
3/4"	"	0.267
7/8" x		
3/4"	EA.	0.267
Union		
1/8"	EA.	0.229
3/16"	"	0.229
1/4"	"	0.229
5/16"	"	0.229
3/8"	"	0.229
3/4"	"	0.229
7/8"	"	0.229
Reducing union		
3/8" x 1/4"	EA.	0.267
5/8" x		
3/8"	EA.	0.267
1/2"	"	0.267
15410.17 Chrome Plated Fittings		
Fittings		
90 ell		
3/8"	EA.	0.200
1/2"	"	0.200
45 ell		
3/8"	EA.	0.200
1/2"	"	0.200
Tee		
3/8"	EA.	0.267
1/2"	"	0.267
Coupling		
3/8"	EA.	0.200
1/2"	"	0.200
Union		
3/8"	EA.	0.200
1/2"	"	0.200
Tee		
1/2" x 3/8" x 3/8"	EA.	0.267
1/2" x 3/8" x 1/2"	"	0.267
15410.18 Glass Pipe		
Glass pipe		
1-1/2" dia.	L.F.	0.160
2" dia.	"	0.178
3" dia.	"	0.200
4" dia.	"	0.229
6" dia.	"	0.267
15410.30 Pvc/cpvc Pipe		
PVC schedule 40		
1/2" pipe	L.F.	0.033
3/4" pipe	"	0.036
1" pipe	"	0.040
1-1/4" pipe	"	0.044

15 MECHANICAL

Plumbing

15410.30 Pvc/cpvc Pipe (Cont.)

Description	UNIT	MAN/HOURS
1-1/2" pipe	L.F.	0.050
2" pipe	"	0.057
2-1/2" pipe	"	0.067
3" pipe	"	0.080
4" pipe	"	0.100
6" pipe	"	0.200
8" pipe	"	0.267
Fittings, 1/2"		
90 deg ell	EA.	0.100
45 deg ell	"	0.100
Tee	"	0.114
Reducing insert	"	0.133
Threaded	"	0.100
Male adapter	"	0.133
Female adapter	"	0.100
Union	"	0.160
Cap	"	0.133
Flange	"	0.160
3/4"		
90 deg elbow	EA.	0.133
45 deg elbow	"	0.133
Tee	"	0.160
Reducing insert	"	0.114
Threaded	"	0.133
1"		
90 deg elbow	EA.	0.160
45 deg elbow	"	0.160
Tee	"	0.178
Reducing insert	"	0.160
Threaded	"	0.178
Male adapter	"	0.200
Female adapter	"	0.200
Union	"	0.267
Cap	"	0.160
Flange	"	0.267
1-1/4"		
90 deg elbow	EA.	0.229
45 deg elbow	"	0.229
Tee	"	0.267
Reducing insert	"	0.267
Threaded	"	0.267
Male adapter	"	0.267
Female adapter	"	0.267
Union	"	0.320
Cap	"	0.267
Flange	"	0.320
1-1/2"		
90 deg elbow	EA.	0.229
45 deg elbow	"	0.229
Tee	"	0.267
Reducing insert	"	0.267
Threaded	"	0.267
Male adapter	"	0.267

15410.30 Pvc/cpvc Pipe (Cont.)

Description	UNIT	MAN/HOURS
Female adapter	EA.	0.267
Union	"	0.400
Cap	"	0.267
Flange	"	0.400
2"		
90 deg elbow	EA.	0.267
45 deg elbow	"	0.267
Tee	"	0.320
Reducing insert	"	0.320
Threaded	"	0.320
Male adapter	"	0.320
Female adapter	"	0.320
Union	"	0.500
Cap	"	0.320
Flange	"	0.500
2-1/2"		
90 deg elbow	EA.	0.500
45 deg elbow	"	0.500
Tee	"	0.533
Reducing insert	"	0.533
Threaded	"	0.533
Male adapter	"	0.533
Female adapter	"	0.533
Union	"	0.667
Cap	"	0.500
Flange	"	0.667
3"		
90 deg elbow	EA.	0.667
45 deg elbow	"	0.667
Tee	"	0.727
Reducing insert	"	0.667
Threaded	"	0.667
Male adapter	"	0.667
Female adapter	"	0.667
Union	"	0.800
Cap	"	0.667
Flange	"	0.800
4"		
90 deg elbow	EA.	0.800
45 deg elbow	"	0.800
Tee	"	0.889
Reducing insert	"	0.800
Threaded	"	0.800
Male adapter	"	0.800
Female adapter	"	0.800
Union	"	1.000
Cap	"	0.800
Flange	"	1.000
PVC schedule 80 pipe		
1-1/2" pipe	L.F.	0.050
2" pipe	"	0.057
3" pipe	"	0.080
4" pipe	"	0.100

15 MECHANICAL

Plumbing

15410.30 Pvc/cpvc Pipe (Cont.)

Description	UNIT	MAN/HOURS
Fittings, 1-1/2"		
90 deg elbow	EA.	0.267
45 deg elbow	"	0.267
Tee	"	0.400
Reducing insert	"	0.267
Threaded	"	0.267
Male adapter	"	0.267
Female adapter	"	0.267
Union	"	0.400
Cap	"	0.267
Flange	"	0.400
2"		
90 deg elbow	EA.	0.320
45 deg elbow	"	0.320
Tee	"	0.500
Reducing insert	"	0.320
Threaded	"	0.320
Male adapter	"	0.320
Female adapter	"	0.320
2-1/2"		
90 deg elbow	EA.	0.500
45 deg elbow	"	0.500
Tee	"	0.667
Reducing insert	"	0.500
Threaded	"	0.500
Male adapter	"	0.500
Female adapter	"	0.500
Union	"	0.667
Cap	"	0.500
Flange	"	0.667
3"		
90 deg elbow	EA.	0.667
45 deg elbow	"	0.667
Tee	"	0.800
Reducing insert	"	0.667
Threaded	"	0.667
Male adapter	"	0.667
Female adapter	"	0.667
Union	"	0.800
Cap	"	0.667
Flange	"	0.800
4"		
90 deg elbow	EA.	0.800
45 deg elbow	"	0.800
Tee	"	1.000
Reducing insert	"	0.800
Threaded	"	0.800
Male adapter	"	0.800
Union	"	1.000
Cap	"	0.800
Flange	"	1.000
CPVC schedule 40		
1/2" pipe	L.F.	0.033

15410.30 Pvc/cpvc Pipe (Cont.)

Description	UNIT	MAN/HOURS
3/4" pipe	L.F.	0.036
1" pipe	"	0.040
1-1/4" pipe	"	0.044
1-1/2" pipe	"	0.050
2" pipe	"	0.057
Fittings, CPVC, schedule 80		
1/2", 90 deg ell	EA.	0.080
Tee	"	0.133
3/4", 90 deg ell	"	0.080
Tee	"	0.133
1", 90 deg ell	"	0.089
Tee	"	0.145
1-1/4", 90 deg ell	"	0.089
Tee	"	0.145
1-1/2", 90 deg ell	"	0.160
Tee	"	0.200
2", 90 deg ell	"	0.160
Tee	"	0.200
Polypropylene, acid resistant, DWV pipe		
Schedule 40		
1-1/2" pipe	L.F.	0.057
2" pipe	"	0.067
3" pipe	"	0.080
4" pipe	"	0.100
6" pipe	"	0.200
Fittings		
1-1/2"		
1/4 bend	EA.	0.200
1/8 bend	"	0.200
Sanitary tee	"	0.400
Cleanout with plug	"	0.400
Wye	"	0.400
Combination wye & 1/8 bend	"	0.400
P-trap	"	0.500
Hub adapter	"	0.200
Coupling	"	0.200
2"		
1/4 bend	EA.	0.229
1/8 bend	"	0.229
Sanitary tee	"	0.471
Cleanout with plug	"	0.471
Wye	"	0.471
Combination wye & 1/8 bend	"	0.471
P-trap	"	0.667
Hub adapter	"	0.267
Mechanical joint adapter	"	0.267
Coupling	"	0.267
3"		
1/4 bend	EA.	0.267
1/8 bend	"	0.267
Sanitary tee	"	0.533
Cleanout with plug	"	0.533
Wye	"	0.533

15 MECHANICAL

Plumbing	UNIT	MAN/HOURS
15410.30 Pvc/cpvc Pipe *(Cont.)*		
Combination wye & 1/8 bend	EA.	0.533
P-trap	"	0.800
Hub adapter	"	0.267
Mechanical joint adapter	"	0.267
Coupling	"	0.267
4"		
1/4 bend	EA.	0.400
1/8 bend	"	0.400
Sanitary tee	"	0.800
Cleanout with plug	"	0.800
Wye	"	0.800
Combination wye & 1/8 bend	"	0.800
P-trap	"	1.333
Hub adapter	"	0.400
Mechanical joint adapter	"	0.400
Coupling	"	0.400
6"		
1/4 bend	EA.	0.667
1/8 bend	"	0.667
Sanitary tee	"	1.333
Cleanout with plug	"	1.333
Wye	"	1.333
Hub adapter	"	0.667
Coupling	"	0.667
Polyethylene pipe and fittings		
SDR-21		
3" pipe	L.F.	0.100
4" pipe	"	0.133
6" pipe	"	0.200
8" pipe	"	0.229
10" pipe	"	0.267
12" pipe	"	0.320
14" pipe	"	0.400
16" pipe	"	0.500
18" pipe	"	0.615
20" pipe	"	0.800
22" pipe	"	0.889
24" pipe	"	1.000
Fittings, 3"		
90 deg elbow	EA.	0.400
45 deg elbow	"	0.400
Tee	"	0.667
45 deg wye	"	0.667
Reducer	"	0.500
Flange assembly	"	0.400
4"		
90 deg elbow	EA.	0.500
45 deg elbow	"	0.500
Tee	"	0.800
45 deg wye	"	0.800
Reducer	"	0.667
Flange assembly	"	0.500
8"		

Plumbing	UNIT	MAN/HOURS
15410.30 Pvc/cpvc Pipe *(Cont.)*		
90 deg elbow	EA.	1.000
45 deg elbow	"	1.000
Tee	"	1.600
45 deg wye	"	1.600
Reducer	"	1.333
Flange assembly	"	1.000
10"		
90 deg elbow	EA.	1.333
45 deg elbow	"	1.333
Tee	"	2.000
45 deg wye	"	2.000
Reducer	"	1.600
Flange assembly	"	1.333
12"		
90 deg elbow	EA.	1.600
45 deg elbow	"	1.600
Tee	"	2.667
45 deg wye	"	2.667
Reducer	"	2.000
Flange assembly	"	1.600
14"		
90 deg elbow	EA.	2.000
45 deg elbow	"	2.000
Tee	"	3.200
45 deg wye	"	3.200
Reducer	"	2.667
Flange assembly	"	2.000
16"		
90 deg elbow	EA.	2.000
45 deg elbow	"	2.000
Tee	"	3.200
45 deg wye	"	3.200
Reducer	"	2.667
Flange assembly	"	2.000
18"		
90 deg elbow	EA.	2.667
45 deg elbow	"	2.667
Tee	"	4.000
45 deg wye	"	4.000
Reducer	"	2.667
Flange assembly	"	2.667
20"		
90 deg elbow	EA.	2.667
45 deg elbow	"	2.667
15410.33 Abs Dwv Pipe		
Schedule 40 ABS		
1-1/2" pipe	L.F.	0.040
2" pipe	"	0.044
3" pipe	"	0.057
4" pipe	"	0.080
6" pipe	"	0.100
Fittings		

15 MECHANICAL

Plumbing	UNIT	MAN/HOURS
15410.33 Abs Dwv Pipe (Cont.)		
1/8 bend		
1-1/2"	EA.	0.160
2"	"	0.200
3"	"	0.267
4"	"	0.320
6"	"	0.400
Tee, sanitary		
1-1/2"	EA.	0.267
2"	"	0.320
3"	"	0.400
4"	"	0.500
6"	"	0.667
Tee, sanitary reducing		
2 x 1-1/2 x 1-1/2	EA.	0.320
2 x 1-1/2 x 2	"	0.333
2 x 2 x 1-1/2	"	0.364
3 x 3 x 1-1/2	"	0.400
3 x 3 x 2	"	0.444
4 x 4 x 1-1/2	"	0.500
4 x 4 x 2	"	0.571
4 x 4 x 3	"	0.615
6 x 6 x 4	"	0.667
Wye		
1-1/2"	EA.	0.229
2"	"	0.320
3"	"	0.400
4"	"	0.500
6"	"	0.667
Reducer		
2 x 1-1/2	EA.	0.200
3 x 1-1/2	"	0.267
3 x 2	"	0.267
4 x 2	"	0.320
4 x 3	"	0.320
6 x 4	"	0.400
P-trap		
1-1/2"	EA.	0.267
2"	"	0.296
3"	"	0.348
4"	"	0.400
6"	"	0.500
Double sanitary, tee		
1-1/2"	EA.	0.320
2"	"	0.400
3"	"	0.500
4"	"	0.667
Long sweep, 1/4 bend		
1-1/2"	EA.	0.160
2"	"	0.200
3"	"	0.267
4"	"	0.400
Wye, standard		
1-1/2"	EA.	0.267

Plumbing	UNIT	MAN/HOURS
15410.33 Abs Dwv Pipe (Cont.)		
2"	EA.	0.320
3"	"	0.400
4"	"	0.500
Wye, reducing		
2 x 1-1/2 x 1-1/2	EA.	0.267
2 x 2 x 1-1/2	"	0.320
4 x 4 x 2	"	0.500
4 x 4 x 3	"	0.533
Double wye		
1-1/2"	EA.	0.320
2"	"	0.400
3"	"	0.500
4"	"	0.667
2 x 2 x 1-1/2 x 1-1/2	"	0.400
3 x 3 x 2 x 2	"	0.500
4 x 4 x 3 x 3	"	0.667
Combination wye and 1/8 bend		
1-1/2"	EA.	0.267
2"	"	0.320
3"	"	0.400
4"	"	0.500
2 x 2 x 1-1/2	"	0.320
3 x 3 x 1-1/2	"	0.400
3 x 3 x 2	"	0.400
4 x 4 x 2	"	0.500
4 x 4 x 3	"	0.500
15410.35 Plastic Pipe		
Fiberglass reinforced pipe		
2" pipe	L.F.	0.062
3" pipe	"	0.067
4" pipe	"	0.073
6" pipe	"	0.080
8" pipe	"	0.133
10" pipe	"	0.160
12" pipe	"	0.200
Fittings		
90 deg elbow, flanged		
2"	EA.	0.800
3"	"	0.889
4"	"	1.000
6"	"	1.333
8"	"	1.600
10"	"	2.000
12"	"	2.667
45 deg elbow, flanged		
2"	EA.	0.667
3"	"	0.800
4"	"	1.000
6"	"	1.333
8"	"	1.600
10"	"	2.000
12"	"	2.667

15 MECHANICAL

Plumbing

15410.35 Plastic Pipe (Cont.)

	UNIT	MAN/HOURS
Tee, flanged		
2"	EA.	1.000
3"	"	1.143
4"	"	1.333
6"	"	1.600
8"	"	2.000
10"	"	2.667
12"	"	4.000
Wye, flanged		
2"	EA.	1.000
3"	"	1.143
4"	"	1.333
6"	"	1.600
8"	"	2.000
10"	"	2.667
12"	"	4.000
Concentric reducer, flanged		
2"	EA.	0.667
4"	"	0.800
6"	"	1.143
8"	"	1.600
10"	"	2.000
12"	"	2.667
Adapter, bell x male or female		
2"	EA.	0.667
3"	"	0.727
4"	"	0.800
6"	"	1.143
8"	"	1.600
10"	"	2.000
12"	"	2.667
Nipples		
2" x 6"	EA.	0.080
2" x 12"	"	0.100
3" x 8"	"	0.123
3" x 12"	"	0.133
4" x 8"	"	0.133
4" x 12"	"	0.160
6" x 12"	"	0.200
8" x 18"	"	0.200
8" x 24"	"	0.229
10" x 18"	"	0.267
10" x 24"	"	0.320
12" x 18"	"	0.364
12" x 24"	"	0.400
Sleeve coupling		
2"	EA.	0.667
3"	"	0.800
4"	"	1.143
6"	"	1.600
8"	"	2.000
10"	"	2.667
Flanges		

Plumbing

15410.35 Plastic Pipe (Cont.)

	UNIT	MAN/HOURS
2"	EA.	0.667
3"	"	0.800
4"	"	1.143
6"	"	1.600
8"	"	2.000
10"	"	2.667
12"	"	2.667

15410.70 Stainless Steel Pipe

	UNIT	MAN/HOURS
Stainless steel, schedule 40, threaded		
1/2" pipe	L.F.	0.114
3/4" pipe	"	0.118
1" pipe	"	0.123
1-1/2" pipe	"	0.133
2" pipe	"	0.145
2-1/2" pipe	"	0.160
3" pipe	"	0.178
4" pipe	"	0.200
Fittings, 1/2"		
90 deg ell	EA.	1.000
45 deg ell	"	1.000
Tee	"	1.333
Cap	"	0.500
Reducer	"	0.667
Union	"	1.000
Flange	"	1.000
3/4"		
90 deg ell	EA.	1.000
45 deg ell	"	1.000
Tee	"	1.333
Cap	"	0.500
Reducer	"	0.667
Union	"	1.000
Flange	"	1.000
1"		
90 deg ell	EA.	1.000
45 deg ell	"	1.000
Tee	"	1.333
Cap	"	0.500
Reducer	"	1.000
Union	"	1.000
Flange	"	1.000
1-1/4"		
90 deg ell	EA.	1.000
45 deg ell	"	1.000
Tee	"	1.333
Cap	"	0.500
Union	"	1.000
Flange	"	1.000
1-1/2"		
90 deg ell	EA.	1.333
45 deg ell	"	1.333
Tee	"	1.600

15 MECHANICAL

Plumbing	UNIT	MAN/HOURS
15410.70 Stainless Steel Pipe *(Cont.)*		
Cap	EA.	0.667
Reducer	"	0.800
Union	"	1.143
Flange	"	1.143
2"		
90 deg ell	EA.	1.600
45 deg ell	"	1.600
Tee	"	2.667
Cap	"	0.800
Reducer	"	1.000
Union	"	1.600
Flange	"	1.600
Type 304, sch 10 pipe		
1" pipe	L.F.	0.100
1-1/4" pipe	"	0.133
1-1/2" pipe	"	0.145
2" pipe	"	0.178
2-1/2" pipe	"	0.200
3" pipe	"	0.229
4" pipe	"	0.267
6" pipe	"	0.320
Fittings, 1"		
90 deg elbow	EA.	1.600
45 deg elbow	"	1.600
Tee	"	2.667
Reducer	"	1.600
Flange	"	1.000
Cap	"	1.000
1-1/4"		
90 deg elbow	EA.	1.600
45 deg elbow	"	1.600
Tee	"	2.667
Reducer	"	1.600
Flange	"	1.000
Cap	"	1.000
1-1/2"		
90 deg elbow	EA.	1.333
45 deg elbow	"	1.333
Tee	"	2.667
Reducer	"	2.000
Flange	"	1.333
Cap	"	1.333
2"		
90 deg elbow	EA.	1.600
45 deg elbow	"	1.600
Tee	"	4.000
Reducer	"	2.667
Flange	"	2.000
Cap	"	2.000
2-1/2"		
90 deg elbow	EA.	1.600
45 deg elbow	"	1.600
Tee	"	4.000

Plumbing	UNIT	MAN/HOURS
15410.70 Stainless Steel Pipe *(Cont.)*		
Reducer	EA.	2.667
Flange	"	2.000
Cap	"	2.000
3"		
90 deg elbow	EA.	2.000
45 deg elbow	"	2.000
Tee	"	5.333
Reducer	"	4.000
Flange	"	2.667
Cap	"	2.667
4"		
90 deg elbow	EA.	2.000
45 deg elbow	"	2.000
Reducer	"	4.000
Flange	"	2.667
Cap	"	2.667
6"		
90 deg elbow	EA.	2.667
45 deg elbow	"	2.667
Tee	"	11.429
Reducer	"	4.000
Flange	"	4.000
Cap	"	4.000
Type 304 tubing		
.035 wall		
1/4"	L.F.	0.044
3/8"	"	0.050
1/2"	"	0.057
5/8"	"	0.067
3/4"	"	0.080
7/8"	"	0.089
1"	"	0.100
.049 wall		
1/4"	L.F.	0.047
3/8"	"	0.053
1/2"	"	0.062
5/8"	"	0.073
3/4"	"	0.089
7/8"	"	0.100
1"	"	0.114
.065 wall		
1/4"	L.F.	0.053
3/8"	"	0.067
1/2"	"	0.073
5/8"	"	0.089
3/4"	"	0.114
7/8"	"	0.133
1"	"	0.160
Type 316 tubing		
.035 wall		
1/4"	L.F.	0.044
3/8"	"	0.050
1/2"	"	0.057

15 MECHANICAL

Plumbing	UNIT	MAN/HOURS
15410.70 Stainless Steel Pipe *(Cont.)*		
5/8"	L.F.	0.067
3/4"	"	0.080
7/8"	"	0.089
1"	"	0.100
.049 wall		
1/4"	L.F.	0.053
3/8"	"	0.067
1/2"	"	0.073
5/8"	"	0.089
3/4"	"	0.114
7/8"	"	0.133
1"	"	0.160
.065 wall		
1/4"	L.F.	0.053
3/8"	"	0.067
1/2"	"	0.073
5/8"	"	0.089
3/4"	"	0.114
7/8"	"	0.133
1"	"	0.160
Fittings, 1/4"		
90 deg elbow	EA.	0.160
Union tee	"	0.267
Union	"	0.267
Male connector	"	0.200
3/8"		
90 deg elbow	EA.	0.200
Union tee	"	0.308
Union	"	0.308
Male connector	"	0.200
1/2"		
90 deg elbow	EA.	0.211
Union tee	"	0.333
Union	"	0.333
Male connector	"	0.200
5/8"		
90 deg elbow	EA.	0.267
Union tee	"	0.400
Union	"	0.400
Male connector	"	0.267
3/4"		
90 deg elbow	EA.	0.267
Union tee	"	0.400
Union	"	0.400
Male connector	"	0.267
7/8"		
90 deg elbow	EA.	0.286
Union tee	"	0.444
Union	"	0.444
Male connector	"	0.286
1"		
90 deg elbow	EA.	0.364
Union tee	"	0.500

Plumbing	UNIT	MAN/HOURS
15410.70 Stainless Steel Pipe *(Cont.)*		
Union	EA.	0.500
Male connector	"	0.400
Type 316 valves		
Gate valves		
1/4"	EA.	0.267
3/8"	"	0.320
1/2"	"	0.348
3/4"	"	0.400
1"	"	0.533
Globe valves		
1/4"	EA.	0.267
3/8"	"	0.320
1/2"	"	0.348
3/4"	"	0.400
1"	"	0.533
Check valves		
1/4"	EA.	0.267
3/8"	"	0.320
1/2"	"	0.348
3/4"	"	0.400
1"	"	0.533
15410.80 Steel Pipe		
Black steel, extra heavy pipe, threaded		
1/2" pipe	L.F.	0.032
3/4" pipe	"	0.032
1" pipe	"	0.040
1-1/2" pipe	"	0.044
2-1/2" pipe	"	0.100
3" pipe	"	0.133
4" pipe	"	0.160
5" pipe	"	0.200
6" pipe	"	0.200
8" pipe	"	0.267
10" pipe	"	0.320
12" pipe	"	0.400
Fittings, malleable iron, threaded, 1/2" pipe		
90 deg ell	EA.	0.267
45 deg ell	"	0.267
Tee	"	0.400
Reducing tee	"	0.400
Cap	"	0.160
Coupling	"	0.320
Union	"	0.267
Nipple, 4" long	"	0.267
3/4" pipe		
90 deg ell	EA.	0.267
45 deg ell	"	0.400
Tee	"	0.400
Reducing tee	"	0.267
Cap	"	0.160
Coupling	"	0.267
Union	"	0.267

15 MECHANICAL

Plumbing	UNIT	MAN/HOURS
15410.80 Steel Pipe (Cont.)		
Nipple, 4" long	EA.	0.267
1" pipe		
90 deg ell	EA.	0.320
45 deg ell	"	0.320
Tee	"	0.444
Reducing tee	"	0.444
Cap	"	0.160
Coupling	"	0.320
Union	"	0.320
Nipple, 4" long	"	0.320
1-1/2" pipe		
90 deg ell	EA.	0.400
45 deg ell	"	0.400
Tee	"	0.571
Reducing tee	"	0.571
Cap	"	0.200
Coupling	"	0.400
Union	"	0.400
Nipple, 4" long	"	0.400
2-1/2" pipe		
90 deg ell	EA.	1.000
45 deg ell	"	1.000
Tee	"	1.333
Reducing tee	"	1.333
Cap	"	0.500
Coupling	"	1.333
Union	"	1.333
Nipple, 4" long	"	1.333
3" pipe		
90 deg ell	EA.	1.333
45 deg ell	"	1.333
Tee	"	2.000
Reducing tee	"	2.000
Cap	"	0.667
Coupling	"	1.333
Union	"	1.333
Nipple, 4" long	"	1.333
4" pipe		
90 deg ell	EA.	1.600
45 deg ell	"	1.600
Tee	"	2.667
Reducing tee	"	2.667
Cap	"	2.667
Coupling	"	0.800
Union	"	2.667
Nipple, 4" long	"	2.667
6" pipe		
90 deg ell	EA.	1.600
45 deg ell	"	1.600
Tee	"	2.667
Reducing tee	"	2.667
Cap	"	0.800
8" pipe		

Plumbing	UNIT	MAN/HOURS
15410.80 Steel Pipe (Cont.)		
90 deg ell	EA.	3.200
45 deg ell	"	3.200
Tee	"	5.000
Reducing tee	"	4.211
Cap	"	1.600
10" pipe		
90 deg ell	EA.	4.000
45 deg ell	"	4.000
Tee	"	5.000
Reducing tee	"	2.000
Cap	"	2.000
12" pipe		
90 deg ell	EA.	5.000
45 deg ell	"	5.000
Tee	"	6.667
Reducing tee	"	6.667
Cap	"	2.667
Butt welded, 1/2" pipe		
90 deg ell	EA.	0.267
45 deg ell	"	0.267
Tee	"	0.400
3/4" pipe		
90 deg ell	EA.	0.267
45 deg. ell	"	0.267
Tee	"	0.400
1" pipe		
90 deg ell	EA.	0.320
45 deg ell	"	0.320
Tee	"	0.444
1-1/2" pipe		
90 deg ell	EA.	0.400
45 deg. ell	"	0.400
Tee	"	0.571
Reducing tee	"	0.571
Cap	"	0.320
2-1/2" pipe		
90 deg. ell	EA.	0.800
45 deg. ell	"	0.800
Tee	"	1.143
Reducing tee	"	1.143
Cap	"	0.400
3" pipe		
90 deg ell	EA.	1.000
45 deg. ell	"	1.000
Tee	"	1.333
Reducing tee	"	1.333
Cap	"	0.667
4" pipe		
90 deg ell	EA.	1.333
45 deg. ell	"	1.333
Tee	"	2.000
Reducing tee	"	2.000
Cap	"	0.667

15 MECHANICAL

Plumbing	UNIT	MAN/HOURS
15410.80 Steel Pipe *(Cont.)*		
6" pipe		
90 deg. ell	EA.	1.600
45 deg. ell	"	1.600
Tee	"	2.667
Reducing tee	"	2.667
Cap	"	0.800
8" pipe		
90 deg. ell	EA.	2.667
45 deg. ell	"	2.667
Tee	"	4.000
Reducing tee	"	4.000
Cap	"	1.600
10" pipe		
90 deg ell	EA.	2.667
45 deg. ell	"	2.667
Tee	"	4.000
Reducing tee	"	4.000
Cap	"	2.000
12" pipe		
90 deg. ell	EA.	3.200
45 deg. ell	"	3.200
Tee	"	5.714
Reducing tee	"	5.714
Cap	"	2.000
Cast iron fittings		
1/2" pipe		
90 deg. ell	EA.	0.267
45 deg. ell	"	0.267
Tee	"	0.400
Reducing tee	"	0.400
3/4" pipe		
90 deg. ell	EA.	0.267
45 deg. ell	"	0.267
Tee	"	0.400
Reducing tee	"	0.400
1" pipe		
90 deg. ell	EA.	0.320
45 deg. ell	"	0.320
Tee	"	0.444
Reducing tee	"	0.444
1-1/2" pipe		
90 deg. ell	EA.	0.400
45 deg. ell	"	0.400
Tee	"	0.571
Reducing tee	"	0.571
2-1/2" pipe		
90 deg. ell	EA.	0.800
45 deg. ell	"	0.800
Tee	"	1.143
Reducing tee	"	1.143
3" pipe		
90 deg. ell	EA.	1.000
45 deg. ell	"	1.000

Plumbing	UNIT	MAN/HOURS
15410.80 Steel Pipe *(Cont.)*		
Tee	EA.	1.600
Reducing tee	"	1.600
4" pipe		
90 deg. ell	EA.	1.333
45 deg. ell	"	1.333
Tee	"	2.000
Reducing tee	"	2.000
6" pipe		
90 deg. ell	EA.	1.333
45 deg. ell	"	1.333
Tee	"	2.000
Reducing tee	"	2.000
8" pipe		
90 deg. ell	EA.	2.667
45 deg. ell	"	2.667
Tee	"	4.000
Reducing tee	"	4.000
15410.82 Galvanized Steel Pipe		
Galvanized pipe		
1/2" pipe	L.F.	0.080
3/4" pipe	"	0.100
1" pipe	"	0.114
1-1/4" pipe	"	0.133
1-1/2" pipe	"	0.160
2" pipe	"	0.200
2-1/2" pipe	"	0.267
3" pipe	"	0.286
4" pipe	"	0.333
6" pipe	"	0.667
90 degree ell, 150 lb malleable iron, galvanized		
1/2"	EA.	0.160
3/4"	"	0.200
1"	"	0.211
1-1/4"	"	0.235
1-1/2"	"	0.267
2"	"	0.320
2-1/2"	"	0.500
3"	"	0.615
4"	"	0.667
5"	"	0.800
6"	"	0.800
45 degree ell, 150 lb m.i., galv.		
1/2"	EA.	0.160
3/4"	"	0.200
1"	"	0.211
1-1/4"	"	0.235
1-1/2"	"	0.267
2"	"	0.320
2-1/2"	"	0.500
3"	"	0.615
4"	"	0.800
5"	"	0.800

15 MECHANICAL

Plumbing	UNIT	MAN/HOURS
15410.82 Galvanized Steel Pipe *(Cont.)*		
6"	EA.	1.000
Tees, straight, 150 lb m.i., galv.		
1/2"	EA.	0.200
3/4"	"	0.229
1"	"	0.267
1-1/4"	"	0.320
1-1/2"	"	0.400
2"	"	0.500
2-1/2"	"	0.667
3"	"	0.800
4"	"	1.000
5"	"	1.143
6"	"	1.333
Tees, reducing, out, 150 lb m.i., galv.		
1/2"	EA.	0.200
3/4"	"	0.229
1"	"	0.267
1-1/4"	"	0.320
1-1/2"	"	0.400
2"	"	0.500
2-1/2"	"	0.667
3"	"	0.800
4"	"	1.000
5"	"	1.143
6"	"	1.333
Couplings, straight, 150 lb m.i., galv.		
1/2"	EA.	0.160
3/4"	"	0.178
1"	"	0.200
1-1/4"	"	0.229
1-1/2"	"	0.267
2"	"	0.320
2-1/2"	"	0.500
3"	"	0.667
4"	"	0.727
5"	"	0.800
6"	"	0.800
Couplings, reducing, 150 lb m.i., galv		
1/2"	EA.	0.160
3/4"	"	0.178
1"	"	0.200
1-1/4"	"	0.229
1-1/2"	"	0.267
2"	"	0.320
2-1/2"	"	0.500
3"	"	0.667
4"	"	0.727
5"	"	0.800
6"	"	0.800
Caps, 150 lb m.i., galv.		
1/2"	EA.	0.080
3/4"	"	0.084
1"	"	0.089

Plumbing	UNIT	MAN/HOURS
15410.82 Galvanized Steel Pipe *(Cont.)*		
1-1/4"	EA.	0.094
1-1/2"	"	0.100
2"	"	0.114
2-1/2"	"	0.145
3"	"	0.200
4"	"	0.250
5"	"	0.308
6"	"	0.400
Unions, 150 lb m.i., galv.		
1/2"	EA.	0.200
3/4"	"	0.229
1"	"	0.267
1-1/4"	"	0.320
1-1/2"	"	0.400
2"	"	0.444
2-1/2"	"	0.533
3"	"	0.667
Nipples, galvanized steel, 4" long		
1/2"	EA.	0.100
3/4"	"	0.107
1"	"	0.114
1-1/4"	"	0.123
1-1/2"	"	0.133
2"	"	0.145
2-1/2"	"	0.160
3"	"	0.200
4"	"	0.267
90 degree reducing ell, 150 lb m.i., galv.		
3/4" x 1/2"	EA.	0.160
1" x 3/4"	"	0.178
1-1/4" x 1"	"	0.200
1-1/4" x 3/4"	"	0.229
1-1/4" x 1/2"	"	0.267
1-1/2" x 1-1/4"	"	0.267
1-1/2" x 1"	"	0.267
1-1/2" x 3/4"	"	0.267
2" x 1-1/2"	"	0.348
2" x 1-1/4"	"	0.348
2" x 1"	"	0.348
2" x 3/4"	"	0.348
2-1/2" x 2"	"	0.500
2-1/2" x 1-1/2"	"	0.500
3" x 2-1/2"	"	0.667
3" x 2"	"	0.667
4" x 3"	"	0.727
Square head plug (C.I.)		
1/2"	EA.	0.089
3/4"	"	0.100
1"	"	0.107
1-1/4"	"	0.114
1-1/2"	"	0.123
2"	"	0.133
2-1/2"	"	0.178

15 MECHANICAL

Plumbing		UNIT	MAN/HOURS
15410.82	**Galvanized Steel Pipe** (Cont.)		
3"		EA.	0.200
4"		"	0.267
5"		"	0.320
6"		"	0.400
Screwed flanges, galv.			
1"		EA.	0.400
1-1/4"		"	0.444
1-1/2"		"	0.500
2"		"	0.500
2-1/2"		"	0.533
3"		"	0.727
4"		"	1.000
5"		"	1.333
6"		"	1.333
15430.23	**Cleanouts**		
Cleanout, wall			
2"		EA.	0.533
3"		"	0.533
4"		"	0.667
6"		"	0.800
8"		"	1.000
Floor			
2"		EA.	0.667
3"		"	0.667
4"		"	0.800
6"		"	1.000
8"		"	1.143
15430.24	**Grease Traps**		
Grease traps, cast iron, 3" pipe			
35 gpm, 70 lb capacity		EA.	8.000
50 gpm, 100 lb capacity		"	10.000
15430.25	**Hose Bibbs**		
Hose bibb			
1/2"		EA.	0.267
3/4"		"	0.267
15430.60	**Valves**		
Gate valve, 125 lb, bronze, soldered			
1/2"		EA.	0.200
3/4"		"	0.200
1"		"	0.267
1-1/2"		"	0.320
2"		"	0.400
2-1/2"		"	0.500
Threaded			
1/4", 125 lb		EA.	0.320
1/2"			
125 lb		EA.	0.320
150 lb		"	0.320
300 lb		"	0.320

Plumbing		UNIT	MAN/HOURS
15430.60	**Valves** (Cont.)		
3/4"			
125 lb		EA.	0.320
150 lb		"	0.320
300 lb		"	0.320
1"			
125 lb		EA.	0.320
150 lb		"	0.320
300 lb		"	0.400
1-1/2"			
125 lb		EA.	0.400
150 lb		"	0.400
300 lb		"	0.444
2"			
125 lb		EA.	0.571
150 lb		"	0.571
300 lb		"	0.667
Cast iron, flanged			
2", 150 lb		EA.	0.667
2-1/2"			
125 lb		EA.	0.667
150 lb		"	0.667
250 lb		"	0.667
3"			
125 lb		EA.	0.800
150 lb		"	0.800
250 lb		"	0.800
4"			
125 lb		EA.	1.143
150 lb		"	1.143
250 lb		"	1.143
6"			
125 lb		EA.	1.600
250 lb		"	1.600
8"			
125 lb		EA.	2.000
250 lb		"	2.000
OS&Y, flanged			
2"			
125 lb		EA.	0.667
250 lb		"	0.667
2-1/2"			
125 lb		EA.	0.667
250 lb		"	0.800
3"			
125 lb		EA.	0.800
250 lb		"	0.800
4"			
125 lb		EA.	1.333
250 lb		"	1.333
6"			
125 lb		EA.	1.600
250 lb		"	1.600
Check valve, bronze, soldered, 125 lb			

15 MECHANICAL

Plumbing	UNIT	MAN/HOURS
15430.60 Valves *(Cont.)*		
1/2"	EA.	0.200
3/4"	"	0.200
1"	"	0.267
1-1/4"	"	0.320
1-1/2"	"	0.320
2"	"	0.400
Threaded		
1/2"		
125 lb	EA.	0.267
150 lb	"	0.267
200 lb	"	0.267
3/4"		
125 lb	EA.	0.320
150 lb	"	0.320
200 lb	"	0.320
1"		
125 lb	EA.	0.400
150 lb	"	0.400
200 lb	"	0.400
Flow check valve, cast iron, threaded		
1"	EA.	0.320
1-1/4"	"	0.400
1-1/2"		
125 lb	EA.	0.400
150 lb	"	0.400
200 lb	"	0.444
2"		
125 lb	EA.	0.444
150 lb	"	0.444
200 lb	"	0.500
2-1/2"		
125 lb	EA.	0.667
250 lb	"	0.800
3"		
125 lb	EA.	0.800
250 lb	"	1.000
4"		
125 lb	EA.	1.143
250 lb	"	1.333
6"		
125 lb	EA.	1.600
250 lb	"	1.600
Vertical check valve, bronze, 125 lb, threaded		
1/2"	EA.	0.320
3/4"	"	0.364
1"	"	0.400
1-1/4"	"	0.444
1-1/2"	"	0.500
2"	"	0.571
Cast iron, flanged		
2-1/2"	EA.	0.800
3"	"	1.000
4"	"	1.333

Plumbing	UNIT	MAN/HOURS
15430.60 Valves *(Cont.)*		
6	EA.	1.600
8"	"	2.000
10"	"	2.667
12"	"	3.200
Globe valve, bronze, soldered, 125 lb		
1/2"	EA.	0.229
3/4"	"	0.250
1"	"	0.267
1-1/4"	"	0.286
1-1/2"	"	0.333
2"	"	0.400
Threaded		
1/2"		
125 lb	EA.	0.267
150 lb	"	0.267
300 lb	"	0.267
3/4"		
125 lb	EA.	0.320
150 lb	"	0.320
300 lb	"	0.320
1"		
125 lb	EA.	0.400
150 lb	"	0.400
300 lb	"	0.400
1-1/4"		
125 lb	EA.	0.400
150 lb	"	0.400
300 lb	"	0.400
1-1/2"		
125 lb	EA.	0.444
150 lb	"	0.444
300 lb	"	0.444
2"		
125 lb	EA.	0.533
150 lb	"	0.533
300 lb	"	0.533
Cast iron flanged		
2-1/2"		
125 lb	EA.	0.800
250 lb	"	0.800
3"		
125 lb	EA.	1.000
250 lb	"	1.000
4"		
125 lb	EA.	1.333
250 lb	"	1.333
6"		
125 lb	EA.	1.600
250 lb	"	1.600
8"		
125 lb	EA.	2.000
250 lb	"	2.000
Butterfly valve, cast iron, wafer type		

15 MECHANICAL

Plumbing — 15430.60 Valves (Cont.)

Description	UNIT	MAN/HOURS
2"		
150 lb	EA.	0.571
200 lb	"	0.667
2-1/2"		
150 lb	EA.	0.667
200 lb	"	0.727
3"		
150 lb	EA.	0.800
200 lb	"	0.889
4"		
150 lb	EA.	1.143
200 lb	"	1.333
6"		
150 lb	EA.	1.600
200 lb	"	1.600
8"		
150 lb	EA.	1.778
200 lb	"	2.000
10"		
150 lb	EA.	2.000
200 lb	"	2.667
Ball valve, bronze, 250 lb, threaded		
1/2"	EA.	0.320
3/4"	"	0.320
1"	"	0.400
1-1/4"	"	0.444
1-1/2"	"	0.500
2"	"	0.571
Angle valve, bronze, 150 lb, threaded		
1/2"	EA.	0.286
3/4"	"	0.320
1"	"	0.320
1-1/4"	"	0.400
1-1/2"	"	0.444
Balancing valve, meter connections, circuit setter		
1/2"	EA.	0.320
3/4"	"	0.364
1"	"	0.400
1-1/4"	"	0.444
1-1/2"	"	0.533
2"	"	0.667
2-1/2"	"	0.800
3"	"	1.000
4"	"	1.333
Balancing valve, straight type		
1/2"	EA.	0.320
3/4"	"	0.320
Angle type		
1/2"	EA.	0.320
3/4"	"	0.320
Square head cock, 125 lb, bronze body		
1/2"	EA.	0.267
3/4"	"	0.320

Plumbing — 15430.60 Valves (Cont.)

Description	UNIT	MAN/HOURS
1"	EA.	0.364
1-1/4"	"	0.400
Radiator temp control valve, with control and sensor		
1/2" valve	EA.	0.500
1" valve	"	0.500
Pressure relief valve, 1/2", bronze		
Low pressure	EA.	0.320
High pressure	"	0.320
Pressure and temperature relief valve		
Bronze, 3/4"	EA.	0.320
Cast iron, 3/4"		
High pressure	EA.	0.320
Temperature relief	"	0.320
Pressure & temp relief valve	"	0.320
Pressure reducing valve, bronze, threaded, 250 lb		
1/2"	EA.	0.500
3/4"	"	0.500
1"	"	0.500
1-1/4"	"	0.571
1-1/2"	"	0.667
Pressure regulating valve, bronze, class 300		
1"	EA.	0.500
1-1/2"	"	0.615
2"	"	0.800
3"	"	1.143
4"	"	1.600
5"	"	2.000
6"	"	2.667
Solar water temperature regulating valve		
3/4"	EA.	0.667
1"	"	0.800
1-1/4"	"	0.889
1-1/2"	"	1.000
2"	"	1.143
2-1/2"	"	2.000
Tempering valve, threaded		
3/4"	EA.	0.267
1"	"	0.320
1-1/4"	"	0.400
1-1/2"	"	0.400
2"	"	0.500
2-1/2"	"	0.667
3"	"	0.800
4"	"	1.143
Thermostatic mixing valve, threaded		
1/2"	EA.	0.286
3/4"	"	0.320
1"	"	0.348
1-1/2"	"	0.400
2"	"	0.500
Sweat connection		
1/2"	EA.	0.286
3/4"	"	0.320

15 MECHANICAL

Plumbing	UNIT	MAN/HOURS
15430.60 Valves *(Cont.)*		
Mixing valve, sweat connection		
1/2"	EA.	0.286
3/4"	"	0.320
Liquid level gauge, aluminum body		
3/4"	EA.	0.320
4125 psi, pvc body		
3/4"	EA.	0.320
150 psi, crs body		
3/4"	EA.	0.320
1"	"	0.320
175 psi, bronze body, 1/2"	"	0.286
15430.65 Vacuum Breakers		
Vacuum breaker, atmospheric, threaded connection		
3/4"	EA.	0.320
1"	"	0.320
Anti-siphon, brass		
3/4"	EA.	0.320
1"	"	0.320
1-1/4"	"	0.400
1-1/2"	"	0.444
2"	"	0.500
Air eliminators, purger, cast iron, threaded		
1"	EA.	0.320
1-1/4"	"	0.400
1-1/2"	"	0.400
2"	"	0.500
2-1/2"	"	1.000
3"	"	1.333
Airtrol fitting, 3/4"	"	0.320
Air eliminator, air vents, 1/4"	"	0.320
Air vent for hot water	"	0.286
15430.68 Strainers		
Strainer, Y pattern, 125 psi, cast iron body, threaded		
3/4"	EA.	0.286
1"	"	0.320
1-1/4"	"	0.400
1-1/2"	"	0.400
2"	"	0.500
250 psi, brass body, threaded		
3/4"	EA.	0.320
1"	"	0.320
1-1/4"	"	0.400
1-1/2"	"	0.400
2"	"	0.500
Cast iron body, threaded		
3/4"	EA.	0.320
1"	"	0.320
1-1/4"	"	0.400
1-1/2"	"	0.400
2"	"	0.500

Plumbing	UNIT	MAN/HOURS
15430.70 Drains, Roof & Floor		
Floor drain, cast iron, with cast iron top		
2"	EA.	0.667
3"	"	0.667
4"	"	0.667
6"	"	0.800
Roof drain, cast iron		
2"	EA.	0.667
3"	"	0.667
4"	"	0.667
5"	"	0.800
6"	"	0.800
15430.80 Traps		
Bucket trap, threaded		
3/4"	EA.	0.500
1"	"	0.533
1-1/4"	"	0.615
1-1/2"	"	0.727
Inverted bucket steam trap, threaded		
3/4"	EA.	0.500
1"	"	0.500
1-1/4"	"	0.444
1-1/2"	"	0.667
With stainless interior		
1/2"	EA.	0.500
3/4"	"	0.500
1"	"	0.500
1-1/4"	"	0.571
Brass interior		
3/4"	EA.	0.500
1"	"	0.533
1-1/4"	"	0.571
Cast steel body, threaded, high temperature		
3/4"	EA.	0.500
1"	"	0.571
1-1/4	"	0.615
1-1/2"	"	0.667
2"	"	0.800
Float trap, 15 psi		
3/4"	EA.	0.500
1"	"	0.533
1-1/4"	"	0.571
1-1/2"	"	0.667
2"	"	0.800
30 psi		
3/4"	EA.	0.500
1"	"	0.533
1-1/4"	"	0.667
1-1/2"	"	0.800
75 psi		
3/4"	EA.	0.500
1"	"	0.533
1-1/4"	"	0.571

15 MECHANICAL

Plumbing	UNIT	MAN/HOURS
15430.80 Traps *(Cont.)*		
1-1/2"	EA.	0.667
125 psi		
3/4"	EA.	0.500
1"	"	0.533
1-1/4	"	0.571
1-1/2	"	0.667
Float and thermostatic trap, 15 psi		
3/4"	EA.	0.500
1"	"	0.533
1-1/4"	"	0.571
1-1/2"	"	0.667
2"	"	0.800
30 psi		
3/4"	EA.	0.500
1"	"	0.533
1-1/4"	"	0.571
1-1/2"	"	0.667
75 psi		
3/4"	EA.	0.500
1"	"	0.533
1-1/4"	"	0.571
1-1/2"	"	0.667
Steam trap, cast iron body, threaded, 125 psi		
3/4"	EA.	0.500
1"	"	0.533
1-1/4"	"	0.571
1-1/2"	"	0.667
Thermostatic trap, low pressure, angle type, 25 psi		
1/2"	EA.	0.500
3/4"	"	0.500
1"	"	0.533
50 psi		
1/2"	EA.	0.500
3/4"	"	0.500
1"	"	0.533
Cast iron body, threaded, 125 psi		
3/4"	EA.	0.500
1"	"	0.571
1-1/4"	"	0.615
1-1/2"	"	0.667
Low pressure, 25 psi, swivel type, 1/2"	"	0.500
Straightway type, 3/4"	"	0.500
Vertical type, 1/2"	"	0.500
Medium pressure, 50 psi, angle type, 1/2"	"	0.500
High pressure, 125 psi, angle type		
1/2"	EA.	0.500
3/4"	"	0.500
1"	"	0.571
Straightway type		
1/2"	EA.	0.500
Thermo disc trap		
3/4"	EA.	0.500
1"	"	0.533

Plumbing	UNIT	MAN/HOURS
15430.80 Traps *(Cont.)*		
Drip pan ell, cast iron		
2-1/2"	EA.	1.000
3"	"	1.143
4"	"	1.333
5"	"	1.600
6"	"	2.000
8"	"	2.667
Steel		
2-1/2"	EA.	1.000
3"	"	1.143
4"	"	1.333
5"	"	1.600
6"	"	2.000
8"	"	2.667

Plumbing Fixtures	UNIT	MAN/HOURS
15440.10 Baths		
Bath tub, 5' long		
Minimum	EA.	2.667
Average	"	4.000
Maximum	"	8.000
6' long		
Minimum	EA.	2.667
Average	"	4.000
Maximum	"	8.000
Square tub, whirlpool, 4'x4'		
Minimum	EA.	4.000
Average	"	8.000
Maximum	"	10.000
5'x5'		
Minimum	EA.	4.000
Average	"	8.000
Maximum	"	10.000
6'x6'		
Minimum	EA.	4.000
Average	"	8.000
Maximum	"	10.000
For trim and rough-in		
Minimum	EA.	2.667
Average	"	4.000
Maximum	"	8.000

15 MECHANICAL

Plumbing Fixtures	UNIT	MAN/HOURS
15440.12 Disposals & Accessories		
Continuous feed		
Minimum	EA.	1.600
Average	"	2.000
Maximum	"	2.667
Batch feed, 1/2 hp		
Minimum	EA.	1.600
Average	"	2.000
Maximum	"	2.667
Hot water dispenser		
Minimum	EA.	1.600
Average	"	2.000
Maximum	"	2.667
Epoxy finish faucet	"	1.600
Lock stop assembly	"	1.000
Mounting gasket	"	0.667
Tailpipe gasket	"	0.667
Stopper assembly	"	0.800
Switch assembly, on/off	"	1.333
Tailpipe gasket washer	"	0.400
Stop gasket	"	0.444
Tailpipe flange	"	0.400
Tailpipe	"	0.500
15440.15 Faucets		
Kitchen		
Minimum	EA.	1.333
Average	"	1.600
Maximum	"	2.000
Bath		
Minimum	EA.	1.333
Average	"	1.600
Maximum	"	2.000
Lavatory, domestic		
Minimum	EA.	1.333
Average	"	1.600
Maximum	"	2.000
Hospital, patient rooms		
Minimum	EA.	2.000
Average	"	2.667
Maximum	"	4.000
Operating room		
Minimum	EA.	2.000
Average	"	2.667
Maximum	"	4.000
Washroom		
Minimum	EA.	1.333
Average	"	1.600
Maximum	"	2.000
Handicapped		
Minimum	EA.	1.600
Average	"	2.000
Maximum	"	2.667
Shower		

Plumbing Fixtures	UNIT	MAN/HOURS
15440.15 Faucets (Cont.)		
Minimum	EA.	1.333
Average	"	1.600
Maximum	"	2.000
For trim and rough-in		
Minimum	EA.	1.600
Average	"	2.000
Maximum	"	4.000
15440.18 Hydrants		
Wall hydrant		
8" thick	EA.	1.333
12" thick	"	1.600
18" thick	"	1.778
24" thick	"	2.000
Ground hydrant		
2' deep	EA.	1.000
4' deep	"	1.143
6' deep	"	1.333
8' deep	"	2.000
15440.20 Lavatories		
Lavatory, counter top, porcelain enamel on cast iron		
Minimum	EA.	1.600
Average	"	2.000
Maximum	"	2.667
Wall hung, china		
Minimum	EA.	1.600
Average	"	2.000
Maximum	"	2.667
Handicapped		
Minimum	EA.	2.000
Average	"	2.667
Maximum	"	4.000
For trim and rough-in		
Minimum	EA.	2.000
Average	"	2.667
Maximum	"	4.000
15440.30 Showers		
Shower, fiberglass, 36"x34"x84"		
Minimum	EA.	5.714
Average	"	8.000
Maximum	"	8.000
Steel, 1 piece, 36"x36"		
Minimum	EA.	5.714
Average	"	8.000
Maximum	"	8.000
Receptor, molded stone, 36"x36"		
Minimum	EA.	2.667
Average	"	4.000
Maximum	"	6.667
For trim and rough-in		
Minimum	EA.	3.636

15 MECHANICAL

Plumbing Fixtures	UNIT	MAN/HOURS
15440.30 Showers (Cont.)		
Average	EA.	4.444
Maximum	"	8.000
15440.40 Sinks		
Service sink, 24"x29"		
Minimum	EA.	2.000
Average	"	2.667
Maximum	"	4.000
Kitchen sink, single, stainless steel, single bowl		
Minimum	EA.	1.600
Average	"	2.000
Maximum	"	2.667
Double bowl		
Minimum	EA.	2.000
Average	"	2.667
Maximum	"	4.000
Porcelain enamel, cast iron, single bowl		
Minimum	EA.	1.600
Average	"	2.000
Maximum	"	2.667
Double bowl		
Minimum	EA.	2.000
Average	"	2.667
Maximum	"	4.000
Mop sink, 24"x36"x10"		
Minimum	EA.	1.600
Average	"	2.000
Maximum	"	2.667
Washing machine box		
Minimum	EA.	2.000
Average	"	2.667
Maximum	"	4.000
For trim and rough-in		
Minimum	EA.	2.667
Average	"	4.000
Maximum	"	5.333
15440.50 Urinals		
Urinal, flush valve, floor mounted		
Minimum	EA.	2.000
Average	"	2.667
Maximum	"	4.000
Wall mounted		
Minimum	EA.	2.000
Average	"	2.667
Maximum	"	4.000
For trim and rough-in		
Minimum	EA.	2.000
Average	"	4.000
Maximum	"	5.333

Plumbing Fixtures	UNIT	MAN/HOURS
15440.60 Water Closets		
Water closet flush tank, floor mounted		
Minimum	EA.	2.000
Average	"	2.667
Maximum	"	4.000
Handicapped		
Minimum	EA.	2.667
Average	"	4.000
Maximum	"	8.000
Bowl, with flush valve, floor mounted		
Minimum	EA.	2.000
Average	"	2.667
Maximum	"	4.000
Wall mounted		
Minimum	EA.	2.000
Average	"	2.667
Maximum	"	4.000
For trim and rough-in		
Minimum	EA.	2.000
Average	"	2.667
Maximum	"	4.000
15440.70 Water Heaters		
Water heater, electric		
6 gal	EA.	1.333
10 gal	"	1.333
15 gal	"	1.333
20 gal	"	1.600
30 gal	"	1.600
40 gal	"	1.600
52 gal	"	2.000
66 gal	"	2.000
80 gal	"	2.000
100 gal	"	2.667
120 gal	"	2.667
Oil fired		
20 gal	EA.	4.000
50 gal	"	5.714
15440.90 Miscellaneous Fixtures		
Electric water cooler		
Floor mounted	EA.	2.667
Wall mounted	"	2.667
Wash fountain		
Wall mounted	EA.	4.000
Circular, floor supported	"	8.000
Deluge shower and eye wash	"	4.000
15440.95 Fixture Carriers		
Water fountain, wall carrier		
Minimum	EA.	0.800
Average	"	1.000
Maximum	"	1.333
Lavatory, wall carrier		

15 MECHANICAL

Plumbing Fixtures	UNIT	MAN/HOURS
15440.95 Fixture Carriers (Cont.)		
Minimum	EA.	0.800
Average	"	1.000
Maximum	"	1.333
Sink, industrial, wall carrier		
Minimum	EA.	0.800
Average	"	1.000
Maximum	"	1.333
Toilets, water closets, wall carrier		
Minimum	EA.	0.800
Average	"	1.000
Maximum	"	1.333
Floor support		
Minimum	EA.	0.667
Average	"	0.800
Maximum	"	1.000
Urinals, wall carrier		
Minimum	EA.	0.800
Average	"	1.000
Maximum	"	1.333
Floor support		
Minimum	EA.	0.667
Average	"	0.800
Maximum	"	1.000
15450.30 Pumps		
In-line pump, bronze, centrifugal		
5 gpm, 20' head	EA.	0.500
20 gpm, 40' head	"	0.500
50 gpm		
50' head	EA.	1.000
100' head	"	1.000
70 gpm, 100' head	"	1.333
100 gpm, 80' head	"	1.333
250 gpm, 150' head	"	2.000
Cast iron, centrifugal		
50 gpm, 200' head	EA.	1.000
100 gpm		
100' head	EA.	1.333
200' head	"	1.333
200 gpm		
100' head	EA.	2.000
200' head	"	2.000
Centrifugal, close coupled, c.i., single stage		
50 gpm, 100' head	EA.	1.000
100 gpm, 100' head	"	1.333
Base mounted		
50 gpm, 100' head	EA.	1.000
100 gpm, 50' head	"	1.333
200 gpm, 100' head	"	2.000
300 gpm, 175' head	"	2.000
Suction diffuser, flanged, strainer		
3" inlet, 2-1/2" outlet	EA.	1.000
3" outlet	"	1.000

Plumbing Fixtures	UNIT	MAN/HOURS
15450.30 Pumps (Cont.)		
4" inlet		
3" outlet	EA.	1.333
4" outlet	"	1.333
6" inlet		
4" outlet	EA.	1.600
5" outlet	"	1.600
6" Outlet	"	1.600
8" inlet		
6" outlet	EA.	2.000
8" outlet	"	2.000
10" inlet		
8" outlet	EA.	2.667
Vertical turbine		
Single stage, C.I., 3550 rpm, 200 gpm, 50'head	EA.	2.667
Multi stage, 3550 rpm		
50 gpm, 100' head	EA.	2.000
100 gpm		
100' head	EA.	2.000
200 gpm		
50' head	EA.	2.667
100' head	"	2.667
Bronze		
Single stage, 3550 rpm, 100 gpm, 50' head	EA.	2.000
Multi stage, 3550 rpm, 50 gpm, 100' head	"	2.000
100 gpm		
100' head	EA.	2.000
200 gpm		
50' head	EA.	2.667
100' head	"	2.667
Sump pump, bronze, 1750 rpm, 25 gpm		
20' head	EA.	10.000
150' head	"	13.333
50 gpm		
100' head	EA.	10.000
100 gpm		
50' head	EA.	10.000
Condensate pump, simplex		
1000 sf EDR, 2 gpm	EA.	6.667
2000 sf EDR, 3 gpm	"	6.667
4000 sf EDR, 6 gpm	"	7.273
6000 sf EDR, 9 gpm	"	7.273
Duplex, bronze		
8000 sf EDR, 12 gpm	EA.	7.273
10,000 sf EDR, 15 gpm	"	10.000
15,000 sf EDR, 23 gpm	"	11.429
20,000 sf EDR, 30 gpm	"	16.000
25,000 sf EDR, 38 gpm	"	16.000
30,000 sf EDR, 45 gpm	"	17.778
40,000 sf EDR, 60 gpm	"	10.000
50,000 sf EDR, 75 gpm	"	11.429
75,000 sf EDR, 112 gpm	"	13.333
100,000 sf EDR, 150 gpm	"	20.000

15 MECHANICAL

Plumbing Fixtures	UNIT	MAN/HOURS
15450.40 Storage Tanks		
Hot water storage tank, cement lined		
10 gallon	EA.	2.667
70 gallon	"	4.000
200 gallon	"	5.714
900 gallon	"	10.000
1100 gallon	"	10.000
2000 gallon	"	10.000
15480.10 Special Systems		
Air compressor, air cooled, two stage		
5.0 cfm, 175 psi	EA.	16.000
10 cfm, 175 psi	"	17.778
20 cfm, 175 psi	"	19.048
50 cfm, 125 psi	"	21.053
80 cfm, 125 psi	"	22.857
Single stage, 125 psi		
1.0 cfm	EA.	11.429
1.5 cfm	"	11.429
2.0 cfm	"	11.429
Automotive, hose reel, air and water, 50' hose	"	6.667
Lube equipment, 3 reel, with pumps	"	32.000
Tire changer		
Truck	EA.	11.429
Passenger car	"	6.154
Air hose reel, includes, 50' hose	"	6.154
Hose reel, 5 reel, motor oil, gear oil, lube, air & water	"	32.000
Water hose reel, 50' hose	"	6.154
Pump, for motor or gear oil, fits 55 gal drum	"	0.800
For chassis lube	"	0.800
Fuel dispensing pump, lighted dial, one product		
One hose	EA.	6.667
Two hose	"	6.667
Two products, two hose	"	6.667

Heating & Ventilating	UNIT	MAN/HOURS
15555.10 Boilers		
Cast iron, gas fired, hot water		
115 mbh	EA.	20.000
175 mbh	"	21.818
235 mbh	"	24.000
940 mbh	"	48.000
1600 mbh	"	60.000
3000 mbh	"	80.000
6000 mbh	"	120.000
Steam		

Heating & Ventilating	UNIT	MAN/HOURS
15555.10 Boilers *(Cont.)*		
115 mbh	EA.	20.000
175 mbh	"	21.818
235 mbh	"	24.000
940 mbh	"	48.000
1600 mbh	"	60.000
3000 mbh	"	80.000
6000 mbh	"	120.000
Electric, hot water		
115 mbh	EA.	12.000
175 mbh	"	12.000
235 mbh	"	12.000
940 mbh	"	24.000
1600 mbh	"	48.000
3000 mbh	"	60.000
6000 mbh	"	80.000
Steam		
115 mbh	EA.	12.000
175 mbh	"	12.000
235 mbh	"	12.000
940 mbh	"	24.000
1600 mbh	"	48.000
3000 mbh	"	60.000
6000 mbh	"	80.000
Oil fired, hot water		
115 mbh	EA.	16.000
175 mbh	"	18.462
235 mbh	"	21.818
940 mbh	"	40.000
1600 mbh	"	48.000
3000 mbh	"	60.000
6000 mbh	"	120.000
Steam		
115 mbh	EA.	16.000
175 mbh	"	18.462
235 mbh	"	21.818
940 mbh	"	40.000
1600 mbh	"	48.000
3000 mbh	"	60.000
6000 mbh	"	120.000
15610.10 Furnaces		
Electric, hot air		
40 mbh	EA.	4.000
60 mbh	"	4.211
80 mbh	"	4.444
100 mbh	"	4.706
125 mbh	"	4.848
160 mbh	"	5.000
200 mbh	"	5.161
400 mbh	"	5.333
Gas fired hot air		
40 mbh	EA.	4.000
60 mbh	"	4.211

15 MECHANICAL

Heating & Ventilating	UNIT	MAN/HOURS
15610.10 **Furnaces** *(Cont.)*		
80 mbh	EA.	4.444
100 mbh	"	4.706
125 mbh	"	4.848
160 mbh	"	5.000
200 mbh	"	5.161
400 mbh	"	5.333
Oil fired hot air		
40 mbh	EA.	4.000
60 mbh	"	4.211
80 mbh	"	4.444
100 mbh	"	4.706
125 mbh	"	4.848
160 mbh	"	5.000
200 mbh	"	5.161
400 mbh	"	5.333

Refrigeration	UNIT	MAN/HOURS
15670.10 **Condensing Units**		
Air cooled condenser, single circuit		
3 ton	EA.	1.333
5 ton	"	1.333
7.5 ton	"	3.810
20 ton	"	4.000
25 ton	"	4.000
30 ton	"	4.000
40 ton	"	5.714
50 ton	"	5.714
60 ton	"	5.000
With low ambient dampers		
3 ton	EA.	2.000
5 ton	"	2.000
7.5 ton	"	4.000
20 ton	"	5.333
25 ton	"	5.333
30 ton	"	5.333
40 ton	"	6.667
50 ton	"	7.273
60 ton	"	7.273
Dual circuit		
10 ton	EA.	4.000
15 ton	"	5.714
20 ton	"	5.714
25 ton	"	5.714
30 ton	"	5.714
40 ton	"	6.667

Refrigeration	UNIT	MAN/HOURS
15670.10 **Condensing Units** *(Cont.)*		
50 ton	EA.	6.667
60 ton	"	6.667
80 ton	"	8.889
100 ton	"	8.889
120 ton	"	8.889
With low ambient dampers		
15 ton	EA.	5.714
20 ton	"	5.714
25 ton	"	5.714
30 ton	"	5.714
40 ton	"	6.667
50 ton	"	6.667
60 ton	"	6.667
80 ton	"	8.889
100 ton	"	8.889
120 ton	"	8.889
15680.10 **Chillers**		
Chiller, reciprocal		
Air cooled, remote condenser, starter		
20 ton	EA.	8.000
25 ton	"	8.000
30 ton	"	8.000
40 ton	"	12.000
50 ton	"	13.333
60 ton	"	14.118
80 ton	"	21.818
100 ton	"	24.000
120 ton	"	26.667
150 ton	"	30.000
180 ton	"	34.286
200 ton	"	40.000
Water cooled, with starter		
20 ton	EA.	8.000
25 ton	"	8.000
30 ton	"	12.000
40 ton	"	12.000
50 ton	"	13.333
60 ton	"	14.118
80 ton	"	21.818
100 ton	"	24.000
120 ton	"	26.667
150 ton	"	30.000
180 ton	"	34.286
200 ton	"	40.000
Packaged, air cooled, with starter		
20 ton	EA.	6.000
25 ton	"	6.000
30 ton	"	6.000
40 ton	"	6.000
50 ton	"	8.000
60 ton	"	8.000
80 ton	"	12.000

15 MECHANICAL

Refrigeration		UNIT	MAN/HOURS
15680.10	**Chillers** (Cont.)		
100 ton		EA.	12.000
120 ton		"	12.000
Heat recovery, air cooled, with starter			
40 ton		EA.	12.000
50 ton		"	12.000
60 ton		"	16.000
75 ton		"	24.000
100 ton		"	24.000
Water cooled, with starter			
40 ton		EA.	12.000
50 ton		"	12.000
60 ton		"	16.000
75 ton		"	24.000
100 ton		"	26.667
Centrifugal, single bundle condenser, with starter			
80 ton		EA.	34.286
130 ton		"	40.000
160 ton		"	43.636
180 ton		"	48.000
230 ton		"	53.333
280 ton		"	60.000
360 ton		"	60.000
460 ton		"	80.000
560 ton		"	85.714
670 ton		"	96.000
15710.10	**Cooling Towers**		
Cooling tower, propeller type			
100 ton		EA.	8.000
200 ton		"	12.000
300 ton		"	20.000
400 ton		"	24.000
600 ton		"	34.286
800 ton		"	48.000
1000 ton		"	60.000
Centrifugal			
100 ton		EA.	8.000
200 ton		"	12.000
300 ton		"	20.000
400 ton		"	24.000
600 ton		"	34.286
800 ton		"	48.000
1000 ton		"	60.000

Heat Transfer		UNIT	MAN/HOURS
15780.10	**Computer Room A/C**		
Air cooled, alarm, high efficiency filter, elec. heat			
3 ton		EA.	6.154
5 ton		"	6.667
7.5 ton		"	8.000
10 ton		"	10.000
15 ton		"	11.429
Steam heat			
3 ton		EA.	6.154
5 ton		"	6.667
7.5 ton		"	8.000
10 ton		"	10.000
15 ton		"	11.429
Hot water heat			
3 ton		EA.	6.154
5 ton		"	6.667
7.5 ton		"	8.000
10 ton		"	10.000
15 ton		"	11.429
Air cooled condenser, low ambient damper			
3 ton		EA.	1.600
5 ton		"	2.000
7.5 ton		"	4.000
10 ton		"	5.714
15 ton		"	4.706
Water cooled, high efficiency filter, alarm, elec. heat			
3 ton		EA.	5.714
5 ton		"	6.667
7.5 ton		"	10.000
10 ton		"	11.429
15 ton		"	13.333
Steam heat			
3 ton		EA.	5.714
5 ton		"	6.667
7.5 ton		"	10.000
10 ton		"	11.429
15 ton		"	13.333
Hot water heat			
3 ton		EA.	5.714
5 ton		"	6.667
7.5 ton		"	10.000
10 ton		"	11.429
15 ton		"	13.333
Chilled water, alarm, high eff. filter, elec. heat			
7.5 ton		EA.	7.273
10 ton		"	8.889
15 ton		"	10.000
Steam heat			
7.5 ton		EA.	7.273
10 ton		"	8.889
15 ton		"	10.000
Hot water heat			
7.5 ton		EA.	7.273
10 ton		"	8.889

15 MECHANICAL

Heat Transfer	UNIT	MAN/HOURS
15780.10 Computer Room A/C *(Cont.)*		
15 ton	EA.	10.000
15780.20 Rooftop Units		
Packaged, single zone rooftop unit, with roof curb		
2 ton	EA.	8.000
3 ton	"	8.000
4 ton	"	10.000
5 ton	"	13.333
7.5 ton	"	16.000
15830.10 Radiation Units		
Baseboard radiation unit		
1.7 mbh/lf	L.F.	0.320
2.1 mbh/lf	"	0.400
Enclosure only		
Two tier	L.F.	0.133
Three tier	"	0.133
Copper element only, 3/4" dia.		
Two tier	L.F.	0.200
Three tier	"	0.267
Fin-tube, 16 ga, sloping cover, 1-1/4" steel		
One tier	L.F.	0.267
Two tier	"	0.320
2" steel		
Two tier	L.F.	0.320
Three tier	"	0.400
1-1/4" copper		
Two tier	L.F.	0.267
18 ga flat cover, 1-1/4" steel		
One tier	L.F.	0.267
Two tier	"	0.320
Three tier	"	0.400
2" steel		
One tier	L.F.	0.267
Two tier	"	0.320
Three tier	"	0.400
1-1/4" copper		
One tier	L.F.	0.267
Two tier	"	0.320
Three tier	"	0.400
15830.20 Fan Coil Units		
Fan coil unit, 2 pipe, complete		
200 cfm ceiling hung	EA.	2.667
Floor mounted	"	2.000
300 cfm, ceiling hung	"	3.200
Floor mounted	"	2.667
400 cfm, ceiling hung	"	3.810
Floor mounted	"	2.667
500 cfm, ceiling hung	"	4.000
Floor mounted	"	3.077
600 cfm, ceiling hung	"	4.420
Floor mounted	"	3.636

Heat Transfer	UNIT	MAN/HOURS
15830.20 Fan Coil Units *(Cont.)*		
800 cfm, ceiling hung	EA.	5.000
Floor mounted	"	3.810
1000 cfm, ceiling hung	"	5.714
Floor mounted	"	4.211
1200 cfm ceiling hung	"	6.667
Floor mounted	"	5.000
15830.70 Unit Heaters		
Steam unit heater, horizontal		
12,500 btuh, 200 cfm	EA.	1.333
17,000 btuh, 300 cfm	"	1.333
40,000 btuh, 500 cfm	"	1.333
60,000 btuh, 700 cfm	"	1.333
70,000 btuh, 1000 cfm	"	2.000
Vertical		
12,500 btuh, 200 cfm	EA.	1.333
17,000 btuh, 300 cfm	"	1.333
40,000 btuh, 500 cfm	"	1.333
60,000 btuh, 700 cfm	"	1.333
70,000 btuh, 1000 cfm	"	1.333
Gas unit heater, horizontal		
27,400 btuh	EA.	3.200
38,000 btuh	"	3.200
56,000 btuh	"	3.200
82,200 btuh	"	3.200
103,900 btuh	"	5.000
125,700 btuh	"	5.000
133,200 btuh	"	5.000
149,000 btuh	"	5.000
172,000 btuh	"	5.000
190,000 btuh	"	5.000
225,000 btuh	"	5.000
Hot water unit heater, horizontal		
12,500 btuh, 200 cfm	EA.	1.333
17,000 btuh, 300 cfm	"	1.333
25,000 btuh, 500 cfm	"	1.333
30,000 btuh, 700 cfm	"	1.333
50,000 btuh, 1000 cfm	"	2.000
60,000 btuh, 1300 cfm	"	2.000
Vertical		
12,500 btuh, 200 cfm	EA.	1.333
17,000 btuh, 300 cfm	"	1.333
25,000 btuh, 500 cfm	"	1.333
30,000 btuh, 700 cfm	"	1.333
50,000 btuh, 1000 cfm	"	1.333
60,000 btuh, 1300 cfm	"	1.333
Cabinet unit heaters, ceiling, exposed, hot water		
200 cfm	EA.	2.667
300 cfm	"	3.200
400 cfm	"	3.810
600 cfm	"	4.211
800 cfm	"	5.000
1000 cfm	"	5.714

15 MECHANICAL

Heat Transfer	UNIT	MAN/HOURS
15830.70 Unit Heaters (Cont.)		
1200 cfm	EA.	6.667
2000 cfm	"	8.889

Air Handling	UNIT	MAN/HOURS
15855.10 Air Handling Units		
Air handling unit, medium pressure, single zone		
1500 cfm	EA.	5.000
3000 cfm	"	8.889
4000 cfm	"	10.000
5000 cfm	"	10.667
6000 cfm	"	11.429
7000 cfm	"	12.308
8500 cfm	"	13.333
10,500 cfm	"	16.000
12,500 cfm	"	17.778
15,500 cfm	"	22.857
17,500 cfm	"	26.667
20,500 cfm	"	32.000
25,000 cfm	"	40.000
31,500 cfm	"	53.333
Rooftop air handling units		
4950 cfm	EA.	8.889
7370 cfm	"	11.429
9790 cfm	"	13.333
14,300 cfm	"	11.429
21,725 cfm	"	11.429
33,000 cfm	"	13.333
15870.20 Exhaust Fans		
Belt drive roof exhaust fans		
640 cfm, 2618 fpm	EA.	1.000
940 cfm, 2604 fpm	"	1.000
1050 cfm, 3325 fpm	"	1.000
1170 cfm, 2373 fpm	"	1.000
2440 cfm, 4501 fpm	"	1.000
2760 cfm, 4950 fpm	"	1.000
3890 cfm, 6769 fpm	"	1.000
2380 cfm, 3382 fpm	"	1.000
2880 cfm, 3859 fpm	"	1.000
3200 cfm, 4173 fpm	"	1.333
3660 cfm, 3437 fpm	"	1.333
4070 cfm, 3694 fpm	"	1.333
5030 cfm, 3251 fpm	"	1.333
5830 cfm, 6932 fpm	"	1.600
6380 cfm, 3817 fpm	"	1.600

Air Handling	UNIT	MAN/HOURS
15870.20 Exhaust Fans (Cont.)		
8460 cfm, 6721 fpm	EA.	1.600
10,970 cfm, 5906 fpm	"	2.000
12,470 cfm, 6620 fpm	"	2.667
7000 cfm, 3449 fpm	"	2.000
13,000 cfm, 5456 fpm	"	2.000
11,250 cfm, 4854 fpm	"	2.000
18,490 cfm, 7405 fpm	"	3.636
11,300 cfm, 3232 fpm	"	3.478
18,330 cfm, 4488 fpm	"	3.478
21,720 cfm, 5131 fpm	"	3.478
31,110 cfm, 6965 fpm	"	4.000
Direct drive fans		
60 to 390 cfm	EA.	1.000
145 to 590 cfm	"	1.000
295 to 860 cfm	"	1.000
235 to 1300 cfm	"	1.000
415 to 1630 cfm	"	1.000
590 to 2045 cfm	"	1.000
805 cfm, 3235 fpm	"	1.000
1455 cfm, 4360 fpm	"	1.000
1385 cfm, 3655 fpm	"	1.000
2260 cfm, 4930 fpm	"	1.000
1720 cfm, 3870 fpm	"	1.000
2700 cfm, 5220 fpm	"	1.000
Terminal blenders and cooling		
400 cfm	EA.	1.600
800 cfm	"	1.600
1200 cfm	"	2.000
2000 cfm	"	2.000

Air Distribution	UNIT	MAN/HOURS
15890.10 Metal Ductwork		
Rectangular duct		
Galvanized steel		
Minimum	Lb.	0.073
Average	"	0.089
Maximum	"	0.133
Aluminum		
Minimum	Lb.	0.160
Average	"	0.200
Maximum	"	0.267
Fittings		
Minimum	EA.	0.267
Average	"	0.400
Maximum	"	0.800

15 MECHANICAL

Air Distribution	UNIT	MAN/HOURS
15890.30 Flexible Ductwork		
Flexible duct, 1.25" fiberglass		
5" dia.	L.F.	0.040
6" dia.	"	0.044
7" dia.	"	0.047
8" dia.	"	0.050
10" dia.	"	0.057
12" dia.	"	0.062
14" dia.	"	0.067
16" dia.	"	0.073
Flexible duct connector, 3" wide fabric	"	0.133
15895.10 Roof Curbs		
8" high, insulated, with liner and raised can		
15" x 15"	EA.	0.400
17" x 17"	"	0.400
19" x 19"	"	0.400
21" x 21"	"	0.400
25" x 25"	"	0.500
28" x 28"	"	0.533
32" x 32"	"	0.571
36" x 36"	"	0.571
40" x 40"	"	0.571
44" x 44"	"	0.615
48" x 48"	"	0.615
52" x 52"	"	0.667
56" x 56"	"	0.667
60" x 60"	"	0.800
64" x 64"	"	0.800
68" x 68"	"	0.889
72" x 72"	"	1.000
15910.10 Dampers		
Horizontal parallel aluminum backdraft damper		
12" x 12"	EA.	0.200
16" x 16"	"	0.229
20" x 20"	"	0.286
24" x 24"	"	0.400
28" x 28"	"	0.444
32" x 32"	"	0.500
36" x 36"	"	0.571
40" x 40"	"	0.667
44" x 44"	"	0.727
48" x 48"	"	0.800
"Up", parallel dampers		
12" x 12"	EA.	0.200
16" x 16"	"	0.229
20" x 20"	"	0.286
24" x 24"	"	0.400
28" x 28"	"	0.444
32" x 32"	"	0.500
36" x 36"	"	0.571
40" x 40"	"	0.667
44" x 44"	"	0.727

Air Distribution	UNIT	MAN/HOURS
15910.10 Dampers (Cont.)		
48" x 48"	EA.	0.800
"Down", parallel dampers		
12" x 12"	EA.	0.200
16" x 16"	"	0.229
20" x 20"	"	0.286
24" x 24"	"	0.400
28" x 28"	"	0.444
32" x 32"	"	0.500
36" x 36"	"	0.571
40" x 40"	"	0.667
44" x 44"	"	0.727
48" x 48"	"	0.800
Fire damper, 1.5 hr rating		
12" x 12"	EA.	0.400
16" x 16"	"	0.400
20" x 20"	"	0.400
24" x 24"	"	0.400
28" x 28"	"	0.571
32" x 32"	"	0.667
36" x 36"	"	0.800
40" x 40"	"	0.889
44" x 44"	"	1.000
48" x 48"	"	1.143
15940.10 Diffusers		
Ceiling diffusers, round, baked enamel finish		
6" dia.	EA.	0.267
8" dia.	"	0.333
10" dia.	"	0.333
12" dia.	"	0.333
14" dia.	"	0.364
16" dia.	"	0.364
18" dia.	"	0.400
20" dia.	"	0.400
Rectangular		
6x6"	EA.	0.267
9x9"	"	0.400
12x12"	"	0.400
15x15"	"	0.400
18x18"	"	0.400
21x21"	"	0.500
24x24"	"	0.500
Lay in, flush mounted, perforated face, with grid		
6x6/24x24	EA.	0.320
8x8/24x24	"	0.320
9x9/24x24	"	0.320
10x10/24x24	"	0.320
12x12/24x24	"	0.320
15x15/24x24	"	0.320
18x6/24x24	"	0.320
18x18/24x24	"	0.320
Two-way slot diffuser with balancing damper, 4'	"	0.800

15 MECHANICAL

Air Distribution	UNIT	MAN/HOURS
15940.20 Relief Ventilators		
Intake ventilator, aluminum, with screen, no curbs		
12" x 12"	EA.	0.667
16" x 16"	"	0.800
20" x 20"	"	0.800
30" x 30"	"	1.143
36" x 36"	"	1.333
42" x 42"	"	1.333
48" x 48"	"	1.600
15940.40 Registers And Grilles		
Lay in flush mounted, perforated face, return		
6x6/24x24	EA.	0.320
8x8/24x24	"	0.320
9x9/24x24	"	0.320
10x10/24x24	"	0.320
12x12/24x24	"	0.320
Rectangular, ceiling return, single deflection		
10x10	EA.	0.400
12x12	"	0.400
14x14	"	0.400
16x8	"	0.400
16x16	"	0.400
18x8	"	0.400
20x20	"	0.400
24x12	"	0.400
24x18	"	0.400
36x24	"	0.444
36x30	"	0.444
Wall, return air register		
12x12	EA.	0.200
16x16	"	0.200
18x18	"	0.200
20x20	"	0.200
24x24	"	0.200
Ceiling, return air grille		
6x6	EA.	0.267
8x8	"	0.320
10x10	"	0.320
Ceiling, exhaust grille, aluminum egg crate		
6x6	EA.	0.267
8x8	"	0.320
10x10	"	0.320
12x12	"	0.400
14x14	"	0.400
16x16	"	0.400
18x18	"	0.400
15940.80 Penthouse Louvers		
Penthouse louvers		
12" high, extruded aluminum, 4" louver		
6' perimeter	EA.	2.000
8' perimeter	"	2.000
10' perimeter	"	2.000

Air Distribution	UNIT	MAN/HOURS
15940.80 Penthouse Louvers *(Cont.)*		
12' perimeter	EA.	2.000
14' perimeter	"	2.667
16' perimeter	"	3.200
18' perimeter	"	4.444
20' perimeter	"	5.333
16" high x 4' perimeter	"	2.000
6' perimeter	"	2.000
8' perimeter	"	2.000
10' perimeter	"	2.000
12' perimeter	"	2.000
14' perimeter	"	2.667
16' perimeter	"	3.200
18' perimeter	"	4.444
20' perimeter	"	5.333
22' perimeter	"	6.667
24' perimeter	"	8.889
20" high x 4' perimeter	"	2.000
6' perimeter	"	2.000
8' perimeter	"	2.000
10' perimeter	"	2.000
12' perimeter	"	2.000
14' perimeter	"	2.667
16' perimeter	"	3.200
18' perimeter	"	4.444
20' perimeter	"	5.333
22' perimeter	"	6.667
24' perimeter	"	8.889
24" high x 4' perimeter	"	2.000
6' perimeter	"	2.000
8' perimeter	"	2.000
10' perimeter	"	2.000
12' perimeter	"	2.000
16' perimeter	"	3.200
18' perimeter	"	4.444
20' perimeter	"	5.333
22' perimeter	"	6.667
24' perimeter	"	8.889

Controls	UNIT	MAN/HOURS
15950.10 Hvac Controls		
Pressure gauge, direct reading gage cock and siphon	EA.	0.500
Control valve, 1", modulating		
2-way	EA.	0.667
3-way	"	1.000
Self contained control valve w/ sensing elmnt, 3/4"	"	0.500

15 MECHANICAL

Controls	UNIT	MAN/HOURS
15950.10 Hvac Controls *(Cont.)*		
Control dampers, round		
6" dia.	EA.	0.320
8" dia	"	0.320
10" dia	"	0.320
12" dia	"	0.320
12" dia	"	0.400
18" dia	"	0.400
20" dia	"	0.400
Rectangular, parallel blade standard leakage		
12" x 12"	EA.	0.400
16" x 16"	"	0.400
20" x 20"	"	0.400
28" x 28"	"	0.500
32" x 32"	"	0.500
36" x 36"	"	0.667
40" x 40"	"	0.800
44" x 44"	"	1.000
48" x 48"	"	1.143
48" x 52"	"	1.333
48" x 56"	"	1.333
48" x 60"	"	1.333
48" x 64"	"	1.333
48" x 68"	"	1.333
48" x 72"	"	1.333
Low leakage		
12" x 12"	EA.	0.400
16" x 16"	"	0.400
20" x 20"	"	0.400
24" x 24"	"	0.400
28" x 28"	"	0.500
32" x 32"	"	0.571
36" x 36"	"	0.667
40" x 40"	"	0.800
44" x 44"	"	1.000
48" x 48"	"	1.143
48" x 56"	"	1.333
48" x 60"	"	1.333
48" x 64"	"	1.333
48" x 68"	"	1.333
48" x 72"	"	1.333
Rectangular, opposed horizontal blade		
12" x 12"	EA.	0.400
16" x 16"	"	0.400
20" x 20"	"	0.400
24" x 24"	"	0.400
28" x 28"	"	0.500
32" x 32"	"	0.533
36" x 36"	"	0.667
40" x 40"	"	0.800
44" x 44"	"	1.000
48" x 48"	"	1.143

Controls	UNIT	MAN/HOURS
15950.10 Hvac Controls *(Cont.)*		
48" x 52"	EA.	1.143
48" x 56"	"	1.333
48" x 60"	"	1.333
48" x 64"	"	1.333
48" x 68"	"	1.333
48" x 72"	"	1.333

16 ELECTRICAL

16050.30 Bus Duct

Basic Materials	UNIT	MAN/HOURS
Bus duct, 100a, plug-in		
10', 600v	EA.	2.759
With ground	"	4.211
10', 277/480v	"	2.759
With ground	"	4.211
Cable tap box	"	2.500
End closure	"	0.400
Edgewise hanger	"	0.727
Flatwise hanger	"	0.727
Outside elbow	"	0.800
Inside elbow	"	0.800
Outside tee	"	1.100
Inside tee	"	1.100
Outlet cover	"	0.400
Wall flange	"	0.400
Circuit breakers, with enclosure		
1 pole		
15a-60a	EA.	1.000
70a-100a	"	1.250
2 pole		
15a-60a	EA.	1.100
70a-100a	"	1.301
3 pole		
15a-60a	EA.	1.159
70a-100a	"	1.509
Bus duct, copper feeder duct, 277/480v, 4 wire		
800a	L.F.	0.400
1000a	"	0.500
1200a	"	0.533
1350a	"	0.615
1600a	"	0.727
2000a	"	0.800
2500a	"	0.851
3000a	"	0.952
Weatherproof		
800a	L.F.	0.444
1000a	"	0.533
1350a	"	0.667
1600a	"	0.727
2000a	"	0.833
2500a	"	0.899
3000a	"	0.976
4000a	"	1.509
5000a	"	1.818
Plug-in feeder duct, 277/480v, 4 wire		
400a	L.F.	0.400
600a	"	0.444
800a	"	0.500
1000a	"	0.500
1200a	"	0.533
1350a	"	0.615
1600a	"	0.727
2000a	"	0.800

16050.30 Bus Duct (Cont.)

Basic Materials	UNIT	MAN/HOURS
2500a	L.F.	0.851
3000a	"	0.952
Copper flanged ends, 277/480v, 4 wire		
225a	EA.	2.500
400a	"	2.759
600a	"	2.963
800a	"	3.077
1000a	"	3.200
1200a	"	3.265
1350a	"	3.333
1600a	"	3.478
2000a	"	3.478
2500a	"	3.636
3000a	"	3.810
4000a	"	4.444
5000a	"	4.706
Bus duct, copper elbows, 277/480v-4w		
225a-1000a	EA.	2.105
1200a-3000a	"	2.500
4000a-5000a	"	2.963
Tees, 277/480v-4w		
225a-1000a	EA.	2.222
1200a-3000a	"	2.581
4000a-5000a	"	2.963
Crosses, 277/480v-4w		
225a-1000a	EA.	2.222
1200a-3000a	"	2.581
4000a-5000a	"	2.963
Copper end closures, 277/480v-4w		
225a-1000a	EA.	0.899
1200a-3000a	"	1.194
4000a-5000a	"	1.667
Tap boxes, 277/480v-4w		
225a	EA.	3.478
400a	"	4.444
600a	"	7.273
800a	"	8.000
1000a	"	10.000
1200a	"	11.004
1350a	"	13.008
1600a	"	14.011
2000a	"	16.985
2500a	"	22.989
3000a	"	27.972
4000a	"	37.915
5000a	"	44.944
Circuit breaker, adapter cubicle		
225a	EA.	1.509
400a	"	1.600
600a	"	1.702
800a	"	1.818
1000a	"	1.905
1200a	"	2.000

16 ELECTRICAL

Basic Materials	UNIT	MAN/HOURS
16050.30 Bus Duct *(Cont.)*		
1600a	EA.	2.105
2000a	"	2.222
Transformer taps, 1 phase 277/480v		
600a	EA.	7.273
800a	"	8.000
1000a	"	10.000
1200a	"	11.004
1350a	"	13.008
1600a	"	14.011
2000a	"	16.985
2500a	"	22.989
3000a	"	27.972
4000a	"	37.915
5000a	"	45.977
3 phase, 480v, 3 wire		
600a	EA.	7.273
800a	"	8.000
1000a	"	10.000
1200a	"	11.004
1350a	"	13.008
1600a	"	14.011
2000a	"	16.985
2500a	"	22.989
3000a	"	27.972
4000a	"	37.915
5000a	"	45.977
3 phase, 4 wire, 277/480v		
600a	EA.	7.273
800a	"	8.000
1000a	"	10.000
1200a	"	11.004
1350a	"	13.008
1600a	"	14.011
2000a	"	16.985
2500a	"	22.989
3000a	"	27.972
4000a	"	40.816
5000a	"	45.977
Transformer connection, 4 wire, 277/480v		
600a	EA.	2.759
800a	"	2.857
1000a	"	2.963
1200a	"	3.077
1350a	"	3.200
1600a	"	3.333
2000a	"	3.478
2500a	"	3.636
3000a	"	3.810
4000a	"	4.444
5000a	"	4.706
Unfused reducers, 3 wire, 480v, 3 phase		
400a	EA.	2.500
600a	"	3.810

Basic Materials	UNIT	MAN/HOURS
16050.30 Bus Duct *(Cont.)*		
800a	EA.	4.706
1000a	"	5.000
1200a	"	5.333
1350a	"	5.714
1600a	"	6.154
2000a	"	6.400
2500a	"	6.667
3000a	"	7.273
4000a	"	8.753
5000a	"	10.796
4 wire, 277/480v, 3 phase, 400a	"	2.759
600a	"	4.000
800a	"	5.000
1000a	"	5.333
1200a	"	5.714
1350a	"	6.154
1600a	"	6.667
2000a	"	6.957
2500a	"	7.273
3000a	"	8.247
4000a	"	9.744
5000a	"	11.994
Circuit breaker reducers, 4 wire, 277/480v		
400a	EA.	2.222
600a	"	3.478
800a	"	4.211
1000a	"	4.706
1350a	"	5.333
1600a	"	5.714
2000a	"	6.154
2500a	"	6.667
3000a	"	6.957
4000a	"	7.273
5000a	"	10.000
Expansion fittings, 4 wire, 277/480v		
225a	EA.	2.500
400a	"	3.810
600a	"	4.706
800a	"	5.000
1000a	"	5.333
1200a	"	5.714
1350a	"	5.926
1600a	"	6.154
2000a	"	6.667
2500a	"	7.273
3000a	"	8.753
4000a	"	10.796
5000a	"	11.994
Wall flanges		
225a-2500a	EA.	4.000
3000a-5000a	"	6.154
Weather seals	"	1.000
Roof flanges	"	4.000

16 ELECTRICAL

Basic Materials	UNIT	MAN/HOURS
16050.30 Bus Duct *(Cont.)*		
Fire barriers	EA.	1.509
Spring hangers	"	1.739
Sway brace collars	"	1.250
Hook sticks		
8'	EA.	
14'	"	
Fusible switches, 240v, 3 phase		
30a	EA.	1.000
60a	"	1.250
100a	"	1.509
200a	"	2.105
400a	"	4.000
600a	"	6.154
208v, 4 wire		
30a	EA.	1.194
60a	"	1.356
100a	"	1.818
200a	"	2.759
400a	"	5.000
600a	"	8.000
600v		
30a	EA.	1.000
60a	"	1.250
100a	"	1.509
200a	"	2.105
400a	"	4.000
600a	"	6.154
800a	"	6.667
1000a	"	8.000
1200a	"	11.004
1600a	"	11.994
480v, 4 wire		
30a	EA.	1.194
60a	"	1.356
100a	"	1.818
200a	"	2.759
400a	"	5.000
600a	"	8.000
800a	"	8.247
1000a	"	11.004
1200a	"	11.994
1600a	"	14.011
Fusible combination starters, 600v, 3 phase		
Size 0	EA.	1.290
Size 1	"	1.600
Size 2	"	1.818
Size 3	"	2.581
Circuit breaker combination starters, 600v, 3 phase		
Size 0	EA.	1.290
Size 1	"	1.600
Size 2	"	1.818
Size 3	"	2.581
Fusible combination contactors, 600v, 3 phase		

Basic Materials	UNIT	MAN/HOURS
16050.30 Bus Duct *(Cont.)*		
30a	EA.	1.290
60a	"	1.600
100a	"	1.818
200a	"	2.581
Circuit breaker, combination contactors, 600v, 3 phase		
30a	EA.	1.290
60a	"	1.600
100a	"	1.818
200a	"	2.581
Fusible contactor electrically held, 480v, 4 wire		
30a	EA.	1.290
60a	"	1.600
100a	"	1.818
200a	"	2.759
Mechanically held		
30a	EA.	1.290
60a	"	1.600
100a	"	1.860
200a	"	2.759
Circuit breakers, 240v, 3 phase		
15a-60a	EA.	1.159
70a-100a	"	1.600
600v, 3 phase		
15a-60a	EA.	1.159
125a-225a	"	2.286
250a-400a	"	4.211
500a-600a	"	5.333
700a-800a	"	8.000
900a-1000a	"	10.000
1200a-1600a	"	11.004
120/208v, 4 wire		
15a-60a	EA.	1.290
70a-100a	"	1.860
277/480v, 4 wire		
15a-60a	EA.	1.290
70a-100a	"	1.905
125a-225a	"	2.759
250a-400a	"	5.000
500a-600a	"	8.000
700a-800a	"	8.000
900a-1000a	"	11.004
1200a-1600a	"	14.011
600v, 3 phase, 65,000 aic.		
60a	EA.	1.159
70a-100a	"	1.600
125a-225a	"	2.286
250a-400a	"	4.211
500a-600a	"	6.154
700a-800a	"	8.000
900a-1000a	"	10.000
277/480v, 4 wire, 65,000 aic.		
15a-60a	EA.	1.290
70a-100a	"	1.905

16 ELECTRICAL

Basic Materials	UNIT	MAN/HOURS
16050.30 Bus Duct *(Cont.)*		
125a-225a	EA.	2.759
250a-400a	"	5.000
500a-600a	"	6.154
700a-800a	"	8.000
900a-1000a	"	10.349
600v, 3 phase, current limiting		
15a-60a	EA.	1.159
70a-100a	"	1.600
125a-225a	"	2.286
250a-400a	"	4.211
500a-600a	"	6.154
700a-800a	"	8.000
900a-1000a	"	10.000
277/480v, 4 wire, current limiting		
15a-60a	EA.	1.290
70a-100a	"	40.000
125a-225a	"	2.759
250a-400a	"	5.000
500a-600a	"	6.154
700a-800a	"	8.000
900a-1000a	"	11.004
Capacitors, 3 phase, 240v		
5 kvar	EA.	5.000
7.5 kvar	"	6.154
10 kvar	"	8.097
15 kvar	"	9.195
480v		
2.5 kvar	EA.	2.759
5 kvar	"	4.706
7.5 kvar	"	5.714
10 kvar	"	8.097
15 kvar	"	8.999
20 kvar	"	11.494
25 kvar	"	13.008
30 kvar	"	14.210
Transformers, 3 phase, 480v		
1.0 kva	EA.	1.739
1.5 kva	"	2.000
2 kva	"	2.500
3 kva	"	2.759
5 kva	"	4.000
7.5 kva	"	5.000
10 kva	"	5.333
16110.12 Cable Tray		
Cable tray, 6"	L.F.	0.059
Ventilated cover	"	0.030
Solid cover	"	0.030
Flat 90	EA.	0.500
Outside 90	"	0.500
Inside 90	"	0.500
Flat 45	"	0.500
Outside 45	"	0.500

Basic Materials	UNIT	MAN/HOURS
16110.12 Cable Tray *(Cont.)*		
Inside 45	EA.	0.500
Adjustable elbow	"	0.500
Support riser	"	0.500
Adjustable riser	"	0.500
Tee	"	1.739
Cross	"	1.818
Blind end	"	0.296
Expansion joint	"	0.500
Box connector	"	2.500
Standard dropout	"	0.500
2"	"	0.615
3"	"	0.800
4"	"	1.000
Cable tray, 9"	L.F.	0.070
Ventilated cover	"	0.040
Solid cover	"	0.040
Flat 90	EA.	0.533
Outside 90	"	0.533
Inside 90	"	0.533
Flat 45	"	0.533
Outside 45	"	0.533
Inside 45	"	0.533
Adjustable elbow	"	0.533
Support riser	"	0.533
Adjustable riser	"	0.533
Tee	"	1.818
Cross	"	1.818
Blind end	"	0.320
Expansion joint	"	0.533
Box connector	"	2.581
Standard dropout	"	0.533
2"	"	0.667
3"	"	0.667
4"	"	1.096
Cable tray, 12"	L.F.	0.080
Ventilated cover	"	0.050
Solid cover	"	0.050
Flat 90	EA.	0.533
Outside 90	"	0.533
Inside 90	"	0.533
Flat 45	"	0.533
Outside 45	"	0.533
Inside 45	"	0.533
Adjustable elbow	"	0.533
Support riser	"	0.533
Adjustable riser	"	0.533
Tee	"	2.000
Cross	"	2.000
Blind end	"	0.348
Expansion joint	"	0.615
Box connector	"	2.759
Standard dropout	"	0.615
2"	"	0.727

16 ELECTRICAL

Basic Materials	UNIT	MAN/HOURS
16110.12 Cable Tray *(Cont.)*		
3"	EA.	0.899
4"	"	1.159
Cable tray, 18"	L.F.	0.100
Ventilated cover	"	0.059
Solid cover	"	0.059
Flat 90	EA.	0.727
Outside 90	"	0.727
Inside 90	"	0.727
Flat 45	"	0.727
Outside 45	"	0.727
Inside 45	"	0.727
Adjustable elbow	"	0.727
Support riser	"	0.727
Adjustable riser	"	0.727
Tee	"	2.105
Cross	"	2.105
Blind end	"	0.400
Expansion joint	"	0.727
Box connector	"	2.963
Standard dropout	"	0.667
2"	"	0.727
3"	"	0.952
4"	"	1.194
Cable tray, 24"	L.F.	0.123
Ventilated cover	"	0.070
Solid cover	"	0.070
Flat 90	EA.	0.727
Outside 90	"	0.727
Inside 90	"	0.727
Flat 45	"	0.727
Outside 45	"	0.727
Inside 45	"	0.727
Adjustable elbow	"	0.727
Support riser	"	0.727
Adjustable riser	"	0.727
Tee	"	2.222
Cross	"	2.222
Blind end	"	0.444
Expansion joint	"	0.727
Box connector	"	3.478
Standard dropout	"	0.667
2"	"	0.800
3"	"	0.976
4"	"	1.250
Cable tray, 36"	L.F.	0.145
Ventilated cover	"	0.080
Solid cover	"	0.080
Flat 90	EA.	0.851
Outside 90	"	0.851
Inside 90	"	0.851
Flat 45	"	0.851
Outside 45	"	0.851
Inside 45	"	0.851

Basic Materials	UNIT	MAN/HOURS
16110.12 Cable Tray *(Cont.)*		
Adjustable elbow	EA.	0.851
Support riser	"	0.851
Adjustable riser	"	0.851
Tee	"	2.286
Cross	"	2.963
Blind end	"	0.471
Expansion joint	"	0.727
Box connector	"	3.810
Standard dropout	"	0.727
2"	"	0.800
3"	"	1.000
4"	"	1.290
Reducers		
9" - 6"	EA.	0.500
12" - 9"	"	0.500
18" - 12"	"	0.615
24" - 18"	"	0.727
36" - 18"	"	0.800
36" - 24"	"	0.899
Conduit dropouts		
3/4"	EA.	0.348
1"	"	0.348
1-1/4"	"	0.400
1-1/2"	"	0.500
2"	"	0.533
2-1/2"	"	0.727
3"	"	0.800
Wall brackets		
6"	EA.	0.145
9"	"	0.145
12"	"	0.200
18"	"	0.250
24"	"	0.296
36"	"	0.400
16110.15 Fiberglass Cable Tray		
Fiberglass cable tray, 6"	L.F.	0.040
Tray cover	"	0.030
Horizontal		
90	EA.	0.276
45	"	0.276
30	"	0.276
Inside		
90	EA.	0.276
45	"	0.276
30	"	0.276
Horizontal tee	"	0.727
Horizontal cross	"	1.096
Splice plate	"	0.145
Floor flange	"	0.348
Panel flange	"	1.250
End plate	"	0.145
Nylon rivet	"	0.050

16 ELECTRICAL

Basic Materials	UNIT	MAN/HOURS
16110.15 Fiberglass Cable Tray (Cont.)		
Barrier strip	EA.	0.050
Hold down clamp	"	0.050
Drop out	"	0.533
Cover stand off	"	0.050
Wall bracket	"	0.200
Sealer	"	
Outside		
90	EA.	0.276
45	"	0.276
30	"	0.276
Fiberglass cable tray, 9"	L.F.	0.050
Tray cover	"	0.040
Horizontal		
90	EA.	0.276
45	"	0.276
30	"	0.276
Inside		
90	EA.	0.276
45	"	0.276
30	"	0.276
Horizontal tee	"	0.727
Horizontal cross	"	1.096
Splice plate	"	0.145
Floor flange	"	0.348
Panel flange	"	1.250
End plate	"	0.145
Nylon rivet	"	0.050
Barrier strap	"	0.050
Hold down clamp	"	0.050
Drop out	"	0.533
Cover stand off	"	0.050
Wall bracket	"	0.200
Sealer	"	
Outside		
90	EA.	0.276
45	"	0.276
30	"	0.276
Fiberglass cable tray, 12"	L.F.	0.059
Tray cover	"	0.050
Horizontal		
90	EA.	0.348
45	"	0.348
30	"	0.348
Inside		
90	EA.	0.348
45	"	0.348
Inside 30	"	0.348
Horizontal tee	"	0.952
Horizontal cross	"	1.290
Splice plate	"	0.160
Floor flange	"	0.364
Panel flange	"	1.600
End plate	"	0.160

Basic Materials	UNIT	MAN/HOURS
16110.15 Fiberglass Cable Tray (Cont.)		
Nylon rivet	EA.	0.050
Barrier strip	"	0.050
Hold down clamp	"	0.050
Drop out	"	0.533
Cover stand off	"	0.050
Wall bracket	"	0.200
Sealer	"	
Outside		
90	EA.	0.348
45	"	0.348
30	"	0.348
Fiberglass cable tray, 18"	L.F.	0.070
Tray cover	"	0.059
Horizontal		
90	EA.	0.400
45	"	0.400
30	"	0.400
Inside		
90	EA.	0.400
45	"	0.400
30	"	0.400
Horizontal tee	"	1.538
Horizontal cross	"	1.356
Splice plate	"	0.160
Floor flange	"	0.381
Panel flange	"	1.702
End plate	"	0.170
Nylon rivet	"	0.050
Barrier strip	"	0.050
Hold down clamp	"	0.050
Drop out	"	0.533
Cover stand off	"	0.050
Wall bracket	"	0.200
Sealer	"	
Outside		
90	EA.	0.400
45	"	0.400
30	"	0.400
Fiberglass cable tray, 24"	L.F.	0.080
Tray cover	"	0.070
Horizontal		
90	EA.	0.500
45	"	0.500
30	"	0.500
Inside		
90	EA.	0.500
45	"	0.500
30	"	0.500
Horizontal tee	"	1.000
Horizontal cross	"	1.509
Splice plate	"	0.160
Floor flange	"	0.400
Panel flange	"	1.818

16 ELECTRICAL

Basic Materials	UNIT	MAN/HOURS
16110.15 Fiberglass Cable Tray *(Cont.)*		
End plate	EA.	0.170
Nylon Rivet	"	0.050
Barrier strip	"	0.050
Hold down clamp	"	0.050
Drop out	"	0.533
Cover stand off	"	0.050
Wall bracket	"	0.200
Sealer	"	
Outside		
90	EA.	0.500
45	"	0.500
30	"	0.500
Fiberglass cable tray, 30"	L.F.	0.089
Tray cover	"	0.080
Horizontal		
90	EA.	0.615
45	"	0.615
30	"	0.615
Inside		
90	EA.	0.615
45	"	0.615
30	"	0.615
Horizontal tee	"	1.096
Horizontal cross	"	1.600
Splice plate	"	0.160
Floor flange	"	0.400
Panel flange	"	1.905
End plate	"	0.200
Nylon rivet	"	0.050
Barrier strip	"	0.050
Hold down clamp	"	0.050
Dropout	"	0.615
Cover stand off	"	0.050
Wall bracket	"	0.200
Sealer	"	
Outside		
90	EA.	0.615
45	"	0.615
30	"	0.615
Fiberglass cable tray, 36"	L.F.	0.100
Tray cover	"	0.089
Horizontal		
90	EA.	0.667
45	"	0.667
30	"	0.667
Inside		
90	EA.	0.667
45	"	0.667
30	"	0.667
Horizontal tee	"	1.194
horizontal cross	"	1.702
Splice plate	"	0.160
Floor flange	"	0.444

Basic Materials	UNIT	MAN/HOURS
16110.15 Fiberglass Cable Tray *(Cont.)*		
Panel flange	EA.	2.105
End plate	"	0.200
Nylon rivet	"	0.050
Barrier strip	"	0.050
Hold down clamp	"	0.050
Drop out	"	0.667
5Cover stand off	"	0.050
Wall bracket	"	0.200
Sealer	"	
Outside		
90	EA.	0.667
45	"	0.667
30	"	0.667
Reducers		
12" - 6"	EA.	0.296
12" - 9"	"	0.296
18" - 6"	"	0.348
18" - 9"	"	0.348
18" - 12"	"	0.400
24" - 6"	"	0.400
24" - 9"	"	0.400
24" - 12"	"	0.400
24" - 18"	"	0.444
30" - 9"	"	0.400
30" - 12"	"	0.400
30" - 18"	"	0.444
30" - 24"	"	0.500
36" - 9"	"	0.500
36" - 12"	"	0.500
36" - 18"	"	0.533
36" - 24"	"	0.615
36" - 30"	"	0.727
16110.20 Conduit Specialties		
Rod beam clamp, 1/2"	EA.	0.050
Hanger rod		
3/8"	L.F.	0.040
1/2"	"	0.050
All thread rod		
1/4"	L.F.	0.030
3/8"	"	0.040
1/2"	"	0.050
5/8"	"	0.080
Hanger channel, 1-1/2"		
No holes	EA.	0.030
Holes	"	0.030
Channel strap		
1/2"	EA.	0.050
3/4"	"	0.050
1"	"	0.050
1-1/4"	"	0.080
1-1/2"	"	0.080
2"	"	0.080

16 ELECTRICAL

Basic Materials	UNIT	MAN/HOURS
16110.20 Conduit Specialties (Cont.)		
2-1/2"	EA.	0.123
3"	"	0.123
3-1/2"	"	0.123
4"	"	0.145
5"	"	0.145
6"	"	0.145
Conduit penetrations, roof and wall, 8" thick		
1/2"	EA.	0.615
3/4"	"	0.615
1"	"	0.800
1-1/4"	"	0.800
1-1/2"	"	0.800
2"	"	1.600
2-1/2"	"	1.600
3"	"	1.600
3-1/2"	"	2.000
4"	"	2.000
Plastic duct bank conduit spacer, 3" separation		
2"	EA.	0.050
3"	"	0.050
4"	"	0.050
5"	"	0.050
6"	"	0.050
Intermediate, 3" separation		
2"	EA.	0.050
3"	"	0.050
4"	"	0.050
5"	"	0.050
6"	"	0.050
Base with 1-1/2" separation		
2"	EA.	0.050
3"	"	0.160
4"	"	0.160
5"	"	0.160
6"	"	0.160
Intermediate, 1-1/2" separation		
2"	EA.	0.160
3"	"	0.160
3-1/2"	"	0.160
4"	"	0.160
5"	"	0.160
6"	"	0.160
OD beam clamp, 1/4"	"	0.200
Threaded rod couplings		
1/4"	EA.	0.050
3/8"	"	0.050
1/2"	"	0.050
5/8"	"	0.050
3/4"	"	0.050
Hex nuts		
1/4"	EA.	0.050
3/8"	"	0.050
1/2"	"	0.050

Basic Materials	UNIT	MAN/HOURS
16110.20 Conduit Specialties (Cont.)		
5/8"	EA.	0.050
3/4"	"	0.050
Square nuts		
1/4"	EA.	0.050
3/8"	"	0.050
3/8"	"	0.050
5/8"	"	0.050
3/4"	"	0.050
Channel closure strip	L.F.	0.133
Channel end cap	EA.	0.133
Li-channel trapeze hangers		
12" long	EA.	0.145
18" long	"	0.145
24" long	"	0.145
30" long	"	0.250
36" long	"	0.250
42" long	"	0.296
Channel spring nuts		
1/4"	EA.	0.059
3/8"	"	0.080
1/2"	"	0.100
Fireproofing, for conduit penetrations		
1/2"	EA.	0.500
3/4"	"	0.500
1"	"	0.500
1-1/4"	"	0.727
1-1/2"	"	0.727
2"	"	0.727
2-1/2"	"	0.899
3"	"	0.899
3-1/2"	"	1.250
4"	"	1.509
16110.21 Aluminum Conduit		
Aluminum conduit		
1/2"	L.F.	0.030
3/4"	"	0.040
1"	"	0.050
1-1/4"	"	0.059
1-1/2"	"	0.080
2"	"	0.089
2-1/2"	"	0.100
3"	"	0.107
3-1/2"	"	0.123
4"	"	0.145
5"	"	0.182
6"	"	0.200
90 deg. elbow		
1/2"	EA.	0.190
3/4"	"	0.250
1"	"	0.308
1-1/4"	"	0.381
1-1/2"	"	0.400

16 ELECTRICAL

Basic Materials	UNIT	MAN/HOURS
16110.21 Aluminum Conduit *(Cont.)*		
2"	EA.	0.444
2-1/2"	"	0.571
3"	"	0.667
3-1/2"	"	0.800
4"	"	0.889
5"	"	1.143
6"	"	2.222
Coupling		
1/2"	EA.	0.050
3/4"	"	0.059
1"	"	0.080
1-1/4"	"	0.089
1-1/2"	"	0.100
2"	"	0.107
2-1/2"	"	0.123
3"	"	0.123
3-1/2"	"	0.145
4"	"	0.160
5"	"	0.160
6"	"	0.190
16110.22 Emt Conduit		
EMT conduit		
1/2"	L.F.	0.030
3/4"	"	0.040
1"	"	0.050
1-1/4"	"	0.059
1-1/2"	"	0.080
2"	"	0.089
2-1/2"	"	0.100
3"	"	0.123
3-1/2"	"	0.145
4"	"	0.182
90 deg. elbow		
1/2"	EA.	0.089
3/4"	"	0.100
1"	"	0.107
1-1/4"	"	0.123
1-1/2"	"	0.145
2"	"	0.190
2-1/2"	"	0.211
3"	"	0.242
3-1/2"	"	0.258
4"	"	0.286
Connector, steel compression		
1/2"	EA.	0.089
3/4"	"	0.089
1"	"	0.089
1-1/4"	"	0.107
1-1/2"	"	0.145
2"	"	0.190
2-1/2"	"	0.250
3"	"	0.286

Basic Materials	UNIT	MAN/HOURS
16110.22 Emt Conduit *(Cont.)*		
3-1/2"	EA.	0.308
4"	"	0.333
Coupling, steel, compression		
1/2"	EA.	0.059
3/4"	"	0.059
1"	"	0.059
1-1/4"	"	0.089
1-1/2"	"	0.107
2"	"	0.145
2-1/2"	"	0.222
3"	"	0.250
3-1/2"	"	0.286
4"	"	0.308
1 hole strap, steel		
1/2"	EA.	0.040
3/4"	"	0.040
1"	"	0.040
1-1/4"	"	0.050
1-1/2"	"	0.050
2"	"	0.050
2-1/2"	"	0.059
3"	"	0.059
3-1/2"	"	0.059
4"	"	0.059
Connector, steel set screw		
1/2"	EA.	0.070
3/4"	"	0.070
1"	"	0.070
1-1/4"	"	0.107
1-1/2"	"	0.145
2"	"	0.182
2-1/2"	"	0.242
3"	"	0.267
3-1/2"	"	0.296
4"	"	0.348
Insulated throat		
1/2"	EA.	0.070
3/4"	"	0.070
1"	"	0.070
1-1/4"	"	0.107
1-1/2"	"	0.145
2"	"	0.182
2-1/2"	"	0.242
3"	"	0.267
3-1/2"	"	0.296
4"	"	0.348
Connector, die cast set screw		
1/2"	EA.	0.059
3/4"	"	0.059
1"	"	0.059
1-1/4"	"	0.089
1-1/2"	"	0.107
2"	"	0.145

16 ELECTRICAL

Basic Materials	UNIT	MAN/HOURS
16110.22 Emt Conduit *(Cont.)*		
2-1/2"	EA.	0.200
3"	"	0.222
3-1/2"	"	0.250
4"	"	0.286
Insulated throat		
1/2"	EA.	0.059
3/4"	"	0.059
1"	"	0.059
1-1/4"	"	0.089
1-1/2"	"	0.107
2"	"	0.145
2-1/2"	"	0.200
3"	"	0.222
3-1/2"	"	0.250
4"	"	0.286
Coupling, steel set screw		
1/2"	EA.	0.040
3/4"	"	0.040
1"	"	0.040
1-1/4"	"	0.050
1-1/2"	"	0.080
2"	"	0.107
2-1/2"	"	0.160
3"	"	0.190
3-1/2"	"	0.222
4"	"	0.250
Diecast set screw		
1/2"	EA.	0.040
3/4"	"	0.040
1"	"	0.040
1-1/4"	"	0.050
1-1/2"	"	0.080
2"	"	0.107
2-1/2"	"	0.160
3"	"	0.186
3-1/2"	"	0.222
4"	"	0.250
1 hole malleable straps		
1/2"	EA.	0.040
3/4"	"	0.040
1"	"	0.040
1-1/4"	"	0.050
1-1/2"	"	0.050
2"	"	0.050
2-1/2"	"	0.059
3"	"	0.059
3-1/2"	"	0.059
4"	"	0.059
EMT to rigid compression coupling		
1/2"	EA.	0.100
3/4"	"	0.100
1"	"	0.150
Set screw couplings		

Basic Materials	UNIT	MAN/HOURS
16110.22 Emt Conduit *(Cont.)*		
1/2"	EA.	0.100
3/4"	"	0.100
1"	"	0.145
Set screw offset connectors		
1/2"	EA.	0.100
3/4"	"	0.100
1"	"	0.145
Compression offset connectors		
1/2"	EA.	0.100
3/4"	"	0.100
1"	"	0.145
Type "LB" set screw condulets		
1/2"	EA.	0.229
3/4"	"	0.296
1"	"	0.381
1-1/4"	"	0.444
1-1/2"	"	0.533
2"	"	0.615
2-1/2"	"	0.727
3"	"	1.000
3-1/2"	"	1.333
4"	"	1.600
Type "T" set screw condulets		
1/2"	EA.	0.296
3/4"	"	0.400
1"	"	0.444
1-1/4"	"	0.533
1-1/2"	"	0.615
2"	"	0.667
Type "C" set screw condulets		
1/2"	EA.	0.250
3/4"	"	0.296
1"	"	0.381
1-1/4"	"	0.444
1-1/2"	"	0.533
2"	"	0.381
Type "LL" set screw condulets		
1/2"	EA.	0.250
3/4"	"	0.296
1"	"	0.381
1-1/4"	"	0.444
1-1/2"	"	0.533
2"	"	0.615
Type "LR" set screw condulets		
1/2"	EA.	0.250
3/4"	"	0.296
1"	"	0.381
1-1/4"	"	0.444
1-1/2"	"	0.533
2"	"	0.615
Type "LB" compression condulets		
1/2"	EA.	0.296
3/4"	"	0.500

16 ELECTRICAL

Basic Materials	UNIT	MAN/HOURS
16110.22 Emt Conduit *(Cont.)*		
1"	EA.	0.500
Type "T" compression condulets		
1/2"	EA.	0.400
3/4"	"	0.444
1"	"	0.615
Condulet covers		
1/2"	EA.	0.123
3/4"	"	0.123
1"	"	0.123
1-1/4"	"	0.123
1-1/2"	"	0.145
2"	"	0.145
2-1/2"	"	0.145
3"	"	0.182
3-1/2"	"	0.182
4"	"	0.182
Clamp type entrance caps		
1/2"	EA.	0.250
3/4"	"	0.296
1"	"	0.400
1-1/4"	"	0.533
1-1/2"	"	0.615
2"	"	0.899
2-1/2"	"	1.000
3"	"	1.509
3-1/2"	"	1.739
4"	"	2.222
Slip fitter type entrance caps		
1/2"	EA.	0.250
3/4"	"	0.296
1"	"	0.400
1-1/4"	"	0.533
1-1/2"	"	0.615
2"	"	0.899
2-1/2"	"	1.000
3"	"	1.509
3-1/2"	"	1.739
4"	"	2.222
16110.23 Flexible Conduit		
Flexible conduit, steel		
3/8"	L.F.	0.030
1/2	"	0.030
3/4"	"	0.040
1"	"	0.040
1-1/4"	"	0.050
1-1/2"	"	0.059
2"	"	0.080
2-1/2"	"	0.089
3"	"	0.107
Flexible conduit, liquid tight		
3/8"	L.F.	0.030
1/2"	"	0.030

Basic Materials	UNIT	MAN/HOURS
16110.23 Flexible Conduit *(Cont.)*		
3/4"	L.F.	0.040
1"	"	0.040
1-1/4"	"	0.050
1-1/2"	"	0.059
2"	EA.	0.080
2-1/2"	"	0.089
3"	"	0.107
4"	"	0.145
Connector, straight		
3/8"	EA.	0.080
1/2"	"	0.080
3/4"	"	0.089
1"	"	0.100
1-1/4"	"	0.107
1-1/2"	"	0.123
2"	"	0.145
2-1/2"	"	0.182
3"	"	0.190
Straight insulated throat connectors		
3/8"	EA.	0.123
1/2"	"	0.123
3/4"	"	0.145
1"	"	0.145
1-1/4"	"	0.182
1-1/2"	"	0.211
2"	"	0.229
2-1/2"	"	0.267
3"	"	0.333
4"	"	0.421
90 deg connectors		
3/8"	EA.	0.148
1/2"	"	0.148
3/4"	"	0.170
1"	"	0.182
1-1/4"	"	0.229
1-1/2"	"	0.250
2"	"	0.267
2-1/2"	"	0.333
3"	"	0.381
4"	"	0.444
90 degree insulated throat connectors		
3/8"	EA.	0.145
1/2"	"	0.145
3/4"	"	0.170
1"	"	0.178
1-1/4"	"	0.229
1-1/2"	"	0.250
2"	"	0.267
2-1/2"	"	0.333
3"	"	0.381
4"	"	0.444
Flexible aluminum conduit		
3/8"	L.F.	0.030

16 ELECTRICAL

Basic Materials	UNIT	MAN/HOURS
16110.23 Flexible Conduit *(Cont.)*		
1/2"	L.F.	0.030
3/4"	"	0.040
1"	"	0.040
1-1/4"	"	0.050
1-1/2"	"	0.059
2"	"	0.080
2-1/2"	"	0.089
3"	"	0.107
3-1/2"	"	0.123
4"	"	0.145
Connector, straight		
3/8"	EA.	0.100
1/2"	"	0.100
3/4"	"	0.107
1"	"	0.123
1-1/4"	"	0.145
1-1/2"	"	0.182
2"	"	0.190
2-1/2"	"	0.222
3"	"	0.276
4"	"	0.348
Straight insulated throat connectors		
3/8"	EA.	0.089
1/2"	"	0.089
3/4"	"	0.089
1"	"	0.100
1-1/4"	"	0.107
1-1/2"	"	0.123
2"	"	0.145
2-1/2"	"	0.182
3"	"	0.190
3-1/2"	"	0.222
4"	"	0.276
90 deg connectors		
3/8"	EA.	0.145
1/2"	"	0.145
3/4"	"	0.145
1"	"	0.170
1-1/4"	"	0.182
1-1/2"	"	0.200
2"	"	0.211
2-1/2"	"	0.229
3"	"	0.267
90 deg insulated throat connectors		
3/8"	EA.	0.145
1/2"	"	0.145
3/4"	"	0.145
1"	"	0.170
1-1/4"	"	0.182
1-1/2"	"	0.200
2"	"	0.211
2-1/2"	"	0.229
3"	"	0.267

Basic Materials	UNIT	MAN/HOURS
16110.23 Flexible Conduit *(Cont.)*		
3-1/2"	EA.	0.333
4"	"	0.421
16110.24 Galvanized Conduit		
Galvanized rigid steel conduit		
1/2"	L.F.	0.040
3/4"	"	0.050
1"	"	0.059
1-1/4"	"	0.080
1-1/2"	"	0.089
2"	"	0.100
2-1/2"	"	0.145
3"	"	0.182
3-1/2"	"	0.190
4"	"	0.211
5"	"	0.286
6"	"	0.381
90 degree ell		
1/2"	EA.	0.250
3/4"	"	0.308
1"	"	0.381
1-1/4"	"	0.444
1-1/2"	"	0.500
2"	"	0.533
2-1/2"	"	0.667
3"	"	0.889
3-1/2"	"	1.000
4"	"	1.333
5"	"	2.222
6"	"	3.333
Couplings, with set screws		
1/2"	EA.	0.050
3/4"	"	0.059
1"	"	0.080
1-1/4"	"	0.100
1-1/2"	"	0.123
2"	"	0.145
2-1/2"	"	0.190
3"	"	0.250
3-1/2"	"	0.286
4"	"	0.308
5"	"	0.444
6"	"	0.500
Split couplings		
1/2"	EA.	0.190
3/4"	"	0.250
1"	"	0.276
1-1/4"	"	0.308
1-1/2"	"	0.381
2"	"	0.571
2-1/2"	"	0.571
3"	"	0.727
3-1/2"	"	1.000

16 ELECTRICAL

Basic Materials	UNIT	MAN/HOURS
16110.24 Galvanized Conduit (Cont.)		
4"	EA.	1.333
5"	"	1.633
6"	"	2.051
Erickson couplings		
1/2"	EA.	0.444
3/4"	"	0.500
1"	"	0.615
1-1/4"	"	0.889
1-1/2"	"	1.000
2"	"	1.333
2-1/2"	"	1.860
3"	"	2.105
3-1/2"	"	2.500
4"	"	2.667
5"	"	2.963
6"	"	3.200
Seal fittings		
1/2"	EA.	0.667
3/4"	"	0.800
1"	"	1.000
1-1/4"	"	1.143
1-1/2"	"	1.333
2"	"	1.600
2-1/2"	"	1.905
3"	"	2.105
3-1/2"	"	2.500
4"	"	2.963
5"	"	4.444
6"	"	5.000
Entrance fitting, (weather head), threaded		
1/2"	EA.	0.444
3/4"	"	0.500
1"	"	0.571
1-1/4"	"	0.727
1-1/2"	"	0.800
2"	"	0.889
2-1/2"	"	1.000
3"	"	1.333
3-1/2"	"	1.739
4"	"	2.500
5"	"	3.478
6"	"	4.444
Locknuts		
1/2"	EA.	0.050
3/4"	"	0.050
1"	"	0.050
1-1/4"	"	0.050
1-1/2"	"	0.059
2"	"	0.059
2-1/2"	"	0.080
3"	"	0.080
3-1/2"	"	0.080
4"	"	0.089

Basic Materials	UNIT	MAN/HOURS
16110.24 Galvanized Conduit (Cont.)		
5"	EA.	0.089
6"	"	0.089
Plastic conduit bushings		
1/2"	EA.	0.123
3/4"	"	0.145
1"	"	0.190
1-1/4"	"	0.222
1-1/2"	"	0.250
2"	"	0.308
2-1/2"	"	0.500
3"	"	0.667
3-1/2"	"	0.800
4"	"	0.889
5"	"	1.143
6"	"	1.600
Conduit bushings, steel		
1/2"	EA.	0.123
3/4"	"	0.145
1"	"	0.190
1-1/4"	"	0.222
1-1/2"	"	0.250
2"	"	0.308
2-1/2"	"	0.500
3"	"	0.667
3-1/2"	"	0.800
4"	"	0.889
5"	"	1.143
6"	"	1.600
Pipe cap		
1/2"	EA.	0.050
3/4"	"	0.050
1"	"	0.050
1-1/4"	"	0.080
1-1/2"	"	0.080
2"	"	0.080
2-1/2"	"	0.089
3"	"	0.089
3-1/2"	"	0.089
4"	"	0.107
5"	"	0.145
6"	"	0.200
GRS elbows, 36" radius		
2"	EA.	0.667
2-1/2"	"	0.808
3"	"	1.053
3-1/2"	"	1.250
4"	"	1.509
5"	"	2.500
6"	"	3.810
42" radius		
2"	EA.	0.808
2-1/2"	"	1.000
3"	"	1.250

16 ELECTRICAL

Basic Materials	UNIT	MAN/HOURS
16110.24 Galvanized Conduit *(Cont.)*		
3-1/2"	EA.	1.509
4"	"	1.739
5"	"	2.857
6"	"	4.000
48" radius		
2"	EA.	0.930
2-1/2"	"	1.127
3"	"	1.429
3-1/2"	"	1.739
4"	"	2.162
5"	"	3.077
6"	"	4.444
Threaded couplings		
1/2"	EA.	0.050
3/4"	"	0.059
1"	"	0.080
1-1/4"	"	0.089
1-1/2"	"	0.100
2"	"	0.107
2-1/2"	"	0.123
3"	"	0.145
3-1/2"	"	0.145
4"	"	0.160
5"	"	0.182
6"	"	0.190
Threadless couplings		
1/2"	EA.	0.100
3/4"	"	0.123
1"	"	0.145
1-1/4"	"	0.190
1-1/2"	"	0.250
2"	"	0.308
2-1/2"	"	0.500
3"	"	0.615
3-1/2"	"	0.808
4"	"	1.000
5"	"	1.250
6"	"	5.333
Threadless connectors		
1/2"	EA.	0.100
3/4"	"	0.123
1"	"	0.145
1-1/4"	"	0.190
1-1/2"	"	0.250
2"	"	0.308
2-1/2"	"	0.500
3"	"	0.615
3-1/2"	"	0.808
4"	"	1.000
5"	"	1.250
6"	"	1.509
Setscrew connectors		
1/2"	EA.	0.080

Basic Materials	UNIT	MAN/HOURS
16110.24 Galvanized Conduit *(Cont.)*		
3/4"	EA.	0.089
1"	"	0.100
1-1/4"	"	0.123
1-1/2"	"	0.145
2"	"	0.190
2-1/2"	"	0.250
3"	"	0.308
3-1/2"	"	0.381
4"	"	0.500
5"	"	0.615
6"	"	0.808
Clamp type entrance caps		
1/2"	EA.	0.308
3/4"	"	0.381
1"	"	0.444
1-1/4"	"	0.500
1-1/2"	"	0.615
2"	"	0.727
2-1/2"	"	0.941
3"	"	1.127
3-1/2"	"	1.379
4"	"	2.424
"LB" condulets		
1/2"	EA.	0.308
3/4"	"	0.381
1"	"	0.444
1-1/4"	"	0.500
1-1/2"	"	0.615
2"	"	0.727
2-1/2"	"	1.000
3"	"	1.379
3-1/2"	"	1.739
4"	"	2.105
"T" condulets		
1/2"	EA.	0.381
3/4"	"	0.444
1"	"	0.500
1-1/4"	"	0.571
1-1/2"	"	0.615
2"	"	0.727
2-1/2"	"	1.127
3"	"	1.509
3-1/2"	"	1.860
4"	"	2.222
"X" condulets		
1/2"	EA.	0.444
3/4"	"	0.500
1"	"	0.571
1-1/4"	"	0.615
1-1/2"	"	0.667
2"	"	0.879
Blank steel condulet covers		
1/2"	EA.	0.100

16 ELECTRICAL

16110.24 Galvanized Conduit *(Cont.)*

Basic Materials	UNIT	MAN/HOURS
3/4"	EA.	0.100
1"	"	0.100
1-1/4"	"	0.123
1-1/2"	"	0.123
2"	"	0.123
2-1/2"	"	0.145
3"	"	0.145
3-1/2"	"	0.145
4"	"	0.200
Solid condulet gaskets		
1/2"	EA.	0.050
3/4"	"	0.050
1"	"	0.050
1-1/4"	"	0.080
1-1/2"	"	0.080
2"	"	0.080
2-1/2"	"	0.100
3"	"	0.100
3-1/2"	"	0.100
4"	"	0.145
One-hole malleable straps		
1/2"	EA.	0.040
3/4"	"	0.040
1"	"	0.040
1-1/4"	"	0.050
1-1/2"	"	0.050
2"	"	0.050
2-1/2"	"	0.059
3"	"	0.059
3-1/2"	"	0.059
4"	"	0.080
5"	"	0.080
6"	"	0.080
One-hole steel straps		
1/2"	EA.	0.040
3/4"	"	0.040
1"	"	0.040
1-1/4"	"	0.050
1-1/2"	"	0.050
2"	"	0.050
2-1/2"	"	0.059
3"	"	0.059
3-1/2"	"	0.059
4"	"	0.080
Bushed chase nipples		
1/2"	EA.	0.059
3/4"	"	0.070
1"	"	0.089
1-1/4"	"	0.100
1-1/2"	"	0.123
2"	"	0.145
2-1/2"	"	0.145
3"	"	0.182

16110.24 Galvanized Conduit *(Cont.)*

Basic Materials	UNIT	MAN/HOURS
3-1/2"	EA.	0.250
4"	"	0.296
Offset nipples		
1/2"	EA.	0.059
3/4"	"	0.070
1"	"	0.089
1-1/4"	"	0.107
1-1/2"	"	0.123
2"	"	0.145
3"	"	0.182
Short elbows		
1/2"	EA.	0.145
3/4"	"	0.200
1"	"	0.250
1-1/4"	"	0.296
1-1/2"	"	0.348
2"	"	0.400
Pulling elbows, female to female		
1/2"	EA.	0.250
3/4"	"	0.296
1"	"	0.400
1-1/4"	"	0.533
1-1/2"	"	0.727
2"	"	0.851
Grounding locknuts		
1/2"	EA.	0.080
3/4"	"	0.080
1"	"	0.080
1-1/4"	"	0.089
1-1/2"	"	0.089
2"	"	0.089
2-1/2"	"	0.100
3"	"	0.100
3-1/2"	"	0.100
4"	"	0.145
Insulated grounding metal bushings		
1/2"	EA.	0.190
3/4"	"	0.222
1"	"	0.250
1-1/4"	"	0.308
1-1/2"	"	0.381
2"	"	0.444
2-1/2"	"	0.667
3"	"	0.808
3-1/2"	"	0.941
4"	"	1.053
5"	"	1.569
6"	"	1.739
Nipples		
1/2" x		
4"	EA.	0.145
6"	"	0.145
8"	"	0.145

16 ELECTRICAL

Basic Materials	UNIT	MAN/HOURS
16110.24 Galvanized Conduit (Cont.)		
10"	EA.	0.145
12"	"	0.145
3/4" x		
4"	EA.	0.145
6"	"	0.145
8"	"	0.145
10"	"	0.145
12"	"	0.145
1" x		
4"	EA.	0.145
6"	"	0.145
8"	"	0.145
10"	"	0.145
12"	"	0.145
1-1/4" x		
4"	EA.	0.250
6"	"	0.250
8"	"	0.250
10"	"	0.250
12"	"	0.250
1-1/2" x		
4"	EA.	0.250
6"	"	0.250
8"	"	0.250
10"	"	0.250
12"	"	0.250
2" x		
4"	EA.	0.250
6"	"	0.250
8"	"	0.250
10"	"	0.250
12"	"	0.250
2-1/2" x		
6"	EA.	0.300
8"	"	0.300
10"	"	0.300
12"	"	0.300
3" x		
6"	EA.	0.300
8"	"	0.300
10"	"	0.300
12"	"	0.300
3-1/2" x		
6"	EA.	0.300
8"	"	0.300
10"	"	0.300
12"	"	0.300
4" x		
8"	EA.	0.400
10"	"	0.400
12"	"	0.400
5" x		
8"	EA.	0.400

Basic Materials	UNIT	MAN/HOURS
16110.24 Galvanized Conduit (Cont.)		
10"	EA.	0.400
12"	"	0.400
6" x		
8"	EA.	0.400
10"	"	0.400
12"	"	0.400
16110.25 Plastic Conduit		
PVC conduit, schedule 40		
1/2"	L.F.	0.030
3/4"	"	0.030
1"	"	0.040
1-1/4"	"	0.040
1-1/2"	"	0.050
2"	"	0.050
2-1/2"	"	0.059
3"	"	0.059
3-1/2"	"	0.080
4"	"	0.080
5"	"	0.089
6"	"	0.100
Couplings		
1/2"	EA.	0.050
3/4"	"	0.050
1"	"	0.050
1-1/4"	"	0.059
1-1/2"	"	0.059
2"	"	0.059
2-1/2"	"	0.059
3"	"	0.080
3-1/2"	"	0.080
4"	"	0.100
5"	"	0.100
6"	"	0.100
90 degree elbows		
1/2"	EA.	0.100
3/4"	"	0.123
1"	"	0.123
1-1/4"	"	0.145
1-1/2"	"	0.190
2"	"	0.222
2-1/2"	"	0.250
3"	"	0.308
3-1/2"	"	0.381
4"	"	0.500
5"	"	0.615
6"	"	0.727
Terminal adapters		
1/2"	EA.	0.100
3/4"	"	0.100
1"	"	0.100
1-1/4"	"	0.160
1-1/2"	"	0.160

16 ELECTRICAL

Basic Materials	UNIT	MAN/HOURS
16110.25 Plastic Conduit *(Cont.)*		
2"	EA.	0.160
2-1/2"	"	0.222
3"	"	0.222
3-1/2"	"	0.222
4"	"	0.381
5"	"	0.381
6"	"	0.381
End bells		
1"	EA.	0.100
1-1/4"	"	0.160
1-1/2"	"	0.160
2"	"	0.160
2-1/2"	"	0.222
3"	"	0.222
3-1/2"	"	0.222
4"	"	0.381
5"	"	0.381
6"	"	0.381
LB conduit body		
1/2"	EA.	0.190
3/4"	"	0.190
1	"	0.190
1-1/4"	"	0.308
1-1/2"	"	0.308
2"	"	0.308
2-1/2"	"	0.444
3"	"	0.533
3-1/2"	"	0.615
4"	"	0.727
Direct burial, conduit		
2"	L.F.	0.050
3"	"	0.059
4"	"	0.080
5"	"	0.089
6"	"	0.100
Encased burial conduit		
2"	L.F.	0.050
3"	"	0.059
4"	"	0.080
5"	"	0.089
6"	"	0.100
"EB" and "DB" duct, 90 degree elbows		
1-1/2"	EA.	0.145
2"	"	0.229
3"	"	0.381
3-1/2"	"	0.444
4"	"	0.533
5"	"	0.667
6"	"	0.899
45 degree elbows		
1-1/2"	EA.	0.229
2"	"	0.229
3"	"	0.381

Basic Materials	UNIT	MAN/HOURS
16110.25 Plastic Conduit *(Cont.)*		
3-1/2"	EA.	0.444
4"	"	0.533
5"	"	0.667
6"	"	0.899
Couplings		
1-1/2"	EA.	0.059
2"	"	0.059
3"	"	0.080
3-1/2"	"	0.080
4"	"	0.100
5"	"	0.100
6"	"	0.160
Bell ends		
1-1/2"	EA.	0.160
2"	"	0.160
3"	"	0.222
3-1/2"	"	0.222
4"	"	0.381
5"	"	0.381
6"	"	0.381
Female adapters, 1-1/2"	"	0.200
5 degree couplings		
1-1/2"	EA.	0.070
2"	"	0.070
3"	"	0.100
4"	"	0.145
5"	"	0.145
6"	"	0.145
45 degree elbows		
1/2"	EA.	0.123
3/4"	"	0.145
1"	"	0.145
1-1/4"	"	0.182
1-1/2"	"	0.229
2"	"	0.267
2-1/2"	"	0.296
3"	"	0.381
3-1/2"	"	0.444
4"	"	0.615
5"	"	0.727
6"	"	0.899
Female adapters		
1/2"	EA.	0.123
3/4"	"	0.123
1"	"	0.123
1-1/4"	"	0.200
1-1/2"	"	0.200
2"	"	0.200
2-1/2"	"	0.267
3"	"	0.267
3-1/2"	"	0.267
4"	"	0.444
5"	"	0.444

16 ELECTRICAL

Basic Materials	UNIT	MAN/HOURS
16110.25 Plastic Conduit *(Cont.)*		
6"	EA.	0.444
Expansion couplings		
1/2"	EA.	0.123
3/4"	"	0.123
1"	"	0.145
1-1/4"	"	0.200
1-1/2"	"	0.200
2"	"	0.200
2-1/2"	"	0.296
3"	"	0.296
3-1/2"	"	0.296
4"	"	0.444
5"	"	0.444
6"	"	0.444
Plugs		
2"	EA.	0.200
3"	"	0.296
3-1/2"	"	0.296
4"	"	0.444
5"	"	0.444
6"	"	0.500
Type "T" condulets		
1/2"	EA.	0.296
3/4"	"	0.296
1"	"	0.296
1-1/4"	"	0.500
1-1/2"	"	0.500
2"	"	0.500
EB & DB female adapters		
2"	EA.	0.250
3"	"	0.381
3-1/2"	"	0.615
4"	"	0.727
5"	"	1.000
6"	"	1.600
16110.27 Plastic Coated Conduit		
Rigid steel conduit, plastic coated		
1/2"	L.F.	0.050
3/4"	"	0.059
1"	"	0.080
1-1/4"	"	0.100
1-1/2"	"	0.123
2"	"	0.145
2-1/2"	"	0.190
3"	"	0.222
3-1/2"	"	0.250
4"	"	0.308
5"	"	0.381
90 degree elbows		
1/2"	EA.	0.308
3/4"	"	0.381
1"	"	0.444

Basic Materials	UNIT	MAN/HOURS
16110.27 Plastic Coated Conduit *(Cont.)*		
1-1/4"	EA.	0.500
1-1/2"	"	0.615
2"	"	0.800
2-1/2"	"	1.143
3"	"	1.333
3-1/2"	"	1.633
4"	"	2.000
5"	"	2.500
Couplings		
1/2"	EA.	0.059
3/4"	"	0.080
1"	"	0.089
1-1/4"	"	0.107
1-1/2"	"	0.123
2"	"	0.145
2-1/2"	"	0.182
3"	"	0.190
3-1/2"	"	0.200
4"	"	0.222
5"	"	0.250
1 hole conduit straps		
3/4"	EA.	0.050
1"	"	0.050
1-1/4"	"	0.059
1-1/2"	"	0.059
2"	"	0.059
3"	"	0.080
3-1/2"	"	0.080
4"	"	0.100
"L.B." condulets with covers		
1/2"	EA.	0.500
3/4"	"	0.500
1"	"	0.615
1-1/4"	"	0.727
1-1/2"	"	0.879
2"	"	1.000
2-1/2"	"	1.379
3"	"	1.739
3-1/2"	"	2.162
4"	"	2.500
"T" condulets with covers		
1/2"	EA.	0.571
3/4"	"	0.615
1"	"	0.667
1-1/4"	"	0.808
1-1/2"	"	0.941
2"	"	1.053
2-1/2"	"	1.509
3-1/2"	"	2.222
4"	"	2.667
5"	"	3.333

16 ELECTRICAL

Basic Materials	UNIT	MAN/HOURS
16110.28 Steel Conduit		
Intermediate metal conduit (IMC)		
1/2"	L.F.	0.030
3/4"	"	0.040
1"	"	0.050
1-1/4"	"	0.059
1-1/2"	"	0.080
2"	"	0.089
2-1/2"	"	0.119
3"	"	0.145
3-1/2"	"	0.182
4"	"	0.190
90 degree ell		
1/2"	EA.	0.250
3/4"	"	0.308
1"	"	0.381
1-1/4"	"	0.444
1-1/2"	"	0.500
2"	"	0.571
2-1/2"	"	0.667
3"	"	0.889
3-1/2"	"	1.143
4"	"	1.333
Couplings		
1/2"	EA.	0.050
3/4"	"	0.059
1"	"	0.080
1-1/4"	"	0.089
1-1/2"	"	0.100
2"	"	0.107
2-1/2"	"	0.123
3"	"	0.145
3-1/2"	"	0.145
4"	"	0.160
16110.32 Flexible Wiring Systems		
Single circuit cables		
5'	EA.	0.059
10'	"	0.100
15'	"	0.145
20'	"	0.200
25'	"	0.267
30'	"	0.296
40'	"	0.400
Two circuit cables		
5'	EA.	0.059
10'	"	0.100
15'	"	0.145
20'	"	0.200
25'	"	0.267
30'	"	0.296
40'	"	0.400
Two wire switch and receptacle cables		
5'	EA.	0.059

Basic Materials	UNIT	MAN/HOURS
16110.32 Flexible Wiring Systems (Cont.)		
10'	EA.	0.100
15'	"	0.145
20'	"	0.200
25'	"	0.267
30'	"	0.296
40'	"	0.348
Three wire switch		
5'	EA.	0.059
10'	"	0.100
15'	"	0.145
20'	"	0.200
25'	"	0.267
30'	"	0.296
40'	"	0.348
Distribution boxes		
2 circuit	EA.	0.533
3 circuit	"	0.667
4 circuit	"	0.800
6 circuit	"	1.096
12 circuit	"	2.222
18 circuit	"	2.963
Tap boxes		
1 single pole switch	EA.	0.400
2 single pole switches	"	0.533
1 3 way switch	"	0.444
1 4 way switch	"	0.533
1 receptacle	"	0.400
2 receptacles	"	0.533
4 receptacles	"	0.800
1 clock	"	0.400
2 clocks	"	0.533
4 clocks	"	0.800
Dust cap	"	0.100
Cable coupler	"	0.200
Reversing connector	"	0.296
16110.35 Surface Mounted Raceway		
Single Raceway		
3/4" x 17/32" Conduit	L.F.	0.040
Mounting Strap	EA.	0.053
Connector	"	0.053
Elbow		
45 degree	EA.	0.050
90 degree	"	0.050
internal	"	0.050
external	"	0.050
Switch	"	0.400
Utility Box	"	0.400
Receptacle	"	0.400
3/4" x 21/32" Conduit	L.F.	0.040
Mounting Strap	EA.	0.053
Connector	"	0.053
Elbow		

16 ELECTRICAL

Basic Materials	UNIT	MAN/HOURS
16110.35 Surface Mounted Raceway *(Cont.)*		
45 degree	EA.	0.050
90 degree	"	0.050
internal	"	0.050
external	"	0.050
Switch	"	0.400
Utility Box	"	0.400
Receptacle	"	0.400
1-1/4" x 7/8" Conduit	L.F.	0.040
Mounting Strap	EA.	0.053
Connector	"	0.053
Elbow		
90 degree	EA.	0.050
internal	"	0.050
external	"	0.050
Switch Box	"	0.400
Receptacle Box	"	0.400
1-29/32" x 7/8" Conduit	L.F.	0.040
Mounting Strap	EA.	0.053
Connector	"	0.053
Elbow		
90 degree	EA.	0.050
internal	"	0.050
external	"	0.050
Switch Box	"	0.400
Receptacle Box	"	0.400
2-3/4" x 1-15/32" Conduit	L.F.	0.040
Mounting Strap	EA.	0.053
Connector	"	0.053
Elbow		
90 degree	EA.	0.050
internal	"	0.050
external	"	0.050
Switch Cover	"	0.400
Receptacle Cover	"	0.400
Double Raceway		
5-1/2" x 2" Conduit	L.F.	0.044
Mounting Strap	EA.	0.067
Connector	"	0.067
Elbow		
90 degree	EA.	0.057
internal	"	0.057
external	"	0.057
Receptacle Cover	"	0.400
16110.40 Underfloor Duct		
Underfloor blank duct, insert duct		
7/8"	L.F.	0.050
1-3/8"	"	0.050
1-7/8"	"	0.050
Box opening plugs	EA.	0.145
Duct end plugs	"	0.145
Sleeve couplings	"	0.348
Expansion couplings	"	0.348

Basic Materials	UNIT	MAN/HOURS
16110.40 Underfloor Duct *(Cont.)*		
Vertical elbow	EA.	0.348
Offset elbow	"	0.348
Horizontal elbow	"	0.348
Adjustable elbow	"	0.348
Cabinet connector	"	1.194
Y-take off	"	0.615
Underfloor duct leveling legs	"	0.145
Conduit adapters		
1/2"	EA.	0.250
3/4"	"	0.250
1"	"	0.250
1-1/4"	"	0.296
2"	"	0.348
Reducer bushings		
1-1/4" x 3/4"	EA.	0.200
1-1/4" x 1"	"	0.200
2" x 1-1/2"	"	0.250
Support couplers		
1 standard	EA.	0.250
2 standard	"	0.296
3 standard	"	0.348
Supports		
1 duct	EA.	0.145
2 duct	"	0.170
3 duct	"	0.190
4 duct	"	0.250
5 duct	"	0.296
Single level junction box		
1 standard	EA.	0.800
2 standard	"	1.509
3 standard	"	2.963
4 standard	"	4.000
Two level junction boxes		
1 standard	EA.	1.000
2 standard	"	2.000
Sealing compound	"	
Insert adapters	"	0.145
Ellipsoids	"	0.348
Insert closing cap	"	0.145
Marker Screws	"	0.145
Access boxes	"	1.000
Closing caps	"	0.145
Afterset markers	"	0.145
Cell markers	"	0.145
Tie down straps	"	0.145
Plastic grommets	"	0.400
Metal grommets	"	0.400
Receptacle		
Duplex, 20a	EA.	0.500
Single		
30a	EA.	0.500
50a	"	0.615
Double duplex	"	0.615

16 ELECTRICAL

Basic Materials	UNIT	MAN/HOURS
16110.40 Underfloor Duct *(Cont.)*		
Single, 20a	EA.	0.444
Double single	"	0.500
Twist lock	"	0.444
1 conduit opening	"	0.444
2 conduit openings	"	0.500
1 bushed opening	"	0.444
2 bushed openings	"	0.500
Amphenol connector		
1"	EA.	0.500
2"	"	0.533
5"	"	0.615
Standpipes		
Aluminum	EA.	0.250
Brass	"	0.250
Abandonment plates		
Aluminum	EA.	0.250
Brass	"	0.250
Split bell caps		
Aluminum	EA.	0.250
Brass	"	0.250
Flush floor receptacles		
Aluminum	EA.	0.727
Brass	"	0.727
Flush floor telephone		
Aluminum	EA.	0.500
Brass	"	0.500
Super underfloor duct blank duct		
1/2"	L.F.	0.059
7/8"	"	0.059
1-3/8"	"	0.059
1-7/8"	"	0.059
Box opening plugs	EA.	0.145
End plugs	"	0.145
Conduit adapters	"	0.727
Sleeve coupling	"	0.444
Expansion coupling	"	0.444
Reducing coupling	"	0.444
Vertical elbow	"	0.444
Offset elbow	"	0.444
Horizontal elbow	"	0.444
Adjustable elbow	"	0.444
Cabinet connector	"	1.509
Super underfloor duct Y-take off	"	0.727
Leveling legs	"	0.145
Support couplers		
1 super	EA.	0.250
2 super	"	0.296
1 super, 1 standard	"	0.296
2 super, 2 standard	"	0.348
Single level junction boxes		
1 super	EA.	1.509
2 super	"	2.000
4 super	"	3.478

Basic Materials	UNIT	MAN/HOURS
16110.40 Underfloor Duct *(Cont.)*		
Double level junction boxes		
1 super	EA.	1.194
2 super	"	1.509
16110.50 Wall Duct		
Lay-in wall duct, 10"	L.F.	0.059
Horizontal elbow	EA.	0.727
Edgewise elbow	"	0.727
Tee	"	0.899
Cross	"	1.096
Cabinet connector	"	1.600
Reverse elbow	"	0.533
Sweep elbow	"	0.727
Partition	"	0.059
Straight tunnel	"	0.276
Elbow tunnel	"	0.348
Tee kit	"	0.400
Ceiling dropout	"	1.000
Coupling device	"	0.145
End cap	"	0.276
Lay-in wall duct, 18"	L.F.	0.080
Horizontal elbow	EA.	0.800
Edgeware elbow	"	1.000
Tee	"	1.096
Cross	"	1.250
Reverse elbow	"	0.727
Sweep elbow	"	0.800
Partition	"	0.100
Straight tunnel	"	0.400
Elbow tunnel	"	0.500
Tee kit	"	0.533
Ceiling dropout	"	1.096
Coupling device	"	0.200
Reducer coupling	"	0.400
Cabinet connector	"	2.000
End cap	"	0.400
16110.60 Trench Duct		
Trench duct, with cover		
9"	L.F.	0.170
12"	"	0.200
18"	"	0.267
24"	"	0.348
30"	"	0.400
36"	"	0.571
Tees		
9"	EA.	1.739
12"	"	2.000
18"	"	2.222
24"	"	2.500
30"	"	2.963
36"	"	3.376
Vertical elbows		

16 ELECTRICAL

Basic Materials	UNIT	MAN/HOURS
16110.60 Trench Duct *(Cont.)*		
9"	EA.	0.800
12"	"	1.096
18"	"	1.356
24"	"	1.667
30"	"	2.000
36"	"	2.500
Cabinet connectors		
9"	EA.	2.000
12"	"	2.105
18"	"	2.424
24"	"	2.500
30"	"	2.759
36"	"	2.963
End closers		
9"	EA.	0.615
12"	"	0.667
18"	"	0.800
24"	"	1.096
30"	"	1.290
36"	"	1.455
Horizontal elbows		
9"	EA.	1.509
12"	"	1.739
18"	"	2.105
24"	"	2.500
30"	"	2.857
36"	"	3.200
Crosses		
9"	EA.	2.000
12"	"	2.222
18"	"	2.500
24"	"	2.759
30"	"	3.200
36"	"	3.478
16110.80 Wireways		
Wireway, hinge cover type		
2-1/2" x 2-1/2"		
1' section	EA.	0.154
2'	"	0.190
3'	"	0.250
5'	"	0.381
10'	"	0.667
4" x 4"		
1'	EA.	0.250
2'	"	0.250
3'	"	0.308
4'	"	0.308
10'	"	0.800
6" x 6"		
1'	EA.	0.381
2'	"	0.381
3'	"	0.444

Basic Materials	UNIT	MAN/HOURS
16110.80 Wireways *(Cont.)*		
4'	EA.	0.444
5'	"	0.571
10'	"	0.889
8" x 8"		
1'	EA.	0.444
2'	"	0.444
3'	"	0.500
4'	"	0.500
5'	"	0.615
12" x 12"		
1'	EA.	0.615
2'	"	0.615
3'	"	0.727
4'	"	0.727
5'	"	0.889
Fittings		
2-1/2" x 2-1/2"		
Drop hanger	EA.	0.123
Bracket hanger	"	0.123
Panel adapter	"	0.500
End plate	"	0.123
U-connector	"	0.123
Tee	"	0.200
Cross	"	0.250
90 degree elbow	"	0.200
Sweep elbow	"	0.200
45 degree elbow	"	0.200
Lay-in adapter	"	0.145
4" x 4"		
Drop hanger	EA.	0.145
Bracket hanger	"	0.145
Panel adapter	"	0.615
End plate	"	0.145
U-connector	"	0.145
Tee	"	0.250
Cross	"	0.348
90 degree elbow	"	0.250
Sweep elbow	"	0.250
45 degree elbow	"	0.250
Lay-in adapter	"	0.200
6" x 6"		
Drop hanger	EA.	0.145
Bracket hanger	"	0.145
Reducing bushing	"	0.145
Panel adapter	"	0.727
End plate	"	0.145
U-connector	"	0.145
Tee	"	0.250
Cross	"	0.348
90 degree elbow	"	0.250
Sweep elbow	"	0.250
45 degree elbow	"	0.250
Lay-in adapter	"	0.250

16 ELECTRICAL

Basic Materials	UNIT	MAN/HOURS
16110.80 Wireways *(Cont.)*		
8" x 8"		
Drop hanger	EA.	0.145
Bracket hanger	"	0.145
Reducing bushing	"	0.145
Panel adapter	"	0.899
End plate	"	0.145
U-connector	"	0.145
Tee	"	0.400
Cross	"	0.444
90 degree elbow	"	0.400
Sweep elbow	"	0.400
45 degree elbow	"	0.400
Lay-in adapter	"	0.250
10" x 10"		
Drop hanger	EA.	0.145
Bracket hanger	"	0.145
Reducing bushing	"	0.145
Panel adapter	"	1.000
End plate	"	0.145
U-connector	"	0.145
Tee	"	0.444
Cross	"	0.500
90 degree elbow	"	0.444
Sweep elbow	"	0.444
45 degree elbow	"	0.444
Lay-in adapter	"	0.296
12" x 12"		
Drop hanger	EA.	0.145
Bracket hanger	"	0.145
Reducing bushing	"	0.145
Panel adapter	"	1.250
End plate	"	0.145
U-connector	"	0.615
Tee	"	0.533
Cross	"	0.615
90 degree elbow	"	0.615
Sweep elbow	"	0.615
45 degree elbow	"	0.615
Lay-in adapter	"	0.533
Raintight wireway, 4" x 4"		
1' section	EA.	0.400
5'	"	0.400
10'	"	1.000
Fittings		
90 degree elbow	EA.	0.400
Tee	"	0.444
Cross	"	0.500
Panel adapter	"	0.727
End plate	"	0.145
Gusset bracket	"	0.145
6" x 6"		
1' section	EA.	0.500
5'	"	0.727

Basic Materials	UNIT	MAN/HOURS
16110.80 Wireways *(Cont.)*		
10'	EA.	1.509
Fittings		
90 degree elbow	EA.	0.500
Tee	"	0.500
Cross	"	0.727
Panel adapter	"	1.000
End plate	"	0.145
Gusset bracket	"	0.145
16120.41 Aluminum Conductors		
Type XHHW, stranded aluminum, 600v		
#8	L.F.	0.005
#6	"	0.006
#4	"	0.008
#2	"	0.009
1/0	"	0.011
2/0	"	0.012
3/0	"	0.014
4/0	"	0.015
300 MCM	"	0.020
350 MCM	"	0.023
400 MCM	"	0.028
500 MCM	"	0.033
600 MCM	"	0.040
700 MCM	"	0.047
750 MCM	"	0.052
THW, stranded		
#8	L.F.	0.005
#6	"	0.006
#4	"	0.008
#3	"	0.009
#1	"	0.010
1/0	"	0.011
2/0	"	0.012
3/0	"	0.012
4/0	"	0.015
250 MCM	"	0.018
300 MCM	"	0.020
350 MCM	"	0.023
400 MCM	"	0.028
500 MCM	"	0.033
600 MCM	"	0.040
700 MCM	"	0.047
750 MCM	"	0.052
XLP, stranded		
#6	L.F.	0.005
#4	"	0.008
#2	"	0.009
#1	"	0.010
1/0	"	0.011
2/0	"	0.012
3/0	"	0.014
4/0	"	0.015

16 ELECTRICAL

Basic Materials	UNIT	MAN/HOURS
16120.41 Aluminum Conductors *(Cont.)*		
250 MCM	L.F.	0.016
300 MCM	"	0.020
350 MCM	"	0.023
400 MCM	"	0.028
500 MCM	"	0.033
600 MCM	"	0.040
700 MCM	"	0.047
750 MCM	"	0.052
1000 MCM	"	0.057
Bare stranded aluminum wire		
#4	L.F.	0.008
#2	"	0.009
1/0	"	0.011
2/0	"	0.012
3/0	"	0.014
4/0	"	0.015
Triplex XLP cable		
#4	L.F.	0.015
#2	"	0.020
1/0	"	0.030
4/0	"	0.048
Aluminum quadruplex XLP cable		
#4	L.F.	0.018
#2	"	0.023
1/0	"	0.032
2/0	"	0.042
4/0	"	0.064
Triplexed URD-XLP cable		
#6	L.F.	0.011
#4	"	0.014
#2	"	0.018
1/0	"	0.028
2/0	"	0.033
3/0	"	0.040
4/0	"	0.047
250 MCM	"	0.055
350 MCM	"	0.057
Type S.E.U. cable		
#8/3	L.F.	0.025
#6/3	"	0.028
#4/3	"	0.035
#2/3	"	0.038
#1/3	"	0.040
1/0-3	"	0.042
2/0-3	"	0.044
3/0-3	"	0.052
4/0-3	"	0.057
Type S.E.R. cable with ground		
#8/3	L.F.	0.028
#6/3	"	0.035
#4/3	"	0.038
#2/3	"	0.040
#1/3	"	0.044

Basic Materials	UNIT	MAN/HOURS
16120.41 Aluminum Conductors *(Cont.)*		
1/0-3	L.F.	0.050
2/0-3	"	0.055
3/0-3	"	0.059
4/0-3	"	0.067
#6/4	"	0.038
#4/4	"	0.044
#2/4	"	0.044
#1/4	"	0.050
1/0-4	"	0.052
2/0-4	"	0.057
3/0-4	"	0.064
4/0-4	"	0.076
16120.43 Copper Conductors		
Copper conductors, type THW, solid		
#14	L.F.	0.004
#12	"	0.005
#10	"	0.006
Stranded		
#14	L.F.	0.004
#12	"	0.005
#10	"	0.006
#8	"	0.008
#6	"	0.009
#4	"	0.010
#3	"	0.010
#2	"	0.012
#1	"	0.014
1/0	"	0.016
2/0	"	0.020
3/0	"	0.025
4/0	"	0.028
250 MCM	"	0.030
300 MCM	"	0.033
350 MCM	"	0.040
400 MCM	"	0.044
500 MCM	"	0.052
600 MCM	"	0.059
750 MCM	"	0.067
1000 MCM	"	0.076
THHN-THWN, solid		
#14	L.F.	0.004
#12	"	0.005
#10	"	0.006
Stranded		
#14	L.F.	0.004
#12	"	0.005
#10	"	0.006
#8	"	0.008
#6	"	0.009
#4	"	0.010
#2	"	0.012
#1	"	0.014

16 ELECTRICAL

Basic Materials	UNIT	MAN/HOURS
16120.43 Copper Conductors *(Cont.)*		
1/0	L.F.	0.016
2/0	"	0.020
3/0	"	0.025
4/0	"	0.028
250 MCM	"	0.030
350 MCM	"	0.040
XHHW		
#14	L.F.	0.004
#10	"	0.006
#8	"	0.008
#6	"	0.009
#4	"	0.009
#2	"	0.011
#1	"	0.014
1/0	"	0.016
2/0	"	0.019
3/0	"	0.025
XLP, 600v		
#12	L.F.	0.005
#10	"	0.006
#8	"	0.008
#6	"	0.009
#4	"	0.010
#3	"	0.011
#2	"	0.012
#1	"	0.014
1/0	"	0.016
2/0	"	0.020
3/0	"	0.026
4/0	"	0.028
250 MCM	"	0.030
300 MCM	"	0.033
350 MCM	"	0.039
400 MCM	"	0.044
500 MCM	"	0.052
600 MCM	"	0.059
750 MCM	"	0.067
1000 MCM	"	0.076
Bare solid wire		
#14	L.F.	0.004
#12	"	0.005
#10	"	0.006
#8	"	0.008
#6	"	0.009
#4	"	0.010
#2	"	0.012
Bare stranded wire		
#8	L.F.	0.008
#6	"	0.010
#4	"	0.010
#2	"	0.011
#1	"	0.014
1/0	"	0.018

Basic Materials	UNIT	MAN/HOURS
16120.43 Copper Conductors *(Cont.)*		
2/0	L.F.	0.020
3/0	"	0.025
4/0	"	0.028
250 MCM	"	0.030
300 MCM	"	0.033
350 MCM	"	0.040
400 MCM	"	0.044
500 MCM	"	0.052
Type "BX" solid armored cable		
#14/2	L.F.	0.025
#14/3	"	0.028
#14/4	"	0.031
#12/2	"	0.028
#12/3	"	0.031
#12/4	"	0.035
#10/2	"	0.031
#10/3	"	0.035
#10/4	"	0.040
#8/2	"	0.035
#8/3	"	0.040
Steel type, metal clad cable, solid, with ground		
#14/2	L.F.	0.018
#14/3	"	0.020
#14/4	"	0.023
#12/2	"	0.020
#12/3	"	0.025
#12/4	"	0.030
#10/2	"	0.023
#10/3	"	0.028
#10/4	"	0.033
Metal clad cable, stranded, with ground		
#8/2	L.F.	0.028
#8/3	"	0.035
#8/4	"	0.042
#6/2	"	0.030
#6/3	"	0.038
#6/4	"	0.044
#4/2	"	0.040
#4/3	"	0.044
#4/4	"	0.055
#3/3	"	0.050
#3/4	"	0.059
#2/3	"	0.057
#2/4	"	0.067
#1/3	"	0.076
#1/4	"	0.084
16120.45 Flat Conductor Cable		
Flat conductor cable, with shield, 3 conductor		
#12 awg	L.F.	0.059
#10 awg	"	0.059
4 conductor		
#12 awg	L.F.	0.080

16 ELECTRICAL

Basic Materials	UNIT	MAN/HOURS
16120.45 Flat Conductor Cable *(Cont.)*		
#10 awg	L.F.	0.080
Transition boxes		
#12 awg	L.F.	0.089
#10 awg	"	0.089
Flat conductor cable communication, with shield		
10 conductor	L.F.	0.059
16 conductor	"	0.070
24 conductor	"	0.100
Power and communication heads, duplex receptacle	EA.	0.800
Double duplex receptacle	"	0.952
Telephone	"	0.800
Receptacle and telephone	"	0.952
Blank cover	"	0.145
Transition boxes		
Surface	EA.	0.727
Flush	"	1.000
Flat conductor cable fittings		
End caps	EA.	0.145
Insulators	"	0.296
Splice connectors	"	0.444
Tap connectors	"	0.444
Cable connectors	"	0.444
Terminal blocks	"	0.615
Tape	"	
16120.47 Sheathed Cable		
Non-metallic sheathed cable		
Type NM cable with ground		
#14/2	L.F.	0.015
#12/2	"	0.016
#10/2	"	0.018
#8/2	"	0.020
#6/2	"	0.025
#14/3	"	0.026
#12/3	"	0.027
#10/3	"	0.027
#8/3	"	0.028
#6/3	"	0.028
#4/3	"	0.032
#2/3	"	0.035
Type U.F. cable with ground		
#14/2	L.F.	0.016
#12/2	"	0.019
#10/2	"	0.020
#8/2	"	0.023
#6/2	"	0.027
#14/3	"	0.020
#12/3	"	0.022
#10/3	"	0.025
#8/3	"	0.028
#6/3	"	0.032
Type S.F.U. cable, 3 conductor		
#8	L.F.	0.028

Basic Materials	UNIT	MAN/HOURS
16120.47 Sheathed Cable *(Cont.)*		
#6	L.F.	0.031
#3	"	0.040
#2	"	0.044
#1	"	0.050
#1/0	"	0.055
#2/0	"	0.064
#3/0	"	0.070
#4/0	"	0.076
Type SER cable, 4 conductor		
#6	L.F.	0.036
#4	"	0.039
#3	"	0.044
#2	"	0.048
#1	"	0.055
#1/0	"	0.064
#2/0	"	0.067
#3/0	"	0.076
#4/0	"	0.084
Flexible cord, type STO cord		
#18/2	L.F.	0.004
#18/3	"	0.005
#18/4	"	0.006
#16/2	"	0.004
#16/3	"	0.004
#16/4	"	0.005
#14/2	"	0.005
#14/3	"	0.006
#14/4	"	0.007
#12/2	"	0.006
#12/3	"	0.007
#12/4	"	0.008
#10/2	"	0.007
#10/3	"	0.008
#10/4	"	0.009
#8/2	"	0.008
#8/3	"	0.009
#8/4	"	0.010
16130.10 Floor Boxes		
Adjustable floor boxes, steel	EA.	0.533
Cast bronze round	"	0.727
1 gang	"	0.800
2 gang	"	0.952
3 gang	"	1.000
Aluminum round	"	0.727
1 gang	"	0.800
2 gang	"	0.952
3 gang	"	1.000
Steel plate single recept	"	0.145
Duplex receptacle	"	0.182
Twist lock receptacle	"	0.182
Plug, 3/4"	"	0.145
1" plug	"	0.145

16 ELECTRICAL

Basic Materials	UNIT	MAN/HOURS
16130.10 Floor Boxes *(Cont.)*		
Carpet flange	EA.	0.145
Adjustable bronze plates for round cast boxes		
1/2" plug	EA.	0.145
3/4" plug	"	0.145
1" plug	"	0.145
1-1/4" plug	"	0.182
2" plug	"	0.200
Combination plug	"	0.200
Duplex receptacle plug	"	0.200
Adjustable aluminum plates for round cast boxes		
1/2" plug	EA.	0.145
3/4" plug	"	0.145
1" plug	"	0.145
1-1/4" plug	"	0.182
2" plug	"	0.200
Combination plug	"	0.200
Duplex receptacle plug	"	0.200
Adjustable bronze plates for gang type boxes		
1/2" plug	EA.	0.145
3/4" plug	"	0.145
1" plug	"	0.145
1-1/4" plug	"	0.182
2" plug	"	0.200
Carpet plate		
1 gang	EA.	0.145
2 gang	"	0.145
3 gang	"	0.200
Adjustable aluminum plates for gang type boxes		
1/2" plug	EA.	0.145
3/4" plug	"	0.145
1" plug	"	0.145
1-1/4" plug	"	0.182
2" plug	"	0.200
Duplex recept	"	0.200
Carpet plate		
1 gang	EA.	0.145
2 gang	"	0.145
3 gang	"	0.200
4 gang carpet plate	"	0.571
Telephone	"	0.500
Floor box nozzles, horizontal		
Duplex recept	EA.	0.533
Single recept	"	0.533
Double duplex recept	"	0.727
Vertical with duplex recept	"	0.615
Double duplex recept	"	0.727
Floor box bell nozzles split bell	"	0.250
One piece bell	"	0.250
Floor box standpipe		
1/2" x 3"	EA.	0.145
1/2" x 1"	"	0.145
Poke thru floor outlets		
2" floor	EA.	1.000

Basic Materials	UNIT	MAN/HOURS
16130.10 Floor Boxes *(Cont.)*		
3" floor	EA.	1.194
4" floor	"	1.290
7" floor	"	1.509
9" floor	"	1.600
11" floor	"	1.818
13" floor	"	2.000
16130.40 Boxes		
Round cast box, type SEH		
1/2"	EA.	0.348
3/4"	"	0.421
SEHC		
1/2"	EA.	0.348
3/4"	"	0.421
SEHL		
1/2"	EA.	0.348
3/4"	"	0.444
SEHT		
1/2"	EA.	0.421
3/4"	"	0.500
SEHX		
1/2"	EA.	0.500
3/4"	"	0.615
Blank cover	"	0.145
1/2", hub cover	"	0.145
Cover with gasket	"	0.178
Rectangle, type FS boxes		
1/2"	EA.	0.348
3/4"	"	0.400
1"	"	0.500
FSA		
1/2"	EA.	0.348
3/4"	"	0.400
FSC		
1/2"	EA.	0.348
3/4"	"	0.421
1"	"	0.500
FSL		
1/2"	EA.	0.348
3/4"	"	0.400
FSR		
1/2"	EA.	0.348
3/4"	"	0.400
FSS		
1/2"	EA.	0.348
3/4"	"	0.400
FSLA		
1/2"	EA.	0.348
3/4"	"	0.400
FSCA		
1/2"	EA.	0.348
3/4"	"	0.400
FSCC		

16 ELECTRICAL

Basic Materials

16130.40 Boxes (Cont.)

Description	UNIT	MAN/HOURS
1/2"	EA.	0.400
3/4"	"	0.500
FSCT		
1/2"	EA.	0.400
3/4"	"	0.500
1"	"	0.571
FST		
1/2"	EA.	0.500
3/4"	"	0.571
FSX		
1/2"	EA.	0.615
3/4"	"	0.727
FSCD boxes		
1/2"	EA.	0.615
3/4"	"	0.727
Rectangle, type FS, 2 gang boxes		
1/2"	EA.	0.348
3/4"	"	0.400
1"	"	0.500
FSC, 2 gang boxes		
1/2"	EA.	0.348
3/4"	"	0.400
1"	"	0.500
FSS, 2 gang boxes		
3/4"	EA.	0.400
FS, tandem boxes		
1/2"	EA.	0.400
3/4"	"	0.444
FSC, tandem boxes		
1/2"	EA.	0.400
3/4"	"	0.444
FS, three gang boxes		
3/4"	EA.	0.444
1"	"	0.500
FSS, three gang boxes, 3/4"	"	0.500
Weatherproof cast aluminum boxes, 1 gang, 3 outlets		
1/2"	EA.	0.400
3/4"	"	0.500
2 gang, 3 outlets		
1/2"	EA.	0.500
3/4"	"	0.533
1 gang, 4 outlets		
1/2"	EA.	0.615
3/4"	"	0.727
2 gang, 4 outlets		
1/2"	EA.	0.615
3/4"	"	0.727
1 gang, 5 outlets		
1/2"	EA.	0.727
3/4"	"	0.800
2 gang, 5 outlets		
1/2"	EA.	0.727
3/4"	"	0.800

16130.40 Boxes (Cont.)

Description	UNIT	MAN/HOURS
2 gang, 6 outlets		
1/2"	EA.	0.851
3/4"	"	0.899
2 gang, 7 outlets		
1/2"	EA.	1.000
3/4"	"	1.096
Weatherproof and type FS box covers, blank, 1 gang	"	0.145
Tumbler switch, 1 gang	"	0.145
1 gang, single recept	"	0.145
Duplex recept	"	0.145
Despard	"	0.145
Red pilot light	"	0.145
SW and		
Single recept	EA.	0.200
Duplex recept	"	0.200
2 gang		
Blank	EA.	0.182
Tumbler switch	"	0.182
Single recept	"	0.182
Duplex recept	"	0.182
3 gang		
Blank	EA.	0.200
Tumbler switch	"	0.200
4 gang		
Tumbler switch	EA.	0.250
Explosion proof boxes type E		
1/2"	EA.	0.348
3/4"	"	0.400
1"	"	0.500
1-1/4"	"	0.571
1-1/2"	"	0.615
Type L.B.		
1/2"	EA.	0.400
3/4"	"	0.500
1"	"	0.571
1-1/4"	"	0.667
1-1/2"	"	0.727
2"	"	0.800
Type C		
1/2"	EA.	0.400
3/4"	"	0.500
1"	"	0.571
1-1/4"	"	0.667
1-1/2"	"	0.727
2"	"	0.800
Type CA		
1/2"	EA.	0.571
3/4"	"	0.727
Type L		
1/2"	EA.	0.400
3/4"	"	0.500
1"	"	0.571
1-1/4"	"	0.667

16 ELECTRICAL

Basic Materials	UNIT	MAN/HOURS
16130.40 **Boxes** (Cont.)		
1-1/2"	EA.	0.727
2"	"	0.800
Type N		
1/2"	EA.	0.400
3/4"	"	0.500
1"	"	0.615
1-1/4"	"	0.667
Type T		
1/2"	EA.	0.533
3/4"	"	0.727
1"	"	0.851
1-1/4"	"	1.000
1-1/2"	"	1.159
2"	"	1.290
Type TA		
1/2"	EA.	0.727
3/4"	"	0.800
Type X		
1/2"	EA.	0.727
3/4"	"	0.851
1"	"	1.000
1-1/4"	"	1.159
1-1/2"	"	1.290
2"	"	1.455
With union hubs		
1/2"	EA.	0.727
3/4"	"	0.800
Box covers		
Surface	EA.	0.200
Sealing	"	0.200
Dome	"	0.200
1/2" nipple	"	0.200
3/4" nipple	"	0.200
16130.45 **Explosion Proof Fittings**		
Flexible couplings with female unions		
1/2" x 18"	EA.	0.200
3/4" x 18"	"	0.276
1" x 18"	"	0.348
1-1/4" x 18"	"	0.421
1-1/2" x 18"	"	0.500
2" x 18"	"	0.571
1/2" x 24"	"	0.250
3/4" x 24"	"	0.296
1" x 24"	"	0.400
1-1/4" x 24"	"	0.444
1-1/2" x 24"	"	0.571
2" x 24"	"	0.615
Female seal-offs		
1/2"	EA.	0.571
3/4"	"	0.667
1"	"	0.727
1-1/4"	"	0.851

Basic Materials	UNIT	MAN/HOURS
16130.45 **Explosion Proof Fittings** (Cont.)		
1-1/2"	EA.	1.000
2"	"	1.159
2-1/2"	"	1.739
3"	"	2.162
4"	"	2.667
Conduit plugs		
1/2"	EA.	0.145
3/4"	"	0.145
1"	"	0.145
1-1/4"	"	0.250
1-1/2"	"	0.250
2"	"	0.296
2-1/2"	"	0.296
3"	"	0.348
4"	"	0.348
Sealing cement		
1 pound	EA.	
5 pound	"	
Fibre		
1 ounce	EA.	
8 ounce	"	
Male unions		
1/2"	EA.	0.200
3/4"	"	0.242
1"	"	0.276
1-1/4"	"	0.296
1-1/2"	"	0.348
2"	"	0.421
2-1/2"	"	0.500
3"	"	0.727
4"	"	0.899
Female unions		
1/2"	EA.	0.200
3/4"	"	0.242
1"	"	0.276
1-1/4"	"	0.296
1-1/2"	"	0.348
2"	"	0.421
2-1/2"	"	0.500
3"	"	0.727
4"	"	0.899
Male elbows		
1/2"	EA.	0.250
3/4"	"	0.296
1"	"	0.348
1-1/4"	"	0.444
Female elbows		
1/2"	EA.	0.250
3/4"	"	0.296
1"	"	0.348
1-1/4"	"	0.444
Pulling elbows		
1/2"	EA.	0.348

16 ELECTRICAL

Basic Materials	UNIT	MAN/HOURS
16130.45 Explosion Proof Fittings (Cont.)		
3/4"	EA.	0.444
1"	"	0.500
1-1/4"	"	0.615
1-1/2"	"	0.727
2"	"	1.905
2-1/2"	"	2.500
3"	"	2.963
3-1/2"	"	3.478
4"	"	4.211
Male expansion couplings		
1/2"	EA.	0.250
3/4"	"	0.296
1"	"	0.444
Female expansion couplings		
1/2"	EA.	0.250
3/4"	"	0.296
1"	"	0.444
16130.60 Pull And Junction Boxes		
4"		
Octagon box	EA.	0.114
Box extension	"	0.059
Plaster ring	"	0.059
Cover blank	"	0.059
Square box	"	0.114
Box extension	"	0.059
Plaster ring	"	0.059
Cover blank	"	0.059
4-11/16"		
Square box	EA.	0.114
Box extension	"	0.059
Plaster ring	"	0.059
Cover blank	"	0.059
Switch and device boxes		
2 gang	EA.	0.114
3 gang	"	0.114
4 gang	"	0.160
Device covers		
2 gang	EA.	0.059
3 gang	"	0.059
4 gang	"	0.059
Handy box	"	0.114
Extension	"	0.059
Switch cover	"	0.059
Switch box with knockout	"	0.145
Weatherproof cover, spring type	"	0.080
Cover plate, dryer receptacle 1 gang plastic	"	0.100
For 4" receptacle, 2 gang	"	0.100
Duplex receptacle cover plate, plastic	"	0.059
4", vertical bracket box, 1-1/2" with		
RMX clamps	EA.	0.145
BX clamps	"	0.145
4", octagon device cover		

Basic Materials	UNIT	MAN/HOURS
16130.60 Pull And Junction Boxes (Cont.)		
1 switch	EA.	0.059
1 duplex recept	"	0.059
4", octagon swivel hanger box, 1/2" hub	"	0.059
3/4" hub	"	0.059
4" octagon adjustable bar hangers		
18-1/2"	EA.	0.050
26-1/2"	"	0.050
With clip		
18-1/2"	EA.	0.050
26-1/2"	"	0.050
4", square face bracket boxes, 1-1/2"		
RMX	EA.	0.145
BX	"	0.145
4" square to round plaster rings	"	0.059
2 gang device plaster rings	"	0.059
Surface covers		
1 gang switch	EA.	0.059
2 gang switch	"	0.059
1 single recept	"	0.059
1 20a twist lock recept	"	0.059
1 30a twist lock recept	"	0.059
1 duplex recept	"	0.059
2 duplex recept	"	0.059
Switch and duplex recept	"	0.059
4-11/16" square to round plaster rings	"	0.059
2 gang device plaster rings	"	0.059
Surface covers		
1 gang switch	EA.	0.059
2 gang switch	"	0.059
1 single recept	"	0.059
1 20a twist lock recept	"	0.059
1 30a twist lock recept	"	0.059
1 duplex recept	"	0.059
2 duplex recept	"	0.059
Switch and duplex recept	"	0.059
4" plastic round boxes, ground straps		
Box only	EA.	0.145
Box w/clamps	"	0.200
Box w/16" bar	"	0.229
Box w/24" bar	"	0.250
4" plastic round box covers		
Blank cover	EA.	0.059
Plaster ring	"	0.059
4" plastic square boxes		
Box only	EA.	0.145
Box w/clamps	"	0.200
Box w/hanger	"	0.250
Box w/nails and clamp	"	0.250
4" plastic square box covers		
Blank cover	EA.	0.059
1 gang ring	"	0.059
2 gang ring	"	0.059
Round ring	"	0.059

16 ELECTRICAL

Basic Materials	UNIT	MAN/HOURS
16130.65 Pull Boxes And Cabinets		
Galvanized pull boxes, screw cover		
4x4x4	EA.	0.190
4x6x4	"	0.190
6x6x4	"	0.190
6x8x4	"	0.190
8x8x4	"	0.250
8x10x4	"	0.242
8x12x4	"	0.250
Screw cover		
10x10x4	EA.	0.308
12x12x6	"	0.444
12x15x6	"	0.444
12x18x6	"	0.500
15x18x6	"	0.571
18x24x6	"	0.615
18x30x6	"	0.727
24x36x6	"	0.727
Cast iron junction box, unflanged		
6x6x4		
3/4" tap	EA.	0.500
1" tap	"	0.500
Two 1/2" taps	"	0.500
3/4" taps	"	0.500
6" adapter plate	"	0.348
6" exterior collar	"	0.348
Screw cover cabinet		
12x12x4	EA.	0.615
12x16x4	"	0.615
12x16x6	"	0.615
12x18x4	"	0.667
12x18x6	"	0.667
18x18x4	"	1.000
18x18x6	"	1.000
18x24x6	"	1.143
24x24x6	"	1.333
24x36x6	"	1.667
36x48x6	"	2.500
NEMA 3R, rain tight screw cover enclosures		
6x6x4	EA.	0.211
8x6x4	"	0.296
8x8x4	"	0.296
10x8x4	"	0.400
10x10x4	"	0.400
12x8x4	"	0.444
12x12x4	"	0.444
15x12x4	"	0.533
8x8x6	"	0.400
10x8x6	"	0.444
10x10x6	"	0.444
12x8x6	"	0.533
12x10x6	"	0.548
12x12x6	"	0.548
18x12x6	"	0.702

Basic Materials	UNIT	MAN/HOURS
16130.80 Receptacles		
Contractor grade duplex receptacles, 15a 120v		
Duplex	EA.	0.200
125 volt, 20a, duplex, standard grade	"	0.200
Ground fault interrupter type	"	0.296
250 volt, 20a, 2 pole, single, ground type	"	0.200
120/208v, 4 pole, single receptacle, twist lock		
20a	EA.	0.348
50a	"	0.348
125/250v, 3 pole, flush receptacle		
30a	EA.	0.296
50a	"	0.296
60a	"	0.348
277v, 20a, 2 pole, grounding type, twist lock	"	0.200
Dryer receptacle, 250v, 30a/50a, 3 wire	"	0.296
Clock receptacle, 2 pole, grounding type	"	0.200
125v, 20a single recept. grounding type		
Standard grade	EA.	0.200
Specification	"	0.200
Hospital	"	0.200
Isolated ground orange	"	0.250
Duplex		
Specification grade	EA.	0.200
Hospital	"	0.200
Isolated ground orange	"	0.250
250v, 20a, duplex, 2 pole, grounding, spec. grade	"	0.200
Combination recepts, 20a, 125v and 250v, duplex	"	0.200
GFI hospital grade recepts, 20a, 125v, duplex	"	0.296
125/250v, 3 pole, 3 wire surface recepts		
30a	EA.	0.296
50a	"	0.296
60a	"	0.348
Cord set, 3 wire, 6' cord		
30a	EA.	0.296
50a	"	0.296
125/250v, 3 pole, 3 wire cap		
30a	EA.	0.400
50a	"	0.400
60a	"	0.444
16198.10 Electric Manholes		
Precast, handhole, 4' deep		
2'x2'	EA.	3.478
3'x3'	"	5.556
4'x4'	"	10.256
Power manhole, complete, precast, 8' deep		
4'x4'	EA.	14.035
6'x6'	"	20.000
8'x8'	"	21.053
6' deep, 9' x 12'	"	25.000
Cast in place, power manhole, 8' deep		
4'x4'	EA.	14.035
6'x6'	"	20.000
8'x8'	"	21.053

16 ELECTRICAL

Basic Materials	UNIT	MAN/HOURS
16199.10 Utility Poles & Fittings		
Wood pole, creosoted		
25'	EA.	2.353
30'	"	2.963
35'	"	3.478
40'	"	3.791
45'	"	6.957
50'	"	7.207
55'	"	7.547
Treated, wood preservative, 6"x6"		
8'	EA.	0.500
10'	"	0.800
12'	"	0.889
14'	"	1.333
16'	"	1.600
18'	"	2.000
20'	"	2.000
Aluminum, brushed, no base		
8'	EA.	2.000
10'	"	2.667
15'	"	2.759
20'	"	3.200
25'	"	3.810
30'	"	4.396
35'	"	5.000
40'	"	6.250
Steel, no base		
10'	EA.	2.500
15'	"	2.963
20'	"	3.810
25'	"	4.520
30'	"	5.096
35'	"	6.250
Concrete, no base		
13'	EA.	5.517
16'	"	7.273
18'	"	8.791
25'	"	10.000
30'	"	12.121
35'	"	14.035
40'	"	16.000
45'	"	17.021
50'	"	18.182
55'	"	19.048
60'	"	20.000
Pole line hardware		
Wood crossarm		
4'	EA.	1.333
8'	"	1.667
10'	"	2.051
Angle steel brace		
1 piece	EA.	0.250
2 piece	"	0.348
Eye nut, 5/8"	"	0.050

Basic Materials	UNIT	MAN/HOURS
16199.10 Utility Poles & Fittings *(Cont.)*		
Bolt (14-16"), 5/8"	EA.	0.200
Transformer, ground connection	"	0.250
Stirrup	"	0.308
Secondary lead support	"	0.400
Spool insulator	"	0.200
Guy grip, preformed		
7/16"	EA.	0.145
1/2"	"	0.145
Hook	"	0.250
Strain insulator	"	0.364
Wire		
5/16"	L.F.	0.005
7/16"	"	0.006
1/2"	"	0.008
Soft drawn ground, copper, #8	"	0.008
Ground clamp	EA.	0.308
Perforated strapping for conduit, 1-1/2"	L.F.	0.145
Hot line clamp	EA.	0.800
Lightning arrester		
3kv	EA.	1.000
10kv	"	1.600
30kv	"	2.000
36kv	"	2.500
Fittings		
Plastic molding	L.F.	0.145
Molding staples	EA.	0.050
Ground wires staples	"	0.030
Copper butt plate	"	0.296
Anchor bond clamp	"	0.145
Guy wire		
1/4"	L.F.	0.030
3/8"	"	0.050
Guy grip		
1/4"	EA.	0.050
3/8"	"	0.050

Power Generation	UNIT	MAN/HOURS
16210.10 Generators		
Diesel generator, with auto transfer switch		
30kw	EA.	30.769
50kw	"	30.769
75kw	"	42.105
100kw	"	47.059
125kw	"	50.000
150kw	"	57.143

16 ELECTRICAL

Power Generation	UNIT	MAN/HOURS
16210.10 Generators *(Cont.)*		
175kw	EA.	66.667
200kw	"	80.000
250kw	"	88.889
300kw	"	100.000
350kw	"	114.286
400kw	"	133.333
450kw	"	145.455
500kw	"	160.000
600kw	"	200.000
750kw	"	200.000
16230.10 Capacitors		
Three phase capacitors		
240v		
1.5 kvar	EA.	2.500
2.5 kvar	"	3.200
3.0 kvar	"	4.000
4 kvar	"	5.000
5 kvar	"	5.333
6 kvar	"	5.714
7.5 kvar	"	6.154
10 kvar	"	8.000
15 kvar	"	9.501
20 kvar	"	11.994
25 kvar	"	13.008
40 kvar	"	18.018
50 kvar	"	20.997
60 kvar	"	21.505
75 kvar	"	25.000
100 kvar	"	29.963
480v		
1.5 kvar	EA.	2.500
2.5 kvar	"	3.200
3 kvar	"	4.000
4 kvar	"	5.000
5 kvar	"	5.333
6 kvar	"	5.714
7.5 kvar	"	6.154
10 kvar	"	8.000
12.5 kvar	"	9.501
15 kvar	"	11.994
18 kvar	"	12.500
20 kvar	"	13.008
22.5 kvar	"	13.491
25 kvar	"	14.842
30 kvar	"	14.842
35 kvar	"	16.000
40 kvar	"	18.018
45 kvar	"	20.000
50 kvar	"	20.997
60 kvar	"	21.978
70 kvar	"	24.024

Power Generation	UNIT	MAN/HOURS
16230.10 Capacitors *(Cont.)*		
75 kvar	EA.	25.000
80 kvar	"	27.027
90 kvar	"	28.986
100 kvar	"	29.963
125 kvar	"	33.058
150 kvar	"	37.037
16320.10 Transformers		
Floor mtd, one phase, int. dry, 480v-120/240v		
3 kva	EA.	1.818
5 kva	"	3.077
7.5 kva	"	3.478
10 kva	"	3.810
15 kva	"	4.301
25 kva	"	7.547
37.5 kva	"	9.412
50 kva	"	10.256
75 kva	"	10.667
100 kva	"	11.594
Three phase, 480v-120/208v		
15 kva	EA.	6.015
30 kva	"	9.412
45 kva	"	10.811
75 kva	"	10.959
112.5 kva	"	12.698
150 kva	"	13.559
225 kva	"	15.385
Single phase, dry type, 2400v		
167 kva	EA.	22.472
250 kva	"	29.963
333 kva	"	37.559
5000v		
167 kva	EA.	22.472
250 kva	"	29.963
333 kva	"	37.559
8660v		
167 kva	EA.	27.491
250 kva	"	34.934
333 kva	"	67.797
1500v		
167 kva	EA.	27.491
250 kva	"	34.934
333 kva	"	42.553
Three phase, dry type transformer, 2400v		
225 kva	EA.	25.000
300 kva	"	27.491
500 kva	"	42.553
750 kva	"	52.632
5000v		
225.0 kva	EA.	25.000
300 kva	"	27.491
500 kva	"	42.553
750 kva	"	52.632

16 ELECTRICAL

Power Generation	UNIT	MAN/HOURS
16320.10 Transformers *(Cont.)*		
8660v		
225.0 kva	EA.	29.963
300 kva	"	32.520
500 kva	"	47.619
750 kva	"	57.554
1500v		
225 kva	EA.	29.963
300 kva	"	32.520
500 kva	"	47.619
750 kva	"	57.554
Buck boost transformers		
.25 kva	EA.	1.000
.50 kva	"	1.250
.75 kva	"	1.509
1.00 kva	"	1.739
1.50 kva	"	2.000
2.00 kva	"	2.500
3.00 kva	"	2.963
16350.10 Circuit Breakers		
Molded case, 240v, 15-60a, bolt-on		
1 pole	EA.	0.250
2 pole	"	0.348
70-100a, 2 pole	"	0.533
15-60a, 3 pole	"	0.400
70-100a, 3 pole	"	0.615
480v, 2 pole		
15-60a	EA.	0.296
70-100a	"	0.400
3 pole		
15-60a	EA.	0.400
70-100a	"	0.444
70-225a	"	0.615
Draw out air circuit breakers		
600a	EA.	16.000
800a	"	18.182
1600a	"	24.242
2000a	"	27.586
3000a	"	32.000
4000a	"	38.095
Load center circuit breakers, 240v		
1 pole, 10-60a	EA.	0.250
2 pole		
10-60a	EA.	0.400
70-100a	"	0.667
110-150a	"	0.727
3 pole		
10-60a	EA.	0.500
70-100a	"	0.727
Load center, G.F.I. breakers, 240v		
1 pole, 15-30a	EA.	0.296
2 pole, 15-30a	"	0.400
Key operated breakers, 240v, 1 pole, 10-30a	"	0.296

Power Generation	UNIT	MAN/HOURS
16350.10 Circuit Breakers *(Cont.)*		
Tandem breakers, 240v		
1 pole, 15-30a	EA.	0.400
2 pole, 15-30a	"	0.533
Bolt-on, G.F.I. breakers, 240v, 1 pole, 15-30a	"	0.348
Enclosed breaker, 120v, 1 pole, 15-50a, NEMA 1	"	0.800
240v, 2 pole		
15-60a, NEMA 1	EA.	1.250
70-100a, NEMA 1	"	1.739
3 pole		
15-60a, NEMA 1	EA.	1.509
70-100a, NEMA 1	"	2.222
Enclosed circuit breakers		
120v, 1 pole, NEMA 3R, 15-50a	EA.	0.899
240v, 2 pole, NEMA 3R		
15-60a	EA.	1.250
70-100a	"	1.739
3 pole, NEMA 3R		
15-60a	EA.	1.509
70-100a	"	2.222
480v, NEMA 1		
1 pole, 15-50a	EA.	0.800
2 pole, 15-60a	"	1.250
70-100a	"	1.509
3 pole, NEMA 1		
15-60a	EA.	1.509
70-100a	"	2.222
480v, 1 pole, 15-50a, NEMA 3R	"	1.000
2 pole		
2 pole, 15-60a, NEMA 3R	EA.	1.250
70-100a, NEMA 3R	"	1.739
3 pole		
15-60a, NEMA 3R	EA.	1.509
70-100a, NEMA 3R	"	2.222
70-100a, NEMA 1	"	1.739
3 pole		
15-60a, NEMA 1	EA.	1.739
70-100a, NEMA 1	"	2.222
Enclosed breakers, 600v, 2 phase, NEMA 3R		
15-60a	EA.	1.250
70-100a	"	1.739
3 phase, NEMA 3R		
15-60a	EA.	1.509
70-100a	"	2.222
600v, 3 phase, NEMA 1		
125a	EA.	2.222
150a	"	2.963
175a	"	2.963
200a	"	2.963
225a	"	2.963
250a	"	6.154
300a	"	6.154
350a	"	6.154
400a	"	6.154

16 ELECTRICAL

Power Generation

16350.10 Circuit Breakers (Cont.)

Description	UNIT	MAN/HOURS
500a	EA.	9.744
600a	"	9.744
700a	"	10.753
800a	"	10.753
900a	"	15.009
1000a	"	15.009
1200a	"	18.519
1400a	"	18.519
1600a	"	24.024
1800a	"	29.963
2000a	"	29.963
600v, 3 phase, NEMA 3R		
125-225a	EA.	2.222
250-400a	"	5.714
500-600a	"	9.744
700-800a	"	11.004
900-1000a	"	15.009
1000-1200a	"	19.002
1400-1600a	"	24.024
1800-2000a	"	29.963

16360.10 Safety Switches

Description	UNIT	MAN/HOURS
Fused, 3 phase, 30 amp, 600v, heavy duty		
NEMA 1	EA.	1.143
NEMA 3r	"	1.143
NEMA 4	"	1.600
NEMA 12	"	1.739
60a		
NEMA 1	EA.	1.143
NEMA 3r	"	1.143
NEMA 4	"	1.600
NEMA 12	"	1.739
100a		
NEMA 1	EA.	1.739
NEMA 3r	"	1.739
NEMA 4	"	2.000
NEMA 12	"	2.500
200a		
NEMA 1	EA.	2.500
NEMA 3r	"	2.500
NEMA 4	"	2.759
NEMA 12	"	3.478
400a		
NEMA 1	EA.	5.517
NEMA 3r	"	5.517
NEMA 4	"	5.755
NEMA 12	"	7.018
600a		
NEMA 1	EA.	8.000
NEMA 3r	"	8.000
NEMA 4	"	8.989
NEMA 12	"	12.308
Non-fused, 240-600v, heavy duty, 3 phase, 30 amp		

16360.10 Safety Switches (Cont.)

Description	UNIT	MAN/HOURS
NEMA 1	EA.	1.143
NEMA 3r	"	1.143
NEMA 4	"	1.739
NEMA 12	"	1.739
60a		
NEMA1	EA.	1.143
NEMA 3r	"	1.143
NEMA 4	"	1.739
NEMA 12	"	1.739
100a		
NEMA 1	EA.	1.739
NEMA 3r	"	1.739
NEMA 4	"	2.500
NEMA 12	"	2.500
200a, NEMA 1	"	2.500
600a, NEMA 12	"	12.308
Bolt-on hubs		
3/4" - 1-1/2"	EA.	0.250
2"	"	0.296
2-1/2"	"	0.296
3"	"	0.348
3-1/2"	"	0.400
4"	"	0.400
Watertight hubs		
1/2"	EA.	0.250
3/4"	"	0.296
1"	"	0.400
1-1/4"	"	0.444
1-1/2"	"	0.471
2"	"	0.500
2-1/2"	"	0.533
3"	"	0.615
3-1/2"	"	0.800
4"	"	0.851
Non-fused, 600v, 3 pole, NEMA 7		
600a	EA.	2.222
100a	"	3.200
225a	"	4.000
NEMA 9		
60a	EA.	2.500
100a	"	3.333
225a	"	4.211
Fusible bolted pressure switches, 600v/3 pole, NEMA 1		
800a	EA.	16.000
1200a	"	21.978
1600a	"	25.000
2000a	"	29.963
2500a	"	34.934
3000a	"	44.944
4000a	"	51.948
Non-fusible		
800a	EA.	14.493
1200a	"	20.000

16 ELECTRICAL

Power Generation	UNIT	MAN/HOURS
16360.10 Safety Switches *(Cont.)*		
1600a	EA.	22.989
2000a	"	27.972
2500a	"	34.934
3000a	"	44.944
4000a	"	51.948
Fusible load interrupter switches, 4.16 kv, NEMA 1		
200a	EA.	29.963
600a	"	70.175
Fusible load interrupter switch, 13.8 kv		
NEMA 1, 600a	EA.	100.000
NEMA 3R, 600a	"	100.000
4.16 kv, NEMA 3R		
200a	EA.	29.963
600a	"	70.175
Non-fused load interrupter switch, 4.16 kv, NEMA 1		
200a	EA.	29.963
600a	"	70.175
13.8 kv, NEMA 1, 600a	"	100.000
4.16 kv, NEMA 3R		
200a	EA.	29.963
600a	"	70.175
13.8 kv, NEMA 3R, 600a	"	100.000
Interrupter switch accessories, strip heater	"	
Cable lugs	"	
Key interlock	"	
Auxiliary switch	"	
Lightning arrester		
5 kva	EA.	
15 kv	"	
16365.10 Fuses		
Fuse, one-time, 250v		
30a	EA.	0.050
60a	"	0.050
100a	"	0.050
200a	"	0.050
400a	"	0.050
600a	"	0.050
600v		
30a	EA.	0.050
60a	"	0.050
100a	"	0.050
200a	"	0.050
400a	"	0.050
Fusetron, 600v		
200a	EA.	0.050
400a	"	0.050
Fuse, amp-trap, K1, 250v		
30a	EA.	0.050
60a	"	0.050
100a	"	0.050
200a	"	0.050
400a	"	0.050

Power Generation	UNIT	MAN/HOURS
16365.10 Fuses *(Cont.)*		
600a	EA.	0.050
600v		
30a	EA.	0.050
60a	"	0.050
100a	"	0.050
200a	"	0.050
400a	"	0.050
K5, 250v		
30a	EA.	0.050
60a	"	0.050
100a	"	0.050
200a	"	0.050
400a	"	0.050
600a	"	0.050
600v		
30a	EA.	0.050
60a	"	0.050
100a	"	0.050
200a	"	0.050
400a	"	0.050
600a	"	0.050
J, 600v		
30a	EA.	0.050
60a	"	0.050
100a	"	0.050
200a	"	0.050
400a	"	0.050
L, 600v		
1200a	EA.	0.400
1600a	"	0.400
2000a	"	0.400
2500a	"	0.400
3000a	"	0.400
4000a	"	0.400
5000a	"	0.400
Fuse cl-ay 250v		
600a	EA.	0.296
1200a	"	0.296
1600a	"	0.296
2000a	"	0.296
600v		
1200a	EA.	0.296
1600a	"	0.296
2000a	"	0.296
Reducers, 600v		
60a-30a	EA.	0.145
100a-30a	"	0.145
100a-60a	"	0.145
200a-60a	"	0.250
200a-100a	"	0.250
400a-100a	"	0.348
400a-200a	"	0.348
600a-100a	"	0.400

16 ELECTRICAL

Power Generation	UNIT	MAN/HOURS
16365.10 Fuses *(Cont.)*		
600a-200a	EA.	0.400
600a-400a	"	0.400
16395.10 Grounding		
Ground rods, copper clad, 1/2" x		
6'	EA.	0.667
8'	"	0.727
10'	"	1.000
5/8" x		
5'	EA.	0.615
6'	"	0.727
8'	"	1.000
10'	"	1.250
3/4" x		
8'	EA.	0.727
10'	"	0.800
Ground rod clamp		
5/8"	EA.	0.123
3/4"	"	0.123
Coupling, on threaded rods, 3/4"	"	0.050
Ground receptacles	"	0.250
Bus bar, copper, 2" x 1/4"	L.F.	0.145
Copper braid, 1" x 1/8", for door ground	EA.	0.100
Brazed connection for		
#6 wire	EA.	0.500
#2 wire	"	0.800
#2/0 wire	"	1.000
#4/0 wire	"	1.143
Ground rod couplings		
1/2"	EA.	0.100
5/8"	"	0.100
Ground rod, driving stud		
1/2"	EA.	0.100
5/8"	"	0.100
3/4"	"	0.100
Ground rod clamps, #8-2 to		
1" pipe	EA.	0.200
2" pipe	"	0.250
3" pipe	"	0.296
5" pipe	"	0.348
6" pipe	"	0.444
#4-4/0 to		
1" pipe	EA.	0.200
2" pipe	"	0.250
3" pipe	"	0.296
3" pipe	"	0.348
8 pipe	"	0.444
8 pipe	"	0.667
10 pipe	"	0.952
12 pipe	"	1.290

Service And Distribution	UNIT	MAN/HOURS
16425.10 Switchboards		
Switchboard, 90" high, no main disconnect, 208/120v		
400a	EA.	7.921
600a	"	8.000
1000a	"	8.000
1200a	"	10.000
1600a	"	11.940
2000a	"	14.035
2500a	"	16.000
277/480v		
600a	EA.	8.163
800a	"	8.163
1600a	"	11.940
2000a	"	14.035
2500a	"	16.000
3000a	"	27.586
4000a	"	29.630
Main breaker sections, 600v		
1200a, GFI	EA.	16.985
1600a, GFI	"	19.512
2000a, GFI	"	20.000
2500a, GFI	"	25.000
3000a, GFI	"	29.963
4000a, GFI	"	34.934
Switchboard meter sections, 600v		
400a	EA.	8.000
600a	"	10.000
800a	"	11.004
1000a	"	13.491
2000a	"	16.000
2500a	"	20.000
3000a	"	25.000
4000a	"	29.963
Insulated case, draw out compartment, 208/120v		
800a	EA.	2.500
1600a	"	2.963
2000a	"	3.478
2500a	"	3.478
3000a	"	4.000
4000a	"	4.790
Accessories for power trip breakers		
Shunt trip	EA.	0.500
Key interlock	"	2.222
Lifting and transport truck	"	4.494
Lifting device	"	1.333
Bus duct connection, 3 phase, 4 wire		
225a	EA.	2.963
400a	"	2.963
600a	"	3.333
800a	"	4.000
2500a	"	6.015
3000a	"	7.477
4000a	"	8.791
Provision for mounting current transformers		

16 ELECTRICAL

Service And Distribution	UNIT	MAN/HOURS
16425.10 Switchboards (Cont.)		
800a & below primary	EA.	2.963
1000 to 1500a primary	"	2.963
2000 to 6000a primary	"	2.963
Provision for mounting potential transformers		
2000a max	EA.	3.810
Switchboard instruments		
Voltmeter	EA.	1.000
Ammeter, incoming line	"	1.000
Wattmeter	"	1.000
Varmeter	"	1.000
Power factor meter	"	1.000
Frequency meter	"	1.000
Recording voltmeter	"	2.000
Wattmeter	"	2.000
Power factor meter	"	2.000
Frequency meter	"	2.000
Instrument phase select switch	"	0.500
Enclosure, 90" high, 3 phase, 4 wire		
1000a	EA.	6.838
1200a	"	7.018
1600a	"	8.602
2000a	"	13.333
5500a	"	15.686
3000a	"	18.182
4000a	"	23.529
Circuit breakers, 600v, 100a, frame		
15-30a, 1 pole	EA.	0.296
15-60a, 2 pole	"	0.348
70-100a, 2 pole	"	0.400
15-60a, 3 pole	"	0.444
70-100a, 3 pole	"	0.500
Bolt on breakers, 600v, 225a frame, 110-225a		
2 pole	EA.	0.615
3 pole	"	1.096
400a frame, 250-400a, 2 pole	"	1.250
800a frame		
450-600a, 2 pole	EA.	1.905
700-800a, 2 pole	"	2.500
450-600a, 3 pole	"	4.211
700-800a, 3 pole	"	4.444
Bolt on branch breakers, 600v		
1000-2000a, 2 pole	EA.	5.333
2500a, 2 pole	"	10.753
1000-2000a, 3 pole	"	8.000
2500a, 3 pole	"	11.004
3000a, 3 pole	"	20.000
Metal clad substation switch board, selector switch		
600a, 5kv	EA.	42.105
15kv	"	47.904
Fused switch, 600a		
5kv	EA.	34.934
15kv	"	34.934
1200a		

Service And Distribution	UNIT	MAN/HOURS
16425.10 Switchboards (Cont.)		
5kv	EA.	40.000
15kv	"	40.000
Oil cutout switch		
5 kv	EA.	15.009
15 kv	"	18.018
Liquid air terminal section	"	8.000
Dry air terminal section	"	8.502
Auxiliary compartment	"	29.963
16430.20 Metering		
Outdoor wp meter sockets, 1 gang, 240v, 1 phase		
Includes sealing ring, 100a	EA.	1.509
150a	"	1.778
200a	"	2.000
Die cast hubs, 1-1/4"	"	0.320
1-1/2"	"	0.320
2"	"	0.320
Indoor meter center, main switch single phase, 240v		
400a	EA.	8.000
600a	"	11.004
800a	"	11.696
Main breaker		
400a	EA.	8.000
600a	"	11.004
800a	"	11.696
1000a	"	16.000
1200a	"	16.495
1600a	"	18.018
Terminal box		
800a	EA.	10.000
1600a	"	18.018
Main switch, three phase, 208v		
400a	EA.	8.502
600a	"	11.994
800a	"	13.491
Main breaker		
400a	EA.	8.502
600a	"	11.994
800a	"	13.491
1000a	"	16.985
1200a	"	18.018
1600a	"	20.997
Terminal box		
800a	EA.	13.008
1600a	"	20.997
Indoor meter center		
2 meters	EA.	5.000
3 meters	"	6.154
4 meters	"	7.273
5 meters	"	8.000
6 meters	"	8.999
Plug on breakers, single phase, 208v		
60a	EA.	0.250

16 ELECTRICAL

Service And Distribution — 16430.20 Metering (Cont.)

Description	UNIT	MAN/HOURS
70a	EA.	0.250
80a	"	0.250
90a	"	0.250
100a	"	0.348
Indoor meter center, single phase, 125a breakers		
3 meters	EA.	6.154
4 meters	"	7.273
5 meters	"	8.000
6 meters	"	8.502
7 meters	"	10.000
8 meters	"	11.004
10 meters	"	11.994
150a breakers		
3 meters	EA.	6.154
4 meters	"	7.273
6 meters	"	8.000
7 meters	"	10.000
8 meters	"	11.004
200a breakers		
3 meters	EA.	6.154
4 meters	"	7.273
6 meters	"	8.000
7 meters	"	10.000
8 meters	"	11.004
Indoor meter center, three phase, 125a breakers		
3 meters	EA.	6.154
4 meters	"	7.273
5 meters	"	8.000
6 meters	"	8.999
7 meters	"	10.000
8 meters	"	11.004
10 meters	"	11.994
150a breakers		
3 meters	EA.	6.154
4 meters	"	7.273
6 meters	"	8.502
7 meters	"	11.004
8 meters	"	11.994
200a breakers		
3 meters	EA.	6.667
4 meters	"	7.273
6 meters	"	8.999
7 meters	"	11.004
8 meters	"	11.994
NEMA 3R, meter center, main switch, 1 phase, 240v		
400a	EA.	8.000
600a	"	10.000
800a	"	11.004
Main breaker		
400a	EA.	8.000
600a	"	10.000
800a	"	12.308
1000a	"	15.009
1200a	EA.	16.000
Terminal box		
225a	EA.	7.273
800a	"	11.494
1600a	"	18.018
NEMA 3R, three phase, 280v		
400a	EA.	8.502
600a	"	11.994
800a	"	13.008
Main breaker		
400a	EA.	8.502
600a	"	11.994
800a	"	13.008
1000a	"	16.985
1200a	"	18.018
Terminal box		
225a	EA.	8.000
800a	"	13.008
1600a	"	20.997
NEMA 3R meter center, single phase, 208v, 100a		
2 meters	EA.	5.000
3 meters	"	6.154
4 meters	"	7.273
5 meters	"	8.000
6 meters	"	8.999
125a, 3 meters	"	6.154
4 meters	"	7.273
6 meters	"	8.239
7 meters	"	10.000
8 meters	"	11.004
150a, 3 meters	"	6.154
4 meters	"	7.273
6 meters	"	8.502
7 meters	"	10.000
8 meters	"	11.004
NEMA 3R center, 3 phase, 208v, 125a breakers		
3 meters	EA.	6.154
4 meters	"	7.273
6 meters	"	8.502
7 meters	"	10.000
8 meters	"	11.004
150a		
3 meters	EA.	6.667
4 meters	"	7.273
6 meters	"	8.999
7 meters	"	10.499
8 meters	"	11.494
200a		
3 meters	EA.	6.667
4 meters	"	7.273
6 meters	"	8.999
7 meters	"	11.004
8 meters	"	11.494

16 ELECTRICAL

Service And Distribution	UNIT	MAN/HOURS
16430.20 Metering *(Cont.)*		
NEMA 3R, center plug-on breakers, 208v, 1 phase		
60a	EA.	0.250
70a	"	0.250
90a	"	0.250
100a	"	0.348
125a	"	0.400
16460.10 Transformers		
Pad mounted, single phase, dry type, 480v-120/240v		
15 kva	EA.	8.000
25 kva	"	8.989
37.5 kva	"	10.000
50 kva	"	10.959
3 phase		
225 kva	EA.	25.000
300 kva	"	30.769
500 kva	"	38.095
750 kva	"	47.059
1000 kva	"	50.000
1500 kva	"	57.143
Substation transformers, outdoor, 5 kv - 208v		
112.5 kva	EA.	21.978
150 kva	"	24.024
225 kva	"	27.972
300 kva	"	29.963
500 kva	"	44.944
750 kva	"	55.172
1000 kva	"	65.041
15 kv, 208v		
112 kva	EA.	27.972
150 kva	"	29.963
225 kva	"	34.934
300 kva	"	40.000
500 kva	"	50.000
750 kva	"	60.150
1000 kva	"	70.175
5kv, 480v		
112kva	EA.	21.978
150 kva	"	24.024
225 kva	"	27.972
300 kva	"	29.963
500 kva	"	44.944
750 kva	"	55.172
1000 kva	"	65.041
1500 kva	"	74.766
2000 kva	"	89.888
2500 kva	"	109.589
15 kv, 480v		
112.5 kva	EA.	27.972
150 kva	"	29.963
225 kva	"	34.934
300 kva	"	40.000
500 kva	"	50.000

Service And Distribution	UNIT	MAN/HOURS
16460.10 Transformers *(Cont.)*		
750 kva	EA.	60.150
1000 kva	"	70.175
1500 kva	"	80.000
2000 kva	"	89.888
2500 kva	"	119.403
Pad mounted 3 phase, 15 kv outdoor		
50 kva	EA.	10.256
75 kva	"	11.765
112 kva	"	12.903
150 kva	"	14.545
225 kva	"	15.385
300 kva	"	17.021
500 kva	"	27.586
750 kva	"	36.364
1000 kva	"	44.444
1500 kva	"	53.333
Dry type, for power gear, 5 kv indoor		
75 kva	EA.	16.000
112.5 kva	"	18.605
150 kva	"	21.053
225 kva	"	23.529
300 kva	"	25.000
500 kva	"	27.586
750 kva	"	36.364
16470.10 Panelboards		
Indoor load center, 1 phase 240v main lug only		
30a - 2 spaces	EA.	2.000
100a - 8 spaces	"	2.424
150a - 16 spaces	"	2.963
200a - 24 spaces	"	3.478
200a - 42 spaces	"	4.000
Main circuit breaker		
100a - 8 spaces	EA.	2.424
100a - 16 spaces	"	2.759
150a - 16 spaces	"	2.963
150a - 24 spaces	"	3.200
200a - 24 spaces	"	3.478
200a - 42 spaces	"	3.636
3 phase, 480/277v, main lugs only, 120a, 30 circuits	"	3.478
277/480v, 4 wire, flush surface		
225a, 30 circuits	EA.	4.000
400a, 30 circuits	"	5.000
600a, 42 circuits	"	6.015
208/120v, main circuit breaker, 3 phase, 4 wire		
100a		
12 circuits	EA.	5.096
20 circuits	"	6.299
30 circuits	"	7.018
225a		
30 circuits	EA.	7.767
42 circuits	"	9.524
400a		

16 ELECTRICAL

Service And Distribution

16470.10 Panelboards (Cont.)

Description	UNIT	MAN/HOURS
30 circuits	EA.	14.815
42 circuits	"	16.000
600a, 42 circuits	"	18.182
120/208v, flush, 3 ph., 4 wire, main only		
100a		
12 circuits	EA.	5.096
20 circuits	"	6.299
30 circuits	"	7.018
225a		
30 circuits	EA.	7.767
42 circuits	"	9.524
400a		
30 circuits	EA.	14.815
42 circuits	"	16.000
600a, 42 circuits	"	18.182
Panelboard accessories		
Grounding bus	EA.	0.348
Handle lock device	"	0.145
Factory assembled panel		
1 pole space	EA.	0.348
2 pole space	"	0.145
3 pole space	"	0.133
Panelboards 1 phase, 240/120v main circuit breaker		
Single phase, 3 wire, 120/240v flush		
100a, 20 circuits	EA.	3.478
225a, 30 circuits	"	4.000
240/120v, main lugs only		
100a		
8 circuits	EA.	2.963
12 circuits	"	2.963
20 circuits	"	2.963
225a		
24 circuits	EA.	3.478
30 circuits	"	3.810
42 circuits	"	3.810
Distribution panelboards, 3 ph, main breaker		
225a	EA.	16.000
400a	"	18.018
600a	"	21.978
800a	"	24.024
1000a	"	27.972
1200a	"	29.963
Single phase		
225a	EA.	14.011
400a	"	16.000
600a	"	20.000
800a	"	24.024
1000a	"	27.972
1200a	"	29.963
Fusible distribution panelboards, 3 phase, 600v		
100a	EA.	14.011
200a	"	16.000
400a	"	20.000

16470.10 Panelboards (Cont.)

Description	UNIT	MAN/HOURS
600a	EA.	24.024
800a	"	27.972
Single phase		
100a	EA.	11.994
200a	"	14.011
400a	"	18.018
600a	"	21.978
800a	"	25.974
Hospital panels, operating room		
3kv - 208v	EA.	6.154
3kv - 277v	"	6.154
5kv - 208v	"	6.154
5kv - 277v	"	6.154
Coronary care		
3kv - 208v	EA.	7.273
3kv - 277v	"	7.273
5kv - 208v	"	7.273
5kv - 277v	"	7.273
Intensive care		
3kv - 208v	EA.	8.000
3kv - 277v	"	8.000
5kv - 208v	"	8.000
5kv - 277v	"	8.000
15kv - 208v	"	11.994
15kv - 277v	"	11.994
25kv - 208v	"	16.000
25kv - 277v	"	16.000
Explosion proof, 240v, m.l.b. 20a, single phase		
6 breakers	EA.	11.004
8 breakers	"	11.747
10 breakers	"	12.500
12 breakers	"	13.245
14 breakers	"	14.011
16 breakers	"	14.011
18 breakers	"	15.504
20 breakers	"	16.260
22 breakers	"	16.985
24 breakers	"	17.738

16480.10 Motor Controls

Description	UNIT	MAN/HOURS
Motor generator set, 3 phase, 480/277v, w/controls		
10kw	EA.	27.586
15kw	"	30.769
20kw	"	32.000
25kw	"	34.783
30kw	"	36.364
40kw	"	38.095
50kw	"	40.000
60kw	"	44.444
75kw	"	50.000
100kw	"	61.538
125kw	"	66.667
150kw	"	66.667

16 ELECTRICAL

Service And Distribution

16480.10 Motor Controls (Cont.)

	UNIT	MAN/HOURS
200kw	EA.	72.727
250kw	"	72.727
300kw	"	80.000
2 pole, 230 volt starter, w/NEMA-1		
1 hp, 9a, size 00	EA.	1.000
2 hp, 18a, size 0	"	1.000
3 hp, 27a, size 1	"	1.000
5 hp, 45a, size 1p	"	1.000
7-1/2 hp, 45a, size 2	"	1.000
15 hp, 90a, size 3	"	1.000
2 pole, w/NEMA-4 enclosure		
2 hp, 18a, size 1	EA.	1.600
5 hp, 45a, size 1p	"	1.600
7-1/2 hp, 45a, size 2	"	1.600
3 pole, 2 hp, 9a, 200-575v starter		
W/NEMA-1, size 00	EA.	1.333
W/NEMA-4 enclosure, size 00	"	1.739
5hp, 18a		
W/NEMA-1 enclosure, size 0	EA.	1.333
W/NEMA-4 enclosure, size 0	"	1.739
7.5-10hp, 27a		
7.5-10hp 27a, w/NEMA-1 enclosure, size 1	EA.	1.333
W/NEMA-4 enclosure size 1	"	1.739
10-25hp, 45a		
W/NEMA-1 enclosure, size 2	EA.	1.333
W/NEMA-4 enclosure, size 2	"	1.739
25-50hp, 90a		
W/NEMA-1 enclosure, size 3	EA.	1.739
W/NEMA-4 enclosure, size 3	"	2.500
40-100hp, 135a		
W/NEMA-1 enclosure, size 4	EA.	2.500
W/NEMA-4 enclosure, size 4	"	3.478
75-200hp, 270a		
W/NEMA-1 enclosure, size 5	EA.	5.517
W/NEMA-4 enclosure, size 5	"	7.018
Magnetic starter accessories		
On-off-auto selector switch kit	EA.	0.320
With pilot light	"	0.348
Control center main lug only, 208v, 3 phase		
600a	EA.	11.994
1200a	"	16.000
Main circuit breakers, 208v, 3 phase		
400a	EA.	10.000
600a	"	14.011
800a	"	16.000
1000a	"	18.018
1200a	"	20.000
Non-reversing starters		
Size 1	EA.	0.727
Size 2	"	1.250
Size 3	"	1.509
Size 4	"	1.739
Reversing starters		

16480.10 Motor Controls (Cont.)

	UNIT	MAN/HOURS
Size 1	EA.	0.727
Size 2	"	1.096
Fusible switch, non-revolving starters		
Size 1	EA.	0.727
Size 2	"	1.250
Size 3	"	1.509
Size 4	"	1.739
Reversing starters		
Size 1	EA.	0.727
Size 2	"	1.096
Two speed, non-reversing starter		
Size 1	EA.	0.727
Size 2	"	1.096
Magnetic starter, 600v, 2 pole, NEMA 3R		
Size 0, 2 hp	EA.	1.000
Size 1, 5hp	"	1.096
NEMA 3R		
Size 2, 7.5 hp	EA.	1.143
Size 3, 15 hp	"	1.194
NEMA 7		
Size 0, 2 hp	EA.	1.739
Size 1, 5 hp	"	2.000
Size 2, 7.5 hp	"	2.222
NEMA 12		
Size 0, 2 hp	EA.	1.509
Size 1, 5 hp	"	1.739
Size 2, 7.5 hp	"	2.000
Size 3, 15 hp	"	2.222
3 pole, NEMA 1		
Size 6	EA.	10.000
Size 7	"	11.994
Size 8	"	16.000
NEMA 4		
Size 6	EA.	14.011
Size 7	"	16.000
Size 8	"	20.000
NEMA 3R		
Size 0	EA.	1.250
Size 1	"	1.356
Size 2	"	1.818
Size 3	"	1.905
Size 4	"	2.759
NEMA 7		
Size 0	EA.	2.000
Size 1	"	2.162
Size 2	"	2.222
Size 3	"	2.759
Size 4	"	4.444
Size 5	"	9.744
Size 6	"	16.000
NEMA 12		
Size 00	EA.	1.739
Size 0	"	1.860

16 ELECTRICAL

Service And Distribution

16480.10 Motor Controls (Cont.)

	Unit	Man/Hours
Size 1	EA.	1.905
Size 2	"	2.000
Size 3	"	2.500
Size 4	"	3.478
Size 5	"	7.273
Size 6	"	12.500
Size 7	"	14.011
Size 8	"	20.000
Reversing magnetic starters, 600v, 3 pole, NEMA 1		
Size 00	EA.	1.250
Size 0	"	1.290
Size 1	"	1.356
Size 2	"	1.509
Size 3	"	1.739
Size 4	"	2.000
Size 5	"	5.333
Size 6	"	9.501
Size 7	"	11.004
Size 8	"	18.018
NEMA 4		
Size 0	EA.	1.739
Size 4	"	1.818
Size 2	"	1.905
Size 3	"	2.000
Size 4	"	2.500
Size 5	"	7.273
Size 6	"	12.500
Size 7	"	15.009
Size 8	"	20.000
NEMA 7		
Size 0	EA.	2.000
Size 1	"	2.222
Size 2	"	2.500
Size 3	"	2.963
NEMA 12		
Size 0	EA.	1.739
Size 1	"	1.905
Size 2	"	2.000
Size 3	"	2.222
Size 4	"	2.500
Size 5	"	7.273
Size 6	"	14.011
Size 7	"	16.000
Size 8	"	20.000
Electrically held lighting contactors, NEMA 1, 20a		
2 pole	EA.	1.000
3 pole	"	1.250
4 pole	"	1.509
6 pole	"	2.000
8 pole	"	2.500
10 pole	"	2.963
12 pole	"	3.478
30a		

16480.10 Motor Controls (Cont.)

	Unit	Man/Hours
2 pole	EA.	2.759
3 pole	"	2.963
4 pole	"	3.200
300a		
2 pole	EA.	4.211
3 pole	"	5.333
400a		
2 pole	EA.	4.211
3 pole	"	5.333
600a		
2 pole	EA.	6.667
3 pole	"	9.249
800a		
2 pole	EA.	8.000
3 pole	"	11.004
Mechanically held lighting contactors, NEMA 1, 20a		
2 pole	EA.	1.000
3 pole	"	1.250
4 pole	"	1.509
6 pole	"	2.000
8 pole	"	2.500
10 pole	"	2.963
30a		
2 pole	EA.	1.000
3 pole	"	1.250
4 pole	"	1.509
5 pole	"	1.739
60a		
2 pole	EA.	1.000
3 pole	"	1.250
4 pole	"	1.509
5 pole	"	1.739
100a		
2 pole	EA.	1.250
3 pole	"	1.739
4 pole	"	2.000
5 pole	"	2.500
200a		
2 pole	EA.	1.739
3 pole	"	2.500
4 pole	"	3.200
300a		
2 pole	EA.	4.211
3 pole	"	5.333
400a		
2 pole	EA.	4.211
3 pole	"	5.333
600a		
2 pole	EA.	6.667
3 pole	"	8.889
800a		
2 pole	EA.	8.000
3 pole	"	11.429

16 ELECTRICAL

Service And Distribution	UNIT	MAN/HOURS
16480.10 Motor Controls (Cont.)		
AC relays, control type open, 15a, 600v		
2 pole	EA.	1.000
3 pole	"	1.250
4 pole	"	1.509
6 pole	"	2.000
8 pole	"	2.500
10 pole	"	2.963
12 pole	"	3.478
16490.10 Switches		
Oil switches, medium voltage, bus components		
Switches, 277/120v, toggle device only	EA.	1.600
With oil 35kv, g&w gram 44, 4 way switch	"	8.000
Weatherproof enclosure		
3 way switch	EA.	10.000
4 way switch	"	10.959
Fused interrupter load, 35kv		
20A		
1 pole	EA.	16.000
2 pole	"	17.021
3 way	"	17.021
4 way	"	18.182
30a, 1 pole	"	16.000
3 way	"	17.021
4 way	"	18.182
Weatherproof switch, including box & cover, 20a		
1 pole	EA.	16.000
2 pole	"	17.021
3 way	"	18.182
4 way	"	18.182
3 way, oil switch, 15kv enclosure	"	11.940
Pedestal for 35kv double breaker switch	"	5.000
Bus terminal connector, 2	"	2.500
2 to 3	"	2.500
Support connector, 3	"	1.600
Tee connector, 2 to 3	"	2.000
Flexible bus stud connector	"	1.739
End cap 3	"	1.333
Weldment connection, 3	"	1.000
Plate switch, 1 gang	"	0.050
Start stop stations, manual motor starters	"	0.727
Lockout switch	"	0.250
Forward-reverse switch	"	0.727
On-off switch	"	0.727
Open-close switch	"	0.727
Forward-reverse-stop switch	"	1.000
Standard 3 button switch any standard legend	"	1.000
Standard 3 button with lockout	"	1.000
Manual motor starters, tog, 115/230v		
Size 1 gp	EA.	1.000
Size 2	"	1.000
Button		
Size 0	EA.	1.000

Service And Distribution	UNIT	MAN/HOURS
16490.10 Switches (Cont.)		
Size 1	EA.	1.000
Size 2	"	1.000
3-phase		
Size 0	EA.	1.333
Size 1	"	1.333
Time & float switches	"	1.600
Astronomical time switch, 40a, 240v	"	1.000
Timer switch 0-5 minute, with box	"	0.500
Single pole/single throw time, 277v, NEMA-1	"	0.727
Single toggle switch, 20a, 120v, with pilot	"	0.250
3-way toggle	"	0.296
Photo electric switches		
1000 watt		
105-135v	EA.	0.727
208-277v	"	0.727
3000 watt, 105-130v		
Double throw	EA.	1.000
Single throw	"	1.000
Double pole/single throw, 210-250v	"	1.333
Dimmer switch and switch plate		
600w	EA.	0.308
1000w	"	0.348
Dimmer switch incandescent		
1500w	EA.	0.702
2000w	"	0.748
Fluorescent		
12 lamps	EA.	0.500
20 lamps	"	0.552
30 lamps	"	0.602
40 lamps	"	0.702
Time clocks with skip, 40a, 120v		
SPST	EA.	0.748
SPDT	"	0.748
DPST	"	0.748
DPDT	"	1.000
SPST	"	1.000
Astronomic time clocks with skip, 40a, 120v		
DPST	EA.	0.748
SPST	"	1.000
SPDT	"	0.748
Raintight time clocks, 40a, 120v		
SPDT	EA.	1.000
DPST	"	1.000
Contractor grade wall switch 15a, 120v		
Single pole	EA.	0.160
Three way	"	0.200
Four way	"	0.267
Specification grade toggle switches, 20a, 120-277v		
Single pole	EA.	0.200
Double pole	"	0.296
3 way	"	0.250
4 way	"	0.296
30a, 120-277v		

16 ELECTRICAL

Service And Distribution	UNIT	MAN/HOURS
16490.10 Switches *(Cont.)*		
Single pole	EA.	0.200
Double pole	"	0.296
3 way	"	0.250
Specification grade key switches, 20a, 120-277v		
Single pole	EA.	0.200
Double pole	"	0.296
3 way	"	0.250
4 way	"	0.296
Red pilot light handle switches, 20a, 120-277v		
Single pole	EA.	0.200
Double pole	"	0.296
3 way	"	0.250
30a, 120-277v		
Single pole	EA.	0.200
Double pole	"	0.296
3 way	"	0.250
Momentary contact switches, 20a		
SPDT, ivory	EA.	0.250
SPDT, locking	"	0.296
Maintained contact switches		
SPDT ivory	EA.	0.250
DPDT ivory	"	0.250
SPDT locking	"	0.296
DPDT locking	"	0.348
Mercury switch, 3 way	"	0.250
Door switches, open on or off	"	0.500
Combination switch and pilot light, single pole	"	0.296
3 way	"	0.348
Combination switch and receptacle, single pole	"	0.296
3 way	"	0.296
Combination two switches, single pole/single pole	"	0.250
3 way	"	0.400
Switch plates, plastic ivory		
1 gang	EA.	0.080
2 gang	"	0.100
3 gang	"	0.119
4 gang	"	0.145
5 gang	"	0.160
6 gang	"	0.182
Stainless steel		
1 gang	EA.	0.080
2 gang	"	0.100
3 gang	"	0.123
4 gang	"	0.145
5 gang	"	0.160
6 gang	"	0.182
Brass		
1 gang	EA.	0.080
2 gang	"	0.100
3 gang	"	0.123
4 gang	"	0.145
5 gang	"	0.160
6 gang	"	0.182

Service And Distribution	UNIT	MAN/HOURS
16490.20 Transfer Switches		
Automatic transfer switch 600v, 3 pole		
30a	EA.	3.478
60a	"	3.478
100a	"	4.762
150a	"	6.015
225a	"	8.000
260a	"	8.000
400a	"	10.000
600a	"	15.094
800a	"	18.182
1000a	"	21.053
1200a	"	22.857
1600a	"	25.000
2000a	"	29.630
2600a	"	42.105
3000a	"	50.000
16490.80 Safety Switches		
Safety switch, 600v, 3 pole, heavy duty, NEMA-1		
30a	EA.	1.000
60a	"	1.143
100a	"	1.600
200a	"	2.500
400a	"	5.517
600a	"	8.000
800a	"	10.526
1200a	"	14.286

Lighting	UNIT	MAN/HOURS
16510.05 Interior Lighting		
Recessed fluorescent fixtures, 2'x2'		
2 lamp	EA.	0.727
4 lamp	"	0.727
2 lamp w/flange	"	1.000
4 lamp w/flange	"	1.000
1'x4'		
2 lamp	EA.	0.667
3 lamp	"	0.667
2 lamp w/flange	"	0.727
3 lamp w/flange	"	0.727
2'x4'		
2 lamp	EA.	0.727
3 lamp	"	0.727
4 lamp	"	0.727
2 lamp w/flange	"	1.000

16 ELECTRICAL

16510.05 Interior Lighting (Cont.)

Lighting	UNIT	MAN/HOURS
3 lamp w/flange	EA.	1.000
4 lamp w/flange	"	1.000
4'x4'		
4 lamp	EA.	1.000
6 lamp	"	1.000
8 lamp	"	1.000
4 lamp w/flange	"	1.509
6 lamp w/flange	"	1.509
8 lamp, w/flange	"	1.509
Surface mounted incandescent fixtures		
40w	EA.	0.667
75w	"	0.667
100w	"	0.667
150w	"	0.667
Pendant		
40w	EA.	0.800
75w	"	0.800
100w	"	0.800
150w	"	0.800
Contractor grade recessed down lights		
100 watt housing only	EA.	1.000
150 watt housing only	"	1.000
100 watt trim	"	0.500
150 watt trim	"	0.500
Recessed incandescent fixtures		
40w	EA.	1.509
75w	"	1.509
100w	"	1.509
150w	"	1.509
Exit lights, 120v		
Recessed	EA.	1.250
Back mount	"	0.727
Universal mount	"	0.727
Emergency battery units, 6v-120v, 50 unit	"	1.509
With 1 head	"	1.509
With 2 heads	"	1.509
Mounting bucket	"	0.727
Light track single circuit		
2'	EA.	0.500
4'	"	0.500
8'	"	1.000
12'	"	1.509
Fittings and accessories		
Dead end	EA.	0.145
Starter kit	"	0.250
Conduit feed	"	0.145
Straight connector	"	0.145
Center feed	"	0.145
L-connector	"	0.145
T-connector	"	0.145
X-connector	"	0.200
Cord and plug	"	0.100
Rigid corner	"	0.145

16510.05 Interior Lighting (Cont.)

Lighting	UNIT	MAN/HOURS
Flex connector	EA.	0.145
2 way connector	"	0.200
Spacer clip	"	0.050
Grid box	"	0.145
T-bar clip	"	0.050
Utility hook	"	0.145
Fixtures, square		
R-20	EA.	0.145
R-30	"	0.145
40w flood	"	0.145
40w spot	"	0.145
100w flood	"	0.145
100w spot	"	0.145
Mini spot	"	0.145
Mini flood	"	0.145
Quartz, 500w	"	0.145
R-20 sphere	"	0.145
R-30 sphere	"	0.145
R-20 cylinder	"	0.145
R-30 cylinder	"	0.145
R-40 cylinder	"	0.145
R-30 wall wash	"	0.145
R-40 wall wash	"	0.145
Explosion proof, incan., surface mounted		
100w - 200w	EA.	1.739
300w	"	1.739
500w	"	1.739
With guard		
100w-200w	EA.	2.222
300w	"	2.222
500w	"	2.222
Reflectors for incan. light fixtures, dome	"	0.250
Angle	"	0.250
Highbay	"	0.296
Explosion proof fluor. fixtures, 800 ms.		
1 lamp	EA.	2.222
2 lamp	"	2.667
3 lamp	"	2.963
4 lamp	"	3.200
Explosion proof hp sodium fixtures		
50w-70w	EA.	2.222
100w	"	2.222
150w	"	2.500
200w	"	2.500
250w	"	2.500
310w	"	2.500
400w	"	2.500
With guard		
50w-70w	EA.	2.500
100w	"	2.500
150w	"	2.759
200w	"	2.759
250w	"	2.759

16 ELECTRICAL

16510.05 Interior Lighting (Cont.)

Lighting	UNIT	MAN/HOURS
310w	EA.	2.759
400w	"	2.759
Explosion proof metal halide fixtures		
175w	EA.	2.500
250w	"	2.500
400w	"	2.500
With guard, 175w	"	2.759
250w	"	2.759
400w	"	2.759
Energy saving rapid start fluor. lamps		
F30 cw	EA.	0.100
F40 cw	"	0.100
F40 cwx	"	0.100
F30 ww	"	0.100
F40 ww	"	0.100
F40 wwx	"	0.100
Slimline		
F48 cw	EA.	0.145
F96 cwx	"	0.145
F48 ww	"	0.145
F96 ww	"	0.145
F96 wwx	"	0.145
High output		
F96 cwx	"	0.145
F96 cw	"	0.145
Power groove, F48 cw	"	0.145
Circle		
Fc6 cw	EA.	0.100
Fc8 cw	"	0.100
Fc12 cw	"	0.100
Fc16 cw	"	0.100
Fc6 ww	"	0.100
Fc8 ww	"	0.100
Fc12 ww	"	0.100
Fc16 ww	"	0.100
Incandescent lamps		
200w	EA.	0.100
300w	"	0.100
500w	"	0.100
750w	"	0.145
1000w	"	0.200
1500w	"	0.200
Energy saving reflector floodlight lamps		
25w	EA.	0.100
30w	"	0.100
50w	"	0.100
75w	"	0.100
120w	"	0.100
150w	"	0.100
200w	"	0.100
300w	"	0.100
500w	"	0.145
750w	"	0.145

16510.05 Interior Lighting (Cont.)

Lighting	UNIT	MAN/HOURS
Reflector spotlight		
75w	EA.	0.100
100w	"	0.100
125w	"	0.100
150w	"	0.100
250w	"	0.100
300w	"	0.100
400w	"	0.145
500w	"	0.145
1000w	"	0.200
Medium par flood lamps		
75w	EA.	0.100
100w	"	0.100
150w	"	0.100
200w	"	0.100
300w	"	0.100
500w	"	0.145
Medium par spot lamps		
75w	EA.	0.100
120w	"	0.100
150w	"	0.100
Tubular quartz lamps		
100w	EA.	0.145
150w	"	0.145
200w	"	0.145
400w	"	0.145
500w	"	0.200
750w	"	0.200
1000w	"	0.250
1250w	"	0.250
1500w	"	0.250
Ballast replacements rapid start fluor		
1f-40-120v	EA.	0.727
1f-40-277v	"	0.727
1f-96-120v	"	0.727
1f-96-277v	"	0.727
2f-40-120v	"	0.727
2f-40-277v	"	0.727
2f-96-120v	"	0.727
2f-96-277v	"	0.727
Circline, 1fc6-1fc16	"	0.727
Very high output, 1500ma		
1f48-120v	EA.	0.727
1f48-277v	"	0.727
1f96-120v	"	0.727
1f96-277v	"	0.727
2f48-120v	"	0.727
2f48-277v	"	0.727
2f96-120v	"	0.727
2f96-277v	"	0.727
Mercury, multi tap		
475w	EA.	1.000
100w	"	1.000

16 ELECTRICAL

Lighting	UNIT	MAN/HOURS
16510.05 Interior Lighting (Cont.)		
175w	EA.	1.000
250w	"	1.000
400w	"	1.000
1000w	"	1.000
Metal halide, multi tap		
175w	EA.	1.000
250w	"	1.000
400w	"	1.000
1000w	"	1.000
1500w	"	1.000
High pressure sodium		
70w	EA.	1.000
100w	"	1.000
150w	"	1.000
250w	"	1.000
400w	"	1.000
1000w	"	1.000
16510.08 Energy Efficient Interior Lighting		
Ballast		
Fluorescent, 12 VDC		
Min.	EA.	0.667
Ave.	"	0.667
Max.	"	0.667
24 VDC		
Min.	EA.	0.667
Ave.	"	0.667
Max.	"	0.667
Pressure sodium, 12 VDC		
Min.	EA.	0.667
Ave.	"	0.667
Max.	"	0.667
24 VDC		
Min.	EA.	0.667
Ave.	"	0.667
Max.	"	0.667
Lamps		
Photovoltaic source, Fluorescent, 12 VDC, 7 Watt		
Min.	EA.	0.267
Ave.	"	0.267
Max.	"	0.267
11 Watt		
Min.	EA.	0.267
Ave.	"	0.267
Max.	"	0.267
15 Watt		
Min.	EA.	0.267
Ave.	"	0.267
Max.	"	0.267
25 Watt		
Min.	EA.	0.267
Ave.	"	0.267
Max.	"	0.267

Lighting	UNIT	MAN/HOURS
16510.08 Energy Efficient Int. Lighting (Cont.)		
30 Watt		
Min.	EA.	0.267
Ave.	"	0.267
Max.	"	0.267
LED, 85-265 V, 300 lumens		
Min.	EA.	0.267
Ave.	"	0.267
Max.	"	0.267
600 lumens		
Min.	EA.	0.267
Ave.	"	0.267
Max.	"	0.267
12-24 V, 2500 lumens		
Min.	EA.	0.667
Ave.	"	0.667
Max.	"	0.667
Compact fluorescent, 13 Watt, 900 lumens		
Min.	EA.	0.267
Ave.	"	0.267
Max.	"	0.267
36 Watt, 2800 lumens		
Min.	EA.	0.267
Ave.	"	0.267
Max.	"	0.267
40 Watt, 3150 lumens		
Min.	EA.	0.267
Ave.	"	0.267
Max.	"	0.267
Low Pressure Sodium, 18 Watt, 1800 lumens		
Min.	EA.	0.267
Ave.	"	0.267
Max.	"	0.267
35 Watt, 5000 lumens		
Min.	EA.	0.267
Ave.	"	0.267
Max.	"	0.267
40 Watt, 3150 lumens		
Min.	EA.	0.267
Ave.	"	0.267
Max.	"	0.267
16510.10 Lighting Industrial		
Surface mounted fluorescent, wrap around lens		
1 lamp	EA.	0.800
2 lamps	"	0.889
4 lamps	"	1.000
Wall mounted fluorescent		
2-20w lamps	EA.	0.500
2-30w lamps	"	0.500
2-40w lamps	"	0.667
Indirect, with wood shielding, 2049w lamps		
4'	EA.	1.000
8'	"	1.600

16 ELECTRICAL

16510.10 Lighting Industrial (Cont.)

Lighting	UNIT	MAN/HOURS
Industrial fluorescent, 2 lamp		
4'	EA.	0.727
8'	"	1.333
Strip fluorescent		
4'		
1 lamp	EA.	0.667
2 lamps	"	0.667
8'		
1 lamp	EA.	0.727
2 lamps	"	0.889
Wire guard for strip fixture, 4' long	"	0.348
Strip fluorescent, 8' long, two 4' lamps	"	1.333
With four 4' lamps	"	1.600
Wet location fluorescent, plastic housing		
4' long		
1 lamp	EA.	1.000
2 lamps	"	1.333
8' long		
2 lamps	EA.	1.600
4 lamps	"	1.739
Parabolic troffer, 2'x2'		
With 2 "U" lamps	EA.	1.000
With 3 "U" lamps	"	1.143
2'x4'		
With 2 40w lamps	EA.	1.143
With 3 40w lamps	"	1.333
With 4 40w lamps	"	1.333
1'x4'		
With 1 T-12 lamp, 9 cell	EA.	0.727
With 2 T-12 lamps	"	0.889
With 1 T-12 lamp, 20 cell	"	0.727
With 2 T-12 lamps	"	0.889
Steel sided surface fluorescent, 2'x4'		
3 lamps	EA.	1.333
4 lamps	"	1.333
Outdoor sign fluor., 1 lamp, remote ballast		
4' long	EA.	6.015
6' long	"	8.000
Recess mounted, commercial, 2'x2', 13" high		
100w	EA.	4.000
250w	"	4.494
High pressure sodium, hi-bay open		
400w	EA.	1.739
1000w	"	2.424
Enclosed		
400w	EA.	2.424
1000w	"	2.963
Metal halide hi-bay, open		
400w	EA.	1.739
1000w	"	2.424
Enclosed		
400w	EA.	2.424
1000w	"	2.963
High pressure sodium, low bay, surface mounted		
100w	EA.	1.000
150w	"	1.143
250w	"	1.333
400w	"	1.600
Metal halide, low bay, pendant mounted		
175w	EA.	1.333
250w	"	1.600
400w	"	2.222
Indirect luminare, square, metal halide, freestanding		
175w	EA.	1.000
250w	"	1.000
400w	"	1.000
High pressure sodium		
150w	EA.	1.000
250w	"	1.000
400w	"	1.000
Round, metal halide		
175w	EA.	1.000
250w	"	1.000
400w	"	1.000
High pressure sodium		
150w	EA.	1.000
250w	"	1.000
400w	"	1.000
Wall mounted, metal halide		
175w	EA.	2.500
250w	"	2.500
400w	"	3.200
High pressure sodium		
150w	EA.	2.500
250w	"	2.500
400w	"	3.200
Wall pack lithonia, high pressure sodium		
35w	EA.	0.889
55w	"	1.000
150w	"	1.600
250w	"	1.739
Low pressure sodium		
35w	EA.	1.739
55w	"	2.000
Wall pack hubbell, high pressure sodium		
35w	EA.	0.889
150w	"	1.600
250w	"	1.739
Compact fluorescent		
2-7w	EA.	1.000
2-13w	"	1.333
1-18w	"	1.333
Handball & racquet ball court, 2'x2', metal halide		
250w	EA.	2.500
400w	"	2.759
High pressure sodium		

16 ELECTRICAL

16510.10 Lighting Industrial (Cont.)

Lighting	UNIT	MAN/HOURS
250w	EA.	2.500
400w	"	2.759
Bollard light, 42" w/found., high pressure sodium		
70w	EA.	2.581
100w	"	2.581
150w	"	2.581
Light fixture lamps		
Lamp		
20w med. bipin base, cool white, 24"	EA.	0.145
30w cool white, rapid start, 36"	"	0.145
40w cool white "U", 3"	"	0.145
40w cool white, rapid start, 48"	"	0.145
70w high pressure sodium, mogul base	"	0.200
75w slimline, 96"	"	0.200
100w		
Incandescent, 100a, inside frost	EA.	0.100
Mercury vapor, clear, mogul base	"	0.200
High pressure sodium, mogul base	"	0.200
150w		
Par 38 flood or spot, incandescent	EA.	0.100
High pressure sodium, 1/2 mogul base	"	0.200
175w		
Mercury vapor, clear, mogul base	EA.	0.200
Metal halide, clear, mogul base	"	0.200
High pressure sodium, mogul base	"	0.200
250w		
Mercury vapor, clear, mogul base	EA.	0.200
Metal halide, clear, mogul base	"	0.200
High pressure sodium, mogul base	"	0.200
400w		
Mercury vapor, clear, mogul base	EA.	0.200
Metal halide, clear, mogul base	"	0.200
High pressure sodium, mogul base	"	0.200
1000w		
Mercury vapor, clear, mogul base	EA.	0.250
High pressure sodium, mogul base	"	0.250

16510.30 Exterior Lighting

Lighting	UNIT	MAN/HOURS
Exterior light fixtures		
Rectangle, high pressure sodium		
70w	EA.	2.500
100w	"	2.581
150w	"	2.581
250w	"	2.759
400w	"	3.478
Flood, rectangular, high pressure sodium		
70w	EA.	2.500
100w	"	2.581
150w	"	2.581
400w	"	3.478
1000w	"	4.494
Round		
400w	EA.	3.478

16510.30 Exterior Lighting (Cont.)

Lighting	UNIT	MAN/HOURS
1000w	EA.	4.494
Round, metal halide		
400w	EA.	3.478
1000w	"	4.494
Light fixture arms, cobra head, 6', high press. sodium		
100w	EA.	2.000
150w	"	2.500
250w	"	2.500
400w	"	2.963
Flood, metal halide		
400w	EA.	3.478
1000w	"	4.494
1500w	"	6.015
Mercury vapor		
250w	EA.	2.759
400w	"	3.478
Incandescent		
300w	EA.	1.739
500w	"	2.000
1000w	"	3.200

16510.40 Energy Efficient Exterior Lighting

Lighting	UNIT	MAN/HOURS
Solar Powered, led area light, 100 Watt, Zone 4		
Min.	EA.	1.333
Ave.	"	1.600
Max.	"	2.000
Zone 2		
Min.	EA.	1.333
Ave.	"	1.600
Max.	"	2.000
Zone 4DD		
Min.	EA.	1.333
Ave.	"	1.600
Max.	"	2.000
Zone 2DD		
Min.	EA.	1.333
Ave.	"	1.600
Max.	"	2.000

16510.90 Power Line Filters

Lighting	UNIT	MAN/HOURS
Heavy duty power line filter, 240v		
100a	EA.	10.000
300a	"	16.000
600a	"	24.242

16600.20 Central Inverter Systems

Lighting	UNIT	MAN/HOURS
Central inverter systems		
500va	EA.	2.963
1000va	"	4.000
1500va	"	5.333
2400va	"	6.667
3000va	"	8.502

16 ELECTRICAL

Lighting	UNIT	MAN/HOURS
16600.20 Central Inverter Systems *(Cont.)*		
4500va	EA.	10.000
6000va	"	11.004
7500va	"	14.011
10,000va	"	16.000
16,600va	"	22.989
25,000va	"	34.934
16610.30 Uninterruptible Power		
Uninterruptible power systems, (U.P.S.), 3kva	EA.	8.000
5 kva	"	11.004
7.5 kva	"	16.000
10 kva	"	21.978
15 kva	"	22.857
20 kva	"	24.024
25 kva	"	25.000
30 kva	"	25.974
35 kva	"	27.027
40 kva	"	27.972
45 kva	"	28.986
50 kva	"	29.963
62.5 kva	"	32.000
75 kva	"	34.934
100 kva	"	36.036
150 kva	"	50.000
200 kva	"	55.172
300 kva	"	74.766
400 kva	"	89.888
500 kva	"	109.589
16670.10 Lightning Protection		
Lightning protection		
Copper point, nickel plated, 12'		
1/2" dia.	EA.	1.000
5/8" dia.	"	1.000

Communications	UNIT	MAN/HOURS
16720.10 Fire Alarm Systems		
Master fire alarm box, pedestal mounted	EA.	16.000
Master fire alarm box	"	6.015
Box light	"	0.500
Ground assembly for box	"	0.667
Bracket for pole type box	"	0.727
Pull station		
Waterproof	EA.	0.500
Manual	"	0.400

Communications	UNIT	MAN/HOURS
16720.10 Fire Alarm Systems *(Cont.)*		
Horn, waterproof	EA.	1.000
Interior alarm	"	0.727
Coded transmitter, automatic	"	2.000
Control panel, 8 zone	"	8.000
Battery charger and cabinet	"	2.000
Batteries, nickel cadmium or lead calcium	"	5.000
CO2 pressure switch connection	"	0.727
Annunciator panels		
Fire detection annunciator, remote type, 8 zone	EA.	1.818
12 zone	"	2.000
16 zone	"	2.500
Fire alarm systems		
Bell	EA.	0.615
Weatherproof bell	"	0.667
Horn	"	0.727
Siren	"	2.000
Chime	"	0.615
Audio/visual	"	0.727
Strobe light	"	0.727
Smoke detector	"	0.667
Heat detection	"	0.500
Thermal detector	"	0.500
Ionization detector	"	0.533
Duct detector	"	2.759
Test switch	"	0.500
Remote indicator	"	0.571
Door holder	"	0.727
Telephone jack	"	0.296
Fireman phone	"	1.000
Speaker	"	0.800
Remote fire alarm annunciator panel		
24 zone	EA.	6.667
48 zone	"	13.008
Control panel		
12 zone	EA.	2.963
16 zone	"	4.444
24 zone	"	6.667
48 zone	"	16.000
Power supply	"	1.509
Status command	"	5.000
Printer	"	1.509
Transponder	"	0.899
Transformer	"	0.667
Transceiver	"	0.727
Relays	"	0.500
Flow switch	"	2.000
Tamper switch	"	2.963
End of line resistor	"	0.348
Printed ckt. card	"	0.500
Central processing unit	"	6.154
UPS backup to c.p.u.	"	8.999
Smoke detector, fixed temp. & rate of rise comb.	"	1.600

16 ELECTRICAL

Communications

16720.50 Security Systems

	UNIT	MAN/HOURS
Sensors		
Balanced magnetic door switch, surface mounted	EA.	0.500
With remote test	"	1.000
Flush mounted	"	1.860
Mounted bracket	"	0.348
Mounted bracket spacer	"	0.348
Photoelectric sensor, for fence		
6 beam	EA.	2.759
9 beam	"	4.255
Photoelectric sensor, 12 volt dc		
500' range	EA.	1.600
800' range	"	2.000
Capacitance wire grid kit		
Surface	EA.	1.000
Duct	"	1.600
Tube grid kit	"	0.500
Vibration sensor, 30 max per zone	"	0.500
Audio sensor, 30 max per zone	"	0.500
Inertia sensor		
Outdoor	EA.	0.727
Indoor	"	0.500
Ultrasonic transmitter, 20 max per zone		
Omni-directional	EA.	1.600
Directional	"	1.333
Transceiver		
Omni-directional	EA.	1.000
Directional	"	1.000
Passive infra-red sensor, 20 max per zone	"	1.600
Access/secure unit, balanced magnetic switch	"	1.600
Photoelectric sensor	"	1.600
Photoelectric fence sensor	"	1.600
Capacitance sensor	"	1.739
Audio and vibration sensor	"	1.600
Inertia sensor	"	1.600
Ultrasonic sensor	"	1.739
Infra-red sensor	"	2.000
Monitor panel, with access/secure tone, standard	"	1.739
High security	"	2.000
Emergency power indicator	"	0.500
Monitor rack with 115v power supply		
1 zone	EA.	1.000
10 zone	"	2.500
Monitor cabinet, wall mounted		
1 zone	EA.	1.000
5 zone	"	1.600
10 zone	"	1.739
20 zone	"	2.000
Floor mounted, 50 zone	"	4.000
Security system accessories		
Tamper assembly for monitor cabinet	EA.	0.444
Monitor panel blank	"	0.348
Audible alarm	"	0.500
Audible alarm control	"	0.348

Communications

16720.50 Security Systems (Cont.)

	UNIT	MAN/HOURS
Termination screw, terminal cabinet		
25 pair	EA.	1.600
50 pair	"	2.500
150 pair	"	5.000
Universal termination, cabinets & panel		
Remote test	EA.	1.739
No remote test	"	0.727
High security line supervision termination	"	1.000
Door cord for capacitance sensor, 12"	"	0.500
Insulation block kit for capacitance sensor	"	0.348
Termination block for capacitance sensor	"	0.348
Guard alert display	"	0.615
Uninterrupted power supply	"	8.000
Plug-in 40kva transformer		
12 volt	EA.	0.348
18 volt	"	0.348
24 volt	"	0.348
Test relay	"	0.348
Coaxial cable, 50 ohm	L.F.	0.006
Door openers	EA.	0.500
Push buttons		
Standard	EA.	0.348
Weatherproof	"	0.444
Bells	"	0.727
Horns		
Standard	EA.	1.000
Weatherproof	"	1.250
Chimes	"	0.667
Flasher	"	0.615
Motion detectors	"	1.509
Intercom units	"	0.727
Remote annunciator	"	5.000

16730.20 Clock Systems

	UNIT	MAN/HOURS
Clock systems		
Single face	EA.	0.800
Double face	"	0.800
Skeleton	"	2.759
Master	"	5.000
Signal generator	"	4.000
Elapsed time indicator	"	0.800
Controller	"	0.533
Clock and speaker	"	1.096
Bell		
Standard	EA.	0.533
Weatherproof	"	0.800
Horn		
Standard	EA.	0.727
Weatherproof	"	0.952
Chime	"	0.533
Buzzer	"	0.533

16 ELECTRICAL

Communications

16730.20 Clock Systems (Cont.)

	UNIT	MAN/HOURS
Flasher	EA.	0.615
Control Board	"	3.478
Program unit	"	5.000
Block back box	"	0.500
Double clock back box	"	0.667
Wire guard	"	0.200

16740.10 Telephone Systems

	UNIT	MAN/HOURS
Communication cable		
25 pair	L.F.	0.026
100 pair	"	0.029
150 pair	"	0.033
200 pair	"	0.040
300 pair	"	0.042
400 pair	"	0.044
Cable tap in manhole or junction box		
25 pair cable	EA.	3.810
50 pair cable	"	7.547
75 pair cable	"	11.268
100 pair cable	"	15.094
150 pair cable	"	22.222
200 pair cable	"	29.630
300 pair cable	"	44.444
400 pair cable	"	61.538
Cable terminations, manhole or junction box		
25 pair cable	EA.	3.756
50 pair cable	"	7.477
100 pair cable	"	15.094
150 pair cable	"	22.222
200 pair cable	"	29.630
300 pair cable	"	44.444
400 pair cable	"	61.538
Telephones, standard		
1 button	EA.	2.963
2 button	"	3.478
6 button	"	5.333
12 button	"	7.619
18 button	"	8.889
Hazardous area		
Desk	EA.	7.273
Wall	"	5.000
Accessories		
Standard ground	EA.	1.600
Push button	"	1.600
Buzzer	"	1.600
Interface device	"	0.800
Long cord	"	0.800
Interior jack	"	0.400
Exterior jack	"	0.615
Hazardous area		
Selector switch	EA.	3.200
Bell	"	3.200
Horn	"	4.211

16740.10 Telephone Systems (Cont.)

	UNIT	MAN/HOURS
Horn relay	EA.	3.077

16740.30 Call Systems

	UNIT	MAN/HOURS
Call systems, single bed station	EA.	0.533
Double bed station	"	0.727
Call-in cord	"	0.200
Pull cord	"	0.200
Pillow speaker	"	0.276
Dome light	"	0.533
Zone light	"	0.533
Stake station	"	0.615
Duty station	"	0.500
Utility station	"	0.615
Nurses station	"	0.533
Surgical station	"	0.727
Master station	"	2.500
Control station	"	8.000
Annunciator	"	2.000
Power supply	"	1.538
Speakers	"	0.800
Foot switch	"	0.296
Code blue systems		
Bed station	EA.	0.727
Dome light	"	0.667
Zone light	"	0.727
Pull cord	"	0.250
Nurses station	"	0.533
Annunciator	"	2.000
Power supply	"	1.455
Nurse station indicator, alarm annunciators, flush		
4 circuit	EA.	4.000
6 circuit	"	8.000
12 circuit	"	12.012
Desktop		
4 circuit	EA.	3.478
6 circuit	"	3.478
12 circuit	"	5.000

16750.20 Signaling Systems

	UNIT	MAN/HOURS
Signaling systems		
4" bell	EA.	0.602
6" bell	"	0.650
10" bell	"	0.748
Buzzer		
Size 0	EA.	0.444
Size 1	"	0.444
Size 2	"	0.500
Size 3	"	0.533
Horn	"	0.615
Chime	"	0.533
Push button		
Standard	EA.	0.400
Weatherproof	"	0.500

16 ELECTRICAL

Communications	UNIT	MAN/HOURS
16750.20 Signaling Systems *(Cont.)*		
Door opener		
Mortise	EA.	0.500
Rim	"	0.400
Transformer	"	0.444
Contractor grade doorbell chime kit		
Chime	EA.	1.000
Doorbutton	"	0.320
Transformer	"	0.500
16770.30 Sound Systems		
Power amplifiers	EA.	3.478
Pre-amplifiers	"	2.759
Tuner	"	1.455
Horn		
Equalizer	EA.	1.600
Mixer	"	2.222
Tape recorder	"	1.860
Microphone	"	1.000
Cassette Player	"	2.162
Record player	"	1.905
Equipment rack	"	1.290
Speaker		
Wall	EA.	4.000
Paging	"	0.800
Column	"	0.533
Single	"	0.615
Double	"	4.444
Volume control	"	0.533
Plug-in	"	0.800
Desk	"	0.400
Outlet	"	0.400
Stand	"	0.296
Console	"	8.000
Power supply	"	1.290
16780.10 Antennas And Towers		
Guy cable, alumaweld		
1x3, 7/32"	L.F.	0.050
1x3, 1/4"	"	0.050
1x3, 25/64"	"	0.059
1x19, 1/2"	"	0.070
1x7, 35/64"	"	0.080
1x19, 13/16"	"	0.100
Preformed alumaweld end grip		
1/4" cable	EA.	0.100
3/8" cable	"	0.100
1/2" cable	"	0.145
9/16" cable	"	0.200
5/8" cable	"	0.250
Fiberglass guy rod, white epoxy coated		
1/4" dia.	L.F.	0.145
3/8" dia	"	0.145
1/2" dia	"	0.200

Communications	UNIT	MAN/HOURS
16780.10 Antennas And Towers *(Cont.)*		
5/8" dia	L.F.	0.250
Preformed glass grip end grip, guy rod		
1/4" dia.	EA.	0.145
3/8" dia.	"	0.200
1/2" dia.	"	0.250
5/8" dia.	"	0.250
Spelter socket end grip, 1/4" dia. guy rod		
Standard strength	EA.	0.500
High performance	"	0.500
3/8" dia. guy rod		
Standard strength	EA.	0.348
High performance	"	0.500
Timber pole, Douglas Fir		
80-85 ft	EA.	19.512
90-95 ft	"	22.222
Southern yellow pine		
35-45 ft	EA.	10.959
50-55 ft	"	14.035
16780.50 Television Systems		
TV outlet, self terminating, w/cover plate	EA.	0.308
Thru splitter	"	1.600
End of line	"	1.333
In line splitter multitap		
4 way	EA.	1.818
2 way	"	1.702
Equipment cabinet	"	1.600
Antenna		
Broad band uhf	EA.	3.478
Lightning arrester	"	0.727
TV cable	L.F.	0.005
Coaxial cable rg	"	0.005
Cable drill, with replacement tip	EA.	0.500
Cable blocks for in-line taps	"	0.727
In-line taps ptu-series 36 tv system	"	1.143
Control receptacles	"	0.449
Coupler	"	2.424
Head end equipment	"	6.667
TV camera	"	1.667
TV power bracket	"	0.800
TV monitor	"	1.455
Video recorder	"	2.105
Console	"	8.502
Selector switch	"	1.379
TV controller	"	1.404

16 ELECTRICAL

Resistance Heating	UNIT	MAN/HOURS
16850.10 Electric Heating		
Baseboard heater		
2', 375w	EA.	1.000
3', 500w	"	1.000
4', 750w	"	1.143
5', 935w	"	1.333
6', 1125w	"	1.600
7', 1310w	"	1.818
8', 1500w	"	2.000
9', 1680w	"	2.222
10', 1875w	"	2.286
Unit heater, wall mounted		
750w	EA.	1.600
1500w	"	1.667
2000w	"	1.739
2500w	"	1.818
3000w	"	2.000
4000w	"	2.286
Thermostat		
Integral	EA.	0.500
Line voltage	"	0.500
Electric heater connection	"	0.250
Fittings		
Inside corner	EA.	0.400
Outside corner	"	0.400
Receptacle section	"	0.400
Blank section	"	0.400
Infrared heaters		
600w	EA.	1.000
2000w	"	1.194
3000w	"	2.000
4000w	"	2.500
Controller	"	0.667
Wall bracket	"	0.727
Radiant ceiling heater panels		
500w	EA.	1.000
750w	"	1.000
Unit heaters, suspended, single phase		
3.0 kw	EA.	2.759
5.0 kw	"	2.759
7.5 kw	"	3.200
10.0 kw	"	3.810
Three phase		
5 kw	EA.	2.759
7.5 kw	"	3.200
10 kw	"	3.810
15 kw	"	4.211
20 kw	"	5.333
25 kw	"	6.400
30 kw	"	8.000
35 kw	"	8.000
Unit heater thermostat	"	0.533
Mounting bracket	"	0.727

Resistance Heating	UNIT	MAN/HOURS
16850.10 Electric Heating *(Cont.)*		
Relay	EA.	0.615
Duct heaters, three phase		
10 kw	EA.	3.810
15 kw	"	3.810
17.5 kw	"	4.000
20 kw	"	6.154

Controls	UNIT	MAN/HOURS
16910.40 Control Cable		
Control cable, 600v, #14 THWN, PVC jacket		
2 wire	L.F.	0.008
4 wire	"	0.010
6 wire	"	0.131
8 wire	"	0.145
10 wire	"	0.160
12 wire	"	0.182
14 wire	"	0.211
16 wire	"	0.222
18 wire	"	0.242
20 wire	"	0.250
22 wire	"	0.286
Audio cables, shielded, #24 gauge		
3 conductor	L.F.	0.004
4 conductor	"	0.006
5 conductor	"	0.007
6 conductor	"	0.009
7 conductor	"	0.011
8 conductor	"	0.012
9 conductor	"	0.014
10 conductor	"	0.015
15 conductor	"	0.018
20 conductor	"	0.023
25 conductor	"	0.027
30 conductor	"	0.030
40 conductor	"	0.036
50 conductor	"	0.042
#22 gauge		
3 conductor	L.F.	0.004
4 conductor	"	0.006
#20 gauge		
3 conductor	L.F.	0.004
10 conductor	"	0.015
15 conductor	"	0.018
#18 gauge		
3 conductor	L.F.	0.004

16 ELECTRICAL

Controls	UNIT	MAN/HOURS
16910.40 Control Cable *(Cont.)*		
4 conductor	L.F.	0.006
Microphone cables, #24 gauge		
2 conductor	L.F.	0.004
3 conductor	"	0.005
#20 gauge		
1 conductor	L.F.	0.004
2 conductor	"	0.004
2 conductor	"	0.004
3 conductor	"	0.006
4 conductor	"	0.007
5 conductor	"	0.009
7 conductor	"	0.011
8 conductor	"	0.012
Computer cables shielded, #24 gauge		
1 pair	L.F.	0.004
2 pair	"	0.004
3 pair	"	0.006
4 pair	"	0.007
5 pair	"	0.009
6 pair	"	0.011
7 pair	"	0.012
8 pair	"	0.014
50 pair	"	0.039
Coaxial cables		
RG 6/u	L.F.	0.006
RG 6a/u	"	0.006
RG 8/u	"	0.006
RG 8a/u	"	0.006
RG 9/u	"	0.006
RG 11/u	"	0.006
RG 58/u	"	0.006
RG 59/u	"	0.006
RG 62/u	"	0.006
RG 174/u	"	0.006
RG 213/u	"	0.006
MATV and CCTV camera cables		
1 conductor	L.F.	0.004
2 conductor	"	0.005
4 conductor	"	0.006
7 conductor	"	0.009
12 conductor	"	0.015
13 conductor	"	0.016
14 conductor	"	0.018
28 conductor	"	0.027
Fire alarm cables, #22 gauge		
6 conductor	L.F.	0.010
9 conductor	"	0.015
12 conductor	"	0.016
#18 gauge		
2 conductor	L.F.	0.005
4 conductor	"	0.007
#16 gauge		
2 conductor	L.F.	0.007

Controls	UNIT	MAN/HOURS
16910.40 Control Cable *(Cont.)*		
4 conductor	L.F.	0.008
#14 gauge		
2 conductor	L.F.	0.008
#12 gauge		
2 conductor	L.F.	0.010
Plastic jacketed thermostat cable		
2 conductor	L.F.	0.004
3 conductor	"	0.005
4 conductor	"	0.006
5 conductor	"	0.008
6 conductor	"	0.009
7 conductor	"	0.012
8 conductor	"	0.013

Supporting Construction Reference Data

This section contains information, text, charts and tables on various aspects of construction. The intent is to provide the user with a better understanding of unfamiliar areas in order to be able to estimate better. This information includes actual takeoff data for some areas and also selected explanations of common construction materials, methods and common practices.

REMODELING / ALLOWANCES — 01000

ADDITIONAL COSTS INVOLVED IN REMODELING CONSTRUCTION

SMALL AMOUNTS OF WORK

<u>Material</u>
Since remodeling work often entails a smaller scope of work, many of the materials that are bought at a discount in large quantities for a new construction project have to be purchased at a lesser discount or at the regular retail price. To take this factor into account, it is prudent to add an additional 10 percent to the cost of materials for small remodeling projects.

<u>Labor</u>
In order to obtain maximum labor efficiency and normal production, quantities of work must be sufficient to keep workers busy for long enough periods of time, usually a full day. Even with a full day's worth of a particular task, the start-up and wind-down/cleanup (non-productive time) of a typical day take a significant amount of time. When the task requires less than a full day, the non-productive time increases as a percentage of the work installed. Therefore, the cost of labor per unit increases to a minimum of 2 to 4 hours for the installing function.

PATCHING TO MATCH
Matching and fitting new materials to the pre-existing construction is usually a costly factor in remodeling work, whether it be the exterior siding of a building or interior paint or plaster. Sometimes creative, and often expensive, methods have to be used to match colors, patterns and materials. Adding 10 to 30 percent to the labor costs will insulate against this.

WORKING AROUND ONGOING OPERATIONS
One of the most time-consuming parts of a remodeling job is the distractions caused by working around people. Just normal traffic flow of people through or near a work area affects the way the work gets done. For example, when a contractor has to take precautions during demolition work for the safety of passersby, the job will take longer and cost more. These costs increase even more when work is being done in an office or retail store when the business must continue to function during the remodeling. Adding 5 to 50 percent (depending upon severity) to the labor estimate will provide for this.

DIFFICULT ACCESS
Contractors often find their access to work areas restricted in any number of ways. In remodeling work, the contractor loses the luxury of a logical building process that a new construction project enjoys. Walls have to be cut and patched for the installation of new mechanical or electrical systems; large building materials do not fit through existing doors or windows; even access to the structure's exterior can be limited by added features such as urban locations, landscaping, fences, garages and swimming pools. Delivery and distribution of materials should be approached on a man-hour basis. Add 5 to 20 percent to the labor process for a realistic delivery allowances.

WORKING AROUND SENSITIVE EQUIPMENT
Doing a remodeling job quickly and efficiently usually means being able to make a mess at first and then cleaning up after. Some job locations, however, only allow a minimum amount of dust and debris. When working in close proximity to sensitive machines, computers especially, a job can become up to 100 percent more expensive for labor. Also to be factored in are the hanging of plastic curtains, longer dumpster rental, and laborers cleaning up on a continual basis.

WORKING OFF PEAK HOURS
When a remodeling job involves accommodating a functioning business it is often to the contractor's advantage to work odd hours. Sometimes it is even a requirement. Many retail operations that are open late, or even 24 hours, schedule remodeling work around the second or third shift. While this is not overtime, it is more costly for the contractor to hire subcontractors and tradesmen. However, the distractions and accommodations necessitated by building occupants are much less than if the remodeling work was done during the normal business day. The greater efficiency and speed allowed by working in a quiet environment helps to offset the higher labor costs and reduced productivity. Still, it is justifiable to add 10 to 15 percent to the estimated cost of labor scheduled during off peak hours.

INEFFICIENT WORK METHODS
Working around people and adverse conditions often make the most efficient work methods and the proper tools unusable. The less efficient the method, the more must be spent on additional labor and equipment rental, such as scaffolding and tools a remodeling contractor may not normally have on hand. Depending on the job, another 10 to 40 percent may be added to the labor estimate.

FINISHES / PAINTING — 09900

Paint Failure. There is a cause for every paint failure, and in most instances the failure can be prevented by observing specific precautions. The most often observed paint failures are: cracking flaking, scaling, alligatoring, bleeding, excessive chalking, blistering, fading, spotting, washing and discoloration. The probable causes, preventive measures and corrective measures are discussed in the following subsections:

Alligatoring. When a rupturing of the top paint coat causes the surface to break up into irregular areas separated by wide cracks in "alligator hide" fashion, the condition is called alligatoring. Its cause is due to applying paint to unseasoned wood or from applying a heavy coat of paint over a relatively soft undercoat before it has thoroughly dried. To correct, remove the entire paint film and repaint.

Bleeding. When the color of a previous coat is absorbed into the top coat, this condition is called bleeding. It is caused by the partial solubility of the pigment of the undercoat in the vehicle of the top coat. Asphalt or bituminous undercoat are particularly susceptible to bleeding. In order to prevent or correct bleeding, care must be exercised to select a vehicle for the top coats which is not a solvent for the undercoat pigments.

Blistering. Blistering is evidenced by blister like irregularities on the film of a painted surface. The most common cause of blistering is the application of paint over a damp or wet surface. Under the action of the sun's rays, the moisture is drawn out of the wood (or vaporized on a metallic surface), raising the paint coating with it in the form of blisters. Blistering can also be caused by using too much dryer in the undercoat. To avoid blistering, do not paint damp or wet surfaces, green lumber or greasy spots. Avoid excessive amounts of dryers in undercoat. To correct, remove the paint film in the blistered area, let the surface dry and repaint.

Blushing. A surface in which blushing has occurred is characterized by a white discoloration in the paint film or a precipitation of an ingredient. It is caused by the condensation of moisture on the paint film or by improper composition of the vehicle or solvent. By applying paint under conditions which do not permit moisture to condense on the applied film, blushing can be avoided. Do not paint cold metal under humid conditions. Blushing can be corrected only by removing the paint film in the affected areas and repainting.

Chalking. Chalking of a painted surface can be detected by gentle rubbing which will disclose loose powder on the paint film. Slow, uniform chalking is a desirable quality in the paint, but excessive chalking is undesirable and indicates a paint film failure. It is caused by excessive rain, fog or high humidity during the application of the paint or during the drying period. Paints low in binder content or high in inert pigments have a tendency toward early and excessive chalking.

To prevent chalking, use a paint which is neither deficient in oil content nor high in inert pigment content and apply only under dry conditions. To correct chalked surfaces, remove all loose chalked substance from the surface with a wire brush and repaint the surface with a paint of good quality.

Checking. Checking of a painted surface can be detected by the appearance in the topcoat of small openings or ruptures which divide the surface into small irregular areas, leaving the undercoat visible through the breaks in the topcoat. It is usually caused by too soft an undercoat or by applying a coat over an underlying coat which has not thoroughly dried. To correct, wire brush or sandpaper to remove the loose film and apply a new coat of paint, making sure it is as elastic as the previous coat.

Cracking. Cracks are breaks in a paint film which extend through to the surface to which the paint was applied. Where cracking is severe, flaking, scaling or peeling usually follows. It is usually caused by improper compounding of the paint, resulting in a hard and brittle paint film lacking in elasticity. Low grade paints are often inelastic because they are deficient in oils or contain too much inert material. Cracking will also result from painting a surface upon which too many coats of paint already exist. To avoid cracking, use paint of proper consistency and elasticity, and remove heavy coats of paint from the surface to be painted prior to repainting.

Crawling or Creeping. This defect can be distinguished by the little drops or islands which are formed by the paint film. It often occurs when paint is applied on an oily, waxy or greasy surface or on a very smooth surface such as glass and polished metal. To correct, remove all greasy or oily spots from the surface. On glossy surfaces, wash with a mild soda ash solution or sand with fine sandpaper.

Dulling. Dulling is characterized by a loss in gloss which develops in high gloss paints and enamels and is caused by improper compounding, use of very old stocks of paints and enamels or through the use of too much turpentine as a thinner. To avoid dulling, use fresh paint stocks of good quality. To correct, sand the dulled coat with fine sandpaper and repaint.

Flaking. Flaking is the dropping off of small pieces of a paint coat which has generally started with checking, developing into cracking, and finally the small cracked sections fail away from the surface. The causes and corrective measures are the same as for cracking.

Mildew. Mildew, a form of plant life, is a fungus frequently found on exposed surfaces in damp, warm locations, particularly on soft paint films. It is caused by soft paint films becoming sticky in the warm location and windblown spores and decayed and dried vegetation adhering to the surfaces. The oil in the paint becomes infected, and breeding of the mildew spores takes place in the damp, warm environment. Mildew can be prevented by using a hard drying paint, applied under dry conditions. Paints which contain zinc oxide are resistant to mildew infections. To correct, remove the paint coat with a blowtorch, wash the area with an abrasive soap and water, or with a water solution of trisodium phosphate, rinsing the surface with clear water and allowing it to dry. The addition of one half ounce of mercuric chloride per gallon of paint or the use of less oils and more turpentine is advisable where mildew is likely to occur. Exercise extreme caution in handling of paints with mercuric chloride or other fungicides to prevent poisoning of the skin.

FINISHES / PAINTING — 09900

Peeling. Peeling of a paint film is evidenced by large scales of the film curling and peeling off the painted surface. It is usually caused by the application of paint in the presence of moisture or from being applied over a faulty priming coat. It can be caused by moisture getting behind the paint film at a corner, or in the case of wood, at knotholes which are not properly sealed. Painting unseasoned lumber will also cause peeling. To avoid peeling, seal knot holes and end grains of lumber with shellac before painting and insure that the edges and corners of steel shapes are thoroughly covered. To correct, remove all loose, peeled paint by wire brush or blowtorch, clean the surface of dust film, and repaint on a dry surface only.

Runs and Sags. Ripples, runs and sags on a vertical surface are caused by the application of too heavy a coat of paint, or by using a paint which has been thinned excessively. Other contributing causes are the use of stiff, inflexible brushes or incomplete brushing of the film. It can be prevented by applying a uniform coat of paint of the correct consistency, using a flexible brush and properly brushing out the paint. When runs or sags appear on spray painted surfaces, it is also indicative that too heavy a coat was applied, or that the paint was excessively thinned. To correct, remove the runs or sags with sandpaper and repaint the area.

Scaling. Scaling is an aggravated form of flaking. It is evidenced by large sections of paint coming loose and falling from the painted surface. Scaling is usually preceded by cracking, and like cracking, is usually caused by the paint drying hard and brittle and being unable to expand or contract as the painted object does with changes in temperature and moisture. Scaling frequently occurs when unseasoned lumber is painted. In some cases, previous coats of paint may have lost their elasticity and become lifeless. This causes poor adhesion in the old coat and, in drying, the new coat shrinks and pulls the old film loose from the painted surface. To avoid, do not paint unseasoned lumber, or over a paint which dries hard and brittle, or over old lifeless paint films. Remove the paint with commercial paint removers, scrapers or blowtorch. Remove dust from the surface and repaint only when the surface is thoroughly dry, using a paint which dries with an elastic film.

Slow Drying. The drying time of a paint film varies with the inherent characteristics of the pigment or of the vehicle. However, certain faulty conditions may prolong the drying period and cause the condition to be termed a paint fault. Paints, which under normal drying conditions are tacky or sticky for 12 hours or longer after application, are likely to catch dust and dirt, promote mildew formations or fail by checking or alligatoring. It is usually caused by using paints to which a small amount of mineral oil has been added through negligence or error. Such paint may never dry thoroughly. Old linseed oil that has become fatty by exposure or inferior dryers and thinners frequently contribute to slow drying of paint coats. This condition can be avoided by painting when the temperature is above 50°F or by applying thin coats of paint in colder weather. Use less oil and allow ample time for each coat to dry before applying subsequent coats. Once paint is applied and fails to dry, the condition can be corrected by allowing sufficient additional time for drying, or by removing the paint with commercial paint removers or by scraping, and then repainting with the right type of paint under proper conditions.

Spotting. The appearance of discolored spots or craters in a painted surface is called spotting. It is usually caused by too few coats, such as painting on new work using only two coats, or painting old work with one coat. The lack of controlled penetration causes uneven fading in the pigments. Sap in wood may affect the paint and cause spotting. Spotting may be caused by improperly sealed nail heads, or rain or hail on a freshly painted surface. On plastered surfaces, an inferior primer or sealer may cause spotting as the alkali if the plaster burns through. To prevent spotting, apply sufficient coats of paint and avoid painting when rain is imminent. Insure that all old paint, plaster surfaces and nail heads are properly sealed with quality paint films. To correct, apply an additional coat of paint. Paint surfaces spotted from rain or hail must be sanded smooth before repainting.

Washing. Washing of paint is evidenced by streaks in the paint surface, the discoloration generally accumulating at the lower edges of boards, panels, girders, etc. Washing is caused by using paints having pigments which form water soluble compounds or by painting under damp conditions. This defect often occurs when paints of inferior quality are used. To prevent washing, use only good quality paints. To correct, remove the paint film completely and repaint.

Wrinkling. Wrinkling of a paint coat is evidenced by the paint film gathering in small wrinkles. Wrinkling may be caused by the application of an excessively thick coat, or by failure to brush out the paint properly. It is frequently caused by too much dryer in the paint or by using paints which have been excessively thinned with oil and applied thick. To avoid wrinkling, do not apply thick coats or use excessive amounts of dryer or oil. To correct, remove the wrinkles with sandpaper and repaint with properly thinned paint which does not have excessive amount of dryer or oil in it. If the coat of paint is excessively wrinkled, strip off the old coat and repaint.

Painting Checklist.

(1) Is all paint material tested or approved?
(2) Should surface preparation samples for sandblasting be required?
(3) Are color samples required?
(4) Is the contractor required to submit a list of paint materials for approval?
(5) Are delivered paint materials as specified?
(6) Prior to painting, check each surface with contractor for preparation, type of paint, number of coats and color.
(7) Is pretreatment required?
(8) Is paint properly mixed? Not excessively thinned?
(9) Is coverage satisfactory? Surface appearance and texture? Color?
(10) Is adequate ventilation provided in confined space? Lighting conditions satisfactory?
(11) Are weather conditions satisfactory for painting?

GENERAL / ALLOWANCES   01024

TYPICAL BUILDING COST BROKEN DOWN BY CSI FORMAT
(Commercial Construction)

Division	New Construction	Remodeling Construction
1. General Requirements	6 to 8%	Up to 30%
2. Sitework	4 to 6%	
3. Concrete	15 to 20%	
4. Masonry	8 to 12%	
5. Metals	5 to 7%	
6. Wood and Plastics	1 to 5 %	
7. Thermal and Moisture Protection	4 to 6%	
8. Doors and Windows	5 to 7%	Up to 30%
9. Finishes	8 to 12 %	
10. Specialties		
11. Architectural Equipment		
12. Furnishings	6 to 10%	
13. Special Construction		
14. Conveying Systems		
15. Mechanical	15 to 25%	Up to 30%
16. Electrical	8 to 12%	
TOTAL COST	100%	

GENERAL / MEASUREMENT

01025

CONVERSION FACTORS

Change	To	Multiply By
ATMOSPHERES	POUNDS PER SQUARE INCH	14.696
ATMOSPHERES	INCHES OF MERCURY	29.92
ATMOSPHERES	FEET OF WATER	34
BARRELS, OIL	GALLONS, OF OIL	42
BARRELS, CEMENT	POUNDS OF CEMENT	376
BOGS OR SACKS, CEMENT	POUNDS OF CEMENT	94
BTU/MIN	FOOT-POUNDS/SEC	12.96
BTU/MIN	HORSE-POWER	0.02356
BTU/MIN	KILOWATTS	0.01757
BTU/MM	WATTS	17.57
CENTIMETERS	INCHES	0.3937
CENTIMETERS OF MERCURY	ATMOSPHERES	0.01316
CENTIMETERS OF MERCURY	FEET OF WATER	0.4461
CUBIC INCHES	CUBIC FEET	0.00058
CUBIC FEET	CUBIC INCHES	1728
CUBIC FEET	CUBIC YARDS	0.03703
CUBIC YARDS	CUBIC FEET	27
CUBIC INCHES	GALLONS	0.00433
CUBIC FEET	GALLONS	7.48
FEET	INCHES	12
FEET	YARDS	0.3333
YARDS	FEET	3
FEET OF WATER	ATMOSPHERES	0.02950
FEET OF WATER	INCHES OF MERCURY	0.8826
GALLONS	CUBIC INCHES	231
GALLONS	CUBIC FEET	0.1337
GALLONS	POUNDS OF WATER	8.33
GALLONS	QUARTS	4
GALLONS PER MIN	CUBIC FEET SEC	0.002228
GALLONS PER MIN	CUBIC FEET HOUR	8.0208
GALLONS WATER PER MIN	TONS WATER/24 HOURS	6.0086
HORSE-POWER	FOOT-LBS./SEC	550
INCHES	CENTIMETERS	2.540
INCHES	FEET	0.0833
INCHES	MILLIMETERS	25.4
INCHES OF WATER	POUNDS PER SQ. INCH	0.0361
INCHES OF WATER	INCHES OF MERCURY	0.0735
INCHES OF WATER	OUNCES PER SQUARE INCH	0.578
INCHES OF WATER	OUNCES PER SQUARE FOOT	5.2
INCHES OF MERCURY	INCHES OF WATER	13.6
INCHES OF MERCURY	FEET OF WATER	1.1333
INCHES OF MERCURY	POUNDS PER SQUARE INCH	0.4914
KILOMETERS	MILES	0.6214
METERS	INCHES	39.37
MILES	FEET	5280
MILLIMETERS	CENTIMETERS	0.1
MILLIMETERS	INCHES	0.03937
OUNCES (FLUID)	CUBIC INCHES	1.805
OUNCES	POUNDS	0.0625
POUNDS	OUNCES	16
POUNDS PER SQUARE INCH	INCHES OF WATER	27.72
POUNDS PER SQUARE INCH	FEET OF WATER	2.310
POUNDS PER SQUARE INCH	INCHES OF MERCURY	2.04
POUNDS PER SQUARE INCH	ATMOSPHERES	0.0681
QUARTS	CUBIC INCHES	67.20
SQUARE INCHES	SQUARE FEET	0.00694
SQUARE FEET	SQUARE INCHES	144
SQUARE FEET	SQUARE YARDS	0.11111
SQUARE YARDS	SQUARE FEET	9
SQUARE MILES	ACRES	640
SHORT TONS	POUNDS	2000
SHORT TONS	LONG TONS	0.89285
TONS OF WATER/24 HOURS	GALLONS PER MINUTE	0.16643
YARDS	FEET	3
YARDS	CENTIMETERS	91.44
YARDS	INCHES	36

GENERAL / MEASUREMENT — 01025

CONVERSION CALCULATIONS

Commercial Measure
16 grams .... = 1 ounce
16 ounces .... = 1 pound
2,000 pounds .... = 1 ton

Long Measure
12 inches .... = 1 foot
3 feet .... = 1 yard
16 ½ feet .... = 1 rod
40 rods .... = 1 furlong
8 furlongs (5,280 ft.) .... = 1 mile
3 miles .... = 1 league

Square Measure
144 square inches .... = 1 square foot
9 square feet .... = 1 square yard
30 ¼ square yards .... = 1 square rod
160 square rods .... = 1 acre
4,840 square yards .... = 1 acre
640 acres .... = 1 square mile
36 square miles .... = 1 township

Surveyor's Measure
7.92 inches .... = 1 link
25 links .... = 1 rod
4 rods (66 ft.) .... = 1 chain
10 chains .... = 1 furlong
8 furlongs .... = 1 mile
1 square mile .... = 1 section

Cubic Measure
1728 cubic inches .... = 1 cubic foot
27 cubic feet .... = 1 cubic yard
128 cubic feet .... = 1 cord (wood/stone)
231 cubic inches .... = 1 U.S. gallon
7.48 U.S. Gallons .... = 1 cubic foot
2150.4 cubic inches .... = 1 U.S. bushel

Liquid Measure
4 fluid ounces .... = 1 gill
4 gills .... = 1 pint
2 pints .... = 1 quart
4 quarts .... = 1 gallon
9 gallons .... = 1 firkin
31 ½ gallons .... = 1 barrel
2 barrels .... = 1 hogshead

Dry Measure
2 pints .... = 1 quart
8 quarts .... = 1 peck
4 pecks .... = 1 bushel
2150.42 cubic inches .... = 1 bushel

GENERAL / MEASUREMENT 01025

SQUARE
$$A = a^2$$

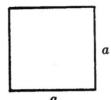

RECTANGLE
$$A = bh$$

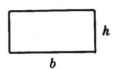

TRIANGLE
$$A = \frac{1}{2}bh$$

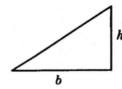

PARALLELOGRAM
$$A = bh = ab\ Sin\phi$$

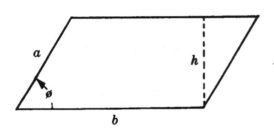

TRAPEZOID
$$A = \left(\frac{a+b}{2}\right)h$$

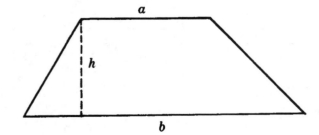

CIRCLE
$$A = \pi r^2 = \frac{\pi d^2}{4}$$

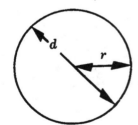

$Circumference = C = 2\pi r = \pi d$

GENERAL / MEASUREMENT 01025

ELLIPSE
$A = \pi ab$

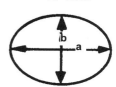

PARABOLA
$A = \frac{2}{3} bh$

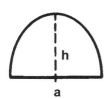

VOLUMES

CUBE
$V = a^3$

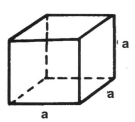

CYLINDER
$V = \pi r^3 h$

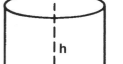

PYRAMID
$V = \frac{1}{3} \text{(Base)} \, h$

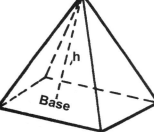

CONE
$V = \frac{1}{3} \pi r^2 h$

SPHERE
$V = \frac{4}{3} \pi r^3 = \frac{1}{6} \pi d^3$

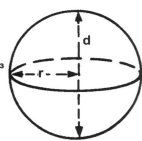

WEDGE
$V = \frac{1}{2} abc$

GENERAL / MEASUREMENT 01295

CONVERSION FACTORS
ENGLISH TO SI (SYSTEM INTERNATIONAL)

To Convert from	To	Multiply by
LENGTH		
Inches	Millimetres	25.4[a]
Feet	Metres	0.3048[a]
Yards	Metres	0.9144[a]
Miles (statute)	Kilometres	1.609
AREA		
Square inches	Square millimetres	645.2
Square feet	Square metres	0.0929
Square yards	Square metres	0.8361
VOLUME		
Cubic inches	Cubic millimetres	16.387
Cubic feet	Cubic metres	0.02832
Cubic yards	Cubic metres	0.7646
Gallons (U.S. liquid)[b]	Cubic metres[c]	0.003785
Gallons (Canadian liquid)[b]	Cubic metres[c]	0.004546
Ounces (U.S. liquid)[b]	Millilitres[c,d]	29.57
Quarts (U.S. liquid)[b]	Litres[c,d]	0.9464
Gallons (U.S. liquid)[b]	Litres[c]	3.785
FORCE		
Kilograms force	Newtons	9.807
Pounds force	Newtons	4.448
Pounds force	Kilograms force[d]	0.4536
Kips	Newtons	4448
Kips	Kilograms force[d]	453.6
PRESSURE, STRESS, STRENGTH (FORCE PER UNIT AREA)		
Kilograms force per sq. centimetre	Megapascals	0.09807
Pounds force per square inch (psi)	Megapascals	6895
Kips per square inch	Megapascals	6.895
Pounds force per square inch (psi)	Kilograms force per square centimetre[d]	0.07031
Pounds force per square foot	Pascals	47.88
Pounds force per square foot	Kilograms force per square metre[d]	4.882
SENDING MOMENT OR TORQUE		
Inch-pounds force	Metre-kilog. force[d]	0.01152
Inch-pounds force	Newton-metres	0.1130
Foot-pounds force	Metre-kilog. force[d]	0.1383
Foot-pounds force	Newton-metres	1.356
Metre-kilograms force	Newton-metres	9.807
MASS		
Ounce (avoirdupois)	Grams	28.35
Pounds (avoirdupois)	Kilograms	0.4536
Tons (metric)	Kilograms	1000[a]
Tons, short (2000 pounds)	Kilograms	907.2
Tons, short (2000 pounds)	Megagrams[e]	0.9072
MASS PER UNIT VOLUME		
Pounds mass per cubic foot	Kilog. per cubic metre	16.02
Pounds mass per cubic yard	Kilog. per cubic metre	0.5933
Pds. mass per gallon (U.S. liquid)[b]	Kilog. per cubic metre	119.8
Pds. mass p/gal. (Canadian liquid)[b]	Kilog. per cubic metre	99.78
TEMPERATURE		
Degrees Fahrenheit	Degrees Celsius	$t_K = (1F - 32)/1.8$
Degrees Fahrenheit	Degrees Kelvin	$t_K = (1F + 459.67)/1.8$
Degree Celsius	Degree Kelvin	$t_K = 1C + 273.15$

[a] The factor given is exact.
[b] One U.S. gallon equals 0.8327 Canadian gallon.
[c] 1 litre = 1000 millilitres = 10,000 cubic centimetres = 1 cubic decimetre = 0.001 cubic metre.
[d] Metric but not SI unit.
[e] Called "tonne" in England. Called "metric ton" in other metric systems.

SITEWORK / SHORING & UNDERPINNING | 01295

TRENCH BRACING

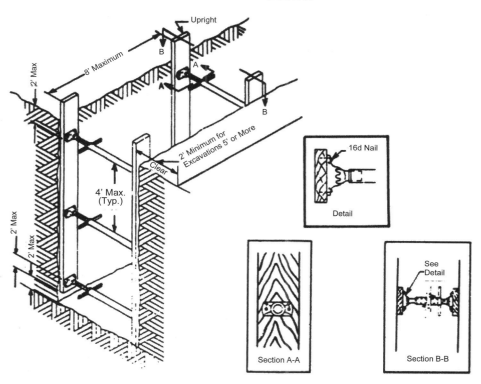

CLOSED VERTICAL SHEETING

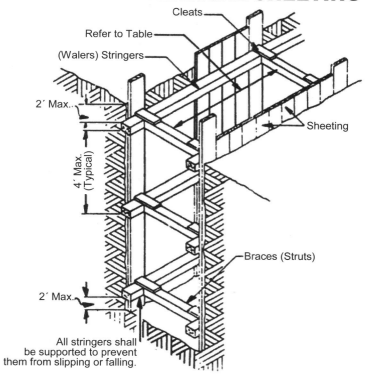

SITEWORK / EARTHWORK 02200

Soil Identification. For most purposes, soils can usually be identified visually and by texture, as described in the chart that follows. For design purposes, however, soils must be formally identified and their performance characteristics determined in a laboratory by skilled soil mechanics.

Classification	Identifying Characteristics
Gravel	Rounded or water-worn pebbles or bulk rock grains. No cohesion or plasticity. Gritty, granular and crunchy underfoot.
Sand	Granular, gritty, loose grains, passing a No. 4 sieve and between .002 and .079 inches in diameter. Individual grains readily seen and felt. No plasticity or cohesion. When dry, it cannot be molded but will crumble when touched. The coarse grains are rounded; the fine grains are visible and angular.
Silt	Fine, barely visible grains passing a No. 200 sieve and between .0002 and .002 inches in diameter. Little or no plasticity and no cohesion. A dried cast is easily crushed. Is permeable and movement of water through the voids occurs easily and is visible. Feels gritty when bitten and will not form a thread.
Clay	Invisible particles under .0002 inches in diameter. Cohesive and highly plastic when moist. Will form a long, thin, flexible thread when rolled between the hands. Does not feel gritty when bitten. Will form hard lumps or clods when dry which resist crushing. Impermeable, with no apparent movement of water through voids.
Muck and Organic Silt	Thoroughly decomposed organic material often mixed with other soils of mineral origin. Usually black with fibrous remains. Odorous when dried and burnt. Found as deposits in swamps, peat bogs and muskeg flats.
Peat	Partly decayed plant material. Mostly organic. Highly fibrous with visible plant remains. Spongy and easily identified.

SITEWORK / EARTHWORK 02200

Classification by Particle Size. Soils can be classified in general terms by the nature of their predominant particle size or the grading of the particle sizes. These particle sizes are usually grouped into gravel, coarse sand, medium sand, fine sand, silt, clay and colloids.

The major divisions of soils are:

Coarse-Grained(Granular)		Fine-Grained		Organic	
Gravel	Sand	Silt	Clay	Muck	Peat

Soils comprised primarily of sand particles are referred to as "granular soils," while fine-grained soils are commonly called "heavy soils." It is accepted practice in the field to refer to a particular soil as a coarse sand, or silt, or by any of the particle size groupings which describe the soil generally from a visual examination.

Classification of Soil Mixtures.

Class	% Sand	% Silt	% Clay
Sandy	80-100	0-20	0-20
Sandy clay loam	50-80	0-30	20-30
Sandy loam	50-80	0-50	0-20
Loam	30-50	30-50	0-20
Silty loam	0-50	50-80	0-20
Silt	0-20	80-100	0-20
Silty clay	0-30	50-80	20-30
Silty clay	0-20	50-70	30-50
Clay	20-50	20-50	20-30
Sandy clay	50-70	0-20	30-50
Clay	0-50	0-50	30-100

Material	Approx. In-Bank Weight (lbs. per cu. yd.)	Percent Swell
Clay, dry	2300	40
Clay, wet	3000	40
Granite, decomposed	4500	65
Gravel, dry	3250	10-15
Gravel, wet	3600	10-15
Loam, dry	2800	15-35
Loam, wet	3370	25
Rock, well blasted	4200	65
Sand, dry	3250	10-15
Sand, wet	3600	10-15
Shale and soft rock	3000	65
Slate	4700	65

SITEWORK / PILE DRIVING 02355

PILES AND PILE DRIVING

General. A pile is a column driven or jetted into the ground which derives its supporting capabilities from end-bearing on the underlying strata, skin friction between the pile surface and the soil, or from a combination of end-bearing and skin friction.

Piles can be divided into two major classes: **Sheet piles** and **load-bearing piles**. Sheet piling is used primarily to restrain lateral forces as in trench sheeting and bulkheads, or to resist the flow of water as in cofferdams. It is prefabricated and is available in steel, wood or concrete. Load-bearing piles are used primarily to transmit loads through soil formations of low bearing values to formations that are capable of supporting the designed loads. If the load is supported predominantly by the action of soil friction on the surface of the pile, it is called a **friction pile**. If the load is transmitted to the soil primarily through the lower tip, it is called an **end-bearing pile**.

There are several load-bearing pile types, which can be classified according to the material from which they are fabricated:
- Timber (Treated and untreated)
- Concrete (Precast and cast in place)
- Steel (H-Section and steel pipe)
- Composite (A combination of two or more materials)

Some of the additional uses of piling are to: eliminate or control settlement of structures, support bridge piers and abutments and protect them from scour, anchor structures against uplift or overturning, and for numerous marine structures such as docks, wharves, fenders, anchorages, piers, trestles and jetties.

Timber Piles. Timber piles, treated or untreated, are the piles most commonly used throughout the world, primarily because they are readily available, economical, easily handled, can be easily cut off to any desired length after driving and can be easily removed if necessary. On the other hand, they have some serious disadvantages which include: difficulty in securing straight piles of long length, problems in driving them into hard formations and difficulty in splicing to increase their length. They are generally not suitable for use as end-bearing piles under heavy load and they are subject to decay and insect attack. Timber piles are resilient and particularly adaptable for use in waterfront structures such as wharves, docks and piers for anchorages since they will bend or give under load or impact where other materials may break. The ease with which they can be worked and their economy makes them popular for trestle construction and for temporary structures such as falsework or centering. Where timber piles can be driven and cut off below the permanent groundwater level, they will last indefinitely; but above this level in the soil, a timber pile will rot or will be attacked by insects and eventually destroyed. In sea water, marine borers and fungus will act to deteriorate timber piles. Treatment of timber piles increases their life but does not protect them indefinitely.

Concrete Piles. Concrete piles are of two general types, precast and cast-in-place. The advantages in the use of concrete piles are that they can be fabricated to meet the most exacting conditions of design, can be cast in any desired shape or length, possess high strength and have excellent resistance to chemical and biological attack. Certain disadvantages are encountered in the use of precast piles, such as:

(a) Their heavy weight and bulk (which introduces problems in handling and driving).
(b) Problems with hair cracks which often develop in the concrete as a result of shrinkage after curing (which may expose the steel reinforcement to deterioration).
(c) Difficulty encountered in cut-off or splicing.
(d) Susceptibility to damage or breakage in handling and driving.
(e) They are more expensive to fabricate, transport and drive.

Precast piles are fabricated in casting yards. Centrifugally spun piles (or piles with square or octagonal cross-sections) are cast in horizontal forms, while round piles are usually cast in vertical forms. With the exception of relatively short lengths, precast piles must be reinforced to provide the designed column strengths and to resist damage or breakage while being transported or driven.

Precast piles can be tapered or have parallel sides. The reinforcement can be of deformed bars or be prestressed or poststressed with high strength steel tendons. Prestressing or prestressing eliminates the problem of open shrinkage cracks in the concrete. Otherwise, the pile must be protected by coating it with a bituminous or plastic material to prevent ultimate deterioration of the reinforcement. Proper curing of the precast concrete in piles is essential.

Cast-in-place pile types are numerous and vary according to the manufacturer of the shell or inventor of the method. In general, they can be classified into two groups: shell-less types and the shell types. The shell-less type is constructed by driving a steel shell into the ground and filling it with concrete as the shell is pulled from the ground. The shell type is constructed by driving a steel shell into the ground and filling it in place with concrete. Some of the advantages of cast-in-place concrete piles are: lightweight shells are handled and driven easily, lengths of the shell may be increased or decreased easily, shells may be transported in short lengths and quickly assembled, the problem of breakage is eliminated and a driven shell may be inspected for shell damage or an uncased hole for "pinching off." Among the disadvantages are problems encountered in the proper centering of the reinforcement cages, in placing and consolidating the concrete without displacement of the reinforcement steel or segregation of the concrete, and shell damage or "pinching-off" of uncased holes.

Shell type piles are fabricated of heavy gage metal or are fluted, corrugated or spirally reinforced with heavy wire to make them strong enough to be driven without a mandrel.

SITEWORK / PILE DRIVING   02355

Other thin-shell types are driven with a collapsible steel mandrel or core inside the casing. In addition to making the driving of a long thin shell possible, the mandrel prevents or minimizes damage to the shell from tearing, buckling, collapsing or from hard objects encountered in driving.

Some shell type piles are fabricated of heavy gauge metal with enlargement at the lower end to increase the end bearing.

These enlargements are formed by withdrawing the casing two to three feet after placing concrete in the lower end of the shell. This wet concrete is then struck by a blow of the pile hammer on a core in the casing and the enlargement is formed. As the shell is withdrawn, the core is used to consolidate the concrete after each batch is placed in the shell. The procedure results in completely filling the hole left by the withdrawal of the shell.

Steel Piles. A steel pile is any pile fabricated entirely of steel. They are usually formed of rolled steel H sections, but heavy steel pipe or box piles (fabricated from sections of steel sheet piles welded together) are also used. The advantages of steel piles are that they are readily available, have a thin uniform section and high strength, will take hard driving, will develop high load-bearing values, are easily cut off or extended, are easily adapted to the structure they are to support, and breakage is eliminated. Some disadvantages are: they will rust and deteriorate unless protected from the elements; acid, soils or water will result in corrosion of the pile; and greater lengths may be required than for other types of piles to achieve the same bearing value unless bearing on rock strata. Pipe pile can either be driven open-end or closed-end and can be unfilled, sand filled or concrete filled. After open-end pipe piles are driven, the material from inside can be removed by an earth auger, air or water jets, or other means, inspected, and then filled with concrete. Concrete filled pipe piles are subject to corrosion on the outside surface only.

Composite Piles. Any pile that is fabricated of two or more materials is called a composite pile. There are three general classes of composite piles: wood with concrete, steel with concrete, and wood with steel. Composite piles are usually used for a special purpose or for reasons of economy.

Where a permanent ground-water table exists and a composite pile is to be used, it will generally be of concrete and wood. The wood portion is driven to below the water table level and the concrete upper portion eliminates problems of decay and insect infestation above the water table. Composite piles of steel and concrete are used where high bearing loads are desired or where driving in hard or rocky soils is expected. Composite wood and steel piles are relatively uncommon.

It is important that the pile design provides for a permanent joint between the two materials used, so constructed that the parts do not separate or shift out of axial alignment during driving operations.

Sheet Piles. Sheet piles are made from the same basic materials as other piling: wood, steel and concrete. They are ordinarily designed so as to interlock along the edges of adjacent piles.

Sheet piles are used where support of a vertical wall of earth is required, such as trench walls, bulkheads, waterfront structures or cofferdams. Wood sheet piling is generally used in temporary installations, but is seldom used where water-tightness is required or hard driving expected. Concrete sheet piling has the capability of resisting much larger lateral loads than wood sheet piling, but considerable difficulty is experienced in securing water-tight joints. The type referred to as "fishmouth" type is designed to permit jetting out the joint and filling with grout, but a seal is not always effected unless the adjacent piles are wedged tightly together. Concrete sheet piling has the advantage that it is the most permanent of all types of sheet piling.

Steel sheet piling is manufactured with a tension-type interlock along its edges. Several different shapes are available to permit versatility in its use. It has the advantages that it can take hard driving, has reasonably watertight joints and can be easily cut, patched, lengthened or reinforced. It can also be easily extracted and reused. Its principal disadvantage is its vulnerability to corrosion.

Types of Pile Driving Hammers. A pile-driving hammer is used to drive load-bearing or sheet piles. The commonly used types are: drop, single-acting, double-acting, differential acting and diesel hammers. The most recent development is a type of hammer that utilizes high-frequency sound and a dead load as the principal sources of driving energy.

Drop Hammers. These hammers employ the principle of lifting a heavy weight by a cable and releasing it to fall on top of the pile. This type of hammer is rapidly disappearing from use, primarily because other types of pile driving hammers are more efficient. Its disadvantages are that it has a slow rate of driving (four to eight blows per minute), that there is some risk of damaging the pile from excessive impact, that damage may occur in adjacent structures from heavy vibration and that it cannot be used directly for driving piles under water. Drop hammers have the advantages of simplicity of operation, ability to vary the energy by changing the height of fall and they represent a small investment in equipment.

Single-Acting Hammers. These hammers can be operated either on steam or compressed air. The driving energy is provided by a free-falling weight (called a ram) which is raised after each stroke by the action of steam or air on a piston. They are manufactured as either open or closed types. Single-acting hammers are best suited for jobs where dense or elastic soil materials must be penetrated or where long heavy timber or precast concrete piles must be driven. The closed type can be used for underwater pile driving. Its advantages include: faster driving (50 blows or more per minute), reduction in skin friction as a result of

SITEWORK / PILE DRIVING — 02355

more frequent blows, lower velocity of the ram which transmits a greater proportion of its energy to the pile and minimizes piles damage during driving, and it has underwater driving capability. Some of its disadvantages are: requires higher investment in equipment (i.e. steam boiler, air compressor, etc.), higher maintenance costs, greater set-up and moving time required, and a larger operating crew.

Double-Acting Hammers. These hammers are similar to the single-acting hammers except that steam or compressed air is used both to lift the ram and to impart energy to the falling ram. While the action is approximately twice as fast as the single-acting hammer (100 blows per minute or more), the ram is much lighter and operates at a greater velocity, thereby making it particularly useful in high production driving of light or medium-weight piles of moderate lengths in granular soils. The hammer is nearly always fully encased by a steel housing which also permits direct driving of piles under water.

Some of its advantages are: faster driving rate, less static skin friction develops between blows, has underwater driving capability and piles can be driven more easily without leads.

Among its disadvantages are: it is less suitable for driving heavy piles in high-friction soils and the more complicated mechanism results in higher maintenance costs.

Differential-Acting Hammers. This type of hammer is, in effect, a modified double-acting hammer with the actuating mechanism having two different diameters. A large-diameter piston operates in an upper cylinder to accelerate the ram on the downstroke and a small-diameter piston operates in a lower cylinder to raise the ram. The additional energy added to the falling ram is the difference in areas of the two pistons multiplied by the unit pressure of the steam or air used. This hammer is a short-stroke, fast-acting hammer with a cycle rate approximately that of the double-acting hammer.

Its advantages are that it has the speed and characteristics of the double-acting hammer with a ram weight comparable to the single-acting type, and it uses 25 to 35 percent less steam or air. It is also more suitable for driving heavy piles under more difficult driving conditions than the double-acting hammer. It is available in the open or closed-type cases, the latter permitting direct underwater pile driving. Its principal disadvantage is higher maintenance costs.

Diesel Hammers. This hammer is a self-contained driving unit which does not require an auxiliary steam boiler or air compressor. It consists essentially of a ram operating as a piston in a cylinder. When the ram is lifted and allowed to fall in the cylinder, diesel fuel is injected in the compression space between the ram and an anvil placed on top of the pile. The continued downstroke of the ram compresses the air and fuel to ignition heat and the resultant explosion drives the pile downward and the ram upward to start another cycle. This hammer is capable of driving at a rate of from 80 to 100 blows per minute. Its advantages are that it has a low equipment investment cost, is easily moved, requires a small crew, has a high driving rate, does not require a steam boiler or air compressor and can be used with or without leads for most work. Its disadvantages are that it is not self-starting (the ram must be mechanically lifted to start the action) and it does not deliver a uniform blow. The latter disadvantage arises from the fact that as the reaction of the pile to driving increases, the reaction to the ram increases correspondingly. That is, when the pile encounters considerable resistance, the rebound of the ram is higher and the energy is increased automatically. The operator is required to observe the driving operations closely to identify changing driving conditions and compensate for such changes with his controls to avoid damaging the pile.

Diesel hammers can be used on all types of piles and they are best suited to jobs where mobility or frequent relocation of the pile driving equipment is necessary.

SITEWORK / PILE DRIVING 02355

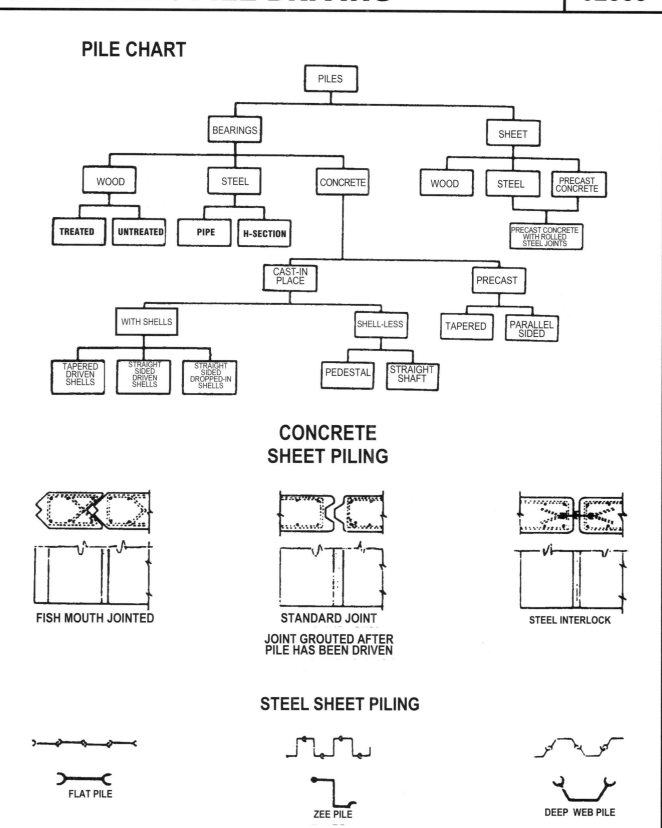

SITEWORK / PILE DRIVING 02355

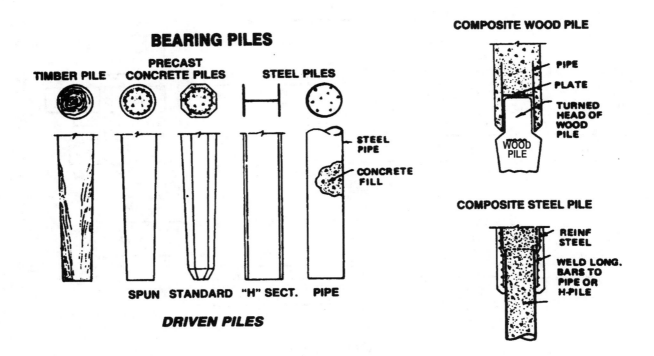
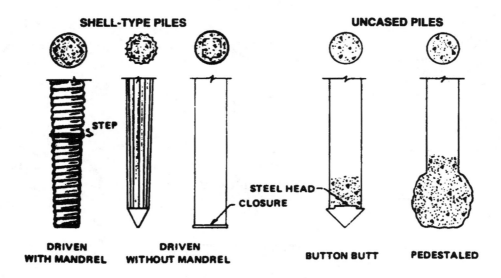

SITEWORK / ROADWORK 02500

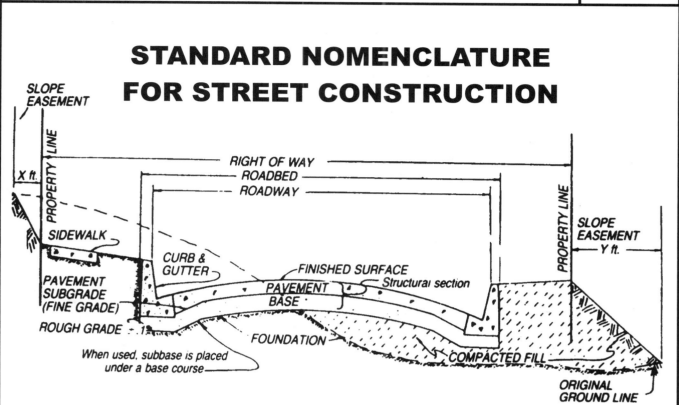

STANDARD NOMENCLATURE FOR STREET CONSTRUCTION

SITEWORK / CONCRETE PAVERS 02515

CONCRETE MASONRY PAVING UNITS

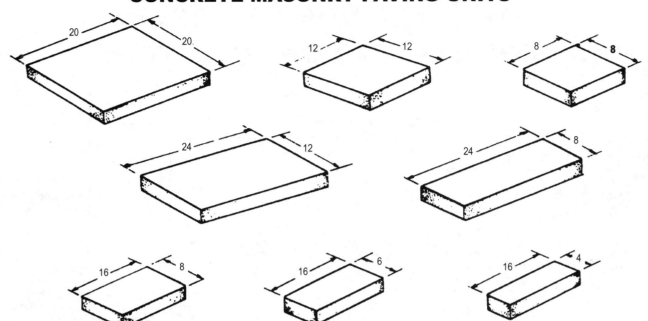

NOTE: Sizes are nominal and will vary by manufacture.

HEXAGON PAVER UNITS
Various Sizes Available

ROUND PAVING UNITS
Various Sizes Available

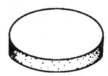

VEHICULAR PAVING UNITS

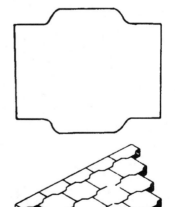

INTERLOCKING PAVER
7¼" x 3" x 8½"

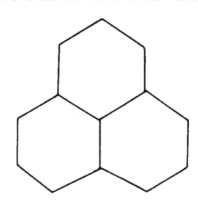

INTERLOCKING PAVER
12" x 3⅝" x 12"

TURF PAVER
24" x 3⅝" x 24"

SITEWORK / PIPE 02610

PIPE

Clay Pipe. Clay pipe is manufactured by blending various clays together, milling, mixing, extruding and firing in a kiln to obtain vitrification. The physical properties of the pipe can be changed by varying the proportions of the several clays used. The pipe is supplied in two basic styles: spigot and socket; and plain end.

Spigot and Socket Pipe has a spigot on one end and a socket on the other, and is commonly referred to as "bell and spigot" pipe. The plans generally specify the type of joint to be used from the several types of jointing methods available. This type of pipe is manufactured with matching polyurethane gaskets molded on the spigot and socket which form a tight seal when the pipe is jointed.

Plain End Pipe is without a socket on either end and is joined with special couplings. The coupling consists of a circular rubber sleeve, two stainless steel compression bands with tightening devices and a corrosion resistant shear ring. Sometimes this joint is supplied with a cardboard form, open at the top, which is filled with portland cement mortar to resist shear and prevent future corrosion of the bands.

Concrete Pipe. Unreinforced and reinforced concrete pipe is manufactured by casting in stationary or revolving metal molds. At the present time, the design practice is to specify reinforced concrete pipe far all purposes.

Unreinforced Concrete Pipe is cast in vertical steel molds, usually in pipe sizes of 21 inches or less, and is of the spigot and socket type. No steel reinforcement is used and the pipe is usually intended for use in irrigation systems and under light loading conditions.

Reinforced Concrete Pipe (RCP) is made in a number of different manufacturing processes and for a wide variety of pressure and non-pressure classes. It is available in standard sizes or it can be made to order to any diameter desired. Some of the larger diameters include diameters of 12 and 14 feet. A large variety of joint details are used with RCP. Tongue and groove joints are used for storm drain pipelines.

Reinforced concrete pipe for wastewater pipeline projects is supplied with gasketed joints and a polyvinyl chloride (PVC) plastic liner cast into the pipe.

(a) **Cast Pipe** is cast vertically in steel forms with the reinforcing cage securely held in place. The reinforcement is generally elliptical in shape to provide the maximum structural strength to resist the loads imposed on the pipe by the backfill and other stresses. Consolidation of the concrete is obtained by the use of external form vibrators.

(b) **Centrifugally Spun Pipe** is manufactured by introducing concrete into a spinning horizontal steel cylinder into which the reinforcement cage has been previously installed and which is equipped with end dams to provide the proper pipe wall thickness. The speed of rotation of the mold is increased and the centrifugal force produces a smooth, dense concrete pipe.

(c) **Pressure Pipe** may be cast or centrifugally spun pipe but it usually has a circular steel reinforcement cage (or cages) designed not only to resist the trench loading, but also the internal pressures exerted on the pipe from the fluid under pressure in the line.

Concrete Cylinder Pipe. This class of pipe is generally used for high pressure water lines and sewer force mains and is available in sizes ranging from 10 inches to 60 inches and larger in special cases.

A sheet steel cylinder is wrapped with the designed steel reinforcement and a concrete lining is centrifugally spun in the interior of the steel cylinder. An exterior coating of concrete is applied generally by the gunite process, while the cylinder is slowly rotated. These coatings vary in thickness from 1/2 to 3/4 of an inch. The joints are commonly of the steel ring and rubber gasket type, but are generally designed for the special purpose for which the pipe line is intended.

Definition of Terms. In general, the terms used to designate types of reinforced concrete pipe refer to the process used in manufacture.

Cast RCP (Cast Reinforced Concrete Pipe). A concrete pipe having one or more cylindrical or elliptical cages of reinforcement steel embedded in it, the concrete for which is cast with the forms in a vertical position.

CSRCP (Centrifugally Spun Reinforced Concrete Pipe). A concrete pipe having one or more cylindrical or elliptical cages of reinforcement steel embedded in it, and cast in a horizontal position while the forms are spinning rapidly. This type of pipe may be designated as Spun RCP or as CCP (Centrifugal Concrete Pipe).

RCP (Reinforced Concrete Pipe). A reinforced concrete pipe manufactured by either the casting or spinning method.

Steel Reinforcement. Steel for reinforcing concrete pipe is generally furnished in large coils which will permit the use of machines to fabricate the "cages." The continuous steel rod is wound spirally at a prescribed pitch on a drum of the proper diameter. Where the rod crosses a longitudinal spacer rod, it is electrically welded to it so that the complete cage is relatively rigid.

Reinforcement cages for pipe designed for external loading are generally elliptical in shape to take full advantage of the steel in tension. Pipe to be used with relatively small external loads or pipe designed for pressure lines will have circular cages.

Reinforcement cages must be rigidly fixed in the forms so that the placement of concrete or the effects of centrifugal spinning will not result in distortion or displacement of the steel The orientation of an elliptical cage must be marked on the forms to assure that the minor axis can be located after the concrete is placed.

SITEWORK / SEPTIC SYSTEMS — 02740

CAPACITIES FOR SEPTIC TANKS SERVING AN INDIVIDUAL DWELLING

No. of bedrooms	Capacity of tank (gals.)
2 or less	750
3	900
4	1,000

CONCRETE / FORMWORK

03110

FORM NOMENCLAURE

FALSEWORK NOMENCLAURE

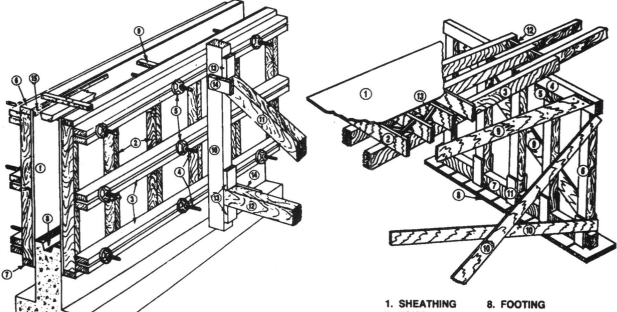

1. SHEATHING
2. STUDS
3. WALES
4. FORM BOLTS
5. NUT WASHER
6. TOP PLATE
7. BOTTOM PLATE
8. KEY-WAY
9. SPREADER
10. STRONGBACK
11. BRACE
12. STRUT
13. CLEATS
14. SCAB
14. POUR STRIP

1. SHEATHING
2. JOIST
3. STRINGER
4. CAP
5. CORBEL
6. POST
7. SILL
8. FOOTING
9. SWAY BRACE
10. LONGITUDINAL BRACE
11. SCAB
12. BLOCKING
13. BRIDGING

TYPICAL PAN-JOIST FORM CONSTRUCTION

TYPICAL WAFFLE SLAB FORM CONSTRUCTION

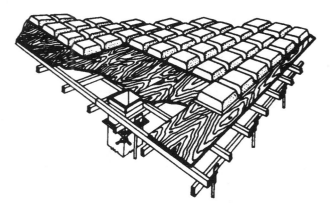

CONCRETE / REINFORCING STEEL — 03210

STANDARD SIZES OF STEEL REINFORCEMENT BARS

STANDARD REINFORCEMENT BARS

Bar Designation Number*	Nominal Weight lb. per ft.	Nominal Dimensions		
		Diameter, in.	Cross Sectional Area, sq. in.	Perimeter, in.
3	0.376	0.375	0.11	1.178
4	0.668	0.500	0.20	1.571
5	1.043	0.625	0.31	1.963
6	1.502	0.750	0.44	2.356
7	2.044	0.875	0.60	2.749
8	2.670	1.000	0.79	3.142
9	3.400	1.128	1.00	3.544
10	4.303	1.270	1.27	3.990
11	5.313	1.410	1.56	4.430
14	7.65	1.693	2.25	5.32
18	13.60	2.257	4.00	7.09

*The bar numbers are based on the number of 1/8 inches included in the nominal diameter of the bar.

Type of Steel & ASTM Specification No.	Size Nos. Inclusive	Grade	Tensile Strength Min., psi.	Yield (a) Min., psi
Billet Steel A 615	3-11	40	70,000	40,000
	3-11 14, 18	60	90,000	60,000
	11, 14, 18	75	100,000	75,000

CONCRETE / WELDED WIRE FABRIC — 03220

Style Designation	Steel Area sq. in per ft.		Weight Approx. lbs. per 100 sq. ft.
	Longit.	Transv.	
Rolls			
6×6 — W1.4×W1.4	.03	.03	21
6×8 — W2×W2	.04	.04	29
6×6 — W2.9×W2.9	.08	.06	42
6×6 — W4×W4	.08	.08	58
4×4 — W1.4×W1.4	.04	.04	31
4×4 — W2×W2	.06	.06	43
4×4 — W2.9×W2.9	.09	.09	62
4×4 — W4×W4	.12	.12	
Sheets			
6×6 — W2.9×W2.9	.06	.06	42
6×6 — W4×W4	.08	.08	58
6×6 — W5.5×W5.5	.11	.11	80
4×4 — W4×W4	.12	.12	86

CONCRETE / REINFORCING STEEL — 03210

COMMON TYPES OF STEEL REINFORCEMENT BARS

ASTM specifications for billet steel reinforcing bars (A 615) require identification marks to be rolled into the surface of one side of the bar to denote the producer's mill designation, bar size and type of steel. For Grade 60 and Grade 75 bars, grade marks indicating yield strength must be show. Grade 40 bars show only three marks (no grade mark) in the following order:

1st — Producing Mill (usually an initial)
2nd — Bar Size Number (#3 through # 18)
3rd — Type (N for New Billet)

NUMBER SYSTEM — GRADE MARKS

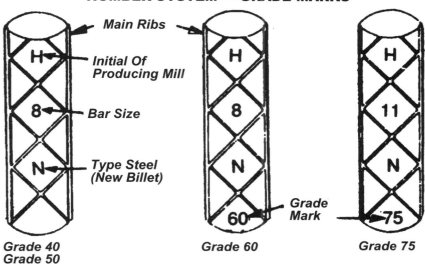

LINE SYSTEM — GRADE MARKS

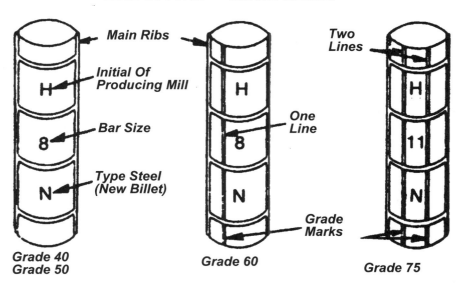

CONCRETE/STRUCTURAL CONCRETE | 03310

Insofar as is possible, the moisture content should be kept uniform to avoid problems in determining the proper amount of water to be added for mixing. Mixing water must be reduced to compensate for moisture in the aggregate in order to control the slump of the concrete and avoid exceeding the specified water-cement ratio.

Handling Concrete by Pumping Methods. Transportation and placement of concrete by pumping is another method gaining increased popularity. Pumps have several advantages, the primary one being that a pump will high-lift concrete without the need for an expensive crane and bucket. Since the concrete is delivered through pipe and hoses, concrete can be conveyed to remote locations in buildings, in tunnels, to locations otherwise inaccessible on steep hillside slopes for anchor walls, pipe bedding or encasement, or for placing concrete for chain link fence post bases. Concrete pumps have been found to be economical and expedient in the placement of concrete, and this has promoted the use and acceptance of this development. The essence of proper concrete pumping is the placement of the concrete in its final location without segregation.

Modern concrete pumps, depending on the mix design and size of line, can pump to a height of 200 feet or a horizontal distance of 1,000 feet. They can handle, economically, structural mixes, standard mixes, low slump mixes, mixes with two-inch maximum size aggregate and light weight concrete. When a special pump mix is required for structural concrete in a major structure, the mix design must be approved by the Engineer and checked and confirmed by the Supervisor of the Materials Control Group. The Inspector should obtain the pump manufacturer's printed information and evaluate its characteristics and ability to handle the concrete mixture specified for the project.

If concrete is being placed for a major reinforced structure, it is important that the placement continue without interruption. The Inspector should be sure that the contractor has ready access to a back-up pump to be used in the event of a breakdown. In order to further insure the success of the concrete placement by the pumping method, the user should be aware of the following points:

(a) A protective grating over the receiving hopper of the pump is necessary to exclude large pieces of aggregate or foreign material.

(b) The pump and lines require lubrication with a grout of cement and water. All of the excess grout is to be wasted prior to pumping the concrete.

(c) All changes in direction must be made by a large radius bend with a maximum bend of 90 degrees. Wye connections induce segregation and shall not be used.

(d) Pump lines should be made of a material capable of resisting abrasion and with a smooth interior surface having a low coefficient of friction. Steel is commonly used for pump lines, because a chemical reaction occurs between the concrete and the aluminum. Aluminum pipe should not be used for pumping concrete and some of the new plastic or rubber tubing is gaining acceptance. Hydrogen is generated which results in a swelling of the concrete, causing a significant reduction in compressive strength. This reaction is aggravated by any of the following: abrasive coarse aggregate, non-uniformly graded sand, low-slump concrete, low sand-aggregate ratio, high-alkali cement or when no air-entraining agent is used.

(e) During temporary interruptions in pumping, the hopper must remain nearly full, with an occasional turning and pumping to avoid developing a hard slug of concrete in the lines.

(f) Excessive line pressures must be avoided. When this occurs, check these points as the probable cause: segregation caused by too low a slump or too high a slump; large particle contamination caused by large pieces of aggregate or frozen lumps not eliminated by the grating; poor gradation of aggregates or particle shape; rich or lean spots caused by improper mixing.

(g) Corrections must be made to correct excessive slump loss as measured at the transit-mixed concrete truck and as measured at the hose outlet. This may be attributable to porous aggregate, high temperature or rapid setting mixes.

(h) Two transit-mix concrete trucks must be used simultaneously to deliver concrete into the pump hopper. These trucks must be discharged alternately to assure a continuous flow of concrete as trucks are replaced.

(i) Samples of concrete for test specimens prepared to determine the acceptance of the concrete quality are to be taken as required for conventional concrete.

Sampling is done before the concrete is deposited in the pump hopper. However, it is suggested that, where possible, the effect of pumping on the compressive strength be checked by taking companion samples, so identified, from the end of the pump line at the same time. The Record of Test must be properly noted as being a special mix used for pumping purposes. This will enable the Materials Control Group to compile a complete history of mix designs and their respective compressive strengths.

The prudent use of pumped concrete can result in economy and improved quality. However, only the control exercised by the operator will assure continued high standards of quality concrete.

Pump lines must be properly fastened to supports to eliminate excessive vibration. Couplings must be easily and securely fastened in a manner that will prevent mortar leakage. It is preferable to use the flexible hose only at the discharge point. This hose must be moved in such a manner as to avoid kinks or sharp bends. The pump line should be protected from excessive heat during hot weather by water sprinkling or shade.

CONCRETE/STRUCTURAL CONCRETE — 03310

COMPRESSIVE STRENGTH FOR VARIOUS WATER-CEMENT RATIOS
(The strengths listed are based on the use of normal portland cement)

WATER/CEMENT RATIO		PROBABLE 28-DAY STRENGTH	
WEIGHT	GALS./100#	PSI	MEGAPASCALS*
.40	4.8	5000	34
.45	5.4	4500	31
.50	6.0	4000	28
.55	6.6	3500	24
.60	7.2	3000	21
.65	7.8	2500	17
.70	8.4	2000	14

* International system equivalent.

APPROXIMATE CONTENT OF SAND, CEMENT AND WATER PER CUBIC YARD OF CONCRETE

Based on aggregates of average grading and physical characteristics in concrete mixes having a water-cement ratio (W/C) of about .65 by weight (or 7.8 gallons) per sack of cement; 3-in, slump; and a medium natural sand having a fineness modulus of about 2.75.

COURSE AGGREGATE MAX. SIZE	WATER		CEMENT	% SAND
	POUNDS	GALLONS		
3/8	385	46	590	57
1/2	365	44	560	50
3/4	340	41	525	43
1	325	39	500	39
1 1/2	300	36	460	37

It can be noted from the above chart that, for a given slump, the amount of mixing water increases as the size of the course aggregate decreases. The size of the course aggregate controls the sand content in the same way; that is, the amount of sand required in the mix increases as the size of the course aggregate decreases.

Other typical examples are contained in the pamphlet published by the Portland Cement Association entitled "Design and Control of Concrete Mixtures."

Effects of Temperature on Concrete. Concrete mixtures gain strength rapidly in the first few days after placement. While the rate of gain in strength diminishes, concrete continues to become stronger with time over a period of many years, so long as drying of the concrete is prevented. Its strength at 28 days is considered to be the compressive strength upon which the Engineer bases his calculations. The temperature of the atmosphere has a significant effect upon the development of strength in concrete. Lower temperatures retard and higher temperatures accelerate the gain in strength.

Most destructive of the natural forces is freezing and thawing action. While the concrete is still wet or moist, expansion of the water as it is converted into ice results in severe damage to the fresh concrete. In situations where freezing may be encountered, high early strength cement may be used. Also, the mixing water or the aggregate (or both) may be preheated before mixing. Covering the concrete, and using steam or salamanders to heat the concrete under the covering, will help prevent freezing. Air-entraining agents help to diminish the effects of freezing of fresh concrete as well as in subsequent freezing and thawing cycles throughout the life of the concrete.

Hot weather will present problems of a different nature in placing concrete. Concrete will set up faster and tend to shrink and crack at the surface. To minimize this problem, the concrete should be placed without delay after mixing. Avoid the use of accelerators (perhaps even use a retarding agent), dampen all subgrade and forms, protect the freshly placed concrete from hot dry winds, and provide for adequate curing. Crushed ice or chilled water can be used as part of the mixing water to reduce the temperature of the mix in extremely hot areas.

CONCRETE/STRUCTURAL CONCRETE — 03310

Admixture	Purpose	Effects on Concrete	Advantages	Disadvantages
Accelerator	Hasten setting.	Improves cement dispersion and increases early strength.	Permits earlier finishing, form removal, and use of the structure.	Increases shrinkage, decreases sulfate resistance, tends to clog mixing and handling equipment.
Air-Entraining Agent	Increase workability and reduce mixing water.	Reduces segregation, bleeding and increases freeze-thaw resistance. Increases strength	Increases workability and reduces finishing time.	Excess will reduce strength and increase slump. Bulks concrete volume.
Bonding Agent	Increase bond to old concrete.	Produces a non-dusting, slip resistant finish,	Permits a thin topping without roughening old concrete, self-curing, ready in one day.	Quick setting and susceptible to damage from fats, oils and solvents.
Densifier	To obtain dense concrete.	Increased workability and strength.	Increases workability and increases waterproofing characteristics, more impermeable.	Care must be used to reduce mixing water in proportion to amount used.
Foaming Agent	Reduce weight.	Increases insulating properties.	Produces a more plastic mix, reduces dead weight loads.	Its use must be very carefully regulated — following instructions explicitly.
Retarder	Retard setting.	Increases control of setting.	Provides more time to work and finish concrete.	Performance varies with cement used — adds to slump. Requires stronger forms.
Water Reducer and Retarder	Increase compressive and flexural strength.	Reduces segregation, bleeding, absorption, shrinkage, and increases cement dispersion.	Easier to place work, provides better control.	Performance varies with cement. Of no use in cold weather.
Water Reducer, Retarder and Air-Entraining Agent	Increases workability.	Improves cohesiveness. Reduces bleeding and segregation.	Easier to place and work.	Care must be taken to avoid excessive air entrainment.

MASONRY / UNIT MASONRY | 04200

ARCHITECTURAL WALL PATTERNS (BONDS)

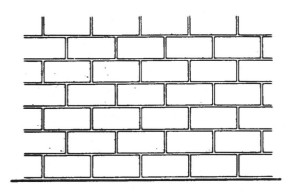

COMMON BOND
8" x 16" UNITS

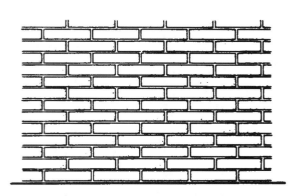

COMMON BOND
4" x 16" UNITS

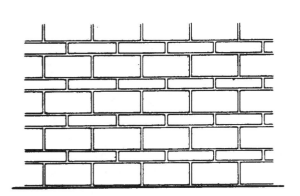

COURSED ASHLER
8" x 16" & 4" x 16" UNITS

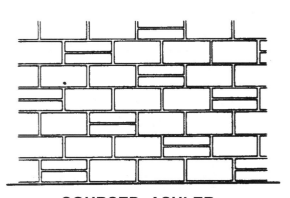

COURSED ASHLER
8" x 16" & 4" x 16" UNITS

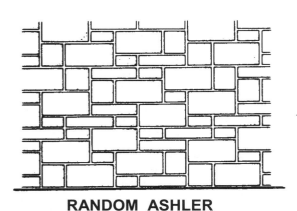

RANDOM ASHLER
8" x 16" & 4" x 16" UNITS
AND 4" x 8" UNITS

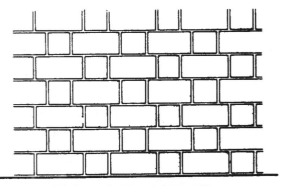

COURSED ASHLER
8" x 16" & 8" x 8" UNITS

MASONRY / UNIT MASONRY 04200

ARCHITECTURAL WALL PATTERNS (BONDS) — (Continued)

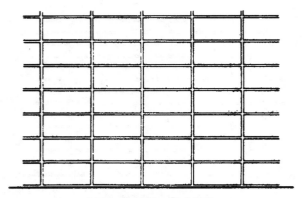

STACKED BOND
8" x 16" UNITS

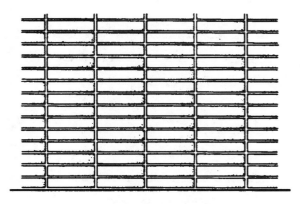

STACKED BOND
8" x 16" & 4" x 16" UNITS

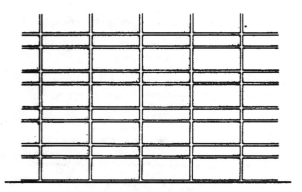

STACKED BOND
8" x 16" & 4" x 16" UNITS

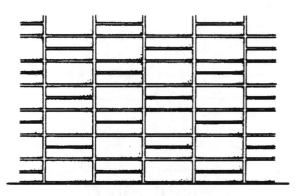

STACKED BOND
8" x 16" & 4" x 16" UNITS

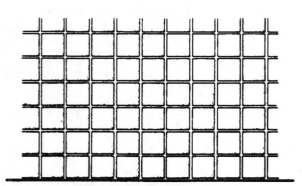

STACKED BOND VERTICAL SCORED UNITS

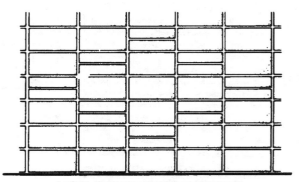

USE OF BLOCK DESIGN IN STACKED BOND

MASONRY / SPLIT FACE CMU   04200

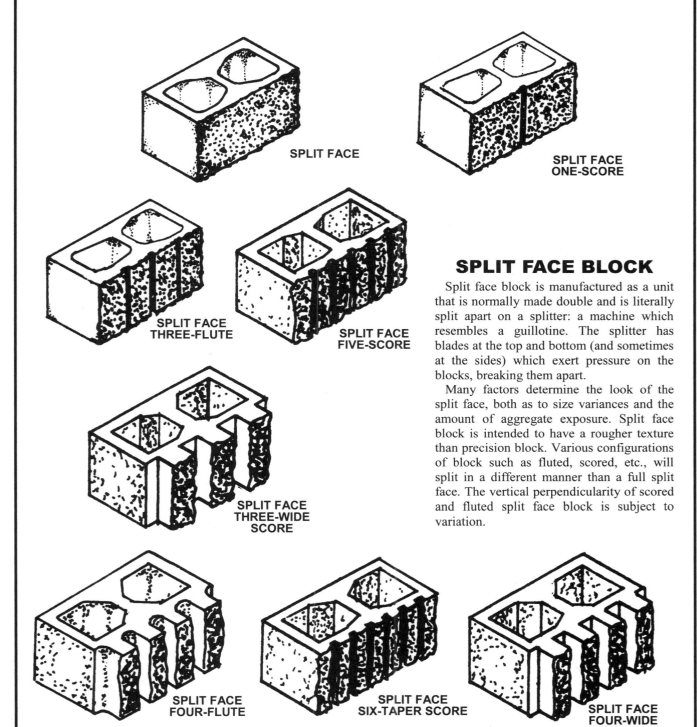

SPLIT FACE BLOCK

Split face block is manufactured as a unit that is normally made double and is literally split apart on a splitter: a machine which resembles a guillotine. The splitter has blades at the top and bottom (and sometimes at the sides) which exert pressure on the blocks, breaking them apart.

Many factors determine the look of the split face, both as to size variances and the amount of aggregate exposure. Split face block is intended to have a rougher texture than precision block. Various configurations of block such as fluted, scored, etc., will split in a different manner than a full split face. The vertical perpendicularity of scored and fluted split face block is subject to variation.

NOTE: Split face units shown in this manual are a small sampling of the broad range of concrete masonry architectural units available from the industry on special order. Depths and widths of scores vary. Consult a local manufacturer for specific information.

MASONRY / BOND BEAMS / LINTELS 04220

TYPICAL DETAILS — LINTELS AND BOND BEAMS

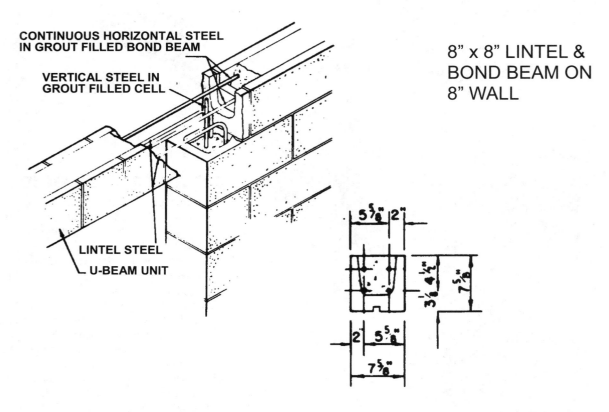

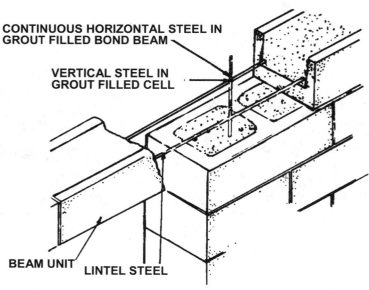

METALS / WELDING 05050

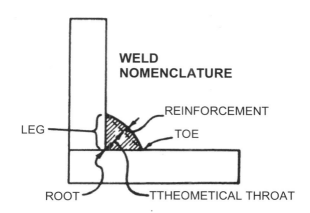

WELD NOMENCLATURE

Labels: REINFORCEMENT, LEG, TOE, ROOT, TTHEOMETICAL THROAT

WELDED JOINTS

  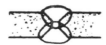 

SQUARE BUTT   SINGLE VEE BUTT   DOUBLE VEE BUTT   SINGLE U BUTT

  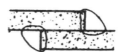 

DOUBLE U BUTT   SINGLE FILLET LAP   DOUBLE FILLET LAP   STRAP JOINT

 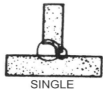 

SINGLE BEVEL TEE   DOUBLE BEVEL TEE   SINGLE J TEE   SQUARE TEE

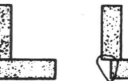

DOUBLE J TEE   CLOSED CORNER (FLUSH) JOINT   HALF OPEN CORNER JOINT

WELDING POSITIONS

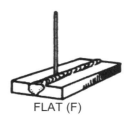  

FLAT (F)   HORIZONTAL (H)   VERTICAL (V)   OVERHEAD (OH)

METALS / FASTENING 05050

BOLTS IN COMMON USAGE

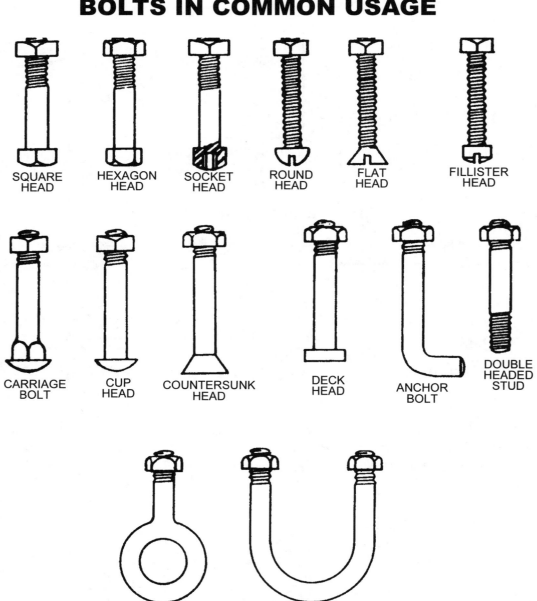
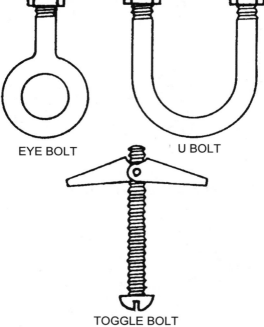

METALS / STRUCTURAL STEEL | 05120

EXAMPLE OF SIMPLIFIED STRUCTURAL STEEL TAKEOFF METHOD

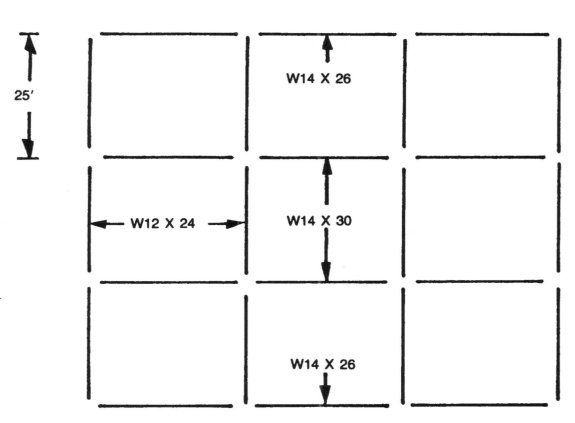

	#		LF. EA.		LBS./LF.			
W14 X 26	6	×	30	×	26	=	4,680	LBS.
W14 X 30	6	×	30	×	30	=	5,400	
W12 X 24	12	×	25	×	24	=	7,200	
							17,280	LBS.
							OR	
							9 ±	TONS

AFTER MAIN MEMBERS ARE ESTIMATED, ADD:
2 TO 3% FOR BASE PLATES
4 TO 5% FOR COLUMN SPLICES
4 TO 5% FOR MISCELLANEOUS COSTS

WOOD & PLASTICS / FASTENERS 06050

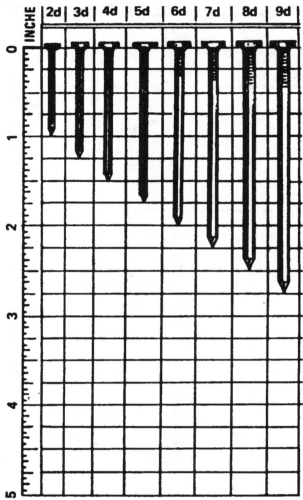

COMMON WIRE NAILS (ACTUAL SIZE)

Cut Nails. Cut nails are angular-sided, wedge-shaped with a blunt point.

Wire Nails. Wire nails are round shafted, straight, pointed nails, and are used more generally than cut nails. They are stronger than cut nails and do not buckle as easily when driven into hard wood, but usually split wood more easily than cut nails. Wire nails are available in a variety of sizes varying from two penny to sixty penny.

Nail Finishes. Nails are available with special finishes. Some are galvanized or cadmium plated to resist rust. To increase the resistance to withdrawal, nails are coated with resins or asphalt cement (called cement coated). Nails which are small, sharp-pointed, and often placed in the craftsman's mouth (such as lath or plaster board nails) are generally blued and sterilized.

WOOD & PLASTICS / FASTENERS 06050

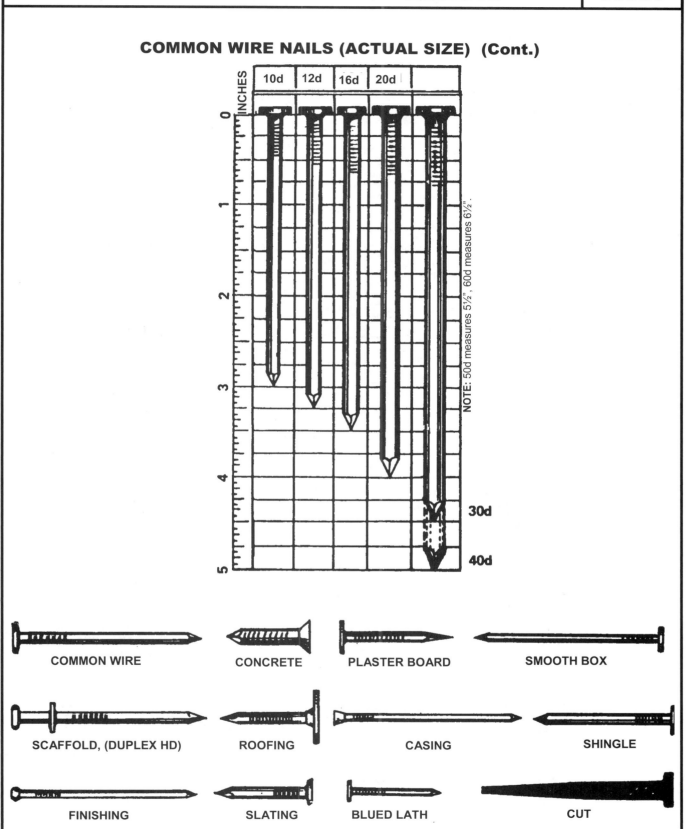

COMMON WIRE NAILS (ACTUAL SIZE) (Cont.)

NOTE: 50d measures 5½", 60d measures 6½".

- COMMON WIRE
- CONCRETE
- PLASTER BOARD
- SMOOTH BOX
- SCAFFOLD, (DUPLEX HD)
- ROOFING
- CASING
- SHINGLE
- FINISHING
- SLATING
- BLUED LATH
- CUT

WOOD & PLASTICS / ROUGH CARPENTRY | 06100

PLYWOOD — BASIC GRADE MARKS
AMERICAN PLYWOOD ASSOCIATION (APA)

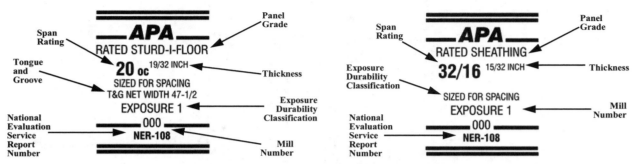

The American Plywood Association's trademarks appear only on products manufactured by APA member mills. The marks signify that the product is manufactured in conformance with APA performance standards and/or U.S. Product Standard PS 1-83 for Construction and Industrial Plywood.

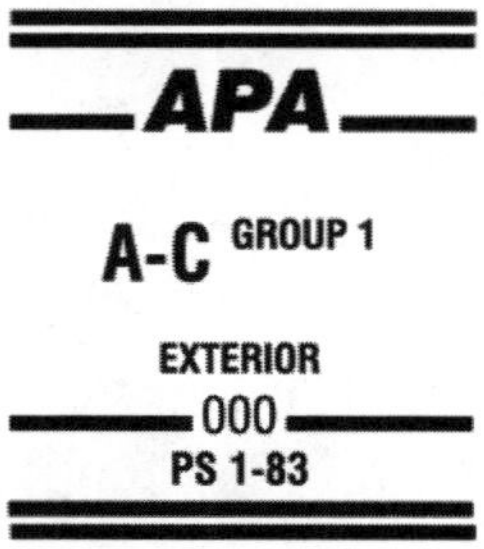

APA A-C
For use where appearance of one side is important in exterior applications such as soffits, fences, structural uses, boxcar and truck linings, farm buildings, tanks, trays, commercial refrigerators, etc. **Exposure Durability Classification: Exterior. Common Thicknesses:** ¼, 11/32, ⅜, 15/32, ½, 19/32, 5/8, 23/32, ¾.

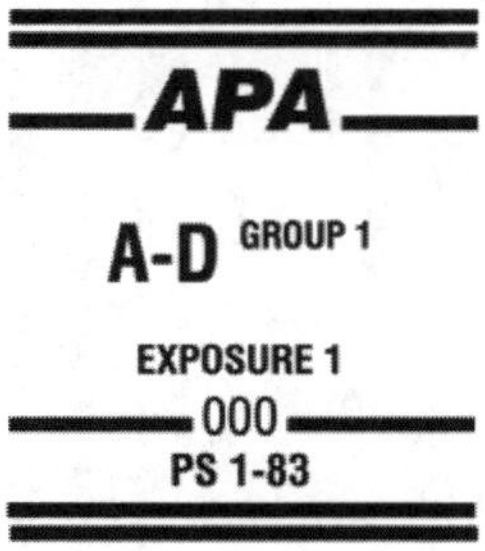

APA A-D
For use where appearance of only one side is important in interior applications, such as paneling, built-ins, shelving, partitions, flow racks, etc. **Exposure Durability Classifications: Interior, Exposure 1. Common Thicknesses:** ¼, 11/32, ⅜, 15/32, ½, 19/32, 5/8, 23/32, ¾.

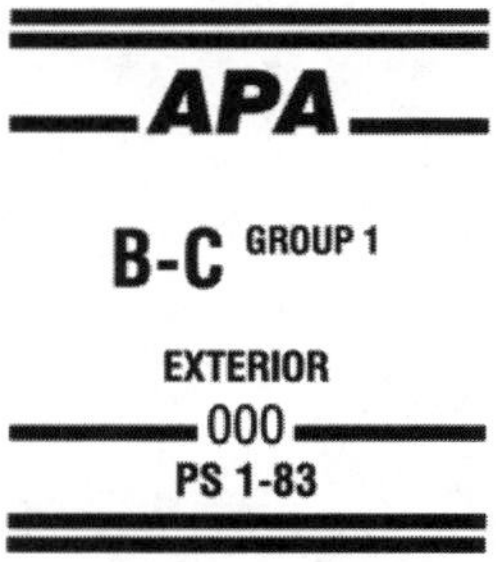

APA B-C
Utility panel for farm service and work buildings, boxcar and truck linings, containers, tanks, agricultural equipment, as a base for exterior coatings and other exterior uses. **Exposure Durability Classification: Exterior. Common Thicknesses:** ¼, 11/32, ⅜, 15/32, ½, 19/32, 5/8, 23/32, ¾.

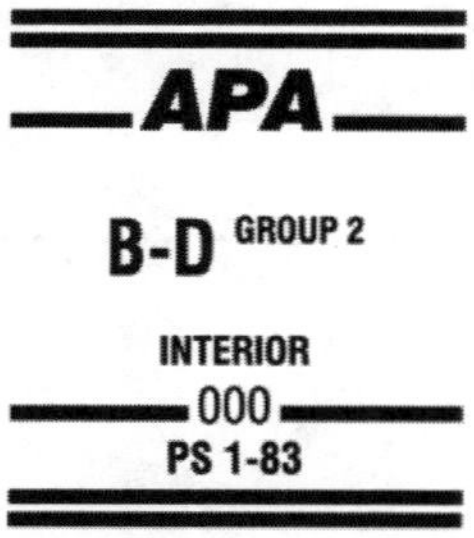

APA B-D
Utility panel for backing, sides or built-ins, industry shelving, slip sheets, separator boards, bins and other interior or protected applications. **Exposure Durability Classifications: Interior, Exposure 1. Common Thicknesses:** ¼, 11/32, ⅜, 15/32, ½, 19/32, 5/8, 23/32, ¾.

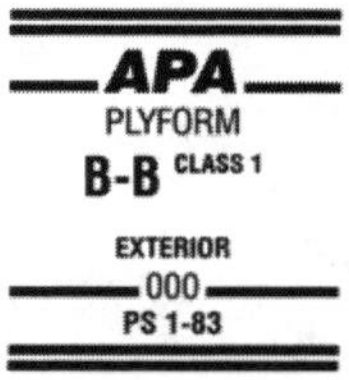

APA proprietary concrete form panels designed for high reuse. Sanded both sides and mill-oiled unless otherwise specified. Class I, the strongest, stiffest and more commonly available, is limited to Group 1 faces, Group 1 or 2 crossbands, and Group 1, 2, 3 or 4 inner plies. Class II is limited to Group 1 or 2 faces (Group 3 under certain conditions) and Group 1, 2, 3 or 4 inner plies. Also available in HDO for very smooth concrete finish, in Structural I, and with special overlays. **Exposure Durability Classification: Exterior. Common Thicknesses:** 19/32, ¾, 23/32, ¼.

Plywood panel manufactured with smooth, opaque, resin-treated fiber overlay providing ideal base for paint on one or both sides. Excellent material choice for shelving, factory work surfaces, paneling, built-ins, signs and numerous other construction and industrial applications. Also available as a 303 Siding with texture-embossed or smooth surface on one side only and Structural I. **Exposure Durability Classification: Exterior. Common Thicknesses:** 11/32, ⅜, 19/32, ½, 19/32, 5/8, 23/32, ¾.

WOOD & PLASTICS / ROUGH CARPENTRY | 06100

SPECIALTY PANELS

```
HDO · A · A · G-1 · EXT – APA · 000 · PS1 – 83
```

Plywood panel manufactured with a hard, semi-opaque resin-fiber overlay on both sides. Extremely abrasion resistant and ideally suited to scores of punishing construction and industrial applications, such as concrete forms, industrial tanks, work surfaces, signs, agricultural bins, exhaust ducts, etc. Also available with skid-resistant screen-grid surface and in Structural I. *Exposure Durability Classification:* **Exterior.** *Common Thicknesses:* 3/8, ½, 5/8, 3/4

```
MARINE · A · A · EXT – APA · 000 · PS1 – 83
```

Specialty designed plywood panel made only with Douglas fir or western larch, solid jointed cores, and highly restrictive limitations on core gaps and faces repairs. Ideal for both hulls and other marine applications. Also available with HDO or MDO faces. *Exposure Durability Classification:* **Exterior.** *Common Thicknesses:* 1/4, 3/8, ½, 5/8, 3/4.

Unsanded and touch-sanded panels, and panels with "B" or better veneer on one side only, usually carry the APA trademark on the panel back. Panels with both sides of "B" or better veneer, or with special overlaid surfaces (such as Medium Density Overlay), carry the APA trademark on the panel edge, like this:

GLOSSARY OF TERMS

Some of the words and terms used in the grading of lumber follow:

Bow. A deviation flatwise from a straight line drawn from end to end of the piece. It is measured at the point of greatest distance from the straight line.
Checks. A separation of the wood which normally occurs across the annual rings and usually as a result of seasoning.
Crook. A deviation edgewise from a straight line drawn from end to end of the piece. It is measured at the point of greatest distance from the straight line.
Cup. A deviation from a straight line drawn across the piece from edge to edge. It is measured at the point of greatest distance from the straight line.
Flat Grain. The annual growth rings pass through the piece at an angle of less than 45 degrees with the flat surface of the piece.

Warp. Any deviation from a true or plane surface, including crook, cup, bow or any combination thereof.
Mixed Grain. The piece may have vertical grain, flat grain or a combination of both vertical and flat grain.
Pitch. An accumulation of resin which occurs in separations in the wood or in the wood cells themselves.
Shake. A separation of the wood which usually occurs between the rings of annual growth.
Splits. A separation of the wood due to tearing apart of the wood cells.
Vertical Grain. The annual growth rings pass through the piece at an angle of 45 degrees or more with the flat surface of the piece.
Wane. Bark or lack of wood from any cause, except eased edges (rounded) on the edge or corner of a piece of lumber.

WOOD & PLASTICS / ROUGH CARPENTRY | 06100

LUMBER GRADING
GRADING-MARK ABBREVIATIONS

GRADES
(Listed alphabetically — not by quality)

COM	Common
CONST	Construction
ECON	Economy
No. 1	Number One
SEL-MER	Select Merchantable
SEL-STR	Select Structural
STAN	Standard
UTIL	Utility

ALSC TRADEMARKS

CLIS	California Lumber Inspection Service
NELMA	Northeastern Lumber Mfrs. Assoc., Inc.
NH&PMA	Northern Hardwood & Pine Mfrs. Assoc., Inc.
PLIB	Pacific Lumber Inspection Bureau
RIS	Redwood Inspection Service
SPIB	Southern Pine Inspection Bureau
TP	Timber Products Inspection
WCLB	West Coast Lumber Inspection Bureau
WWP	Western Wood Products Association

SPECIES GROUPINGS

AF	Alpine Fir
DF	Douglas Fir
HF	Hem Fir
SP	Sugar Pine
PP	Ponderosa Pipe
LP	Lodgepole Pine
IWP	Idaho White Pine
ES	Engelmann Spruce
WRC	Western Red Cedar
INC CDR	Incense Cedar
L	Larch
LP	Lodgepole Pine
MH	Mountain Hemlock
WW	White Wood

MOISTURE CONTENT

S-GRN	Surfaced at a moisture content of more than 19%.
S-DRY	Surfaced at a moisture content of 19% or less.
MC-15	Surfaced at a moisture content of 15% or less.

WOOD & PLASTICS / ROUGH CARPENTRY — 06100

FRAMING ESTIMATING RULES OF THUMB

For 16" O.C. stud partitions figure 1 stud for every L.F. of wall; add for top and bottom plates.

For any type of framing, the quantity of basic framing members (in L.F.) can be determined based on spacing and surface area (S.F.):

12" O.C.	1.2 L.F./S.F.
16" O.C.	1.0 L.F./S.F.
24" O.C.	0.8 L.F./S.F.

(Doubled-up members, bands, plates, framed openings, etc., must be added.)

Framing accessories, nails, joist hangers, connectors, etc., should be estimated as separate material costs. Installation should be included with framing. Rule of thumb allowance is 0.5 to 1.5% of lumber cost for rough hardware. Another is 30 to 40 pounds of nails per M.B.F.

BOARD FEET/LINEAR FEET FOR LUMBER

Nominal Size	Actual Size	Board Feet Per Linear Foot	Linear Feet Per 1000 Board Feet
1 x 2	¾ x 1 ½	.167	6000
1 x 3	¾ x 2 ½	.250	4000
1 x 4	¾ x 3 ½	.333	3000
1 x 6	¾ x 5 ½	.500	2000
1 x 8	¾ x 7 ¼	.666	1500
1 x 10	¾ x 9 ¼	.833	1200
1 x 12	¾ x 11 ¼	1.0	1000
2 x 2	1 ½ x 1 ½	.333	3000
2 x 3	1 ½ x 2 ½	.500	2000
2 x 4	1 ½ x 3 ½	.666	1500
2 x 6	1 ½ x 5 ½	1.0	1000
2 x 8	1 ½ x 7 ¼	1.333	750
2 x 10	1 ½ x 9 ¼	1.666	600
2 x 12	1 ½ x 11 ¼	2.0	500

WOOD & PLASTICS / ROUGH CARPENTRY — 06100

Redwood. Redwood is a fairly strong and moderately lightweight material. The heartwood is red but the sapwood is white. One of the principal advantages of redwood is that the heartwood is highly resistant (but not entirely immune) to decay, fungus and insects. Standard Specifications require that all redwood used in permanent installations shall be "select heart." Grade marking shall be in accordance with the standards established in the California Redwood Association. Grade marking shall be done by, or under the supervision of the Redwood Inspection Service.

Redwood is graded for specific uses as indicated in the following table:

REDWOOD GRADING

Type of Lumber	Grade	Typical Use
Grades for Dimension Only Listed Here	Clear All Heart	Exceptionally fine, knot free, straight-grained timbers. This grade is used primarily for stain finish work of high quality.
	Clear	Same as Clear All Heart except that this grade may contain sound sapwood and medium stain.
	Select Heart	**This grade only is to be used in Agency work, unless otherwise specified in the plans or specifications.** It is sound, live heartwood free from splits or streaks with sound knots. It is generally used where the timber is in contact with the ground, as in posts, mudsills, etc.
	Select Construction Heart	Slightly less quality than Select Heart. It may have some sapwood in the piece. Used for general construction purposes when redwood is needed.
	Construction Common	Same requirement as Construction Heart except that it will contain sapwood and medium stain. Its resistance to decay and insect attack is reduced.
	Merchantable	Used for fence posts, garden stakes, etc.
	Economy	Suitable for crating, bracing and temporary construction.

DOUGLAS FIR GRADING

Type of Lumber	Grade	Typical Use
Select Structural Joists and Planks	Select Structural	Used where strength is the primary consideration, with appearance desirable.
	No. 1	Used where strength is less critical and appearance not a major consideration.
	No. 2	Used for framing elements that will be covered by subsequent construction.
	No. 3	Used for structural framing where strength is required but appearance is not a factor.
Finish Lumber	Superior	For all types of uses as casings, cabinet, exposed members, etc., where a fine appearance is desired.
	Prime	
	E	
Boards (WCLIB)* * Grading is by West Coast Lumber Inspection Bureau rules, but sizes conform to Western Wood Products Assn. rules. These boards are still manufactured by some mills.	Select Merchantable	Intended for use in housing and light construction where a knotty type of lumber with finest appearance is required.
	Construction	Used for sub-flooring, roof and wall sheathing, concrete forms, etc. Has a high degree of serviceability.
	Standard	Used widely for general construction purposes, including subfloors, roof and wall sheathing, concrete forms, etc. Seldom used in exposed construction because appearance.
	Utility	Used in general construction where low cost is a factor and appearance is not important. (Storage shelving, crates, bracing, temporary scaffolding etc.)

WOOD & PLASTICS / ROUGH CARPENTRY | 06100

BOARD FEET CONVERSION TABLE

Nominal Size (In.)	ACTUAL LENGTH IN FEET								
	8	10	12	14	16	18	20	22	24
1 x 2		1 2/3	2	2 1/3	2 2/3	3	3 ½	3 2/3	4
1 x 3		2 ½	3	3 ½	4	4 ½	5	5 ½	6
1 x 4	2 ¾	3 1/3	4	4 2/3	5 1/3	6	6 2/3	7 1/3	8
1 x 5		4 1/6	5	5 5/6	6 2/3	7 ½	8 1/3	9 1/6	10
1 x 6	4	5	6	7	8	9	10	11	12
1 x 7		5 5/8	7	8 1/6	9 1/3	10 ½	11 2/3	12 5/6	14
1 x 8	5 1/3	6 2/3	8	9 1/3	10 2/3	12	13 1/3	14 2/3	16
1 x 10	6 2/3	8 1/3	10	11 2/3	13 1/3	15	16 2/3	18 1/3	20
1 x 12	8	10	12	14	16	18	20	22	24
1¼ x 4		4 1/6	5	5 5/6	6 2/3	7 ½	8 1/3	9 1/6	10
1¼ x 6		6 ¼	7 ½	8 ¾	10	11 ¼	12 ½	13 ¾	15
1¼ x 8		8 1/3	10	11 2/3	13 1/3	15	16 2/3	18 1/3	20
1¼ x 10		10 5/12	12 ½	14 7/12	16 2/3	18 ¾	20 5/6	22 11/12	25
1¼ x 12		12 ½	15	17 ½	20	22 ½	25	27 ½	30
1½ x 4	4	5	6	7	8	9	10	11	12
1½ x 6	6	7 ½	9	10 ½	12	13 ½	15	16 ½	18
1½ x 8	8	10	12	14	16	18	20	22	24
1½ x 10	10	12 ½	15	17 ½	20	22 ½	25	27 ½	30
1½ x 12	12	15	18	21	24	27	30	33	36
2 x 4	5 1/3	6 2/3	8	9 1/3	10 1/3	12	13 1/3	14 2/3	16
2 x 6	8	10	12	14	16	18	20	22	24
2 x 8	10 2/3	13 1/3	16	18 2/3	21 1/3	24	26 2/3	29 1/3	32
2 x 10	13 1/3	16 2/3	20	23 1/3	26 2/3	30	33 1/3	36 2/3	40
2 x 12	16	20	24	28	32	36	40	44	48
3 x 6	12	15	18	21	24	27	30	33	36
3 x 8	16	20	24	28	32	36	40	44	48
3 x 10	20	25	30	35	40	45	50	55	60
3 x 12	24	30	36	42	48	54	60	66	72
4 x 4	10 2/3	13 1/3	16	18 2/3	21 1/3	24	26 2/3	29 1/3	32
4 x 6	16	20	24	28	32	36	40	44	48
4 x 8	21 1/3	26 2/3	32	37 1/3	42 2/3	48	53 1/3	58 2/3	64
4 x 10	26 2/3	33 1/3	40	46 2/3	53 1/3	60	66 2/3	73 1/3	80
4 x 12	32	40	48	56	64	72	80	88	96

MOISTURE PROTECTION / ROOFING — 07000

SLOPE AREA CALCULATIONS

Rise and Run	Multiply Flat Area by	LF of Hips or Valleys per LF of Common Run
2 in 12	1.014	1.424
3 in 12	1.031	1.436
4 in 12	1.054	1.453
5 in 12	1.083	1.474
6 in 12	1.118	1.500
7 in 12	1.158	1.530
8 in 12	1.202	1.564
9 in 12	1.250	1.600
10 in 12	1.302	1.641
11 in 12	1.357	1.685
12 in 12	1.413	1.732

MOISTURE PROTECTION / DOWNSPOUTS — 07600

DOWNSPOUT/VERTICAL LEADER CALCULATIONS

Roof Type	Slope	S.F. Roof/ Sq. In. Leader
Gravel	Less than ¼" per foot	300
Gravel	Greater than ¼" per foot	250
Metal or Shingle	Any	200

Alternate calculations:

$$\text{Diameter of downspout/leader} = 1.128 \sqrt{\frac{\text{Area of drainage}}{\text{SF Roof/Sq. Inch}}}$$

MOISTURE PROTECTION / ROOFING — 07600

TYPICAL MINIMUM SIZE OF VERTICAL CONDUCTORS AND LEADERS

Size of leader or conductor (Inches)	Maximum projected roof area (Square feet)
2	544
2 ½	987
3	1,610
4	3,460
5	6,280
6	10,200
8	22,000

TYPICAL MINIMUM SIZE OF ROOF GUTTERS

Diameter gutter (Inches)	MAXIMUM PROJECTED ROOF AREA FOR GUTTERS OF VARIOUS SLOPES			
	1/16 in. Ft. slope (Sq. ft.)	1/8 in. per Ft. slope (Sq. ft.)	¼ in. per Ft. slope (Sq. ft.)	1/2 in. per Ft. /slope (Sq. ft.)
3	170	240	340	480
4	360	510	720	1,020
5	625	880	1,250	1,770
6	960	1,360	1,920	2,770
7	1,380	1,950	2,760	3,900
8	1,990	2,800	3,980	5,600
10	3,600	5,100	7,200	10,000

FINISHES / METAL FRAMING SYSTEMS | 09110

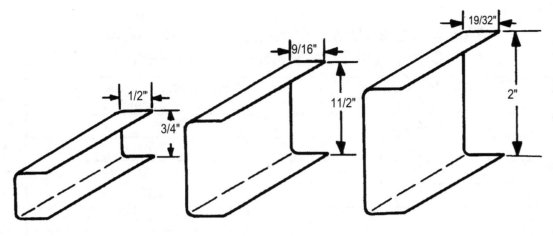

COLD ROLLED CHANNELS

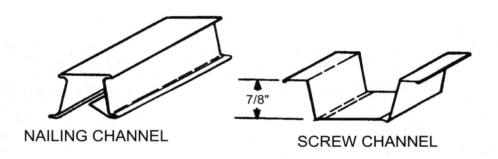

NAILING CHANNEL   SCREW CHANNEL

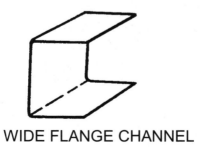

CHANNEL STUD

WIDE FLANGE CHANNEL   CEE STUD

FINISHES / LATH & PLASTER — 09200

STUDLESS SOLID PARTITION

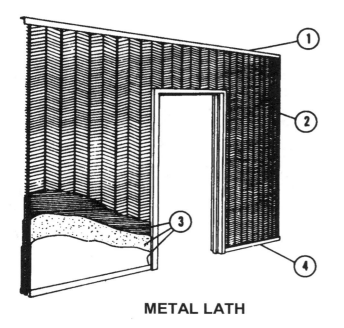
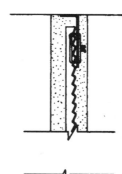

(1) Ceiling Runner
(2) Rib Metal Lath
(3) Plaster
(4) Combination Floor Runner and Screed

METAL LATH

STUDLESS SOLID PARTITION

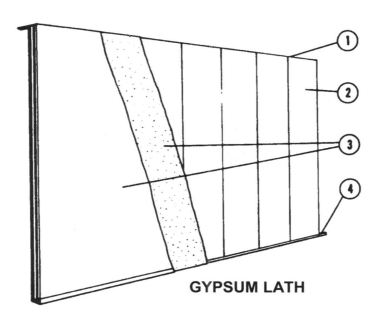
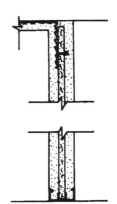

(1) Ceiling Runner
(2) Long length Gypsum Lath
(3) Plaster
(4) Combination Floor Runner and Screed

GYPSUM LATH

FINISHES / LATH & PLASTER 09200

SUSPENDED CEILINGS

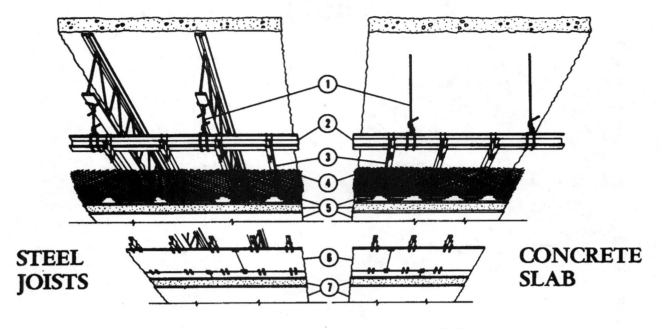

STEEL JOISTS **CONCRETE SLAB**

(1) Hanger
(2) Main Runner Channel
(3) Furring Channel
(4) Metal or Wire Fabric Lath
(5) Plaster
(6) Gypsum Lath
(7) Plaster

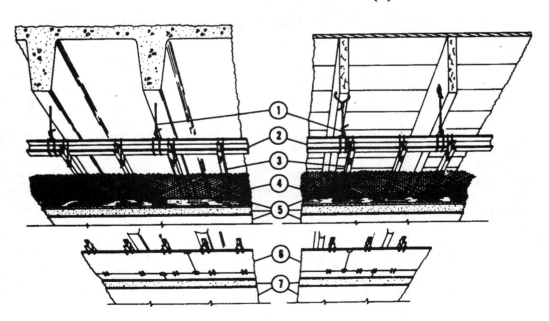

CONCRETE JOISTS **WOOD JOISTS**

FINISHES / LATH & PLASTER 09200

STEEL STUD
Hollow Partition

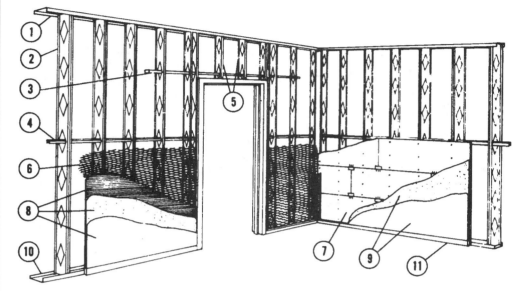

(1) Ceiling Runner Track
(2) Nailable Stud
(3) Door Opening Stiffener
(4) Partition Stiffener
(5) Jack Studs
(6) Metal or Wire Fabric Lath (screwed or wire tied)
(7) Gypsum Lath (nailed, clipped or screwed)
(8) Three Coats of Plaster (Scratch, Brown, Finish)
(9) Two Coats of Plaster (Brown, Finish)
(10) Floor Runner Track
(11) Flush Metal Base

SCREW STUD
Hollow Partition

(1) Ceiling Runner Track
(2) Screwed Stud
(3) Door Opening Stiffener
(4) Partition Stiffener
(5) Jack Studs
(6) Metal or Wire Fabric Lath (screwed on)
(7) Gypsum Lath (screwed on)
(8) Three Coats of Plaster (Scratch, Brown, Finish)
(9) Two Coats of Plaster (Brown, Finish)
(10) Floor Runner Track
(11) Flush Metal Base

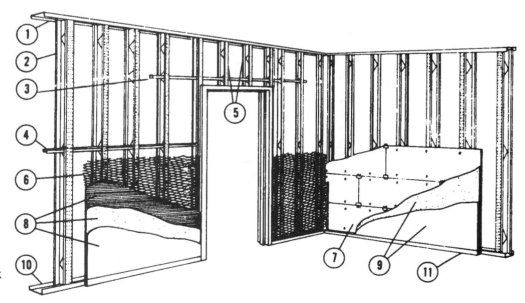

FINISHES / LATH & PLASTER — 09200

LOAD-BEARING HOLLOW PARTITION
Structural Stud

(1) Ceiling Runner Track
(2) Structural Studs (prefabricated)
(3) Structural Stud (nailable)
(4) Jack Studs
(5) Partition Stiffener (bridging)
(6) Metal or Wire Fabric Lath (wired-tied, nailed or stapled)
(7) Gypsum Lath (nailed or stapled)
(8) Three Coats of Plaster (Scratch, Brown, Finish)
(9) Two Coats of Plaster (Brown, Finish)

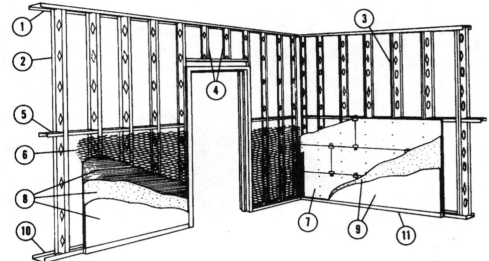

(10) Floor Runner Track (11) Flush Metal Base

VERTICAL FURRING
With Studs

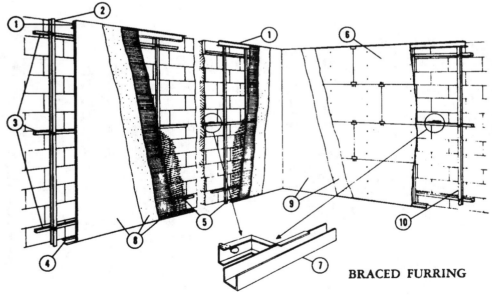

(1) Ceiling Runner Track
(2) Channel Studs
(3) Horizontal Stiffener
(4) Floor Runner
(5) Metal or wire Fabric Lath
(6) Gypsum Lath
(7) Bracing
(8) Three Coats of Plaster (Scratch, Brown, Finish)
(9) Two Coats of Plaster (Brown, Finish)
(10) Screw Channel Studs

FREE STANDING FURRING   BRACED FURRING

FINISHES / LATH & PLASTER　　09200

DOUBLE CHANNEL STUD
Hollow Partition

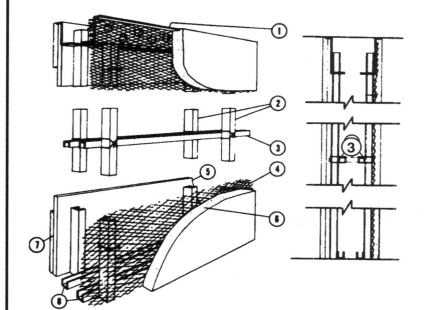

(1) Ceiling Runner
(2) Channel Studs
(3) Partition Stiffener and Channel Spacer
(4) Metal or Wire Fabric Lath (wired-tied)
(5) Gypsum Lath (clipped on)
(6) Three Coats of Plaster (Scratch, Brown, Finish)
(7) Two Coats of Plaster (Brown, Finish)
(8) Floor Runners (channel)

SINGLE CHANNEL STUD
Solid Partition

(1) Ceiling Runner
(2) Channel Stud
(3) Metal or Wire Fabric Lath (wired-tied)
(4) Plaster
(5) Combination Floor Runner and Screed

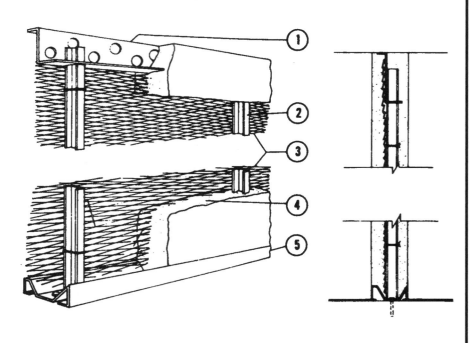

FINISHES / LATH & PLASTER — 09200

FURRED CEILINGS

STEEL JOISTS   CONCRETE JOISTS

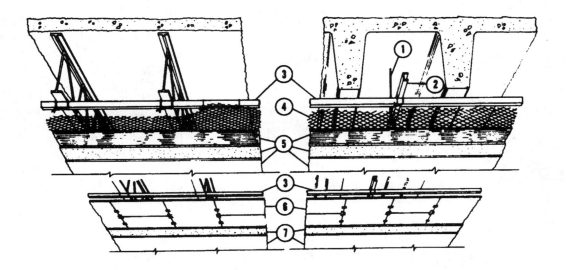

WOOD JOISTS

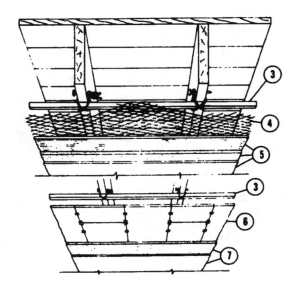

(1) Hanger
(2) ¾-inch Channel
(3) Cross Furring
(4) Metal or Wire Fabric Lath
(5) Plaster
(6) Gypsum Lath
(7) Plaster

FINISHES / LATH & PLASTER — 09200

CEILINGS

STEEL JOISTS   CONCRETE JOISTS

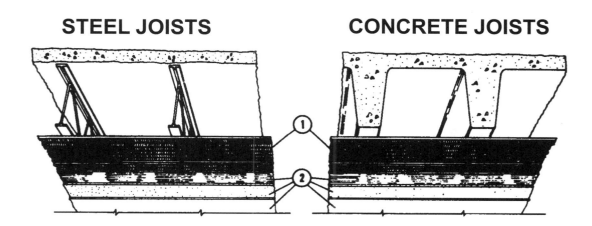

WOOD JOISTS

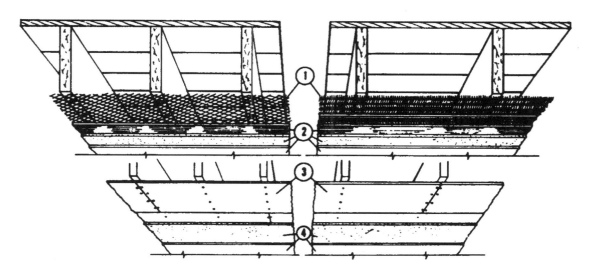

(1) Metal or Wire Fabric Lath
(2) Plaster
(3) Gypsum Lath
(4) Plaster

FINISHES / PLASTER ACCESSORIES 09205

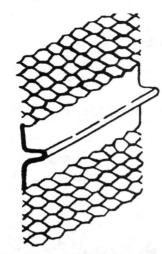 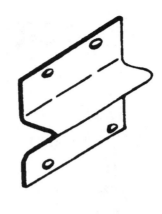

BASE OR PARTING SCREEDS

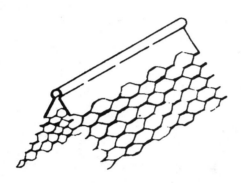 

SMALL NOSE CORNER BEADS

WIRE BULL NOSE CORNER BEADS

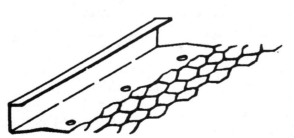 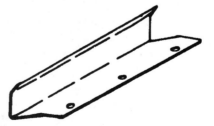

SQUARE CASING BEADS

FINISHES / METAL LATH 09205

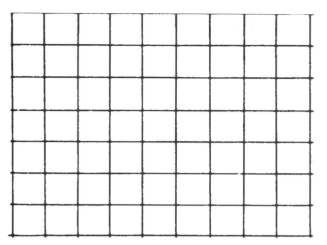

PLAIN WIRE FABRIC LATH

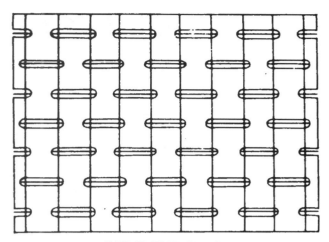

SELF-FURRING WIRE FABRIC LATH

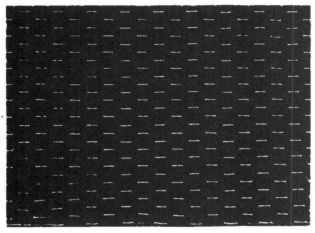

PAPER BACKED WOVEN WIRE FABRIC LATH

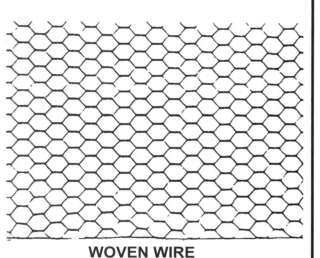

WOVEN WIRE FABRIC LATH
(Also Available Self-Furred)

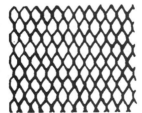

FLAT DIAMOND MESH METAL LATH

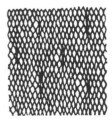

SELF-FURRING METAL LATH

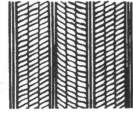

FLAT RIB METAL LATH

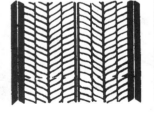

RIB METAL LATH

RIB METAL LATH

FINISHES / FURRING 09205

VERTICAL FURRING
Studless

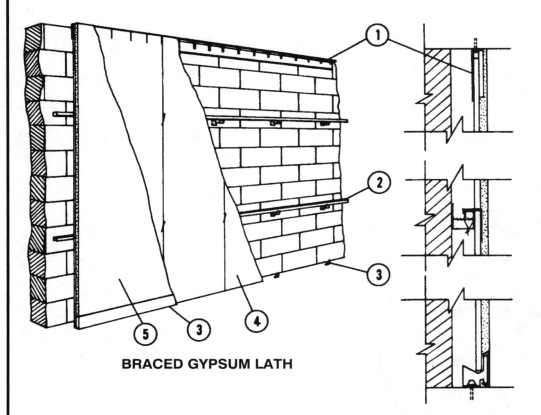

BRACED GYPSUM LATH

(1) Ceiling Runner
(2) Horizontal Stiffener (secured to bracing attachment)
(3) Metal Base and Clips
(4) Gypsum Lath
(5) Three Coats of Plaster (Scratch, Brown, Finish) (Minimum plaster thickness is ¾ inch.)

COLUMN FURRING

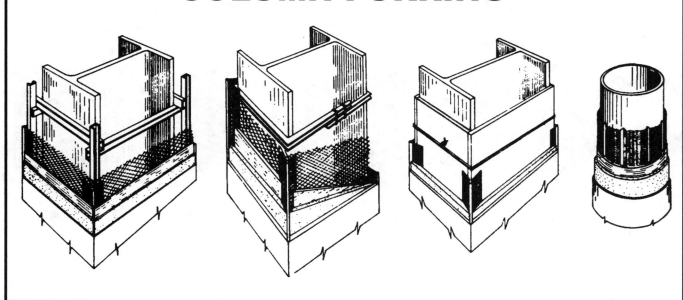

FINISHES / FURRING　　09205

CONTACT FURRING

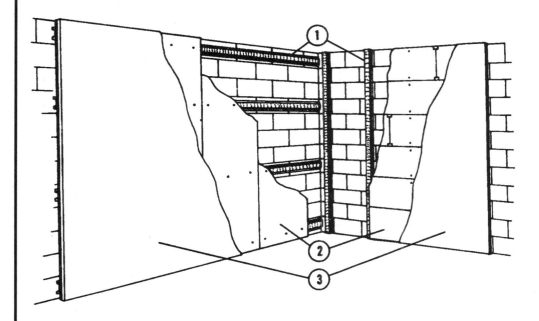

(1) Screw Channel
(2) Gypsum Lath (screwed on)
(3) Two Coats of Plaster (Brown, Finish)

FALSE BEAMS

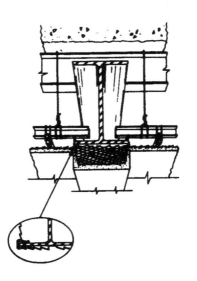

LATH DIRECT TO UNDERSIDE OF BEAM

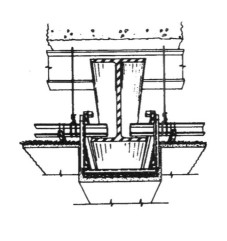

CHANNEL BRACKETS TO MAIN RUNNERS

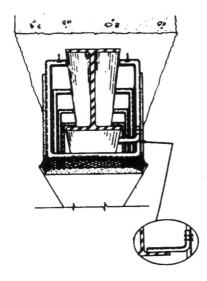

CHANNEL BRACKETS TO SLAB

FINISHES / LATHING 09205

MAXIMUM SPACING OF SUPPORTS FOR METAL LATH (Inches)

Type Of Lath	Weight of Lath Lb. Per Sq. Yd.	WALLS AND PARTITIONS			CEILINGS	
		Wood Studs	Solid Partitions	Steel Studs Wall Furring, Etc.	Wood or Concrete	Metal
Diamond Mesh (flat expanded)	2.5	16	16	13 1/2	12	12
	3.4	16	16	16	16	
Flat Rib	2.75	16	16	16	16	16
	3.4	19	24 (3)	19	19	19
3/8" Rib (1) (2)	3.4	24	(4)	24	24	24
	4.0	24	(4)	24	24	24
3/4" Rib	5.4	-	(4)	24 (5)	36 (6)	36 (6)
Sheet Lath	4.5	24	(4)	24	24	24

NOTE: Weights are exclusive of paper, fiber or other backing.
(1) 3.4 lb. 3/8" Rib Lath is permissible under Concrete Joists at 27" c.c.
(2) These spacings are based on a narrow bearing surface for the lath. When supports with a relatively wide bearing surface are used, these spacings may be increased accordingly, and still assure satisfactory work.
(3) This spacing permissible for Solid Partitions not exceeding 16' in height. For greater heights, permanent horizontal stiffener channels or rods must be provided on channel side of partitions, every 6' vertically, or else spacing shall be reduced 25%.
(4) For studless solid partitions, lath erected vertically.
(5) For interior wall furring or for application over solid surfaces for stucco.
(6) For contact or ceilings only.

FINISHES / LATHING   09205

TYPES OF LATH-ATTACHMENT TO WOOD AND METAL SUPPORTS

TYPE OF LATH	NAILS Type & Size	NAILS Max Spacing Vertical (In)	NAILS Max Spacing Horizontal (In)	SCREWS Max Spacing Vertical (In)	SCREWS Max Spacing Horizontal (In)	STAPLES Wire Gauge No.	STAPLES Crown	STAPLES Leg	STAPLES Max Spacing Vertical (In)	STAPLES Max Spacing Horizontal (In)
1. Diamond Mesh Expanded Metal Lath and Flat Rib Metal Lath	4d blued smooth box 1 ½ No. 14 gauge 7/32" head (clinched) 1" No.11 gauge 7/16" head, barbed 1 ½" No.11 gauge 7/16" head, barbed	6 6 6	– – 6	6	6	16	3/4	7/8	6	6
2. 3/8" Rib Metal Lath and Sheet Lath	1 ½" No. 11 ga. 7/16" head, barbed	6	6	6	6	16	3/4	1 1/2	At Ribs	At Ribs
3. ¾" Rib Metal Lath	4d common 1 ½" No.12 ½ gauge ¼" head 2" No.11 gauge 7/16" head, barbed	– At Ribs	– At Ribs	At Ribs	At Ribs	16	3/4	1 5/8	At Ribs	At Ribs
4. Wire Fabric Lath	4d blued smooth box (clinched) 1" No.11 gauge 7/16" head, barbed 1 ½" No.11 gauge 7/16" head, barbed 1 ¼" No. 12 ga. 3/8" head, furring 1" No.12 gauge 3/8" head	6 6 6 6 6	– – 6 6	6	6	16	7/16	7/8	6	6
5. 3/8" Gypsum Lath	1 1/8" No.13 gauge 12/61" head, blued	8	8	8	8	16	3/4	7/8	8	8
6. ½" Gypsum Lath	1 ¼" No.13 gauge 12/61" head, blued	8	8 6	8 6	8 6	16	3/4	1 1/8	8	8 6

FINISHES / PLASTER ACCESSORIES | 09205

STRESS RELIEF (CONTROL JOINTS)

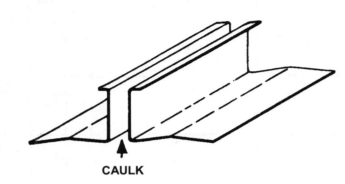
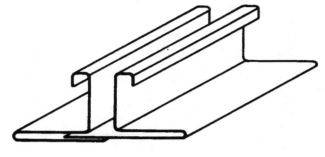

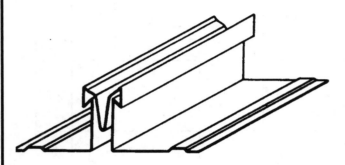

REVEALS

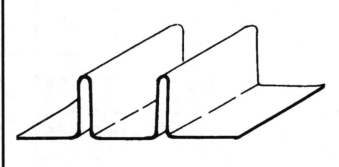
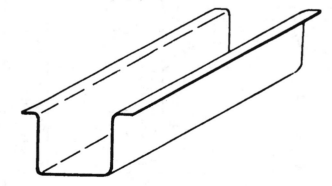

FINISHES / PLASTER ACCESSORIES | 09205

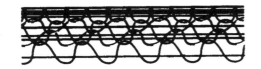
CORNER REINFORCEMENT (EXTERIOR) WIRE

CORNER REINFORCEMENT (EXTERIOR) EXPANDED METAL

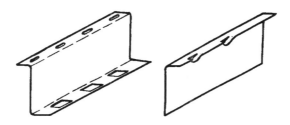
PARTITION RUNNERS (Z AND L SHAPE)

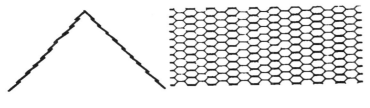
EXPANDED METAL CORNERITE

WIRE CORNERITE

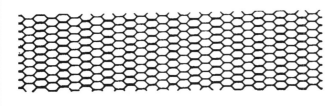
STRIP REINFORCEMENT (EXPANDED METAL)

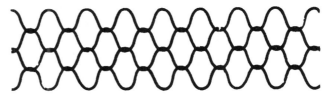
STRIP REINFORCEMENT (WIRE)

FINISHES / PLASTER ACCESSORIES | 09205

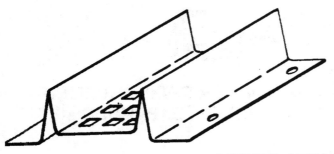 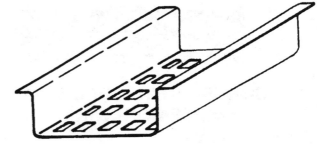

VENTILATING SCREEDS

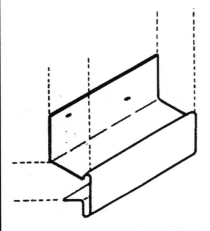 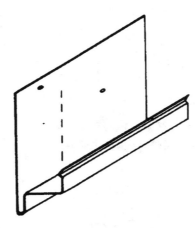 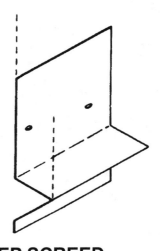

DRIP SCREEDS

WEEP SCREED
(Also Available with Perforations)

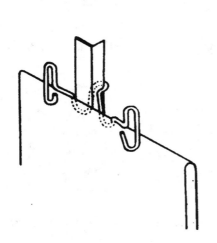 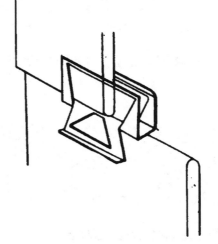

GYPSUM LATH ATTACHMENTS CLIPS

FINISHES / FURRING & LATHING — 09205

VERTICAL FURRING

VERTICAL FURRING MEMBER	UNBRACED STUD SPACING				BRACED STUD SPACING			
	24"	19"	16"	12"	24"	19"	16"	12"
	Maximum Furring Heights				Maximum Vertical Distance Between Braces			
3/4" Channel	6'	7'	8'	9'	5'	5'	6'	7'
1 1/2" Channel	8'	9'	10'	12'	6'	7'	8'	9'
2" Channel	9'	10'	11'	13'	7'	8'	9'	10'
2" Prefab. Stud	8'	9'	10'	11'	6'	7'	8'	9'
2 1/2" Prefab. Stud	10'	11'	12'	14'	8'	9'	10'	11'
3 1/4" Prefab. Stud	14'	16'	17'	20'	11'	13'	14'	16'

TYPES OF LATH—MAXIMUM SPACING OF SUPPORTS

TYPE OF LATH	Minimum Weight (psy), Gauge & Mesh Size	VERTICAL			HORIZONTAL	
		WOOD	METAL		Wood or Concrete	Metal
			Solid Plaster Partitions	Other		
Expanded Metal Lath (Diamond Mesh)	2.5 3.4	16" 16"	16" 16"	12" 16"	12" 16"	12" 16"
Flat rib Expanded Metal Lath	2.75 3.4	16" 19"	16" 24"	16" 19"	16" 19"	16" 19"
Stucco Mesh Expanded Metal Lath	1.8 and 3.6	16"	-	-	-	-
3/8" Rib Expanded Metal Lath	3.4 4.0	24" 24"	-	24" 24"	24" 24"	24" 24"
Sheet Lath	4.5	24"	-	24"	24"	24"
3/4" Rib Expanded Metal Lath (Not manufactured in West)	5.4	-	-	-	36"	36"
Wire Fabric Lath — Welded	1.95 lbs., 11 ga., 2"x2" 1.4 lbs., 16 ga., 2"x2" 1.4 lbs., 18 ga., 1"x1"	24" 16" 16"	24" 16" -	24" 16" -	24" 16" -	24" 16" -
Wire Fabric Lath — Woven	1.4 lbs., 17 ga., 1 1/2" Hex. 1.4 lbs., 18 ga., 1" Hex.	24" 24"	16" 16"	16" 16"	24" 24"	16" 16"
3/8" Gypsum Lath (plain)	-	16"	-	16"	16"	16"
(Large Size)	-	16"	-	16"	16"	16"
1/2" Gypsum Lath (plain)	-	24"	-	24"	24"	24"
(Large Size)	-	24"	No supports; Erected vertically	24"	24"	16"
5/8" Gypsum Lath (Large Size)	-	24"	No supports; Erected vertically	24"	24"	16"

FINISHES / PLASTER — 09210

PLASTERING TABLES

THICKNESS OF PLASTER

PLASTER	FINISHED THICKNESS OF PLASTER FROM FACE OF LATH, MASONRY, CONCRETE	
	Gypsum Plaster	Portland Cement Plaster
Expanded Metal Lath	5/8" minimum	5/8" minimum
Wire Fabric Lath	5/8" minimum	3/4" minimum (interior) 7/8" minimum (exterior)
Gypsum Lath	1/2" minimum	
Gypsum Veneer Base	1/16" minimum	1/2" minimum
Masonry Walls	1/2" minimum	7/8" maximum
Monolithic Concrete Walls	5/8" maximum	1/2" maximum
Monolithic Concrete Ceilings	3/8" maximum	

GYPSUM PLASTER PROPORTIONS

NUMBER OF COATS	COAT	PLASTER BASE OR LATH	MAXIMUM VOLUME AGGREGATE PER 100# NEAT PLASTER (CUBIC FEET)	
			Damp Loose Sand	Perlite or Vermiculite
Two-Coat Work	Basecoat	Gypsum Lath	2 1/2	2 1/2
	Basecoat	Masonry	3	3
Three-Coat Work	First Coat	Lath	2	2
	Second Coat	Lath	3	3
	First & Second Coat	Masonry	3	3

PORTLAND CEMENT PLASTER

COAT	VOLUME CEMENT	MAXIMUM WEIGHT (OR VOLUME) LIME PER VOLUME CEMENT	MAXIMUM VOLUME SAND PER VOLUME CEMENT	APPROXIMATE MINIMUM THICKNESS	MINIMUM PERIOD MOIST CURING	MINIMUM INTERVAL BETWEEN COATS
First	1	20 lbs.	4	3/8"	48 Hours	48 Hours
Second	1	20 lbs.	5	1st & 2nd Coats total 3/4"	48 Hours	7 Days
Finish	1	1	3	1st, 2nd & Finish Coats total 7/8"	-	

PORTLAND CEMENT - LIME PLASTER

COAT	VOLUME CEMENT	MAXIMUM WEIGHT (OR VOLUME) LIME PER VOLUME CEMENT	MAXIMUM VOLUME SAND PER VOLUME CEMENT	APPROXIMATE MINIMUM THICKNESS	MINIMUM PERIOD MOIST CURING	MINIMUM INTERVAL BETWEEN COATS
First	1	1	4	3/8"	48 Hours	48 Hours
Second	1	1	4 1/2	1st & 2nd Coats total 3/4"	48 Hours	7 Days
Finish	1	1	3	1st, 2nd & Finish Coats total 7/8"	-	

FINISHES / VENEER PLASTER 09215

METAL STUD CONSTRUCTION

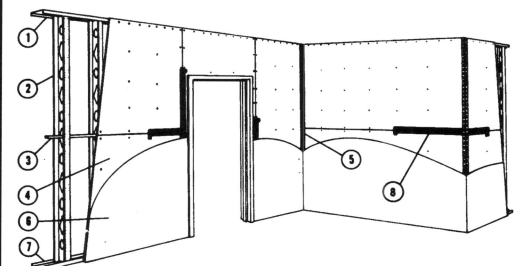

(1) Ceiling Runner Track
(2) Metal Stud (nailable or screw)
(3) Horizontal Stiffener
(4) Large Size Lath
(5) Angle Reinforcement
(6) Veneer Plaster 1/16 to 1/8 inch thick)
(7) Floor Runner Track
(8) Joint Reinforcement

WOOD STUD CONSTRUCTION

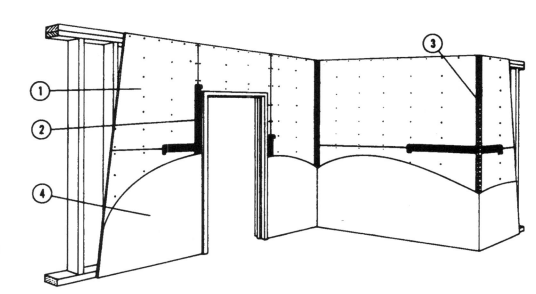

(1) Large Size Lath
(2) Joint Reinforcement
(3) Corner Bead
(4) Veneer Plaster (1/16 to 1/8 inch thick)

FINISHES / PLASTER — 09220

EXTERIOR LATH AND PLASTER

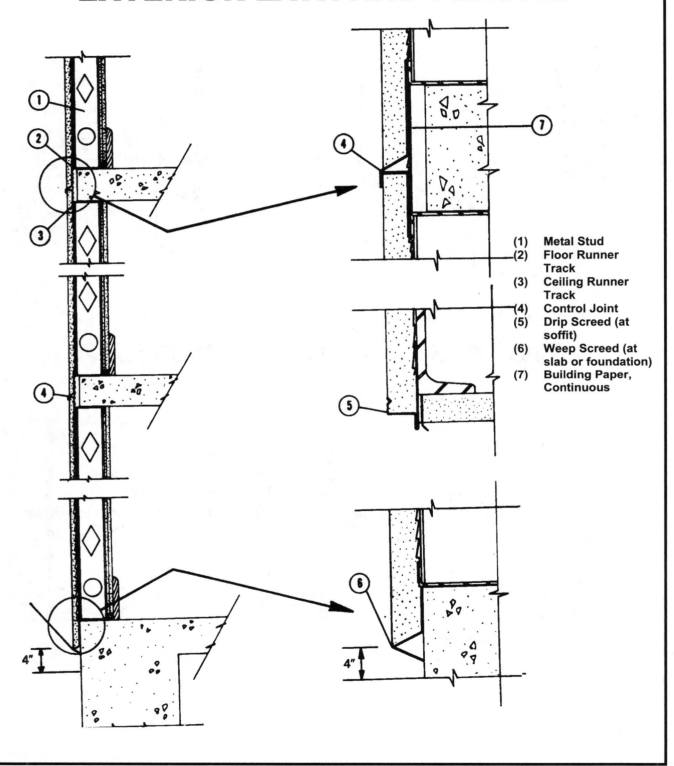

(1) Metal Stud
(2) Floor Runner Track
(3) Ceiling Runner Track
(4) Control Joint
(5) Drip Screed (at soffit)
(6) Weep Screed (at slab or foundation)
(7) Building Paper, Continuous

FINISHES / PLASTER — 09220

EXTERIOR LATH AND PLASTER

OPEN WOOD FRAME CONSTRUCTION

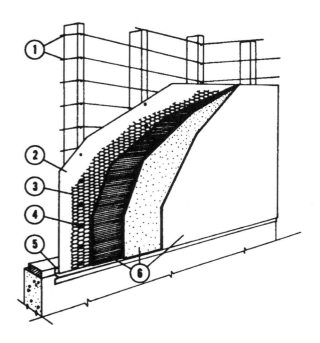

(1) Wire Backing
(2) Building Paper
(3) Wire Fabric Lath
(4) Approved Fasteners
(5) Weep Screed
(6) Three Coats of Plaster (Scratch, Brown, Finish)

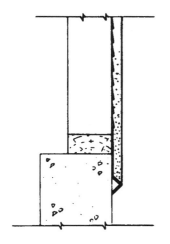

SHEATHED WOOD FRAME CONSTRUCTION

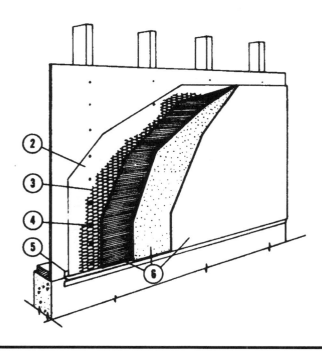

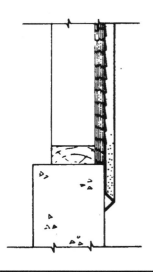

MECHANICAL / PIPE   15060

PIPE WEIGHTS

CAST IRON PIPE

SERVICE WEIGHT

Size, Inches	Weight of Pipe	Weight of Water	Total Weight— Lbs.
2"	3.8	1.45	5.3
3"	5.6	3.2	8.8
4"	7.5	5.5	13.0
5"	9.8	8.7	18.5
6"	12.4	12.5	24.9
8"	18.5	21.7	40.2

EXTRA HEAVY

Size, Inches	Weight of Pipe	Weight of Water	Total Weight — Lbs.
2"	4.3	1.45	5.8
3"	8.3	3.2	11.5
4"	10.8	5.5	16.3
5"	13.3	8.7	22.0
6"	16.0	12.5	28.5
8"	26.5	21.7	48.2

STEEL PIPE

Pipe Size	W/40	H_2O/Lbs.	Total Lbs./L.F.
2"	3.65	1.45	5.1
2 1/2"	5.79	2.07	7.86
3"	7.57	3.2	10.77
3 1/2"	9.11	4.28	13.39
4"	10.8	5.51	16.31
5"	14.6	8.66	23.26
6"	18.0	12.5	30.5
8"	28.6	21.66	50.26
10"	40.5	34.15	74.65
12"			

MECHANICAL / SPRINKLER SYSTEMS | 15300

SPRINKLER AREA CALCULATIONS

Typical maximum floor area allowed per system riser:

Light Hazard	**Ordinary Hazard**	**Extra Hazard**
52,000 S.F.	40,000-52,000 S.F.	25,000 S.F.

Typical maximum floor area coverage allowed per sprinkler head:

Light Hazard	**Ordinary Hazard**	**Extra Hazard**
130-200 S.F.	100-130 S.F.	90 S.F.

Typical maximum spacing between lines and sprinkler heads:

Light Hazard	**Ordinary Hazard**	**Extra Hazard**
12-15 feet	12-15 feet	12 feet

Note: This data is for estimating purposes only. Check all applicable codes and regulations for specific requirements.

MECHANICAL / FIRE PROTECTION — 15300

SPRINKLER HEAD CALCULATIONS
Typical maximum quantity of sprinkler heads allowed by pipe size.

Light Hazard:

For sprinklers below ceiling:

<u>Steel</u>
- 1 in. pipe .... 2 sprinklers
- 1 ¼ in. pipe .... 3 sprinklers
- 1 ½ in. pipe .... 5 sprinklers
- 2 in. pipe ....10 sprinklers
- 2 ½ in. pipe ....30 sprinklers
- 3 in. pipe ....60 sprinklers
- 3 ½ in. pipe ....100 sprinklers

<u>Copper</u>
- 1 in. tube ....2 sprinklers
- 1 ¼ in. tube ....3 sprinklers
- 1 ½ in. tube ....5 sprinklers
- 2 in. tube ....12 sprinklers
- 2 ½ in. tube ....40 sprinklers
- 3 in. tube ....65 sprinklers
- 3 ½ in. tube ....115 sprinklers

For sprinklers above and below ceiling:

<u>Steel</u>
- 1 in. .... 2 sprinklers
- 1 ¼ in. .... 4 sprinklers
- 1 ½ in. .... 7 sprinklers
- 2 in. ....15 sprinklers
- 2 ½ in. ....50 sprinklers

<u>Copper</u>
- 1 in. ....2 sprinklers
- 1 ¼ in. ....4 sprinklers
- 1 ½ in. ....7 sprinklers
- 2 in. ....18 sprinklers
- 2 ½ in. ....65 sprinklers

Ordinary Hazard:

For sprinklers above ceiling:

<u>Steel</u>
- 1 in. pipe .... 2 sprinklers
- 1 ¼ in. pipe .... 3 sprinklers
- 1 ½ in. pipe .... 5 sprinklers
- 2 in. pipe ....10 sprinklers
- 2 ½ in. pipe ....20 sprinklers
- 3 in. pipe ....40 sprinklers
- 3 ½ in. pipe ....65 sprinklers
- 4 in. pipe ....100 sprinklers
- 5 in. pipe ....160 sprinklers
- 6 in. pipe ....275 sprinklers

<u>Copper</u>
- 1 in. tube ....2 sprinklers
- 1 ¼ in. tube ....3 sprinklers
- 1 ½ in. tube ....5 sprinklers
- 2 in. tube ....12 sprinklers
- 2 ½ in. tube ....25 sprinklers
- 3 in. tube ....45 sprinklers
- 3 ½ in. tube ....75 sprinklers
- 4 in. tube ....115 sprinklers
- 5 in. tube ....180 sprinklers
- 6 in. tube ....300 sprinklers

For sprinklers above and below ceiling:

<u>Steel</u>
- 1 in. .... 2 sprinklers
- 1 ¼ in. .... 4 sprinklers
- 1 ½ in. .... 7 sprinklers
- 2 in. ....15 sprinklers
- 2 ½ in. ....30 sprinklers
- 3 in. ....60 sprinklers

<u>Copper</u>
- 1 in. ....2 sprinklers
- 1 ¼ in. ....4 sprinklers
- 1 ½ in. ....7 sprinklers
- 2 in. ....18 sprinklers
- 2 ½ in. ....40 sprinklers
- 3 in. ....65 sprinklers

Extra Hazard:

For sprinklers below ceiling:

<u>Steel</u>
- 1 in. pipe .... 1 sprinkler
- 1 ¼ in. pipe .... 2 sprinklers
- 1 ½ in. pipe .... 5 sprinklers
- 2 in. pipe .... 8 sprinklers
- 2 ½ in. pipe ....15 sprinklers
- 3 in. pipe ....27 sprinklers
- 3 ½ in. pipe ....40 sprinklers
- 4 in. pipe ....55 sprinklers
- 5 in. pipe ....90 sprinklers
- 6 in. pipe ....150 sprinklers

<u>Copper</u>
- 1 in. tube .... 1 sprinkler
- 1 ¼ in. tube ....2 sprinklers
- 1 ½ in. tube ....5 sprinklers
- 2 in. tube ....8 sprinklers
- 2 ½ in. tube ....20 sprinklers
- 3 in. tube ....30 sprinklers
- 3 ½ in. tube ....45 sprinklers
- 4 in. tube ....65 sprinklers
- 5 in. tube ....100 sprinklers
- 6 in. tube ....170 sprinklers

Note: This data is for estimating purposes only. Check all applicable codes and regulations for specific requirements.

MECHANICAL / FIRE PROTECTION — 15300

SPRINKLER HAZARD OCCUPANCIES

Typical Light Hazard Occupancies:

Churches
Clubs
Eaves and overhangs, if combustible construction with no combustible beneath
Educational
Hospitals
Institutional
Libraries, except large stack rooms
Museums
Nursing or convalescent homes
Office, including data processing
Residential
Restaurant seating areas
Theaters seating areas
Theaters and auditoriums excluding stages and prosceniums
Unused attics

Typical Ordinary Hazard Occupancies (Group 1):

Automobile parking garages
Bakeries
Beverage manufacturing
Canneries
Dairy products manufacturing and processing
Electronic plants
Glass and glass products manufacturing
Laundries
Restaurant service areas

Typical Ordinary Hazard Occupancies (Group 2):

Cereal mills
Chemical plants - ordinary
Cold Storage warehouses
Confectionery products
Distilleries
Leather goods mfg.
Libraries-large stack room areas
Mercantiles
Machine shops
Metal working
Printing and publishing
Textile mfg.
Tobacco Products mfg.
Wood product assembly

Typical Ordinary Hazard Occupancies (Group 3):

Feed mills
Paper and pulp mills
Paper process plants
Piers and wharves
Repair garages
Tire manufacturing
Warehouses (having moderate to higher combustibility of content, such as paper, household furniture, paint, general storage, whiskey, etc.)[1]
Wood machining

Typical Extra Hazard Occupancies (Group 1):

Combustible hydraulic fluid use areas
Die casting
Metal extruding
Plywood and particle board manufacturing
Printing (using inks with below 100°F [37.8°C] flash points)
Rubber reclaiming, compounding, drying, milling, vulcanizing
Saw mills
Textile picking, opening, blending, garnetting, carding, combining of cotton, synthetics, wool shoddy or burlap
Upholstering with plastic foams

Typical Extra Hazard Occupancies (Group 2):

Asphalt saturating
Flammable liquids spraying
Flow coating
Mobile Home or Modular Building assemblies (where finished enclosure is present and has combustible interiors)
Open Oil quenching
Solvent cleaning
Varnish and paint dipping

MECHANICAL / PLUMBING — 15400

TYPICAL MINIMUM SIZE OF HORIZONTAL BUILDING STORM DRAINS AND BUILDING STORM SEWERS

Diameter of drain	Maximum projected area in square feet for various slopes		
	1/8 inch per feet slope	1/4 inch per feet slope	1/2 inch per feet slope
3	822	1160	1644
4	1880	2650	3760
5	3340	4720	6680
6	5350	7550	10700
8	11500	16300	23000
10	20700	29200	41400
12	33300	47000	66600
15	59500	84000	119000

MECHANICAL / PLUMBING 15400

PLUMBING BUDGETS

Budget estimates for plumbing can be determined as a percentage of total building costs depending on building type. For example:

- Apartments 9 to 12 percent
- Assembly 4 to 7 percent
- Banks 3 to 6 percent
- Dormitories 7 to 10 percent
- Factories 4 to 8 percent
- Hospitals 8 to 12 percent
- Motels 9 to 12 percent
- Office Buildings 4 to 7 percent
- Retail (small) 4 to 7 percent
- Retail (large) 3 to 6 percent
- Schools 3 to 6 percent
- Warehouses 3 to 7 percent

MECHANICAL / PLUMBING VENTS 15410

TYPICAL SIZE & LENGTH OF PLUMBING VENTS

Diameter of soil or waste stack (in.)	Total fixture units connected to stack (dfu)	DIAMETER OF VENT PIPE										
		1-1/4	1	2	2-1/2	3	4	5	6	8	10	12
1-1/4	2	30										
1-1/2	8	50	150									
1-1/2	10	30	100									
2	12	30	75	200								
2	20	26	50	150								
2-1/2	42		30	100	300							
3	10		42	150	360	1040						
3	21		32	110	270	810						
3	53		27	94	230	680						
3	102		25	86	210	620						
4	43			35	85	250	980					
4	140			27	65	200	750					
4	320			23	55	170	640					
4	540			21	50	150	580					
5	190				28	82	320	990				
5	490				21	63	250	760				
5	940				18	49	210	670				
5	1400				16	33	190	590				
6	500					26	130	400	1000			
6	1100					22	100	310	780			
6	2000					20	84	260	660			
6	2900						77	240	600			
8	1800						31	95	240	940		
8	3400						24	73	190	720		
8	5600						20	62	160	610		
8	7600						18	56	140	560		

MECHANICAL / PLUMBING PIPING — 15410

TYPICAL SIZES OF FIXTURE WATER SUPPLY PIPES

Fixture	Nominal pipe size (inches)
Bath tubs	1/2
Combination sink and tray	1/2
Drinking fountain	3/8
Dishwasher (domestic)	1/2
Kitchen sink, residential	1/2
Kitchen sink, commercial	3/4
Lavatory	3/8
Laundry tray, 1, 2 or 3 compartments	1/2
Shower (single head)	1/2
Sinks (service, slop)	1/2
Sinks flushing rim	3/4
Urinal (flash tank)	3/8
Urinal (direct flush valve)	1
Water closet (tank type)	3/8
Water closet (flush valve type)	1
Hose bibs	1/2
Wall hydrant	1/2

MECHANICAL / HVAC — 15500

TYPICAL VENTILATION AIR REQUIREMENTS FOR SPECIAL USES

Occupancy Classification	Required ventilation air in cfm per human occupant
Special areas	
Lockers	2 * (or 30 per locker)
Wardrobes	2 *
Public bathrooms	40 **
Private bathrooms	25 **
Swimming pools	15 (per occupant)
Water closet (flush valve type)	1
Exitways and corridors	1 1/2 *

*Per square foot floor area.
**Per water closet or urinal.

TYPICAL VENTILATION AIR REQUIREMENTS FOR RETAIL USES

Occupancy Classification	Required ventilation air in cfm per human occupant
Mercantile	
Sales floors and showrooms (basement & grade floors)	7
Sales and showrooms (upper floors)	7
Storage areas	5
Dressing rooms	7
Malls	7
Shipping areas	15
Elevators	7
Supermarkets	
Meat processing rooms	5
Drugs stores	
Pharmacists' work rooms	20
Specialty shops	
Pet shops	1.0*
Florists	5
Greenhouses	5

*cfm per sq. ft. floor area.

MECHANICAL / HVAC   15500

TYPICAL VENTILATION AIR REQUIREMENTS FOR RESIDENTIAL USES

Occupancy Classification	Required ventilation air in cfm per human occupant
Residential	
General living areas	5
Bedrooms	5
Kitchens	20
Basements, utility rooms	5
Mobile homes	5
Hotels, motels	
Bedrooms (single, double)	7
Living rooms (suites)	10
Corridors	5
Lobbies	7
Conference rooms (small)	20
Assembly rooms (large)	15

TYPICAL VENTILATION AIR REQUIREMENTS FOR STORAGE USES

Occupancy Classification	Required ventilation air in cfm per human occupant
Storage	
Garages, service stations, parking garages (enclosed)	1.5*
Auto repair shops	1.5**
Warehouses	
General	7

*cfm/s.f. floor area
**Must have positive engine exhaust system.

MECHANICAL / HVAC — 15500

TYPICAL VENTILATION AIR REQUIREMENTS FOR FACTORY AND INDUSTRIAL USES

Occupany Classification	Required ventilation air in cfm per human occupant
Factory and industrial	
Metalworking & finishing	35
Automotive engine test	*Require Special Exhaust Systems*
Paint spray booths	
Picking, etching & plating lines	
Degreasing booths	
Sandblasting booths	
Chemicals and pharmaceuticals	
Dusty operations	30
Rooms containing potential gas emitters	20
Drying oven rooms	15
Fermentation rooms	15
Pillmaking booths	10
Packaging areas	10
Utility rooms	7
Computer rooms	7
Textiles-clothes manufacturer	15
Electronics & aerospace circuit board & soldering rooms	20
Wood products, papermaking	20
Brewing, distilling, wineries, bottling	20*
Food processing	20
Tobacco processing	20
Power plants	
Control rooms	10
Boiler rooms	35
Generator rooms	20
Sewage treatment plants	
Control rooms	10
Compressor/blower motor rooms	20
Glass & ceramic manufacturer	20
Agricultural	20

MECHANICAL / HVAC   15500

TYPICAL VENTILATION AIR REQUIREMENTS FOR BUSINESS USES

Occupany Classification	Required ventilation air in cfm per human occupant
Business	
Banks	
(see offices)	
Vaults	5
Barber, beauty and health services	
Beauty shops (hair dressers)	25
Reducing salons	25
Sauna baths, steam rooms	5
Barber shops	7
Photo studios	
Camera rooms, stages	5
Dark rooms	
Shoe repair shops	
Workrooms/trade areas	10
Offices	
General office space and showrooms	15
Conference rooms	25
Drafting/art rooms	7
Doctor's consultation rooms	10
Waiting rooms	10
Lithographing rooms	7
Diazo printing rooms	7
Computer rooms	5
Keypunch rooms	7
Communication	
TV/radio broadcasting booths, studios	30
Motion picture and TV stages	30
Pressrooms	15
Composing rooms	7
Engraving rooms	7
Telephone switchboard rooms (manual)	7
Telephone switchgear rooms (automatic)	7
Teletypewriter/facsimile rooms	5
Research institutes	
Laboratories:	
Light duty; non-chemical	15
Chemical	15
Heavy-duty	15
Radioisotope, chemical & biologically toxic	15
Machine shops	15
Dark rooms, spectroscopy rooms	10
Animal rooms	40
Veterinary hospitals	
Kennels, stalls	25
Operating rooms	25
Reception rooms	10

MECHANICAL / HVAC | 15500

TYPICAL VENTILATION AIR REQUIREMENTS FOR INSTITUTIONAL USES

Occupancy Classification	Required ventilation air in cfm per human occupant
Institutional	
Prisons	
Cell blocks	7
Eating halls	15
Guard stations	7

MECHANICAL / DUCTWORK — 15890

AIR CONDITIONING
RECOMMENDED SHEET METAL GAUGES AND CONSTRUCTION FOR RECTANGULAR DUCT

LOW PRESSURE — LOW VELOCITY = 2" W.G. MAX

PLATE NO.	DIMENSION OF LONGEST SIDE OF DUCT	Steel Metal Gauges - Steel	Steel Metal Gauges - Aluminum	Plain "S" Slip (B) / Pocket Lock (K) / Drive Slip (A)	Hemmed "S" Slip (C) / Bar Slip (E) / Seam (1)	Reinforced Bar Slip (G)	Angle Slip (H) / Alternate Bar Slip (F) / Angle RFD Pocket (L)	Companion Angles (M) / Angle Reinforced Standing Seam (J)	Reinforcing Between Joints
6	Thru 12"	26	24 (.020)	A-B-K	—	—	—	—	
6	13" thru 18"	24	22 (.025)	A-B-K	—	—	—	—	
7 / 7A	19" thru 30"	24	22 (.025)	K @ 5' cc / A	C-E- @ 5' cc / C-E- @ 10' cc	—	—	—	1" x 1" x 1/8" @ 5' cc
8	31" thru 42"	22	20 (.032)	K @ 5' cc	E-G-K @ 5' cc / E-G-K @ 10' cc	—	—	—	1" x 1" x 1/8" @ 5' cc
9	43" thru 54"	22	20 (.032)	K @ 4' cc / K @ 8' cc	E-@ 4' cc / E-@ 8' cc	G- @ 4' cc / G- @ 8' cc	—	—	1½" x 1½" x 1/8" @ 4' cc
9	55" thru 60"	20	18 (.040)	K @ 4' cc / K @ 8' cc	E-@ 4' cc / E-@ 8' cc	G- @ 4' cc / G- @ 8' cc	—	—	1½" x 1½" x 1/8" @ 4' cc
10	61" thru 84"	20	18 (.040)	—	—	G- @ 4' cc / G- @ 5' cc	H- @ 4' cc / F- @ 4' cc / L- @ 4' cc / H- @ 5' cc / F- @ 5' cc / L- @ 5' cc	J- @ 2' cc	1½" x 1½" x 1/8" @ 2' cc / 1½" x 1½" x 1/8" @ 2'-6" cc
11	85" thru 96"	18	16 (.051)	—	—	—	H- @ 4' cc / L- @ 4' cc / H- @ 5' cc / L- @ 5' cc	M- @ 4' cc / M- @ 5' cc / J- @ 2' cc	1½" x 1½" x 3/16" @ 2' cc / 1½" x 1½" x 3/16" @ 2'-6" cc / 1½" x 1½" x 3/16" @ 2' cc
12	Over 96"	18	16 (.051)	—	—	—	H- @ 4' cc / L- @ 4' cc / H- @ 5' cc / L- @ 5' cc	M- @ 4' cc / M- @ 5' cc / J- @ 2' cc	2" x 2" x ¼ @ 2' cc / 2" x 2" x ¼ @ 2'-6" cc / 2" x 2" x ¼ @ 2' cc

H (height dimension) — up to 42" = 1"
H (height dimension) — 43" to 96" = 1½"
H (height dimension) — over 96" = 2"

MECHANICAL / DUCTWORK

15890

AIR CONDITIONING (Cont.)
LONGITUDINAL SEAMS
FOR SHEET METAL DUCTWORK

Fig. "N"
PITTSBURGH LOCK

Fig. "Z"
BUTTON PUNCH SNAP LOCK

Fig. "O"
ACME LOCK-GROOVED SEAM

Fig. "T"
DOUBLE SEAM

Approximately 2" Spacing
Between "Buttons"

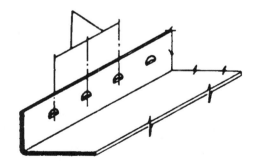

DETAIL NO.1
MALE PIECE-SNAP LOCK

MECHANICAL / DUCTWORK 15890

AIR CONDITIONING
TYPICAL DUCT CONNECTIONS
CROSS JOINTS FOR SHEET METAL DUCTWORK
(NOT TO SCALE)

H - HEIGHT REFERRED TO IN DIMESIONS

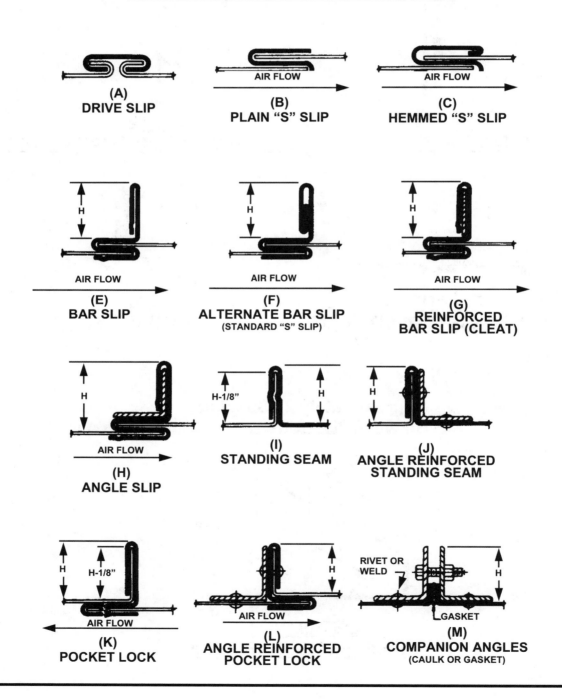

MECHANICAL / DUCTWORK 15890

TYPICAL DUCT CONSTRUCTION SHEET METAL GAGES IN ONE– AND TWO– FAMILY DWELLINGS

Metal Gauges (duct not enclosed in partitions)		
ROUND DUCTS		
Diameter, Inches	Minimum thickness galvanized sheet gage	Minimum thickness aluminum B&S gage
Less than 12	30	26
12-14	28	26
15-18	26	24
Over 18	24	22
RECTANGULAR DUCTS		
Width, Inches	Minimum thickness galvanized sheet gage	Minimum thickness aluminum B&S gage
Less than 14	28	24
14-24	26	22
25-30	24	22
Over 30	22	20
Metal Gauges (duct enclosed in partitions)		
Width, Inches	Minimum thickness galvanized sheet gage	Minimum thickness aluminum B&S gage
14 or less	30	26
Over 14	28	24

TYPICAL DUCT CONSTRUCTION SHEET METAL GLASS
(All uses except 1– and 2– family dwellings)

RECTANGULAR DUCTS		
Maximum side inches	**Steel min. Galv. Sheet Gage**	**Aluminum Min. B&S Gage**
Through 12	26 (0.022 in.)	24 (0.020 in.)
13 through 30	24 (0.028 in.)	22 (0.025 in.)
31 through 54	22 (0.034 in.)	20 (0.032 in.)
55 through 84	20 (0.040 in.)	18 (0.040 in.)
Over 84	18 (0.052 in.)	16 (0.051 in.)

ROUND DUCTS			
Diameter Inches	Spiral seam duct	Longitudinal seam duct	Fittings
	Steel min. Galv. Sht. Gage	Steel min. Galv. Sht. Gage	Steel min. Galv. Sht. Gage
Through 12	28 (0.019 in.)	26 (0.022 in.)	26 (0.022 in.)
13 through 18	26 (0.022 in.)	24 (0.028 in.)	24 (0.028 in.)
19 through 28	24 (0.028 in.)	22 (0.034 in.)	22 (0.034 in.)
29 through 36	22 (0.034 in.)	20 (0.040 in.)	20 (0.040 in.)
37 through 52	20 (0.040 in.)	18 (0.052 in.)	18 (0.052 in.)

ELECTRICAL / WIRES & CABLES

16120

APPLICATIONS FOR CONDUCTORS USED FOR GENERAL WIRING

	AMBIENT TEMPERATURE								FEATURES
	60°C 140°F	75°C 167°F	85°C 185°F	90°C 194°F	110°C 230°F	200°C 392°F	Dry	Dry or Wet	
R	X						X		Code Rubber
RH		X					X		Heat Resistant
RHH				X			X		More Heat Resistant
RW	X							X	Moisture Resistant
RH-RW	X							X	Moisture and Heat Resistant
		X					X		Moisture and Heat Resistant
RHW		X						X	Moisture and Heat Resistant
RU	X						X		Latex Rubber
RUH		X					X		Heat Resistant
RUW	X							X	Moisture Resistant
T	X						X		Thermoplastic
TW	X							X	Moisture Resistant
THHN				X			X		Heat Resistant
THW		X						X	Moisture and Heat Resistant
THWN		X						X	Moisture and Heat Resistant
MI			X					X	Mineral Insulated Metal Sheathed
V		X					X		Varnished Cambric
AVA					X		X		With Asbestos
AVB			X				X		With Asbestos
AVL					X			X	With Asbestos

This table does not include special condition conductors, thickness of conductor insulation, or reference to all other protective coverings.

GENERAL CLASSIFICATION OF INSULATIONS:

A .... Asbestos
H .... Heat Resistant
MI .... Mineral Insulation
R .... Rubber
RU .... Latex Rubber
V .... Varnished Cambric
T .... Thermoplastic
W .... (Water) Moisture Resistant

ELECTRICAL / WIRES & CABLES 16120

WIRE AND SHEET METAL GAGES
(In Decimals of an Inch)

Name of Gage	American Wire Gage (A.W.G.) (Corresponds to Brown & Sharpe Gage)	Birmingham Iron Wire Gage (B.W.G.)	United States Standard Gage (U.S.S.G.)	
Principal Use	Electrical Wire & Non-Ferrous Sheet Metal	Iron or Steel Wire	Ferrous Sheet Metal	
Gage No.				Gage No.
00 00000				00 00000
0 00000	.5800			0 00000
00000	.5165	.500		00000
0000	.4600	.454		0000
000	.4096	.425		000
00	.3648	.380		00
0	.3249	.340		0
1	.2893	.300		1
2	.2576	.284		2
3	.2294	.259	23.91	3
4	.2048	.238	.2242	4
5	.1819	.220	.2092	5
6	.1620	.203	.1943	6
7	.1443	.180	.1793	7
8	.1285	.165	.1644	8
9	.1144	.148	.1495	9
10	.1019	.134	.1345	10
11	.0907	.120	.1196	11
12	.0808	.109	.1046	12
13	.0720	.095	.0897	13
14	.0641	.083	.0747	14
15	.0571	.072	.0673	15
16	.0508	.065	.0598	16
17	.0453	.058	.0538	17
18	.0403	.049	.0478	18
19	.0359	.042	.0418	19
20	.0320	.035	.0359	20
21	.0285	.032	.0329	21
22	.0253	.028	.0299	22
23	.0226	.025	.0269	23
24	.0201	.022	.0239	24
25	.0179	.020	.0209	25
26	.0159	.018	.0179	26
27	.0142	.016	.0164	27
28	.0126	.014	.0149	28
29	.0113	.013	.0135	29
30	.0100	.012	.0120	30
31	.0089	.010	.0105	31
32	.0080	.009	.0097	32
33	.0071	.008	.0090	33
34	.0063	.007	.0082	34
35	.0056	.005	.0075	35
36	.0050	.004	.0067	36
37	.0045		.0064	37
38	.0040		.0060	38
39	.0035			39
40	.0031			40

BNi *Building News*

Geographic Cost Modifiers

The costs as presented in this book attempt to represent national averages. Costs, however, vary among regions, states and even between adjacent localities.

In order to more closely approximate the probable costs for specific locations throughout the U.S., this table of Geographic Cost Modifiers is provided. These adjustment factors are used to modify costs obtained from this book to help account for regional variations of construction costs and to provide a more accurate estimate for specific areas. The factors are formulated by comparing costs in a specific area to the costs as presented in the Costbook pages. An example of how to use these factors is shown below. Whenever local current costs are known, whether material prices or labor rates, they should be used when more accuracy is required.

Cost Obtained from Costbook Pages **X** Location Cost Adjustment Factor Divided by 100 **= Adjusted Cost**

For example, a project estimated to cost $125,000 using the Costbook pages can be adjusted to more closely approximate the cost in Los Angeles:

$$\$125,000 \times \frac{119}{100} = \$148,750$$

BNi. Building News

GEOGRAPHIC COST MODIFIERS 01025

State	Metropolitan Area	Multiplier
AK	ANCHORAGE	132
AL	ANNISTON	81
	AUBURN	82
	BIRMINGHAM	82
	DECATUR	84
	DOTHAN	83
	FLORENCE	84
	GADSDEN	82
	HUNTSVILLE	84
	MOBILE	86
	MONTGOMERY	81
	OPELIKA	82
	TUSCALOOSA	81
AR	FAYETTEVILLE	79
	FORT SMITH	79
	JONESBORO	78
	LITTLE ROCK	82
	NORTH LITTLE ROCK	82
	PINE BLUFF	80
	ROGERS	79
	SPRINGDALE	79
	TEXARKANA	79
AZ	FLAGSTAFF	94
	MESA	94
	PHOENIX	95
	TUCSON	93
	YUMA	94
CA	BAKERSFIELD	116
	CHICO	118
	FAIRFIELD	120
	FRESNO	118
	LODI	117
	LONG BEACH	119
	LOS ANGELES	119
	MERCED	118
	MODESTO	114
	NAPA	120
	OAKLAND	124
	ORANGE COUNTY	118
	PARADISE	114
	PORTERVILLE	116
	REDDING	114
	RIVERSIDE	116
	SACRAMENTO	118
	SALINAS	120
	SAN BERNARDINO	116
	SAN DIEGO	117
	SAN FRANCISCO	129

GEOGRAPHIC COST MODIFIERS 01025

State	Metropolitan Area	Multiplier
CA	SAN JOSE	126
	SAN LUIS OBISPO	113
	SANTA BARBARA	116
	SANTA CRUZ	120
	SANTA ROSA	121
	STOCKTON	117
	TULARE	118
	VALLEJO	120
	VENTURA	116
	VISALIA	118
	WATSONVILLE	118
	YOLO	118
	YUBA CITY	118
CO	BOULDER	103
	COLORADO SPRINGS	100
	DENVER	101
	FORT COLLINS	110
	GRAND JUNCTION	99
	GREELEY	108
	LONGMONT	103
	LOVELAND	110
	PUEBLO	105
CT	BRIDGEPORT	113
	DANBURY	113
	HARTFORD	112
	MERIDEN	113
	NEW HAVEN	113
	NEW LONDON	110
	NORWALK	117
	NORWICH	110
	STAMFORD	117
	WATERBURY	112
DC	WASHINGTON	105
DE	DOVER	105
	NEWARK	106
	WILMINGTON	106
FL	BOCA RATON	80
	BRADENTON	80
	CAPE CORAL	78
	CLEARWATER	81
	DAYTONA BEACH	75
	FORT LAUDERDALE	83
	FORT MYERS	78
	FORT PIERCE	81
	FORT WALTON BEACH	76
	GAINESVILLE	80
	JACKSONVILLE	78
	LAKELAND	78

GEOGRAPHIC COST MODIFIERS 01025

State	Metropolitan Area	Multiplier
FL	MELBOURNE	75
	MIAMI	81
	NAPLES	79
	OCALA	79
	ORLANDO	77
	PALM BAY	75
	PANAMA CITY	77
	PENSACOLA	76
	PORT ST. LUCIE	81
	PUNTA GORDA	78
	SARASOTA	80
	ST. PETERSBURG	80
	TALLAHASSEE	75
	TAMPA	80
	TITUSVILLE	75
	WEST PALM BEACH	80
	WINTER HAVEN	78
GA	ALBANY	86
	ATHENS	89
	ATLANTA	92
	AUGUSTA	86
	COLUMBUS	79
	MACON	83
	SAVANNAH	87
HI	HONOLULU	138
IA	CEDAR FALLS	91
	CEDAR RAPIDS	102
	DAVENPORT	106
	DES MOINES	104
	DUBUQUE	95
	IOWA CITY	97
	SIOUX CITY	91
	WATERLOO	91
ID	BOISE CITY	102
	POCATELLO	102
IL	BLOOMINGTON	113
	CHAMPAIGN	109
	CHICAGO	125
	DECATUR	107
	KANKAKEE	113
	NORMAL	113
	PEKIN	111
	PEORIA	111
	ROCKFORD	113
	SPRINGFIELD	108
	URBANA	109
IN	BLOOMINGTON	102
	ELKHART	96

GEOGRAPHIC COST MODIFIERS 01025

State	Metropolitan Area	Multiplier
IN	EVANSVILLE	99
	FORT WAYNE	100
	GARY	107
	GOSHEN	96
	INDIANAPOLIS	103
	KOKOMO	101
	LAFAYETTE	101
	MUNCIE	101
	SOUTH BEND	102
	TERRE HAUTE	100
KS	KANSAS CITY	120
	LAWRENCE	109
	TOPEKA	96
	WICHITA	87
KY	LEXINGTON	91
	LOUISVILLE	102
	OWENSBORO	101
LA	ALEXANDRIA	89
	BATON ROUGE	93
	BOSSIER CITY	90
	HOUMA	93
	LAFAYETTE	91
	LAKE CHARLES	93
	MONROE	89
	NEW ORLEANS	95
	SHREVEPORT	90
MA	BARNSTABLE	124
	BOSTON	128
	BROCKTON	118
	FITCHBURG	120
	LAWRENCE	121
	LEOMINSTER	120
	LOWELL	124
	NEW BEDFORD	118
	PITTSFIELD	118
	SPRINGFIELD	119
	WORCESTER	120
	YARMOUTH	124
MD	BALTIMORE	95
	CUMBERLAND	98
	HAGERSTOWN	90
ME	AUBURN	87
	BANGOR	87
	LEWISTON	87
	PORTLAND	88
MI	ANN ARBOR	119
	BATTLE CREEK	111
	BAY CITY	116

GEOGRAPHIC COST MODIFIERS — 01025

State	Metropolitan Area	Multiplier
MI	BENTON HARBOR	111
	DETROIT	120
	EAST LANSING	117
	FLINT	116
	GRAND RAPIDS	112
	HOLLAND	112
	JACKSON	107
	KALAMAZOO	111
	LANSING	117
	MIDLAND	115
	MUSKEGON	112
	SAGINAW	116
MN	DULUTH	107
	MINNEAPOLIS	112
	ROCHESTER	107
	ST. CLOUD	105
	ST. PAUL	112
MO	COLUMBIA	114
	JOPLIN	103
	KANSAS CITY	118
	SPRINGFIELD	96
	ST. JOSEPH	117
	ST. LOUIS	115
MS	BILOXI	79
	GULFPORT	79
	HATTIESBURG	79
	JACKSON	79
	PASCAGOULA	79
MT	BILLINGS	96
	GREAT FALLS	90
	MISSOULA	91
NC	ASHEVILLE	73
	CHAPEL HILL	79
	CHARLOTTE	82
	DURHAM	81
	FAYETTEVILLE	75
	GOLDSBORO	80
	GREENSBORO	81
	GREENVILLE	79
	HICKORY	72
	HIGH POINT	81
	JACKSONVILLE	72
	LENOIR	72
	MORGANTON	72
	RALEIGH	80
	ROCKY MOUNT	72
	WILMINGTON	72
	WINSTON SALEM	77

GEOGRAPHIC COST MODIFIERS — 01025

State	Metropolitan Area	Multiplier
ND	BISMARCK	84
	FARGO	98
	GRAND FORKS	81
NE	LINCOLN	84
	OMAHA	91
NH	MANCHESTER	106
	NASHUA	106
	PORTSMOUTH	111
NJ	ATLANTIC CITY	126
	BERGEN	129
	BRIDGETON	125
	CAPE MAY	125
	HUNTERDON	128
	JERSEY CITY	130
	MIDDLESEX	129
	MILLVILLE	125
	MONMOUTH	129
	NEWARK	129
	OCEAN	130
	PASSAIC	130
	SOMERSET	128
	TRENTON	128
	VINELAND	125
NM	ALBUQUERQUE	91
	LAS CRUCES	91
	SANTA FE	91
NV	LAS VEGAS	109
	RENO	97
NY	ALBANY	119
	BINGHAMTON	116
	BUFFALO	118
	DUTCHESS COUNTY	119
	ELMIRA	118
	GLENS FALLS	120
	JAMESTOWN	112
	NASSAU	137
	NEW YORK	148
	NEWBURGH	119
	NIAGARA FALLS	121
	ROCHESTER	118
	ROME	109
	SCHENECTADY	119
	SUFFOLK	137
	SYRACUSE	118
	TROY	119
	UTICA	109

GEOGRAPHIC COST MODIFIERS 01025

State	Metropolitan Area	Multiplier
OH	AKRON	112
	CANTON	107
	CINCINNATI	105
	CLEVELAND	114
	COLUMBUS	115
	DAYTON	115
	ELYRIA	114
	HAMILTON	105
	LIMA	115
	LORAIN	114
	MANSFIELD	115
	MASSILLON	107
	MIDDLETOWN	115
	SPRINGFIELD	109
	STEUBENVILLE	115
	TOLEDO	109
	WARREN	111
	YOUNGSTOWN	111
OK	ENID	86
	LAWTON	86
	OKLAHOMA CITY	85
	TULSA	80
OR	ASHLAND	109
	CORVALLIS	112
	EUGENE	112
	MEDFORD	109
	PORTLAND	114
	SALEM	112
	SPRINGFIELD	112
PA	ALLENTOWN	118
	ALTOONA	110
	BETHLEHEM	118
	CARLISLE	113
	EASTON	118
	ERIE	112
	HARRISBURG	113
	HAZLETON	118
	JOHNSTOWN	104
	LANCASTER	93
	LEBANON	115
	PHILADELPHIA	134
	PITTSBURGH	116
	READING	119
	SCRANTON	116
	SHARON	112
	STATE COLLEGE	98
	WILKES BARRE	116
	WILLIAMSPORT	97

GEOGRAPHIC COST MODIFIERS 01025

State	Metropolitan Area	Multiplier
PA	YORK	113
PR	MAYAGUEZ	73
	PONCE	74
	SAN JUAN	75
RI	PROVIDENCE	122
SC	AIKEN	89
	ANDERSON	71
	CHARLESTON	76
	COLUMBIA	76
	FLORENCE	73
	GREENVILLE	76
	MYRTLE BEACH	73
	NORTH CHARLESTON	81
	SPARTANBURG	73
	SUMTER	76
SD	RAPID CITY	81
	SIOUX FALLS	85
TN	CHATTANOOGA	84
	CLARKSVILLE	83
	JACKSON	83
	JOHNSON CITY	83
	KNOXVILLE	80
	MEMPHIS	84
	NASHVILLE	83
TX	ABILENE	88
	AMARILLO	92
	ARLINGTON	87
	AUSTIN	89
	BEAUMONT	88
	BRAZORIA	88
	BROWNSVILLE	73
	BRYAN	86
	COLLEGE STATION	86
	CORPUS CHRISTI	84
	DALLAS	89
	DENISON	87
	EDINBURG	73
	EL PASO	81
	FORT WORTH	87
	GALVESTON	93
	HARLINGEN	73
	HOUSTON	88
	KILLEEN	77
	LAREDO	78
	LONGVIEW	78
	LUBBOCK	91
	MARSHALL	87
	MCALLEN	73

GEOGRAPHIC COST MODIFIERS

01025

State	Metropolitan Area	Multiplier
TX	MIDLAND	87
	MISSION	73
	ODESSA	87
	PORT ARTHUR	88
	SAN ANGELO	87
	SAN ANTONIO	90
	SAN BENITO	73
	SAN MARCOS	89
	SHERMAN	87
	TEMPLE	77
	TEXARKANA	79
	TEXAS CITY	93
	TYLER	84
	VICTORIA	74
	WACO	77
	WICHITA FALLS	87
UT	OGDEN	95
	OREM	93
	PROVO	93
	SALT LAKE CITY	92
VA	CHARLOTTESVILLE	86
	LYNCHBURG	83
	NEWPORT NEWS	88
	NORFOLK	91
	PETERSBURG	78
	RICHMOND	90
	ROANOKE	76
	VIRGINIA BEACH	91
VT	BURLINGTON	97
WA	BELLEVUE	119
	BELLINGHAM	111
	BREMERTON	113
	EVERETT	117
	KENNEWICK	101
	OLYMPIA	113
	PASCO	100
	RICHLAND	101
	SEATTLE	119
	SPOKANE	98
	TACOMA	116
	YAKIMA	104

GEOGRAPHIC COST MODIFIERS 01025

State	Metropolitan Area	Multiplier
WI	APPLETON	113
	BELOIT	117
	EAU CLAIRE	113
	GREEN BAY	112
	JANESVILLE	117
	KENOSHA	118
	LA CROSSE	114
	MADISON	116
	MILWAUKEE	118
	NEENAH	113
	OSHKOSH	113
	RACINE	118
	SHEBOYGAN	112
	WAUKESHA	118
	WAUSAU	113
WV	CHARLESTON	113
	HUNTINGTON	113
	PARKERSBURG	113
	WHEELING	113
WY	CASPER	85
	CHEYENNE	85

Square Foot Tables

The following Square Foot Tables list hundreds of actual projects for dozens of building types, each with associated building size, total square foot building cost and percentage of project costs for total mechanical and electrical components. This data provides an overview of construction costs by building type. These costs are for actual projects. The variations within similar building types may be due, among other factors, to size, location, quality and specified components, materials and processes. Depending upon all such factors, specific building costs can vary significantly and may not necessarily fall within the range of costs as presented. The data has been updated to reflect current construction costs.

BUILDING CATEGORY	PAGE
Commercial	774
Residential	775
Educational	776
Hotel/Motel	777
Industrial	778
Medical	778
Public Facilities	779
Offices	780
Recreational	782
Religious	783

SQUARE FOOT TABLES

COMMERCIAL

AUTO DEALERSHIP

Project Size Gross S.F.	Project Cost $/S.F.	% Cost Mechanical	% Cost Electrical
7,700	158.50	16.7	9.4
16,100	90.00	10.2	15.9
20,000	94.60	12.9	23.4
26,300	96.20	12.5	22.0
43,600	76.40	19.4	13.2
53,600	127.10	12.5	11.5

BUSINESS CENTER

Project Size Gross S.F.	Project Cost $/S.F.	% Cost Mechanical	% Cost Electrical
3,900	95.00	12.0	9.2
9,900	91.80	9.1	7.6
54,400	61.40	3.4	12.2
135,000	66.70	8.2	1.5

CINEMA

Project Size Gross S.F.	Project Cost $/S.F.	% Cost Mechanical	% Cost Electrical
18,000	217.10	10.9	6.7
22,500 A	133.60	6.6	4.2

MALL/PLAZA

Project Size Gross S.F.	Project Cost $/S.F.	% Cost Mechanical	% Cost Electrical
9,700	67.10	15.0	13.3
10,500	107.40	8.0	13.5
16,300	97.90	9.4	10.6
26,900	97.50	15.0	7.0
36,000	82.40	10.0	11.0
36,300	93.70	12.4	8.2
44,720	97.90	18.3	12.4
59,100	95.80	9.8	9.5
60,000 R	98.30	10.0	9.5
64,100	98.80	22.4	18.4
66,000	113.20	13.5	11.5
67,400	82.20	21.0	15.0
73,500	118.90	14.9	6.9

MALL/PLAZA (Cont.)

Project Size Gross S.F.	Project Cost $/S.F.	% Cost Mechanical	% Cost Electrical
142,000	74.70	7.1	8.0
220,000	197.40	11.0	6.4
223,700	58.10	9.0	9.3
321,200	69.10	7.3	7.4
379,900	90.10	11.2	6.2
405,100	92.80	13.9	6.0
482,000	154.70	10.5	9.4
630,000	103.30	12.2	12.4

RESTAURANT

Project Size Gross S.F.	Project Cost $/S.F.	% Cost Mechanical	% Cost Electrical
4,300 R	170.80	6.9	8.7
4,400 R	251.20	14.6	8.5
5,800	204.20	28.0	10.6
6,800 A	238.70	7.0	11.1
7,360 R	257.20	16.0	6.5
9,600	255.20	24.7	13.1
10,000 R	265.50	21.0	10.0
10,100	229.10	28.6	18.4
10,600	421.90	20.4	6.4
22,900 R	274.20	15.8	16.9

RETAIL STORE

Project Size Gross S.F.	Project Cost $/S.F.	% Cost Mechanical	% Cost Electrical
1,000	257.20	12.8	6.7
3,000 R	230.00	14.3	10.5
12,300	315.40	14.0	10.0
30,000	159.40	15.6	26.2
61,300	82.40	13.3	13.0
115,000	114.00	14.6	11.3
154,700	161.60	11.2	12.4
314,700 R	136.30	13.8	9.4

A = Addition   R = Remodel

SQUARE FOOT TABLES

RESIDENTIAL

APARTMENTS

Project Size Gross S.F.	Project Cost $/S.F.	% Cost Mechanical	% Cost Electrical
3,700	95.50	14.9	4.4
13,900	130.70	7.9	7.2
19,200 R	207.80	45.3	7.5
19,700	121.80	7.4	10.4
23,700	133.70	10.6	4.3
26,500	126.60	25.2	12.8
35,100	103.40	16.4	5.6
54,000	179.10	23.3	13.1
62,700	130.20	17.0	9.0
67,300	118.90	13.8	8.4
70,600	66.90	18.1	7.8
75,300	147.70	13.2	8.1
75,600	143.80	14.5	8.9
72,200	124.50	18.4	10.9
77,600	161.20	26.9	14.3
88,100	149.80	15.3	9.3
89,500	140.30	10.7	11.1
94,100	84.60	8.5	6.8
96,000	105.70	17.0	13.3
102,000	183.20	17.8	12.5
103,200	83.30	12.1	8.9
103,600	134.10	19.1	9.0
105,200	170.80	14.9	8.9
106,200	122.90	12.8	9.3
110,900	118.20	15.6	8.4
111,800	167.70	17.3	7.4
115,900	125.80	12.5	8.6
117,200	78.60	12.9	8.0
119,000	107.90	15.6	8.4

APARTMENTS (Cont.)

Project Size Gross S.F.	Project Cost $/S.F.	% Cost Mechanical	% Cost Electrical
119,400	70.70	18.6	8.7
144,300	157.20	15.0	8.6
176,300	133.70	19.1	9.0
192,300	78.60	10.1	6.0
210,900	154.80	19.1	9.5
220,200	159.30	14.8	7.5
253,900	206.80	20.6	7.7
369,500	151.20	15.8	9.0

CONDOS/TOWNHOUSE

Project Size Gross S.F.	Project Cost $/S.F.	% Cost Mechanical	% Cost Electrical
8,600	135.70	8.5	4.4
16,700	109.40	13.2	6.5
18,000	230.00	14.6	9.7
18,400	118.10	12.9	5.8
74,800	107.00	13.8	5.2
111,700	135.00	9.2	7.1
150,300	150.90	15.9	7.9
278,800	235.80	14.1	7.9
1,109,900	108.50	9.8	4.7

SINGLE-FAMILY HOMES

Project Size Gross S.F.	Project Cost $/S.F.	% Cost Mechanical	% Cost Electrical
600 R	97.70	16.3	2.0
900	126.90	33.0	3.0
2,100	146.90	8.8	4.0
2,200	314.70	23.8	3.4
2,500	154.40	6.1	7.1
2,900	148.80	7.7	2.8
3,000	104.40	9.5	6.8
3,100	233.60	15.5	4.6
3,600	146.40	8.5	3.0
3,700	204.60	9.8	3.4

A = Addition   R = Remodel

SQUARE FOOT TABLES

RESIDENTIAL (Cont.)

SINGLE-FAMILY HOMES (Cont.)

Project Size Gross S.F.	Project Cost $/S.F.	% Cost Mechanical	% Cost Electrical
4,200	177.90	15.5	5.7
4,600	282.40	9.0	4.7
5,200	367.10	8.3	6.0
5,700	215.00	7.4	3.7
5,700	146.30	8.7	12.6
21,300*	104.20	26.0	4.0
22,700*	96.90	27.0	4.6
45,000*	147.40	7.8	2.5
51,458*	106.80	11.0	5.0

*TOWNHOUSES

EDUCATIONAL

ADMINISTRATION (OFFICES)

Project Size Gross S.F.	Project Cost $/S.F.	% Cost Mechanical	% Cost Electrical
53,700	251.20	15.4	9.0

ATHLETIC FACILITY

Project Size Gross S.F.	Project Cost $/S.F.	% Cost Mechanical	% Cost Electrical
38,100	249.20	16.8	9.8
44,100	230.00	16.8	8.4
100,000	157.20	19.2	5.3
160,000	284.80	13.7	7.9
247,500	238.90	11.2	9.0
271,000	224.60	13.2	7.6
283,100	270.40	14.6	6.0

AUDITORIUM/PERFORMING ARTS

Project Size Gross S.F.	Project Cost $/S.F.	% Cost Mechanical	% Cost Electrical
9,900	402.70	17.5	29.2
17,800	368.30	11.4	16.1
29,200	364.50	16.2	10.5
62,700	265.00	13.0	12.9

CLASSROOM

Project Size Gross S.F.	Project Cost $/S.F.	% Cost Mechanical	% Cost Electrical
35,400	365.40	10.2	6.8
70,000	165.00	24.3	13.1
78,900	306.90	13.3	14.3
80,100	249.20	16.9	10.5
100,000	248.40	21.1	9.8
166,000	171.60	16.1	11.2
298,400	161.00	11.3	11.9

COMPLETE COLLEGE FACILITIES

95,300	258.90	19.4	11.7
450,000	308.90	18.6	10.8

ELEMENTARY SCHOOL

18,000	189.40	19.1	13.8
30,800	174.30	16.0	10.0
31,600	145.70	12.1	11.3
35,700	213.10	17.3	8.7
40,000	185.10	18.5	13.4
40,500	162.30	22.1	11.3
57,000	140.20	22.1	7.5
69,700	218.60	20.6	9.3
91,400	176.30	18.2	7.6

HIGH SCHOOL

116,400	193.50	18.1	12.9
133,000	166.70	17.8	10.7
184,000	293.60	23.0	10.0
217,200 R	158.20	26.4	10.6
254,000 R	122.10	17.2	14.6
431,700	186.00	13.1	9.6

A = Addition   R = Remodel

SQUARE FOOT TABLES

EDUCATIONAL (Cont.)

JUNIOR HIGH SCHOOL

Project Size Gross S.F.	Project Cost $/S.F.	% Cost Mechanical	% Cost Electrical
26,000	243.70	9.5	9.3
28,100	151.20	11.9	9.5
52,800	195.60	18.3	8.6
91,600	240.50	21.2	11.0
123,700	196.80	29.1	9.1

LABORATORY/RESEARCH

Project Size Gross S.F.	Project Cost $/S.F.	% Cost Mechanical	% Cost Electrical
9,200	421.90	25.4	4.5
80,300	331.80	20.2	15.5

LIBRARY

Project Size Gross S.F.	Project Cost $/S.F.	% Cost Mechanical	% Cost Electrical
6,900	221.20	17.9	10.3
8,200	203.10	18.8	10.6
12,000	263.20	19.4	15.9
15,000	230.00	15.7	18.7
16,300	194.90	11.3	7.4
28,600	174.60	15.7	9.0
30,100	236.80	13.0	11.0
37,700	186.00	14.6	8.0
43,500	157.20	16.6	6.7
47,900	275.50	29.7	9.0
51,400	281.50	13.1	12.4
63,400 A	211.70	13.5	8.3
64,000	198.20	10.9	11.4
74,000	224.60	17.2	8.0
75,600	298.00	12.5	16.7
176,000	165.10	11.2	9.8

SPECIAL NEEDS FUNCTION

Project Size Gross S.F.	Project Cost $/S.F.	% Cost Mechanical	% Cost Electrical
15,200	187.20	16.6	8.4
27,900	225.50	19.4	11.2

STUDENT CENTER/MULTIPURPOSE

Project Size Gross S.F.	Project Cost $/S.F.	% Cost Mechanical	% Cost Electrical
90,000	284.60	16.2	10.0
49,600	210.20	18.2	9.2
187,700	316.50	15.3	8.8
194,800	167.10	17.2	8.5

HOTEL/MOTEL

CONVENTION/CONFERENCE CENTER

Project Size Gross S.F.	Project Cost $/S.F.	% Cost Mechanical	% Cost Electrical
8,600 A	249.90	20.7	16.3
71,900	277.10	12.5	13.8
433,800	144.80	22.1	8.3

HOTEL

Project Size Gross S.F.	Project Cost $/S.F.	% Cost Mechanical	% Cost Electrical
19,900 A	143.00	16.8	5.8
25,875 R	128.10	11.8	10.5
48,400 A	256.10	23.6	8.2
64,300 R	345.10	20.3	10.1
104,200 A	160.40	15.0	8.8
108,040	136.90	13.0	7.0
110,100	177.30	18.4	10.5
132,000	331.50	16.4	5.4
135,900 A	204.80	15.2	7.7
144,100 A	230.00	19.3	11.5
231,000	248.80	15.2	8.4
449,800 A	143.80	13.1	7.0

HOTEL/INN

Project Size Gross S.F.	Project Cost $/S.F.	% Cost Mechanical	% Cost Electrical
57,400	166.30	13.0	10.0
73,000	109.00	24.4	18.1
75,900	144.80	16.5	7.6
162,000	174.30	17.5	8.0
197,000	166.80	15.7	7.7
277,900	122.50	18.8	9.4

A = Addition   R = Remodel

SQUARE FOOT TABLES

INDUSTRIAL

MANUFACTURING

Project Size Gross S.F.	Project Cost $/S.F.	% Cost Mechanical	% Cost Electrical
14,300	141.50	13.0	7.0
18,500	213.50	20.5	13.5
26,600	76.40	6.1	12.1
31,400	168.60	18.9	17.7
33,400	97.50	23.1	15.8
37,300	72.80	3.0	24.0
43,400	96.90	14.7	11.7
45,400	157.20	15.9	15.0
79,800	163.30	16.0	14.5
81,100	174.60	8.2	8.4
137,400	109.60	41.6	13.6
179,600	126.40	26.3	13.6
186,000	118.20	19.5	11.7

RESEARCH AND DEVELOPMENT

Project Size Gross S.F.	Project Cost $/S.F.	% Cost Mechanical	% Cost Electrical
89,140	225.70	20.8	9.0
100,400	299.10	36.1	25.1
114,200	292.70	21.4	9.8
125,000	199.20	20.8	8.4
140,000	278.30	20.1	11.2

WAREHOUSE W/OFFICE

Project Size Gross S.F.	Project Cost $/S.F.	% Cost Mechanical	% Cost Electrical
14,000	65.00	6.5	9.2
19,000	58.30	2.3	1.8
19,700	76.60	9.5	7.0
31,200	65.70	7.0	7.0
40,500	83.90	6.6	10.5
62,000	106.70	10.8	10.0
96,200	61.80	2.2	6.3
105,000	60.00	5.3	11.1
149,800	62.70	14.5	8.6

WAREHOUSE W/OFFICE (Cont.)

Project Size Gross S.F.	Project Cost $/S.F.	% Cost Mechanical	% Cost Electrical
168,600	69.10	11.4	6.3
209,600	63.70	4.9	10.3
402,400	86.50	14.9	8.3

MEDICAL

EDUCATION CENTER

Project Size Gross S.F.	Project Cost $/S.F.	% Cost Mechanical	% Cost Electrical
35,400	387.20	10.2	6.8

HOSPITALS

Project Size Gross S.F.	Project Cost $/S.F.	% Cost Mechanical	% Cost Electrical
9,300 R	382.60	29.5	12.0
15,900 A	282.20	27.4	15.3
16,600 R	101.80	14.7	9.6
22,000	642.50	31.6	17.5
39,100	320.10	23.3	7.9
63,800	257.90	23.4	7.8
98,000 A	370.90	20.5	17.0
100,200 A	591.80	22.2	9.6
103,900 A	328.00	31.1	11.7
109,300	324.90	20.7	15.7
148,700	255.30	25.2	11.8
154,700	414.70	19.3	10.5
165,484	360.50	21.5	14.8
165,700 A	351.70	26.2	17.7
179,400 A	214.30	28.9	20.3
182,800	319.00	19.1	14.0
265,000	380.60	31.2	14.1
281,100	364.70	24.1	14.6
435,000	377.60	21.5	13.7
694,300	209.20	28.9	9.7
772,300	385.50	31.5	13.4

A = Addition   R = Remodel

SQUARE FOOT TABLES

MEDICAL (Cont.)

MEDICAL OFFICES/CENTERS

Project Size Gross S.F.	Project Cost $/S.F.	% Cost Mechanical	% Cost Electrical
3,000	161.70	13.7	11.5
5,500	168.40	8.5	18.7
10,000	203.60	13.9	9.4
10,600	257.80	13.2	8.9
16,300	213.40	21.8	13.2
18,300	116.40	7.1	13.4
20,600	167.10	14.3	6.4
24,900	294.40	21.0	15.1
27,000	297.00	19.7	10.1
28,400	207.20	13.6	9.3
30,500	224.60	17.2	12.7
32.000	177.90	18.1	10.5
44,300	179.40	24.4	16.2
50,200	103.40	9.8	4.9
51,200	262.30	21.0	12.0
64,600	124.80	13.2	6.4
66,000	116.40	9.8	8.8
80,000	99.80	8.2	8.3
137,175	166.80	12.8	6.6

NURSING HOMES

Project Size Gross S.F.	Project Cost $/S.F.	% Cost Mechanical	% Cost Electrical
11,600 A	488.90	53.2	7.9
16,800	318.30	33.3	7.8
31,900 A	239.30	22.0	11.0
64,100	205.20	20.6	11.1
290,000	287.70	16.1	13.6

RESEARCH

Project Size Gross S.F.	Project Cost $/S.F.	% Cost Mechanical	% Cost Electrical
34,600	272.40	18.1	3.0

PUBLIC FACILITIES

ANIMAL CENTER

Project Size Gross S.F.	Project Cost $/S.F.	% Cost Mechanical	% Cost Electrical
20,000	345.10	20.7	4.6
39,100	257.90	22.9	6.8
44,300	245.90	8.1	5.8

AUTO DEALERSHIP

Project Size Gross S.F.	Project Cost $/S.F.	% Cost Mechanical	% Cost Electrical
7,700	160.90	16.7	9.4

BROADCASTING

Project Size Gross S.F.	Project Cost $/S.F.	% Cost Mechanical	% Cost Electrical
20,000	470.50	20.0	13.0
29,500	363.00	16.6	15.0
45,000 R	263.20	15.0	13.0

CIVIC CENTER

Project Size Gross S.F.	Project Cost $/S.F.	% Cost Mechanical	% Cost Electrical
6,000	259.20	9.2	2.8
23,900	325.60	12.3	10.9
34,400	148.40	3.5	17.8
69,800 A	388.80	17.7	8.8
206,500	210.30	15.0	11.0

CORRECTION FACILITIES

Project Size Gross S.F.	Project Cost $/S.F.	% Cost Mechanical	% Cost Electrical
44,600	314.40	20.9	13.1
66,000	192.00	15.0	23.0
257,800 A	332.00	20.9	10.7
360,000	223.80	32.7	13.2

FIRE STATION

Project Size Gross S.F.	Project Cost $/S.F.	% Cost Mechanical	% Cost Electrical
6,900	280.60	12.4	9.7
7,600	215.60	16.0	9.5
8,430	296.40	12.8	9.6
9,600	258.50	13.5	11.8

A = Addition   R = Remodel

SQUARE FOOT TABLES

PUBLIC FACILITIES (Cont.)

GOVERMENT BUILDINGS

Project Size Gross S.F.	Project Cost $/S.F.	% Cost Mechanical	% Cost Electrical
12,300	201.10	13.5	10.7
23,500	329.60	11.1	15.6
27,300	260.10	19.5	9.3
31,600	292.90	27.1	11.3
46,600	275.70	17.4	13.8
72,100	300.20	24.4	10.7
78,200	249.10	19.7	15.2
332,900	241.70	16.2	14.2
364,100	316.60	14.6	12.9
771,000	349.80	17.6	11.3

MUSEUM

Project Size Gross S.F.	Project Cost $/S.F.	% Cost Mechanical	% Cost Electrical
27,600	241.50	17.8	14.1
30,100	277.50	18.6	7.9
43,264	258.50	11.1	8.3
63,000	279.80	8.8	18.1

PARKING GARAGE

Project Size Gross S.F.	Project Cost $/S.F.	% Cost Mechanical	% Cost Electrical
66,000	73.90	2.8	3.1
169,000	54.40	10.3	3.7
562,700	49.10	2.4	6.2

TRANSPORTATION

Project Size Gross S.F.	Project Cost $/S.F.	% Cost Mechanical	% Cost Electrical
7,300	443.80	15.1	3.2
14,300	356.70	13.3	16.6
23.000	246.50	9.6	13.7
35,500	205.20	1.0	19.0
49,100	306.10	35.5	11.6

TRANSPORTATION (Cont.)

Project Size Gross S.F.	Project Cost $/S.F.	% Cost Mechanical	% Cost Electrical
288,100	176.40	8.3	13.3
2,160,000	285.70	23.3	11.5

OFFICES

BANKS

Project Size Gross S.F.	Project Cost $/S.F.	% Cost Mechanical	% Cost Electrical
2,900	410.40	5.3	4.3
3,100	183.20	5.9	6.7
3,300	266.40	10.3	16.4
3,600	200.70	10.3	12.2
4,000	228.60	8.0	9.0
4,100	226.10	21.4	13.0
4,200	244.00	8.6	14.21
4,400	281.90	12.2	12.7
4,500	175.10	12.4	13.5
4,900	276.50	11.7	11.2
5,900	205.50	9.3	13.8
6,000	231.20	11.6	7.3
6,100	310.80	11.0	8.0
7,000	418.10	6.0	9.0
7,300	255.60	11.9	11.3
7,700	286.50	8.0	7.5
7,800	314.00	11.0	11.7
8,000	159.60	10.0	14.0
9,200	253.20	9.9	12.1
9,400	183.30	11.7	11.8
10,200	382.30	12.6	12.6
12,600	136.60	7.0	18.0

A = Addition   R = Remodel

SQUARE FOOT TABLES

OFFICES (Cont.)

BANKS (Cont.)

Project Size Gross S.F.	Project Cost $/S.F.	% Cost Mechanical	% Cost Electrical
13,300	240.80	9.5	8.3
13,800	215.60	10.00	9.7
15,000	184.50	15.4	12.4
15,200	147.50	9.2	12.9
15,500	188.10	9.8	10.3
16,000	123.50	13.4	23.1
20,100	116.20	13.0	11.0
21,700	245.90	8.8	11.3
44,800	210.00	13.0	8.2
53,200	379.30	14.9	7.2
62,100	217.70	10.1	7.9
95,100	286.40	13.5	4.3

OFFICE BUILDINGS

Project Size Gross S.F.	Project Cost $/S.F.	% Cost Mechanical	% Cost Electrical
2,600	257.60	17.2	9.4
3,400	210.50	10.5	11.3
3,800	206.60	16.4	11.8
4,400	204.20	12.8	8.5
4,500	175.70	13.0	7.0
5,100	127.70	16.6	10.2
5,200	167.40	8.0	5.7
6,700	249.20	16.8	10.4
7,500	268.30	10.1	8.0
7,900	209.00	17.4	9.0
8,100	323.00	10.6	11.0
10,600	186.90	9.9	9.7
10,900	107.50	13.8	10.8

OFFICE BUILDINGS (Cont.)

Project Size Gross S.F.	Project Cost $/S.F.	% Cost Mechanical	% Cost Electrical
11,300	212.10	17.0	6.0
13,000	144.50	15.0	9.0
14,400	189.50	19.8	12.9
14,500	145.20	17.3	12.5
17,000	207.50	14.7	7.9
17,800 A	112.10	10.4	11.0
18,100	217.30	22.5	11.6
19,300	118.10	11.1	8.1
24,600	109.10	18.5	14.1
27,700	131.10	19.6	5.5
27,800	237.90	12.7	5.1
27,800	128.10	17.8	10.3
32,500 R	219.20	13.9	6.4
35,400	128.00	15.0	12.0
36.500	110.90	10.2	10.2
42,300	137.60	10.3	7.7
44,400	187.40	23.5	14.7
44,400	98.50	11.0	5.0
44,500	135.70	11.5	3.1
45,400	117.20	19.1	13.2
47,300	120.50	18.5	8.0
49,700	202.50	22.1	7.2
50,000	193.00	19.4	15.6
50,400	209.30	23.2	7.8
52,200	136.00	18.3	7.8
52,900	158.50	4.4	3.9
53,700	242.60	15.4	9.0

A = Addition   R = Remodel

SQUARE FOOT TABLES

OFFICES (Cont.)

OFFICE BUILDINGS (Cont.)

Project Size Gross S.F.	Project Cost $/S.F.	% Cost Mechanical	% Cost Electrical
54,000	98.20	14.4	2.6
56,000	111.30	10.6	6.2
56,500	153.10	19.1	11.0
72,000 R	53.80	12.9	20.2
74,000	106.70	13.5	6.8
80,800	107.80	12.0	6.0
81,800	148.40	21.1	9.5
81,900	124.80	19.9	9.0
82,000	173.50	14.0	4.2
83,100	204.60	16.0	8.6
85,400	179.40	18.7	8.6
86,200	140.50	14.7	11.0
99,900	162.60	16.6	6.6
100,000	181.70	10.6	12.1
100,000	107.10	16.3	10.4
116,400	171.60	12.6	10.0
134,500	350.00	14.1	12.4
140,000	271.20	20.1	11.2
155,700	337.90	11.1	6.1
171,000	158.40	24.0	7.7
174,300	195.00	12.3	9.1
203,300	263.60	19.0	13.2
265,800	365.00	11.0	10.0
287,300	101.10	10.7	4.1
319,800	158.40	13.4	5.6
350,000	233.90	20.0	13.0
360,900	141.60	14.2	11.1
394,000	112.60	22.5	8.7

OFFICE BUILDINGS (Cont.)

Project Size Gross S.F.	Project Cost $/S.F.	% Cost Mechanical	% Cost Electrical
430,000	168.60	21.8	9.3
490,000	179.30	11.2	7.5
588,400	368.60	15.0	9.0
606,000	147.40	9.4	8.8
620,000	406.60	12.6	12.8
733,500	112.60	10.8	4.2

RECREATIONAL

ARENA

Project Size Gross S.F.	Project Cost $/S.F.	% Cost Mechanical	% Cost Electrical
315,200	421.90	10.4	8.00
385,800	241.50	13.2	8.80
727,000	218.40	14.9	7.40

HEALTH CLUB

15,900	118.10	8.7	9.7
21,800	167.00	10.3	10.3
30,100	258.20	11.4	19.4
66,400	127.10	10.7	8.5

RECREATIONAL CENTER

9,900	229.40	6.6	9.8
14,000	149.80	11.2	5.0
14,000	179.40	11.3	16.7
15,700	136.60	17.8	14.3
20,000	309.90	19.1	8.9
21,200	174.40	16.0	9.6

A = Addition   R = Remodel

SQUARE FOOT TABLES

RECREATIONAL (Cont.)

RECREATIONAL CENTER (Cont.)

Project Size Gross S.F.	Project Cost $/S.F.	% Cost Mechanical	% Cost Electrical
26,000	203.10	9.3	7.0
53,400 A	242.90	11.7	6.4
69,800	393.80	17.7	8.8

RELIGIOUS

CHURCH

Project Size Gross S.F.	Project Cost $/S.F.	% Cost Mechanical	% Cost Electrical
4,100	276.50	8.8	13.5
10,400 R	376.40	12.8	8.1
11,100	165.80	10.9	12.7
13,400	248.40	5.5	6.0
14,500	169.60	12.8	7.4
15,200	219.70	18.3	8.0
15,700	191.30	14.4	8.4

CHURCH (Cont.)

Project Size Gross S.F.	Project Cost $/S.F.	% Cost Mechanical	% Cost Electrical
16,000	311.20	14.1	9.6
20,900	194.10	16.0	14.0
21,500	213.20	7.3	8.0
22,900	192.50	12.1	9.5
30,600	145.30	15.5	8.0
42,700	164.30	18.6	7.7

MULTI-PURPOSE

4,400	169.90	11.5	17.5
5,800 A	216.70	8.9	7.5
6,400	295.70	15.6	15.8
9,000	144.80	7.7	5.9
9,000	205.50	16.0	6.6
10,100	137.90	11.1	12.0
12,000	272.70	13.1	10.7
18,400	149.20	10.8	10.1
19,500	263.20	16.0	17.3

A = Addition   R = Remodel

For more information subscribe to **Design Cost & Data**

BNi Building News

INDEX

- A -

A-BAND HANGER 240, 570
ABANDON CATCH BASIN 12, 457
ABOVE GROUND TANK 230, 564
ABRASIVE SURFACE 114, 507
    SURFACE TILE 176, 177, 539
ABS PIPE 37, 294, 469, 595
AC RELAY 427, 662
ACCESS CONTROL 205, 552
    DOOR 149, 524
    FLOOR 203, 551
    STAIR 219, 559
ACCESS/SECURE CONTROL 443, 670
ACCESSORY REINFORCING 82, 492
ACCORDION FOLDING DOORS 159, 530
    PARTITION 208, 553
ACID ETCH 83, 185, 492, 542
ACID-PROOF COUNTER 226, 563
ACOUSTICAL BLOCK 102, 501
    CEILING 196, 548
    PANEL 179, 540
    TILE 180, 540
ACRYLIC CARPET 184, 542
AD PLYWOOD 131, 515
ADAPTER CUBICLE 342, 619
    TERMINAL 372, 634
ADHESIVE ROOF 145, 522
ADHESIVE-BED TILE 175, 538
ADJUSTABLE BAR HANGER 399, 648
    FLOOR BOX 392, 644
    ROLLER 239, 569
    SHELF 131, 515
    VALVE BOX 32, 467
ADMIXTURE 83
AIR COMPRESSOR 5, 324, 611
    CONDENSER 326, 612
    CONDITIONING 326, 612
    DISTRIBUTION PANE 180, 540
    ELIMINATOR 314, 606
    ENTRAINING AGENT 83
    MONITORING TEST 230
    POWERED HOIST 236, 568
    TOOL 5
    VENT 314, 606
AIRFIELD CATCH BASIN 49, 475
    MARKING 56, 478
ALARM LOCK 165, 532
    VALVE 252, 575
ALL THREAD ROD 354, 625
ALUMINUM BOX 392, 644
    CABLE 45, 473
    CONDUIT 362, 629
    DOOR 170, 535
    DOWNSPOUT 148, 524
    EXPANSION JOINT 148, 524
    FACED PANEL 179
    FACED TILE 180, 540
    FLASHING 146, 523
    GRATING 115, 507
    GUTTER 148, 524
    HATCH 149, 524
    JACKET 248, 573
    LOUVER 117, 508
    PIPE 50, 476
    PLAQUE 205, 551
    POLE 403, 650
    RAILING 117, 508
    ROOF 145, 522
    ROOF PANEL 140, 141, 520
    SHINGLE 139, 519
    SIDING PANEL 142, 521
    STOREFRONT 161, 530
    THRESHOLD 167, 533
    TILE 180, 540
    TREAD 114, 507
    VENT 202, 550
AMBIENT DAMPER 326, 612
AMMETER 415, 656
AMPLIFIER 447, 672
AMP-TRAP FUSE 411, 654
ANCHOR 109
    BOLT 109
    BRICK 98, 499
    DOVETAIL 98, 499
    PIPE 33, 467
    RAFTER 119, 509
    SLOT 98, 499
    WOOD 119, 509
ANGLE BAR 52, 477
    GUARD 116, 508
    STEEL 104, 110, 502, 505
    STEEL BRACE 403, 650
    SUPPORT 240, 570
    VALVE 312, 313, 605
ANNUNCIATOR 446, 671
    PANEL 442, 669
ANODIZED TILE 180, 540
ANTENNA 448, 672
ANTI-SIPHON BREAKER 314, 606
APRON 129, 514
ARCH CULVERT 51, 476
ARCHITECTURAL FEE 1
ARMOR PLATE 166, 533
ARRESTER LIGHTNING 404, 411, 650, 654
ASBESTOS REMOVAL 231, 565
    VACUUM 230
ASH RECEIVER 210, 554
ASHLAR VENEER 106, 503
ASPHALT CURB REMOVAL 13, 457
    DAMPPROOFING 135, 517
    DEMOLITION 14, 458
    EXPANSION JOINT 80, 491
    MEMBRANE 134, 517
    PRIME COAT 54, 478
ASPHALTIC PAINT 135, 517
ASTRAGAL WEATHERSTRIPPING 167, 533
ASTRONOMIC TIME CLOCK 428, 662
ATTENUATION BLANKET 181, 540
ATTIC INSULATION 138, 519
AUDIO CABLE 450, 673
    SENSOR 443, 670
AUGER FENCE 61, 481
AUTO GRIP HANGER 241, 570
    TRANSFER SWITCH 404, 650
AUTOMOTIVE EQUIPMENT 324, 611
    HOIST 235, 567
AUTOPSY TABLE 224, 561
AUXILIARY SWITCH 411, 654
AWNING WINDOW 163, 532

- B -

BACK MOUNTED TILE 176, 538
    SPLASH 131, 515
BACKBOARD 63, 482
    BASKETBALL 221, 559
BACKER ROD 152, 526
BACKFILL 25, 463
    HAND 23, 462
BACKHOE 6
BACKHOE/LOADER 6
BACKSTOP 63
    BASEBALL 64
BACK-UP BLOCK 102, 501
    BRICK 99, 500
BAKE OVEN 217, 558
BALANCE LABORATORY 223, 560
BALANCING VALVE 312, 605
BALL TREE 68, 484
    VALVE 312, 313, 605
BALLAST 53, 477
    REPLACEMENT 434, 665
    ROOF 144, 522
BALL-BEARING HINGE 164
BALUSTER 130, 514
BAND SAW 222, 560
    STAGE 214, 556
BANDING GRATING 115, 507
BANK COUNTER 213, 556
    DUCT 354, 626
    RUN GRAVEL 20, 461
BAR HANGER 399, 648
    PARALLEL 221, 560
    REINFORCING 97, 499
BARBED WIRE 56, 57, 478, 479
BARE SOLID WIRE 389, 643
    STRANDED WIRE 389, 643
BARGE CONSTRUCTION 20, 461
BARRICADE 3
BARRIER VAPOR 82, 135, 492, 517
BASE CABINET 132, 226, 515, 563
    COLONIAL 129, 514
    COLUMN 106, 119, 503, 509
    FLASHING 146, 523
    GRANITE 106, 503
    GROUTING 94, 498
    MANHOLE 41, 471
    TERRAZZO 178, 539
    WALL 183, 541
BASEBALL BACKSTOP 64
BASEBOARD HEATER 449, 673
    RADIATION 330, 614
BASKET LOCKER 206, 552
BASKETBALL BACKBOARD .. 63, 221, 482, 559
BATCH DISPOSAL 318, 608
BATH FAUCET 318, 608
BATHROOM LOCK 165, 532
BATTEN SIDING 143, 521
BEAD MOLDING 130, 514
    PARTING 130, 514
    PLASTER 173, 175, 537, 538
    SASH 130, 514
BEAM 110, 505
    BOLSTER 82, 492
    BOND 84, 97, 103, 493, 499, 502
    CHAIR 82, 492
    CLAMP 243, 571
    FIREPROOFING 138, 519
    FURRING 112, 171, 506, 536
    GRADE 87, 494
    HANGER 119, 509
    PLASTER 173, 537
    PRECAST 90, 495
    PRESTRESSED 89, 495
    REINFORCING 78, 490
    SHEETROCK 174, 537
BEARING PAD 93, 497
BELL AND SPIGOT 49, 475
    END 372, 373, 635
    STANDARD 444, 670
BELLOWS COPPER 149, 524
BENCH 62, 481
    FLOOR MOUNTED 206, 552
    SAW 222, 560
BENCHES 186, 543
BEND SOIL 259, 579
BENTONITE MEMBRANE 134, 517
BEVELED CONCRETE 83, 492
    SIDING 143, 521
BHMA HINGE 164
BIBB HOSE 308, 603
BICYCLE EXERCISE 223, 561
BI-FOLD DOOR 157, 529
BIKE RACK 63, 482
BIN ICE STORAGE 219, 559
BINDER COURSE 55, 478
BI-PASSING DOOR 157, 529
BIRCH DOOR 156, 528
    PANELING 133, 515
    VENEER 133, 515
BIRDSCREEN 229
BITUMINOUS MEMBRANE 134, 517
    SIDEWALK 55, 478
BLACK IRON FITTING 301, 599
    RISER HANGER 241, 570
    STEEL 301, 599
    STEEL HANGER 239, 569
    WIRE HANGER 244, 571

INDEX

BLEACHER	62, 481
GYM	221, 560
BLOCK CONCRETE	89, 101, 495, 501
DAMPPROOFING	135, 517
DEMOLITION	9, 455
GLASS	105, 503
GRANITE	101, 501
GROUT	97, 499
REMOVAL	11, 456
SPLASH	92, 497
TERMINAL	48, 474
VENT	202, 550
BLOOD ANALYZER	222, 560
REFRIGERATOR	223, 560
BLOWN-IN INSULATION	138, 519
BLUESTONE	101, 501
BOARD CEMENT	134, 517
COMPOSITE	136, 518
CONCRETE FINISH	83, 492
DIVING	229, 564
DOCK	215, 557
PROTECTIVE	134, 517
RAFTER	124, 511
RIDGE	124, 511
SIDING	143, 521
SUB-FLOORING	126, 512
BOLLARD PIPE	61, 481
BOLSTER SLAB	82, 492
BOLT CARRIAGE	119, 509
DEAD	165, 532
DOOR	166, 533
ON BREAKER	415, 656
WOOD	119, 509
BOLTED PIPE	39, 470
STEEL	110, 505
TRUSS	111, 505
BOLT-ON HUB	410, 653
BOND BEAM	97, 103, 499, 502
BEAM PLACEMENT	84, 493
ROOF	144
BONDERIZED FLAGPOLE	204, 551
BOOKKEEPER	2
BOOTH SPRAY	221, 560
BORED LOCK	165, 532
BOTTLE COOLER	219, 559
BOTTOM BEAM	70, 486
PIVOT HINGE	164
PLATE	125, 512
BOUNDARIES	3
BOWL TOILET	321, 609
BOX BEAM RAIL	61, 481
CABINET	400, 649
DEVICE	399, 648
DISTRIBUTION	377, 637
EXPLOSION PROOF	396, 646
FLOOR	392, 644
JUNCTION	399, 648
PLASTER	400, 648
PLASTIC	400, 648
PULL	400, 649
RECEIVING	207, 553
RIB ROOF	141, 520
SCREW COVER	400, 649
TAP	341, 378, 619, 637
TERMINAL	416, 418, 656, 657
WEATHERPROOF	395, 646
BOXING RING	221, 560
BOX-RIB PANEL	142, 521
BRACKET BOX	399, 648
WALL	349, 623
BRANCH BREAKER	415, 656
BRASS FITTING	278, 279, 281, 285, 588, 589, 591
BRAZED CONNECTION	413, 655
BREAKER CIRCUIT	340-345, 407-415, 619-621, 652-656
MAIN	416, 656
VACUUM	314, 606
BRICK ANCHOR	98, 499
CHIMNEY	100, 500
FIRE RATED	100, 500
FIREWALL	100, 500
BRICK MANHOLE	42, 471
PAVER	101, 501
REMOVAL	10, 11, 107, 456, 504
BRICKWORK	186, 543
BRIDGE CRANE	236, 568
BRIDGING	119, 509
BROILER KITCHEN	217, 558
BRONZE PLAQUE	204, 551
RAILING	117, 508
VALVE	309, 603
BROOM FINISH	83, 492
BRUSH	186, 543
CUTTING	18, 460
BUCK BOOST TRANSFORMER	407, 652
BUCKET TRAP	315, 606
BUFF BRICK	100, 500
BUILDING METAL	228, 564
PAPER	135, 517
BUILT-IN OVEN	220, 559
SHELVING	210, 554
BUILT-UP PLAQUE	204, 551
ROOF	144, 522
ROOF REMOVAL	10, 456
BULK MAIL SLOT	207, 553
BULKHEAD FORMWORK	75, 488
BULLDOZER	6
BULLETIN BOARD	204, 551
BULLETPROOF GLASS	169, 534
PARTITION	213, 556
BULLNOSE GLAZED	103, 502
BUMPER DOCK	215, 557
DOOR	166, 532
BURLAP CONCRETE	84, 493
RUB	83, 492
BUS BAR	413, 655
DUCT	340, 619
DUCT CONNECTION	414, 655
GROUND	341
BUSH HAMMER	83, 492
BUSHED NIPPLE	369, 633
BUSHING BRASS	279, 589
CONDUIT	365, 631
FLUSH COPPER	271, 584
BUTT HINGE	164
PILE	29, 465
WELDED PIPE	303, 600
BUTTERFLY VALVE	312, 605
BUTYL CAULK	107, 504
CAULKING	151, 525
MEMBRANE	134, 517
BUZZER	446, 671
BX CABLE	389, 643
CLAMP	399, 648

- C -

C CONDULET	360, 628
CAB ELEVATOR	232
CABINET BASE	132, 515
EXTINGUISHER	207, 553
HEATER	332, 614
KITCHEN	226, 563
LABORATORY	222, 560
MONITOR	443, 670
CABLE AUDIO	450, 673
BX	389, 643
CAMERA	451, 674
COAXIAL	451, 674
COMMUNICATION	445, 671
COMPUTER	451, 674
CONNECTOR	391, 644
COPPER	45, 473
COUPLER	378, 637
ELECTRIC	390, 643
FIRE ALARM	451, 674
FITTING	391, 644
GUY	447, 672
LUG	411, 654
MAN-HOLE	445, 671
MICROPHONE	451, 674
CABLE SER	387, 642
SEU	387, 642
SUPPORT	47, 474
TAP BOX	340, 619
THERMOSTAT	452, 674
THHN-THWN	388, 642
TRAY	347-352, 353, 622-625
TRAY FITTING	353, 625
UNGROUNDED	45, 473
CAGE LADDER	114, 507
CAMERA CABLE	451, 674
CANT STRIP	125, 144, 512, 522
CAP BLACK IRON	302, 599
BRASS	280, 589
CONDUIT	366, 631
COPPER	271, 276, 277, 584, 587
ENTRANCE	360, 629
PILE	72, 87, 487, 494
PILE REINFORCING	78, 490
PVC	288, 289, 290, 593, 594
STAINLESS	297, 597
STEEL	307, 602
WELDED	303, 600
CAPACITANCE SENSOR	443, 670
CAPACITOR	346, 405, 622, 651
CARBORUNDUM	83, 492
CARD ACCESS CONTROL	205, 552
CARPENTRY ROUGH	120, 509
CARPET CLEAN	184, 542
PLATE	393, 645
TILE	184, 203, 542
CARRIAGE BOLT	119, 509
CARRIER CHANNEL	181
FIXTURE	322, 609
CART LAUNDRY	215, 557
CARTRIDGE EXTINGUISHER	207, 553
CARVED DOOR	157, 529
CASE REFRIGERATED	219, 559
CASED BORING	8, 455
CASEMENT WINDOW	161, 163, 531
CASING	129, 514
BEAD	175, 538
RANCH	129, 514
TRIM	131, 515
CASSETTE PLAYER	447, 672
CAST ALUMINUM BOX	395, 646
BOX	393, 645
IRON BOILER	325, 611
IRON FITTING	304, 601
IRON STEP	116, 508
IRON STRAINER	315, 606
IRON TREAD	113, 506
IRON VALVE	309, 603
CATCH BASIN	48, 475
CAULK MASONRY	107, 504
CAULKING	151, 525
CAVITY WALL	97, 99, 499, 500
WALL ANCHOR	98, 499
C-CLAMP	240, 243, 570, 571
CCTV CABLE	451, 674
CDX	126, 512
CEDAR CLOSET	133, 515
DECKING	127, 513
SHINGLE	140, 520
SIDING	142, 143, 521
CEILING ACCESS DOOR	149, 524
BLOCKING	120, 509
DIFFUSER	335, 616
EXPANSION	117, 508
FURRING	122, 171, 510, 536
GRILLE	337, 617
HEATER	219, 559
INSULATION	135, 138, 517, 519
JOIST	120, 509
LATH	172, 173, 536, 537
REMOVAL	10, 456
CEILINGS	196, 548
CELLULAR DECKING	111, 505

INDEX

CEMENT BOARD 134, 517
    FIBER 94, 498
    FIBER BOARD 93, 497
    KEENES 173, 537
    PLASTER 174, 537
    SEALING 398, 647
CENTER BULB 81, 491
CENTRIFUGAL PUMP 323, 610
CENTRIFUGE LABORATORY 223, 561
CHAIN HOIST DOOR 159, 530
    LINK 13, 457
    LINK FENCES 187, 543
    TRENCHER 24, 463
CHAIR BAND 214, 556
    HYDROTHERAPY 223, 561
    LIFEGUARD 229, 564
    RAIL 129, 514
    REINFORCING 82, 492
CHAMFER STRIP 75, 488
CHANGE TRAILER 230
CHANNEL CLOSURE STRIP 355, 626
    DOOR FRAME 110
    FURRING 112, 171, 506, 536
    MANHOLE 42, 471
    SLAB 94, 497
    SPRING NUT 355, 626
    STRAP 354, 625
CHECK VALVE 252, 310, 575, 603
CHEMICAL EXTINGUISHER 207, 552
    PREPARATION 185, 542
    STRIPPING 185, 186, 542, 543
CHERRY VENEER 133, 515
CHESTNUT VENEER 133, 515
CHICKEN WIRE 65, 482
CHIME 442, 446, 669, 671
CHIMNEY BRICK 100, 500
    FLUE 108, 504
    MASONRY 108, 504
CHOPPER FOOD 218, 558
CHUTE MAIL 207, 553
    TRASH 216, 236, 557, 567
CIRCUIT BREAKER 340-345, 407-415, 619-621, 652-656
CLADDING PANEL 92, 497
CLAMP BEAM 355, 626
    GROUND ROD 413, 655
    ROD BEAM 354, 625
    TYPE CAP 367, 632
CLASSROOM LOCK 165, 532
    LOCKSET 165, 532
CLAY BRICK FLOOR 182, 541
    COURT 63
CL-AY FUSE 412, 654
CLAY PIPE 41, 42, 51, 471, 472, 476
    TILE FLOOR 105, 503
CLEANING MASONRY 107, 504
CLEANOUT 291, 594
    DANDY 262, 580
    PVC 42, 472
    STORM DRAIN 51, 476
CLEARING TREE 19, 460
CLEFT NATURAL 101, 501
CLEVIS HANGER 238, 569
CLOCK RECEPTACLE 378, 402, 637, 649
CLOCK-TIMER HOSPITAL 225, 561
CLOSED CELL CAULKING 151, 525
CLOSET CEDAR 133, 515
    DOOR 157, 529
    FLANGE 276, 587
    POLE 129, 514
CLOSURE STRIP 355, 626
CLOTH FACED FIBERGLASS 179, 540
CMU 101, 501
    GROUT 97, 499
CO2 EXTINGUISHER 206, 552
COAT RACK 211, 555
COATED ROOF 144, 522
    STEEL PIPE 50, 476
COAX CABLE 448, 672
COAXIAL CABLE 444, 451, 670, 674
COBRA HEAD FIXTURE 440, 668

CODE BLUE SYSTEM 446, 671
COFFEE URN 217, 558
COILED PIPE 43, 472
COILING DOOR 158, 529
    GRILLE 159, 529
COIN DRYER 215, 557
    WASHER 215, 557
COLONIAL BASE 129, 514
    MOLDING 130, 514
COLOR GROUP TILE 176
COLUMN 110, 505
    BASE 106, 119, 503, 509
    BASE PLATE 110, 505
    COVER 170, 535
    FIREPROOFING 138, 519
    FOOTING 72, 487
    FURRING 112, 506
    PIER 74, 488
    PLASTER 173, 537
    PRECAST 90, 496
    SHEETROCK 174, 537
    TIMBER 128, 513
COMBINATION DOUBLE WY 262, 580
    RECEPTACLE 402, 649
    STARTER 345, 621
    SWITCH 429, 663
    WYE 291, 594
    WYE SOIL 261, 580
COMMERCIAL CARPET 184, 542
    DOOR 155, 161, 528, 530
COMMON BRICK 99, 500
COMPACT BASE 23, 462
    BORROW 21, 461
    FLUORESCENT 438, 667
    KITCHEN 216, 557
COMPACTED CONCRETE 55, 478
COMPACTOR INDUSTRIAL 216, 557
    RESIDENTIAL 219, 559
COMPANION FLANGE 272, 585
COMPARTMENT SHOWER 202, 550
COMPOSITE BOARD 136, 518
    DECKING 112, 506
COMPRESSION COUPLING 359, 628
    ELBOW 285, 591
    FITTING 284, 357, 591, 627
    LUG 46, 473
    SPLICE 48, 474
COMPRESSOR 5
    AUTOMOTIVE 221, 560
COMPROMISE SPLICE BAR 53, 477
COMPUTER CABLE 451, 674
    FLOOR 203, 551
CONCEALED Z BAR 181, 540
CONCENTRIC REDUCER 296, 597
CONCRETE BLOCK 187, 543
    BLOCK REMOVAL 11, 456
    CATCH BASIN 48, 475
    COLUMN 90, 496
    CROSS TIE 53, 477
    CUTTING 17, 459
    DEMOLITION 9, 455
    DOWEL 78, 490
    FENCE HOLE 61, 481
    FIBROUS 55, 478
    FILLED PILE 28, 465
    FLOORS 185, 542
    HARDENER 83, 492
    HEADWALL 51, 476
    MANHOLE 42, 471
    PARKING BARRIER 62, 481
    PILING 29, 465
    PIPE 49, 475
    PLACEMENT 84, 493
    PLANK 93, 497
    PLANTER 68, 484
    POLE 403, 650
    POST 61, 481
    PRECAST 89, 495
    PRESTRESSED 89, 495
    PUMP 84, 493

CONCRETE RECEPTOR 201, 550
    REINFORCEMENT 76, 489
    REMOVAL 11, 18, 456, 460
    SHEET PILING 30, 465
    SLEEPER 125, 512
    TESTING 3
    TOPPING 85, 493
    WATERPROOFING 135, 517
CONDENSATE PUMP 323, 610
CONDUCTIVE FLOOR 182, 541
    TERRAZZO 178, 539
    TILE 176, 538
CONDUCTOR ALUMINUM 385, 641
    COPPER 388, 642, 643
    STRANDED 388, 642
CONDUIT ALUMINUM 356, 362, 626, 629
    BODY 372, 635
    BUSHING 365, 631
    CAP 366, 631
    DROPOUT 349, 623
    EMT 357, 627
    FLEXIBLE 361, 629
    GALVANIZED 363, 630
    LOCKNUT 365, 631
    PENETRATION 354, 356, 626
    PLASTIC COATE 375, 377, 636, 637
    PLUG 397, 647
    PVC 371, 634
    STEEL 361, 629
CONDULET 359, 374, 628, 635, 636
    COVER 360, 368, 629, 632
    GASKET 368, 633
    LB 368, 632
CONNECTION BUS DUCT 414, 655
CONNECTOR COMPRESSION 357, 627
    DIE CAST 358, 627
    SET SCREW 358, 627
CONTACT SWITCH 429, 663
CONTINUOUS DISPOSAL 318, 608
    FOOTING 86, 494
    HIGH CHAIR 82, 492
CONTROL DAMPER 338, 618
    JOINT 99, 500
    VALVE 338, 617
CONVECTION OVEN 217, 558
COOLER BOTTLE 219, 559
    WATER 322, 609
COPING CLAY 105, 503
    LIMESTONE 106, 503
    PRECAST 92, 497
COPPER BELLOWS 149, 524
    BRAID 413, 655
    CABLE 45, 473
    DOWNSPOUT 147, 523
    FLASHING 146, 523
    GRAVEL STOP 147, 523
    HANGER 241, 570
    PIPE FITTING 276, 587
    PLAQUE 204, 551
    VENT 148, 524
    WIRE HOOK 245, 572
CORD ELECTRIC 392, 644
    SET 402, 649
CORED PLUG 281, 590
    SLAB 91, 496
CORING CONCRETE 12, 457
CORK RUNNING TRACK 63, 482
    WALL COVERING 199, 549
CORNER BEAD 173, 175, 537, 538
    CABINET 226, 563
    GUARD 110, 199, 549
    POST 56, 58, 479
    PROTECTION 116, 508
CORNICE MOLDING 131, 515
CORRUGATED DOWNSPOUT 147, 523
    METAL PIPE 50, 51, 475, 476
    PANEL 142, 521
    ROOF 141, 145, 520, 522
COUNTER BANK 213, 556
    DOOR 158, 529

INDEX

COUNTER FLASHING 146, 523
   GRIDDLE 217, 558
   LABORATORY 222, 560
   LAVATORY 319, 608
COUNTERSUNK PLUG 281, 590
COUNTERTOP 131, 515
COUNTER-TOP RANGE 220, 559
COUPLING 292, 594
   ABS 38, 469
   BLACK IRON 302, 599
   BRASS 280, 589
   COMPRESSION 357, 627
   COPPER 276, 277, 587
   EMT 359, 628
   ERICKSON 364, 631
   NO-HUB 253
   PLASTIC 37, 469
   PVC 39, 288, 289, 290, 470
   SET SCREW 358, 628
   SPLIT 364, 630
   STEEL 306, 602
   THREADED 366, 632
   THREADLESS 367, 632
   WITH STOP 264, 581
COURSE BINDER 55, 478
COURT LIGHTING 438, 667
   RECREATIONAL 63
   TENNIS 63, 482
COVE BASE 182, 541
   BASE GLAZED 103, 502
   BASE TILE 177, 539
   MOLDING 129, 514
COVED STAIR RISER 183, 541
COVER CATCH BASIN 49, 475
   CONDULET 360, 368, 629, 632
   MANHOLE 116, 507, 508
   POOL 229, 564
CPVC FITTING 291, 594
   PIPE 287, 291, 593, 594
CRACK PATCH 174, 537
   REPAIR 95, 498
   SEALING 54, 478
CRANE 7
   JIB 237, 568
   MOBILIZATION 4
   TRAVELING 237, 568
CRAWL SPACE INSULATION ... 135, 517
CRAWLER CRANE 7
CREOSOTE 132
CREOSOTED POLE 402, 650
CREW TRAILER 4
CROSS CAST IRON 35, 468
   NO-HUB 254, 576
   TIE 53, 477
CROSSARM POLE 403, 650
CROWN MOLDING 130, 514
CRUSHED MARBLE 101, 501
CRUSHER RUN 25
CUBICLE OFFICE 208, 553
CULVERT FORMWORK 70, 486
   REINFORCEMENT 76, 489
CURB FORM 72, 487
   FORMWORK 75, 488
   GRANITE 106, 503
   INLET 42, 472
   INLET FRAME 49, 475
   REMOVAL 12, 457
   ROOF 334, 616
   TERRAZZO 177, 178, 539
CURING PAPER 84, 492
CURTAIN GYM 221, 560
   WALL 170, 534
CURVED CURB FORM 71, 486
   STAIR TERRAZZO 178, 539
   WALL FORM 75, 488
CUSTOM DOOR 158, 529
CUT AND FILL 25, 463
   STONE 105, 503
   TREE 19, 460

CUT-OUT 10, 456
   COUNTERTOP 131, 515
CUTTING GRATING 115, 507
CYLINDER PILING 29, 465

- D -

DAIRY PRODUCT 219, 559
DAMPER 338, 618
DANDY CLEANOUT 275, 587
   SOIL 259, 578
DARBY 83, 492
DB CONDUIT 373, 635
   DUCT 373, 635
DEAD BOLT 165, 532
DECK INSULATION 137, 518
DECKING CELLULAR 111, 505
   METAL 111, 505
   OPEN 111, 505
   WOOD 127, 513
DECKS, METAL 189, 544
   WOOD, STAINED 189, 544
DELICATESSEN CASE 219, 559
DELIVERY CART 215, 557
DELUGE SHOWER 322, 609
DENTAL EQUIPMENT 225, 562
   LIGHT 225, 562
DEPOSITORY NIGHT 213, 556
DESANCO FITTING COPPER ... 275, 586
DETECTOR SMOKE 167, 533
DEVICE COVER 399, 648
DIAPHRAGM PUMP 6
DIECAST SET SCREW 359, 628
DIESEL GENERATOR 404, 650
DIFFUSER SUCTION 323, 610
DIMMER SWITCH 428, 662
DIRECT BURIAL CONDUIT 373, 635
   DRIVE FAN 333, 615
DIRECTIONAL ARROW 56, 478
DIRECTORY BOARD 204, 551
   SHELF 210, 554
DISAPPEARING STAIR 220, 559
DISC TRAP 317, 607
DISH SOAP 177, 538
DISHWASHER 218, 558
   RESIDENTIAL 219, 559
DISINFECT PIPE 252, 301, 575
DISPENSER TISSUE 211, 555
DISPOSAL COMMERCIAL 218, 558
   RESIDENTIAL 219, 559
   SUIT 230
DISTRIBUTION 414, 655
   BOX 377, 637
   PANELBOARDS 421, 659
DITCH EXCAVATION 23, 462
DIVIDER HOSPITAL 202, 550
   STRIP TERRAZZO 177
DIVING BOARD 229, 564
DLH JOIST 111, 505
DOCK BOARD 215, 557
   LEVELER 215, 557
DOME LIGHT 446, 671
DOMESTIC MARBLE 106, 503
   WELL 31, 466
DOOR ACCESS 149, 524
   ALUMINUM 170, 535
   BUMPER 166, 532
   CHAIN HOIST 159, 530
   CLOSER 166, 532
   COMMERCIAL 155, 528
   COORDINATOR 166, 533
   FLUSH 157, 529
   FOLDING 159, 530
   FRAME 153, 527
   FRAME GROUT 97, 499
   FRAMES, METAL 190, 545
   HEADER 125, 512
   HOLDER 166, 532
   MOTOR 159
   OPENER 444, 447, 670, 672

DOOR OVERHEAD 158, 529
   PLATE 166, 533
   REMOVAL 10, 456
   REVOLVING 161, 530
   ROLL UP 228
   SEAL 215, 557
   SHOWER 201, 550
   SLIDING GLASS 160, 530
   SOLID CORE 156, 528
   SPECIAL 213, 556
   SWITCH 429, 443, 663, 670
   VAULT 213, 556
   VERTICAL LIFT 160, 530
DOORS 185, 542
   AND WINDOWS 185, 542
DOORS, METAL 189, 197, 544, 548
   WOOD 190, 197, 545, 548
DORMITORY WARDROBE 211, 555
DOUBLE BARREL LUG 47, 474
   HUB PIPE 258, 578
   HUB SOIL 259, 578
   HUNG WINDOW 162, 531
   OVEN 217, 558
   TEE 91, 92, 496, 497
   WALLED TANK 230, 564
   WYE NO-HUB 256, 577
   WYE SOIL 258, 261, 578, 580
DOUGLAS FIR 127, 513
DOVETAIL ANCHOR SLOT 98, 499
DOWEL REINFORCING 78, 490
DOWN DAMPER 335, 616
DOWNSPOUT 147, 523
   ALUMINUM 148, 524
DOZER 7
   EXCAVATION 21, 461
DRAG LINE 21, 461
DRAIN FIELD 43, 472
   TILE 51, 52, 476
DRAINAGE BOOT 146, 523
   PIPE 37, 469
   UNDERSLAB 52, 476
DRAPERY TRACK 227, 563
DRAW OUT BREAKER 407, 652
   OUT COMPARTMENT 414, 655
DRAWER BASE 226, 563
DRILLED WELL 31, 466
DRIP PAN ELL 317, 607
DRIVE SCREW 243
DRIVE-UP WINDOW 213, 556
DRIVING CAP 28, 464
   STUD 413, 655
DROP EAR ELL 270, 584
   PANEL 71, 486
DROPCLOTHS 185, 542
DROPOUT CONDUIT 349, 623
DRUM TRAP 276, 587
DRY CLEANER 215, 557
   RUB 83, 492
   SET TILE 176, 538
   STONE WALL 105, 503
   TYPE EXTINGUISHER 207, 552
   TYPE SPRAY BOOTH 221, 560
   TYPE TRANSFORMER 406, 420, 651, 658
DRYER COIN 215, 557
   DARKROOM 220, 559
   RECEPTACLE 402, 649
DRYVIT PANEL 141, 520
DRYWALL 174, 537
   REMOVAL 10, 456
DUCT BANK 354, 626
   BUS 340, 619
   DETECTOR 442, 669
   FEEDER 340, 619
   HEATER 450, 673
   INSULATION 251, 575
   LAY-IN 381, 382, 639
   UNDERFLOOR 379, 380, 638
DUCTWORK 198, 549
   METAL 334, 615
DUMBELL JOINT 81, 491

INDEX

DUMP FEE 230	EXPANSION SHIELD 109	FINK TRUSS 128, 513
TRUCK 6	EXPLOSION PROOF BOXES 396, 646	FIN-TUBE RADIATOR 330, 614
DUMPSTER 11, 230, 456	PROOF BREAKER 422, 659	FIR DECKING............................. 127, 513
DUPLEX BOX 392, 644	PROOF LIGHTING 432, 664	FLOOR 181, 541
PUMP 324, 610	EXPOSED AGGREGATE 101, 501	SIDING 142, 521
DUST CAP 378, 637	EXTENSION BOX 399, 648	FIRE ALARM CABLE 451, 674
DUTCH DOOR 153	EXTERIOR DOOR 157, 158, 529	ALARM SYSTEM 442, 669
DWV COPPER 263, 581	TILE 176	BARRIER 344, 621
FITTING COPPER 272, 585	EXTINGUISHER CABINET................... 207, 553	CONNECTION 252, 575
	FIRE 206, 552	DAMPER 335, 616
- E -	EXTRA HEAVY PIPE 258, 578	DETECTION ANNUNCIATOR 442, 669
	EXTRACTOR LAUNDRY 215, 557	DOOR 158, 529
EB CONDUIT 373, 635	WASHER 214, 556	EXTINGUISHER.................... 206, 552
DUCT 373, 635	EYE NUT 403, 650	RATED BRICK 100, 500
EDGE FORM............................... 73, 487		RATED DOOR...................... 153
FORMS 55, 478	- F -	RATED FRAME..................... 154
FORMWORK.......................... 75, 488		RATED PANEL...................... 179
REPAIR 95, 498	F&M PLATE 242, 571	RESISTANT SHEETROCK..................... 174
EDGEWISE HANGER..................... 340, 619	FABRIC CHAIN LINK 57, 479	RETARDANT 132
ELAPSED TIME INDICATOR 444, 670	FORM 95	RETARDENT ROOF 140
ELASTIC SHEET ROOF 144, 522	FABRICATION METAL 116, 508	FIREPROOF VAULT 213, 556
ELASTOMERIC FLASHING 99, 500	FACING PANEL 106, 503	FIREPROOFING 180, 355, 626
ELBOW ABS 37, 469	PANELS 105, 503	FIREWALL BRICK................... 100, 500
CAST IRON............................ 34, 468	FACTORY FINISH FLOOR 181	FIRING RANGE 214, 556
PLASTIC 36, 469	FAN EXHAUST 333, 615	FITTING CONDULET 374, 636
POLYETHYLENE 293, 595	FASCIA BOARD 124, 511	CPVC 291, 594
PREINSULATED 33, 467	PLASTER 173, 537	NO-HUB 253, 256, 576
PULLING 369, 398, 633, 647	FASTENER INSULATION................... 136, 518	POLYETHYLENE 292, 595
PVC 38, 289, 470, 593	FAUCET 318, 608	POLYPROPYLENE 291, 594
ELECTRIC BOILER 325, 611	FEE PROFESSIONAL 1	PVC 290, 594
ELEVATOR 232, 566	FEEDER DUCT 340, 619	REDUCER COPPER............... 275, 586
FURNACE 326, 611	FELT CONTROL JOINT....................... 81, 491	SOIL 258, 578
HOIST 236, 567	FEMALE ADAPTER COPPER266, 275, 582, 586	STAINLESS 300, 599
LOCKSET 165, 532	ADAPTER PVC 288, 289, 290, 593, 594	WELDED 303, 600
ELECTRICAL FEE 2	CONNECTOR 284, 591	FIXED SASH 161, 531
ELEVATED FLOOR 203, 551	DROP EAR ELL.................... 270, 584	TYPE LOUVER 117, 508
SLAB 75, 488	FLANGED ELL..................... 270, 584	WINDOW 162, 531
STAIRS 73, 487	FLUSH BUSHING 272, 585	FIXTURE LIGHTING 430, 663
ELEVATOR 232, 233, 234, 566	FENCE 56, 478	PLUMBING 317, 607
ENAMEL PAINT 200	POST HOLE 61, 481	SQUARE 432, 664
ENCASED BURIAL CONDUIT 373, 635	REUSE 13, 457	FLAKER ICE 219, 559
ENCLOSED BREAKER 407, 652	SENSOR 443, 670	FLAMEPROOF TILE 180
ENCLOSURE TELEPHONE 210, 554	TENNIS 63	FLAMEPROOFING 180
END BELL 372, 635	FENCES, WOOD OR MASONRY 188, 544	FLANGE 297, 597
CAP 427, 662	FIBER BOARD 93, 497	PVC 288, 289, 290, 593, 594
CLEANOUT 276, 587	CANT 125, 512	FLANGED PIPE 39, 470
CLOSURE 340, 619	FORM CONCRETE............... 71, 486	VALVE 31, 311, 466, 604
GRIP 447, 672	PANEL 179, 540	FLARE END PIPE 50, 475
ENGINEER 2	FIBERBOARD INSULATION 137, 518	FITTING BRASS 284, 591
ENGINEERING FEE 2	SHEATHING 127, 513	FLASHING 144, 146, 522, 523
ENGRAVED PLAQUE................. 204, 205, 551	FIBERGLASS BENCH 62, 481	MASONRY............................ 99, 500
ENTRANCE CAP 360, 367, 629, 632	CABLE TRAY 350, 623	FLAT APRON 129, 514
DOOR 161, 530	DIVING BOA 229, 564	CONDUCTOR CABLE 390, 643
FITTING 365, 631	INSULATION 135-137, 246, 517-518, 572	CONDUIT CABLE 340, 619
MAT 227, 563	PANEL 179, 540	DUMBELL 81, 491
ENTRY LOCK 165, 532	PIPE INSUL 248, 573	RIB 172, 536
LOCKSET 165, 532	PLANTER 68, 484	SLAB 91, 496
EPOXY CONCRETE 95, 498	REINFORCED 295, 596	WASHER 355
FLOORING 184, 542	RIGID 251, 575	FLATWISE HANGER 340, 619
GROUT 95, 498	SEAT 62, 481	FLEXIBLE CONDUIT 362, 630
TERRAZZO 178, 184, 539, 542	TANK 229, 564	CORD 392, 644
EQUALIZER 447, 672	FIBROUS CONCRETE 55, 478	COUPLING 397, 647
EQUIPMENT CURB............................. 72, 487	DAMPPROOFING 135, 517	FLOAT 83, 492
MOBILIZATION 4	FIELD ENGINEER................................. 2	FINISH 83, 492
PAD 85, 493	WELDED TRUSS 110, 505	GLASS 168, 534
PAD REINFORCING................ 77, 489	FILE ROOM DOOR 213, 556	TRAP 315, 606
RACK 447, 672	FILL BORROW 21, 461	FLOOD LAMP 434, 665
ERICKSON COUPLING 364, 631	CAP 230, 565	FLOODLIGHT 439, 668
ESTABROOK TY 276, 587	GYPSUM 93, 497	FLOOR ACCESS 203, 551
ETCH ACID 83, 492	FILLED BLOCK 89, 102, 495, 501	BASE 183, 541
EXERCISE BICYCLE 223, 561	FILLER JOINT 99, 500	BOX 393, 645
EXHAUST HOOD 218, 558	PAINT 200	BUMPER 166, 532
EXIT LIGHT 431, 664	FILM VIEWER 224, 561	CLAY TILE 105, 503
LOCK 165, 532	FILTER BARRIER 26, 463	CLEANOUT 308, 603
EXPANDED LATH 172, 536	CLOTH 26, 463	DRAIN 315, 606
EXPANSION COUPLING 398, 648	LINE 440, 668	FINISH 83, 492
FITTING 343, 620	FINISH PAVING 55, 478	FINISHING.......................... 182, 541
JOINT 80, 173, 491, 537	PLASTER 173, 537	GYM 182, 541
JOINT COVER 117, 508	SHOTCRETE 83, 492	HARDENER......................... 83, 492
JOINT ROOF 148, 524	FINISHING SHEETROCK................... 175, 537	INDUSTRIAL WOOD................ 182, 541

INDEX

FLOOR JOIST 121, 510
   MASONRY 182, 541
   MOUNTED BENCH 206, 552
   MOUNTED TRANSFORMER 406, 651
   PATCH 95
   PEDESTAL 203, 551
   PLATE 115, 507
   REMOVAL 10, 456
   SAFE 228, 564
   SHEATHING 126, 512
   SLAB 71, 91, 486, 496
   TERRAZZO 177, 539
   TILE 175, 177, 538, 539
FLOORING MARBLE 106, 503
FLOORS 198, 549
FLOW CHECK VALVE 310, 604
FLUE LINER 108, 504
FLUID APPLIED FLOOR 182, 541
   APPLIED MEMBRAN 134, 517
   APPLIED ROOF 145, 522
   TYPE GROUT 94, 498
FLUORESCENT FIXTURE ... 430, 436, 663, 666
FLUSH BOLT 166, 533
   BUSHING COPPER 271, 584
   DOOR 153, 157, 527, 529
   PANEL DOOR 170, 535
FLUTED COLUMN 106, 503
   STEEL CASING 28, 465
FOAM EXTINGUISHER 207, 553
   POLYETHYLENE 81, 491
FOAMGLASS CANT 144, 522
   INSULATION 137, 518
FOIL BACKED INSULATION 136, 518
FOLDING DOOR 159, 530
   STAIR 220, 559
FOOD PREPARATION 218, 558
FOOTING EXCAVATION 23, 462
FOREMAN 2
FORM FABRIC 95
   LINER 83, 492
   SLAB 112, 506
FORMBOARD 93, 497
FORMICA 131, 515
   PARTITION 201, 550
FORMWORK OPENINGS 73, 487
FOUNDATION BACKFILL 22, 25, 462
   BLOCK 102, 501
   MAT 87, 494
   VENT 97, 499
FOUNTAIN WASH 322, 609
FRAME CATCH BASIN 49, 475
   COVER 42, 471
   DOOR 157, 529
   MANHOLE 116, 508
FRAMING ANCHOR 119, 509
   CEILING 120, 509
   DOOR 126, 512
   ROOF 123, 511
   SOFFIT 125, 512
   TIMBER 127, 513
   WALL 125, 512
   WINDOW 126, 512
FREIGHT ELEVATOR 233, 566
FREQUENCY METER 415, 656
FRICTION HANGER 241, 570
   PILING 28, 465
FRONT DOOR LOCK 165, 532
FRONT-END LOADER 22, 462
FROZEN FOOD CASE 219, 559
FRYER KITCHEN 217, 558
FS BOX 394, 645
FSA BOX 394, 645
FSC BOX 394, 645
FSCA BOX 394, 645
FSCD BOX 394, 646
FSL BOX 394, 645
FSR BOX 394, 645
FULL MORTISE BUTT 164
   VISION GLASS 168, 533

FURRING CEILING 171, 536
   CHANNEL 112, 506
   STEEL 171, 536
FUSABLE PRESSURE SWITCH 410, 653
   STARTER 345, 621
FUSE 412, 654
FUSED REDUCER 343, 620
   SAFETY SWITCH 409, 653
FUSIBLE SWITCH 344, 411, 424
                           621, 654, 660

- G -

GABION 26, 464
GABLE LOUVER 202, 550
   TRUSS 128, 513
GABLE-END RAFTER 123, 511
GALVANIZED ACCESSORY 82, 492
   DECKING 111, 505
   DUCTWORK 334, 615
   GRATING 115, 507
   MESH 79, 490
   METAL DOOR 153, 527
   PIPE 50, 476
   RAILING 114, 507
   REGLET 75, 488
   REINFORCEMENT 76, 489
   SHINGLE 139, 519
   STEEL 109, 505
   WALL TIE 97, 499
   WIRE FENCE 65, 482
GARBAGE HANDLING 216, 557
GAS FURNACE 326, 611
   METER 43, 472
   PUMP 324, 611
   STERILIZER 224, 561
GASKET CONDULET 368, 633
   GLASS 169, 534
GATE FENCE 59, 480
   POST 56, 479
   SWING 58, 479
   VALVE 309, 603
GEL EPOXY 95, 498
GENERATOR 5
   SET 422, 659
GFI BREAKER 407, 652
GIRDER 110, 505
   PRECAST 90, 495
GIRT 110, 505
GLASS 168, 533, 534
   BACKBOARD 221
   BEAD 130, 514
   CLOTH TILE 180, 540
   DOOR 160, 530
   ENCASED BOARD 204, 551
   FIBER BOARD 93, 497
   FRAMING 170, 534
   WIRE 169, 534
GLASSWARE WASHER 223, 560
GLAZED BLOCK 103, 502
   BRICK 100, 500
   FLOOR 182, 541
   TILE 100, 500
   WALL TILE 175, 538
GLAZING ACCESSORY 169, 534
GLOBE VALVE 311, 604
GLOVE BAG 230
GOLF SHELTER 63, 482
GRAB BAR 210, 554
GRADALL 23, 462
GRADING 25, 463
GRANDSTAND 62, 481
GRANITE BLOCK 101, 501
   CURB REMOVAL 13, 457
   VENEER 106, 503
GRANOLITHIC TOPPING 83, 492
GRANULAR WATERPROOFING 134, 517
GRASS CLOTH 199, 549
GRATE CUTTING 115, 507
   METAL 115, 507

GRATINGS, METAL 193, 546
GRAVEL FILL 52, 476
   MULCH 69, 484
   ROOF 144, 522
   STOP 147, 523
GRAVITY VENTILATOR 150, 524
GRAY SPONGE 81, 491
GREEN CONCRETE 83, 492
GRIDDLE KITCHEN 217, 558
GRILLE DOOR 159, 529
   METAL 117, 508
   VENTILATION 336, 617
GRINDER ELECTRIC 222, 560
GROUND BUS 341
   FAULT 407, 652
   HYDRANT 319, 608
   ROD 413, 655
   SLAB 75, 488
   STRAP 400, 648
GROUNDING LOCKNUT 370, 633
GROUT BEAM 97, 499
   BLOCK 89, 97, 102, 495, 499, 501
   CRACK SEAL 95, 498
   EPOXY 95
   MASONRY 97, 107, 499, 504
   WALL 97, 499
GRS CONDUIT 363, 630
   FITTING 366, 631
GRUBBING 18, 460
GUARD ALERT DISPLAY 444, 670
GUNITE 83, 492
GUTTER 148, 524
   COPPER 147, 523
   REMOVAL 12, 457
   STEEL 148, 524
GUTTERS AND DOWNSPOUT 191, 545
GUY CABLE 447, 672
   GRIP 403, 650
   ROD 447, 672
   TREE 68, 484
GYM CURTAIN 221, 560
   FLOOR 181, 182, 541
   LOCKER 206, 552
GYPSUM FILL 93, 497
   MASONRY 103, 502
   PLANK 94, 498
   PLASTER 173, 537
   SHEATHING 127, 513

- H -

HALF ROUND GUTTER 148, 523
   ROUND MOLDING 130, 514
   SURFACE BUTT 164
HALIDE FIXTURE 433, 437, 665, 667
   LIGHT 435, 666
HAMMERED GLASS 169, 534
HAND BUGGY 84, 493
   DRYER 211, 554
   EXCAVATION 24, 463
   LOADING 19, 460
   SEEDING 66, 483
   SPLIT SHAKE 140, 520
   SPREADING 65, 483
HANDBALL COURT 63
HANDHOLE 402, 649
HANDICAPPED PARKING 62, 481
   PARTITION 201, 550
   PLUMBING 319, 608
   SYMBOL 56, 478
HANDLE LOCK DEVICE 421, 659
HANDLING RUBBISH 11, 456
   WASTE 216, 557
HANDY BOX 399, 648
HANGAR DOOR 158, 529
HANGER BUS DUCT 340, 619
   CHANNEL 354, 625
   JOIST 119, 509
   ROD 354, 625

INDEX

HARDBOARD DOOR ... 156, 528
    UNDERLAYMENT ... 126, 512
HARDENER CONCRETE ... 83, 492
HASP ASSEMBLY ... 166, 533
HAT RACK ... 211, 555
HATCH ROOF ... 149, 524
HAULING RUBBISH ... 11, 456
H-BEAM FENCE ... 57, 479
HEADER ... 125, 512
    CEILING ... 121, 510
    PIPE ... 19, 460
HEADWALL ... 51, 476
    REMOVAL ... 18, 460
HEAT DETECTION ... 442, 669
    RECOVERY ... 328, 613
HEAT/SMOKE VENT ... 149, 524
HEATER ... 5
    DUCT ... 450, 673
    ELECTRIC ... 219, 449, 559, 673
    UNIT ... 331, 614
    WATER ... 321, 609
HEAVY CASTING ... 116, 507
    DUTY DOOR ... 153, 527
    DUTY TERRAZZO ... 177
    STEEL ... 110, 505
    WASHER ... 243
HERRINGBONE FLOOR ... 182
HEX BUSHING ... 279, 589
    NUT ... 240, 243, 355, 626
    TRAP COPPER ... 275, 586
HIGH CHAIR REINFORCING ... 82, 492
    PRESSURE SODIUM ... 435, 437, 666, 667
    SEAM ROOF ... 141, 520
    STRENGTH BLOCK ... 102, 501
HIGHBAY LIGHTING ... 432, 664
HINGE ... 165
    COVER WIREWAY ... 383, 640
HIP RAFTER ... 124, 511
HOIST INDUSTRIAL ... 236, 567
HOLDER DOOR ... 166, 532
HOLE CLIP ... 242, 571
    CONDUIT ... 354, 626
    PLASTER ... 173, 537
HOLLOW BLOCK ... 101, 501
    CLAY TILE ... 104, 502
    COLUMN ... 133, 516
    CORE DOOR ... 156, 157, 528
    METAL DOOR ... 153, 527
    METAL FRAME ... 153, 527
HOOD EXHAUST ... 218, 558
    RANGE ... 219, 559
HOOK STICK ... 344, 621
    WIRE ... 245, 572
HORIZONTAL DAMPER ... 335, 616
    JACKING ... 26, 464
    REINFORCING ... 80, 97, 491, 499
    SIDING ... 142, 521
    WINDOW ... 162, 531
HORN ... 442, 669
    STANDARD ... 444, 670
    WATERPROOF ... 442, 669
HOSE REEL ... 324, 611
HOSPITAL BED ... 224, 561
    CUBICLE ... 202, 550
    EQUIPMENT ... 223, 561
    FAUCET ... 318, 608
    PANEL ... 422, 659
    RECEPTACLE ... 402, 649
    SYSTEM ... 446, 671
    TRACK ... 202, 550
    WARDROBE ... 211, 555
HOT LINE CLAMP ... 404, 650
    PLATE LABORATORY ... 222, 560
    WATER DISPENSER ... 318, 608
    WATER STORAGE TAN ... 324, 611
HP SODIUM ... 432, 664
H-SECTION PILE ... 27, 464
HUB ADAPTER ... 291, 594
    BOLT-ON ... 410, 653
    UNION ... 397, 647

HUNG SLAB ... 71, 486
HYDRANT WALL ... 252, 575
HYDRAULIC ELEVATOR ... 233, 566
    EXCAVATOR ... 21, 23, 461, 462
    LIFT ... 222, 560

- I -

ICE CUBE MAKER ... 218, 558
    MAKER ... 219, 559
    STORAGE BIN ... 219, 559
IMC CONDUIT ... 376, 637
IMPACT BARRIER ... 61, 481
IN LINE PUMP ... 322, 610
    LINE SPLICE ... 46, 473
    LINE SPLITTER ... 448, 672
    LINE TURBINE ... 44, 473
INCANDESCENT FIXTURE ... 431, 664
    LAMP ... 433, 665
INCINERATOR LABORATOR ... 222, 560
    MEDICAL ... 216, 557
INCREASING FEMALE ADA ... 266, 582
    MALE ADAPT ... 267, 583
INCUBATOR ... 224, 561
    LABORATORY ... 222, 560
INDIRECT LUMINARE ... 438, 667
INDOOR METER CENTER ... 416, 656
    TERMINATION ... 46, 473
INDUSTRIAL COMPACTOR ... 216, 557
    FLOURESCEN ... 437, 667
    WINDOW ... 161, 531
    WOOD FLOOR ... 182, 541
INERTIA SENSOR ... 443, 670
INFRA-RED SENSOR ... 443, 670
INJECTION GROUT ... 95, 498
INLAID TAPE ... 56, 478
INLET ... 42, 472
    CONNECTION ... 252, 575
INSULATED BLOCK ... 102, 501
    GLASS ... 168, 533
    GLASS DOOR ... 160, 530
    ROOF ... 145, 522
    SIDING ... 142, 521
    THROAT ... 358, 361, 627, 628, 629
    VAULT DOOR ... 213, 556
INSULATION BLOWN-IN ... 138, 519
    CALCIUM ... 248, 573
    EQUIPMENT ... 251, 575
    FASTENERS ... 136, 518
    MASONRY ... 138, 519
    RIGID ... 136, 518
    SPRAYED ... 138, 519
INTERIOR SIGN ... 205, 552
    WALL FORM ... 74, 488
INTERMEDIATE CONDUIT ... 376, 637
    PIVOT HINGE ... 164
INTERRUPTER SWITCH ... 410, 411, 654
INVERT MANHOLE ... 42, 471
INVERTED ROOF SYSTEM ... 145, 522
    TRAP ... 315, 606
INVERTER SYSTEM ... 440, 668
IONIZATION DETECTOR ... 442, 669
IRON BLACK ... 301, 599
    MISCELLANEOUS ... 109, 505
    TREAD ... 114, 507
IRREGULAR STONE ... 101, 501
IRRIGATION LAWN ... 56
ISOLATION MONITOR ... 224, 561
ISOLATOR VIBRATION ... 246, 572

- J -

JACK RAFTER ... 124, 511
JACKET ALUMINUM ... 248, 573
JACKING PIPE ... 26, 464
JALOUSIE WINDOW ... 162, 531
JOB BUILT FORM ... 70, 486
    SCHEDULING ... 2
    TRAILER ... 4
JOINT CLIP ... 173, 537
    EXPANSION ... 80, 117, 491, 508

JOINT FILLER ... 99, 500
    REINFORCING ... 97, 499
JOIST BRIDGING ... 119, 509
    FLOOR ... 121, 510
    HANGER ... 119, 509
    METAL ... 111, 505
    SISTER ... 121, 122, 510
JUMBO BRICK ... 100, 500
JUNCTION BOX ... 401, 445, 649, 671
JUTE PADDING ... 183, 542

- K -

K COPPER ... 263, 581
    TUBE ... 264
KEENES CEMENT ... 173, 537
KETTLE STEAM ... 217, 558
KEY LOCK ... 208, 553
    SWITCH ... 429, 663
KEYED JOINT ... 81, 491
KEYWAY FORM ... 75, 488
KICK PLATE ... 116, 166, 508, 533
KILN DRIED ... 132
KING POST TRUSS ... 129, 514
KITCHEN CABINET ... 132, 226, 515, 563
    COMPACT ... 216, 557
    FAUCET ... 318, 608
    SINK ... 320, 609
KNOTTY PINE ... 133, 515

- L -

L TUBE ... 263, 264, 581
LADDER POOL ... 229, 564
    REINFORCING ... 97, 499
LADDERS ... 193, 546
LAG ROD ... 243
LAMINATED GLASS ... 168, 534
    PANEL ... 208, 553
    PLASTIC ... 131, 515
    RAILING ... 117, 508
    RUBBER ... 215, 557
LAMP INCANDESCENT ... 433, 665
LANDING STEEL ... 113, 506
    TERRAZZO ... 178, 539
LANDSCAPE TIMBER ... 69, 484
LATCHSET ... 165, 532
LATEX PAINT ... 200
LATH EXPANDED ... 172, 536
    STUCCO ... 173, 537
LATHE SHOP ... 222, 560
LATTICE MOLDING ... 130, 514
LAUAN DOOR ... 156, 528
    VENEER ... 133, 515
LAVATORY CARRIER ... 322, 609
    FAUCET ... 318, 608
LAY-IN DUCT ... 381, 639
LB BOX ... 396, 646
    CONDULET ... 359, 360, 368, 376, 628, 632, 636
LEACHING PIT ... 43, 472
LEAD COATED COPPER ... 147, 523
    LINED DOOR ... 153
    LINED FRAME ... 154
LETTER BOX ... 208, 553
    SLOT ... 207, 553
LEVEL FLOOR ... 183, 541
    GAUGE ... 314, 606
    INDICATOR ... 230, 565
LEVELER DOCK ... 215, 557
LH JOIST ... 111, 505
LI-CHANNEL ... 355, 626
LIFEGUARD CHAIR ... 229, 564
LIFT AUTO ... 235, 567
    HYDRAULIC ... 222, 560
LIGHT CASTING ... 116, 507
    DENTAL ... 225, 562
    FIXTURE LAMP ... 439, 668
    GAGE FRAMING ... 112, 506
    POOL ... 229, 564
    SHIELD ... 61, 481
    STEEL ... 110, 505

791

INDEX

LIGHT SURGICAL..................................224, 561
    TRACK..431, 664
LIGHTING...431, 664
    CONTRACTOR...............................425, 661
LIGHTNING ARRESTER...... 404, 411, 650, 654
LIGHTWEIGHT BLOCK........................102, 501
LIMESTONE AGGREGATE..................101, 501
    PANEL..106, 503
LIMING..66, 483
LINE FILTER..440, 668
    MARKING..56, 478
    POST.......................................57, 58, 479
LINEN CHUTE..............................235, 236, 567
LINER FINISH..92, 496
    FLUE..108, 504
    PANEL.........................141, 229, 520, 564
LINOLEUM REMOVAL...............................10, 456
LINTEL LIMESTONE.............................106, 503
    STEEL...104, 502
LIQUID CHALKBOARD..........................201, 550
    LEVEL GAUGE...............................314, 606
    PENETRATION..............................238, 569
LITHONIA LIGHT.....................................438, 667
LL CONDULET...360, 628
LOAD BEARING STUD............................171, 536
    CENTER BREAKER.......................407, 652
    CENTER INDOOR.........................420, 658
LOADER..6, 7
LOCKNUT BRASS..................................282, 590
    CONDUIT..365, 631
    GROUNDING.................................370, 633
LOCKOUT SWITCH................................427, 662
LOCKSET..165, 532
LOCKWASHER...355
LONG FORGED NUT..............................285, 591
    SWEEP ABS..................................295, 596
    SWEEP SOIL.................................258, 578
    TURN TEE......................................274, 586
LOT LINES..3
LOUVER DOOR...................157, 203, 529, 551
    METAL...117, 508
    PENTHOUSE..................................337, 617
    VENT...202, 550
LOW PRESSURE SODIUM....................436, 666
LR CONDULET..360, 628
LUBE EQUIPMENT................................324, 611
LUMINARE............................438, 444, 667, 670

- M -

M TUBE...264, 581
MAGNETIC DOOR HOLDER.................167, 533
    STARTER...........................423, 424, 660
MAHOGANY VENEER............................133, 515
MAIL CHUTE...207, 553
MAIN BREAKER.....................................416, 656
MALE ADAPTER COPPER.. 267, 275, 583, 587
    ADAPTER PVC........ 288, 289, 290, 593, 594
    CONNECTOR.................................284, 591
    CONNECTOR STAINLESS..............300, 599
MALLEABLE IRON..................................301, 599
    STRAP.....................................359, 368, 628, 633
MAN HOLE COMMUNICATION...........445, 671
MANHOFF ADAPTER COPPER.............274, 586
MANHOLE..41, 471
    COVER....................................116, 507, 508
    ELECTRIC.......................................402, 649
    WATERTIGHT................................116, 508
MAP RAIL CHALKBOARD.....................201, 550
MAPLE FLOOR.......................................181, 541
MARBLE CHIP..69, 484
    IMPORTED....................................105, 503
MASKING PAPER AND TAP..................185, 542
MASONRY BASE....................................106, 503
    CHIMNEY..................100, 108, 500, 504
    CLEANING.....................................107, 504
    DEMOLITION......................................9, 455
    FACING PANEL.............................105, 503
    FENCE..13, 457
    FURRING......................................122, 510

MASONRY GLASS..................................105, 503
    GYPSUM...103, 502
    INSULATION..................................138, 519
    LIMESTONE...................................106, 503
    MANHOLE..12, 457
    PLAQUE..205, 551
    PLASTER..105, 503
    REINFORCING................................80, 491
    SANDBLAST...................................107, 504
    SLATE...106, 503
    WATERPROOFING.......................135, 517
MAT FLOOR..227, 563
    FORMWORK....................................73, 487
    FOUNDATION.................................88, 494
    REINFORCING................................79, 490
MEAT CASE..219, 559
MECHANICAL FEE..2
    JOINT..40, 471
    JOINT PIPE......................................34, 467
    SEEDING...66, 483
MEDICAL EQUIPMENT.........................222, 560
    WASTE..216, 557
MEDICINE CABINET..............................211, 554
MEMBRANE CURING..............................84, 492
    NEOPRENE...................................134, 517
    ROOF..144, 522
    WATERPROOFING.......................134, 517
MERCURY FIXTURE...............................434, 665
    SWITCH..429, 663
    VAPOR..439, 668
MESH SLAB..79, 490
    TIE...99, 500
    WIRE..88, 495
MESSAGE TUBE.....................................235, 567
METAL BASE...173, 537
    BUILDING......................................228, 564
    CEILING SYSTEM.........................180, 540
    CLAD CABLE.................................390, 643
    CLAD DOOR..................................158, 529
    DOOR REMOVAL............................10, 456
    FABRICATION...............................113, 506
    FENCE..56, 478
    HALIDE......433, 435, 437, 438, 665, 666, 667
    LADDER...114, 507
    LINER PANELS.............................141, 520
    LINTEL...104, 502
    LOUVER...153
    PAN STAIR.....................................113, 506
    PAN TILE..180, 540
    PIPE......................................50, 51, 475, 476
    PLAQUE..204, 551
    POLE...403, 650
    RAILING...............................114, 117, 507, 508
    ROOF REMOVAL.............................10, 456
    SHELVING.....................................209, 554
    STAIR...113, 506
    STUD..112, 506
    TOILET PARTITIO.........................201, 550
    WINDOW REMOVAL.......................11, 456
METALLIC COLOR TILE................................176
    GROUT...94, 498
    HARDENER.....................................83, 492
    OXIDE MEMBRA............................134, 517
METER CENTER....................................417, 657
    CENTER INDOOR.........................416, 656
    GAS...43, 472
    SECTION..414, 655
    SERUM...222, 560
MICROPHONE.......................................447, 672
    CABLE..451, 674
MICROSCOPE LABORATORY..............223, 561
MILDEW ERADICATION.......................186, 543
MILFORD HANGER...............................244, 571
MILL FRAMED STRUCTURE................127, 513
MILLED WINDOW..................................164, 532
MINERAL FIBER BOARD.........................93, 497
    FIBER PANEL................................179, 540
    FIBER TILE....................................180, 540
    WOOL...138, 519
    WOOL INSULATION.....................136, 518

MINI PIPE...258, 578
MIRROR..211, 554
    PLATE...169, 534
    TILE..169, 534
MISCELLANEOUS CASTING................116, 507
    IRON...109, 505
    LOCK..165, 532
MIXER...447, 672
    KITCHEN.......................................218, 558
MIXING VALVE................................314, 605, 606
MODIFIER EPOXY...95
MOGUL BASE..439, 668
MOISTURE RELIEF VENT.....................148, 524
MOLDED CASE BREAKER....................407, 652
    RUBBER BUMPER........................215, 557
MOMENTARY SWITCH CONTACT.......429, 663
MONITOR RACK....................................443, 670
MONKEY LADDER...................................63, 482
MONOLITHIC TERRAZZO.....................177, 539
MOP SINK...320, 609
MORTISE LOCKSET..............................165, 532
MOSAIC TILE...176
MOSS PEAT...69, 484
MOTION DETECTOR.............................444, 670
MOTOR OPERATOR......................................61
    STARTER......................................427, 662
MOTORIZED CHECKROOM..................214, 556
MOUNTING BRACKET HEAT................449, 673
    BUCKET...431, 664
MOVABLE LOUVER...............................117, 508
MOVING SHRUB......................................68, 484
    TREE..68, 484
MULCH...68, 484
MULLION SECTION..............................169, 534
    STOREFRONT..............................155, 527
    WINDOW.......................................161, 531
MULTIPLE TEE..91, 496
MUSHROOM STATIONARY..................150, 525

- N -

NAILER CEILING...................................121, 510
NAMEPLATE PLAQUE...........................204, 551
NAPKIN DISPENSER.............................211, 555
NARROW STILE DOOR........155, 170, 528, 535
NATURAL CLEFT SLATE......................101, 501
    GAS METER....................................43, 472
NEMA SWITCH......................................409, 653
NEOPRENE FLASHING........145, 146, 522, 523
    JOINT..80, 491
    MEMBRANE...................................134, 517
    ROOF..144, 522
NESTABLE PIPE......................................51, 476
NET SAFETY..4
    TENNIS..63, 482
NIGHT DEPOSITORY............................213, 556
NIPPLE BLACK IRON............................302, 600
    BUSHED..369, 633
    FIBERGLASS................................296, 597
    OFFSET...369, 633
    STEEL..307, 602
NM CABLE..391, 644
NO-HUB ADAPTER COPPER...............274, 586
    PIPE..........................253, 256, 257, 576, 577
NON-CELLULAR DECKING..................112, 506
NON-DESTRUCTIVE TESTING...........238, 569
NON-DRILLING ANCHOR.............................109
NON-FUSABLE SWITCH......................410, 653
NON-FUSED SWITCH..........................409, 653
NON-METALLIC GROUT........................94, 498
    HARDENER....................................83, 492
NONSHRINK GROUT..............................94, 498
NON-SLIP TERRAZZO..........................178, 539
NORMAN BRICK....................................100, 500
NOSING RUBBER..................................183, 541
NURSE STATION..................................446, 671
    STATION INDICAT.......................446, 671
NUT HEX..355, 626
NYLON CARPET....................................184, 542

INDEX

- O -

O RING 151, 525
OAK DOOR 158, 529
   FLOOR 181, 541
   VENEER 133, 515
OBSCURE GLASS 169, 534
OD BEAM CLAMP 355, 626
OFFICE CUBICLE 208, 553
   PARTITION 208, 553
   SAFE 213, 228, 556, 564
   TRAILER 4
OFFSET CONNECTOR 359, 628
   NIPPLE 369, 633
OGEE MOLDING 130, 514
OIL BASE CAULKING 151, 525
   FIRED BOILER 325, 611
   FURNACE 326, 612
   SWITCH 427, 662
   TANK 229, 564
ONE-HOLE STRAP 368, 633
OPEN TYPE DECKING 111, 505
OPPOSED BLADE 339, 618
ORGANIC BED TILE 176, 538
OS&Y VALVE 309, 603
OUTDOOR FLUORESCENT 437, 667
OUTRIGGER FLAGPOLE 204, 551
OVAL ARCH CULVERT 51, 476
OVEN BAKE 217, 558
   CONVECTION 217, 558
OVERHEAD 1
   DOOR 158, 159, 228, 529, 530
OVERSIZED BRICK 100, 500

- P -

PACKAGED CHILLER 328, 612
   HEATER 330, 614
PAD BEARING 93, 497
   EQUIPMENT 72, 487
   REINFORCING 77, 489
   TRANSFORMER 419, 658
PADDING CARPET 183, 542
PAINT ASPHALTIC 135, 517
   SPRAY BOOTH 221, 560
PALM TREE 68, 484
PAN METAL STAIR 113, 506
   SLAB 71, 486
PANEL ACOUSTICAL 179, 540
   CLADDING 92, 497
   LIMESTONE 106, 503
   LINER 141, 520
   SANDWICH 92, 497
   SIDING 142, 521
PANELBOARD 421, 659
PANIC DEVICE 166, 532
PAPER BACKED INSULATION 136, 518
   BACKED LATHE 173, 537
   BUILDING 135, 517
   CURING 84, 492
PARABOLIC TROFFER 437, 667
PARALLEL BAR 221, 560
   BLADE 338, 618
   DAMPER 335, 616
PARGING 105, 503
PARK BENCH 62, 481
PARKING STALL PAINTING 56, 478
PARQUET FLOOR 182, 541
PARTICLE BOARD INSULATION 137, 518
PARTING BEAD 130, 514
PARTITION CLAY TILE 104, 502
   PANEL 92, 497
   REMOVAL 9, 455
   TOILET 201, 550
   WIRE MESH 209, 553
PASSENGER ELEVATOR 232, 566
PASS-THROUGH WASHER 214, 556
PATCH CONCRETE 83, 492
   HOLE 173, 537
PATIO BLOCK 101, 501

PATTERN STRAINER 314, 606
PAVEMENT CUTTING 17, 459
   MARKING 15, 458
   REMOVAL 14, 458
PAVER GRANITE 101, 501
   MARBLE 106, 503
   STONE 101, 106, 501, 503
PAVING CONCRETE 55, 478
   FINISHES 55, 478
PEA GRAVEL 25
PEAT MOSS 69, 484
PECAN VENEER 133, 515
PEDESTRIAN BARRICADE 3
PEGBOARD 133, 515
PENETRATION SEAL 151, 525
PENNSYLVANIA SLATE 139, 519
PERFORATED CLAY PIPE 52, 476
   PVC PIPE 43, 472
   STRAP 242, 571
PERFORMANCE BOND 7
PERLITE 93, 497
   INSULATION 138, 519
   ROOF INSULATION 136, 518
PHOTO EQUIPMENT 220, 559
PHOTOELECTRIC SENSOR 443, 670
PHOTOGRAPHS 3
PHOTOVOLTAIC FLUORESCENT 435, 666
PHYSICAL THERAPY 223, 561
PICKUP TRUCK 6
PICTURE WINDOW 161, 163, 531
PILASTER 74, 89, 488, 495
PILE REPAIR 96, 498
   SPALL 96, 498
PILING DRIVING CAP 28, 464
   MOBILIZATION 4
   POINT 28, 464
PINE DECKING 127, 513
   DOOR 158, 529
   RISER 133, 516
   SIDING 143, 521
PIPE ABS 294, 595
   ALUMINUM 50, 476
   BOLLARD 61, 481
   CAP 366, 631
   CLAY 41, 471
   CLEANOUT 42, 472
   COLUMN 110, 505
   COPPER 263, 581
   CPVC 291, 594
   EXTRA HEAVY 44, 472
   FIBERGLASS 295, 596
   FLASHING 144, 522
   GALVANIZED 50, 305, 476, 601
   GLASS 287, 592
   HANGER 239, 240, 241, 569, 570
   IDENTIFICATION 301
   INSULATION 247, 250, 573, 574
   NO-HUB 253, 256, 576, 577
   PILE POINT 28, 465
   POLYETHYLENE 292, 595
   POLYPROPYLENE 291, 594
   PVC 290, 593
   RAILING 114, 507
   ROLL STAND 239, 240, 569, 570
   STAINLESS 297, 597
   STRUCTURAL 110, 505
   TAP HOLE 32, 467
PIPES 198, 549
PITCH POCKET 147, 523
PIVOT HINGE 164
PLACEMENT SLAB 85, 493
PLANT BED PREPARATION 66, 483
PLAQUE BRONZE 204, 551
   MASONRY 205, 551
PLASTER MASONRY 105, 503
   PATCH 173, 537
   REMOVAL 10, 456
   RING 400, 648
PLASTERBOARD 174, 537

PLASTIC CONDUIT BUSHING 365, 631
   DUCT BANK 354, 626
   FILTER FABRIC 52, 476
   SHEET 134, 517
   SIGN 205, 552
   SQUARE BOX 400, 648
PLATE 125, 512
   ARMOR 166, 533
   DUPLEX 393, 645
   GLASS 167, 168, 533
   MIRROR 169, 534
   SHEAR 119, 509
   STEEL 110, 505
   SWITCH 427, 429, 662, 663
   TOE 115, 507
PLAYER BENCH 63, 482
PLAYGROUND EQUIPMENT 63, 482
PLENUM CEILING 181, 540
PLEXIGLASS 167, 533
PLUG CAST IRON 308, 602
   CONDUIT 397, 399, 647, 648
   NO-HUB 256
   SOIL PIPE 259, 579
PLUNGER DOOR HOLDER 166, 532
PLYWOOD 126, 512
   FINISH 131, 515
   SHEATHING 127, 513
   SHELVING 131, 515
   STRUCTURAL 127, 513
   UNDERLAYMENT 126, 512
PNEUMATIC HOIST 235, 567
   TOOL 5
POCKET PITCH 147, 523
POINT PILING 28, 464
POINTING MASONRY 107, 504
POLE CLOSET 129, 514
   ELECTRIC 448, 672
   LINE HARDWARE 403, 650
   UTILITY 402, 650
POLYACRYLATE FLOOR 178, 539
POLYESTER FLOOR 179, 539
POLYETHYLENE PIPE 43, 292, 472, 595
   VAPOR BARRIER 135, 517
   WRAP 96, 498
POLYPROPYLENE PIPE 291, 594
POLYSTYRENE INSULATION ... 137-138, 518-519
POLYSULFIDE ACRYLIC 107, 504
POLYURETHANE CAULKING 151, 525
   FLOOR 182, 541
   JOINT 80, 491
   MEMBRANE 134, 517
POOL EQUIPMENT 229, 564
   LIGHT 229, 564
PORCELAIN SHINGLE 139, 519
   SINK 320, 609
POROUS CONCRETE PIPE 50, 51, 475, 476
   FILL 82, 492
PORTABLE BLEACHER 62, 481
   STAGE 214, 556
PORTLAND CEMENT GROUT 94, 498
POST CORNER 58, 479
   HOLE 61, 481
   LINE 58, 479
POSTAL ACCESSORIES 207, 553
POSTS TREATED 128, 513
POTENTIAL TRANSFORMER 415, 656
POURED DECK 93, 497
   INSULATION 138, 519
   VERMICULITE 138, 519
POWER FACTOR METER 415, 656
   MANHOLE 402, 649
   TOOL 222, 560
   TRIP BREAKER 414, 415, 655, 656
PRECAST BEAM 90, 495
   HANDHOLE 402, 649
   LINTEL 104, 502
   MANHOLE 41, 471
   PANEL 92, 497
   SEPTIC TANK 43, 472
   TERRAZZO 178, 539

INDEX

PREFINISHED SHELVING 131, 515
PREFORMED TAPE 56, 478
PREINSULATED FITTING 33, 467
PREMOLDED JOINT 80, 491
PRESSER CLOTHING 215, 557
PRESSURE GAUGE 338, 617
    PIPE 38, 470
    REDUCING VALVE 313, 605
    REGULATORS 44, 472
    SWITCH 410, 653
PRESTRESSED CONCRETE 89, 495
PRIMED DOOR 157, 529
    SIDING 143, 521
PRIMER PAINT 200
PRIVACY LOCKSET 165, 532
    SLAT 61, 481
PROCESSOR PHOTO 220, 559
PRODUCE CASE 219, 559
PROFESSIONAL FEE 1
PROFIT 1
PROJECTING SASH 161, 531
    WINDOW 162, 531
PROTECTION SADDLE 250, 574
PROTECTIVE BOARD 134, 517
P-TRAP 276, 291, 587, 594
    ABS 294, 596
    NO-HUB 254, 257, 576, 577
    SOIL 263, 580
PULL BOX 399, 648
    CABINET 400, 649
    ELBOW 369, 398, 633, 647
    STATION 441, 669
PUMP 5
    GAS 324, 611
    SUMP 323, 610
PUMPED CONCRETE 84, 89, 493, 495
PURGER AIR 314, 606
PURLIN 110, 505
PUSH PLATE 166, 533
PUTTYING 185, 542
PVC CAULKING 151, 525
    CEMENT 374
    CHAMFER STRIP 75, 488
    COATED HANGER 244, 571
    CONDUIT 371, 634
    CONTROL JOINT 99, 500
    FITTING 290, 594
    FOAM 81, 491
    FORM LINER 75, 488
    PIPE 36-43, 52, 289, 469, 470-476, 593
    ROOF 144, 522
    SEWER PIPE 42, 472
    WATERSTOP 81, 491

- Q -

QUADRUPLEX CABLE 387, 642
QUARTER ROUND 130, 514
QUARTZ LAMP 434, 665

- R -

RACEWAY 378, 637
RACK-TYPE DISHWASHER 218, 558
RADIAL WALL FORM 74, 488
RADIATOR VALVE 313, 605
RADIOLOGY EQUIPMENT 223, 561
RADIUS CURB 106, 503
    ELBOW 366, 631
    FORMWORK 75, 488
RAFTER 123, 511
    ANCHOR 119, 509
    HIP 124, 511
RAIL ANCHOR 53, 477
    CHAIR 129, 514
    RELAY 52, 477
RAILING 130, 514
    METAL 117, 508
RAILROAD CROSSING 54, 477
RAINTIGHT ENCLOSURE 401, 649
    TIME CLOCK 428, 662

RANCH CASING 129, 514
    MOLDING 131, 515
RANDOM WIDTH FLOOR 182
RANGE ELECTRIC 220, 559
    FIRING 214, 556
    HOOD 219, 559
    KITCHEN 218, 558
RAPID START LAMP 433, 665
RECEIVING BOX 207, 553
RECEPTACLE CABLES 377, 637
    COMBINATION 402, 649
    GROUND 413, 655
RECEPTOR SHOWER 201, 320, 550, 608
RECESS FILLER STRIP 180
RECESSED LIGHT FIXTURE 181
RECIPROCAL CHILLER 327, 612
RECORD PLAYER 447, 672
RECORDING VOLTMETER 415, 656
RECTANGULAR BOX 395, 646
    WALL TIE 97, 499
RED BRICK 99, 500
    LEAD PAINT 200
    NIPPLE 282, 590
    PILOT LIGHT 429, 663
REDUCED PRESSURE ASSEMBLY 238, 569
REDUCER 349, 623
    CABLE TRAY 353, 625
    CAST IRON 41, 471
    COPPER 265, 582
    FUSE 412, 654
    NO-HUB 257, 577
    POLYETHLENE 293, 595
    PREINSULATED 33, 467
    SOIL 262, 580
    STAINLESS 297, 597
    UNFUSED 343, 620
REDUCING COUPLING COP 264, 581
    ELL STEEL 307, 602
    FEMALE ADAPT 267, 582
    INSERT PVC 288, 289, 290, 593, 594
    LONG SWEEP 260, 579
    MALE ADAPTER 268, 583
    ROD 243, 571
    TEE 302, 599
REDWOOD SIDING 143, 144, 521, 522
REEL HOSE 324, 611
REFLECTIVE PAINT 56, 478
REFLECTOR LAMP 433, 665
    LIGHTING 432, 664
REFLECTORIZED SIGN 62, 481
REFRIGERATED CASE 219, 559
REFRIGERATION TUBING 263, 581
REFRIGERATOR BLOOD 223, 560
REFUSE HOPPER 216, 557
REGISTER WALL 336, 617
REGLET 75, 488
    ROOF 147, 523
REINFORCING ACCESSORY 82, 492
    MASONRY 97, 103, 499, 502
    STEEL REPAIRS 95, 498
RELAY AC 427, 662
    RAIL 52, 477
RELIEF VALVE 313, 605
RELINE PIPE 69, 484
REMOVE CATCH BASIN 12, 457
    DUCT INSULATIO 231, 565
    FENCING 13, 457
    GAS PIPING 15, 458
    HEADWALL 18, 460
    HYDRANT 13, 457
    LOOSE PAINT 186, 543
    ROOF 144, 522
    SEPTIC TANK 8, 455
    SEWER PIPE 16, 459
    TANK 8, 455
    TOPSOIL 65, 483
    VALVE 17, 459
    WALL 18, 460
    WATER PIPE 16, 459
REPLACE VALVE 69, 485

REPLACEMENT GLASS BLOCK 105, 503
    SLATE 139, 519
REPOINT BLOCK 107, 504
RESEED 66, 483
RESET HYDRANT 13, 457
    INLET 42, 472
    MANHOLE 12, 457
RESIDENTIAL CARPET 184, 542
    LOUVER 117, 508
RESPIRATOR MASK 230
RETAINING WALL 74, 488
RETARDANT FIRE 132
REVERSING CONNECTOR 378, 637
    STARTER 423, 660
REVOLVING DOOR 161, 530
RIB LATH 172, 536
    ROOF 141, 520
RIBBED BLOCK 102, 501
    SIDING 143, 521
    WATERSTOP 81, 491
RIDGE BOARD 124, 511
    ROOF 145, 522
    VENT 149, 524
    VENTILATOR 228
RIFLE RANGE 214, 556
RIGID STEEL CONDUIT 363, 630
    URETHANE INSULA 137, 518
RING SPLIT 119, 509
    TOOTHED 120, 509
RISER BAND 214, 556
    FRICTION HANGER 241, 570
    MARBLE 106, 503
    RESILIENT 183, 541
    TERRAZZO 178, 539
    WOOD 133, 516
ROCK CAISSON 30, 466
    DRILL FENCE 61, 481
    DRILLING 8, 455
ROCKWOOL INSULATION 138, 519
ROD BEAM CLAMP 354, 625
    COUPLING 355, 626
    GROUND 413, 655
    GUY 447, 672
    HANGER 354, 625
    THREAD 354, 625
    TRAVERSE 227, 563
ROLL ROOF 139, 146, 519, 523
ROLLER COMPACTED CONCRETE 55, 478
ROLL-UP DOOR 158, 159, 228, 529
ROOF ASPHALT 144, 522
    BALLAST 145, 522
    BOND 144
    BUILT-UP 144, 522
    DECKING 111, 505
    DRAIN 315, 606
    EXPANSION 117, 508
    FLANGE 344, 620
    FLUID-APPLIED 145, 522
    HATCH 149, 524
    HIP 124, 511
    INSULATION 136, 137, 518
    PANEL 140, 229, 520, 564
    SCUPPER 146, 523
    SHINGLE 139, 519
    SIPHON 150, 525
    SLAB 91, 496
    WALKWAY 144, 522
ROOFING REMOVAL 10, 456
ROOFTOP AC 332, 615
ROSEWOOD VENEER 133, 515
ROUGH-IN PLUMBING 318, 607
ROUGH-SAWN SIDING 142, 521
ROUND CAST BOX 393, 395, 645, 646
    COLUMN 71, 486
    QUARTER 130, 514
    TRAP COPPER 275, 586
RUBBER BUMPER 215, 557
    LAMINATED 215, 557
    MAT 167, 533
    NOSING 183, 541

INDEX

RUBBERIZED ASPHALT 54, 478
   TRACK 63, 482
RUBBISH HANDLING 216, 557
   REMOVAL 11, 456
RUBBLE STONE 105, 503
RUNNING TRACK 63, 482
RURAL LETTER BOX 207, 553
RUSTICATED CONCRETE 83, 492

- S -

SADDLE TAPPING 31, 466
SAFE OFFICE 213, 228, 556, 564
SAFETY GLASS 168, 533
   NET 4
   TREAD 183, 541
SALAMANDER 5
SALES TAX 2
SALT PRESERVATIVE 132
SAMPLE SOIL 9, 455
SAND BED PAVER 101, 501
   BORROW 21, 461
   FINISH STUCCO 174, 537
SANDBLAST 83, 492
   FINISH 92, 497
   MASONRY 107, 504
SANDING 185, 542
SANDWICH PANEL 92, 141, 497, 520
SANITARY CROSS NO-HUB 253, 576
   CROSS SOIL 262, 580
   T 291, 594
   T ABS 294, 596
   T COPPER 274, 586
   T NO-HUB 253, 576
   T SOIL 258, 260, 578, 579
SASH BEAD 130, 514
   TRANSOM 154, 527
SATIN FINISH LETTER 205
SAW BAND 222, 560
SCAFFOLDING 4
SCHEDULE-40 PIPE 36, 469
SCHEDULE-80 PIPE 37, 469
SCHEDULING SOFTWARE 3
SCOREBOARD GYM 221, 560
SCORED CLAY TILE 104, 502
SCRATCH COAT 174, 537
SCREED 83, 492
SCREEN MOLDING 130, 514
   VENT 202, 550
SCREW COVER BOX 400, 649
   COVER CABINET 401, 649
SCRUB SURGICAL 224, 561
SCUPPER ROOF 146, 523
SEAL CAULKING 151, 525
   COAT 54, 478
   FITTING 364, 631
SEALING CEMENT 398, 647
SEAL-OFF 397, 647
SEAT BACK 62, 481
   GRANDSTAND 62, 481
   RESTORATION 62, 481
SECTION METER 414, 655
SECTIONAL DOOR 160, 530
SECURITY SASH 161, 531
   SYSTEM 443, 670
SEDIMENT FENCE 26, 463
SEEDING 65, 483
SEE-SAW 63, 482
SEH BOX 393, 645
SEHC BOX 393, 645
SEHL BOX 394, 645
SEHT BOX 394, 645
SEHX BOX 394, 645
SELECT COMMON BRICK 99, 500
   SWITCH 415, 656
SELF-DRILLING ANCHOR 110
SENSOR PHOTOELECTRIC 443, 670
SER CABLE 387, 391, 642, 644
SERVICE DOOR 159, 529
   HEAD 390, 644

SERVICE SINK 320, 609
   WINDOW 213, 556
   WYE SOIL 261, 580
SET RETARDER 83
   SCREW 358, 627
   SCREW CONDULET 359, 628
   SCREW COUPLING 358, 359, 628
SEU CABLE 387, 642
SEWER CLEANOUT 42, 472
   CONNECTION 42, 472
S-FORM CONCRETE 76, 489
SFU CABLE 391, 644
SHAG CARPET 184, 542
SHAKE ROOF 140, 520
SHAPER SHOP 222, 560
SHEAR PLATE 119, 509
SHEATHING ROOF 127, 513
   WALL 127, 513
SHEET FLASHING 146, 523
   FLOOR 182, 541
   GLASS 167, 533
   LEAD 181, 540
   METAL ROOF 145, 522
   MIRROR 169, 534
   PILING 20, 30, 461, 465
   PLASTIC 134, 517
   ROOF 144, 522
SHEETROCK 174, 537
   FINISHING 175, 537
SHELF 131, 515
   CHECKROOM 214, 556
   UNIT 210, 554
SHELTER GOLF 63, 482
   TRUCK 216, 557
SHIELD EXPANSION 109
SHINGLE CEDAR 140, 520
   METAL 139, 519
   REMOVAL 10, 456
   WOOD 140, 520
SHOE MOLDING 130, 514
SHORT ELBOW 369, 633
   FORGED NUT 285, 591
   PATTERN CLAMP 241, 570
SHORTY SOIL PIPE 259, 579
SHOTCRETE 83, 492
SHOWER COMPARTMENT 202, 550
   DOOR 201, 550
   FAUCET 319, 608
   ROD 211, 555
SHRUB 66, 483
   MAINTENANCE 68, 484
SHUNT TRIP 414, 655
SHUTTERS AND LOUVRES 194, 547
SIDE BEAM 70, 486
   BEAM CONNECTOR 243, 571
   COILING GRILLE 159, 530
SIDELIGHT 154, 155, 527
SIDEWALK REMOVAL 15, 458
SIDING 140, 142, 520, 521
   STEEL 143, 521
SIDING, METAL 191, 545
SIGN INTERIOR 205, 552
   PLASTIC 205, 552
SIGNAL GENERATOR 444, 670
SILICON CAULKING 151, 525
SILICONE CAULK 107, 504
   DAMPPROOFING 135, 517
   ROOF 145, 522
SILL ANCHOR 119, 509
   LIMESTONE 106, 503
   MARBLE 106, 503
   SLATE 106, 503
   WINDOWWALL 170, 534
SIMULATED PEG FLOOR 182
SINGLE BARREL LUG 47, 474
   GATE 58, 479
   HUB PIPE 258, 578
   HUB SOIL 259, 579
   TEE 91, 496

SINK CARRIER 322, 610
   DARKROOM 220, 559
   ELL 270, 584
SIPHON ROOF 150, 525
SIREN 442, 669
SISTER JOIST 121, 122, 510
   RAFTER 124, 511
SITE GRADING 25, 463
SKYLIGHT 229, 564
SLAB 91, 496
   BOLSTER 82, 492
   CHANNEL 94, 497
   CONCRETE 85, 87, 88, 493, 494, 495
   DRAINAGE 52, 476
   FORM 112, 506
   FORMWORK 71, 486
   MESH 79, 490
   PRECAST 91, 496
   REINFORCING 77, 79, 489, 490
SLATE 101, 501
   PANEL 106, 503
SLEEPER PIPE 32, 467
SLEEVE COUPLING 296, 597
   TAPPING 32, 466
SLIDE 63, 482
   WATER 229, 564
SLIDING FIRE DOOR 158, 529
   GLASS DOOR 160, 530
   WINDOW 163, 531
SLIP COUPLING COPPER ... 265, 272, 581, 585
   FITTER 361, 629
   JOINT ADAPTER 275, 587
   JOINT COPPER 275, 586
SLOT ANCHOR 98, 499
SLURRY SEAL 54, 478
SMOKE DETECTOR 167, 442, 533, 669
   VENT 149, 524
SMOKESTACK MASONRY 108, 504
SNAPPED PAVER 101, 501
SOAP DISH 177, 211, 538, 555
SOCCER GOAL POST 63, 482
SOCIAL SECURITY 2
SOCKET METER 416, 656
SODIUM FIXTURE 435, 666
   LIGHT 437, 667
SOFFIT REPAIR 95, 498
   TILE 176
   VENT 149, 524
SOFTBALL BACKSTOP 64
SOFTENER WATER 220, 559
SOFTWARE SCHEDULING 3
SOIL BORING 8, 455
   FILTER 26, 463
   PIPE 258, 259, 578
   TESTING 3
SOLAR LIGHT 440, 668
SOLAR SCREEN BLOCK 103, 502
   VALVE 313, 605
SOLID ARMORED CABLE 389, 643
   BLOCK 101, 501
   CORE 156, 528
   WALL 99, 500
   WIRE 389, 643
SOUND ABSORPTION WALL 180, 540
   ATTENUATION 181, 540
   DEADENING BOARD 173, 537
SPALL PILE 96, 498
SPANDREL GLASS 168, 533
SPEAKER 447, 672
SPECIAL DOOR 213, 556
SPELTER SOCKET 448, 672
SPIGOT ADAPTER COPPER 274, 586
SPINNER ROOF 150, 525
SPIRAL REINFORCING 77, 489
SPLASH BLOCK 92, 497
SPLICE COMPRESSION 46, 48, 474
   CONNECTOR 391, 644
   IN LINE 46, 473
   IN-LINE 46, 473
SPLINE CEILING TILE 180, 540

INDEX

SPLIT BOLT CONNECTOR 47, 474
   COUPLING 364, 630
   GROUND FACE 102, 501
   RIB PROFILE 102, 501
   RING 119, 509
   RING HANGER 242, 570
SPONGE NEOPRENE 81, 491
   RUBBER PADDING 183, 542
SPOOL INSULATOR 403, 650
SPOT LAMP 434, 665
SPOTLIGHT 434, 665
SPRAY BOOTH 221, 560
SPREAD FOOTING 86, 494
   TOPSOIL 25, 65, 463, 483
SPREADER FEEDER 215, 557
SPREADING TOPSOIL 65, 483
SPRING HANGER 344, 621
   NUT 355, 626
SPRINKLER HEAD 252, 575
SQUARE COLUMN 70, 486
   FACE BRACKET 399, 648
   FIXTURE 432, 664
   HEAD COCK 313, 605
   HEAD PLUG 281, 308, 589, 602
   MANHOLE 116, 508
   NUT 355, 626
   TUB 317, 607
SQUASH COURT 63
STAGE BAND 214, 556
   THEATER 214, 556
STAGING 4
STAINLESS COUNTER 226, 563
   DOOR 159
   EXTINGUISHER 207, 553
   FITTING 300, 599
   FLASHING 146, 523
   GRAB BAR 210, 554
   PARTITION 201, 550
   PIPE 298, 598
   RAILING 117, 508
   REGLET 147, 523
   SHELVING 210
   TILE 180, 540
   TUBING 299, 598
STAIR ACCESS 220, 559
   ACCESSORY 183, 541
   LANDING 113, 506
   MARBLE 106, 503
   PAN 113, 506
   RAIL 114, 507
   TERRAZZO 177, 539
   TREAD 178, 539
   TREAD CERAMIC 177, 539
STALL SHOWER 201, 550
STANDING-SEAM ROOF 141, 145, 520, 522
STARTER 427, 662
   MAGNETIC 423, 424, 660
   REVERSING 424, 660
STATIONARY SIPHON 150, 524
STEAM BOILER 325, 611
   CLEAN 186, 543
   CLEAN MASONRY 107, 504
   GENERATOR MEDICAL 224, 561
   HEAT 329, 613
   KETTLE 217, 558
   STERILIZER 223, 561
   TRAP 315, 316, 606, 607
STEAMER ELECTRIC 218, 558
STEEL ANGLE 104, 502
   BOX BEAM 61, 481
   BRACE 403, 650
   COFFERDAM 20, 461
   EDGING 68, 484
   FORMS 55, 478
   GRATE 115, 507
   GUTTER 148, 524
   HATCH 149, 524
   LINTEL 110, 505
   PIPE 43, 305, 472, 601
   POLE 403, 650

STEEL REPAIR 95, 498
   ROOF PANEL 141, 520
   SHINGLE 139, 519
   SIDING PANEL 142, 521
   STAIR 113, 506
   STRAP 369, 633
   STRUCTURAL 110, 505
   STUD 171, 536
   TANK 229, 246, 564, 572
   TOILET PARTITION 201, 550
   TRANSOM 154, 527
   TREAD 114, 506
   WELDING 109, 505
STEEL, MEDIUM TO HEAVY 195, 547
STEP LIMESTONE 106, 503
   MANHOLE 42, 116, 472, 508
STEPPING STONE 68, 484
STERILIZER GAS 224, 561
   MEDICAL 223, 561
STICK HOOK 344, 621
STILL WATER 224, 561
STIRRUP POLE 403, 650
STONE 25
   CHEMICAL CLEANING 185, 542
   PAVER 101, 501
STOOL MARBLE 106, 503
   MOLDING 131, 515
   SLATE 107, 504
STOP GRAVEL 147, 523
   MOLDING 130, 514
STORAGE SHELVING 209, 554
   TANK REMOVAL 8, 455
   TANKS 188, 544
STOREROOM LOCK 165, 532
STORM DOOR 160, 530
   WINDOW 162, 531
STRAIGHT CROSS SOIL 261, 580
   CURB FORM 71, 486
   T SOIL 258, 260, 262, 578, 579, 580
STRANDED CONDUCTOR 386, 641
   WIRE 388, 389, 642, 643
STRAP CHANNEL 354, 625
   CONDUIT 376, 636
   MALLEABLE 359, 369, 628, 633
   STEEL 357, 369, 627, 633
   TIE 119, 509
STRAW BALE 26, 463
STREET ELL COPPER 269, 273, 583, 585
   REMOVAL 14, 458
STRETCHER HOSPITAL 224, 561
STRINGER TERRAZZO 178, 539
STRIP CANT 125, 512
   CHAMFER 75, 488
   FLOORING 181, 541
   FLOURESCENT 437, 667
   SHINGLE 139, 519
STROBE LIGHT 442, 669
STRUCTURAL BACKFILL 25
   EXCAVATION 22, 462
   FEE 1
   PIPE 110, 505
   PLYWOOD 127, 513
   SHAPE 110, 505
   TUBE 110, 505
STUCCO 174, 537
   LATH 173, 537
STUD CLIP 173, 537
   METAL 112, 171, 506, 536
   REMOVAL 9, 455
STUMP 19, 460
STYROFOAM INSERT 102, 501
SUB-BASE 20, 461
SUB-FLOORING 126, 512
SUBMERSIBLE PUMP 5
SUBSTATION SWITCH BOARD 415, 656
   TRANSFORMER 419, 658
SUCTION DIFFUSER 323, 610
SUMP PUMP 323, 610
SUN SCREEN 92, 497
SUPERINTENDENT 2

SUPPORT CONNECTOR 427, 662
SURFACE BOLT 166, 532
   COVER 400, 648
   MOUNTED CLOSE 166, 532
SURGICAL CHRONOMETER 225, 561
   LIGHT 224, 561
   SCRUB 224, 561
   STATION 446, 671
   TABLE 224, 561
SURVEILLANCE SYSTEM 213, 556
SUSPENDED CEILING 135, 517
   CEILING REM 10, 456
   HEATER 449, 673
SUSPENSION SYSTEM 181, 540
SWALE EXCAVATION 23, 462
SWAY BRACE COLLAR 344, 621
SWEEP SOIL 260, 579
SWING 63, 482
   CHECK VALVE 252, 575
   DOOR 209, 554
   DOWN DOOR 149, 524
   GATE 58, 59, 479, 480
SWITCH BOX 399, 648
   DIMMER 428, 662
   FUSABLE 344, 411, 621, 654
   OIL 427, 662
   PLATE 429, 663
   PRESSURE 410, 653
   SAFETY 409, 653
   SELECT 415, 656
   TIE 53, 477
   TRANSFER 430, 663
SWITCHBOARD INSTRUMENT 415, 656
SWIVEL HANGER BOX 399, 648
   JOINT 276, 587
   TRAP 276, 587
SYNTHETIC LATEX MASTIC 179, 539
SY-PINE DECKING 127, 513
   SIDING 143, 521
   TIMBER 128, 513

- T -

T BOX 396, 647
   CONDULET 360, 368, 375, 628, 632, 636
T&G SIDING 143, 521
TA BOX 397, 647
TABLE LABORATORY 222, 560
TACK BOARD 204, 551
   COAT 54, 478
TANDEM BOX 395, 646
   BREAKER 407, 652
TANK TOILET 321, 609
TANK-TYPE EXTINGUISHER 206, 552
TAP BOX 341, 378, 619, 637
   CONNECTOR 391, 644
   HOLE 32, 467
   TRANSFORMER 342, 620
TAPE RECORDER 447, 672
TAPERED FRICTION PILE 28, 465
TAPPED SANITARY T 258, 578
   T NO-HUB 254, 576
   T SOIL 262, 580
TAPPING SLEEVE 32, 466
   VALVE 32, 467
TARPAULIN 4
TAX 2
T-BAR SYSTEM 181, 540
TEAK VENEER 133, 515
TEE ABS 38, 294, 469, 596
   BLACK IRON 302, 599
   BRASS 278, 284, 588, 591
   BUS DUCT 341, 619
   CAST IRON 34, 40, 468, 471
   CONNECTOR 427, 662
   COPPER 271, 276, 277, 584, 587
   FIBERGLASS 296, 597
   MEMBER 91, 496
   PLASTIC 37, 469
   POLYETHLENE 293, 595

INDEX

TEE PREINSULATED	33, 467
PVC	38, 288, 289, 290, 470, 593, 594
STAINLESS	297, 597
STEEL	306, 602
WELDED	303, 600
TELEPHONE STANDARD	445, 671
TELESCOPING BLEACHER	221, 560
TELLER WINDOW	213, 556
TEMPERED PEGBOARD	133, 515
TEMPERING VALVE	313, 605
TENNIS COURT	63, 482
TERMINAL ADAPTERS	372, 634
BLOCK	48, 391, 474, 644
BOX	416, 418, 656, 657
SCREW	444, 670
TERMINATION INDOOR	46, 473
TERMITE CONTROL	26, 203, 464, 551
TERRA COTTA	105, 503
TERRAZZO EPOXY	184, 542
RECEPTOR	201, 550
REMOVAL	10, 456
TEST CAP COPPER	271, 584
CONNECTION	252, 575
PILE	29
PIT	9, 455
T NO-HUB	253, 576
TESTING WELD	238, 569
TETHER BALL	63, 482
TEXTURE 1-11	142, 521
THERMAL DETECTOR	442, 669
THERMOPLASTIC TAPE	56, 478
THERMOSTAT	449, 673
CABLE	452, 674
THERMOSTATIC VALVE	314, 605
THHN-THWN CONDUCTOR	388, 642
THIN-SET TERRAZZO	178, 539
THREAD ROD	354, 625
THREADED BOLT	119, 509
COUPLING	366, 632
ROD	240, 242, 570, 571
ROD COUPLING	355, 626
VALVE	309, 603
THREADLESS COUPLING	367, 632
THRESHOLD DOOR	167, 533
MARBLE	106, 503
THROAT INSULATED	358, 627
THROUGH-WALL FLASHING	99, 500
THWN CONDUCTOR	388, 642
TIE MESH	99, 500
PLATE	53, 477
PLUG	53, 477
STRAP	119, 509
TILE ACOUSTICAL	180, 540
CLAY	104, 502
CONDUCTIVE	176, 538
QUARRY	177, 539
REMOVAL	10, 456
RESILIENT	182, 541
STRUCTURAL	100, 500
TERRAZZO	178, 539
TILT-UP BRACE	92, 497
TIMBER FRAMING	122, 127, 510, 513
GUARD RAIL	61, 481
LANDSCAPING	69, 484
PARKING BARRIER	62, 481
POLE	448, 672
TIME CLOCK	428, 662
SWITCH	428, 662
TIMEKEEPER	2
TINTED GLASS	168, 533, 534
TISSUE DISPENSER	211, 555
TOE PLATE	115, 507
TOGGLE SWITCH	428, 662
TOILET	321, 609
ACCESSORY	211, 555
CARRIER	322, 610
TOOTHBRUSH HOLDER	211, 555
TOOTHED RING	120, 509
TOP BEAM CLAMP	243, 571
COILING GRILLE	159, 529
TOP PLATE	125, 512
TOPDRESS SOIL	65, 483
TOPPING CONCRETE	85, 493
GRANOLITHIC	83, 492
TOWEL BAR	211, 555
BAR CERAMIC	177, 538
DISPENSER	211, 555
TOWER	447, 672
COOLING	328, 613
TRACK ACCESSORY	53, 477
BOLT	53, 477
LIGHTING	431, 664
MOUNTED LOADER	22, 462
RUNNING	63, 482
SPIKE	53, 477
TRAFFIC SIGN	62, 481
TRAILER	4
TRANSFER SWITCH	404, 650
TRANSFORMER	346, 403-420, 622, 650-658
POTENTIAL	415, 656
TAP	342, 620
TRANSITION BOX	390, 644
BOXES	390, 644
TRANSLUCENT PANEL	150, 525
TRANSMITTER AUTOMATIC	442, 669
ULTRASONIC	443, 670
TRANSOM	154
SASH	154, 527
TRAP FLOAT	316, 606
GREASE	308, 603
TRAPEZE HANGER	355, 626
TRASH CHUTE	216, 236, 557, 567
HANDLING	216, 557
TRAVELING CRANE	236, 568
TRAVERSE ROD	227, 563
TRAVERTINE FLOOR	182, 541
FLOORING	106, 503
TRAY CABLE	347-349, 350-353, 622-625
TREAD ABRASIVE	114, 506
MARBLE	106, 503
METAL	113, 506
RESILIENT	183, 541
SAFETY	183, 541
STEEL	114, 506
WOOD	133, 516
TREATED LUMBER	69, 484
POLE	403, 650
POST	128, 513
TREE MAINTENANCE	68, 484
REMOVAL	19, 460
SMALL	66, 483
TRIMMING	19, 460
TRELLIS PRECAST	93, 497
TRENCH PAVEMENT	14, 458
TRENCHER	24, 25, 463
TRIANGULAR PILE CAP	72, 487
TRIM	185, 192, 198, 542, 546, 549
ALUMINUM	146, 523
BANKS	23, 462
CARPENTRY	129, 514
CASING	129, 131, 514, 515
TRIMMING TREE	19, 460
TRIPLE OVEN	217, 558
TROFFER	437, 667
TROWEL FINISH	173, 537
FINISH STUCCO	174, 537
TRUCK	6
CRANE	7
SHELTER	216, 557
TRUSS	110, 505
REINFORCING	97, 499
STEEL	111, 505
WOOD	128, 513
T-SECTION BUMPER	215, 557
WATERSTOP	146, 523
T-SERIES TILE	100, 500
T-SPLICE	46, 473
TUB BATH	317, 607
TUBE STRUCTURAL	110, 505
TUMBLER HOLDER	211, 555
SWITCH	396, 646
TUMBLESTONE AGGREGATE	101, 501
TURBINE IN LINE	44, 473
VERTICAL	323, 610
TURNOUT	53, 477
TWIN ELL COPPER	272, 585
TWIST LOCK RECEPTACLE	401, 649

- U -

U BOLT	238, 569
LAMP	437, 667
UF CABLE	391, 644
ULTRASONIC TRANSMITTER	443, 670
UNDER SLAB SPRAYING	203, 551
UNDERFLOOR DUCT	379, 380, 638
UNDERGROUND TANK	229, 564
UNDERLAYMENT	126, 512
UNDERPINNING	20, 461
UNDERSLAB DRAINAGE	52, 476
UNEMPLOYMENT TAX	2
UNFACED INSULATION	136, 518
UNFUSED REDUCER	343, 620
UNGLAZED FLOOR	182, 541
FLOOR TILE	175, 538
WALL TILE	176, 538
UNGROUNDED CABLE	45, 473
UNION BLACK IRON	302, 599
BRASS	279, 588
COPPER	272, 276, 277, 585, 587
ELBOW BRASS	284, 591
HUB	397, 647
PVC	288, 289, 290, 593, 594
STAINLESS	297, 300, 597, 599
STEEL	307, 602
T STAINLESS	300, 599
UNIT HEATER	449, 673
KITCHEN	216, 557
MASONRY	103, 502
UNIVERSAL TERMINATION	444, 670
UNRATED DOOR	153, 527
UP DAMPER	335, 616
UPS BACKUP	442, 669
SYSTEM	441, 669
URD-XLP CABLE	387, 642
URETHANE BOARD	251, 575
INSULATION	137, 138, 518, 519
PADDING	184, 542
RIGID	251, 575
ROOF	145, 522
URINAL CARRIER	322, 610
SCREEN	201, 550
URN COFFEE	217, 558
UTENSIL WASHER MEDICAL	224, 561

- V -

VACUUM CARPET	184, 542
SYSTEM	213, 556
VALLEY FLASHING	146, 523
VALVE TAPPING	32, 467
TRIM	252, 575
VACUUM	213, 556
VANITY BATH	226, 563
VAPOR BARRIER	81, 134, 251, 491, 517, 575
VARMETER	415, 656
VARNISH PAINT	200
VAULT DOOR	158, 213, 529, 556
VAULTED ROOF PANEL	150, 525
V-BEAM ROOF	145, 522
VEGETATION CONTROL	68, 484
VENEER	133, 515
ASHLAR	106, 503
BRICK	99, 500
GRANITE	106, 503
VENETIAN BLIND	227, 563
TERRAZZO	178
VENT CAP	230, 565
FOUNDATION	97, 499
MASONRY	202, 550

INDEX

VENT ROOF 148, 524
   SMOKE 149, 524
   SOFFIT 149, 524
VENTILATOR GRAVITY 150, 525
   RELIEF 336, 617
VERMICULITE 93, 497
   INSULATION 138, 519
   PLASTER 173, 537
VERMONT SLATE 139, 519
VERTICAL CHECK VALVE 311, 604
   CUTTER 218, 558
   GRAIN FLOOR 181, 541
   LIFT DOOR 160, 530
   LOUVER 202, 550
   REINFORCING 80, 97, 103, 491, 499, 502
   TURBINE 323, 610
VINYL COATED FENCE 57, 479
   FACED PANEL 179
   FENCE 64, 482
   SHEET 182, 541
   SHEET FLOOR 182, 541
   SHEETROCK 174, 537
   SIDING 142, 521
   TILE 182, 541
   WALL COVERING 199, 549
VISION GLASS 153
   PANEL 213, 556
VISUAL AID BOARD 204, 551
VOLTMETER 415, 656

- W -

WAFERBOARD SHEATHING 127, 513
WAINSCOT TERRAZZO 178, 539
WALK 88, 495
WALKWAY ROOF 144, 145, 522
WALL BASE 183, 541
   BLOCKING 120, 509
   BRACKET 349, 623
   BRICK 99, 500
   CABINET 132, 226, 515, 563
   CAVITY 89, 97, 495, 499
   CLEANOUT 308, 603
   EXPANSION 117, 508
   FABRIC 199, 549
   FINISH 83, 492
   FLANGE 344, 620
   FOOTING 72, 487
   FORMWORK 75, 488
   FURRING 112, 122, 506, 510
   GRILLE 336, 617
   HEATER 219, 559
   HUNG SINK 319, 608
   HYDRANT 252, 319, 575, 608
   INSULATION 136, 137, 518
   LATH 172, 536
   LOUVER 202, 550
   MASONRY REINFORCING 80, 491
   MOUNTED FLAGPOLE 204, 551
   MOUNTED HEATER 449, 673
   PANEL 92, 170, 229, 496, 535, 564
   PENETRATION 238, 569
   PLASTER 173, 537
   RAILING 114, 507
   REMOVAL 11, 456
   RETAINING 74, 488
   SHEETROCK 174, 537
   STONE 105, 503
   TIE 97, 499
   TILE 175, 177, 538, 539
WALLS 192, 199, 546, 549
   AND FLAT SURFACE 185, 542
WALNUT DOOR 158, 529
   VENEER 133, 515
WARDROBE 226, 563
   LOCKER 206, 552
WASH FOUNTAIN 322, 609
WASHER 243
   COIN 215, 557
   EXTRACTOR 214, 556

WASHER FLAT 355
WASHING (GENERAL) 186, 543
   BRICK 107, 504
   MACHINE BOX 320, 609
WASHROOM FAUCET 319, 608
WASTE RECEPTACLE 211, 555
WATCHMAN 2
WATER BATH 223, 560
   BATH LABORATORY 222, 560
   CLEANING/PREPAR 186, 543
   COOLED CHILLER 327, 612
   COOLER 322, 609
   MOTOR ALARM 252, 575
   PIPE CHILLED 32, 467
   PIPE RESTORATION 69, 484
   REDUCING ADMIXTURE 83
   SLIDE 229, 564
   SOFTENER 220, 559
   STILL 224, 561
   TIGHT HUB 410, 653
   VALVE 313, 605
WATER-RESISTANT SHEET 174
WATERSTOP 81, 491
   T-SECTION 146, 523
WATERTIGHT MANHOLE 42, 116, 472, 508
WATTMETER 415, 656
WAX FLOOR 182, 541
WEARING SURFACE 55, 478
WEATHER SEAL 344, 620
WEATHERHEAD 365, 631
WEATHERPROOF BELL 442, 669
   BOX 395, 646
   COVER 399, 648
   DUCT 340, 619
   INSULATION 251, 575
   SWITCH 427, 662
WELD TESTING 238, 569
   X-RAY 238, 569
WELDED PIPE 303, 600
   RAILING 114, 507
   TRUSS 111, 505
WELDING 109, 505
   TEST 3
WELDMENT CONNECTION 427, 662
WELLPOINT 19, 460
WET BED TILE 175, 538
   LOCATION FLUORESCENT 437, 667
   PIPE VALVE 252, 575
   RUB 83, 492
   VALVE 252, 575
WHEEL BARROW 24, 463
   CHAIR PARTITION 201, 550
WHEELED EXTINGUISHER 206, 552
WHITE CEDAR SHINGLE 140, 520
   CEMENT 92, 497
WIDE STILE DOOR 155, 170, 528, 535
WIND DRIVEN SPINNER 150, 525
WINDOW 162, 163, 531, 532
   DRIVE-UP 213, 556
   FRAME 164, 532
   HEADER 125, 512
   REMOVAL 11, 456
   SERVICE 213, 556
   SILL 100, 500
   SILL MARBLE 106, 503
   STEEL 161, 531
   STILL CERAMIC 177, 539
   STOOL MARBLE 106, 503
WINDOWS 192, 546
WINDOWWALL 170, 534, 535
WIRE ALUMINUM 386, 642
   CHICKEN 65, 482
   GLASS 169, 534
   GUARD 437, 667
   HOOK HANGER 244, 571
   MESH 79, 88, 490, 495
   MESH REMOVAL 14, 458
   MESH WALL 209, 553
   STRANDED 388, 642
WIREWAY FITTING 383, 385, 640, 641

WIRING FLEXIBLE 377, 637
WOOD ANCHOR 119, 509
   BENCH 62, 481
   BLEACHING 185, 542
   CANT 125, 144, 512, 522
   CHIP MULCH 68, 484
   COLUMN 133, 516
   CUTTING 18, 460
   DECKING 127, 513
   DOOR REMOVAL 10, 456
   FENCE REMOVAL 13, 457
   FIBER INSULATION 137, 518
   FIBER PANEL 179, 540
   FIBER TILE 180, 540
   FINISH CONCRETE 83, 492
   FLOOR REMOVAL 10, 456
   FRAME 157, 529
   HANDRAIL 117, 508
   INDUSTRIAL FLOOR 182, 541
   PILE 29, 465
   PLATE 125, 512
   POLE 402, 650
   RISER 133, 516
   SCREW 243
   SEAT 62, 481
   SHELVING 131, 515
   SLEEPER 125, 512
   SOFFIT 125, 512
   STUD REMOVAL 10, 456
   WALL 125, 512
   WINDOW REMOVAL 11, 456
WOODWORK ARCHITECTURAL ... 133, 515
WOOL CARPET 184, 542
   MINERAL 138, 519
WYE 291, 594
   ABS 295, 596
   CAST IRON 41, 471
   FIBERGLASS 296, 597
   NO-HUB 253, 576
   POLYETHLENE 293, 595
   SOIL 258, 261, 578, 579

- X -

X BOX 397, 647
   CONDULET 368, 632
XHHW WIRE 388, 643
XLP CABLE 386, 642
   CONDUCTOR 386, 641
   COPPER 45, 473
   WIRE 389, 643
X-RAY 225, 562
   EQUIPMENT 224, 561
   WELD 238, 569

- Y -

Y STRAINER 314, 606

- Z -

Z-TIE WALL 98, 499

Other Estimating References from BNi Building News

2013 BNi General Construction Costbook

Over 12,000 unit prices provide you with cost estimates for all aspects of construction -- from sitework and concrete, to doors and painting.

The BNi General Construction Costbook 2013 is broken down into material and labor costs, to allow for maximum flexibility and accuracy in estimating. You'll find this costbook data invaluable when preparing detailed estimates, bids, checking prices, and calculating the impact of change orders. You also get detailed man-hour tables that let you see the basis for the labor costs based on standard productivity rates. There are also extensive reference tables, providing supporting information, cost derivations, man-hours, estimating tips, and other related information.

Includes a Free PDF version of this book that you can customize by entering you own rates for 13 different labor classifications, and have every single cost item completely re-calculated. **568 pages, 8½ x 11, $99.95**

2013 BNi Facility Manager's Costbook

Quickly and easily estimate the cost of renovations, repairs and new construction for all types of facilities and commercial buildings. The BNi Facilities Managers Costbook 2013 is the first place to turn, whether you're preparing a preliminary estimate, evaluating a contractor's bid, or submitting a formal budget proposal. It provides accurate and up-to-date material and labor costs for thousands of cost items. Provides material costs for thousands of items. Labor costs based on the prevailing rates for each trade and type of work, PLUS man-hour tables tied to each unit costs, so you can clearly see exactly how the labor cost were calculated and make any necessary adjustments. Equipment costs -- including rental and operating costs. Square-foot tables based on the cost-per-square-foot of hundreds of actual projects -- invaluable data for quick, ballpark estimates. **798 pages, 8½ x 11, $99.95.**

2013 BNi Public Works Costbook

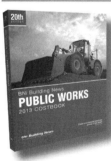

Now you can quickly and easily estimate the cost of all types of public works projects involving roads, excavation, drainage systems and much more. The BNi Public Works Costbook 2013 is the first place to turn, whether you're preparing a preliminary estimate, evaluating a contractor's bid, or submitting a formal budget proposal. It provides accurate and up-to-date material and labor costs for thousands of cost items, based on the latest national averages and standard labor productivity rates. It features material costs for thousands of items based on current national averages (including allowances for transport, handling and storage).

Labor costs based on the prevailing rates for each trade and type of work, equipment costs -- including rental and operating costs.

Square-foot tables based on the cost-per-square-foot of hundreds of actual projects -- invaluable data for quick, ballpark estimates. **478 pages, 8½ x 11, $99.95.**

2013 BNi Mechanical/Electrical Costbook

From air conditioning units to mechanical apparatus, this unique cost guide provides extensive coverage of mechanical/electrical components. Easily estimate the cost of all types of projects involving new construction and renovations in both commercial and residential buildings.

Includes accurate and up-to-date material and labor costs for thousands of mechanical and electrical cost items based on national averages and standard labor productivity rates.

Includes regional cost modifiers and material costs for thousands of items based on current national averages (including allowances for transport, handling and storage) Labor costs based on the prevailing rates for each trade and type of work, PLUS man-hour tables tied to each unit costs, so you can see how the cost was calculated and make any adjustments. Equipment costs: including rental and operating costs. **506 pages, 8½ x 11, $99.95**

2013 BNi Electrical Costbook

From meter to duct, conduit to receptacle, this unique guide provides extensive coverage of the most technical aspects of building construction. Quickly and easily estimate the cost of all types of electrical projects involving new construction and renovations in both commercial and residential buildings.

Provides accurate, up-to-date material and labor costs for thousands of cost items, based on the latest national averages and standard labor productivity rates. Includes detailed regional cost modifiers for adjusting your estimate to your local conditions. Features material costs for thousands of items based on current national averages (including allowances for transport, handling and storage). Labor costs based on the prevailing rates for each trade and type of work, PLUS man-hour tables tied to each unit costs. Includes equipment rental and operating costs. Square-foot tables are included for quick, ballpark estimates.
354 pages, 8½ x 11, $91.95

2013 BNi Home Builder's Costbook

Here's the easy way to estimate the cost of all types of residential construction projects! Accurate and up-to-date material and labor costs for thousands of cost items, based on the latest national averages and standard labor productivity rates.

Includes detailed regional cost modifiers for adjusting your estimate to your local conditions. Material costs are included for thousands of items based on current national averages (including allowances for transport, handling and storage). Labor costs based on the prevailing rates for each trade and type of work, PLUS man-hour tables tied to each unit costs, so you can clearly see exactly how the labor cost was calculated and make any necessary adjustments. **336 pages, 8½ x 11, $81.95**

...deler's Costbook

...ped in cooperation with the National ...ation of Home Builders, this guide lets ...uickly and easily estimate the cost of all ...s of home remodeling projects, including ...itions, new kitchens and baths, and much ...re. It provides accurate and up-to-date ...aterial and labor costs for thousands of ...ost items, based on the latest national averages and standard labor productivity rates. Includes detailed regional cost modifiers for adjusting your estimate to your local conditions. Includes material costs for thousands of remodeling items as well as labor costs for each trade and type of work. Includes equipment costs including rental and operating costs and square-foot tables for quick, ballpark estimates. **497 pages, 8½ x 11, $85.95**

Standard Estimating Practice

A complete guideline on estimation practice by the American Society of Professional Estimators. This practical "how-to" reference presents a standard set of practices and procedures proven to create consistent construction estimates. every step of estimating is covered in detail -- from spec and plan review to what to expect on bidding day. With the procedures in this book you'll see your estimating results become more accurate and more consistent. It includes practical checklists and forms to help you include everything in your bids -- including insurance, outside services, taxes, equipment rental and much more. **526 pages, 8½ x 11, $89.00**

2013 ACE Guide to Construction Costs

Since 1969, architects, contractors and engineers alike have looked to the ACE (Architects, Contractors, Engineers) Guide to Construction Costs as a practical resource for all their construction needs. Whether the job is for general construction, remodeling, building maintenance, or repair, the ACE Guide to Construction Costs provides the most accurate and up-to-date data for material and installation costs, labor and equipment rates, and even adjusted allowances for overhead and profit. It also breaks down all unit and summary costs for every type of structure -- all organized in the 16-Division CSI Masterformat. It also includes prevailing wage rates for over 400 U.S. Metropolitan areas, square foot costs, Americans with Disabilities costs (ADA), production and demolition rates, energy factors, equipment rental rates and much more! **160 pages, 8½ x 11, $62.95**

2013 Architect's Square Foot Costbook

This manual presents detailed square foot costs for 65 buildings tailored specifically to meet the needs of today's architect. For each project you get a complete cost breakdown of the included systems, so you can easily calculate the impact of modifications and enhancements on your own projects. Several of the case studies presented feature significant green building strategies as indicated by the LEED Rating system, encompassing recycled construction waste, recycled material content and re-use, recycled rainwater, bio-retention and wetland storm management, indoor environmental air quality, energy efficient lighting, hybrid HVAC systems, green roofs, and native or adaptive vegetation.

This manual also features detailed costs to enhance the included case studies and allow the design professional to "mix and match" components to more closely follow their own design. **210 pages, 8½ x 11, $64.95**

ORDER ON A FULL 30-DAY MONEY-BACK GUARANTEE

BNi Building News
990 Park Center Dr.
Suite E
Vista, CA 92081

☎ **Phone Order Line**
1-888-264-2665
Fax (760) 734-1540

In A Hurry?
We accept phone orders charged to your
○ Visa, ○ MasterCard, or ○ American Express

Card#_____

Exp. date_____ CVV2 _____

Signature_____

Tax Deductible: Treasury regulations make these references tax deductible when used in

Order online   www.bnibooks.com
Find hundreds of construction references, forms and contracts to help you with your construction business.
Use promo code 841Y

Name_____
e-mail address (for order tracking and special offers)
Company_____
Address_____
City/State/Zip   ○ This is a residence
Total enclosed _____ (In California add 8.25% tax)

Shipping via Fed Ex, or UPS by dollar amount of order: $1 - $100 = $9.75, $101 - $150 = $15.25, $150 - $200 = $18.75.
Find Updated Davis Bacon wages and FREE estimating tools at www.constructionworkzone.com

Prices subject to change without notice

○ 62.95 2013 ACE Guide to Construction Costs	○ 85.95 2013 BNi Home Remodeler's Costbook
○ 64.95 2013 Architect's Square Foot Costbook	○ 91.95 2013 BNi Electrical Costbook
○ 99.95 2013 BNi Facilities Manager's Costbook	○ 99.95 2013 BNi Mechanical/Electrical Costbook
○ 99.95 2013 BNi General Costbook	○ 99.95 2013 BNi Public Works Costbook
○ 81.95 2013 BNi Home Builder's Costbook	○ 89.00 Standard Estimating Practice